Caribbean Amphibians and Reptiles

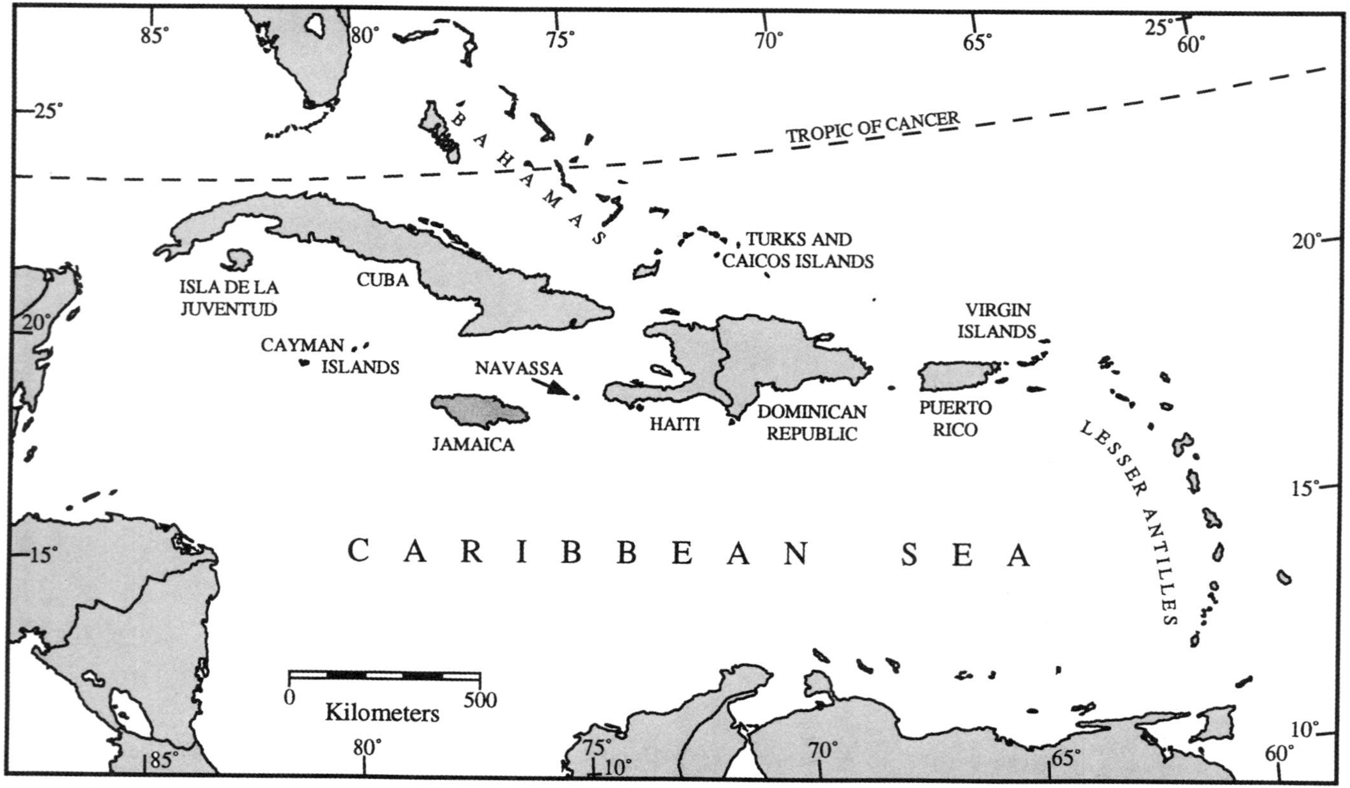
85°
80°
75°
70°
65°
25°
60°
25°
TROPIC OF CANCER
B A H A M A S
TURKS AND
CAICOS ISLANDS
20°
ISLA DE LA
JUVENTUD
CUBA
20°
VIRGIN
ISLANDS
CAYMAN
ISLANDS
NAVASSA
HAITI
DOMINICAN
REPUBLIC
PUERTO
RICO
JAMAICA
LESSER ANTILLES
15°
15°
C A R I B B E A N   S E A
0
500
Kilometers
10°
80°
75°
10°
70°
65°
60°
85°

Caribbean Amphibians and Reptiles

Edited by

Brian I. Crother
Department of Biology
Southeastern Louisiana University

Academic Press
San Diego London Boston New York Sydney Tokyo Toronto

Front cover photograph: *Anolis smallwoodi* from Guantanamo Bay.
(For more details, see Color Plate 2, Figure D.)

This book is printed on acid-free paper.

Academic Press
a division of Harcourt Brace & Company
525 B Street, Suite 1900, San Diego, California 92101-4495, USA
http://www.apnet.com

Academic Press
24-28 Oval Road, London NW1 7DX, UK
http://www.hbuk.co.uk/ap/

Library of Congress Catalog Card Number: 98-87235

International Standard Book Number: 0-12-197955-5

PRINTED IN THE UNITED STATES OF AMERICA
99 00 01 02 03 04 QW 9 8 7 6 5 4 3 2 1

Contents

3 Jamaica

Ronald I. Crombie

4 The Hispaniolan Herpetofauna: Diversity, Endemism, and Historical Perspectives, with Comments on Navassa Island

Robert Powell, Jose A. Ottenwalder, and Sixto J. Inchaustegui

5 The Puerto Rico Area

Richard Thomas

6 The Lesser Antillean Fauna

Ellen J. Censky and Hinrich Kaiser

7 West Indian Herpetoecology

Robert W. Henderson and Robert Powell

8 Evolutionary Relationships

Brian I. Crother

9 Ecological and Historical Biogeography

Gregory K. Pregill and Brian I. Crother

Contributors

Numbers in parentheses indicate the pages on which the authors' contributions begin.

Ellen J. Censky (*181*), *Section of Amphibians and Reptiles, Carnegie Museum of Natural History, Pittsburgh, Pennsylvania 15213*

Ronald I. Crombie (*63*), *Division of Amphibians and Reptiles, National Museum of Natural History, Washington, DC 20560*

Brian I. Crother (*269, 335*), *Department of Biology, Southeastern Louisiana University, Hammond, Louisiana 70402*

William E. Duellman (*357*), *Division of Herpetology, Museum of Natural History, University of Kansas, Lawrence, Kansas 66045*

Alberto R. Estrada (*31*), *Department of Biology, University of Puerto Rico, San Juan, Puerto Rico 00937*

Robert W. Henderson (*223*), *Section of Vertebrate Zoology, Milwaukee Public Museum, Milwaukee, Wisconsin 53233*

Sixto J. Inchuastegui (*93*), *Grupo Jaragua, El Vergel, Santo Domingo, Dominican Republic*

Hinrich Kaiser (*181*), *Institut für Humangenetik, Universität Würzburg, Biozentrum, am Hubland, 97074 Würzburg, Germany*

José A. Ottenwalder (*93*), *Conservation and Management of Biodiversity Project, United Nations Development Programme, Santo Domingo, Dominican Republic*

Robert Powell (*93, 223*), *Department of Natural Sciences, Avila College, Kansas City, Missouri 64145*

Gregory K. Pregill (*335*), *Department of Biology, The University of San Diego, San Diego, California 92110*

Rudolfo Ruibal (*31*), *Department of Biology, University of California, Riverside, California 92521*

Richard Thomas (*169*), *Biology Department, University of Puerto Rico, Rio Piedras, Puerto Rico 00931*

Ernest Williams[†] (*1*), *Museum of Comparative Zoology, Harvard University, Cambridge, Massachusetts 02138*

†Deceased.

Foreword[1]

In honor of Albert Schwartz.

"Just when I get you to a point that you can intelligently discuss biology, you are gone!" That or a similar statement was Al's common lament. However, if you listened carefully, it was said with a note of pride. Al never had a graduate student and more than half of his teaching career was at a two-year college. Yet his imprint as an intellectual mentor is borne by dozens, perhaps hundreds, of us—biologists, medical doctors, dentists, architects, librarians, to name a few professions of his students.

As herpetologists, we cannot fail to recognize Al's legacy to West Indian Herpetology, even though there were some of us who mumbled disparaging remarks about his constant description and splitting taxa. Whereas James Bond provided the foundation for ecological and behavioral studies of the bird fauna, Al's studies provided an even broader framework for the study of amphibians and reptiles; his published studies and his voucher collections serve us and future generations of systematic herpetologists. Al's meticulous attention to details make his collections and publications gold mines of information. There was nothing casual about his collecting; he was intent upon obtaining large samples and geographically closely spaced so that he could better discern speciation through an intimate knowledge of geographic variation. This understanding was gained by intense examination, a remarkable memory for details, and an ability to bring all data-bits into a coherent interpretation. He relied little on statistical analysis; his "taxonomic eye" was amazingly capable. Similarly, his herpetofaunal surveys are models of how biodiversity research needs to be pursued—thorough and repeated sampling (with retention of voucher specimens) of a geographic area and PROMPT study and publication of the results.

As herpetologists, many of you are unaware of Al as a student magnet. Oh yes, at the beginning of each school term, he would grumble something about staring out at all those blank faces. He would then "gird his loins" and like a gladiator march into the lecture arena to challenge

[1]This contribution is a U.S. government work in the public domain.

those faces and minds to think. An imposing and demanding teacher, he excited your imagination, largely I believe through his enthusiasm for biology. For some of us, this enthusiasm attracted us to his lab and research. For me, he shook me out of my sophomore lethargy and impending dropout status. Amphibians and reptiles had had little fascination for me but that was to change and stay changed. I was neither the first nor the last to be so challenged and invigorated by Al. However, I may be unique in being his only academic student to retain a herpetological focus. But of his Albrightian students of the fifties, I was not the only one to owe my continuing fascination with biology to Al. Several of my dentist and M.D. classmates continued to join him in his West Indies research visits, until Al's legs could no longer tolerate field work.

Owing to the thriftiness of an uncle, Al was able to escape the northern winters in 1960 and, to the best of my knowledge, never again ventured north of Florida. His shift to Miami gave him ready access and importantly short flight time to the Antilles. For a number of years, he devoted himself entirely to field work and writing. Then among the student revolts of the late sixties, he returned to full-time teaching at Miami-Dade Community College and part time at Florida Atlantic University, again sharing his enthusiasm for biology with students, some who would continue in biology but most who would not. This book is an intellectual tribute from Al's herpetological colleagues. If I may be a bit presumptuous, I offer Al a tribute of appreciation and love from his students.

September 1997

George R. Zug, Department of Vertebrate Zoology,
National Museum of Natural History, Washington DC 20560

Preface

In a small token of appreciation for Dr. Schwartz's help and friendship, and in recognition of his incredible scientific accomplishments, this book reviews herpetology in the Caribbean. A letter was sent to Dr. Schwartz informing him of the plans and to ask him if he would like to write a short autobiography (a true biography of Dr. Schwartz would be a book unto itself). He never saw the letter. He never knew a tribute was in the offing. The letter had been mailed on a Friday and Dr. Schwartz peacefully passed away that weekend. I somehow suspect he now knows and I hope that he is pleased.

This book presents an overview of the herpetology of the West Indies–Caribbean region. Authoritative reviews have been prepared by herpetologists who have worked extensively in the Caribbean. Chapters cover particular islands and amphibian and reptile ecology, phylogeny, and biogeography. The final chapter is a perspective that places the West Indian herpetofauna in relation to other geographic regions.

The islands of the Caribbean continue to lure scientists interested in testing hypotheses on the manageable ecosystems (not to mention the balmy climate, beaches, rum, etc.). In 10 years another review of Caribbean herpetofauna could include separate chapters on ethology, physiology, conservation biology, and micro-evolution at the molecular level, as well as entirely new chapters on phylogeny, biogeography, and ecology. The latter three, while receiving much attention here, are far from well understood. Only a small fraction of the taxa have been examined in explicit phylogenetic studies, and of those, only a few can be considered well corroborated (stable). To add to the difficulty in estimating stable phylogenies, it seems that more species remain to be discovered and described, as exemplified by continued discoveries of Blair Hedges, Richard Thomas, and Orlando Garrido. Biogeographic explanations, both historical and ecological, remain in debate.

This incomplete understanding of the evolutionary history of the taxa and the areas has naturally resulted in differences in opinion on matters of taxonomy and origins of the herpetofauna in the West Indies. Even the casual reader will note differences among the authors in this book. Empirically, it can be argued that much of this is due to the lack of stable phylogenies (and consistent philosophies), and with increasing numbers of stable phylogenies, differences of opinion on taxonomy and even origins should dwindle.

As for general ecological work, much has been accomplished (see Chapter 7), but far more needs to be done, especially at the most fundamental level of natural history. The natural history of the vast majority of the islands' herpetofauna remains completely unknown, or at best based on museum ecology (stomach contents studies). Given these gaps, I hope information will continue to accumulate at a high rate over the next decades.

Herpetology in the Caribbean region remains in a dynamic state. Opportunities for research on the West Indian herpetofauna abound, but rapid losses of forests and undisturbed areas in these natural laboratories are of great concern. So to all current and future researchers of the island herpetofauna, and to those engaged in preserving the beauty and nature of the islands, godspeed.

Brian I. Crother

Acknowledgments

EDITOR

In 1985 I had decided upon a dissertation project that required field work throughout the West Indies. To help the project along, my major professor, Jay Savage, arranged a meeting for me with Dr. Albert Schwartz. I was on time (*that* was important) for my meeting at Dr. Schwartz's home/library/lab/collection. He showed me his collection (I had no idea he kept such a collection in his garage), and then we sat in his library to talk. Throughout the conversation I prodded him about helping me contact Richard Thomas for help with field work. Finally, he stopped me and said, "Is there something wrong with me?" At first I had no idea what he was talking about and then it sunk in that I had essentially been slowly sticking both feet in my mouth. Here was the most famous Caribbean herpetologist to have ever lived sitting before me, kindly talking to a green graduate student/budding scientist, and I never thought to ask *him* for advice, help, etc.! The embarrassment of the moment will never be forgotten, but I will also never forget the tremendous help he provided over the years. And so, I gladly acknowledge Dr. Schwartz's (Al's) friendship and advice as the impetus for editing this volume in his honor.

Because of the goal and format of this volume, I had to solicit specific topics from the contributing authors: a proposition that turned out be challenging. I am greatly indebted to the contributing authors for their expertise. My gratitude for their understanding and patience is inexpressible: it made this volume a reality. Jonathan Campbell, Charles Crumly, Roy McDiarmid, and Rich Seigel all responded to my questions concerning advice on assembling this volume. I am most grateful, even if I neglected to heed it.

Charles Crumly and Academic Press are thanked for publishing this book. Charles Crumly proved most patient and persistent throughout the project.

A number of colleagues provided reviews of chapters (some multiple chapters): Darrel Frost, Craig Guyer, Robert Henderson, Jonathan Losos, Thomas Ostertag, Kirsten Nicholson, Gregory Pregill, Phil Stouffer, Travis Taggart, Mary White. Their generosity of time and expertise increased the quality of the work herein.

Mary White showed great understanding and support of this project, even when it clearly conflicted with other research. I could not have made this volume a reality without her. "Thank you" is not quite enough a reciprocation. I am in her debt.

CHAPTER 1. History (Williams)

During the preparation of this volume, the most esteemed contributor to the work, Ernest Williams, passed away. I, and hopefully everyone who reads it, will cherish the experience and wisdom to be found in his chapter. Dr. Williams left no acknowledgments, but I do understand from a letter to me that John Cadle read and commented on his paper.

CHAPTER 2. Cuba (Estrada and Ruibal)

We thank Rafael de la Vega for the assistance he provided with publications and information that were crucial for the completion of this article.

CHAPTER 3. Jamaica (Crombie)

In the course of my Jamaican field work I enjoyed the company of many individuals who contributed materially to the success of our ventures in many ways. Alphabetically they are Frank Davis, Dick Franz, Linda (Crout) Gordon, Susan Hansen, Barb Harvey, Blair Hedges, Bill Hilgartner, George Jacobs, Alan Jaslow, Fran McCullough (now Irish), Dyer Moore, Greg Pregill, Miles Roberts, Dave Steadman, Richard Thomas, Fred Thompson, and Bob Tuck. Special appreciation is due Jeremy F. Jacobs, who was along on nearly every trip even though there are no salamanders in Jamaica. He endured the long hikes through the Cockpits and an even more perilous stroll through downtown Kingston, demonstrating great dedication and endurance, if not much common sense.

In Jamaica we received cooperation from and imposed on the hospitality of a great many people. Patrick Fairbairn and Ann Haynes (Department of Natural Resources, Kingston) facilitated our permits and made our visits to Kingston much more endurable. Tom Farr (Institute of Jamaica), Ivan Goodbody (University of the West Indies, Mona), and Jeremy Woodley (UWI and later Discovery Bay Marine Lab) helped in too many ways to list. For hospitality above and beyond the call of duty, Menocal Stephenson and family of Quick Step, the Sutton family of Marshall's Pen, and Patrick Tennison of Windsor deserve special mention. Early in my field work, Tom Jenssen and his family provided much-needed sanctuary in Mandeville, for which I am most grateful. The management and staff of several hotels tolerated many strange beasts (including us) in their rooms with equanimity if not enthusiasm. Most notable are the Malvern in Montego Bay, Mayfair Guest House in Mandeville, and the Port of Call in Black River.

I have swapped notes, data, and general chat on Jamaica with so many people that I apologize to any I have neglected to mention here, but those that distinguish themselves in my admittedly feeble mind are Kurt Auffenberg, Alex Cruz, Dick Franz, Les Garrick, Glen Goodfriend, Blair Hedges, Jeff Lang, Ross MacPhee, Gary Morgan, Storrs Olson, Greg Pregill, Al Schwartz, Dave Steadman, Richard Thomas, Becki Thompson, and Fred Thompson.

Brian Crother has demonstrated both the patience of Job and an unexpected streak of masochism in editing this volume, a commendable if puzzling combination. Thanks dude, but I still can't believe you wouldn't buy the story about evil space aliens stealing my manuscript.

I dedicate this chapter to all those who "went before" in Jamaica, with thanks for the contributions that made this summary possible. I am particularly grateful that I was able to meet and swap Jamaica stories with Gardner Lynn, Al Schwartz, Garth Underwood, and Ernest Williams.

Last but not least, I express my appreciation and admiration to George Proctor, botanist of the Institute of Jamaica, who I met briefly in Kingston many years ago. No matter where I went in Jamaica, from the deepest Cockpits to the peaks in the Blue Mountains, Dr. Proctor had been there first and was fondly remembered by everyone I encountered. I can only hope I have done half as well, George.

CHAPTER 4: Hispaniola (Powell *et al.*)

We thank Brian I. Crother for the invitation to contribute to this volume. The late Albert Schwartz helped each of us immeasurably and on numerous occasions over the years. Robert W. Henderson, S. Blair Hedges, and Brian I. Crother commented critically on early drafts of this chapter and kindly provided access to considerable data and unpublished manuscripts; without their help this chapter could not have been written. Kraig Adler helped with citations to the historical literature. John S. Parmerlee, Jr., assisted with figures. Directors and staff at the Departamento de Vida Silvestre and the Dirección Nacional de Parques have been supportive and helpful and have made available the permits necessary for us to engage in our studies. Andreas Schubert and Gloria Santana, both of the Departamento de Vida Silvestre, have facilitated opportunities for field work by R.P. in the Dominican Republic. We also thank the many colleagues and students who have accompanied us in the field over the years.

Powell's field work was supported by Avila College and via funding by a series of National Science Foundation Research Experiences for Undergraduates (REU) programs.

CHAPTER 6: Lesser Antilles (Censky and Kaiser)

We thank Gregory Pregill for his comments on our manuscript. Kaiser's Caribbean research was funded by grants from the Hans-Krueger-

Stiftung and the Studienstiftung des Deutschen Volkes, by Boehringer Ingelheim Fonds, and by the Deutsche Forschungsgemeinschaft.

CHAPTER 7: Ecology (Henderson and Powell)

We thank Brian Crother for inviting us to write this chapter, and Craig Guyer and Jonathan Losos for constructive, thought-provoking reviews. Blair Hedges and Hinrich Kaiser answered many questions about West Indian frogs, and Susan Otto (librarian, Milwaukee Public Museum [MPM]) was instrumental in procuring literature. We also are indebted to the many kind officials on many islands who granted permission to conduct research in their respective nations.

Funding for Henderson's field work in the West Indies came from a variety of sources: MPM, Zoology Trust Fund (MPM), the late Albert Schwartz, Institute of Museum Services, the Central Florida Herpetological Society, the Chicago Zoological Society, Gulf + Western Dominicana, and the Museo Nacional de Historia Natural in Santo Domingo.

Powell's field work has been made possible largely through opportunities presented by teaching a series of field biology classes at Avila College and via funding provided by the National Science Foundation Research Experience for Undergraduates (REU) Program.

CHAPTER 8: Phylogeny (Crother)

I thank the several people who responded to my inquiries about various groups: Aaron Bauer, James Dixon, Carl Gans, Ronald Heyer, Ulrich Joger, Ulrich Kuch, Jonathan Losos, John Lynch, William Presch, Steven Werman, and Wolfgang Wüster. I thank Tom Ostertag, Travis Taggart, Kevin Toal, and Brian Warren for making many library and copying runs. I am especially grateful to Travis Taggart for making all the figures. Mary White kindly read, edited, and discussed drafts of the manuscript. The research was partly funded by the National Science Foundation (Award DEB-9207751).

CHAPTER 9: Biogeography (Pregill and Crother)

We extend our gratitude to George Zug and Ronald Crombie for their hospitality and free access to the herpetological resources of the U.S. National Museum. We also thank Mary White and David Steadman for their insightful comments on the manuscript.

CHAPTER 10: Perspective (Duellman)

I am grateful to Brian I. Crother for inviting the manuscript and thereby providing me an opportunity to express my views on the subject. I am indebted to B. I. Crother, C. Guyer, and S. B. Hedges for sharing with me copies of manuscripts that they had submitted elsewhere and to J. A. Campbell for the same courtesy, as well as providing up-to-date information on the herpetofauna of Guatemala. But foremost, I am indebted to the late Albert Schwartz for many inspiring discussions; although we frequently entertained the idea of each of us accompanying the other to the West Indies and to Middle America, we never did so. What a pity!

Legends to Color Plates*

The phenomenal diversity of the Antillean herpetofauna precludes illustrating even a small fraction of the many species of amphibians and reptiles worthy of inclusion. Instead, Plates 2–6 were designed to provide a glimpse of the total picture through a sampling of representative types from each of the major regions covered in this volume. Habitats illustrated in Plate 1 were chosen also as representative of the region, and we purposefully selected in nearly every instance habitats altered by human activities, both because these represent the reality of the modern era and because these tend to be less frequently illustrated than the increasingly rare and fragmented pristine environments that often attract the attention of biologists working in the region. In Plate 7, we illustrate diagramatically the ecomorph concept as applied to Greater Antillean anoles, using representative species that also expand the coverage of diversity presented in previous plates. Plate 8 illustrates species whose very existence has become tenuous due to human activities in the region, although only some of the taxa included are officially recognized as "rare," "threatened," or "endangered."

Captions list information about habitats and specimens illustrated, credits the photographers who so graciously provided slides, and summarize some pertinent aspects of natural history, with much of the latter extracted from Schwartz and Henderson (1991).

Plate 1: Representative Antillean Habitats. Unfortunately, few West Indian habitats remain pristine. Although hurricanes and other natural events often wreak havoc, it is primarily the need to sustain the rapidly growing human populations and the accompanying exploitation of resources, introduction of alien species, and alterations of all but the most "worthless" areas that have converted the islands into a very different reality than that experienced by the first Europeans over 500 years ago. "With this in mind, biologists must be willing to examine interactions in severely altered habitats (Powell *et al.,* 1996a), contrary to our inclination to focus on the remaining pristine areas" (Henderson and

---

*Compiled by Robert Powell.

Powell, Chapter 7). (A). Cactus scrub northeast of Gonaïves, Haiti. The removal of woody plants for charcoal production has resulted in many areas being dominated by cacti; also note the deforested hillside in the background. (Insert) Impenetrable cactus scrub near Lago Enriquillo, Provincia de Baoruco, Dominican Republic (photograph by Thomas A. Jenssen; insert by Robert Powell). (B) A recent (1995) clear-cut through formerly virgin cloud forest at Loma Remigio, Sierra de Baoruco, Dominican Republic. Even pristine stands of primary forest are vulnerable to clearing for subsistence agriculture and charcoal production (photograph by Robert Powell). (C) Damage on Guadeloupe from Hurricane Hugo, September 1989 (photograph taken March 1990 by Richard A. Sajdak). (D) The Cockpit Country in Trelawny Parish, Jamaica. To date, such "inhospitable" areas have largely escaped extensive development, but growing human populations are intruding on these areas to greater and greater extents [see, for example, comments by Vogel *et al.,* (1996) on the Hellshire Hills] (photograph by Richard A. Sajdak). (E) Secondary forest mixed with cultivated coconut, mango, cacao, and banana trees at Westerhall Estaste, St. David Parish, Grenada. This composite habitat is extensively exploited by *Corallus grenadensis* (Henderson *et al.,* 1996) (photograph by R. Allan Winstel). (F) The dry, high-elevation pine forests of the Sierra de Baoruco, Provincia de Barahona, Dominican Republic. These pine forests are almost absent from Haiti and many Dominican stands are being destroyed by fire, both inadvertent and intentional (photograph by Richard A. Sajdak). (G) Mangrove swamp near Cabo Rojo, Puerto Rico. Mangrove forests throughout the Antilles are disappearing in order to expose beaches for tourist developments (photograph by Thomas A. Jenssen). (H) Sea grape beach at Cabo Rojo, Puerto Rico. Many beaches used historically by sea turtles and a number of terrestrial lizards have become devoid of reptiles largely as the result of tourist development (photograph by Thomas A. Jenssen).

Plate 2: Selected Amphibians and Reptiles of Cuba. ". . . today's Cuba still treasures much of its original, wild nature. More, without a doubt, than any other Caribbean island. The *mogotes* and steep-sided mountains, swamps and distant or inaccessible estuaries, the scattering of low and muddy keys, and the desert of the southeastern region have proven hard to surmount barriers to settlement so far. More accessible highlands were indeed invaded, a long time ago. Their main use, however, has been coffee and cacao cultivation, which requires deep shadows: This has allowed them to survive dressed up like a densely packed arboretum. The tall vegetation attenuates the impact of rainwater, provides havens to many minor life-forms, and pleases the eye" (Silva Lee, 1996). (A) *Bufo* (formerly *Peltophryne*) *peltocephalus* from Guantanamo Bay, Provincia de Guantánamo. One of seven native Cuban toads, this species is at home in a wide variety of habitats. Males call from April through December from flooded ditches and slowly moving streams (photograph by Richard A. Sajdak). (B) *Eleutherodactylus atkinsi* from Guantanamo Bay, Provincia de Guantánamo. Although largely mesophilic, this widely distributed species, one of more than nearly 40 native Cuban species in this genus, is sufficiently opportunistic to utilize a variety of specific habitats. Males usually call from ground litter or vegetation up to 1.5 m high (photograph by Richard A. Sajdak). (C) *Anolis* (formerly *Chamaeleolis*) *chamaeleonides* from the Península de Guanahacabibes, Provincia de Pinar del Río. These slow-moving lizards rely on crypsis and are also like true chamaeleons in that they can move their eyes independently (photograph courtesy of the late Albert Schwartz). (D) *Anolis smallwoodi* from Guantanamo Bay. One of about 55 native Cuban anoles, this crown giant is usually found at height ≥3 m but will descend to the ground on occasion (photograph by Richard A. Sajdak). (E) *Leiocephalus carinatus,* from Guantanamo Bay, Provincia de Guantánamo. One of six endemic Cuban species in this genus, this xerophilic sit-and-wait forager is common in xeric coastal areas. Functionally omnivorous, these lizards frequently consume flowers, buds, fruits, and seeds along with smaller lizards and a variety of invertebrates, including large quantities of ants (photograph by Richard A. Sajdak). (F) *Alsophis cantherigerus* from Guantanamo Bay, Provincia de Guantánamo. Like many of the other 11 endemic West Indian species, the Cuban Racer has become considerably less common after introduction of the mongoose. Adults are quite catholic in diet, consuming lizards, frogs, birds, mammals (including bats), and even turtles (photograph by Richard A. Sajdak). (G) *Epicrates angulifer* from Guantanamo Bay, Provincia de Guantánamo. This species is the largest West Indian snake, with snout-vent lengths (SVL) approaching 4 m, and there are unconfirmed reports of much larger specimens. The diet changes ontogenetically from lizards to birds and mammals (photograph by Richard A. Sajdak). (H) *Tropidophis melanurus* from Guantanamo Bay, Provincia de Guantánamo. One of nine native species in this genus, these snakes feed primarily on frogs and lizards, but will take birds and small mammals on occasion (photograph by Richard A. Sajdak).

Plate 3: Selected Amphibians and Reptiles of Jamaica. "The list of Jamaican species is by no means complete, however, and the rugged limestone terrain continues to hide surprises and make a mockery of "definitive" statements on the rarity or even extinction of some species" (Crombie, Chapter 3). (A) *Osteopilus brunneus* from near Clarkstown, Trelawney Parish. As in the other four endemic species of Jamaican hylids, both terrestrial and arboreal bromeliads are used for daytime retreats, calling sites, egg deposition, and larval development (photograph by Karen Anderson). (B) *Calyptahyla crucialis* (possibly *Osteopilus;* Hedges, 1996b; see also Chapter 3) from a limestone hillock at Marshall's Pen, Mandeville Parish (700 m). These large tree frogs occur in mesic broadleaf forests. Males usually call late at night, after other frogs. Eggs and tadpoles are in bromeliads (photograph by Jeffrey W. Lang). (C) *Hyla wilderi* (possibly *Osteopilus;* Hedges, 1996b) from Barbecue Bottom, Trelawney Parish. These frogs inhabit mesic woodlands where they are frequently associated with bromeliads, in which eggs are laid and tadpoles develop (photograph by Karen Anderson). (D) *Anolis grahami* from Discovery Bay, St. Ann Parish. This trunk–crown anole, one of eight native Jamaican species, has taken full advantage of habitats provided by humans, frequenting buildings, walls, and fences. The diet is dominated by invertebrates but may occasionally include smaller anoles (photograph by Jonathan B. Losos). (E) *Sphaerodactylus semasiops* from the karst regions in the Cockpit Country. These diminuitive lizards, representing one of seven native Jamaican species, are associated with bromeliads on trees, rocks, or the ground (photograph by Richard Thomas). (F) *Sphaerodactylus parkeri* from Clarendon near Freetown. These generally mesophilic lizards may survive habitat alterations rendering areas more xeric. Individuals have been found in association with caves (photograph by R. G. Tuck, Jr. Courtesy of the National Museum of Natural History). (G) *Celestus crusculus* from Montego Bay, St. James Parish. Like *Sphaerodactylus parkeri,* these mesophiles may be able to tolerate altered xeric conditions. Like most Antillean lizards, these anguids are primarily insectivorous (photograph by Ronald I. Crombie. Courtesy of the National Museum of Natural History). (H) *Typhlops jamaicensis* from Worthy Park. These scolecophidian burrowers are associated with a variety of habitats, including termitaria (photograph by R. G. Tuck, Jr. Courtesy of the National Museum of Natural History).

Plate 4: Selected Amphibians and Reptiles of Hispaniola. Once described by Albert Schwartz as an "island of islands," Hispaniola is a mosaic of mountain ranges separated by rain-shadowed lowlands (see Chapter 4). Its known diversity is astounding and new discoveries are ongoing. (A) *Hyla heilprini* from along the road from Jarabacoa to Manabao, Provincia de La Vega, Dominican Republic. One of only four endemic Hispaniolan hylids, this species is associated with rapidly flowing or torrential streams in mesic broadleaf forests, and the tadpoles have thick, short, muscular tails (photograph by Karen Anderson). (B) *Anolis* (formerly *Chamaelinorops*) *barbouri* from the Sierra de Baoruco, Provincia de Barahona, República Dominicana. An unusual and exceedingly cryptic terrestrial anole, these lizards do not bask in their deeply shaded forest habitats, but nevertheless maintain above-ambient body temperatures (photograph by Thomas A. Jenssen). (C) *Anolis whitemani* from

Cayo Monte Grande, Siete Hermanos, Provincia de Monte Cristi, Dominican Republic. This trunk–ground anole is the most xerophilic member of the cybotoid group and typically is found in desert scrub. Animals from the northernmost part of the species' range are unusual in having yellow dewlaps (photograph by Robert Powell). (D) *Anolis bahorucoensis* from near Paraiso, Sierra de Baoruco, Dominican Republic. One of about 40 native Hispaniolan species, these lizards represent a variant of the bush–grass ecomorph. The apparently bright pattern effectively serves to break up the lizards' outlines in the shaded mesic forests where they occur. The dewlap is diminutive and even males display only rarely (photograph by Robert W. Henderson). (E) *Sphaerodactylus ladae* from arid foothills of the Sierra Martín Garcia, Provincia de Barahona, Dominican Republic. Most specimens of this poorly known species have been taken under dead *Agave* in *Agave*–cactus associations. About 30 endemic Hispaniolan species of *Sphaerodactylus* are known (photograph by James D. Forester). (F) *Leiocephalus schreibersii* from near Monte Cristi, Provincia de Monte Cristi, Dominican Republic. One of 10 endemic Hispaniolan species, this terrestrial sit-and-wait forager effectively regulates its body temperature by shuttling between sun and shade (photograph by Thomas A. Jenssen). (G) *Uromacer catesbyi* from Santo Domingo, Distrito Nacional, Dominican Republic. One of three endemic Hispaniolan "vine snakes" that are highly arboreal but willingly and frequently forage on the ground. These snakes feed primarily on *Osteopilus* and anoles but will take other frogs and lizards as opportunities arise. Generally inoffensive, one reported bite from one of these rear-fanged snakes produced an instant "burning" pain that rapidly subsided (photograph by Robert Powell). (H) *Epicrates striatus* from near Paraiso, Sierra de Baoruco, Provincia de Barahona, Dominican Republic. The largest Hispaniolan snake with maximum known SVL of 233 cm, these snakes inhabit a variety of habitats but are most common in mesic forests. As in many of its West Indian congeners, the diet shifts ontogenetically from lizards and frogs to birds and mammals (photograph by Suzanne Collins. Courtesy of the North American Center for Amphibians and Reptiles).

Plate 5: Selected Amphibians and Reptiles of the Puerto Rico Bank. "Of the 80 or so species in the entire area, 57 occur on Puerto Rico, and, of these, 37 are confined to that island and its immediate offshore cays. As one progresses east of Puerto Rico into the Passage Islands and the Virgin Islands, there is a marked drop-off in species, no doubt due largely to the reduction in the area of the eastern islands but also due to the absence of high-elevation, moist habitat" (Thomas, Chapter 5). (A) *Eleutherodactylus coqui* from El Verde, Río Grande. Named for its distinctive two-note call, this little frog has become a "mascot" of Puerto Rico. Males call from elevated perches and may "parachute" to the ground (Stewart, 1985). Nearly 20 species of *Eleutherodactylus* are known from the Puerto Rico Bank (photograph by Pamela T. Lopez). (B) *Eleutherodactylus locustus* from Pico del Este, El Yunque. Diurnal retreats of this frog are primarily underground litter. Males call from elevated perches (<1.5 m) on vegetation. At El Yunque, where *E. locustus* is sympatric with *E. eneidae,* the latter calls later at night than elsewhere in its range. Rivero (1978) suggested that this shift may have resulted from acoustic competition with *E. locustus,* which has a similar call (photograph by Rafael L. Joglar). (C) *Sphaerodactylus nicholsi* from Isabela. One of about a dozen native Puerto Rican Bank species, these xerophilic geckos are frequently found on beaches where they hide in *Cocos* trash in relatively exposed areas (Schwartz and Henderson, 1991) (photograph by Manuel Leal). (D) *Anolis cristatellus* from Aguirre Forest Reserve, Guayama. This common trunk–ground anole favors sunnier perches than most of the nearly 20 native species of Puerto Rican anoles. Introduced into the Dominican Republic, these lizards have displaced the ecologically similar *A. cybotes* in urban areas (Fitch *et al.,* 1989) (photograph by Manuel Leal). (E) *Anolis stratulus* from El Verde, Río Grande. One of the few trunk–crown anoles that are not primarily green, these well-camouflaged lizards remain primarily on the trunk and seldom venture onto smaller branches and leaves; however, they do wander onto the ground more frequently than most trunk–crown ecomorphs. Relying heavily on crypsis, mean approach distance is 50 cm (photograph by Manuel Leal). (F) *Diploglossus pleii* from Río Abajo Forest Reserve, Arecibo. The only anguid known from Puerto Rico, these small, mesophilic lizards (maximum known SVL, 125 mm) give live birth to two to four young (Greer, 1967) (photograph by Manuel Leal). (G) Hatchling *Epicrates inornatus* from Río Grande. This ecologically versatile snake occupies a variety of habitats and is "not uncommon in urban and suburban areas" (Schwartz and Henderson, 1991). Individuals have been known to aggregate at the mouths of caves in order to capture emerging bats (Rodríguez and Reagan, 1984) (photograph by Rafael L. Joglar). (H) *Arrhyton exiguum* from Cerro de las Cuevas. The presence of this genus on the Puerto Rico Bank as well as on Cuba and Jamaica lends credence to the possibility that Hispaniolan *Darlingtonia* should be included in *Arrhyton.* These small active foragers prey heavily on *Eleutherodactylus* (eggs, hatchlings, and adults), *Sphaerodactylus,* and *Anolis* (photograph by Howard W. Campbell. Courtesy of the National Museum of Natural History).

Plate 6: Selected Amphibians and Reptiles of the Lesser Antilles. These islands form two distinct arcs, the inner of which still has active volcanos. Diversity varies considerably from island bank to bank and is related to island size, diversity of habitat, and distance from the mainland or another large island (see Chapter 6). (A) *Eleutherodactylus barlagnei* from Guadeloupe. One of the many endemic West Indian species in this most diverse of all vertebrate genera, this aquatic species is associated with boulder-strewn mountain torrents (Schwartz, 1967) (photograph by Richard A. Sajdak). (B) *Eleutherodactylus martinicensis* from Dominica. This ecologically diverse species often is exceedingly common in a variety of habitats, including many that have been substantially altered. Nevertheless, some populations have been adversely affected by the recent introductions of *E. johnstonei* on some islands (photograph by Richard A. Sajdak). (C) *Anolis marmoratus* from Capesterre-Belle-Eau, Guadeloupe. These habitat generalists exhibit remarkable pattern diversity in on Guadeloupe and its satellites (Lazell, 1964a) (photograph by Richard A. Sajdak). (D) *Anolis sabanus* from Saba. The distinct disruptive color pattern of this highly saxicolous lizard (Lazell, 1972) is very effective (photograph by Richard A. Sajdak). (E) *Ameiva griswoldi* from Great Bird Island, Antigua. These actively foraging terrestrial lizards are less common on the main island where the mongoose is quite common. In contrast, individuals on Great Bird Island are almost tame (R. W. Henderson, personal communication) (photograph by Richard A. Sajdak). (F) *Corallus grenadensis* from Mt. Hartman (sea level), Grenada. These snakes exhibit the highest population densities in disturbed habitats but require contiguous canopies (Henderson *et al.,* 1996) (photograph by Robert W. Henderson). (G) *Sphaerodactylus elegantulus* from Barbuda. Eggs of this "dwarf gecko" have been found in moist debris under palm fronds and rocks (Schwartz and Henderson, 1991) (photograph by Richard A. Sajdak). (H) The fate of *Alsophis antillensis manselli,* from Montserrat, may not be determined for some time after the cessation of current pyroclastic events (photograph by Richard A. Sajdak).

Plate 7: Ecomorphs of Greater Antillean *Anolis* Lizards. Ecomorphs are species of different phyletic origins with similar morphological adaptations to similar niches (Williams, 1983). Crown giants (A, B) are large (SVL >100 mm), generally green, sometimes patterned lizards with large, often casqued heads and vertebral crests. These active and aggressive foragers are generally found high in the canopies of large trees. Twig dwarfs (C, D) are small (<50 mm), gray, or lichenate anoles with long heads and short bodies and limbs. Slow "crawlers," these lizards forage on small-diameter twigs in the canopy, relying on crypsis and immobility to protect them from predators. Trunk–crown anoles (E, F) are moderately large (>70 mm), generally green lizards with large heads, long bodies, and proportionately short limbs. Foraging on leaves and branches in the canopy and along the upper portion of the trunk, these lizards almost invariably flee upward when threatened. Trunk anoles (G) are small (<50 mm) and range from green to variously patterned with browns and/or grays. Head and body are both short. Foraging almost exclusively on the trunk below perches favored by trunk–crown and above those frequented by trunk–ground ecomorphs, these lizards "squirrel" (run around to the opposite side of the trunk) when threatened. Trunk–ground anoles (H, I) are of moderate size (>60 mm) and have relatively short heads, short, stout bodies, and long limbs. These lizards typically face downward from perches low on tree trunks, searching for prey, often on the ground. When threatened, they will usually flee downward. Grass–bush anoles (J, K) are small (<50 mm), slender lizards with relatively long heads and very long tails. Although ground colors vary, patterns inevitably include lateral and/or dorsal stripes. These adept jumpers forage in bushes and grasses, fleeing toward the ground to denser vegetation when threatened. Additional habitat partitioning occurs between sun- and shade-tolerant species within ecomorph assemblies. (A) *Anolis cuvieri,* a crown giant from Puerto Rico (photograph by Jonathan B. Losos). (B) *Anolis luteogularis,* a crown giant from Cuba (photograph by Jonathan B. Losos). (C) *Anolis angusticeps,* a twig anole from Cuba and the Bahamas (photograph by Jonathan B. Losos). (D) *Anolis insolitus,* a twig anole from Hispaniola (photograph by Jonathan B. Losos). (E) *Anolis porcatus,* a trunk–crown anole from Cuba (and introduced to Hispaniola) (photograph by Robert Powell). (F) *Anolis evermanni,* a trunk–crown anole from Puerto Rico (photograph by Jonathan B. Losos). (G) *Anolis caudalis,* a trunk anole from Hispaniola (photograph by Thomas A. Jenssen). (H) *Anolis cybotes,* a trunk–ground anole from Hispaniola (photograph by Jonathan B. Losos). (I) *Anolis mestrei,* a trunk–ground anole from Cuba (photograph by Jonathan B. Losos). (J) *Anolis olssoni,* a grass–bush anole from Hispaniola (photograph by Jonathan B. Losos). (K) *Anolis pulchellus,* a grass–bush anole from Puerto Rico (photograph by Jonathan B. Losos).

Plate 8: Some Threatened and Endangered Amphibians and Reptiles of the West Indies. "West Indian environments have been subjected to the influence of nonnative Americans, almost always negatively, longer than any other land masses in the New World. Habitat destruction is rampant and forests are rapidly disappearing in order to produce charcoal and to plant crops; Haiti, admittedly the most extreme example, retains less than 1% of its original forest cover. The introduction of the mongoose is correlated with the decline or disappearance of several snake taxa and populations; the entire known range of several species is now less than 0.5 km^2. The introduction of cats, dogs, and rats has had a catastrophic effect on some *Cyclura* populations. Human numbers are booming and development (largely for the tourist trade) continues unabated. Although some versatile frogs and reptiles respond favorably to habitat modification, far more species fare less well'' (Powell and Henderson, 1996a). (A) *Cyclura cornuta,* from Isla Cabritos, Provincia de Independencia, Dominican Republic. This large Hispaniolan iguanid has seen its populations fragmented and many extirpated by human exploitation for food, habitat destruction, and introduced predators and competitors. (Insert) The Mona Island population, *C. cornuta stejnegeri,* considered by some to represent a distinct species, is similarly threatened. Males compete for mating territories by means of highly formalized tests of strength (photograph by Robert Powell; insert by Manuel Leal). (B) *Cyclura nubila* from Guantanamo Bay, Cuba. The Cuban Iguana has been similarly expoited throughout its range. (Insert) *Cyclura collei,* the Jamaican Iguana, was thought to be extinct until rediscovered in the Hellshire Hills in 1970 (Vogel *et al.,* 1996). (photograph by Robert W. Henderson; insert by Peter Vogel). (C) *Eleutherodactylus richmondi* from Monte del Estado, Maricao, Puerto Rico. Populations of this and other forest-dwelling species have declined or have been extirpated at El Yunque, although this area is considered to be the ''best-conserved forest reserve in Puerto Rico'' (Joglar and Burrowes, 1996). Why *E. richmondi* is almost extinct at El Yunque, but are still present in the Sierra de Cayey and Cordillera Central, is difficult to explain. (D) *Alsophis antiguae sajdaki* is restricted to tiny Great Bird Island off Antigua (Henderson, 1990); the nominate subspecies has long been extirpated from the main island of Antigua (Henderson *et al.,* 1996a) (photograph by Richard A. Sajdak). (E) *Celestus* (formerly *Diploglossus*) *carraui,* from the Dominican Republic, is known from only three localities (Henderson, 1988a). Its Hispaniolan congener, *C. anelpistus,* has not been found since it was formally described in 1979 (Schwartz *et al.,* 1979; Henderson, 1988) (photograph by Richard A. Sajdak). (F) Carcass of a *Crocodylus acutus* from Lago Enriquillo, Provincia de Pedernales, Dominican Republic. This relict population has been historically exploited for food, hides, and pharmaceuticals (Schubert and Santana, 1996) (photograph by Andreas Schubert). (G) A *Chelonia mydas* awaiting slaughter near the beach of Isla Saona, Dominican Republic (photograph by Robert W. Henderson). (H) ''Jicotecas'' (*Trachemys stejnegeri*) from Laguna de Saladillo, Parque Nacional Montecristi, Provincia de Monte Cristi, Dominican Republic, being sold by a roadside vender. All of the endemic freshwater sliders of the West Indies are being heavily exploited for food (photograph by Robert Powell). (I) *Epicrates monensis* from Isla Mona, Puerto Rico. A number of populations have been extirpated and others are threatened by introduced predators, habitat destruction, and stochastic processes resulting from population fragmentation and isolation (Tolson, 1996) (photograph by Peter J. Tolson). (J) The mongoose, *Herpestes javanicus,* is implicated in the extirpation of many West Indian species and populations of terrestrial snakes and lizards (Henderson, 1992) (photograph by Richard A. Sajdak).

Color Plates

Plate 1

Plate 2

Plate 3

Plate 4

Plate 5

Plate 6

Plate 7

Plate 8

1

Over 300 Years of Collecting in the Caribbean

Ernest E. Williams
Museum of Comparative Zoology
Harvard University
Cambridge, Massachusetts 02138

Introduction: Where We Are

With Schwartz and Henderson's massive 1991 compendium on the West Indian herpetofauna, an epoch primarily devoted to collection and description has come to an end. The goals will be quite different in the future.

In 1914 Thomas Barbour published a comparable work titled "A Contribution to the Zoogeography of the West Indies, With Special Reference to the Amphibians and Reptiles." His effort at zoogeography was, in fact, primitive and elementary, but he did provide an overview that was quite accurate in regard to the knowledge of its time. He included *Spelerpes infuscata* Peters, allegedly from Haiti, which was a bolitoglossine salamander that Dunn (1926a) would regard only as a badly mislabeled *Oedipus* [now *Lineatriton*] *lineolus* (Cope), a species from Mexico; although anomalous – no other salamander had ever been collected in the West Indies – it was a record attested to by Wilhelm Peters, Curator at Berlin, a very reputable source. He listed only 22 frogs: 17 *Eleutherodactylus*, six *Hyla*, six *Bufo*, one or two *Leptodactylus* (depending what is counted as endemic), and *Phyllobates limbatus*, so identified by Cope.

In lizards he listed 158 species, of which seven were *Mabuya*, and 73 anoles, one of which was *Chamaeleolis* (*chamaeleonides*), one *Xiphocercus* (*valencienni*), and one *Norops* (*ophiolepis*). In non-anoline lizards he reported 29 *Sphaerodactylus*, 15 *Ameiva*, 11 *Leiocephalus*, nine *Celestus*, six *Cyclura*, three *Aristelliger*, a *Gymnodactylus* that is probably a mislabeled specimen, and a smattering of genera of which there was only one endemic example in the West Indies

(*Tarentola*, *Gymnopthalmus*, and *Cricosaura*). Of amphisbaenians he knew *Amphisbaena punctata* Bell. In snakes he was aware of six *Typhlops*, only two *Leptotyphlops*, eight *Epicrates*, six *Tropidophis*, and one "*Lachesis*." In colubrid snakes he recorded 37 species.

It is inevitable to compare this list of the West Indian herpetofauna with that of Schwartz and Henderson (1991) (Table 1.1). Their compendium is an accurate picture of the state of the art today. There are some problems of name changes, but these are remarkably few, except in snakes.

Counting only endemic species, there are a total of 152 species of frogs, 125 of them *Eleutherodactylus*; one of these is the single species supposed to represent a mainland genus, the *Phyllobates* of Barbour (1914). An odd story is concealed behind this apparent suppression of a genus and species. The species *limbatus* was discovered by Cope (1863a) and was referred by him to the genus *Phyllobates*. When Barbour and Noble (1920) in the course of a revision of *Phyllobates* discovered that the shoulder girdle of Cuban *limbatus* did not resemble the shoulder girdle of *Phyllobates* they erected the genus *Sminthillus* for it. Later, G. K. Noble (1921) and then H. W. Parker (1926a) erected mainland species, respectively, *peruvianus* and *brasilensis*. Several authors (including Noble) considered the Cuban animal had eleutherodactyline affinities. Finally, Hedges (1989a) submerged the species in an expanded definition of *Eleutherodactylus* without examining the Cuban animal's shoulder girdle or preparing allozymes from it. He relied primarily on Bogart's statement that its karyotype resembled the *ricordii* group of *Eleutherodactylus* in referring it to that group and also on (his own character and observation) the shape of its liver. *Sminthillus peruvianus* had already been referred to *Phrynopus* by Heyer, as had *S. brasiliensis*. *Euparkerella* was erected by Griffiths (1959).

There are 294 species of lizards, and 128 of these are *Anolis* [*Eleutherodactylus* is close to overtaking *Anolis* as the most numerous genus of the West Indian herpetofauna, but despite new (post-Schwartz and Henderson) descriptions of species of both genera, I believe *Anolis* is still slightly ahead.] Two genera, *Xiphocercus* and *Norops*, have been synonymized with *Anolis* (Etheridge, 1960). (However, see Guyer and Savage, 1986 and the resulting literature: Cannatella and de Queiroz, 1988; Williams, 1989a; and Guyer and Savage, 1992.)

One new endemic anoline genus, *Chamaelinorops*, has been recorded. This is ecologically (Flores *et al.*, 1994) and morphologically

Table 1.1. A Comparison of the Known Herpetofauna in the West Indies as Listed by Barbour (1914) and Schwartz and Henderson (1991).

Barbour (1914) Species	No.	Schwartz and Henderson (1991) Species	No.
		Amphibia	
Spelerpes infuscata X See text			
Hyla	6	*Calyptahyla* [a]	1
		Osteopilus [a]	3
		Hyla	5
Bufo	6	*Peltaphryne* [a]	10
Eleutherodactylus	17	*Eleutherodactylus*	125
Leptodactylus	2	*Leptodactylus*	3
Phyllobates See text			
		Sauria	
Gymnodactylus fasciatus See text			
Gonatodes	1	[as subspecies of *G. albogularis*]	
Phyllodactylus	1	*Phyllodactylus*	2
Aristelliger	3	*Aristelliger*	6
Tarentola	1	*Tarentola*	1
Sphaerodactylus	29	*Sphaerodactylus*	71
Chamaeleolis	1	*Chamaeleolis*	3
Xiphocercus	1	[now a synonym of *Anolis*]	
Anolis	70	*Anolis*	128
Norops	1	[now a synonym of *Anolis*]	
		Chamaelinorops [a]	
Iguana	1	*Iguana*	1
Cyclura	6	*Cyclura* [a]	8
Leiocephalus	11	*Leiocephalus* [a]	14
Celestus	9	*Celestus*	13
		Diploglossus	6
		Sauresia [a]	2
		Wetmorena [a]	1
Ameiva	15	*Ameiva*	18
		Cnemidophorus	1
Gymnopthalmus	1	*Gymnopthalmus*	1

Mabuya	7	*Mabuya* See text	1
Amphisbaena	7	*Amphisbaena*	10
		Cadea	2
	Serpentes		
Typhlops	6	*Typhlops*	20
Leptotyphlops	2	*Leptotyphlops*	8
Epicrates	8	*Epicrates*	9
Chironius	1	*Chironius*	1
Hypsirhynchus	1	*Hypsirhynchus* [a]	1
Ialtris	1	*Ialtris* [a]	3
Tretanorhinus	1	*Tretanorhinus*	1
Uromacer	3	*Uromacer* [a]	3
[*Mastigodryas* as *Alsophis*]		*Mastigodryas*	1
		Darlingtonia [a]	1
Alsophis	15	*Alsophis* [a]	11
		Antillophis [a]	2
Arrhyton	3	*Arrhyton* [a]	10
Leimadophis	12		
		Liophis	5
Lachesis	1	*Bothrops*	2

[a]Endemic genus fide Schwartz and Henderson.

(Forsgaard, 1983) a very peculiar anole; it is both montane and cryptic, and primarily confined to the south island of Hispaniola, the area south of the Cul de Sac-Valle de Neiba trough. All these characteristics are reasons that is was not recorded in the pre-1900s. A single specimen is known from above Constanza in the Cordillera Central of the Dominican Republic.

One genus and species of gecko is missing in the new listing: *Gymnodactylus fasciatus*. Kluge (1964) discussed the relevant problem. Duméril and Bibron (1836) described the species on a single specimen with the locality "Martinique" from the Plee collection. It is not a *Gymnodactylus*, as now understood, but instead a *Homonota*, a genus known only from between 15 and 48°S latitude in Argentina. In view of the fact that the genus has not been re-collected in Martinique and that the known distribution is so remote from that island, the locality is regarded as erroneous.

Seventy-one of the lizards are *Sphaerodactylus* and 18 are *Ameiva*. One is a surprising discovery of a Greater Antillean species of a gecko, *Phyllodactylus*, previously known in the West Indies only from far south in the Lesser Antilles. Only one species (three subspecies) is reported despite the fact that the animal is known from two of the Greater Antilles, from both Haiti and the Dominican Republic in Hispaniola, and also from the dry south central coast of Puerto Rico.

Thirteen West Indian anguids are *Celestus* and six are *Diploglossus*; one of the latter was described from a single specimen by Underwood from Montserrat in the middle of the Lesser Antillean chain. Two are in a genus (*Sauresia*) erected by Gray because instead of being pentadactyle like the previous two genera, these had only four toes on each foot. The other is a montane pentadactyle but earless genus (*Wetmorena*) named for Alexander Wetmore, who searched the mountains for birds. All species of *Celestus* and *Diploglossus* are distinguished on the basis of the structure of the osteoscutes (Strahm and Schwartz, 1977; but see Savage and Lips, 1993). *Sauresia* and *Wetmorena* resemble *Celestus* in "reduced and absent radices."

Twelve amphisbaenians are listed, two of them in the genus *Cadea*, which Gray had erected for *punctata* Bell. Stejneger (1916) casually mentioned the generic change in a note indicating that *punctata* was preoccupied and he substituted *blanoides* for it. [He mentioned another preoccupied name, a generic name for a South American taxon; his excuse for publishing the two names together was a curious one ("it is necessary to use the names in a forthcoming lecture")]. Mary Dickerson provided in the same year the second species in the identical genus (and also a better explanation for the recognition of the genus). All other amphisbaenians remain in the genus *Amphisbaena*.

There are 99 snakes recognized: 21 *Typhlops*, eight *Leptotyphlops*, nine *Epicrates*, and 13 *Tropidophis*. The genera of colubrids, apart from *Hypsirhynchus* (one species) and *Uromacer* (three species), have been changed or added to. Most of them have shifted generic names on the basis of Maglio's (1970) review of cranial osteology and hemipenial morphology. *Leimadophis* has totally disappeared. Sense can be made of the other names only on the basis of Maglio's discussion. *Antillophis* is the only new Maglio genus.

However, one new genus, the distinctive montane genus *Darlingtonia* was added earlier by Cochran (1935). It was named for P. J. Darlington, the second of the first two recorded herpetological explorers of the mountains of Hispaniola. (His herpetological efforts, like those of Wetmore, were peripheral to his main interests. Darlington's major concern was with carabid beetles.)

In place of "*Chrysemys palustris*" there are four freshwater turtles of the genus *Trachemys*; this is not a great change since Stejneger (1904) confessed that Barbour relied on not enough specimens to allow him to estimate the number of species in the total fauna. There is still just one endemic crocodile in the West Indies.

No more than Barbour in 1914 did Schwartz and Henderson (1991) achieve a zoogeography of the herpetofauna of the West Indies. Discreetly, they did not attempt it. They have produced another overview, possibly nearly complete.

Periods in the History of Knowledge of the West Indian Herpetofauna

How did we get where we are? There has been a long history of observation and collecting in the West Indies. The history of knowledge of the herpetofauna of the West Indies can be divided into six periods. (I can only comment on a few workers that stand out in each period and perhaps exemplify the period in which they did their work.)

In these six periods there has been a very visible progression from essentially folk knowledge to scientific knowledge. Initially there was a great deal of carelessness about locality and a great ignorance of the features (characters) that distinguished the kinds (the species) that were recognized. In every period there has been a very evident solidification of standards and a truly steady increase in the individual application of clarity, consistency, and rigor in systematic procedure and example. To be sure, there have been in each period some people ahead of their time and others who were definitely behind their time; most were sometimes ahead and sometimes behind. There was obviously a learning process in which some learned part of the totality of what needed to be learned and some did not.

Schwartz was part of that learning process, and he was very good at what he did. He, like everyone, will be superseded. This is unavoidable and part of the progression of Science.

Period I

The first period was a pre-Linnaean period which I will characterize, using Hans Sloane as the example. (I confess that my choice of

Sloane is not solely on his virtues but also because he wrote about Jamaica, where I had my first experience of the West Indies.)

Hans Sloane (1660-1753) can be considered as the real founder of the British Museum. His collections provided the base for the British Museum (Günther, 1980; de Beer, 1953; Brooks, 1954). (The collections were both botanical and zoological, birds and mammals, but, not on verifiable evidence, herpetological.)

Long before his bequest, he spent 15 months (in another account 20 months) in Jamaica beginning in 1687. In 1725 he reported and figured the amphibians and reptiles that he had seen but apparently not collected. They are all recognizable (certainly to genus) from the figures, except for one lizard and a snake which is only recognizable by its English name, the Yellow Boa of Jamaica, and by its purported size. (The figure is definitely a caricature of a snake; crisscrossing lines presumably indicate scales. The artist must not have seen a Yellow Boa.) The lizards at least show genuine observation and sometimes careful observation.

Sloane was primarily a physician, employed while he was in Jamaica as the personal physician of the Governor-General of Jamaica. By avocation he was a botanist and not in any way a specialist in amphibians and reptiles, but even so his figures show a very good sample of the species that were then conspicuous elements of the herpetofauna, including some animals that may now be extinct. He dissected some of the specimens but succeeded in bringing none of them back to England. (That story is told below.) There were certainly no specimens available when John Edward Gray started the herpetological collections long after the museum was founded (Boulenger, 1906, the internal evidence of Gray's 1845 "Catalogue of the Lizards in the British Museum").

All of Sloane's (1725) figured sample's of the Jamaican herpetofauna have been identified, most of them correctly (Ahrenfeldt, 1954). I comment on Sloane's or Ahrenfeldt's observations and, incidentally, modernize the nomenclature. (In quoting Sloane, I have faithfully reproduced his spelling and punctuation, eliminating only the capitalizations which will distract the reader.)

"Rana arborea maxima" (p. 331) = *Osteopilus brunneus* Gosse, 1851

Sloane's remarks are these: "These tree frogs are very frequent in the woods of Jamaica. They are larger than ordinary frogs, slenderer, and of a light brown or grey colour." The figure is quite good for this period, and there seems no doubt of its identity. Note the date of the name it now bears.

"The land tortoise" (p. 331)

"Tis common in the woods between Guanaboa and town every where." Whether the land tortoise was endemic or imported cannot be determined without a specimen. If it was imported, whether it was *Geochelone denticulata* or *G. carbonaria*, the same observation applies. The population is now extinct. If there was ever an endemic land tortoise, its discovery and naming now belongs to paleontology.

"Lacertus Indicus senembi & Iguana dictus" (p. 333, Plate 273) = *Cyclura collei* Gray, 1845

"Gwanas are very fat and good meat." This large endemic lizard was thought extinct until very recently. It has recently been rediscovered on the Hellshire Hills. It is still very much an endangered species.

"Lacertus major e viridi cinereus, dorso crista breviori donato" (p. 333, Plate 273) = *Anolis garmani* Stejneger, 1899

Sloane says, "This is found frequently in the woods of Jamaica, it differs little from the Guana, but being greener, less, and having a shorter crest or comb along the back. It lays eggs less than a pigeon's egg." The characteristic toe pads of *Anolis* are at least suggested, but the figured dorsal crest and described greenish coloration confirms the identity of the species. It is still common in the relict woods of Jamaica and sometimes occurs along roadsides (E. E. Williams, personal observation).

"Lacertus major cinereus maculatus" (p. 333, Plate 273) = *Ameiva dorsalis* Gray, 1838

Sloane calls it "The Greatest Spotted Lizard" and describes it in detail: "This is about 8 inches long, of the shape of other lizards, reddish under the chaps, light brown on the back and tail, with one yellowish green line along it, it is blue all under in the abdomen, & c." Duméril and Bibron (1839) described it as *Ameiva sloanei* the year after Gray provided the name *Ameiva* dorsalis from a specimen "donated by R. Heward, Esq." In his 1845 "Catalogue of the Specimens of Lizards in the Collection of the British Museum" Gray recognized the reality of Sloane's priority by placing the English name "Sloane's *Ameiva*" alongside his own formal Latin name.

"Lacertus cinereus minor" (p. 333, Plate 273) = ??

Sloane calls this "the least light brown or grey lizard" and describes it as "two inches and a half long, of which the body is in measure but an inch, and not much bigger than a goose-quill, it has four legs, the hindermost pair is the longest, it is all over smooth, of a`light brown or gray color, except the tail, which has brown marks on it, the belly and under side of the body is whitish and the tail very small. It loves moist places and stones, where it is to be found leaping from one to the other, not only here but through the Caribes." This was identified by Ahrenfeldt, perhaps on the advice of Malcolm Smith, whom he consulted, as *Anolis opalinus*. I doubt this identification. The animal is the correct size, if it were a young specimen, although I hesitate because the phrase "not much bigger than a goose-quill" does not accord with the build of the animal (nor the figure!). It lacks the light line along the sides which enables the identification of *A. opalinus* in life and, more important, lacks the suggestion of adhesive pads seen in the figure of *A. garmani*. The behavior as observed is also anomalous. *Anolis opalinus* is arboreal on trees or bushes. It is not terrestrial, not fond of stones or moist places, and not "leaping from one to the other." It is endemic, not found "not only here, but through the Caribes." Indeed, on the basis of behavior, I know of nothing alive today that fits this description. (Perhaps the figure and the text confuse two animals, but I can make nothing of Sloane's text.)

"Lacertus minor laevis" (333, Plate 273) = *Mabuya* sp.

Sloane cites this as "bigger than the former, having a great many brown spots, otherwise much the same, laying a very small white hard shel'd egg...and nesting in rotten holed trees, leaping from one bough to another. Tis very common among old Palisadoes, & c." This is clearly a West Indian *Mabuya*, so much so that it has been the source of confusion: In describing *Scincus sloanei* Daudin (1802a) not only intended to honor Sloane but also quoted the latter's short Latin description and praised his figure. Stejneger (1904), however, was emphatic that *S. sloanei* could only have been based on the same specimen in the Paris Museum that Duméril and Bibron, as they themselves asserted, described as *Eumeces sloanei*. Daudin apparently recognized this specimen as being very similar to Sloane's "Lacertus minor laevis." This specimen, twice described under the trivial name "*sloanei*," unfortunately, was from St. Thomas and, by general agreement, the taxa of the two islands are at least subspecifically different. Fortunately, Schwartz and Henderson (1991) decided that the same subspecies name is applicable from the Puerto Rico Bank all the way to Jamaica. *Mabuya mabouia sloanei* (Daudin, 1802a), in consequence, cannot be the animal from Jamaica.

"Salamandra minima, fusca, maculis albis notata" (p. 334, Plate 273) = *Sphaerodactylus argus*; "The Wood Slave"

"This is about an inch long from the end of the snout, to that of the tail. It has two small eyes, four legs, each of which three quarters of an inch long, with two joints of and five toes, and a thick, blunt tail, of reddish brown colour. The body is not near so thick as one's little finger, 'tis all above of a brown colour, with white little spots like stars, and a white in the belly, and is all over cover'd with scarce discernible scales. The figures are different, one being taken from the lizard when dry'd, and the other from one preserv'd in spirit of wine." The toes in the two figures do not show the characteristic toe structure of the genus, but, given that the animal was so small, the artist might have missed it, as he did not in the much larger *Anolis garmani* and as he should not have in the larger "Lacertus cinereus minor" – if it is an *Anolis*.

"Scincus maximus fuscus" (pp. 334-335, Plate 273) = *Celestus occiduus* Shaw, 1802

Sloane called this "a Galliwasp," the first usage in English of this name for the genus now called *Celestus*. He described it as follows: "it was 11 inches long from head to tail, and six inches about the middle, where biggest, it was almost round every where from a sharp round snout, increasing to the middle, and from thence decreasing to the tail, which ended in a point; the back was hard and a little compress'd, and so was the belly, it had two round spiracula or nostrils in the two corners of the snout. About half an inch behind them were the eyes, and near an inch further two oblong holes for the ears; about three inches from the beginning of the snout towards the tail were the fore feet, which had two joints, and were not over an inch and a half long, with five toes like those of a lizard, the hinder legs were of the same length, & c. and beyond them the anus, covered with a transverse flap, the back or upper parts were all covered over with rhomboidal small rows of scales of a brown colour, with spots of orange color, and belly with the latter only...It is found in marsh grounds, in several parts of the island."

Sloane described three snakes, but only figured one. Given the absurdity of the figure, the artist must not have seen any snake at all.

"Serpens major subflavus" (pp. 336-337, Plate 274) = *Epicrates subflavus* Stejneger, 1901

Sloane called this "the Yellow Snake" and described it as follows: "This is in length about seven or eight feet, its head is not very large, having nostrils and phangs tho' not long, the neck is small, being near two inches about, rather less then his body, which growes bigger, till it be about as big as ones wrist and continues so large to the anus, whence it diminishes by degrees to the tail; its head is of a dark brown colour, and the scales all over the body are of a dark brown, with some yellow streaks here and there, the belly is all yellow, and cover'd with larger and greater scales than the back....It is for the most part in the woody mountains, quoiled up in the paths as ropes in a ship....They feed on birds, rats, & which they swallow whole, and therefore Nature has given them such a folded or rugous tunicle of the stomach, that it may eat things of large dimensions. Many of them have been kill'd with thirteen or fourteen rats in their bellies."

"Serpens major nigricans" (p. 337) = *Alsophis ater* Gosse, 1851

"The black snake is only smaller, else in everything the same, although not venemous." There are two black snakes in Jamaica, one large, *Alsophis ater* Gosse, one small *Arrhyton funereum* Cope. Sloane's adjective "major" decides the issue. The species *ater* is probably extinct.

"Serpens major cinereus" (p. 337)

"There is a sort likewise of a light grey colour." Color is reputed to be variable in *A. ater*; this is surely a color phase of the latter.

Sloane did, in fact, attempt to bring some these "uncommon Creatures" home to England. He intended to bring the Yellow Snake ("seven feet long"), the Guana, and also a crocodile back from Jamaica alive. There was a tragic conclusion to his effort. While all the animals survived for a while on the ship peaceably and successfully, the snake was shot by servants when it invaded their quarters, the Guana was frightened by a seaman and fell into the sea, and the crocodile died of unknown causes (de Beer, 1953, pp. 47-48). Sloane did not preserve any of these animals nor any of the small ones. The only record of all of these animals are the figures and the text appended.

If at least the snake had arrived alive in England, if it had been preserved or even skinned, and if it had been part of collections that Sloane gave to the British Museum, the Yellow Snake of Jamaica would not have had to wait for Stejneger (1901) to give it its formal

Latin name, *Epicrates subflavus*. Stejneger conspicuously did not mention Sloane's note on the Yellow Boa, although he usurped the very epithet that Sloane used.

Two of these figures were sufficiently appreciated that they were recognized and named from the figures. *Scincus* (= *Mabuya*) *sloanei* Daudin 1802 and *Ameiva sloanei* Duméril and Bibron were in obvious honor of the man who discovered them and had them drawn.

Two other species were adequately figured: the Guana and the Crested Anole. The formal naming of these had to wait. The Guana was named by Gray (1845) *Cyclura collei*, not from Sloane's figure but from a stuffed specimen "Presented by the Admiralty, from Haslar Hospital." (It is called "Colley's Cyclura" at the beginning of the description; who Colley was is not explained.)

"The Great Crested Anole of Jamaica," Stejneger's English name for this animal in the title page of his formal description, was not named until the end of the nineteenth century (1899; Stejneger did mention Sloane's figure) and was rescued from a misnomer under which it suffered for many long years. Stejneger (1899) expressly says: "Herpetological writers have shown a curious unanimity in mis-naming the large crested *Anolis* of Jamaica *Anolis edwardsii* Merrem." He was able to demolish the relevance of this name for the Jamaican species very effectively and named the species after West Indian spe-cialist Samuel Garman whom one of the same "herpetological writers" had done an injustice.

Period II

The next period begins with the Systema Naturae of Linnaeus (1758) and ends with the first volume of the "Erpétologie généralee" by Duméril and Bibron (1834).

Not many people have examined even the reprint of the 10th edition of the Systema Naturae, especially for items from the West Indies. All cold-blooded tetrapods were called "Amphibia." Under "Reptilia" there were just four genera: *Testudo*, *Draco*, *Lacerta*, and *Rana*. "Serpentes" included *Crotalus*, *Boa*, *Coluber*, *Anguis*, *Amphis-baena*, and *Caecilia*. The single West Indian representative that Lin-naeus knew is surprisingly the Bahaman *Typhlops lumbricalis*, re-corded under the genus *Anguis*.

The entry is characteristic of this early time:

lumbricalis- 230-7. Gron. mus. 2. p. 52 n. 3

Brown. jam. 460. t. 44. f. 1
Seb. mus. 1. p. 137, t. 86. f. 2
Habitat in America.
Color ex albido flavescens.

The numbers to the right of the name itself are respectively counts of ventrals and subcaudals, testifying to the use of the counts for snakes at so early a period.

This and the preceding period were epochs of random discovery, collection, and description. Many aspects contributed to this phase. Frequently the collector and the describer were quite different individuals. Often the exact locality was never given or, if given, was soon lost. Often the collector and the describer never saw each other. Many specimens had long been in the Cabinets of Curiosities. Many figures or descriptions in travel books that were often much worse than Hans Sloane's were interpreted, regardless of geographical plausibility, as identical with the usually badly preserved specimens that were in hand. Indeed, the localities of many specimens were "unknown," wildly wrong, or extraordinarily imprecise as "America" in Linnaeus.

Laurenti was the first to describe a West Indian species after Linnaeus. It was *Iguana delicatissima*, the second and more obscure of the two species of *Iguana* and the only one endemic to the Lesser Antilles. (Laurenti adopted for the generic name the second part of Linnaeus' name, *Lacerta iguana*, and because he avoided use of tautonomous names, he called the Linnaean iguana, *I. tuberculata*.) In the matter of locality, both species were labeled, typically for the period, "Indiis" (the Indies).

It is clear from the many *Iguana* species that Laurenti cites, many of which are superficially similar to agamids, that he did not possess the restricted modern concept of the genus *Iguana*. Not until the next period did Gray (1845) introduce the modern concept of just two species.

Anders Sparrman described, unwittingly, the first West Indian *Anolis*, *Lacerta bimaculatus*, 1784 and the first West Indian *Sphaerodactylus*, *Lacerta sputator*, 1784. His Plate IV portrays both unmistakably.

The Paris Museum began to be active about this time; it was called at this time the Cabinet du Roi. Only in 1793 did the assembly pass a law that transformed the Cabinet du Roi into the Museum National d' Histoire Naturelle. Previously, the Comte de Lacépède, among other duties, was Keeper of the Cabinet du Roi and published two volumes in

1788 and 1789 (i.e., just before and in the first year of the French Revolution) that described four West Indian species, one from Patrick Browne's (1789) book on Jamaica and three from "Martinique," on the basis of specimens in the Cabinet du Roi. The one from Browne's book was a freshwater turtle from Jamaica now called *Trachemys terrapen*. The other three species are as follows: the anole *Anolis roquet*, the colubrid *Liophis cursor*, and the crotalid *Bothrops lanceolatus*. (I put "Martinique" in quotation marks because the Paris collection, despite the fact that three of these species coincidentally came from Martinique, has a poor record of accuracy in claiming species that actually came from "Martinique." I previously noted that "*Gymnodactylus*" *fasciatus*, which is *Homonota*, was inferred to be from far south in South America.)

It is a curious fact that the holotype of *L. cursor* may be in the collection of the Academy of Natural Sciences in Philadelphia. Lacépède mentions a specific specimen that he describes in detail, including color pattern, and in another part of the book he includes a table giving ventral and subcaudal counts. There is no explicit mention of syntypes. There is therefore no credible reason to doubt the statement of Dixon (1981, p. 300) that ANSP 5580 is the holotype of Lacépède's *cursor*. (Cope described this exact specimen as his new species *Liophis putnamii* but obscurely admits that the specimen is from Paris "Habitat. Martinique. Mus. Acad. Phila.; Mus. Plants in ex." I presume "in ex" means "in exchange.")

Even 30 years after the 10th edition of the Systema Naturae, Lacépède did not use the binomial consistently. Even in his new West Indian species, *terrapen* or *roquet*, and *cursor*, he used just the single name. Only for his new *lanceolatus* does he use *C.* for *Coluber*, the Linnaean genus in which he categorized this very venomous snake. Even among the snakes, in which he is much more consistent, for *cursor* he uses the single name, latinized and italicized, but without the prefatory *C.* Because of this inconsistency Lacépède's names have been rejected as not binomial by the International Commission, and those names not expressly conserved are at least temporarily invalid, including all three of those discussed previously.

Daudin (1802a), in contrast, published during the French Revolution. His eight volume Histoire Naturelle, Generale et Particuliere, des Reptiles bears the odd date: an X for Year 10 of the revolution. He is the inventor of the genus *Anolis*, a native name which has survived many years but only recently has received its generotype (by fiat of the International Commission). He included both *Lacerta bimaculatus* Sparrman and *Lacerta sputator* in *Anolis*. He described only one West

Indian reptile species: *Ameiva erythrocephala*. He described it from a very detailed letter sent by M. Badier to Lacépède. The letter included measurements and diet as well as colors in life. The species is mentioned more briefly just as "Tete Rouge," without the Latin name, in the Additions a l'Histoire Naturelle des Quadrepedes Ovipares beginning on p. 487 of the second volume of Histoire Naturelle des Serpents of Lacépède. The requirement for an actual specimen in the hands of the describer was not understood at this time, certainly not by Daudin, nor, apparently, in the case of "Terrapen", by Lacépède.

As mentioned previously, at long last the galliwasp of Sloane was described as *Lacerta occidua* by George Shaw in his 1802 General Zoology, the first book in English to use the Linnaean system.

That Sloane's was the animal named *L. occidua*, which translates to the Western *Lacerta*, by Shaw is certain. He does not refer explicitly to Sloane but rather to Browne. However, he uses the term galliwasp, a local name in Jamaica which was first brought by Sloane into more general currency. It is also certain that *C. occiduus* was misnamed. Shaw was under the impression that the lizard he called the Australasian Galliwasp, his *Lacerta scincoides*, actually a true giant skink, was just a variety of the anguid *L. occidua*. This is an example of both the extreme confusion of species that was prevalent during this period and especially the total lack of realization that the range of a species might be restricted to a single island. I shall discuss two other examples of this taxonomic and zoogeographic blindness that especially apply to Jamaica.

Shaw also described as *Anguis jamaicensis* the *Typhlops* Patrick Browne (1805) cited as "The Silver Snake." Again, as with Shaw's name *L. occidua*, the name is based entirely on Browne's description; there is no hint of a type.

Neither the *Cyclura* nor the *Anolis* which Hans Sloane had figured, were described during this period. This was partly because there was great and very natural confusion as to the real species, reflected in the synonymies which plague every description or discussion of any species. The characters were not yet known, inevitably, in this early period to distinguish the authentic species.

Nowhere is this more evident than in the West Indian names which consensus attributes to Blasius Merrem (1820): *Anolis equestris* and *Anolis cuvieri*. The characters that are supposed to distinguish the species cause any anole expert to stop and stare: they are definitely not the ones anyone would choose today. The first trivial name identifies it as the Knight anole, which presumably distinguishes the species as having a helmeted head, i.e., a casque. None of the other presumed distinguishing characters listed on facing pages in Latin and in German could be understood from the text itself. There is, however, a footnote to Cuvier's "Le grand *Anolis* á Echarpe" in his La Regne Animale and to a figure that is distinctly more useful. Cuvier's epithet (echarpe = sling in French) translates to the equivalent of "bandirt" (= "in a sling") in German on the left-facing page of Merrem's book. A light streak above each shoulder does look vaguely like a sling.

The name *cuvieri* conveys no information except the intention to honor Cuvier. However, there is a footnote reference to Cuvier's "Le Grand *Anolis* á Crete" and a figure that shows a high tail crest. There is again no indication of types other than the figures, which are excellent and which explain the characters of the tail by which Merrem (1820) chose to define his species.

The phenomenon of providing a formal Latin name is the reason that the generotype of *Anolis* is *Anolis carolinensis* Voight, 1832. In all the several French editions, Cuvier in his Regne Animale used only the French "l'*Anolis* de la Caroline." Friedrich Siegmund Voight was responsible for a German translation as "Das Thierreich." In the course of a mostly literal translation, Voight transformed the French name used by Cuvier into a formal Latin name!

Only two species have been ascribed to Cuvier, the Cuban endemic crocodile, *Crocodylus rhombifer* 1807, three syntypes initially of unknown locality, and the first West Indian amphisbaenid, named *Amphisbaena caeca* Cuvier, 1829 and distinguished by being "totally blind" and erroneously originally attributed to Martinique. Later, *A. caeca* was determined to have come from Puerto Rico and to have eyes after all. The latter is named in barely a phrase, but it did have a type; the *rhombifer* syntypes are at the very least misplaced, probably lost.

The paper in which Cuvier described *C. rhombifer* provides a thorough discussion of the species of crocodiles then known, with figures of the skulls and also the patterns of the dorsal scutes that are still used as taxonomic characters. It was certainly extraordinarily valuable in its time. It is the "time" I complain about, and it is the reason I am making disrespectful comments about Cuvier and Merrem. The neglect of all the data that we now know to be indispensable was not only permissible but all but inevitable. Both were able men who entered herpetology and systematics when these "disciplines" were in their naive very first stage. It was a time when providing a recognizable Latin name, whether from a picture without a physical type or a type without a picture, or merely a unambiguous type locality was itself a service.

Richard Harlan named the first *Cyclura*, *Cyclura carinata* Harlan, 1824, a southern Bahaman lizard. He also provided a long overlooked name, *Hyla crucialis* Harlan, 1829 for the very distinctive Jamaican species for which, long after, Trueb and Tyler (1974) erected the generic name *Calyptahyla* not on externals at all but by utilizing skull characters. (Actually, the generic name was proposed for *Hyla lichenata* Gosse, 1851, the name long used for the earlier *H. crucialis*. There was, in fact, a rather heated dispute between the advocates of the respective names.)

The period between the 10th edition of the Systema Naturae and the Erpétologie générale was unpredictable and, indeed, random regarding the description of species from the West Indies. Species large and small are commingled, as are species that would be expected to be cryptic or fossorial and those that would be expected to be conspicuous. Confusion regarding to identity of species is the most evident feature of this period. Local naturalists are the tale-tellers and museum men, if they are even that, provide formal names – binomial if they were to be accepted by future generations.

Period III

The third period is the entire time span (20 years) of the Erpétologie Générale by Constant Duméril and Gabriel Bibron from 1834 to 1854. (After Bibron's death in 1848 the title pages of the later volumes had three names: The third was Auguste Duméril, the son of Constant.)

This is the very first book, or set of books, which is a complete work with full synonymies, elaborate descriptions (not in Latin), and even dichotomous keys. (I discuss two – one to genera and one to species – by way of illustration; each includes at least one West Indian taxon.) The authors tried to include summaries of aspects of the taxa they treated. Not only the systematics but also what was then known (although it was not as much information as exists today) of the anatomy and physiology and even the habits and behavior of the animals were described. It was an immense step forward and it was a considerable advance over the previous Synopsis Methodica or Table Methodique of Lacépède or the pages of listings provided by Merrem (1820), who merely followed Linnaeus in this respect. In some respects it has not been superseded. No subsequent set of books has even attempted to summarize all the literature. This is no longer possible.

During this period, there was general agreement that it was no longer permissible to describe a species without a specimen in hand. Museums and their collections became more important. It is not especially relevant to the West Indies, but it is significant for the progress of herpetology that Duméril and Bibron (1834, volume 1, p. xi) included a tabulation of the total species known from the time of Lacépède to 1834. I reproduce it here in translation:

	Lacépède in 1790	Daudin in 1805	Merrem in 1820	In the Museum collection in 1834
Turtles	24	62	62	97
Lizards	56	88	83	168
Snakes	172	315	348	391
Amphibians	40	91	87	190

Thus, the total number of species in the Paris Museum in 1834 was 846. At the end of the nine volumes in 1854, Schmidt (1955, p. 601) estimated that 1393 species were treated.

The British Museum started its herpetological collections during this period. John Edward Gray was their first curator.

Regarding Jamaica, in 1845 John Edward Gray began cataloging the lizards of the collections of the British Museum – sans any from Hans Sloane. He included *Cyclura collei*, *Ameiva dorsalis*, *Anolis grahami*, and *A. lineatopus*, all named by himself. He did acknowledge Hans Sloane's priority by printing the English name "Sloane's *Ameiva*" in front of his own formal Latin *A. dorsalis* in the 1845 catalog. He also used Daudin's *M. sloanei* but attributed it to Jamaica.

Gray's 1845 catalog is the first of a series in which a list of the herpetological specimens actually in the collection of the British Museum was intended to be authentically a checklist of the herpetofauna of the whole world. It lists the desiderata of the British Museum (Natural History) by italicizing the names of the species wanted.

Gray was completely a museum man (as were all his successor curators at the British Museum until Grandison and Arnold, recent curators).

In 1851 Philip Henry Gosse, not a museum man, wrote an extraordinarily pleasant book: A Naturalist's Soujourn in Jamaica. The first paragraph of its preface deserves to be quoted often:

> Natural History is far too much a science of dead things; a necrology. It is mainly conversant with dry skins furred or feathered, blackened, shrivelled, and hay-stuffed; with objects, some admirably beautiful, some hideously ugly, impaled on pins, and arranged

> in rows in cork drawers; with uncouth forms, disgusting to sight and smell, bleached and shrunken, suspended by threads and immersed in spirit (in defiance of the aphorism, that "he is born to be hanged will never be drowned") in glass bottles. These distorted things are described; their scales, plates, feathers counted; their forms copied, all shriveled and stiffened as they are; their colours, changed and modified by death or partial decay, carefully set down; their limbs, members, and organs measured, and the results recorded in thousandths of an inch; two names are given to every one; the whole is enveloped in a mystic cloud of Graeco-Latino-English phraseology (often barbaric enough); –and this is Natural History!

Gosse partly retracts this vivid diatribe immediately on the second succeeding page:

> The author would not be misunderstood. He is far from despising the labours of those who describe and catalogue the specimens that travellers send to the cabinets of Europe. Careful and minute descriptions, accurate admeasurements, and distinctive names are absolutely indispensable to science; but they must not be confounded with science itself. Valuable as these details are,...they are the cumbrous machinery by which that knowledge is preserved and communicated to the world.

Gosse could not afford wholly to denigrate the "necrology" of biological collections. I have myself seen his syntypes of *Anolis iodurus* and *Placopsis ocellata* and the holotype of *Draconura catenata* and participated in the discovery that the alleged type of *A. opalinus* was not the true type but was *A. grahami* (Underwood and Williams, 1959). [Apparently, if any type of *A. opalinus* was ever sent to the British Museum, the confusion occurred early; possibly it was even Gosse's mistake. Cope, who in the next period visited the British Museum (1863b) and examined the so-called type of *A. opalinus*, described much later (1895) *Anolis flabellatus*, a synonym of that species. Boulenger in his Catalogue of the Specimens of Lizards in the Collection of the British Museum (1885, vol. 2) considered *A. opalinus* a synonym of *A. grahami* Cope and redescribed *opalinus* years later as *A. flabellatus* Cope, 1895. Everyone who has examined authentic or live *A. opalinus* has agreed that the species is valid. This "type" has only caused confusion.]

Gosse (1850) named *Sphaerodactylus argus* (thus ending the formal anonymity of Sloane's "*Salamandra minima*") and another *Sphaerodactylus*, *S. oxyrhinus*. Both are still valid today. His *A. iodurus* was a synonym of *A. grahami*. His *A. maculatus*, his Zebra Lizard, cited, "especially common around Kingston", (p. 491) is the equivalent of *A. lineatopus*. (Of the two synonymous Gray names Boulenger chose the latter.) His *Placops ocellata* is *A. valencienni* Duméril and Bibron, and his *D. catenata* is *A. sagrei* of the same authors. He is certainly not to be praised for his care in his anoline taxonomy, but he is to be praised for his careful observations. His *A. opalinus* and two *Sphaerodactylus*, and also his one snake *Alsophis ater* and two frogs (*Osteopilus brunneus* and *Eleutherodactylus luteolus*) of his ventures into herpetological taxonomy alone bear his names. The *A. opalinus* survives, despite the absence of a verified type, because he was able to recognizably characterize the species in his description.

Gosse (1850) did not name either the giant anole of Jamaica, which he cites informally as the Venus lizard, insisting that this is (like, in fact, *Anolis*) an Indian name "used by the negroes", (p. 142) or the Yellow Boa. He called them by previously described names *A. edwardsii* Merrem, 1820 and *Chilabothrus inornatus* Reinhardt, 1843. Both are incorrect attributions to Jamaica of animals named from other islands, the anole from Nevis in the Lesser Antilles and the snake from Puerto Rico. In the case of the Yellow Boa, Gosse felt compelled on the inappropriateness of the species name:

> The trivial name *inornatus*, which MM Duméril and Bibron have selected to designate the species, must be considered as comparative, for this boa, when seen alive, in its black and yellow livery, I think, is far from unadorned, the contrast of colours being fine, and the purple iridescent glow that is reflected in the playing light from the dark parts of its polished armor is very rich and brilliant.

Stejneger (late in the next period, 1899 and 1901) had to correct these errors.

Period IV

The fourth period is the latter half of the nineteenth century and the very beginning of the twentieth century and its primary exemplar was Edward Drinker Cope. There continued to be a tendency to separation between the "expert" in the museum and the collector in the field. Cope, while an enthusiastic collector regarding the United States and certain parts of Mexico (Orizaba, Lake Xochimilco, and Hidalgo; Osborn, 1931:489) is the absolute extreme of the museum describer regarding the West Indies. He never saw any West Indian island, but his names for West Indies species are almost competition for Schwartz. They far exceed all other authors except Schwartz. He was brilliant and so obviously so that he impressed every one he met. He soon had the collections of the Philadelphia Academy and the Smithsonian at his disposal and readily borrowed specimens from Louis Agassiz at the Museum of Comparative Zoology. When he visited the British Museum in 1863, he was given free access to the collections, and so it was in the European museums, which he visited the next year. (When he described the British material in August 1864, he felt compelled to thank them for their courtesies: "Of the preceding species of *Anolis*, sixteen have been derived from the British Museum collections. My particular acknowledgments are due to Drs. Gray and Günther, the directors, for the ample facilities afforded me in the examination of these and other objects of interest under their care.")

Most of his herpetological collections are from his early years; actually, most of his new West Indian species are from the first years before he was preoccupied with paleontology. In the earliest years most of his specimens came from Charles Wright from Monte Verde in Cuba (these were deposited in the Philadelphia Academy or the Smithsonian) or from D. F. Weinland from near Jeremie in Haiti (these specimens were the property of the Museum of Comparative Zoology). He had to make his descriptions by external characters that were modified by preservation. His comment on Gosse's descriptions is characteristic of all workers who have not had the opportunity to observe the color of the animals in life: "they scarcely assist the student in separating the species when rendered unnatural by preservation in alcohol." Despite difficulties of this sort, Osborn (1931) quotes Theodore Gill (who knew Cope at the National Museum) as saying; "He found herpetology an art; he left it a science; he found it a device for the naming of specimens, he left it the expression of the coordination of all structural features." This is hyperbole, but there is also a measure of truth. It was in the broader aspects of herpetology that Cope excelled, in the delimitation of larger groups, and in the use of structures that had proved valuable for these larger groups, for example, the pectoral girdle of frogs and the lungs and hemipenes of snakes.

Albert Günther, a German (Gray's successor at the British Museum), described one Hispaniolan taxon as *Hypsirhynchus ferox* (explaining the trivial name of this rather small and inoffensive snake by the observation: "The ferocious physiognomy of the snake, produced by the peculiar form of the snout, in combination with the strong dentition, has caused the denomination.") He also provided the prior trivial name, *dorsalis*, of a taxon (also Hispaniolan) that Cope recognized as the genus *Ialtris*. These deserve mention as the only two generic names of West Indian colubrid snakes from this period that have survived unchanged since the generic revisions of Maglio (1970). However, there is general consensus that Günther's greatest contribution was his choice of a successor, a Belgian named Boulenger.

George Albert Boulenger, who in some sense popularized Cope's ideas by adopting most of them in his "Catalogues of the British Museum" (1882-1896) that are still useful today. They have survived not only because he lists, following the Gray and Günther precedents, all the species that the British Museum did not have, and thus provided a checklist of the world herpetofauna, but also because the useful characters for each group began to be known. Also, he used a standard format for each group that he described. (This was an element of description that Cope did not achieve consistently.) Boulenger's volumes are the superior successor reference volumes that definitively replace the Erpétologie Générale. The Catalogues, in contrast to Duméril and Bibron, do not discuss previous classifications. They do not report distributions except in the briefest of terms; they also do not give any natural history notes. Again they are unlikely to be duplicated, not by any single author or pair or trio of authors.

Boulenger never went into the field. He too never saw an Antillean island; in contradiction to Cope, he is not reported collecting salamanders or snakes. (When he retired, he did research on roses.) During all of his active career, he sat in London, very usefully describing the herpetological spoils of the British Empire. However, Boulenger's contributions to the description of the West Indian herpetofauna are meager, only *Chironius vincenti*, *Sphaerodactylus vincenti* and *Anolis wattsi* were added, all species of the Lesser Antilles.

Samuel Garman was also a collector who was a describer. He had an extremely varied life before he became a museum man, associated with Louis and Alexander Agassiz and eventually the Curator of both Ichthyology and Herpetology at the Museum of Comparative Zoology. He took the opportunity of being Alexander Agassiz's assistant on the ship Blake to make the splendid collections that enriched our knowledge of the Lesser Antilles and other Caribbean islands as well.

It was these collections that were the occasion of serious disagreement between one museum expert and one man in the field. Boulenger, who was the Recorder for the Zoological Record in 1887, made the following comment in an entry on Garman's new names for anoles in the West Indies, primarily the Lesser Antilles:

> *Anolis*, Daud. S. GARMAN, Bull. Essex Inst. xix, has greatly multiplied the binomials of the West Indian forms of this genus. The Recorder would refer the new species and varieties to the following more comprehensive terms: -- 1, *Anolis cristatellus*, D & B: *A. scriptus*, n. sp., p. 28, Silver and Lena Keys, Florida. 2, *A. gingivinus*, Cope: *A. virgatus*, n. sp., p. 41, St. Bart's. 3, *A. leachii*, D & B: *A. asper*, n. sp., p. 31, Marie Galante, *nubilus*, n. sp., p. 32, Redonda, *sabanus*, n. sp., p. 39, *speciosus*, n. sp., p. 42, Marie Galante, *lividus*, n.sp., Montserrat. 4, *A. alligator* D & B: *A. gentilis*, n. var., p, 34, Petit Martinique, *cinereus*, n. var., p. 35, Grenada, *extremus*, n. var., p. 35, Barbados, *luciae*, n. var., p. 44, St. Lucia, *vicentii*, n. var., p. 46, St. Vincent. 5, *A. richardii*, D & B: *griseus*, n. sp., St. Vincent, *A. trossulus*, n. sp., Grenada. Also copious notes on other known spe-

cies. *A. conspersus*, n. sp., P. Am. Phil. Soc., xxiv, p. 273 Grand Cayman, W. Indies [= *A. grahamii*], Gray. -- REC.

Boulenger was able to evaluate these species from preserved specimens (Boulenger, 1906). Garman had sent him examples of his recently described species.

Barbour (1914) and Lazell (1972) revalidated most of these species, and Lazell's is the revision of the Lesser Antillean anoles accepted today. The non-Lesser Antillean Garman species are also accepted as full species. Boulenger was, in my judgment, the best of the museum men, but he had not been able to see the animals alive. There is, in fact, no substitute for experience in the field, but there is also no substitute for a museum's opportunity for careful comparison. Boulenger did accurately identify the species groups to which all these species belong.

Stejneger, long Curator of Herpetology at the United States National Museum, by no means made as great a global herpetological contribution as Boulenger; he did not attempt any herpetological monographs or catalogs. (His original field was ornithology; he only took the job of Curator of Herpetology because it was available. The latter part of his long life was spent in administration.) He did have a legal background, having graduated in law from the University of Christiana in his native Norway, and so he was well prepared to be a member of the International Commission on Zoological Nomenclature in 1898.

He was not, however, merely legalistic; his strong point was accuracy. He was, therefore, bitterly opposed to the (in his view) casual behavior of Cope and Boulenger in employing series of syntypes instead of holotypes. He also instituted, for the same reason, possible confusion of two more species, the detailed description of a holotype, instead of a series.

It was undoubtedly as part of his search for accuracy that he provided names for the Great Crested Anole of Jamaica, his own English name (*Anolis garmani*, 1899), and for the Yellow Boa (*Epicrates subflavus*, 1901) of that island.

It is clear that there was never any justification for the use of *A. edwardsii* for the giant anole of Jamaica. Stejneger was able to demolish the use of this name. First, it was mentioned as "brought from the Nevis in the West Indies, by a young gentleman who came to London for education"; second, while the figure was life size, it is not nearly large enough to be the giant anole of Jamaica; third, it has no trace of the saw tooth dorsal crest that distinguishes it from all other anoles on the island.

Epicrates subflavus presented more difficulty for Stejneger (1901). There is a difference in color (so vividly described by Gosse), unmentioned by Stejneger except in the (borrowed?) epithet, *subflavus*. Stejneger could find only one constant scale difference between *inornatus* and *subflavus*, the contact in *subflavus* of the prefrontals with the preocular rather than the separation by small scales.

Period V

The fifth period begins with Stejneger's "Herpetology of Porto Rico" (1904) and is the period typified by "island surveys." It is the first period in which there was a solid effort to collect the total herpetofauna of the individual large islands of the Antilles. Museum people frequently also collected. The museum man had "the big picture"; he was therefore anxious to fill in the gaps, geographic and taxonomic. He did so by collecting himself, fusing the roles of museum man and the field man.

The Greater Antilles were all surveyed. Puerto Rico was surveyed twice, by Schmidt (1928) and Stejneger (1904). Barbour and Ramsden (1919) did Cuba, Cochran (1941) did Hispaniola, and Lynn and Grant (1940) did Jamaica. The Lesser Antilles never had a genuine summary for all groups. Perhaps this is because the diversity is so special.

Of the successive surveys, Stejneger's is my favorite. He describes individual specimens in hand and when possible color in life. Stejneger (1904), on the first page of "The Herpetology of Porto Rico," he confesses that his own collecting experience was for a relatively frivolous reason:

> The trip that I undertook in conjunction with Dr. C. W. Richmond, Assistant Curator of Birds, was primarily for the purpose of collecting specimens characteristic of the West Indian fauna for exhibition at the Pan-American Exposition in Buffalo. It was originally our intention to proceed from Porto Rico to the smaller on the east and south, but an attack of fever to which we both fell victims prevented us from carrying out our plans. The fever and the long convalescence also seriously interfered with our work in Porto Rico, so we lost nearly three weeks' collecting.

Barbour and Ramsden's (1919) Herpetology of Cuba is impeccable except that, as Schwartz and Cuban workers now active on the island have been able to show, it is woefully inadequate.

Schmidt's production of 1926 is mostly a copy of Stejneger, to the extent of quoting verbatim Stejneger's notes on color in life and repeating his figures, but Schmidt does include "original descriptions." He cites even Cuvier's description of *Amphisbaena caeca*, which he describes as "evidently quite useless." He then presents, not Duméril and Bibron's redescription of the typical material, but Stejneger's account, which he declares "much the most useful."

Cochran's The Herpetology of Hispaniola (1941) is, in some ways, worse. She also fills up space with original descriptions, but she describes the hues and tints provided by specimens in alcohol at inordinate length. Her figures are acceptable, but her plates (those which show patterns a little obscurely) are photos of very obviously dead specimens that add no information at all.

However, she and Lynn and Grant (1940) deserve the faint praise of "useful." I have used them and have usually not been misled.

I am sure that Thomas Barbour was the most conspicuous West Indian herpetologist of the period. I use conspicuous in its invidious sense. I quote entire the paper by Anthony Curtis (1947:224) titled, "Prevalence of Snakes in Haiti":

> Some authors have stated that snakes are practically unknown in Haiti. The late Dr. Barbour states that *Uromacer catesbyi* is widespread but rare. It is really very common, even in the gardens of this town, as many specimens to Dr. Cochran attest. Dr. Barbour stated that *Epicrates striatus* seems to be really uncommon. Every large dead tree which we have opened in the low lands, even in local gardens, contained one or more of these snakes. Many unopened trees are from time to time festooned with their shed skins. We find large boas stretched on branches after rainy nights or in our houses pursuing rats. I have found no boas in the mountains.
>
> In some places the boa is respected as being the emblem of one of their jinn; hence Dr. Barbour may not have seen it. He used to send out blacks he met along the shore, retire to his boat, and buy what they brought in late that day. Many blacks here keep snakes in captivity, but seldom show them to strangers.

Barbour was a special mixture of the professional and the dilettante. He was indeed both a collector and a describer. Perhaps his early work, before the arrival of Arthur Loveridge, was his best. Loveridge was a specialist on Africa, particularly East Africa. He resented taking care of anything else. Loveridge trained Benjamin Shreve to take care of the non-African material. Barbour found this particularly useful. As Shreve complained to me, the several West Indian papers by Barbour and Shreve always took the same course: Shreve did the spade work, and Barbour did the florid introductions; Barbour was always the first author.

Barbour was independently wealthy, and his greatest service was to fund collectors, in particular Noble, Dunn, and Darlington, and also to buy collections. Loveridge pointed out to me that the Great Depression hit him hard, that after 1930 there were no acquisitions comparable to those before that year and far fewer African trips for Loveridge.

Barbour composed three lists of Antillean reptiles and amphibians, in 1930, 1935, and 1937. All are limited to distributional comments of the sort that Anthony Curtis justly rebuked, and the third list must be stigmatized as erroneous because he used trinomials to show "races that are obviously closely related." Unfortunately, in the group that I know most about, he got the "obvious relationships" wrong. *Anolis krugi* is not obviously close to *acutus*, *wattsi*, *forresti*, nor *gingivinus*, all of which Barbour labeled as subspecies of *krugi*. Schwartz and Henderson (1991), following Lazell (1972), treat these "subspecies" as three full species (*krugi*, *acutus*, and *gingivinus*) and two subspecies (*wattsi wattsi* and *wattsi forresti*). *Anolis krugi* is more closely related to *pulchellus* and *poncensis*, as both Stejneger (1904) and Schmidt (1926) had already realized.

Period VI

The sixth period begins with Schwartz's first trip to Cuba. Schwartz is the towering figure of this phase: His legendary NSF proposal (it was refused) to collect all the West Indies herpetofauna, like his near achievement of that goal, were the signal and significant events of this phase.

He did what he pleased and he did it well. He found the area inadequately collected, partly because there were fallacious predictions that there was little to collect. He had found in that first trip to Cuba that the West Indian fauna was very different from the mainland fauna: animals tended to be abundant there, not only the common widespread species but also local populations. He deliberately went off the beaten track, examined areas that had not been seen before, looked for at least local variation within species, found it, and described it. This self-imposed task avoided the vicious presumption that islands were devoid of interest because they were relatively depauperate; he demonstrated the old adage – not for the continent of Africa but for the West Indies – that from the islands, there comes always something new.

He could be critical of others. I experienced some of his criticism. He noticed that I had misspelled *bahorucoensis* throughout one paper and recorded that fact with a (justified) sic after each repetition of the misspelling *baharucoensis* in all the multiple editions of his coauthored checklists. He further complained that I mispronounced *cybotes* with a hard c and took issue with my pronunciation of *Wetmor-ena* for that genus, pointing out that the genus *Wetmore-na* was supposed to honor Alexander Wetmore.

His mode of operation was intended to be professional in all that he did: i.e., (1) to be precise about localities, (2) to take detailed notes of color in life, (3) to collect large series whenever possible, (4) to select a single specimen as holotype, (5) to describe the scales and other external features of all specimens of the taxa carefully and in detail, and (6) to spell exactly, and to ensure that pronunciations honored whomsoever they were intended to honor. He succeeded.

He was not always the discoverer, but he was certainly the inspiration. Let's call the role of some startling surprises which the West Indies has provided: Schwartz himself described three anoles that can only be described as startling, the aquatic *eugenegrahami*, the primitive *sheplani*, and the especially problematic *fowleri*. The ovoviviparous *Eleutherodactylus jasperi* is another species evoking incredulity. Astounding is the fact that it was not until 1966 that Richard Thomas described *Eleutherodactylus coqui*, the most common frog in Puerto Rico, the call of which gives it its name. Also amazing is the distribution of *Phyllodactylus* in the West Indies, Barbados (*pulcher*), both Haiti and the Dominican Republic, and Puerto Rico. [Schwartz (1979) described both of the disjunct Haitian (*sommeri*) and Dominican Republic populations (*hispaniolae*) as distinct species; the Puerto Rican population is the nominate one, *Phyllodactylus wirshingi* Kerster and Smith, 1955.]

Schwartz was primarily a collector and describer, not a jealous collector nor a jealous describer but rather he was genuinely interested in presenting what he believed to be true to the world at large.

He was a collector par excellence. He loved collection for its own sake, so he collected large series. There were larger samples for any problems that came to his notice. He was fortunate also in his choice of the West Indies; the West Indies had fewer taxa than the mainlands but usually greater abundances of the fewer taxa. For him, the West Indies was heuristic and enabling.

He published. Oh how, he published! He published alone or with (usually) one collaborator. He never published just for publication's sake; he was "getting the information out." He was fanatic about this. His ambitions were great, he wanted to collect every West Indian species, and it appears as though he nearly succeeded. Certainly the species he and his cohorts missed are very local (except in Cuba, for extraneous political reasons) and/or have a very special microhabitat.

As noted previously, he did not accomplish these feats alone. His collaborators comprise a formidable list. More than 50 are listed by name in Schwartz and Henderson (1991). The list almost included me, or perhaps it was Rodolfo Ruibal that was the attraction. When Ruibal and I were exploring the Cuban members of the *Anolis homolechis* group, Schwartz at one time orally suggested that he had some ideas and information about that and might join us. Almost immediately he withdrew, suggesting that it might prove complicated. (It is noticeable that he almost always collaborated with one author at a time.)

While Schwartz and I overlapped at times in the field and though, at one time we corresponded almost daily, we never did join forces. My primary concern with anoles was too limited for him.

His assistants and collaborators were diverse. He included some curators who were his students at first (e.g., Zug and Thomas). Zug was eventually distracted by the concerns of the National Museum; Thomas, who had teaching duties at the University of Puerto Rico, remained active in the West Indies, and alone or with Schwartz, has vied in publication numbers with his mentor. He also collaborated with McCoy at Carnegie (on subspecies of *Ameiva auberi*) and with Günther Peters of the Berlin Museum (on *Sphaerodactylus bromeliarum*; his only paper in German). His assistants in the field were mostly funded by himself. (He, like Barbour, was wealthy in his own right. Fortunately, and also fortunately for the West Indies, he was never a dilettante.)

If he specialized at all, he did so by genera and species complexes. Thus, from 1956 through 1967 he described (literally) an incredible and incredibly varied number of species and subspecies. He also reviewed and discussed alone or with one collaborator *Ameiva lineolata*, *A. taeniura*, and *A. chrysolaema*, the Hispaniolan snakes then called *Dromicus*, the *pardalis-maculatus* complex of *Tropidophis*, *Leiocephalus* of the *carinatus*, *cubensis*, and *raviceps* groups, the *scaber* group of Cuban *Sphaerodactylus*; and the *vincenti* group of the Lesser Antilles; the sphaerodactyls of the Greater Puerto Rico region; the large toads of Cuba; and numerous other topics as well.

He never discussed the phylogeny of any group. The nearest that he came to doing so was his paper with Strahm on the characters of anguid osteoderms, and this for him was essentially a question of definition of genera – of what genus to which to assign a particular species. It was not a problem of relationship at all; it was solely a naming problem.

The other paper in which I remember a dendrogram (not a cladogram) is Schwartz and Thomas (1983) "The *difficilis* complex of

Sphaerodactylus (Sauria, Gekkonidae) of Hispaniola." I am impressed by the fact that the paper is divided into two parts: Part I, authored solely by Schwartz on three species of the complex; and Part II authored by Thomas as first author and Schwartz as second author on five species of the complex and the conclusions, including the dendrogram. I assume that Schwartz was at least mildly uncomfortable with branching patterns. I point out that there are two unresolved terminal polytomies on the two branches on the dendrogram. Schwartz was clearly not one to regard precise branching patterns as essential.

He was also not primarily a biogeographer. He wrote only two specifically biogeographical papers: "Some Aspects of the Herpetogeography of the West Indies" in the 1975 Leidy Medal Symposium and "The Herpetogeography of Hispaniola, West Indies" (Schwartz, 1980b). In the latter he documented the concept of north and south island herpetofaunas relying not only on Mertens' suggestion (1939) and Williams' hypothesis (1961), but also on a geological reference (Khudoley and Meyerhoff, 1971) which asserts a fusion of two islands as a demonstrable fact.

He was fond of subspecies, inordinately some would say. He described many species, but the number of subspecies to species is about four to one. There are 40 described subspecies of *Ameiva auberi*, 29 of them Cuban (or the close satellite islands) and the rest Bahaman; and 17 subspecies of *Anolis distichus*, 12 of these Hispaniolan or on close satellite islands. Most of these were named by Schwartz alone or in collaboration.

There was always some flavor of reality to Schwartz's subspecies. At the least, he described geographic variation within species elaborately and recognized the utility of naming "entities." My current bias is that if there is not free exchange of genes, the populations should be called species, not subspecies (i.e., binomials and not trinomials). This might result in an enormous (excessive?) increase in the number of binomials, but I think that this complexity might reflect reality. I favor a case by case reexamination. If subspecies are not species, they are probably clines or morphs, not recognizable geographic units.

I have myself nibbled away at some of his taxa. With Susan Case, I have investigated certain of the distichoid anoles. Schwartz described approximately 20 subspecies in this complex. In one case – *distichus dominicensis* x *d. ignigularis* – there was a distinct hybrid zone with rare alleles and also relatively far away (near the north coast, in fact) there was an area of polymorphism in which dewlap color, the most conspicuous difference between the two "races," ran the whole range between the two "defining" conditions. In the other case in the Barahona region the absence of "interaction" between *distichus ravitergum* and *distichus favillarum* was accomplished by the intervention of a third member of the distichoid anoles, the indubitable species *brevirostris* between them. (This also occurs in another potential contact area. To the east, where *distichus ravitergum* should meet *distichus dominicensis*, there is a band of *brevirostris* between the two.) I do not present these cases as solved but as evidence of the presence of "complications."

The Seventh Period?

What comes after Schwartz? What will the seventh period in the history of the knowledge of the herpetofauna of the West Indies be like? It will concern itself more with history, and with cladogenesis and systematics. It will endeavor to date and document the events that produced the herpetofauna. My view is that this will be a more laborious process than the current enthusiasms may expect.

There will more species recognized in the future, but fewer will be discovered. Thus, the real legacy of Albert Schwartz is to leave the basic data complete enough to make unmistakable the incompleteness of our knowledge of the phylogeny and biogeography of the herpetofauna and raise the consciousness of the difficulties which we face in its analysis.

2

A Review of Cuban Herpetology

Alberto R. Estrada[1] *and Rodolfo Ruibal*[2]

[1]*Department of Biology, University of Puerto Rico, P.O. Box 38030, San Juan, Puerto Rico 00937-1030*

[2]*Department of Biology, University of California, Riverside, California 92521-0427*

History

The first written account of Cuban herpetology, according to Poey (1865), was in the reports of Christopher Columbus in which he mentioned iguanas and crocodiles in the newly discovered islands. Gonzalo Oviedo in 1526 (1959) published his *Historia Natural de las Indias* and also wrote of crocodiles and iguanas, though it is unclear whether he was referring to Cuba or Hispaniola. Oviedo does mention Cuban sea turtles when describing their method of capture, by the natives, with remoras.

The formal taxonomy of Cuban herpetology starts with the description of *Typhlops lumbricalis* by Linnaeus in 1758. However, the type locality of the specimen was listed as "America." In 1807 Cuvier described *Crocodilus rhombifer*. The first major contribution to Cuban herpetology was in Ramon de la Sagra's *Historia Física, Política, y Natural de la Isla de Cuba*, published over a period of years (1838-1853). The herpetological section was written by Cocteau and Bibron and included 34 species, of which 14 were new. Much of this publication was based on specimens brought to Paris by de la Sagra. Separate printings were made in French and Spanish and the actual dates of publication are ambiguous (see Smith and Grant, 1958). Gundlach (1880) specifically states that the herpetological part was first published in volume 6 in Spanish in 1843. Duméril and Bibron had also published Cuban species in their 1836-1841 Erpétologie Générale. Ramon de la Sagra was born in Spain but resided in Cuba for most of his life. He was director of the botanical gardens in Habana between 1824 and 1836 and was the founder of the first Cuban scientific publication; *Anales de Ciencias, Agricultura, Comercio y Artes de la Habana*, in 1827.

Alexander Humbolt traveled in Cuba during 1800 and 1801. In his *Island of Cuba* (1856) Gundlach discusses the differences between the "caiman" (*C. acutus*) and the "cocodrilo" (*C. rhombifer*).

One of the most prominent figures in Cuban natural history was Felipe Poey y Aloy. Between 1851 and 1858 Felipe Poey published in Cuba his *Memorias Sobre la Historia Natural de la Isla de Cuba*. This was the first Cuban publication devoted exclusively to zoological matters. The main herpetological component of this publication was a detailed description by Poey of the circulatory system of *C. acutus* and *C. rhombifer* and a discussion of the aortic circulation and the fact that the head of crocodiles receives a more oxygenated blood than the body. In the same article he also denies the existence of a "peritoneal respiration" that had been proposed by St. Hilaire to be present in crocodiles. Poey was the first director of the Museo de Historia Natural in Habana that was founded in 1838. These collections now reside in the Museo Felipe Poey de la Universidad de la Habana.

Charles Wright, a United States resident in Guantánamo, Cuba, during the middle of the nineteenth century was a collector of botanical and zoological specimens. He sent his herpetological material to the United States National Museum. E. D. Cope, between 1861 and 1864, described 6 new species of amphibians and 10 new species of lizards based on Wright's specimens and other specimens sent to Philadelphia by Poey.

During 1865-1868, Poey also published the *Repertorio Físico Natural de la Isla de Cuba*. It is in this publication that Juan Gundlach (1867) authored his first checklist of the Cuban herpetofauna. Gundlach was born and educated in Germany but came to Cuba as a young man (Adler, 1989). Gundlach and Poey collected and published on Cuban natural history, Poey concentrating on fishes and Gundlach on herpetology and other vertebrates. Gundlach's 1868 list contains 67 species of amphibians and reptiles. He provided distributional and ecological data for many of the species, having observed most of them in the field. He comments on the fact that by this date *Cyclura* had already become rare where once it was abundant along the coast and keys.

The checklist includes the common names for many of the species. He lists 16 species of *Anolis* and his comments on the ecology and color indicate that he had observed all of them in the field. Gundlach traveled extensively throughout Cuba. *Cricosaura typica*, which he and W. Peters described in 1863, was collected by Gundlach in Cabo Cruz, an isolated area in eastern Cuba. He apparently lived in Cabo Cruz with the turtle fishermen. He lists four sea turtle species as known and observed by him or Felipe Poey: *Chelonia mydas, Eretmochelys fimbriata, Caretta caretta* and *Dermochelys coriacea*. The fifth species, *Chelonia virgata* (currently a synonym of *C. mydas*), he lists but states that he has never seen it.

In 1880 Gundlach publishes his book *Contribucion a la Erpetologia Cubana* in Habana with a total of 71 species of amphibians and reptiles. Barbour and Ramsden (1919) provides a detailed listing of the species in Gundlach. Gundlach's opus is more than a checklist and contains extensive information about the natural history of many of the species and detailed descriptions of the color patterns of species that he collected. Gundlach kept live specimens of *Epicrates angulifer, C. rhombifer,* and *C. acutus* and gave detailed information about their behavior in captivity. Barbour and Ramsden (1919) translate and quote a long section from Gundlach detailing the Cuban crocodiles.

After Gundlach's 1880 publication 39 years, lapsed before the Cuban herpetofauna was again reviewed. In 1919 Thomas Barbour and Charles T. Ramsden published *The Herpetology of Cuba* through Harvard's Museum of Comparative Zoology. Barbour had traveled and collected extensively in Cuba and Ramsden was also a collector and lived in Guantánamo. Barbour and Ramsden carefully reviewed Gundlach's work and the other publications referring to Cuban species since 1880. During this time the total number of species had changed only slightly: Gundlach listed 71 species, Barbour and Ramsden listed 73 (see Fig. 2.1). Barbour's contribution to Cuban herpetology included numerous other papers. The most important were the 1914 "A Contribution to the Zoogeography of the West Indies, with Special Reference to Amphibians and Reptiles" and the 1921 "*Sphaerodactylus*," both publications having a substantive Cuban component. In 1937 he published his "Third List of Antillean Reptiles and Amphibians" raising the number of Cuban species to 89, and introducing trinomials for a number of the species for the first time.

During the 1930s and 1940s a total of 11 new species were added to the Cuban herpetofauna. Barbour and Shreve were the major contributors with a total of 6 new species (4 *Eleutherodactylus* and 2 *Anolis*).

Albert Schwartz's contribution to West Indian herpetology started with his 1956 paper with L. H. Ogren that was based on a collection of amphibians and reptiles that he and Ogren collected in Cuba. For the next 35 years, Albert Schwartz was the major taxonomist of West Indian herpetology (see Bourgeois, 1995). In his papers on Cuban herpetology he added 23 new species to the fauna. Schwartz's papers were on taxonomy and distribution and covered all groups of reptiles and amphibians. After the Cuban revolution of 1960 he collaborated with the Cuban herpetologist Orlando Garrido and between 1967 and 1985 they coauthored 14 papers.

Ernest Williams and Rodolfo Ruibal also contributed to Cuban herpetology during the 1960s with publications on the anoline lizards (see Rodriquez-Schettino, 1995, for these and other references).

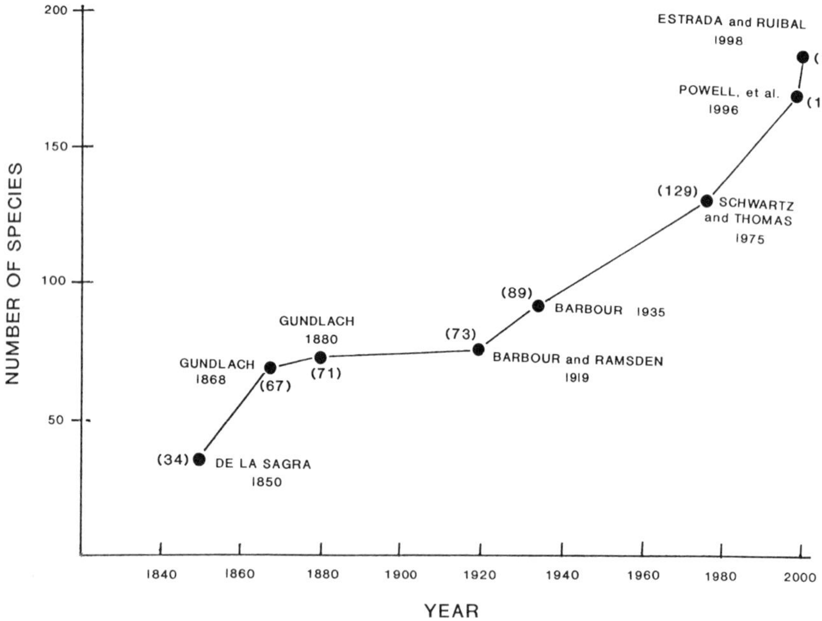

Figure 2.1. The number of recognized species in the Cuban herpetofauna. The totals include marine turtles.

The Cuban Revolution of 1959 ushered in a new era for Cuban science. The revolution restructured the academic and scientific institutions of Cuba (Margulis and Kunz, 1984). During the 1960s changes were initiated in the universities and all the other scientific entities (see Pruna Goodgall, 1994). The restructuring of the Academia de Ciencias de Cuba was accompanied by a surprising growth in scientific research and publication, including herpetology. Much of the research has been published in the new Cuban scientific journals published by the academy: *Ciencias Biológicas*, *Poeyana*, *Reporte de Investigacion del Instituto de Ecología y Systemática*, and *Miscelania*

Zoológica. Publications of other Cuban institutions include *Revista Cubana de Biología*, *Garciana*, *Memorias de la Facultad de Ciencias* and *Torreia*. These publications, and others, contain much herpetological research carried out by the new and large cadre of Cuban biologists produced by the revolution.

The magnitude of the change in Cuban herpetology since 1960 can be gauged by tabulating the first authors of the articles cited in the iguanid bibliography of Rodríguez Schettino (1995). The bibliography lists 68 papers by non-Cuban first authors and 66 papers by Cuban first authors. With few exceptions the Cuban-authored articles are post-1960. Orlando Garrido and Alberto R. Estrada are major contributors, but many other Cuban herpetologists have made substantive contributions. Lourdes Rodríguez-Schettino has made numerous ecological contributions, and Octavio Pérez-Beato, Vincente Berovides, and Georgina Espinosa López have provided electrophoretic analyses of numerous species groups. Many of the Cuban-authored papers represent a new emphasis on ecology and the evolutionary relationship of the species rather than the description of new forms.

A further example of the increased participation of Cuban scientists can be seen by comparying the number of species in the 1996 checklist (Powell *et al.*, 1996a) described by Cubans (authored or coauthored) prior to 1970 with the number post 1969. A total of 8 species were by Cuban biologists prior to 1970, whereas 36 new species were by Cubans after 1969.

During the 1990s major contributions to the Cuban herpetofauna were made by S. Blair Hedges. In collaboration with Orlando Garrido and Alberto Estrada, he added more than 15 new species to the Cuban fauna.

The Species

Geographic Regions

With some simplifications of the map of Mateo and Acevedo (1989), Cuba can be divided into four geographic regions: Western, Central, Camagüey-Maniabón and Eastern (Fig. 2.2). In totality the regions contain more than 20 subregions.

The Western section is subdivided into seven subregions. The Plains of Guanahacabibes, which includes the peninsula with the same nature, and is a karst plains. The Cordillera de Guaniguanico, composed of two main ranges, the Organos and the Rosario ranges, has a maximum elevation of 699 m at the Pan de Gujaibón. The Sierra de los Organos is formed by a chain of large isolated blocks of limestone called "mogotes," and it has numerous interior valleys and is bordered to the north and south by the clay hills the Alturas de Pizarras del Norte and Sur. The Sierra del Rosario occupies the eastern half of the Cordillera de Guaniguanico and is composed of conical hills. The cordillera is surrounded by the southern and northern plains of Pinar del Río. The eastern part of the region contains the subregions (north to south) of the Alturas de La Habana-Matanzas (up to 381m), the Alturas Bejucal-Coliseo, the Llanura de Antemisia-Colón, and the Llanura de Zapata. The first of these is a range of coastal hills, marine terraces and limestone sierras. The Alturas de Bejucal-Coliseo are formed by groups of isolated, and primarily limestone, elevations that extend from the center of the region to the east between the territories of La Habana and Matanzas, with a maximum elevation of 312 m. The Llanura de Antemisia-Colón flank this subregion on the south and is a continuation of the Llanura Sur of Pinar de Río. Currently 80% of this part of the Western region is devoted to agriculture. The Llanura de Zapata is formed by a karst plain that forms the Zapata Peninsula. This is covered by extensive forests, swamp vegetation, and mangrove and constitutes the major moist region of the archipelago and one of the major forest reserves of Cuba.

The Central region is the smallest and is divided into three subregions: The Llanura Corralillo-Yaguajay to the north, the Alturas y Llanos de Cubanacán in the central portion, and to the south the mountains of Guamuhaya, also known as the Escambray. This last subregion contains the Sierra de Trinidad, which occupies most of the western portion, and the Sierra de Sancti Spíritus, which occupies the most eastern portion. Both are separated by the basin of the Río Agabama-Manatí. The maximum elevation of this region is located in the Sierra de Trinidad at Pico San Juan (1140 m).

The Camagüey-Maniabón region has in its central portion a chain of low elevations and valleys called Alturas del Centro de Camagüey. The most prominent is the Sierra de Cubitas, a range of low (maximum elevation 330 m) limestone and serpentine hills, located in the northwest. The central southern area contains the Sierra de Najasa (301 m). In the extreme east is the Alturas de Maniabón-Banes (347 m). These hills are surrounded by the ample Llanura de Norte and Llanura de Sur, which meet in west to form the Llanura de Júcaro-Morón.

The Eastern region in the most mountainous in Cuba. Its western limit is the Llanura del Cauto, which is an alluvial zone with the largest river in Cuba, the Cauto. To the south and to the east along the length of the coast extends the Sierra Maestra, with the highest peak in Cuba, Pico Turquino, at 1972 m. The Cuenca de Santiago is a depression that separates the Sierra Maestra from the Sierra de la Gran Piedra (1214 m) and continues to the east to the Cuenca de Guantánamo. A narrow valley to the north separates the eastern portion of the Sierra Maestra, the Cuenca de Santiago, and the Sierra de la Gran Piedra from the Nipe mountains (995 m) and the Sierra de Cristal (1231 m). Located on the eastern tip of the island are the Sagua-Baracoa mountains (1175 m), which contain the ridges of Moa, Toa, and Baracoa, and the fluvial basins of Jaguaní-Toa and Duaba.

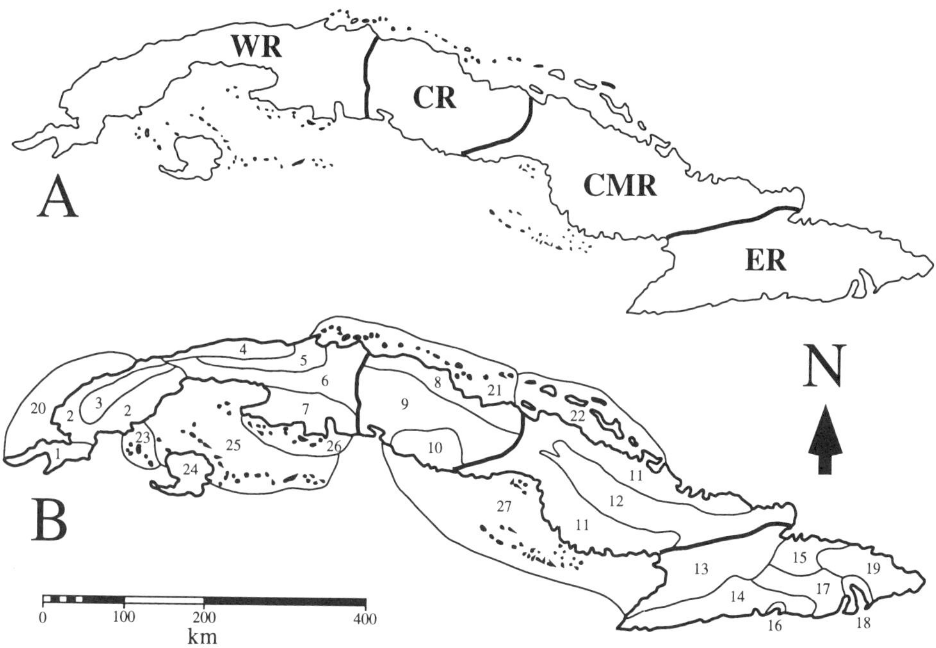

Figure 2.2. **(A)** Geographic regions of Cuba. WR, western region; CR, central region; CMR, Camagüey-Maniobón region; ER, eastern region. **(B) Subregions**: 1, Llanura y Península de Guanahacabibes; 2, Llanuras norte y sur de Pinar de Río; 3, Cordillera de Guaniguanico; 4, Alturas de La Habana-Matanzas; 5, Alturas de Bejucal-Coliseo; 6, Llanura Artemisa-Colón; 7, Llanura y Península de Zapata; 8, Llanura Corralillo-Yaguajay; 9, Alturas y Llanos de Cubancán; 10, Sierra de Guamuhaya; 11, Llanuras de norte y sur de Camagüey; 12, Alturas y Llanos de Camagüey-Maniabón; 13, Llanura de Cauto; 14, Sierra Maestra; 15, Alturas de Nipe-Cristal; 16, Cuenca de Santiago de Cuba; 17, Valle Central; 18, Cuenca de Guantánamo; 19, Montañas de Sagua-Baracoa; 20, Archipiélago de los Colorados; 21, Archipiélago de Sabana; 22, Archipiélago de Camagüey; 23, Cayos de San Felipe; 24, Isla de la Juventud; 25, Archipiélago de los Canarreos; 26, Cayos Blancos del Sur; 27, Archipiélago de Jardines de la Reina (Mateo and Acevedo, 1989).

The Eastern region also includes the marine terraces of the Meseta de Maisí (633 m), the Sierra de Purial (1059 m), Imías (1176 m) and the Meseta de Guaso (890 m). Other subregions are the Isla de la Juventud in the south and along the north coast the archipelagos of the Colorados, Sabana, and Camagüey. Along the south coast the abundant keys of the Canarreos and the Jardines de la Reina. Other keys are located to the northwest of the Isla de la Juventud (the Cayos de San Felipe), and to the south of the Zapata Peninsula (the Cayos Blancos de Sur).

Vegetation

The map of the natural vegetation of Cuba provided by Capote *et al.* (1989) recognizes 30 different botanical associations. In a simpler form the vegetation can be classified into four major groups: (i) broad-leaf tropical forests; (ii) tropical coniferous forests; (iii) tropical thickets; and (iv) herbaceous communities. The broad-leaf tropical forests contain two major divisions: (i) evergreen forests and (ii) semi-deciduous forests. The coniferous forests are characterized by four species of pines. In the west and Isla de la Juventud contain *Pinus caribea* and *P. tropicalis*, whereas the eastern portion of Cuba contain *P. cubensis* and *P. maestrensis*. The Central regions of Camagüey-Maniobón do not have natural pines. The thickets include the dry coastal associations, the vegetation on serpentine soils, and on the limestone mogotes. The herbaceous communities include the swamp vegetation formed by *Cladium jamaicense* and *Typha dominguensis* .

During the nineteenth and twentieth centuries, Cuba was transformed from a predominantly forested island into arable agricultural land. Smith (1954) estimates that pre-Columbian Cuba was 60% forest and that only 15% of the forest survived. Similarly, Seifriz (1943) concluded that “little, if any, of Cuba’s arable land has been left uncultivated”. Since 1989 deforestation has increased due to the economic crisis and the lack of fossil fuels on the island. How this ecological transformation has affected the herpetofauna has gone unstudied. It is obvious that such extreme deforestation has modified the distribution of many species.

Amphibians

Of the 57 know species, 7 are bufonids and there is 1 hylid, 46 leptodactyls, and 1 introduced ranid (Table 2.1; Fig. 2.3). Two other species have been reported as introduced into Cuba, the bufonid *Bufo marinus* and the hylid *Pseudacris crucifer*.

Table 2.1. Cuban Amphibians[a]

Species	Status	Distribution
CLASS AMPHIBIA, ORDER ANURA		
Family Bufonidae		
Bufo cataulaciceps	Endemic	WR, IJ
Bufo empusa	Endemic	IW, IJ
Bufo fustiger	Endemic	WR
Bufo gundlachi	Endemic	IW, IJ
Bufo longinasus	Endemic	
B. l. longinasus		WR
B. l. cajalbanensis		WR
B. l. dunni		CR
B. l. ramsdeni		ER
Bufo peltocephalus		CR-ER, IJ, ASC
Bufo taladai	Endemic	CR-ER
Family Leptodactylidae		
Eleutherodactylus acmonis	Endemic	ER
Eleutherodactylus albipes	Endemic	ER
Eleutherodactylus atkinsi	Endemic	
E. a. atkinsi		IW, IJ, ASC
E. a. estradai		ER
Eleutherodactylus auriculatus	Endemic	IW, IJ
Eleutherodactylus bartonsmithi	Endemic	ER
Eleutherodactylus bresslerae	Endemic	ER
Eleutherodactylus casparii	Endemic	CR
Eleutherodactylus cubanus	Endemic	ER
Eleutherodactylus cuneatus	Endemic	ER
Eleutherodactylus dimidiatus	Endemic	
E. d. dimidiatus		IW
E. d. amelasma		ER
Eleutherodactylus eileenae	Endemic	WR-CMR
Eleutherodactylus emiliae	Endemic	CR
Eleutherodactylus etheridgei	Endemic	ER
Eleutherodactylus glamyrus	Endemic	ER
Eleutherodactylus goini	Endemic	WR
Eleutherodactylus greyi	Endemic	CR
Eleutherodactylus guanahacabibes	Endemic	WR
Eleutherodactylus guantanamera	Endemic	ER
Eleutherodactylus gundlachi	Endemic	ER
Eleutherodactylus iberia	Endemic	ER
Eleutherodactylus intermedius	Endemic	ER

Table 2.1 cont.

Eleutherodactylus ionthus	Endemic	ER
Eleutherodactylus jaumei	Endemic	ER
Eleutherodactylus klinikowskii	Endemic	WR
Eleutherodactylus leberi	Endemic	ER
Eleutherodactylus limbatus	Endemic	IW
Eleutherodactylus orientalis	Endemic	ER
Eleutherodactylus mariposa	Endemic	ER
Eleutherodactylus melacara	Endemic	ER
Eleutherodactylus pezopetrus	Endemic	ER
Eleutherodactylus pinarensis	Endemic	WR, IJ
Eleutherodactylus planirostris	Cuba and Bahamas	
E. p. planirostris		IW, IJ, ASC
Eleutherodactylus principalis	Endemic	ER
Eleutherodactylus ricordii	Endemic	ER
Eleutherodactylus sp. n.	Endemic	WR-CMR, IJ
Eleutherodactylus ronaldi	Endemic	ER
Eleutherodactylus symingtoni	Endemic	WR
Eleutherodactylus tetajulia	Endemic	ER
Eleutherodactylus thomasi	Endemic	WR-CMR
E. t. thomasi		CMR
E. t. trinidadensis		CR
E. t. zayasi		WR
Eleutherodactylus toa	Endemic	ER
Eleutherodactylus tonyi	Endemic	ER
Eleutherodactylus turquinensis	Endemic	ER
Eleutherodactylus varians	Endemic	IW
E. v. varians		CR-ER
E. v. olibrus		WR
E. v. staurometopon		IJ
Eleutherodactylus varleyi	Endemic	IW
Eleutherodactylus zeus	Endemic	WR
Eleutherodactylus zugi	Endemic	WR
E. z. zugi		WR
E. z. erythroproctus		WR
Family Hylidae		
Osteopilus septentrionalis	Cuba and Bahamas	IW
Family Ranidae		
Rana catesbeiana	Introduced	IW

[a]The letters in the distribution column refer to the following geographic regions: WR, Western region; CR Central region; CMR, Camagüey-Maniabón región; ER Eastern region; ASC Archipié-

lago de Sabana-Camagüey; IW, Islandwide and IJ, Isla de la Juventud (Mateo and Acevedo, 1989).

Bufo marinus was introduced in 1946 (Buide, 1967) and, unlike on the other Antillean islands, the introduction was unsuccessful in Cuba. Currently, there is no evidence of a resident population of *B. marinus* on Cuba. The existence of the populations of *P. crucifer* reported in Cuba by Schwartz and Thomas (1975) and by Schwartz and Henderson (1988, 1991) have never been confirmed. For more than 10 years attempts to locate the frogs near Canasí and Marianao have failed. Consequently, neither of these species have been included in the list of Cuban amphibians.

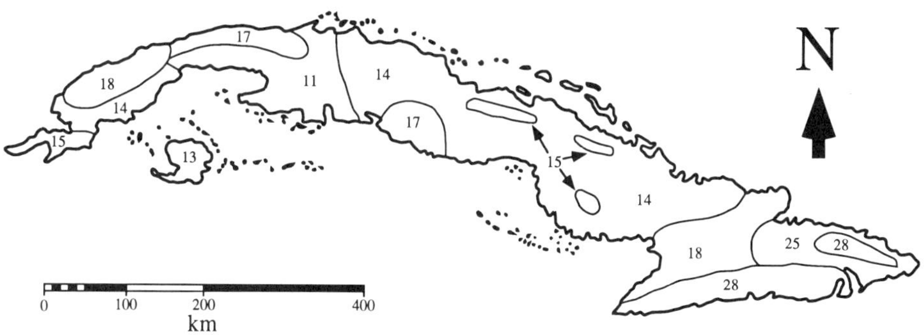

Figure 2.3. Estimated number of amphibian species in the different regions of Cuba. The numbers indicate the number of sympatric species and exclude species known only from a single locality.

The seven bufonid species are endemic to Cuba. *Bufo gundlachi* and *B. empusus* have a wide distribution in the lowlands and also on Isla de la Juventud and on the Coco, Romano, and Guajaba keys of the Sabana-Camagüey archipelago. *Bufo longinasus* is a polytypic species with four disjunct populations: two in the west, *B. l. longinasus* in the highlands of Pizarras del Sur, between the city of Pinar del Río and the Valle de Viñales, and *B. l. cajalbanensis* in the Cajálbana plateau (Valdés and Ruíz, 1980). *Bufo l. dunni* inhabits the mountains of the Sierra de Guamuhaya in the areas of Trinidad as well as Sancti Spíritu (Valdés and Ruíz, 1980; Estrada, 1994). *Bufo l. ramsdeni* is the eastern-most form and inhabits the areas of Monte Verde and Monte Líbano in the Guaso plateau (Valdés and Ruíz, 1980). All the populations of *B. longinasus* are associated with rivers and streams. *B. cataulaciceps* lives in the sandy plains in the west and south of the province of Pinar del Río and the northern part of the Isla de Juventud.

Bufo fustiger has a broad distribution in the lowlands and mountains of western Cuba, from the Guanahacabibes Peninsula to the borders of the Llanura de Zapata. *Bufo peltocephalus* resides in lowlands and moderate elevations from the Central region to extreme eastern Cuba, including the Santa María, Coco, Parédon Grande, Romano, Guajaba and Sabinal keys (Schwartz and Henderson, 1991). *Bufo taladai* is found in the lowlands and mountains of the Eastern region and some populations are known in the Central region in the Sierra Guamuhaya (Schwartz and Henderson, 1991).

The only members of the Hylidae is *Osteopilus septentrionalis*. Although it is not endemic to Cuba, it is widely distributed in Cuba and is found in lowlands, mountains, forests, savannas, mangroves, and agricultural areas. It is present on the Isla de la Juventud and the keys of Sabana-Camagüey and Canarreos. Outside of Cuba it is indigenous to the Bahamas and Cayman Islands and has been introduced to Puerto Rico, the Virgin Islands, and the keys and mainland of Florida.

The bulk of the amphibian species are leptodactylids in the genus *Eleutherodactylus*. Dunn (1926a) divided the Cuban *Eleutherodactylus* into four groups: *auriculatus*, *dimidiatus*, *ricordi*, and *varleyi*. Schwartz (1957) added a fifth group, *symingtoni*. Hedges (1989) performed a phylogenetic synthesis of Antillean *Eleutherodactylus* and proposed three subgeneric categories: *Euhyas*, *Pelorius*, and *Eleutherodactylus*. The Cuban species fall within two subgenera: 12 species in *Eleutherodactylus* and 28 species in *Euhyas*. The species show four patterns of distribution within Cuba: islandwide (7 species), western (7 species), central (3 species), from the Western region to Camagüey-Maniabón (3 species), and the Eastern region (26 species).

Some of the western species are characterized by having local narrow distributions. *Eleutherodactylus guanahacabibes* is endemic to the Guanahacabibes Peninsula (Estrada and Novo, 1985b); *E. zeus* is only found in the Sierra de los Organos (Estrada *et al.*, 1989); *E. klinikowski* is known only from the Sierra de los Organos and a few localities in the Sierra del Rosario (Schwartz and Henderson, 1991); and *E. symingtoni* is only known from five localities from the Sierra de los Organos to the Alturas La Habana-Matanzas (Estrada *et al.*, 1989). *Eleutherodactylus zugi* has two disjunct subspecies (Schwartz, 1960) with populations in the Cordillera de Guaniguanico (*E. z. zugi*) and in the Alturas La Habana-Matanzas (*E. z. erythroproctus*). *Eleutherodactylus pinarensis* exhibits an interesting pattern of distribution: Llanura de Guanahacabibes, Isla de la Juventud, and the rest of the Western region excluding the Llanura de Pinar del Río and the Cordillera de Guaniguanico (Estrada and Novo, 1985a). *E. goini*, a species once considered a subspecies of *E. planirostris* (Estrada and Hedges, 1997b) is found in the forest spurs of the Cordillera de Guaniguanico.

The three species with an exclusively Central region distribution are: *E. caspari*, *E. emiliae* (both endemic to the Sierra de Guamuhaya), and *E. greyi*, which also has some populations in Camagüey-Maniabón region in the Sierra de Cubitas and Sierra de Najasa (Schwartz and Henderson, 1991). From the Western region to Camagüey-Maniabón are found *E. eileenae*, *E. sp. n. and E. thomasi*. *Eleutherodactylus sp. n.* was long confused with *E. cuneatus*, a species with similar riparian habits (Estrada and Hedges, 1998). *Eleutherodactylus thomasi* has three subspecies: *E. t. thomasi* in the Cubitas and Najasa mountains of Camagüey-Maniabón, *E. t. trinidadensis* on the south slopes of the Sierra de Guamuhaya, and *E. t. zayasi* from the Pan de Matanzas in the Alturas de La Habana-Matanzas. Additionally, some populations that have not been subspecifically identified are known from north of the Central region and from Cayo Coco (Schwartz and Henderson, 1991; Estrada, 1993b).

Most species of *Eleutherodactylus* have an Eastern region distribution and 18 of these (*E. albipes*, *E. bartonsmithii*, *E. bresslerae*, *E. cubanus*, *E. etheridgei*, *E. glamyrus*, *E. iberia*, *E. intermedus*, *E. jaumei*, *E. leberi*, *E. orientalis*, *E. mariposa*, *E. melacara*, *E. pezopetrus*, *E. principalis*, *E. tetajulia*, *E. tonyi*, and *E. turquinensis*) are known from a single locality or from a very limited area (Estrada and Alonso, 1997; Estrada and Hedges, 1996a, 1997a; Hedges *et al.*, 1992a, 1995; Schwartz and Henderson, 1991). The other 8 Eastern region species have a broader distribution in the region (*E. acmonis*, *E. cuneatus*, *E. guantanamera*, *E. gundlachi*, *E. ionthus*, *E. ricordii*, *E. ronaldi*, and *E. toa*).

Seven species have a pan-Cuban distribution: *E. atkinsi*, *E. auriculatus*, *E. dimidiatus*, *E. limbatus*, *E. planirostris*, *E. varians*, and *E. varleyi* (Schwartz and Henderson, 1991). There are two subspecies of *E. atkinsi*, one having a distribution throughout Cuba, Isla de la Juventud, and the keys of Sabana-Camagüey (*E. a. atkinsi*) and the other (*E. a. estradai*) located in the Sagua-Baracoa mountains in the Eastern region (Schwartz and Henderson, 1991; Lynch, 1991). Three subspecies are recognized for *E. varians*; *E. v. varians* from the eastern portion of the Llanura de Zapata to the Camagüey-Maniabón region, *E. v. olibrus* from Cordillera de Guaniguanico, and *E. v. staurometopon* from the Isla de la Juventud. The populations of *E. varians* from the Llanura de Guanahacabibes and the Alturas de La Habana-Matanzas have not been assigned to any of the described subspecies (Schwartz and Henderson, 1991).

The only ranid species, *Rana catesbeiana* was introduced at the beginning of the nineteenth century and currently is widely distributed throughout Cuba and the Isla de la Juventud (Schwartz and Henderson, 1991).

Reptiles

There are 122 recognized species of reptiles in Cuba (Table 2.2; Fig. 2.4). There is a single pond turtle, *Trachemys decussata*, that is widely distributed in the swamps, rivers, ponds and dams of the entire archipelago, including most of the keys.

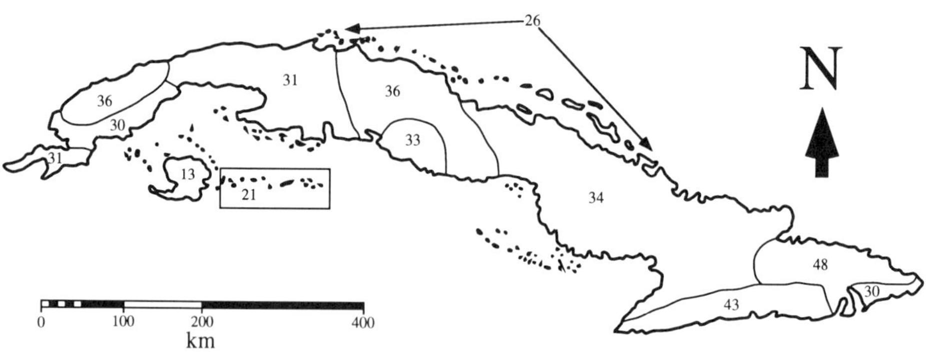

Figure 2.4. Estimated number of reptilian species in the different regions of Cuba. The numbers indicate the number of sympatric species and exclude species known only from a single locality.

Table 2.2. Cuban Reptiles[a]

Species	Status	Distribution
CLASS REPTILIA, ORDER TESTUDINES		
Family Emydidae		
Trachemys decussata	Cuba, Cayman Is.	
T. d. decussata	Endemic	WR-ER
T. d. angusta		WR, IJ
ORDER SQUAMATA, SUBORDER AMPHISBAENIA		
Family Amphisbaenidae		
Amphisbaena blanoides	Endemic	ER
Amphisbaena cubana	Endemic	ER
A. c. cubana		CR-ER, IJ
A. c.barbouri		
Amphisbaena palirostrata	Endemic	IJ
SUBORDER LACERTILIA		
Family Anguidae		
Diploglossus delasagra	Endemic	IW,IJ
Diploglossus nigropunctatus	Endemic	ER
Diploglossus garridoi	Endemic	ER
Family Gekkonidae		
Gonatodes albogularis	Cuba, Jamaica, Cayman Is., and Hispaniola	
G. a. fuscus		IW
Hemidactylus brooki	Cuba, Hispaniola and Puerto Rico	
H. b. brooki		IW
Hemidactylus maboui	Cuba, Hispaniola and Puerto Rico	WR, ER
Hemidactylus turcicus	Introduced	
H. t. turcicus		WR, CR, ER
Sphaerodactylus armasi	Endemic	ER
Sphaerodactylus argus	Cuba and Jamaica	
S. a. argus		WR-CMR, AJR
Sphaerodactylus bromeliarum	Endemic	ER
Sphaerodactylus celicara	Endemic	ER
Sphaerodactylus cricoderus	Endemic	ER
Sphaerodactylus docimus	Endemic	ER
Sphaerodactylus elegans	Endemic	IW, IJ
Sphaerodactylus intermedius	Endemic	WR
Sphaerodactylus nigropunctatus	Cuba and Bahama Is.	
S. n. alayoi		ER

Table 2.2 cont.

S. n. granti		ASC
S. n. lissodesmus		CMR
S. n. ocujal		ER
S. n. strategus		ER
Sphaerodactylus notatus	Cuba and Bahama Is.	
S. n. atactus		IW, IJ
Sphaerodactylus oliveri	Endemic	CR
Sphaerodactyllus ramsdeni	Endemic	ER
Sphaerodactylus richardi	Endemic	WR
Sphaerodactylus ruibali	Endemic	ER
Sphaerodactylus scaber	Endemic	CR-CMR, AS
Sphaerodactylus storeyae	Endemic	IJ, ACA
Sphaerodactylus schwartzi	Endemic	ER
Sphaerodactylus torrei	Endemic	
S. t. torrei		ER
S. t. spielmani		ER
Tarentola americana	Cuba and Bahama Is.	
T. a. americana		IW, ASC
Family Iguanidae		
Cyclura nubila	Cuba and Cayman Is.	
C. n. nubila	Endemic	IW, IJ
Family Polychrotidae		
Anolis ahli	Endemic	CR
Anolis alayoni	Endemic	ER
Anolis alfaroi	Endemic	ER
Anolis allisoni	Cuba and	IW
Anolis allogus	Endemic	IW
Anolis alutaceus	Endemic	IW, IJ
Anolis anfiloquioi	Endemic	ER
Anolis angusticeps	Cuba and Bahama Is.	
A. a. angusticeps		IW, IJ
Anolis argenteolus	Endemic	CMR-ER
Anolis argillaceus	Endemic	ER
Anolis baracoae	Endemic	ER
Anolis barbatus	Endemic	WR
Anolis bartchi	Endemic	WR
Anolis birama	Endemic	ER
Anolis bremeri	Endemic	
A. b. bremeri		WR

Table 2.2 cont.

A. b. insulsepinorum		IJ
Anolis centralis	Endemic	
A. c. centralis		CR-CMR
A. c. litoralis		ER
Anolis chamaelionides	Endemic	IW
Anolis clivicola	Endemic	ER
Anolis confusus	Endemic	ER
Anolis cupeyalensis	Endemic	ER
Anolis cyanopleurus	Endemic	
A. c. cyanopleurus		ER
A. c. orientalis		ER
Anolis delafuentei	Endemic	CR
Anolis equestris	Endemic	
A. e. equestris		WR
A. e. buidei		WR
A. e. cincoleguas		AS
A. e. juraguensis		CR
A. e. persparsus		CR
A. e. potior		AC
A. e. thomasi		CMR
A. e. verreonensis		ER
Anolis fugitivus	Endemic	ER
Anolis garridoi	Endemic	CR
Anolis guafe	Endemic	ER
Anolis guamuhaya	Endemic	CR
Anolis guazuma	Endemic	ER
Anolis homolechis	Endemic	
A. h. homolechis		IW, IJ
A. h. turquinensis		ER
Anolis imias	Endemic	ER
Anolis inexpectatus	Endemic	ER
Anolis isolepis	Endemic	
A. i. isolepis		CR-ER
A. I. altitudinalis		ER
Anolis juangundlachi	Endemic	WR
Anolis jubar	Endemic	
A. j. jubar		CMR
A. j. albertschwartzi		ER
A. j. balaenarum		AC
A. j. cocoensis		AC

Table 2.2 cont.

A. j. cuneus		CMR
A. j.gibarensis		CMR-ER
A. j. maisensi		ER
A. j. oriens		ER
A. j. santamariae		AS
A. j. yaguajayensis		CR
Anolis loysianus	Endemic	IW
Anolis lucius		IW
Anolis luteogularis	Endemic	
A. l. luteogularis		WR
A. l. calceus		WR
A. l. coctilis		ACA
A. l. delacruzi		IJ
A. l. hassleri		IJ
A. l. jaumeai		WR
A. l. nivevultus		WR
A. l. sanfelipensis		CSF
A. l. sectilis		IJ
Anolis macilentus	Endemic	ER
Anolis mestrei	Endemic	ER
Anolis mimus	Endemic	ER
Anolis noblei	Endemic	
A. n. noblei		ER
A. n. galeifer		ER
Anolis ophiolepis	Endemic	IW
Anolis paternus	Endemic	
A. p. paternus		IJ
A. p. pinarensis		WR
Anolis pigmaequestris	Endemic	AS
Anolis porcatus	Endemic	IW, IJ
Anolis porcus	Endemic	IW
Anolis pumilus	Endemic	WR-CR-AS-AC
Anolis quadriocellifer	Endemic	WR
Anolis rubribarbus	Endemic	ER
Anolis sagrei	Cuba, Jamaica, Bahama Is., NA	
A. s. sagrei		IW
A. s. greyi		CMR
Anolis smallwoodi	Endemic	ER
A. s. smallwoodi		

Table 2.2 cont.

A. s. palardis		
A. s. saxuliceps		
Anolis spectrum	Endemic	WR
Anolis vanidicus	Endemic	
A. v. vanidicus		CR
A. v. rejectus		ER
Anolis vermiculatus	Endemic	WR
Anolis vescus	Endemic	ER
Family Teiidae		
Ameiva auberi	Cuba and Bahama Is.	
A. a. auberi	Endemic	WR
A. a. abducta	Endemic	WR
A. a. atrothorax	Endemic	CR
A. a. cacuminus	Endemic	WR
A. a. citra	Endemic	CMR
A. a. denticola	Endemic	WR
A. a. extorris	Endemic	ASC
A. a. extraria	Endemic	ASC
A. a. galbiceps	Endemic	AJR
A. a. garridoi	Endemic	CR
A. a. gemmea	Endemic	CR
A. a. granti	Endemic	ER
A. a. hardyi	Endemic	ER
A. a. llanensis	Endemic	CMR
A. a. marcida	Endemic	IJ
A. a. nigriventris	Endemic	IJ
A. a. orlandoi	Endemic	ASC
A. a. paulsoni	Endemic	WR
A. a. peradusta	Endemic	CR
A. a. procer	Endemic	WR
A. a. pullata	Endemic	WR
A. a. sabulicolor	Endemic	ER
A. a. sanfelipensis	Endemic	CSF
A. a. schwartzi	Endemic	WR
A. a. secta	Endemic	IJ
A. a. sublesta	Endemic	ASC
A. a. ustulata	Endemic	ER
A. a. zugi	Endemic	ACA, WR

Table 2.2 cont.

Family Tropiduridae		
Leiocephalus carinatus	Cuba, Cayman and Bahama Is.	
L. c. carinatus	Endemic	WR
L. c. acuarius	Endemic	ER
L. c. cayensis	Endemic	AJR
L. c. labrossytus	Endemic	WR-CR
L. c. microcyon	Endemic	IJ
L. c. mogotensis	Endemic	WR
L. c. zayasi	Endemic	WR
Leiocephalus cubensis	Endemic	
L. c. cubensis		IW
L. c. gigas		IJ
L. c. minor		CSF
L. c. pambasileus		ACA
L. c. paraphrus		AJR
Leiocephalus macropus	Endemic	
L. m. macropus		ER
L. m. aegialus		CMR
L. m. asbolomus		ER
L. m. felinoi		WR
L. m. hoplites		CR-CMR
L. m. hyacinthurus		CR
L. m. immaculatus		ER
L. m. koopmani		WR
L. m. lenticulatus		CMR
L. m. phylax		ER
L. m. torrei		WR
Leiocephalus onaneyi	Endemic	ER
Leiocephalus raviceps	Endemic	
L. r. raviceps		ER
L. r. delavarai		ER
L. r. jaumei		WR
L. r. klinikowskii		WR
L. r. uzzelli		ER
Leiocephalus stictigaster	Endemic	
L. s. stictigaster		WR
L. s. astictus		IJ
L. s. celeustes		ER
L. s. exothetus		IJ

Table 2.2 cont.

L. s. gibarensis		CMR-ER
L. s. lipomator		CR
L. s. Lucianus		CMR
L. s. naranjoi		CR
L. s. ophioplacodes		CMR
L. s. parasphex		AC
L. s. septentrionalis		ASC
L. s. sierrae		WR
Family Xantusiidae		
Cricosaura typica	Endemic	ER
ORDEN SERPENTES		
Family Boidae		
Epicrates angulifer	Endemic	IW
Family Colubridae		
Alsophis cantherigerus	Cuba and Cayman Is.	
A. c. cantherigerus	Endemic	WR-CR, IJ, ACA
A. c. adspersus	Endemic	ER
A. c. pepei	Endemic	ER
A. c. schwartzi	Endemic	CR, CMR, ER AC
Antillophis andreai	Endemic	
A. a. andreai		WR-CMR
A. a. melopyrrha		ACA
A. a. morenoi		ASC
A. a. nebulatus		IJ
A. a. orientalis		CMR-ER
A. a. peninsulae		WR
Arrhyton ainictum	Endemic	ER
Arrhyton dolichura	Endemic	WR
Arrhyton landoi	Endemic	ER
Arrhyton procerum	Endemic	WR
Arrhyton supernum	Endemic	ER
Arrhyton taeniatum	Endemic	WR, CR, ER, IJ
Arrhyton tanyplectum	Endemic	WR
Arrhyton vittatum	Endemic	IW, IJ
Nerodia clarki	Cuba	
N. c. compressicauda	Gulf Coast, Cuba and Florida Keys	WR-CR
Tretanorhinus variabilis	Cuba and Cayman Is.	

Table 2.2 cont.

T. v. variabilis	Endemic	WR, CR, ER
T. v. binghami	Endemic	ER
T. v. insulaepinorum	Endemic	IJ
T. v. wagleri	Endemic	WR
Family Tropidophidae		
Tropidophis feicki	Endemic	WR
Tropidophis fuscus	Endemic	ER
Tropidophis haetianus	Cuba and Haiti	
T. h. haetianus		ER
Tropidophis maculatus	Endemic	WR
Tropidophis melanurus	Cuba and Cayman Is.	
T. m. melanurus	Endemic	IW
T. m. dysodes	Endemic	WR
T. m. ericksoni	Endemic	IJ
Tropidophis nigriventris	Endemic	
T. n. nigriventris		CMR
T. n. hardyi		CR
Tropidophis pardalis	Endemic	IW, IJ
Tropidophis pilsbryi	Endemic	
T. p. pilsbryi		ER
T. p. galacelidus		CR
Tropidophis semicinctus	Endemic	WR
Tropidophis wrighti	Endemic	CR-ER
Family Typhlopidae		
Typhlops biminiensis	Cuba and Bahama Is.	
T. b. biminiensis		WR, CR, ER
Typhlops lumbricalis	Cuba and Bahama Is.	IW, IJ
ORDER CROCODYLIA		
Family Alligatoridae		
Caiman crocodilus	Introduced	IJ
Family Crocodylidae		
Crocodylus acutus		IW, IJ
Crocodylus rhombifer	Endemic	WR

[a]The letters in the Distribution columns refer to the following geographic regions; WR, Western region; CR, Central region; CMR, Camagüey-Maniabón region; ER, Eastern region; AS, Archipiélago de Sabana; AC, Archipiélago de Camagüey; ASC, Archipiélago de Sabana-Camagüey; CSF, Cayos de San Felipe; ACA, Archipiélago de los Canarreos; AJR, Archipiélago de Jardines de la Reina; IW, Islandwide; IJ, Isla de la Juventud (Mataeo and Acevedo, 1989).

There are two subspecies: *T. d. decussata*, extending from the east end to the western end of Cuba, including the keys of Sabana-Camagüey, and *T. d. angusta* is found in the extreme east end of Cuba, on the Isla de la Juventud, and also on the Cayman Islands (Seidel, 1988a,b, 1990).

Of the remaining 121 reptiles, 93 are saurians. There are three species of *Amphisbaena*. *Amphisbaena cubana* has two subspecies: *A. c. cubana* found from the Ciénaga de Zapata to the Eastern region and Isla de la Juventud and *A. c. barbouri* distributed in the Western region (excluding the western Ciénaga de Zapata). *Amphisbaena blanoides* is common in the Cordillera de Guaniguanico and *A. palirostrata* is only known from the northern portion of the Isla de la Juventud. These last two species were previously included in the genus *Cadea*, however, Hedges (1996b) considered *Cadea* synonymous with *Amphisbaena*.

Three species of anguids are present and all belong to the genus *Diploglossus*. The distribution of *D. delasagra* includes the Western and Central regions, the Alturas del Centro de Camagüey, and the southern part of the Eastern region. *Diploglossus nigropunctatus* is found in the northern part of the Camagüey-Maniabón region and the mountains of Sagua-Baracoa. The third species, *D. garridoi* is from the northern slopes of the Sierra Maestra (Hedges and Thomas, 1998).

The Gekkonidae is represented by four genera: *Gonatodes*, *Hemidactylus*, *Sphaerodactylus*, and *Tarentola*. There is a single species of *Gonatodes*: *G. albogularis fuscus* which is found in some urban areas in Cuba and is a species widely distributed in Central and South America. *Hemidactylus brooki haitianus* is found throughout Cuba, including the keys of Sabana-Camagüey, and also occurs in Hispaniola and Puerto Rico (Schwartz and Henderson, 1991). Another member of this genus, *H. mabouia*, is known from Guantánamo and the city of La Habana, and is also found in Hispaniola, Bahamas, Virgin Islands, Puerto Rico, Mona, Culebra, and Vieques (Schwartz and Henderson, 1991). The third species, *H. turcicus turcicus*, presumably introduced into Cuba, is found in the cities of La Habana, Matanzas, Santa Clara, Cienfuegos, Santiago de Cuba, and Guantánamo.

The genus *Sphaerodactylus* is represented by 18 species, but 4 of these are not endemic to Cuba. The Cuban *Sphaerodactylus* have been classified into seven morphological groups (Thomas *et al.*, 1992; Hedges and Garrido, 1993): (1) the *nigropunctatus* complex contains *S. nigropunctatus* and *S. torrei*. Both are large species with granular scales and a pattern of bands and are sexually dimorphic in color; (2)

S. elegans is similar to the prior complex, but does not have color differences between the sexes; (3) the *intermedius* complex has four species, *S. intermedius*, *S. armasi*, *S. docimus*, and *S. ruibali*, all characterized by imbricate scales and a less prominent pattern of bands; (4) the *scaber* complex has four species, *S. scaber*, *S. oliveri*, *S. richardi*, and *S. storeyae*, all are medium sized, with bulky dorsal scales and a middorsal zone of small granular scales, and all are sexually dichromatic; (5) the *notatus* complex has three species, *S. notatus*, *S. bromeliarum*, and *S. celicara*, are all medium-size lizards with flat imbricate dorsal scales (there are various species of this complex in Hispaniola); (6) the *ramsdeni* complex has three species, *S. ramsdeni*, *S. cricoderus*, and *S. schwartzi*, and all have dark dorsal scales that do not overlap and more than one internasal scale; and (7) *S. argus* is a species with small imbricate dorsals and an ocellated color pattern.

Sphaerodactylus argus argus is known from south coastal localities of the Central, Camagüey-Maniabón, and Eastern regions, including the keys of the Jardines de la Reina. This subspecies is also found in the northern Bahamas and is broadly distributed on Jamaica. *Sphaerodactylus a. andresensis* is found on the island of San Andrés in the Gulf of Honduras (Schwartz and Henderson, 1991). *Sphaerodactylus elegans elegans* has a wide distribution on Cuba, the keys of Sabana-Camagüey, Canarreos, and Jardines de la Reina, as well as on Isla de la Juventud. There are also presumably introduced populations on the Florida Keys (Schwartz and Henderson, 1991). Another subspecies, *S. e. punctatissimus* inhabits the western part of Hispaniola and the island of Gonave. *Sphaerodactylus nigropunctatus* is a polytypic species with at least five subspecies in Cuba and another five in the Bahamas. The Cuban subspecies are: *S. n. alayoi* is found on the edge of the Guaso plateau and the eastern slopes of the Cuenca de Guantánamo; *S. n. granti* is found in the Sabana-Camagüey keys, the Alturas de Maniabón-Banes, the mountains of Nipe-Cristal, and the northeast of the Cauto Valley; *S. n. lissodesmus* is known only from the Sierra de Cubitas in Camagüey; *S. n. ocujal* is found on the southern slopes of the Sierra Maestra and the Meseta de Cabo Cruz; and *S. n. strategus* is found on the coastal slopes of the Guantánamo valley (Schwartz and Henderson, 1991). Of the four subspecies recognized for *S. notatus*, only *S. n. atactus* is found at isolated localities in the Western, Central, Camagüey-Maniabón, and Eastern regions. It is most common in the Eastern region. The populations in the San Felipe keys, Isla de la Juventud, Canarreos keys, and various localities in the Eastern region show marked differences from the *atactus* morphology. *Sphaerodactylus intermedius* is distributed across the Alturas de La Habana-Matanzas; *S. richardi* is found along the length of the coastal forests between the bays of Cochinos and Cienfuegos (Hedges and Garrido, 1993). *Sphaerodactylus oliveri* is located on the spurs of the Sierra de Trinidad. *Sphaerodactylus storeyae* was described by Grant (1944a) but Schwartz (1961), Schwartz and Garrido (1981a), and Schwartz and Henderson (1991) treated it as a subspecies of *S. oliveri*. Hedges and Garrido (1993) proposed considering it a separate species. It is restricted to the Isla de la Juventud and the Canarreos keys. *Sphaerodactylus scaber* is distributed through the Llanura Corralillo-Yaguajay and the Llanura Sur de Camagüey. The rest of the species are found in the Eastern region: *S. armasi*, *S. bromeliarum*, and *S. celicara* in the mountains of Sagua-Baracoa; *S. cricoderus* on the north and south slope of the Sierra Maestra; and *S. docimus* on the South slopes of the neseta de Cabo Cruz and the eastern end of the Sierra Maestra (Schwartz and Henderson, 1991; Thomas *et al.*, 1992). *Sphaerodactylus schwartzi* and *S. ruibali* are distributed to the west and east of the Bahia de Guantánamo. *Sphaerodactylus ramsdeni* has two populations, one in the Sierra de Gran Piedra and another more than 40 km to the northeast at the Meseta de Guaso (Schwartz and Garrido, 1985; Thomas *et al.*, 1992). *Sphaerodactylus torrei* has two subspecies: *S. t. torrei* in the vicinity of Santiago de Cuba and *S. t. spielmani* on the west of Guantánamo bay (Schwartz and Garrido, 1985).

Cyclura nubila has a wide distribution throughout Cuba along the coastal regions and on the keys, including the Isla de la Juventud. The populations of *Cyclura* have decreased due to development and construction along the coast. There are some interior populations in the Sierra de los Organos and the Alturas de La Habana-Matanzas. Three subspecies are recognized: *C. n. nubila* on Cuba and *C. n. caymanensis* and *C. n. lewisi* on the Cayman Islands (Schwartz and Carey, 1977; Schwartz and Henderson, 1991).

The Polychrotidae is the most numerous and best represented family in Cuba with 55 anoline species, including the 4 species of *Chamaeleolis* included in the synonymy of *Anolis* by Hass *et al.* (1993). There are four patterns of distribution: Species that are exclusively western, central or eastern and species that are broadly distributed throughout the island. The broadly distributed species are: *Anolis alutaceous*, *A. allisoni*, *A. angusticeps*, *A. chamaeleonides*, *A. equestris*, *A. homolechis*, *A. loysiana*, *A. lucius*, *A. ophiolepis*, *A. porcatus* and *A. sagrei* (Schwartz and Henderson, 1991). The distribution of *A. allisoni* extends from the eastern edge of the Western region to the rest of the island. Outside of Cuba it is known from populations on the Islas de la Bahia off the north coast of Honduras and Half Moon Cay off the coast of Belize (Ruibal and Williams, 1961). *Anolis allogus* is not present in the eastern part of the Western region and is absent in the Central region (Schwartz and Henderson, 1991). *Anolis angusticeps* is represented in Cuba by the nominative subspecies *A. a. angusticeps*, whereas *A. a. ologapsi* is restricted to the Bahamas (Garrido, 1975d, Hardy, 1967). Eight subspecies have been described for *A. equestris*, but some populations in the Cordillera de Guaniguanico in western Cuba have not been identified subspecifically (Schwartz and Garrido, 1972; Garrido, 1981; Schwartz and Henderson, 1991). Garrido (1973a) described *A. homolechis turquinensis* as an altitudinal subspecies from the Pico Turquino. With respsect to *A. lucius*, populations have been reported in the Cordillera de Guaniguanico, but they require confirmation. *Anolis lucius* has also been accidentally introduced to Cayo Largo del Sur (Garrido and Jaume, 1984). *Anolis sagrei* is the most common of the Cuban anoles, is found throughout the island and keys, and has a single subspecies, *A. s. greyi*, from the Camagüey-Maniabón region (Schwartz and Henderson, 1991).

Eleven species have a distribution in the Western region: *Anolis barbatus* is in the Sierra del Rosario (Garrido, 1982a), and *A. mestrei* and *A. bartschi* are in the Cordillera de Guaniguanico (Schwartz and Henderson, 1991). *Anolis bremeri* has two subspecies, *A. b. bremeri* in the Llanura Sur of Pinar del Río and *A. b. insulaepinarum* on the Isla de la Juventud (Garrido, 1972). *Anolis juangundlachi* is known only from the vicinity of Carlos Rojas in the Llanura Artemisa-Colón (Garrido, 1975c). *Anolis luteogularis* has nine subspecies in the region, including the Isla de la Juventud and the Archipelago de los Canarreos (Schwartz and Garrido, 1972; Schwartz and Henderson, 1991). *Anolis paternus* has two subspecies, one on the Isla de la Juventud (*A. p. paternus*) and the second, *A. p. pinarensis*, from the Llanura Sur of Pinar del Río (Garrido, 1975a). The distribution of *A. pumilus* extends into some localities in the Central region and to the keys of Sabana and Camagüey on the north and also the Isla de la Juventud (Garrido, 1988). *Anolis quardiocellifer* is restricted to the Llanura de Guanahacabibes (Garrido, 1973b). *Anolis spectrum* is known from various localities in the Sierra de los Organos, a single locality near La Habana, and also in the vicinity of Carlos Rojas in the eastern part of the Llanura Artemisa-Colón (Garrido and Schwartz, 1972). *Anolis vermiculatus* is found in the vegetation along the banks of rivers and streams that arise in the Cordillera de Guaniguanico (Schwartz and Henderson, 1991).

The species restricted to the Central region are all endemics of the Sierra de Guamuhaya: *Anolis ahli*, *A. delafuentei*, *A. garridoi*, and *A. guamuhaya* (Garrido *et al.*, 1991; Diaz *et al.*, 1996).

Most species of Cuban anoles have an eastern distribution: *Anolis alayoni*, *A. alfaroi*, *A. anfiloquioi*, *A. argenteolus*, *A. argillaceus*, *A. baracoae*, *A. birama*, *A. clivicola*, *A. cupeyalensis*, *A. fugitivus*, *A. guafe*, *A. guazuma*, *A. imias*, *A. inexpectatus*, *A. macilentus*, *A. mimius*, *A. noblei*, *A. porcus*, *A. rubribarbus*, *A. smallwoodi*, and *A. vescus*. Some of these species stand out because of their restricted distibution. For example, *A. alfaroi* inhabits the grasses and ferns below the pines of La Municion in the Cuchillas de Toa, in the mountains of Sagua-Baracoa (Garrido and Hedges, 1992). *Anolis birama* is restricted to the environs of the Ciénaga de Birama to the northwest of the mouth of the Río Cauto (Garrido, 1990). *Anolis clivicola* is found in the Sierra Maestra above 1000 m in elevation (Schwartz and Henderson, 1991). *Anolis fugitivus* is known from the northeastern slopes of Monte Iberia in the mountains of Sagua-Baracoa (Garrido, 1975c). Garrido (1983) described *A. guazuma* from Pico Turquino and the species is found on the north and south slopes of the Sierra Maestra. *Anolis macilentus* is known only from the vicinity of the Río Pai in the Maseta de Guaso (Garrido and Hedges, 1992). The extreme eastern end of the Sierra Maestra is the locality for *A. mimus* (Garrido, 1975c; Schwartz and Thomas, 1975), whereas *A. vescus* has a similar narrow distribution in the vicinity of Los Calderos in the Sierra de Imías (Garrido and Hedges, 1992). In the Meseta of Cabo Cruz there are two species found exclusively in the dry forests of the zone: *A. guafe* and *A. confusus* (Estrada and Garrido, 1991). Some of the species with an eastern distribution contain various subspecies in different subregions. For example, in the case of *A. cyanopleurus*, one of the subspecies (*A. c. cyanopleurus*) is found in the northwestern part of the Sagua-Baracoa subregion while the other subspecies (*A. c. orientalis*) is restricted to the extreme southeastern portion of the subregion in the Meseta de Maisí (Garrido, 1975c). *Anolis noblei* has two subspecies, *A. n. noblei* in the Sierra Maestra and *A. n. galeifer* in the Sierra Nipe-Cristal (Schwartz and Garrido, 1972).

Other species with patterns of distribution different form those mentioned are *A. centralis*, *A. isolepis*, and *A. jubar*. In the case of *A. centralis* it has a subspecies in the Camagüey-Maniabón region, *A. c. centralis*, and another, *A. c. litoralis* ranging from the southern part of the Sierra Maestra, Cuenca de Santiago de Cuba, Valle Central, Cuenca de Guantánamo, to the southern part of Sagua-Baracoa (Garrido, 1975b). *Anolis isolepis* has two subspecies: *A. i. isolepis* is found in the Guamuhaya mountains and various localities in the Eastern region, and no differences have been documented distinguishing these separate populations. *Anolis i. altitudinalis* is on Pico Turquino in the Sierra Maestra (Garrido, 1985). Of the nine subspecies of *Anolis jubar*, one (*A. j. yaguajayenis*) is from the Central region in the Llanura Corralillo-Yaguajay; two subspecies are found in the keys off the north coast, *A. j. cocoensis* on the Coco, Paredón Grande and Romano keys, and *A. j. santamariae* on Cayo Santa Maria; the nominative subspecies, *A. j. jubar,* is found in part of the Llanura Norte de Camagüey; *A. j. balaenarum* is found on the Los Ballenatos keys in the Bahía de Nuevitas; *A. j. cuneus* is in the eastern half of the Llanura Norte of Camagüey; and in the same region *A. j. gibarense* is found in the Alturas de Maniobón and Gibara extending to the east of Bahía de Nipe (Schwartz and Henderson, 1991; Estrada and Garrido, 1990). The remaining three subspecies are distributed along the coastal zone from Cabo Cruz to the Punta de Maisí; *A. j. oriens* from Cabo Cruz to Santiago de Cuba, *A. j. albertschwartzi* from the Cuenca de Guantánamo to near the Meseta de Maisí, and *A. j. maisensis* on the terraces of Punta de Maisí, *Anolis pigmaequestris* is only known from the Francés and Santa Maria keys on the north coast (Garrido, 1975a).

The family Tropiduridae has 6 species and, with the exception of *Leiocephalus carinatus*, all are endemic to Cuba. *Leiocephalus carinatus* has 13 subspecies, 6 of which are not Cuban but are found in the Bahamas, Cayman Islands and Swan Island. The seven Cuban subspecies are: *L. c. carinatus* along the north coast in the subregion of Alturas de La Habana-Matanzas; *L. c. aquarius* along the entire south coast of the Eastern region, from Cabo Cruz to Maisí; *L. c. cayensis* is in the keys of the Jardines de la Reina; *L. c. labrossytus* is located near Bahía de Cochinos along the coast and the south slopes of the Sierra de Trinidad; *L. c. microcyon* on the Isla de la Juventud; *L. c. mogotensis* in the Sierra de los Organos; and *L. c. zayasi* in the Llanura de Guanahacabibes (Schwartz and Henderson, 1991). Other populations, not identified subspecifically, are known from Cayo Coco, Villa Clara, Nuevitas, Holguín, Inglés and Rosario keys in the Archipiélago de los Canarreos, and the Cayos Blancos south of the Zapata Peninsula (Estrada, 1992, 1993a,b).

Leiocephalus cubensis has 5 described subspecies: *L. c. cubensis* is widely distributed on Cuba, though the records from Guanahacabibes and the Cordillera Guaniguanico are questionable (Schwartz and Henderson, 1991); *L. c. gigas* is exclusively on the Isla de la Juventud; *L. c. minor* is only found on the San Felipe keys to the northwest of

Isla de la Juventud; *L. c. pambasilous* is found on the Hicacos and Campo keys at the extreme western end of the Archipiélago de los Canarreos; and *L. c. paraphrus* is restricted to the keys of the Jardines de la Reina. One population not assigned subspecies status is on Cayo Coco on the north coast in the Sabana-Camagüey keys (Estrada, 1993b). The third species of *Leiocephalus* is *L. macropus*, which has 11 subspecies, of which six are restricted to the subregions of the Eastern region: *L. m. macropus*, *L. m. asbolomus*, *L. m. immaculatus*, *L. m. lenticulatus*, and *L. m. phylax*. One subspecies, *L. m. aegialus*, is restricted to Playa Santa Lucía on the north coast of Camagüey. Other subspecies with narrow local distributions are: *L. m. hoplites* in the northern part of the Llanura Júcaro-Morón, *L. m. hyacinthurus* in the mountains of Guamuhaya, *L. m. felinoi* in the Alturas de La Habana-Matanzas, *L. m. koopmani* in the Llanura de Guanahacabibes, and *L. m. torrei* from San Miguel de los Baños in the Alturas Bejucal-Coliseo (Schwartz and Henderson, 1991). There are also other populations not identified subspecifically in the Sierra de Rosario, Isla de la Juventud, Archipiélago de Sabana, and others in the Eastern region (Schwartz and Henderson, 1991).

Leiocephalus onaneyi is a monotypic species from the summit of the Loma de Macambo in a coastal dry forest to the east of the Cuenca de Guantánamo (Garrido, 1973a). *Leiocephalus raviceps* has five subspecies *L. r. raviceps* along the coast of the Cuenca de Guantánamo; *L. r. delavarai* in the vicinity of Gibara in the Alturas de Mamiabón; *L. r. jaumei* on the western end of the Llanura Sur de Pinar del Río; *L. r. klinikowskii* on the Hicacos Peninsula, and *L. r. uzzelli* along the coastal area between Santiago de Cuba and Guantánamo (Schwartz and Henderson, 1991).

Leiocephalus stictigaster is one of the most diverse species of the Cuban tropidurids with twelve described subspecies. Some of the subspecies are restricted to dry coastal areas: *L. s. stictigaster* on the Llanura de Guanahacabibes, *L. s. astictus* on the southern coast of Isla de la Juventud, *L. s. naranjoi* on the coastal southern slopes of the Sierra de Guamuhaya, *L. s. parasphex* and *L. s. septentrionalis* on the Archipiélago de Sabana-Camagüey, *L. s. lucíanus* in the Playa Santa Lucía on the north coast of Camagüey, and *L. s. gibarensis* of the coastal zone of the Alturas de Maniabón (Schwartz and Henderson, 1991). Other subspecies are located in interior areas such as *L. s. sierrae* in the Alturas de Pizarras surrounding the Sierra de los Organos, *L. s. lipomator* in the Alturas de Cubanacán near Santa Clara, *L. s. ophiplacodes* in the western part of the Llanura Norte de Camagüey, *L. s. exothetus* in the northern area of Isla de la Juventud, and *L. s.*

celeustes on the border between the Llanura de Cauto and the northern slopes of the Sierra Maestra (Schwartz and Henderson, 1991). There are also other populations of this species without subspecific determination along the southeast coast of the Península de Zapata, on the Cayos Blancos south of the peninsula, and in the Sierra de Gran Piedra (Schwartz and Henderson, 1991).

The family Teiidae has only one species in Cuba, *Ameiva auberi*, and it is probably one of the most diverse species of the genus, judging by the fact that 40 subspecies have been described, of which 28 are restricted to Cuba (Schwartz and Henderson, 1991). The distribution of these subspecies covers the entire island and the surrounding keys, ranging from coastal zone to mountains and in different types of forests.

The only member of the Xantusiidae in the Caribbean in *Cricosaura typica*, a species that for many years was only known from a few localities in the plateau near Cabo Cruz in eastern Cuba. Recently the distribution has been expanded and the species is known from populations in the dry forest of the Meseta de Cabo Cruz and the coastal zone along the south slope of the Sierra Maestra extending to the vicinity of Uvero (Estrada and de Armas, 1998). The origin of this species has been the subject of considerable debate, which is ongoing (Crother *et al.*, 1986; Crother and Guyer, 1996; Crother and Presch, 1992, 1994; Hedges *et al.*, 1991; Hedges and Bezy, 1993, 1994; Hedges, 1996b).

The 25 Cuban species of the Serpentes include the families Boidae, Colubridae, Tropidophidae, and Typhlopidae. The Boidae contains an endemic species, *Epicrates angulifer*, known throughout Cuba and the keys of Sabana-Camagüey, Canarreos, Cayos Blanco del Sur, and Isla de la Juventud. The colubrids are represented by 10 species and the genera *Alsophis*, *Antillophis*, *Arrhyton*, *Nerodia*, and *Tretanorhinus*.

The only species of *Alsophis* is *A. cantherigerus* with eight subspecies, of which three are restricted to the Caiman Islands and another to Swan Island (Schwartz and Henderson, 1991). Of the five cuban endemic subspecies the one with the most extensive distribution is *A. c. cantherigerus* found in all of the Western region including the Canarreos keys and Isla de la Juventud, and extends to the northwestern part of the Central region including the Sabana keys (Schwartz and Henderson, 1991). *Alsophis c. schwartzi* is found in the rest of the Central region, the Camagüey-Maniobón region, including the Archipiélago de Camagüey, the southwestern portion of the Llanura del Cauto, the Meseta de Cabo Cruz, Sierra Maestra, Valle Central, the basins of Santiago de Cuba and Guantánamo, and part of the Meseta de Guaso (Schwartz and Henderson, 1991). The subspecies *A. c. adspersus* and *A. c. pepei* extend along the northeast and southeast of the Eastern region. The populations in the extreme eastern end of Cuba are considered intermediates between the *adspersus* and *pepei* subspecies, and those in the southwestern Sierra Maestra are considered intermediates between *pepei* and *schwartzi* (Schwartz and Henderson, 1991).

Antillophis contains one species *A. andreai* with six subspecies. *Antillophis a. andreai* is found throughout the west (except for the Llanura de Guanahacabibes) to the western half of the Central region; *A. a. orientalis* is distributed from the eastern half of Camagüey-Maniabón to the entire Eastern region. There are at least three subspecies with localized distribution: *A. a. peninsulae* in Guanahacabibes, *A. a. nebulatus* on the Isla de la Juventud, *A. a. melopyrrha* on Cantiles key in the Canarreos and *A. a. morenoi* on the keys of Sabana-Camagüey. The populations between the eastern part of the Central region and the western half of Camagüey-Maniabón show intergradation between *andreai* and *orientalis* (Schwartz and Henderson, 1991).

The genus *Arrhyton* has six species in Cuba, *A. ainictum* is known only from the type locality at Cueva del 18 in the extreme east of the Llanura Sur of Camagüey; *A. dolichura* is another species with a poorly documented distribution and is known only from the vicinity of the city of La Habana. *Arrhyton landoi* is restricted to the southern part of the Eastern region. Two species, *A. taeniatum* and *A. vittatum*, have been reported from different localities of the Western, Central, and Eastern regions and Isla de la Juventud (Schwartz and Henderson, 1991). Another species *A. tanyplectum*, is known from a few localities in the Sierra de los Organos. Recently two new species have been described, *A. supernum* in the mountains of Sagua-Baracoa in eastern Cuba, and *A. procerum* from the vicinity of Playa Girón in the Llanura de Zapata (Hedges and Garrido, 1992a).

The two remaining especies of colubrids, include *Nerodia clarki compressicauda*, a water snake of swamps, estuaries and bays of marine or brackish waters, and found at various coastal sites in the Western and Central regions and the keys of Sabana-Camagüey. *Tretanorhinus variabilis*, in constrast, has five subspecies, four endemic to Cuba and the fifth to the Cayman Islands. *Tretanorhinus v. variabilis* is present throughout most of Cuba, except for the Isla de la Juventud where *T. v. insulaepinorum* is found, and in the western half of the Western region where *T. v. wagleri* is present, and the Cabo Cruz plateau in the western Sierra Maestra where *T. v. binghami* is resident.

The populations to the west of La Habana and those from the Canarreos keys have not been identified subspecifically (Schwartz and Henderson, 1991).

The family Tropidophidae is well represented on Cuba with 11 species of *Tropidophis*, of which 9 are endemic. *Tropidophis melanurus* has four subspecies of which only one is found outside of Cuba on Navassa Island. *Tropidophis m. melanurus* is found throughout Cuba, while *T. m. dysodes* is known from a small area in the center of the Llanura Sur de Pinar del Río and *T. m. ericksoni* from the Isla de la Juventud. This species has populations on the Canarreos, Sabana-Camagüey, and San Felipe keys that have no subspecific designation (Schwartz and Henderson, 1991).

Three of the *Tropidophis* species have a western distribution: *T. feicki*, *T. maculatus* and *T. semicinctus*. Another species, *T. nigriventris*, has two disjunct populations, considered subspecies: *T. n. nigriventris* from the Alturas de Camagüey and *T. n. hardyi* in the southern part of the Sierra de Guamuhaya. *Tropidophis pilsbryi* is another species with disjunct populations; *T. p. pilsbryi* in the Eastern region and *T. p. galacelidus* in the Sierra de Guamuhaya (Shwartz and Henderson, 1991). *Tropidophis pardalis* is a common species in western Cuba, but isolated populations are found throughout the main island and Isla de la Juventud and the keys of Sabana-Camagüey. *Tropidophi wrighti* has a distribution from the Central to the Eastern region. *Tropidophis haetianus haetianus* is not endemic to Cuba and has been reported from the northern part of the Eastern region (Schwartz and Henderson, 1991). *Trophidophis fuscus* is known from two localities in the Sagua-Baracoa mountains (Hedges and Garrido, 1992b).

The Family Typhlopidae has two species of *Typhlops* in Cuba: *T. biminiensis*, a Bahaman species that has been reported from some localities in the Western, Central and Eastern regions of Cuba and *T. lumbricalis* widely distributed on Cuba, Isla de la Juventud, and in the southern and northern Bahamas (Schwartz and Henderson, 1991).

The order Crocodylia has three species on Cuba, one an introduced member of the Alligatoridae, and two native species of Crocodylidae: *Caiman crocodylus* was introduced in 1959 in the Cíenaga de Lanier, on Isla de la Juventud, and today the species is found throughout the wet areas of the island (Varona, 1976; Schwartz and Henderson, 1991); *Crocodylus acutus* is widely distributed in all the coastal and swampy areas including the keys and *C. rhombifer*, an endemic species, today found only in the western part of the Cíenaga de Zapata.

3

Jamaica

Ronald I. Crombie

Division of Amphibians and Reptiles, National Museum of Natural History, Smithsonian Institution, Washington, D.C. 20560

Introduction

Jamaica is the second smallest of the major Greater Antillean islands, a bit more than 11,000 km^2 in area, with a maximum elevation of 2256 m. It lies on its own bank, separated by very deep water from both the mainland and the other Antilles, with few, mostly close in-shore surrounding islets (Steers 1940a,b; Steers *et al.*, 1940; Crombie *et al.*, 1984). Consequently it could be considered the smallest major island because the Greater Puerto Rican Bank is slightly larger. Despite its modest size, Jamaica has a diverse (65+ species) and distinctive herpetofauna (Table 3.1). The fauna in general is striking as much by the elements that are absent as by the significant radiations of the taxa that are present, reflecting the island's isolation from its Antillean neighbors and (perhaps) by its geological history relative to the Middle American mainland (Buskirk, 1985; Hedges, 1989a,b). The list of Jamaican species is by no means complete, however, and the rugged limestone terrain continues to hide surprises and make a mockery of "definitive" statements on the rarity or even extinction of some species.

Because this book is a well-deserved tribute to Albert Schwartz, it is interesting to note that Jamaica was not one of his favorite places in the West Indies. For example, compared to the multitude of his papers on most of the rest of the Antilles, Schwartz published only two (Schwartz, 1971a; Schwartz and Fowler, 1973) exclusively on Jamaica and one other (Thomas and Schwartz, 1974) partly on a Jamaican species. Shortly after my first trip to the island in 1970 I visited with Schwartz in Miami, where he encouraged my interest in the Jamaican fauna, candidly admitting that he did not relish working there much more but recognizing that extensive investigations were needed.

Over a beverage or two, we discussed what seemed to me a peculiar prejudice, because I found the place utterly entrancing. As usual, Schwartz had a perfectly pragmatic and understandable basis for his mild aversion to Jamaica. His field trips were strongly automobile oriented and at that time Jamaica had a road system that was generally passable between the major population centers and the tourist resorts of the north coast. The small rural roads that provided access to the "good places" on many other islands were relatively uncommon on Jamaica and important collecting sites could often only be reached by hard overland trek through unforgiving terrain, decidedly not Schwartz's style at that point of his career. He also objected to the "touristy" aspect of Jamaica, which did not require many out-of-the-way pensiones or small eateries. The "tourist prices" for everything may have struck Schwartz, ever thrifty with field money, the hardest and he hated to pay full price for anything. However, Jamaicans rarely bargained, extrapolating the frequent tourist dim-wittedness to all foreigners. Schwartz believed that even the people well out of the tourist mainstream lacked the warmth and friendliness he valued so much on other islands. I did not entirely agree and being young and stupid (a trait some would deny I ever outgrew), I continued my Jamaican field work for the next fifteen years.

As I gained more experience on other Antillean islands, I could understand Schwartz's points far more clearly, and Jamaica was indeed never an "easy" place to work. The proud, aloof, almost arrogant Maroons never totally warmed to my presence in their forests but we shared many trials together and their unparalleled bush knowledge was enormously helpful to my projects. The activities of the U.S. Drug Enforcement Agency in attempting to stem the flow of drugs from Jamaica to the United States often made innocent foreigners suspect and unwelcome in isolated areas. However, the kindness, generosity, and inherent charm of the Jamaican people more than overwhelmed any negative encounters. The culture, the fauna, and all things Jamaican are the product of an environment that can oscillate from idyllic to fierce and hostile, seemingly without effort. The result is a unique kaleidoscopic effect that can seem far more foreign than Jamaica's proximity to the rest of the Antilles would warrant.

Table 3.1. Checklist of the Jamaican Herpetofauna

Anura: Bufonidae

Bufo marinus (Linnaeus, 1758) — Introduced

Anura: Hylidae

"*Calyptahyla*" *crucialis* (Harlan, 1826)
Hyla marianae Dunn, 1926
Hyla wilderi Dunn, 1925
Osteopilus brunneus (Gosse, 1851)
Osteopilus sp. nov.

Anura: Leptodactylidae

Eleutherodactylus alticola Lynn 1937
Eleutherodactylus andrewsi Lynn 1937
Eleutherodactylus cavernicola Lynn, 1954
Eleutherodactylus cundalli Dunn, 1926
Eleutherodactylus fuscus Lynn and Dent, 1943
Eleutherodactylus glaucoreius Schwartz and Fowler, 1973
Eleutherodactylus g. gossei Dunn, 1926
Eleutherodactylus gossei oligaulax Schwartz and Fowler, 1973
Eleutherodactylus grabhami Dunn, 1926
Eleutherodactylus griphus Crombie, 1986
Eleutherodactylus jamaicensis Barbour, 1910
Eleutherodactylus johnstonei Barbour, 1914 – Introduced
Eleutherodactylus junori Dunn, 1926
Eleutherodactylus luteolus (Gosse, 1851)
Eleutherodactylus nubicola Dunn, 1926
Eleutherodactylus orcutti Dunn, 1928
Eleutherodactylus p. pantoni Dunn, 1926
Eleutherodactylus pantoni amiantus Schwartz and Fowler, 1973
Eleutherodactylus pentasyringos Schwartz and Fowler, 1973
Eleutherodactylus p. planirostris Cope, 1863 - Introduced
Eleutherodactylus sisyphodemus Crombie, 1977

Anura: Ranidae

Rana catesbeiana Shaw, 1802 - Introduced

Sauria: Diploglossidae

Celestus barbouri Grant, 1940
Celestus c. crusculus Garman, 1888

Table 3.1 cont.

Celestus crusculus cundalli Grant, 1940
Celestus crusculus molesworthi Grant, 1940
Celestus duquesneyi Grant, 1940
Celestus fowleri Schwartz, 1971
Celestus hewardi Gray, 1845
Celestus microblepharis Underwood, 1959
Celestus occiduus Shaw, 1802
Sauria: Gekkonidae
Aristelliger praesignis Hallowell, 1857
Gonatodes albogularis notatus Reinhardt and Luetken, 1863
Sphaerodactylus a. argus Gosse, 1850
Sphaerodactylus dacnicolor Barbour, 1910
Sphaerodactylus gilvitorques Cope, 1862
Sphaerodactylus goniorhynchus Cope, 1895
Sphaerodactylus oxyrhinus Gosse, 1850
Sphaerodactylus parkeri Grant, 1939
Sphaerodactylus r. richardsoni Gray, 1845
Sphaerodactylus richardsoni gossei Grant, 1939
Sphaerodactylus semasiops Thomas, 1975
Sphaerodactylus sp.
Sauria: Iguanidae
Cyclura collei Gray, 1845
Sauria: Polychrotidae
Anolis garmani Stejneger, 1899
Anolis g. grahamii Gray, 1845
Anolis grahamii aquarum Underwood and Williams, 1959
Anolis l. lineatopus Gray, 1840
Anolis lineatopus ahenobarbus Underwood and Williams, 1959
Anolis lineatopus merope Underwood and Williams, 1959
Anolis lineatopus neckeri Underwood and Williams, 1959
Anolis opalinus Gosse, 1850
Anolis reconditus Underwood and Williams, 1959
Anolis s. sagrei Dumeril and Bibron, 1837 – Introduced ?
Anolis valencienni Dumeril and Bibron, 1837
Sauria: Scincidae
Mabuya sp.

Table 3.1 cont.

Sauria: Teiidae	
	Ameiva dorsalis Gray, 1838
Serpentes: Boidae	
	Epicrates subflavus Stejneger, 1901
Serpentes: Colubridae	
	Alsophis ater (Gosse, 1851)
	Arrhyton callilaemum (Gosse, 1851)
	Arrhyton funereum (Cope, 1862)
	Arrhyton polylepis (Buden, 1966)
Serpentes: Tropidophiidae	
	Tropidophis jamaicensis Stull, 1928
	Tropidophis stejnegeri Grant, 1940
	Tropidophis stullae Grant, 1940
Serpentes: Typhlopidae	
	Typhlops jamaicensis (Shaw, 1802)
Testudines: Emydidae	
	Trachemys terrapen (Lacepede, 1788)
Crocodilia	
	Crocodylus acutus Cuvier, 1807

Location - Topography - Geology

Jamaica is located about 200 km south of the eastern end of Cuba and about the same distance west of the tip of the Tiburon Peninsula of Haiti. The island is approximately 230 km long and 80 km wide at its broadest (central) point, tapering at both ends, with a total area of 11,500+ km^2 (Fig. 3.1). Its topography is very simple compared to the larger Greater Antillean islands, consisting primarily of a large, heterogeneous limestone block atop an igneous and metamorphic core, sloping slightly upward from south to north. The southern alluvial plain is fairly extensive and includes several ridges or uplifts (Portland Point and Hellshire Hills) that were probably islands during periods of elevated sea levels. The dominant feature of the limestone block is the extensive flat or rolling central plateau, mostly 2000-2500 ft in elevation, with a variety of associated ridges (Don Figuerero, Dry Harbour, and Santa Cruz Mountains) and karst areas, including the extremes of the Cockpit Country and the more weathered limestones of the western Round Hill District. Unlike the other Greater Antilles, Jamaica has a paucity of offshore islands, most of them associated with the south coast peninsulas of Portland Point and the Palisadoes (Steers, 1940a,b; Steers *et al.*, 1940; Lazell, 1996). Cabarita Island, off Port Maria, St. Mary, is one of the few north coast islets (Crombie *et al.*, 1984). The more distant Pedro (65 km south) and Morant Cays (95 km southeast) are not on the Jamaican Bank but are currently claimed by Jamaica.

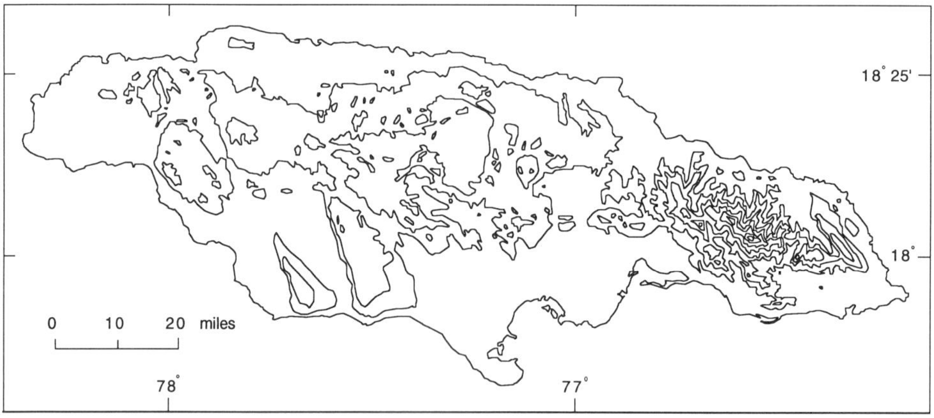

Figure 3.1. Topographic map of Jamaica. The contours mark 1000 ft. Most of the island's elevational features are associated with the limestone block which sits on top of the island. The greatest elevational changes occur on the eastern end as the Blue and John Crow Mountains.

The eastern tip of the island is a largely granitic-shale conglomerate that rudely contacted the limestone block by movement along a fault. The resulting compression and slip faulting caused the uplift of the eastern mountains (Blue and John Crow ranges), which reach an elevation of more than 2200 m. This uplift was a relatively recent event, generally thought to be Pliocene (Buskirk, 1985).

The geological history of Jamaica, including its tectonic movements, submergence, and reemergence, was thoroughly reviewed by Buskirk (1985) and Hedges (1989a,b). Various aspects of Caribbean geology (particularly regarding Jamaica) remain controversial, as does the debate on the role of vicariance versus dispersal in the herpetological colonization of the islands. I believe that more data are required before I would speculate on the "definitive" geological history of Jamaica since I do not agree totally with any of the proposals presented thus far. Rather than greatly expanding the literature cited herein by citing the same literature already summarized in several recent reviews, I suggest interested readers make their own conclusions from these more detailed discussions: Crother and Guyer (1996), Guyer and Crother (1996), and Hedges (1996a,b,c).

Rivers

Despite the porous limestone base of most of the island, Jamaica has a multitude of rivers draining to both coasts but with more emptying to the north. The Negril River and associated Morass is the major stream at the western end of the island and (moving eastward) other important waterways of the northern part of the limestone block include the Lucea, Great, Montego, Martha Brae, Rio Bueno, Roaring, White, Rio Nuevo, and the Wag Water. The southwest coast, considerably lusher than the south-central area, has numerous smaller streams and the Cabarita River system, but the single most extensive wetlands area in Jamaica is the Black River and its tributaries. With its large estuary and shallow bay, in addition to broad inland freshwater swamps, this area remains very poorly collected and barely explored. East of the Black River many river tracks are dry at least part of the year and the Milk River system is the primary drainage on the south-central coast. Other rivers meander through the central plateau and either vanish underground or join other rivers, usually flowing south.

The eastern mountains are also well supplied with rivers. The main systems draining north are the Buff Bay, Spanish, Rio Grande, and Drivers. South of the mountains are the Plantain Garden, Negro-Morant, Yallahs, and Cane-Mammee.

This abundance of surface water is misleading since there is considerable evidence that suggests Jamaica was considerably drier and perhaps even devoid of surface water for long periods after its emergence, possibly until very recently in geological time. Most compelling is the absence of native frogs with larvae that develop in ponds or streams. All other Greater Antillean islands have xeric-adapted bufonids that can breed in ephemeral pools. Another Greater Antillean species, *Leptodactylus albilabris*, copes with xeric habitats by producing a foam nest which can provide food, insulate against extreme temperatures, and deter desiccation for the larvae. Although this species is ubiquitous on the Puerto Rican Bank, including some harshly dry areas, it is restricted to a small mesic enclave on extreme eastern Hispaniola and is absent from all islands to the west. Since Heyer (1978, personal communication, 1996) suggests South American rather than Middle American relationships for *albilabris*, it seems most likely that it is a recent adventive to Hispaniola and never reached the more western islands (Cuba and Jamaica). Jamaica lacks native bufonids and *Leptodactylus* but has 17 endemic *Eleutherodactylus* (direct development) and five hylids, even greater than the hylid radiation (four) on much larger Hispaniola. Whereas most other Antillean hylids are either very generalized pond breeders, including ephemeral pools (*Hyla pulchrilineata*, *Osteopilus dominicensis*, and *O. septentrionalis*) or stream-adapted (*Hyla heilprini*, and *H. vasta*), all the Jamaican species have evolved highly specialized phytotelmatous larvae (Dunn, 1926a; Lannoo *et al.*, 1987; Thompson 1996), that usually develop in bromeliad axils but may use tree holes or other small cavities. The larvae of the undescribed *Osteopilus* are not definitely known but variation in the few known samples of supposed "*Calyptahyla*" *crucialis* larvae suggests that both species may be represented therein. Although there is some dispute whether the Jamaican hylids represent a single radiation (Hedges, 1996b) or convergence in larval morphology (Lannoo *et al.*, 1987; Anderson 1996), it is clear that they do not (and probably cannot) use ponds or streams for reproduction. In addition, some Jamaican *Eleutherodactylus* occasionally or routinely (*E. jamaicensis*) use bromeliads and other cavities as egg deposition sites, also suggesting an extended drier regime on the forest floor. Some genera of normally terrestrial reptiles such as anguid lizards and *Sphaerodactylus* have evolved obligate bromeliad species on Jamaica (*Celestus fowleri*, *Sphaerodactylus oxyrhinus*, and *S. semasiops*), as have a variety of invertebrates (carabid beetles, crabs, scorpions, and worms). In addition, Pregill *et al.* (1992) found Holocene remains of xeric or at least open-formation birds and lizards (*Ameiva dorsalis*, *Leiocephalus* cf. *jamaicensis*) in a cave deep within the mesic Cockpit Country. While hardly conclusive, it is hard to imagine why such extensive adaptation for life in the mesic enclaves of bromeliads would have taken place if moist terrestrial refuges and more typical reproductive sites were available (Fig. 3.2).

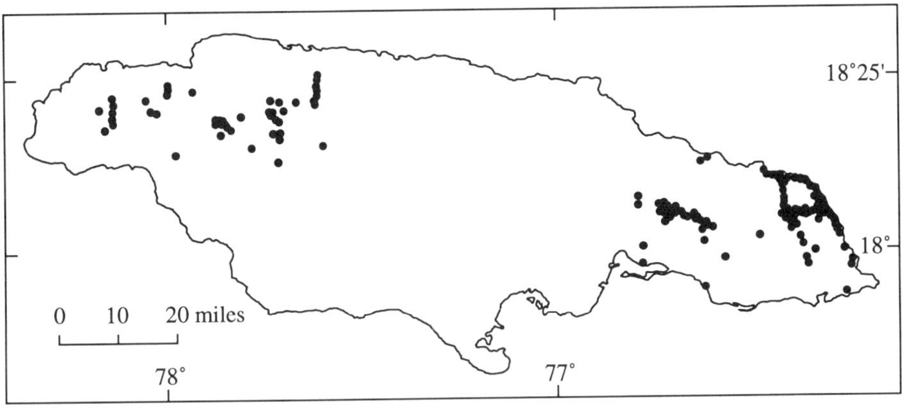

Figure 3.2. Combined superimposed distributions of 15 taxa which exhibit geographic ranges associated with the two major areas of endemism. The western distributions are represented by *Eleutherodactylus fuscus*, *E. griphus*, *E. sisyphodemus*, *Celestus fowleri*, *Sphaerodactylus oxyrhinus*, and *S. semasiops*. The eastern distributions are represented by *E. alticola*, *E. andrewsi*, *E. glaucoreius*, *E. nubicola*, *E. orcutti*, *E. pentasyringos*, *Anolis reconditus*, *Arrhyton polylepis*, and *S. dacnicolor*.

Climate and Vegetation

Jamaica has a seasonal tropical maritime climate, dominated by north or northeast trade winds which bring substantial rainfall to most of the northern part of the island. The northern and eastern slopes of the Blue and John Crow Mountains can receive more than 7 m of rain annually and the Cockpit Country only slightly less. Most of the south coast is considerably more arid and the limited rainfall rapidly percolates away in the limestone areas. A 91-year survey of rainfall in the *Handbook of Jamaica* for 1963 reported a mean rainfall for the combined 14 parishes of 84.83 in., with a rainy season from May to October (6.92-12.15 in.) and a slightly drier one from November to April (2.91-7.74 in.). The mean maximum-minimum temperatures at sea level are 87.5 and 70.8°F. On the central plateau at around 2000 ft this drops to 78.6 and 66.1°F. The highest elevation for which climatological data are available is Hill Gardens, Cinchona, St. Thomas, at about 4400 ft, where mean highs are in the high 60s and overnight lows in the mid-50s. Higher in the Blue Mountains, overnight temperatures near 40°F are not uncommon.

A discussion of Jamaican floristics is almost a moot point since estimates of natural vegetation loss range from 75+% (Downer and Sutton, 1990) to more than 90% (Hedges, manuscript in preparation). Furthermore, most of this habitat destruction (at least 50%) took place during the early period of colonization, when forest was extensively cleared for agricultural and grazing land. Although this habitat alteration continues today, primarily in the higher elevations which were not suitable for early exploitation, it is relatively minor compared to the early clearing. Asprey and Robbins (1953: 361, Fig. 3) mapped their major vegetational types and even at that time the Wet Limestone Forest was centered in the Cockpit Country with disjunct patches in the western Parishes (Hanover and Westmoreland) and additional outliers as far east as St. Ann-St. Catherine. Many of these isolated patches were greatly reduced or nonexistent by the 1970s (R. Crombie, personal observation).

Similarly, Dry Limestone Forest was concentrated along the south coast, particularly the Hellshire Hills-Portland Point area, with isolates in the western parishes, east of Kingston in the rain shadow of the Blue Mountains, and a few small patches along the north coast. All of these have been severely reduced by development and overgrazing since Asprey and Robbins' report.

Montane Forest is restricted to the higher elevations of the Blue and John Crow Mountains. The Lower Montane Rain Forest and Montane Sclerophyll Forest have been severely depleted by development, as Kingston residents seek to escape to the cooler temperatures of the mountains. Still more restricted in area are the Mist Forest and Elfin Woodland, progressively higher on the peaks. However, it should be noted that these "upper" montane communities can extend to surprisingly low elevations in some areas (e.g., Elfin Woodland at 2500 ft in the John Crow Mountains, Asprey and Robbins, 1953: 402). For more detailed descriptions of these communities, the reader is referred to Asprey and Robbins, whom devote more than 50 pages and 20 figures to it.

The obvious question is, "What effect has this extensive deforestation had on the herpetofauna?" Since the most significant clearing of midelevation forest took place 200 years ago when most of the herpetofauna had not even been described, we have no "before" to compare with the "after." Forest-associated species such as many of the current Cockpit endemics may well have been widespread in forest throughout the Central Plateau and the current patterns of distribution are probably totally artifacts of early deforestation. Certainly most of the herpetofauna is ecologically very versatile, occurring in disturbed as well as pristine habitats, but distributional data on the Jamaican herpetofauna are so vestigial that even current range maps are only an approximation of the actual distribution for most species. Looking for historical changes with such paltry data would be pure speculation.

Historical Summary

I have divided this section into two parts: the historical literature, up to 1959, and the recent literature, 1960 to the present. Although this is admittedly arbitrary, there is a rationale behind it. Rather than follow a strictly chronological summary of the literature, it is more cohesive and space efficient to discuss the contributions of a given institution or individual as "eras" with a general chronological arrangement. In some cases (e.g., the Harvard era) the contributions span nearly a century and rather than subsume the more recent contributions with those from the late 1800s, I chose the division of "older" versus "recent" because it approximately corresponds to the time when Schwartz became involved in Jamaican herpetology; it also coincides with the departure of the most famous herpetologist resident in Jamaica, Garth Underwood. The contributions of some individuals (e.g., Coleman J. Goin) slightly span this arbitrary point, in which case I have included those individuals in the period that their publications began. In other cases (e.g., the Harvard era), the publications in both the older and more recent literature are so voluminous that I have separated them. This summary of the literature is not intended to be a bibliography of Jamaican herpetology, which would greatly exceed the space constraints of this chapter, but it is more a "highlights reel" of significant advances.

Historical Literature

Specimens from the New World colonies probably began arriving in Europe soon after trade ships began plying the West Indian waters. Precise locality data rarely were associated with the early specimens, at best only the island where the material allegedly originated, and even these minimal data were unreliable. For example, Gray's (1839)

Tiliqua jamaicensis (type locality "Jamaica") is a synonym of *Diploglossus monotropis*, a species found in lower Middle and northwest South America. The type likely was not even collected in a British colony, evidence that the colonial powers did not respect each others territorial claims in the New World. Jamaican material even made its way to Paris, devoid of **any** locality (which might have been incriminating) and was studied by Daudin, Duméril and Bibron (e.g. *Anolis valencienni* Duméril and Bibron, 1837). The American colonies were not left out and Jamaican specimens in Philadelphia and Washington were described by Harlan, Hallowell, and Cope.

The earliest records of Jamaican reptiles consisted of anecdotal notes in voyage reports or travelogs, usually by surgeons who served as physician to the colonial governor. Grant (1940a: 146-148) adequately summarized most of them and pointed out the fanciful nature of the information therein. The most notable (or at least most extensive) of the early reports was the two-volume epic of Sir Hans Sloane (1725) on his extensive Antillean travels, including a 15 month stay in Jamaica in 1688 and 1689. Ahrenfeldt (1953, 1954) discussed Sloane's herpetological contributions (all appearing in the 1725 volume) in some detail, but Sloane's descriptions were all nonbinomial and not nomenclaturally available. Although some of his natural history information is historically significant, I agree with Gosse's (1848c: 61) comments on Sloane's observations: "like most of his zoological notes, ... full of confusion and error." Although Sloane apparently made a significant collection of plants (Ahrenfeldt, 1954), the status of his herpetological specimens is unclear. Gray (1845) did not list Sloane as a donor of material to the British Museum, the most logical depository. However, the French, whose relations with Britain were strained by conflicts in the New World, were uncharacteristically lavish in their praise of Sloane, even dedicating several species to this "learned traveler and friend of the Sciences" (Duméril and Bibron 1839: 107, as translated by Ahrenfeldt, 1954: 639). Perhaps Sloane's specimens were indeed deposited in Paris but without even general locality.

Following the Sloane tradition, Patrick Browne included some natural history in his three-part treatise on Jamaica (mostly in Part II), but his Latin descriptions were also nonbinomial, and some of his observations appear to be taken almost verbatim (including errors in spelling and logic) from Sloane. There has never been any suggestion that Browne actually collected any herpetological specimens, but his comments and descriptions were used by Lacépède, Shaw, and other later authors.

There is a tendency that the largest, most obvious, and edible species are usually the first taxa described from new territory, with the most unusual ones a close second. Following this trend, the first species described validly and expressly from Jamaica was the fresh-water turtle, *Trachemys terrapen* (Lacépède, 1789), although the description is marginal and the locality not as clear-cut as subsequent treatment would indicate. The page generally cited for the species (p. 129) does indeed contain a description for La Terrapen or Terrapene, nowhere therein are the names latinized or used in binomial form. Lacépède's name *Testudo terrapen* is nomenclaturally validated, somewhat debatably, only by the fold-out table that provides a latinized binomial (see discussion in Barbour and Carr, 1940:393). Barbour and Carr (1940) also point out that the species was described from "aux Antilles" and "particulierement a la Jamaique." However, since Lacépède clearly used Browne's (1789) comments on the species in his description, restriction of the locality to Jamaica is reasonable.

Also drawing from the works of Sloane (1725) and Browne (1789), Shaw (1802) described the largest Jamaican diploglossid and the endemic *Typhlops*.

Aside from these descriptions, most of the early Jamaican material was studied by John Edward Gray at the British Museum. Gray's descriptions were often peremptory, and the specimens poorly preserved, which compounded the taxonomic problems due to unreliable locality data. Still, Gray is responsible for six currently valid Jamaican species, including two of the most common and widespread anoles (*Anolis grahamii*, 1845, and *A. lineatopus*, 1840), a galliwasp (*Celestus hewardii*, 1845), the ground lizard (*Ameiva dorsalis*, 1838), the iguana (*Cyclura collei*, 1845), and the largest species of *Sphaerodactylus* (*richardsoni*, 1845). He also provided a significantly larger number of synonyms and species incorrectly attributed to Jamaica.

One of the American contributions preceding Gray's pioneering studies was Harlan's (1826b) description of the largest Jamaican treefrog. Although *Hyla crucialis* was well and accurately described and the type specimen remains in the Academy of Natural Sciences (ANSP 2180, Crombie, 1973), the description was long overlooked, probably due to it being obscurely attached as a separately titled note to Harlan's larger paper (1826a).

All of the earlier work notwithstanding, the true father of Jamaican herpetology (and many other aspects of Jamaican biology) was Philip Henry Gosse, who lived on the island for 18 months in 1844 and 1845. Gosse was a careful observer and a talented writer who traveled extensively throughout Jamaica. His "A Naturalist's Sojourn in Jamaica" (1851) remains an indispensable classic for anyone interested in the natural history of the island. Unlike many early visitors, who were confined to Kingston and areas easily accessible from it, Gosse was based in the Bluefields/Content region, now in Westmoreland Parish, in the far western limestone districts. Clearing of forests for plantations and lumber was common even at this early stage, but Gosse had easy access to the wet limestone forest that was less suitable for early development. Many significant localities such as the Black River Morass were not far away, but Gosse traveled even to the distant Blue Mountains and was probably the first European to see the full diversity of Jamaican habitats. He certainly was the first to describe it in print. As a result of his travels and meticulous observations, Gosse (1850, 1851) described 13 new species, seven of which are currently recognized as valid species or subspecies. Of his six synonymous names, *Placopsis ocellata* was described by Duméril and Bibron (1839) as *Anolis* (not *Xiphosurus*, as stated in Schwartz and Henderson, 1988) *valencienni* with no locality data, *Trachycephalus lichenatus* is Harlan's (1826b) *Hyla crucialis* (but see Problems section), and *Draconura catenata* is the Cuban species *Anolis sagrei*, obviously a very early introduction or natural dispersant. Gosse's descriptions of *Trachycephalus anochloros* (currently a synonym of *Calyptahyla crucialis*) and *Natrix capistrata* (= *Alsophis ater*) are so distinctive that I am not convinced the species are invalid, as discussed in the Problems section. Gosse also drew extensively from the notes of Dr. Anthony Robinson, a surgeon and botanist who never published his Jamaican observations. However, a copy of his manuscript was preserved in the Institute of Jamaica Library and many of the herpetological notes have since been published (Cockerell, 1894). Richard Hill of Spanish Town also contributed greatly to Gosse's work, particularly "The Birds of Jamaica," and Hill is often credited as being the first permanently resident naturalist of Jamaica.

An interesting adjunct of Gosse's work was the discovery of an unpublished color plate of his Jamaican amphibians and reptiles nearly 100 years after the publication of his "Sojourn" (Underwood, 1949). Chapman Grant disagreed with some of Underwood's identifications (Anonymous, 1950) and a revised identification key appeared in Lewis (1950).

After Harlan's (1826b) early note, the American contributions continued with Hallowell's (1857) description of *Aristelliger praesignis* and Cope's (1862a,b, 1895) additional new species (*Sphaerodactylus gilvitorques* and *goniorhynchus* and *Arrhyton funereus*). Although most new species to this time had only "Jamaica" or an inferred more specific type locality, the two Jamaican species described in Cope (1895), based on material from the University of Pennsylvania West Indian Expedition of 1890-1891, actually had specific localities: *S. goniorhynchus* from Port Antonio (Portland Parish) and *Anolis flabellatus* from Port Morant (St. Thomas) and Port Lucea (Hanover), at opposite ends of the island. Cope included data on coloration, ecology, and behavior from the notes of the expedition. The two *Eleutherodactylus* species Cope reported from the expedition (*Hylodes martinicensis* from Port Lucea and Blue Peak; *Lithodytes lentus* from Port Lucea) are not identifiable (but clearly not the Lesser Antillean *martinicensis* or the St. Croix endemic *lentus*), but the description (handsome black and orange tree-toad ... clinging to a shrub overhanging a stream) of the latter could be *E. jamaicensis*. Although *A. flabellatus* is a synonym of *Anolis opalinus*, the description warrants additional discussion because it contains much of the confusing and contradictory information that typifies so many Cope descriptions. Cope clearly had more than one specimen of *flabellatus*, but his description and measurements were from one individual and variation in the other(s) was briefly noted. The material from this expedition was deposited at the Academy of Natural Sciences, Philadelphia (ANSP) but Malnate (1971) did not list any types of *flabellatus* in the collection. By way of (brief) explanation, Cope traveled frequently between Philadelphia and Washington, usually carrying specimens from both institutions to facilitate study at whichever museum was his current destination. Although perhaps authorized by the authorities of United States National Museum (USNM) and ANSP, these transfers were NOT invoiced or documented in any way and seemed entirely at Cope's discretion. Unfortunately, his organization of research materials was negligible and after his death utter chaos reigned in attempting to return specimens to their proper institutions, which was never really achieved. As a result, many ANSP specimens (including some types) were catalogued at USNM and vice versa, obviously an attempt to simplify what must have been a monumental task. Cope also removed tags when illustrating specimens (or for some other unfathomable reason) and after his death USNM received a sizable batch of these tagless specimens, which were recataloged as "returned by Cope's estate." They were usually completely dataless, but occasionally a label associated with a specimen (or specimens) provided a shred of locality information. Because of these informal "exchanges" between ANSP and USNM, researchers should be aware that Cope's types should be sought at both institutions, regardless of any specific comments on their deposition in Cope's descriptions. Additional missing types may well be included (but unrecognized) in the now dataless "returned by Cope's estate" material in USNM. The types of *A. flabellatus*, listed as "unlocated" by Schwartz and Henderson (1988: 158), are a good example of the "unintentional exchange" due to Cope's transport of specimens between the two institutions. Although not listed in Cochran's (1961) USNM type catalog, two specimens of Cope's *flabellatus* remained on a "problem shelf" in USNM until April 1968, when they were finally recognized as missing syntypes and cataloged as USNM 164931 (from Port Morant, ledger remarks "type") and 164932 (Port Lucea, "type of drawing").

The end of the nineteenth century also marked the beginning of the "Harvard era" in West Indian herpetology, a notable period that continues to the present day. Samuel Garman made collections for the Museum of Comparative Zoology (MCZ) throughout the Antilles, including a brief visit to Jamaica. Although most of the other islands he visited yielded a treasure trove of novelties, Garman (1887a) described only a single currently valid new species from Jamaica, the most common small diploglossid, *Celestus crusculus*. Garman's activities appear to have been limited to the Kingston, Spanish Town, and Moneague area, but his successor, Thomas Barbour, was far more successful in reaching more distant parts of the island through his contacts with major plantation owners. Barbour (1910) added two more valid taxa to the faunal list, including the first indication that Jamaica was inhabited by more than one widespread, exceedingly variable species of *Eleutherodactylus*. With the exception of Harlan (1826b) and Gosse (1851), Jamaican amphibians were neglected by the early writers and Gosse's *Litoria luteola* was the only species of *Eleutherodactylus* recognized on the island. Barbour's (1910) description of *Eleutherodactylus jamaicensis*, an arboreal, bromeliad dwelling species with expanded digital tips and sticky bluish skin secretions, was particularly noteworthy and the species so distinctive that some subsequent authors (Goin, 1954; Crombie, 1977; Flores, 1984) regarded it as unique among Jamaican and even West Indian *Eleutherodactylus*. Barbour's other new species was *Sphaerodactylus dacnicolor*, also an arboreal bromeliad dweller. Barbour (1922) later produced the first list of the Jamaican herpetofauna and discussed Jamaican species in his West Indian checklists and taxonomic reviews. However, his later reviews of Jamaican species (particularly *Anolis* and *Sphaerodactylus*) lacked the initial clarity of his 1910 paper.

Although he taught at Haverford College, Emmett Reid Dunn could be included in the Harvard era due to his strong institutional affiliation with MCZ and his long friendship with Barbour. Dunn single-handedly revamped the Jamaican amphibian fauna in just 10 days of fieldwork in August and September of 1925. He worked primarily at Spaldings, on the Clarendon-Manchester Parish border, near where the weathered central plateau meets the karst of the Cockpits and Dry Harbour Mountains. Dunn (1925) had previously described *Hyla wilderi*, named after and based on specimens collected at Moneague, St. Ann, by Harris Hawthorne Wilder and Inez Whipple Wilder, colleagues from Smith College. The Wilder's interesting discovery may have prompted Dunn to stop in Jamaica on his return from Cuban fieldwork in 1925, but whatever his motivation, the 10 days of frog-oriented fieldwork yielded all seven species previously known from the island and seven new species. Except for *Eleutherodactylus nubicola*, based on older material from two localities in the Blue Mountains (which Dunn did not visit), the new *Hyla* (*marianae*) and five new *Eleutherodactylus* were all from Dunn's Spaldings collections. In more than tripling the number of endemic *Eleutherodactylus*, Dunn conclusively demonstrated that Jamaica did indeed have a significant radiation of the genus, like all other Greater Antillean islands. He later (1928) described the Blue Mountain aquatic frog from Arntully, St. Thomas as *Eleutherodactylus orcutti* and *cunctator*, the latter a minor variant and synonym of the former. The material Dunn (1928) reported as *Eleutherodactylus grabhami* from Arntully (not cited by number in the text but USNM 73862-64) was later reidentified as *E. pantoni* by Lynn (probably = *E. pentasyringos*). Other contributions included an update and key for Jamaican frogs (Dunn 1927) and valuable natural history notes on what is now known as *Calyptahyla crucialis* (*Hyla lichenata*, 1929). Dunn's patronyms for some of his new species acknowledged several prominent Jamaican naturalists (E. Stuart Panton of Mandeville, Frank Cundall of the Institute of Jamaica, and Philip Henry Gosse) and his Jamaican hosts (Percy Junor of Spaldings and the unidentified Mr. Grabham). Dunn's etymology (or lack thereof) for *Hyla marianae*, a female patronym for "Marian", presented an interesting sidelight of mystery. The enigmatic "Marian" is neither mentioned nor identified in ANY of Dunn's works, and queries to Dunn's wife and friends (including Roger Conant and H. G. Dowling) yielded no clue to the identity of the "mystery woman" commemorated in the description of this very distinctive and attractive little treefrog.

Following a short hiatus, the MCZ contributions to West Indian herpetology continued with the work of Ernest E. Williams and his multitudes of students, primarily on *Anolis*. Their taxonomic contributions on Jamaica included the descriptions of four new subspecies of *Anolis grahamii* and *lineatopus* and the Blue Mountain endemic, *Anolis reconditus* (Underwood and Williams, 1959). Williams also maintained close contact with colleagues at the American Museum of Natural History (AMNH), Max Hecht and Karl Koopman, often traveling in the field with them. Hecht's (1952) work on *Aristelliger* contained information on the Jamaican *A. praesignis* and the description of a giant fossil species, *A. titan*. Etheridge (1965) synonymized *A. titan* with the Hispaniolan *A. lar*, but I disagree with this allocation.

The USNM Era

Although Leonhard Stejneger and Doris Cochran were doing extensive work elsewhere in the West Indies, their contributions to Jamaican herpetology were minimal. Stejneger's impeccable nomenclatoral and historical background compelled him to correct the common usage of names for the largest Jamaican anole (1899) and the Jamaican boa (1901), the latter a clear acknowledgment of Sloane's "Serpens major subflavus". Although USNM collectors (including E. A. Chapin, H. L. Clark, W. Harris, W. R. Maxon, G. S. Miller, Jr. and C. R. Orcutt) and collections provided resources for other researchers on Jamaica, the USNM staff added nothing further to the literature until my field work began in 1970.

Lynn and Grant Period

The fieldwork and collections of W. Gardner Lynn and Chapman Grant produced another spate of taxonomic advances and a great increase in our knowledge of the basic biology of the Jamaican herpetofauna. Lynn's extensive travels in the summers of 1932, 1936, and 1941 yielded two new Blue Mountain endemics (*Eleutherodactylus alticola* and *andrewsi*, Lynn 1937), the Portland Ridge cave frog (*E. cavernicola* Lynn, 1954), and the cryptic western limestone species *E. fuscus* (Lynn and Dent, 1943). Lynn and Dent (1942) also described a very distinctive little frog from western Jamaica as *Eleutherodactylus lewisi*, in honor of C. B. Lewis of the Institute of Jamaica. At this time, Gosse's name *luteolus* was being misapplied to the most common and widespread Jamaican *Eleutherodactylus*, but Lynn and Dent (1942) actually redescribed real *luteolus*, as pointed out by Goin (1953). Lynn (1937) also noted the introduction of the Cuban species *Eleutherodactylus ricordi* based on some specimens from Montego Bay. Schwartz (1965c) later demonstrated that *ricordi* and *planirostris* were distinct species and that *planirostris* was the common and widespread taxon that had been introduced in Jamaica

(1973c). In addition to his taxonomic contributions, Lynn (1940) provided substantial ecological information on Jamaican amphibians and a useful historical summary of the literature. Both he and his students used Jamaican material in embryological studies (Lynn, 1936, 1942, 1944; Lynn and Peadon, 1955; Schreckenberg, 1956) and papers on the descriptive morphology of the thyroid (e.g., Lynn and Walsh, 1957). The embryological studies were important additions to our understanding of clutch size and other life history parameters. Lynn (1936) also pointed out that the Sampson's (1904) paper was based on two and possibly three Jamaican species, none of which was actually *Hylodes* (= *Eleutherodactylus*) *martinicensis*. Lynn and all other authors believed that *E. martinicensis* had been introduced and established in Jamaica from the Lesser Antilles, however. Schwartz (1967c) documented that *martinicensis* and *johnstonei* were valid Lesser Antillean species and that the latter had been widely introduced elsewhere.

Grant's Jamaican fieldwork was performed in March-May 1937, March 1938, and April 1946. In searching for reptiles, he visited more xeric areas than Lynn, and their collections complemented each other nicely. Grant named 12 new taxa (three species and nine subspecies) that have received varying degrees of acceptance. In reviewing the large-scaled *Sphaerodactylus*, Grant (1939b) noted the striking differences between the north and south coast populations, corrected Barbour's (1921) confused treatment of them, and proposed *S. parkeri* for the south coast species. He also described a subspecies of the north coast *S. richardsoni* from the mouth of the Roaring River, St. Ann, but variation in *richardsoni* has never been carefully analyzed (Schwartz and Henderson, 1988: 190) and the status of *S. r. gossei* remains questionable.

To his credit, Grant was the first to recognize that Jamaica had a radiation of diploglossine lizards second only to Hispaniola and he did much to clarify the many synonyms and species erroneously reported from Jamaica. He first (1940a) described *Celestus duquesneyi* from the xeric Portland Ridge, Clarendon, a *hewardii*-like species notable not only for its blue-banded tail but also for its habitat (all other larger Jamaican *Celestus* are decided mesophiles). The much more widely distributed *Celestus barbouri* (Grant, 1940b) probably escaped detection and description because earlier authors often regarded the smaller species as juveniles or subadults of *occiduus* or *hewardii*; for those who recognized that these "juveniles" were adults, Cope's *Celestus impressus* (1868) was a convenient catch-all name for moderate sized species. Grant (1940b) discovered that *impressus* was actually based on one specimen of *hewardii* and one of *crusculus*, clearly not applicable to the intermediate-sized *barbouri*. Schwartz (1964: 56) later resolved the problem by designating the larger syntype (ANSP 9225, clearly the specimen Cope described) as the lectotype of *impressus*, firmly placing it in the synonymy of *hewardii*. Had the other syntype (ANSP 9226) been a justifiable lectotype selection, the name *impressus* would have antedated and replaced *crusculus* Garman (1887a).

Other aspects of Grant's diploglossine arrangement were far less rational. He regarded *Celestus occiduus* and *hewardii* as subspecifically related, which Cousens (1956) dismissed as ludicrous. His two new subspecies of *C. crusculus* were also rejected by Cousens, but she admitted that *molesworthi* was possibly worthy of recognition. Schwartz (1964: 55) disagreed and felt that there was definite subspecific differentiation of *crusculus* on Jamaica and that, "likely some or all of the races recognized by Grant are valid." He later (Schwartz and Henderson 1988: 97) concluded that "geographic variation in this species is more complex than considered by Grant." However, in the absence of more detailed analysis, Grant's arrangement still stands (see Problems).

Continuing his preoccupation with subspecies, Grant described two new dwarf boas, *Tropidophis maculatus stulli* and *T. pardalis stejnegeri*, and recognized Stull's *T. m. jamaicensis* as yet another Jamaican subspecies. Schwartz and Marsh (1960) transferred all three as subspecies of the Hispaniola *T. haetianus*, concluding that, "certainly there is no question of the distinctness of the three Jamaican forms from one another." S. Blair Hedges (personal communication, 1996) informs me that all three taxa are valid species, more closely allied to the Cuban radiation, thus confirming Grant's suspicions on their relationships if not his subspecific arrangement. Grant's other subspecies (three of *Anolis lineatopus*, one of *Sphaerodactylus argus*) were all very weakly defined and not placed in the perspective of variation throughout the species distribution. Thomas (1975) synonymized *S. argus henriquesi* with the nominate subspecies. Underwood and Williams (1959) retained *A. lineatopus neckeri* but synonymized *coxi* and *lynni* with the nominate subspecies.

It should be noted that Grant's Jamaican descriptions were published in such an idiosyncratic and inconsistent way that they had potential nomenclatural consequences, which fortunately never arose. The first and simplest problem involves the date of publication of the Institute of Jamaica *Science Series Bulletins* on the herpetology of Jamaica and the Cayman Islands (Grant, 1940b, 1940c). Grant consistently listed the date of both as 1941 in his personally issued bibliographies, although he cited the dates as 1940 in some of his later publications (e.g., Grant, 1951). Lynn's bibliography (USNM reprint files) lists *The Herpetology of Jamaica* as published in 1940. Grant greatly respected Leonhard Stejneger and maintained close contact with both him and Doris Cochran throughout his career, so it is likely that Grant sent Stejneger copies of his papers as soon as possible. The USNM copy of the Jamaican monograph is marked (in Stejneger's hand) "received 30 Aug 1941," and the *Natural History Notes of the Natural History of Jamaica* for April 1941 (Vol. 1, No. 1) announces that the publication was "now on sale at the Institute bookshop." I was unable to resolve the date problem at the library of the Institute of Jamaica in the 1970s, but things were somewhat in disarray at that time, so I continue to use the "traditional" date of 1940 for the only species actually described in the 1940b monograph, *Celestus barbouri*. It is clear that the "preliminary descriptions" in *Jamaica Today* (Grant, 1940a) appeared first since Grant cited it in 1940b issue (including pagination) as the original description for most of the new taxa. However, both *Tropidophis maculatus stulli* and *T. pardalis stejnegeri* were listed as "subsp. nov." in the 1940b issue, with no mention of their previous description in the 1940a issue. It could be argued that in the *Jamaica Today* article's the descriptions were not nomenclaturally valid descriptions, published in a journal that was not "generally available," but all new taxa described therein were accompanied by type localities, numbers for the holotypes, and diagnostic characters, albeit brief. It is unclear why Grant considered seven of the nine taxa as validly described in the 1940a issue but specifically proposed the two *Tropidophis* as "new" in the 1940b issue, but the original description of all must stand as in the 1940a issue. It should also be noted that although he clearly dedicated the taxon to "Mrs. Olive Stull Davis" (1940a,b) he used a masculine ending for his patronym. Schwartz and Fowler (1973: 131) properly corrected it to *stullae*, but without comment.

Goin/University of Florida

Coleman Jett Goin made his primary contribution in describing and quantifying the pattern of polymorphism in Jamaican *Eleutherodactylus*, with preliminary data on the genetic basis of this variation (Goin, 1950, 1954, 1958, 1960). His summer fieldwork in 1948, 1950, 1952, and 1957 resulted in sizable herpetological collections that formed the basis of his genetic studies and were also used in his work on the maxillary dentition of frogs (Goin, 1959a). However, taxonomic progress was (and still is) the inevitable result of fresh field collections and Goin's were no exception. Until Goin's work, few researchers had attached much importance to the lack of agreement between Gosse's (1851) description of *Litoria luteola* and the common *Eleutherodactylus* found throughout Jamaica to which the name had been applied. Dunn perhaps had an inkling when he described *E. gossei* (1926a) but he still mistakenly referred other *gossei* to *luteolus*. Lynn (1940) synonymized *gossei* with *luteolus* to return to the status quo. Goin collected specimens that matched Gosse's description and convincingly demonstrated that *E. luteolus* was a western Jamaican species with a fairly restricted distribution in that limestone region (1953). Although *Eleutherodactylus lewisi* (Lynn and Dent, 1942) was clearly a synonym of *luteolus*, Goin never suggested it and the synonymy first appeared in Cochran (1961: 45), possibly on the advice of Dr. Lynn. Appropriately, the most widespread and abundant Jamaican frog was now named *Eleutherodactylus gossei*, a fitting tribute to Gosse's pioneering contributions. Goin also made the first attempt at a phylogeny of Jamaican frogs with his comments on the relationships of the *gossei* group.

Goin described only one new species, *Eleutherodactylus lynni* (Goin and Cooper, 1950), which Schwartz and Fowler (1973) synonymized with *E. cundalli* (but see Problems). Goin also noticed that Taylor (1952) described a Dunn specimen of *Hyla wilderi* (erroneously cataloged with Panamanian locality data) as a new species, *Hyla shrevei*. Ironically, Taylor (1952: 1) stated of his supposed new species, "It is presumed that it is a rivulet species, living in the neighborhood of small streams, rather than a bromeliad species." Not only was Taylor's locality information for this species incorrect, but also his assumptions about its ecology were also 100% wrong, since *Hyla wilderi* is a bromeliad obligate. Although the University of Florida was (and continues to be) very involved in various aspects of Jamaican biology, the only other herpetologically notable paper of this period was Laessle (1961), which contains considerable information on hylid frog larvae and has import for other Jamaican bromeliadicoles.

The Underwood Era

After his Army service in India during World War II, Garth Underwood accepted a position at the University of the West Indies, teaching at both the Jamaican and Trinidadian campuses. His remarkably diverse reptilian interests utilized a variety of Jamaican species for broader evolutionary studies, including, for example, papers on the visual system (1951), digital structure (1954a), systematics

(1954b), the circulatory system (Adams *et al.*, 1957), and syntheses of evolutionary processes (1977). Underwood was active in the Natural History Society of Jamaica, becoming president in 1953, and he published several popular and educational papers in their *Natural History Notes* (1949, 1950,1954c,d,e), including his multipart "Introduction to the Study of Jamaican Reptiles" (1951-52). His taxonomic contributions include the classic study of Jamaican anoles with Ernest Williams (1959), which resolved the status of older names by examination of type specimens, synonymized most of the subspecies Grant (1940b) recognized, and reanalyzed variation of each species throughout Jamaica, resulting in a redefined *Anolis grahamii* (with two subspecies) and *lineatopus* (with four subspecies). They also described the Blue Mountain endemic, *Anolis reconditus*. Later, Underwood (1959a) described *Diploglossus microblepharis*, which remains known only from the type, and reraised the question of whether the diploglossine genera *Celestus* and *Diploglossus* are indeed distinct, a debate that continues unresolved (see Problems). Underwood's departure from the West Indies in 1960 marked the end of a productive and important period of contributions by a resident biologist. However, it is noted with pleasure that Dr. Underwood has recently (1993) continued his Antillean studies with a paper on the *Clelia* of the Lesser Antilles, long an area of special import to him.

In addition to Underwood, other members of the Natural History Society also published herpetological information in the Natural History Notes (nearly 50 titles by my count). Some of these were trivial, but others contained valuable ecological (e.g., Lewis, 1941, 1943; Panton, 1952), behavioral, or distributional data, including information on the spread of *Eleutherodactylus martinicensis* (= *johnstonei*) from the area of original introduction in Kingston (Perkins, 1942; Anonymous, 1943; Jeffrey-Smith, 1946).

Modern Era

It is only appropriate that the discussion of the "modern literature" should begin with the Harvard contributions. Under the tutelage of Ernest E. Williams, Harvard students and postdocs produced an enormous number of papers on virtually all aspects of anole biology, many of them involving Jamaican species only obliquely if at all. A representative (but by no means complete) selection of those most important for Jamaica follows, arranged alphabetically by author: Blake (1986), Floyd and Jenssen (1983), Gorman and Atkins (1968a), Hicks (1973), Jenssen (1973, 1977, 1979), Jenssen and Andrews (1984), Jenssen and Nunez (1994), Lazell (1966), Licht and Gorman

(1970), Rand (1964a, 1967a,b), Schoener and Schoener (1971a), Taylor and Gorman (1975), Trivers (1976), Wyles and Gorman (1980a). Although most of the data for the papers were collected on short visits to Jamaica, both A. S. Rand (August 1961 - June 1962 in Kingston) and Thomas A. Jenssen (early 1970s in Mandeville) were resident while conducting their ecological studies. Certainly the major thrust of the Harvard studies was *Anolis*, but other Jamaican contributions included an important paper on xenodontine snakes (Maglio, 1970) and an insightful thesis on relationships of *Eleutherodactylus* (Flores, 1984).

Harvard has not had a monopoly on *Anolis*, however, and several recent authors discussed Jamaican species, e.g., Landwer *et al.* (1995), Losos (1990a,b,c, 1992), Losos and Sinervo (1989), Powell and Russell (1992). Peter Vogel at the University of the West Indies, Mona, has studied *Anolis lineatopus* in some detail (Vogel, 1983, 1984; Vogel and Brockhusen-Holzer, 1984; Vogel and Curio, 1983; Vogel *et al.*, 1986). He also published an interesting paper on *Cyclura* (Vogel *et al.*, 1996), adding substantially to Woodley's (1971, 1980) "rediscoery" of the species.

Frogs were much less popular than anoles and only a few papers on ecology and morphology appeared during this period: Garrick *et al.* (1985), Lannoo *et al.* (1987), Pough *et al.* (1977), Stewart (1979), Stewart and Martin (1980), and Trueb (1970). Mahon and Aiken (1977) in the "first" report of *Rana catesbeiana* from Jamaica noted that 22 pairs were intentionally introduced in the Black River in March 1967 and were well established by the time of their paper. Unfortunately, as early as April 1946 Chapman Grant reported sight records of ranids (which he suspected were *R. catesbeiana*) from the "Baldwin, Main, or Parnassus Rivers" (Grant 1946). Taxonomic papers included some by Crombie, Hedges, and Schwartz, Trueb (1972), and Trueb and Tyler (1974). Trueb (1972) unsuccessfully attempted to have Harlan's *Hyla crucialis* suppressed for nomenclatural purposes. Trueb and Tyler (1974) ambitiously attempted to resolve higher systematics of Antillean hylids but suffered from a lack of field experience in the area. The genus *Osteopilus* was resurrected for most of the casque-headed species (including the Jamaican *Hyla brunnea*), but *lichenata* (= *crucialis*) was placed in a monotypic genus, *Calyptahyla* (see Problems).

Mike Seidel included the Jamaican pond-turtle in his revisions of Antillean emydids (1988a, 1996; Seidel and Atkins, 1987).

In terms of cumulative field time, comprehensiveness of collections, and geographic scope, three groups are notable in the "Modern" period: Al Schwartz, Blair Hedges, Ron Crombie and their respective associates. Each of these will be discussed in greater detail.

Albert Schwartz

As noted previously, Albert Schwartz did not publish extensively on exclusively Jamaican topics but this is somewhat misleading in a number of ways. Schwartz was a master of including nuggets of information (often substantial ones) on topics quite unrelated to the title of the paper. For example, Schwartz (1964b) included very significant taxonomic information on Jamaican "*Diploglossus*" (including designation of a lectotype for *D. impressus*) in a paper on the Hispaniolan *costatus* group. I learned very early that each and every Schwartz paper should be read carefully and thoroughly.

Schwartz's Jamaican fieldwork began in 1961 (June-August) with additional collections in 1967 (June-September), 1969, and August 1970 (as summarized in Schwartz and Fowler, 1973). The earliest collection was the basis for Buden's (1966) revision of Jamaican *Dromicus* (now *Arrhyton*, sensu Maglio, 1970), including the description of *D. polylepis* from the northeast coast. With the exception of Thomas's (1975) *Sphaerodactylus argus* group review, the rest of the reptile collections were never summarized. However, Schwartz and Fowler (1973) used the frog collections to produce the humbly titled "Progress Report," with the stated purpose of bringing the, "knowledge of Jamaican frogs closer to that of the frogs of other Greater Antillean islands." It succeeded admirably. With his characteristic thoroughness, Schwartz (and his associates) collected fresh material of every Jamaican species except *Eleutherodactylus alticola*, and the wealth of associated ecological data presented in the 1973 paper far surpassed everything accumulated by others to that point. The keen Schwartzian eye for variation resulted in the description of four new subspecies (one each of *Eleutherodactylus cundalli* and *gossei*, and two of *E. pantoni*), two of which were later elevated to full species. Schwartz and Fowler (1973) also synonymized *E. lynni* (Goin and Cooper, 1950) with *E. c. cundalli*, but I remain unconvinced that this was justified (see Problems).

Other contributions included description of the bromeliad-dwelling *Diploglossus fowleri* (Schwartz, 1971a), a thorough summary of information on the Jamaican iguana in a review of the genus *Cyclura* (Schwartz and Carey, 1977), and a re-evaluation of the distinctive and unique holotype of *Sphaerodactylus gilvitorques* (Thomas and Schwartz, 1974). Thomas's (1975) review of the *Sphaerodactylus argus* group included the description of the striking, ocellate

Sphaerodactylus semasiops. (I've tried hard to forgive Richard for getting to this amazing beast first.) Thomas also considered *Sphaerodactylus dacnicolor* and *oxyrhinus* to be conspecific (but recognizable subspecies), despite their thoroughly different habitat (terrestrial vs bromeliads) and broadly allopatric distributions at opposite ends of the island. They are once again considered full species.

S. Blair Hedges

Blair Hedges first visited Jamaica in 1986 with one of my field teams. With National Science Foundation support during his tenure at the University of Maryland and later at Penn State University he and his associates have made numerous trips to the island to continue their pursuit of tissue for molecular analysis. An impressive number and variety of papers have emerged from the Jamaican side of Blair's Antillean-wide project, some of which I even agree with. Among the topics addressed are *Anolis* relationships and biogeography (Burnell and Hedges, 1990; Hass *et al.*, 1993; Hedges and Burnell, 1990); *Eleutherodactylus* relationships (Bogart and Hedges, 1995; Hedges, 1989a,b); natural history data (Hedges, 1987, Hedges *et al.*, 1994); protein variation in *Typhlops jamaicensis* (Hedges, 1989c); and general Caribbean biogeography (Hedges, 1982, 1996 a,b,c; Hedges *et al.*, 1994). Hedges' primary interest was relationships of species throughout the Antilles and his interpretations of Jamaican species have been controversial. His Jamaican *Eleutherodactylus* phylogeny is so totally opposed to earlier gut-feeling concepts (Goin, 1954; Schwartz and Fowler, 1973; Crombie, 1977, 1985) and more carefully tested arrangements (Flores, 1984; Joglar, 1989) that it requires additional analysis. The few alpha-taxonomic conclusions (*Eleutherodactylus glaucoreius* as a valid species and confirmation of *pentasyringos* as distinct, Hedges, 1989a) are valid.

Ronald I. Crombie

My Jamaican fieldwork began with a 6-week trip in the summer of 1970 and continued mostly in the summer months in increments of 1-3 months, until 1985, with more than 15 months total field time. Collections were made in every month of the year except November and December. Virtually every stretch of passable road was traveled and overland hikes were made across every area of significant topography, including transects of the Cockpit Country from south to north (twice), west to east, and east to west. Collections were made from every parish and 95% of the herpetofauna was sampled. Taxonomic advances included description of the two minute Cockpit endemic *Eleutherodactylus* (Crombie, 1977, 1985) and elevation of *pentasyringos* to a full species (1985). General vertebrate surveys of Cabarita Island (Crombie *et al.*, 1984) and the Cockpit Country (Pregill *et al.*, 1992) were also published. Additional descriptive work on larval morphology of hylids and acoustic analyses of frog calls is planned, in addition to description of the new *Osteopilus*.

Problems

Throughout this chapter I have mentioned "problems" involving the taxonomic status or other aspects of some species; I will discuss them in taxonomic order. However, I should preface this section by stating that in general, data on the distribution, systematics, and natural history of Jamaican amphibians and reptiles remain so basic that very few "definitive" statements can be made. The following assessment is very much my personal opinion, but it is based on a substantial amount of data and experience. Some may disagree (and I know some do), and others may fault my "objectivity". I tend to expect an author's opinions on controversial subjects, not an egalitarian discourse.

Hylidae

The Jamaican hylid frogs remain a problem on several levels. The undescribed *Osteopilus* is represented by only four specimens from three localities, despite its broad distribution and apparent abundance (based on call surveys). Schwartz and Fowler's (1972) record of "*Calyptahyla*" *crucialis* from eight miles south Seaman's Valley, Portland is probably this species. *Osteopilus* sp. nov. is about the same size as *O. brunneus*, with the cranial ornamentation of *crucialis*, and its call is somewhat intermediate between the two. It is clearly not a hybrid (confirmed by molecular data, S. B. Hedges, personal communication), but it does negate the morphological distinction between the genera *Osteopilus* and *Calyptahyla*, hence my use of "*Calyptahyla*" *crucialis* in this chapter. My analysis of the osteology of *crucialis* based on two additional specimens (MCZ 11251 dry and USNM 252374, cleared and stained) indicates that most of Trueb and Tyler's osteological characters distinguishing *Osteopilus* and *Calyptahyla* are either equivocal (characters 1, 2, 19, 20) or aberrant (6) in the one skeleton they used. Both of my skeletons have a normal (= *Osteopilus*-like) cultriform process of the parasphenoid, very unlike that illustrated for BMNH 52.12.1 by Trueb and Tyler. The only differences between the skulls of *crucialis* and *brunneus* are minor details of decoration seen within other genera of casque-headed hylids (e.g. *Trachycephalus*). When a skeleton of *Osteopilus* sp. nov. becomes available I anticipate that the osteological distinctions between the two genera will be even more negligible. Morphologically, the major difference is the lack of nuptial pads in males of *Calyptahyla* and their presence in the four *Osteopilus* (including sp. nov.). Hedges (1996c) suggested that *Calyptahyla* was a synonym of *Osteopilus* based on molecular data, and I agree completely.

However, Hedges (1996c) also concluded that the two smaller Jamaican hylids (as well as the Hispaniolan *Hyla pulchrilineata* and *vasta*) were also referable to *Osteopilus*. This is not a novel idea, since Dunn (1926a) first suggested that *wilderi* and *marianae* might be "neotenic" forms of *brunneus* and Pregill (1981a) described a similar form of paedomorphosis in the Antillean bufonids. Although the skulls of *wilderi* and *marianae* do resemble juvenile *Osteopilus* in some respects, because of major differences in larval morphology (i.e., Lannoo *et al.*, 1987) I am skeptical of Hedges' arrangement. Clearly the generic level relationships of Jamaican (and other Antillean) hylids needs careful analysis.

Leptodactylidae

The roster of Jamaican *Eleutherodactylus* may not be complete. Zoogeography would argue for a high-elevation endemic in the John Crow Mountains and I have heard unfamiliar calls there. Call variation in *E. jamaicensis* suggests that east-west speciation may have occurred in this species. *Eleutherodactylus cundalli* also has unprecedented variation in morphology, call, and ecology, particularly in the startling reproductive behavior of the Windsor Cave "*cundalli*" reported by Diesel *et al.* (1995). Early in my fieldwork I noted differences in size, skin fragility, limb proportions, and calls of cave-associated versus forest *cundalli* in the western limestone regions, similar to the characters Goin and Cooper (1950) used to diagnose *E. lynni*. Schwartz and Fowler (1973) discounted this as intraspecific variation but I remain unconvinced.

Diploglossidae

Much remains to be done on the relationships and biology of the Jamaican galliwasps. As noted above, the subspecies of *Celestus crusculus* remain controversial. Grant's subspecies *molesworthi*, from the highly endemic northeast coast, is very distinctive and almost surely a valid species, but *cundalli* from the central plateau is poorly defined from the nominate, xerophitic subspecies. *Celestus barbouri*, *duquesneyi*, *fowleri*, and *microblepharis* are generally very poorly represented in collections, with two of them (*fowleri* and *microblepharis*) known only from the types. The validity of these species is not in question but their relationships and distributions would be greatly clarified by additional material. However, Dips (even large ones) can be extremely difficult to see, much less catch on a limestone substrate. I have visited all the type localities at least twice (and spent considerable time at or near that of *fowleri*), with very little success. Blair Hedges and Richard Thomas have had similar experiences with these taxa.

I agree with Cousens (1956) that Grant's subspecific arrangement of *hewardi* and *occiduus* is unrealistic. Certainly the giant *occiduus* is in serious decline from observations by Gosse (1851) and particularly from the astounding abundance in the recent Cockpit Cave remains reported by Pregill *et al.* (1992), but I think declarations that the species is extinct are premature. As noted above, dips in general and giant species in particular can be most elusive. Schwartz's discoveries of the huge *Diploglossus anelpistus* (Schwartz *et al.*, 1978) and *carraui* (Incháustegui *et al.*, 1985) in "well-collected" areas of the Dominican Republic are but two examples. Part of the problem may be that, other than size, the characters used to distinguish *occiduus* from *hewardi* are unreliable and nobody has yet critically examined the available material of the two taxa. Variation in *C. hewardi* has also never been analyzed and there are a few specimens (USNM and CM) from an uncharacteristically high elevation (Clydesdale) in the Blue Mountains that may be a distinct species.

Gekkonidae

Although I list *Aristelliger praesignis* as a monotypic species, there is a subspecies from the Swan Islands (*A. p. nelsoni*) that has never been formally synonymized (Bauer and Russell, 1993). The genus is remarkably conservative and characters to distinguish between unquestionably distinct species are few, so analysis of intraspecific variation is meaningless using traditional meristic and morphological characters. Jamaican, Cayman, and Swan Islands *praesignis* "look" different, and molecular data might reveal them to be distinct taxa. As noted previously, Etheridge (1965) synonymized the fossil species *Aristelliger titan* (Hecht, 1951) with the Hispaniolan *A. lar*. Although we have not found any additional fossils of giant *Aristelliger* in Jamaica, the presence of an Hispaniolan species in the Pleistocene of Jamaica would be a zoogeographic anomaly. Etheridge took a commendably conservative approach, but we have since prepared considerably more skeletal material of extant *Aristelliger* than was available to Etheridge and we find that other than size, there are few osteological characters that will distinguish between obviously distinct species either. Consequently, I prefer to recognize *titan* as a valid species on purely zoogeographic grounds.

The insular subspecies *Gonatodes albogularis notatus* is also problematical. Its distribution around major centers of human habitation in both Jamaica and Hispaniola smacks of an introduction, but the status of *albogularis* in northern South America and its supposed Middle American subspecies *fuscus* are also chaotic.

Jamaican *Sphaerodactylus* are in reasonably good shape, thanks to Thomas' (1975) careful review of the *argus* group. However, the very distinctive *S. gilvitorques* has never been rediscovered in Jamaica or elsewhere (Thomas and Schwartz, 1974), but there is no reason to doubt that the type actually came from Jamaica. As noted previously, Grant's (1940b) study of variation in *S. richardsoni* is suspect and the status of *S. r. gossei* needs verification. *Sphaerodactylus* sp. in my list is a diminutive taxon related to *goniorhynchus*, under study by Richard Thomas. It is a xeric, coastal species (as opposed to the mesophilic *goniorhynchus*) but with some seemingly "intermediate" populations near Kingston.

4

The Hispaniolan Herpetofauna: Diversity, Endemism, and Historical Perpectives, with Comments on Navassa Island

Robert Powell[1], *José A. Ottenwalder,*[2] *and Sixto J. Incháustegui*[3]

[1] *Department of Natural Sciences, Avila College, 11901 Wornall Road, Kansas City, Missouri 64145*

[2] *Conservation and Management of Biodiversity Project, United Nations Development Programme, P.O. Box 1424, Santo Domingo, República Dominicana*

[3] *Grupo Jaragua, El Vergel 33, El Vergel, Santo Domingo, República Dominicana*

Introduction

Once, in a casual conversation with Albert Schwartz, he spoke of Hispaniola as an "island of islands" in reference to the myriad pockets of dramatically different habitats often separated by only a few kilometers. The discussion had focused on the tremendous diversity of the Hispaniolan herpetofauna and the difficulty experienced by those who had never been there in accepting as valid the surprisingly vast number of species and subspecies that had been described. However, the diversity is real, amazing for an island so relatively small in area (ca. 76,500 km^2), and a visiting herpetologist quickly appreciates the distinctiveness of geographically proximate forms separated solely by habitat.

Geography and Physiography

The variety of amphibians and reptiles may be attributed largely to three factors: (1) the rugged and mountainous terrain of which the dominant relief features are parallel ranges that run primarily from the northwest and west in a generally easterly direction (Weil *et al.*, 1982; Lewis and Draper, 1990), (2) a variety of satellite islands of various sizes and with exceedingly different topographies and habitats, and (3) the peculiar geological history of the island, which is unique among the Greater Antilles.

The mountainous terrain of parallel ranges (Fig. 4.1), along with an extended complex of satellite ranges, results in an inordinately diverse topography. Extensive valleys lie between the primary ranges, which are often impressive in their elevation. Satellite ranges create narrow and often broken coastal lowlands, and a lowland plain covers much of the eastern end of the island. This structural complexity, in combination with the very high elevations and resultant rainshadows, produces a juxtaposition of, among other habitat types, harsh deserts, dry scrub forests, rainforests, and high-elevation pine savannas.

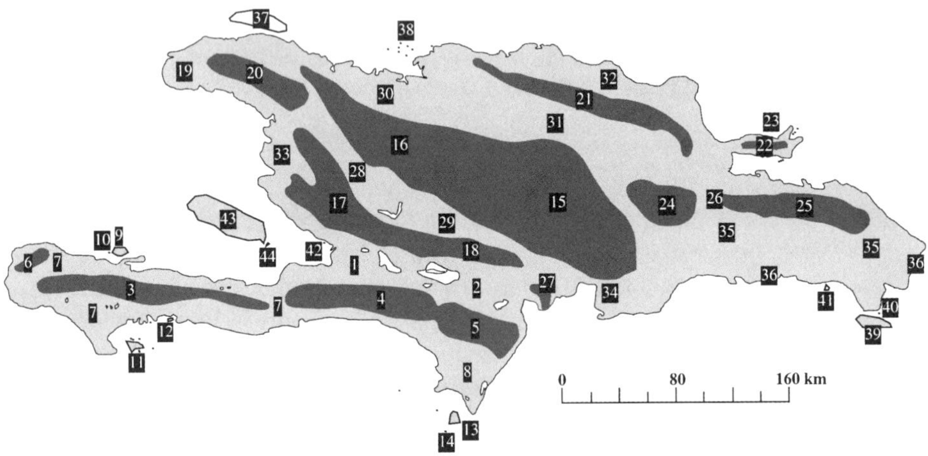

Figure 4.1. Biogeographic and physiographic features of Hispaniola (modified from Schwartz, 1980b; Schwartz and Henderson, 1984; Hedges, 1998). Darker shading indicates the approximate extent of major mountain ranges. **Key:** 1, Plaine du Cul de Sac; 2, Valle de Neiba; **South Island:** 3, Massif de la Hotte; 4, Massif de la Selle; 5, Sierra de Baoruco; 6, Monts Cartaches; 7, Peninsula de la Tiburon; 8, Península de Barahona; 9, Île Grande Cayemite; 10, Île Petite Cayemite; 11, Île à Vache; 12, Île Grosse Caye; 13, Isla Beata; 14, Isla Alto Velo; **North Island:** 15, Cordillera Central; 16, Massif du Nord; 17, Montagnes du Trou-d'Eau and Montagnes Noires; 18, Sierra de Neiba; 19, Pointe de Nord Ouest; 20, Montagnes du Nord Ouest; 21, Cordillera Septentrional; 22, Sierra de Samaná; 23, Península de Samaná; 24, Sierra de Yamasá; 25, Cordillera Oriental; 26, Los Haitises; 27, Sierra de Martín García; 28, Plateau Central; 29, Valle de San Juan; 30, Plaine du Nord; 31, Valle de Cibao; 32, northern coastal plain; 33, Vallée de l'Artibonite; 34, Llanos de Azua; 35, Llanura del Este; 36, Costera del Caribe; 37, Île de la Tortue; 38, Cayos Siete Hermanos; 39, Isla Saona; 40, Isla Catalinita; 41, Isla Catalina; 42, Île à Cabrit; **Île de la Gonâve:** 43, Île de la Gonâve; 44, Île Petite Gonâve.

Important satellite islands (Fig. 4.1), many with endemic taxa, range in size from Île de la Gonâve, with an area of approximately 650 km^2 and a maximum elevation of over 700 m near its southeasternmost point (Schwartz, 1980b), to individual cays in the Cayos Siete Hermanos, the smallest of which (Cayo Muertos) has an area measuring less than 0.02 km^2 and none are more than a few meters above sea level (Burns *et al.*, 1992). Other satellites of herpetological significance include Île de la Tortue, Grande Cayemite, Île à Vache, and Île Grosse Caye on the Haitian side of the main island, whereas Isla Saona, Isla Catalina, Isla Beata, and Isla Alto Velo lie on the Dominican side. In addition, Navassa Island, situated between Hispaniola and Jamaica, has a unique and apparently disproportionately large herpetofauna (Thomas, 1966a). With an area of < 4 km^2 (Proctor, 1959), this island supported, at least historically, eight species of reptiles, most of which demonstrate Hispaniolan affinities.

Although "Considerable controversy exists over the interpretation of the early tectonic evolution of Hispaniola" (Lewis and Draper, 1990), the peculiar history of the island lies largely in its origin as two paleoislands (Schwartz, 1978a, 1980b), although as many as four elements may have combined to form the present island (Lewis and Draper, 1990). The two major elements (the north and south paleoislands) ultimately were joined via the lateral translocation of southern Hispaniola; in effect, the south paleoisland "caught" the north island after the latter collided with the Bahama Platform (Lewis and Draper, 1990). The juncture of the paleoislands most likely occurred during the Miocene (Huebeck and Mann, 1991) and maybe as early as the Eocene (Khudoley and Meyerhoff, 1971); however, (correlated primarily with rising or falling sea levels) partial or complete separation may have occurred at intervals throughout the Pleistocene. Reef limestones deposited during the Pleistocene are currently exposed in coastal areas and in the former marine channel (Lewis and Draper, 1990). In the herpetological literature (e.g., Williams, 1961), the paleoislands have become known as the North and South islands, respectively (Fig. 4.1; subsequently used parenthetical numbers in the text correspond to those used in the figure to designate biogeographic and/or physiographic entities). The former marine strait is now a barren valley known as the Plaine de Cul-de-Sac in Haiti (1) and the Valle de Neiba in the Dominican Republic (2). Much of this valley still lies below sea level and it is characterized by four remaining large lakes, two of which are saline: Etang Saumâtre in Haiti and nearby Lago Enriquillo in the Dominican Republic.

Further complicating the historical biogeography of Hispaniola is the sketchy pattern of emergence and submergence affecting both paleoislands (Hedges, 1996b). The present Cordillera Central, Cordillera Oriental, and possibly a portion of the La Selle – Baoruco range were emergent during the mid-Tertiary (Bowin, 1975; Lewis, 1980; Maurrassee, 1982; McLaughlin *et al.*, 1991). The remainder of the South Island probably began to emerge during the Miocene, in correlation with the collision between the paleoislands, and the entire island took shape during the Pliocene and Pleistocene (Lewis and Draper, 1990). The Sierra de Neiba and the Sierra de Martín García, along the southern shore of the North Island, apparently emerged in the late Miocene or Pliocene (McLaughlin *et al.*, 1991).

The South Island is the smaller of the two paleoislands, with an area of over 9500 km^2, and is much less complex than the North Island. The principal relief features constitute a series of three major mountain ranges (3-5), which run from west to east, and a smaller, independent range, the Monts Cartaches (6), lies near the northwesternmost tip of the Tiburon Peninsula (7). The westernmost of the three major ranges is the Massif de la Hotte (3), the highest peak of which (Pic Macaya) has an elevation of 2347 m and lies near the western end of the range. The central range is the Massif de la Selle (4), with several peaks exceeding 2100 m. The easternmost of these ranges is the Sierra de Baoruco (5), with a maximum elevation of 2367 m (Loma del Toro). The first two of these ranges form the backbone of the narrow Tiburon Peninsula, the width of which ranges from 30 to 68 km. Extending south of the Sierra de Baoruco for approximately 85 km is the Península de Barahona (8). The Loma Gran Sabana, a limestone ridge with a maximum elevation of nearly 1100 m, vegetated by dry forest, and running essentially from north to south, effectively divides the peninsula into an extremely xeric western region and a somewhat less xeric eastern plain. Important satellite islands associated with the South Island are Île Grand Cayemite (9) and Île Petite Cayemite (10) off the northern shore of the Tiburon Peninsula, Île-à-Vache (11) and Île Grosse Caye (12) off the southern shore of the peninsula, and Isla Beata (13) and Isla Alto Velo (14) off the southern tip of the Península de Barahona.

The structurally much more complex North Island has an area of about 67,700 km^2. The most dominant relief feature is the Cordillera

Central (15), which extends to the northwest as the Massif du Nord (16). Its highest peak, Pico Duarte, with an elevation of 3087 m, is the highest point in the entire West Indies. Essentially paralleling the southern shore of the North Island and to the south of the Cordillera Central lie the Montagnes du Trou-d'Eau and Montagnes Noires (17), which extend to the east as the Sierra de Neiba (18). The highest peak, Pico Neiba, has an elevation of 2279 m. The northwesternmost part of the island, Pointe de Nord Ouest (19), is generally dry. Uplands in that area form the Montagnes du Nort Ouest (20). To the east and along the northern coast, parallel to the Cordillera Central is the Cordillera Septentrional (21), the greatest elevation of which (Pico Diego de Ocampo) is 1250 m. Smaller, more or less isolated ranges largely to the east include the Sierra de Samaná (22) (highest point, Monte Mesa, 606 m) on the Península de Samaná (23) and the Sierra de Yamasá (24), essentially an eastern isolate of the Cordillera Central which blends into the Cordillera Oriental (25), situated to the south of the Bahía de Samaná. The maximum elevation in these ranges is 856 m at the Picos Siete Cabezas. The western extent of the Cordillera Oriental extends into the karst landscape of the Los Haitises region (26), which is sometimes considered separately as the Sierra de El Seibo. The Sierra de Martín García (27) is an eastern isolate of the Sierra de Neiba. Its highest peak is Loma del Curro, with an elevation of 1343 m. This small range is literally a mesic island surrounded by desert.

Important lowlands of the North Island include the Plateau Central (28) – Valle de San Juan (29), which lies between the Cordillera Central – Massif du Nord to the north and the Montagnes du Trou-d'Eau – Montagnes Noires – Sierra de Neiba to the South. This valley is locally mesic, but largely xeric as a result of rainshadows from the northeasterly trade winds. The Plaine du Nord (30), north of the Cordillera Central – Massif du Nord, is largely mesic near the western coast, but becomes increasingly xeric as it extends to the east and becomes the Valle de Cibao (31). Farther to the east, the area becomes gradually more mesic until the easternmost portions are among the most mesic lowlands on the Dominican side of the island. North of the Cordillera Septentrional lies the northern coastal plain (32). South of the Massif du Nord is the Vallée de l'Artibonite (33), which is mesic along the coast and increasingly more xeric in inland areas. The arid Llanos de Azua (34) lie to the south and in the rainshadow of the Cordillera Central. Much of the eastern portion of the island is flat to somewhat hilly. This Llanura del Este (35) is moderately mesic. Craggy dogtooth limestone and pocket beaches constitute most of the Costera del Caribe (36).

Important satellite islands include Île de la Tortue (37) in the northwest. The Cayos Siete Hermanos (38) form an archipelago of small islets off the north central coast. Isla Saona (39) and the much smaller Isla Catalinita (40) are situated off the southeastern tip of the island, Isla Catalina (41) lies a few kilometers to the west off the southern coast, and tiny Île à Cabrit (42) is situated off the southwestern coast of the North Island.

Île de la Gonâve (43) and its minute satellite, Île Petite Gonâve (44), lie slightly closer to the North than to the South Island, but a currently submerged bank extending to the southeast provided a historical connection with both paleoislands (Schwartz, 1980b). In addition to several insular endemics, its herpetofauna is composed of a combination of both North and South island species.

Navassa Island (not shown in Fig. 4.1) is somewhat of an anomaly. Although more proximate to Hispaniola than Jamaica (approximately 65 km west of Cap de Irois, Haiti), it cannot be considered a true satellite, because of distance and faunal affinities (Thomas, 1966a). The island itself is a small limestone mass characterized by steep cliffs, about 10-15 m in height, outlining a central plateau with a maximum elevation of nearly 70 m (Proctor, 1959; Turner, 1960).

Vegetation

Because of its very diverse physiography, Hispaniola supports a diversity of floral communities. Although difficult to summarize, Hedges (1999), in describing West Indian vegetation patterns, provided a useful overview (paraphrased herein). Characterization of forest types is largely from SEA/DVS (1990), Hager and Zanoni (1993), and Hedges (1999).

Prevailing winds from the northeast carry moisture which typically precipitates on northern and eastern regions. Southern areas usually are dry. Vegetation patterns generally follow rainfall patterns. Well-developed moist forests occur on northern and eastern slopes and dry, xerophytic vegetation is found in southern regions or in rainshadows.

Before human alterations, Hispaniola was largely forested (Hedges, 1999). Lowland rainforests, with high canopies and tall buttressed trees, graded into montane rainforests on lower slopes. These forests have high, dense canopies and occur on slopes up to about 1500 m elevation receiving more than 2 m of rainfall annually. Cloud forests occur at cloud level (usually > 1000 m). These forests have low canopies (10-15 m) and are subject to the highest humidity of any forest type. Pine forests and elfin woodlands can be found at higher elevations (from about 1000 m to >2400 m). Pine forests tend to be relatively dry, whereas elfin woodlands are characterized by windblown tangles of mossy trees and shrubs up to 5 m. Wet limestone forests, with thin soils and canopies up to 15-25 m in height, cover karst regions, usually in areas receiving less than 2 m of annual precipitation. Dry scrub forests, often associated with coastal areas, have low canopies (to 10 m), many thorny shrubs and cacti, and grow on exposed rock or very thin soil in areas receiving less than 1 m rainfall annually. These occur in the drier south and those areas shielded by mountain ranges from the prevailing winds.

Essentially all the historically widespread lowland rainforests have been destroyed, and were the first to disappear after colonization of the island (Powell and Henderson, 1996a). Most other forest types are disappearing but may still be found in patches throughout the island. Wet limestone and dry scrub forests are being destroyed at rates slower than those of most other types, mainly because of difficulties posed for human access and exploitation. Still, many areas, such as those in the vicinity of Monte Cristi, have been sufficiently altered that the subsequent secondary growth produces an impenetrable tangle of cacti and thorny shrubs. Montane rainforests remain largely on only the steepest mountain slopes, and even in regions where the forest superficially appears to be natural many components consist of cultivated species (Lenart *et al.*, 1997). Haiti has been almost entirely deforested. Some recent estimates of remaining forested areas are as low as 1% (e.g., Hedges, 1996b; see Chapter 7). The only remaining extensive tracts of natural forest in the Dominican Republic are in the Cordillera Central, but patches, some of moderate size, exist in other areas (e.g., Sierra de Baoruco).

Navassa is less xeric than most small West Indian islands, although it is comparable to the moderately xeric scrub forests on Hispaniola, and supports a relatively luxuriant flora dominated by *Ficus* trees and occasional savanna-like areas (Proctor, 1959; Turner, 1960), although the latter may be attributable to human activities (Thomas, 1966a).

Table 4.1. Amphibians of Hispaniola.[a]

Species	Status	Origin or other distribution
	Class Amphibia, Order Anura	
	Family Bufonidae	
Bufo fluviaticus	Endemic	North Island
Bufo fractus	Endemic	North Island
Bufo guentheri	Endemic	North Island invader of South Island
Bufo marinus	Introduced	Central/South America (widely introduced)
	Family Hylidae	
Hyla heilprini	Endemic	Islandwide
Hyla pulchrilineata	Endemic	Islandwide
Hyla vasta	Endemic	Islandwide
Osteopilus dominicensis	Endemic	Islandwide
	Family Leptodactylidae	
Eleutherodactylus abbotti	Endemic	Islandwide
Eleutherodactylus alcoae	Endemic	South Island
Eleutherodactylus amadeus	Endemic	South Island
Eleutherodactylus apostates	Endemic	South Island
Eleutherodactylus armstrongi	Endemic	South Island
Eleutherodactylus audanti	Endemic	South Island invader of North Island
E. a. audanti		South Island
E. a. melatrigonum		North Island
E. a. notidodes		North Island
Eleutherodactylus auriculatoides	Endemic	North Island
Eleutherodactylus bakeri	Endemic	South Island
Eleutherodactylus brevirostris	Endemic	South Island
Eleutherodactylus caribe	Endemic	South Island
Eleutherodactylus chlorophenax	Endemic	South Island
Eleutherodactylus corona	Endemic	South Island
Eleutherodactylus counouspeus	Endemic	South Island
Eleutherodactylus darlingtoni	Endemic	South Island
Eleutherodactylus dolomedes	Endemic	South Island
Eleutherodactylus eunaster	Endemic	South Island
Eleutherodactylus flavescens	Endemic	North Island
Eleutherodactylus fowleri	Endemic	South Island
Eleutherodactylus furcyensis	Endemic	South Island
Eleutherodactylus glandulifer	Endemic	South Island
Eleutherodactylus glanduliferoides	Endemic	South Island
Eleutherodactylus glaphycompus	Endemic	South Island
Eleutherodactylus grahami	Endemic	North Island
Eleutherodactylus haitianus	Endemic	North Island
Eleutherodactylus heminota	Endemic	South Island

Table 4.1 cont.

Eleutherodactylus hypostenor	Endemic	South Island
Eleutherodactylus inoptatus	Endemic	Islandwide
Eleutherodactylus jugans	Endemic	South Island
Eleutherodactylus lamprotes	Endemic	South Island
Eleutherodactylus leoncei	Endemic	South Island
Eleutherodactylus lucioi	Endemic	North Island
Eleutherodactylus minutus	Endemic	North Island
Eleutherodactylus montanus	Endemic	North Island
Eleutherodactylus sp. "N"	Endemic	North Island
Eleutherodactylus nortoni	Endemic	South Island
Eleutherodactylus oxyrhynchus	Endemic	South Island
Eleutherodactylus parabates	Endemic	North Island
Eleutherodactylus parapelates	Endemic	South Island
Eleutherodactylus patriciae	Endemic	North Island
Eleutherodactylus paulsoni	Endemic	South Island
*Eleutherodactylus pictissimus**	Endemic	South Island invader of North Island
E. p. pictissimus		South Island
E. p. apantheatus		Valle de Neiba / Plaine de Cul-de-Sac
E. p. eremus		North Island
Eleutherodactylus pituinus	Endemic	North Island
Eleutherodactylus poolei	Endemic	North Island
Eleutherodactylus probolaeus	Endemic	North Island
Eleutherodactylus rhodesi	Endemic	North Island
Eleutherodactylus rufifemoralis	Endemic	South Island
Eleutherodactylus ruthae *	Endemic	Islandwide
E. r. ruthae		North Island
E. r. aporostegus		South Island
E. r. bothroboans		North Island
E. r. tychathrous		North Island
Eleutherodactylus schmidti	Endemic	North Island
E. s. schmidti		North Island
E. s. limbensis		North Island
E. s. rucillensis		North Island
Eleutherodactylus sciagraphus	Endemic	South Island
Eleutherodactylus semipalmatus	Endemic	South Island
Eleutherodactylus thorectes	Endemic	South Island
Eleutherodactylus ventrilineatus	Endemic	South Island
Eleutherodactylus warreni	Endemic	North Island
Eleutherodactylus weinlandi *	Endemic	North Island
E. w. weinlandi		North Island
E. w. chersonesodes		North Island
E. w. paralius		North Island

Table 4.1 cont.

Eleutherodactylus wetmorei	Endemic	South Island invader of North Island
E. w. wetmorei		South Island
E. w. ceraemerus		South Island
E. w. diplasius		South Island
E. w. sommeri		North Island
Leptodactylus dominicensis	Endemic	North Island

*Some North Island isolates are not assigned to subspecies.

Family Ranidae

Rana catesbeiana	Introduced	Eastern North America (widely introduced)

[a] The species list is taken from Powell et al. (1996a), but with the addition of a species currently being described (Hedges, 1997). Origins of endemic species are from Schwartz (1980b), except for taxa subsequently described or when the taxonomy has changed. Non-Hispaniolan distributions are from Schwartz and Henderson (1991). Please note that footnotes are inserted after each family.

Diversity and Endemism

The Hispaniolan herpetofauna consists of 65 currently recognized species of amphibians (Table 4.1) and 146 recognized species of reptiles (Table 4.2). The taxonomy used is that of Powell *et al.* (1996a), with exceptions noted in the tables. Anurans are placed in only four families and six genera. Twenty subspecies in six species are recognized. Reptiles are placed in three orders, 15 families, and 24 genera [but see comments regarding iguanian lizard families *sensu* Frost and Etheridge (1989), in Powell *et al.* (1996a) and Hedges (1996a)]. Two hundred and thirty-one subspecies in 55 species are recognized, but six of these represent situations in which a single subspecies of a polytypic species inhabits Hispaniola. Of these, one subspecies is introduced (*Anolis c. cristatellus*), three are endemic (*Trachemys stejnegeri vicina*, *Sphaerodactylus elegans punctatissimus*, and *Cyclura c. cornuta*), one clearly is native (*Gonatodes albogularis notatus*), and whether the last (*Mabuya bistriata sloanei*) is native or introduced is uncertain.

The relative paucity of families and genera compared to the diversity of species on Hispaniola is consistent with the general pattern seen throughout the Greater Antilles, in large part because both the insular and the regional faunas are dominated by three genera: *Eleutherodactylus*, *Sphaerodactylus*, and *Anolis* (see Chapter 7). One likely cause of this preponderance of so few genera is the difficulty and subsequent rarity of overwater colonization of an island [Hedges, 1996a, and references therein; but also see comments in Crother and Guyer (1996a,b) and in rebuttals by Hedges (1996d,e)]. This generalization may apply, even though West Indian *Eleutherodactylus* appar ently arose by proto-Antillean vicariance (Hedges, 1989a,b, 1996a,b,c), and Guyer and Savage (1996a,b) have proposed an alternative scenario. Nevertheless, that ancestral Hispaniolan *Anolis* and *Sphaerodactylus* underwent spectacular radiations is apparent from evidence suggesting closer affinities among Hispaniolan forms than among more morphologically similar species from different islands (see discussion of the ecomorph concept in Chapter 7).

The possible differences in origins may also explain to some degree the obvious differences in the patterns seen among the species densities of amphibians (Fig. 4.2) and reptiles (Fig. 4.3). Amphibian distribution evidently and logically is correlated to a large degree with moisture (which, in turn, is correlated to some extent with elevation) (Hedges, 1997). Increased species densities, indicated by darker shades in Fig. 4.2, easily can be superimposed on the uplands indicated in Fig. 4.1. Historical factors, however, may complicate this issue and apparently are largely responsible for the fact that the highest elevations (e.g., the Cordillera Central) do not support a greater number of species than lower ranges on the South Island. Conversely, low, xeric regions, less hospitable to amphibians, exhibit consistently lower densities. The greatest reptilian species densities, in contrast, are in many instances associated with areas along the edges of biogeographic regions (Fig. 4.3). This ecotonal effect is most obvious where junctions between regions coincide with the greater boundary between the two paleoislands. Reptiles may be more likely than amphibians to move readily into areas bordering their historical ranges, creating concentrations of diversity along boundary zones.

The dramatically greater number of reptilian subspecies is testament to the remarkable adaptability of many reptiles when faced with the myriad habitats on Hispaniola, and testament to the pronounced differences between even geographically proximate habitats – differences even the most mobile reptiles frequently fail to overcome. The relatively few amphibian subspecies (all in the genus *Eleutherodactylus*), in contrast, speaks to the confinement of many anurans largely to moist upland regions (Fig. 4.2) (Hedges, 1999). In addition, the many ecologically versatile reptilian species with broad distributions contrasts sharply with the larger percentage of amphibian species endemic to ranges which are small and often restricted to one type of habitat (Schwartz and Henderson, 1991; see Chapter 7).

Table 4.2. Reptiles of Hispaniola.[a]

Species	Status	Origin or other distribution
	CLASS REPTILIA, ORDER TESTUDINES	
	Family Emydidae	
Trachemys decorata	Endemic	Uncertain (see discussion in Schwartz, 1980b)
Trachemys stejnegeri	West Indian	North Island, Bahamas, Puerto Rico (widely introduced)
T. s. vicina	Endemic	North Island
	ORDER SQUAMATA, SUBORDER AMPHISBAENIA	
	Family Amphisbaenidae	
Amphisbaena caudalis	Endemic	South Island
Amphisbaena gonavensis[b]	Endemic	Île de la Gonâve
Amphisbaena innocens	Endemic	South Island invader of North Island
Amphisbaena hyporissor[b]	Endemic	South Island
A. h. hyporissor		South Island
A. h. leberi		South Island
Amphisbaena manni	Endemic	North Island invader of South Island
	SUBORDER LACERTILIA	
	Family Anguidae	
Celestus agasepsoides	Endemic	South Island invader of North Island
Celestus anelpistus	Endemic	North Island
Celestus carraui	Endemic	North Island
Celestus costatus	Endemic	Islandwide; Navassa
C. c. costatus		South Island
C. c. aenetergum		North Island - Isla Catalinita
C. c. chalcorhabdus		North Island
C. c. emys		North Island - Île de la Tortue
C. c. leionotus		North Island
C. c. melanchrous		North Island
C. c. neiba		North Island
C. c. nesobous		South Island - Île-à-Vache

Table 4.2 cont.

C. c. oreistes		South Island
C. c. psychonothes		North Island
C. c. saonae		North Island - Isla Saona
Celestus curtissi	Endemic	North Island invader of South Island
C. c. curtissi		North Island, Île de la Gonâve
C. c. aporus		South Island
C. c. diastatus		North Island
C. c. hylonomus		North Island
Celestus darlingtoni	Endemic	North Island
Celestus haetianus	Endemic	South Island
C. h. haetianus		South Island
C. h. mylicus		South Island
C. h. surdus		South Island
Celestus macrotus	Endemic	South Island
Celestus marcanoi	Endemic	North Island
Celestus sepsoides	Endemic	Islandwide
Celestus stenurus	Endemic	Islandwide
C. s. stenurus		South Island
C. s. alloeides		North Island
C. s. rugosus		North Island
C. s. weinlandi		North Island, Valle de Neiba/Plaine de Cul-de-Sac
Celestus warreni	Endemic	North Island
	Family Gekkonidae	
Aristelliger expectatus	Endemic	Islandwide
Aristelliger lar	Endemic	Islandwide
Gonatodes albogularis	Native	Central and South America, Cuba
G. a. notatus	West Indian	Jamaica, Cayman Islands
Hemidactylus haitianus	West Indian	Cuba, Puerto Rico
Hemidactylus mabouia	Native (?)	North and South islands; Africa, Neotropics
Phyllodactylus wirshingi	West Indian	Puerto Rico
P. w. hispaniolae	Endemic	North Island, Valle de Neiba
P. w. sommeri	Endemic	North Island

Table 4.2 cont.

Sphaerodactylus altavelensis	Endemic	North Island invader of South Island
S. a. altavelensis		South Island, Isla Alto Velo
S. a. brevirostratus		North and South islands, Plaine de Cul-de-Sac
S. a. enriquilloensis		North and South islands, Valle de Neiba
S. a. lucioi		North Island
Sphaerodactylus armstrongi	Endemic	South Island
S. a. armstrongi		South Island
S. a. hypsinephes		South Island
Sphaerodactylus asterulus	Endemic	North Island
Sphaerodactylus callocricus	Endemic	North Island
Sphaerodactylus cinereus	Endemic	North Island invader of South Island
S. c. cinereus		Plaine de Cul-de-Sac
S. c. stejnegeri		North Island
Sphaerodactylus clenchi	Endemic	North Island
S. c. clenchi		North Island
S. c. apocoptus		North Island
Sphaerodactylus cochranae	Endemic	North Island
Sphaerodactylus copei	Endemic	South Island invader of North Island
S. c. copei		South Island, Plaine de Cul-de-Sac
S. c. astreptus		South Island
S. c. cataplexis		South Island, Bahamas (introduced)
S. c. deuterus		Île de la Gonâve
S. c. enochrus		South Island
S. c. pelates		South Island
S. c. picturatus		South Island
S. c. polyommatus		South Island, Île Grande Cayemite
S. c. websteri		South Island
Sphaerodactylus cryphius	Endemic	South Island, Valle de Neiba
Sphaerodactylus darlingtoni	Endemic	North Island
S. d. darlingtoni		North Island
S. d. bobilini		North Island

Table 4.2 cont.

S. d. mekistus		North Island
S. d. noblei		North Island
Sphaerodactylus difficilis	Endemic	North Island invader of South Island
S. d. difficilis		North Island
S. d. anthracomus		South Island
S. d. diolenius		North Island
S. d. euopter		North Island - Île de la Tortue
S. d. lycauges		North Island
S. d. peratus		North Island
S. d. typhlopous		North Island
Sphaerodactylus elasmorhynchus	Endemic	South Island
Sphaerodactylus elegans	West Indian	Cuba
S. e. punctatissimus	Endemic	Islandwide, Île de la Gonâve
Sphaerodactylus epiurus	Endemic	North Island
Sphaerodactylus ladae	Endemic	North Island
Sphaerodactylus lazelli	Endemic	North Island
Sphaerodactylus leucaster	Endemic	North Island
Sphaerodactylus nycteropus	Endemic	South Island
Sphaerodactylus ocoae	Endemic	North Island
Sphaerodactylus omoglaux	Endemic	South Island, Plaine de Cul-de-Sac
Sphaerodactylus perissodactylius	Endemic	North Island
Sphaerodactylus plummeri	Endemic	South Island
Sphaerodactylus randi	Endemic	South Island
S. r. randi		South Island
S. r. methorius		South Island
S. r. strahmi		South Island
Sphaerodactylus rhabdotus	Endemic	North Island invader of South Island
Sphaerodactylus samanensis	Endemic	North Island
Sphaerodactylus savagei	Endemic	North Island
S. s. savagei		North Island
S. s. juanilloensis		North Island
Sphaerodactylus shrevei	Endemic	North Island
Sphaerodactylus sommeri	Endemic	North Island
Sphaerodactylus streptophorus	Endemic	South Island

Table 4.2 cont.

S. s. streptophorus		South Island
S. s. sphenophanes		South Island
Sphaerodactylus thompsoni	Endemic	South Island
Sphaerodactylus williamsi	Endemic	North Island
Sphaerodactylus zygaena	Endemic	South Island
	Family Iguanidae	
Cyclura cornuta	West Indian	Navassa, Isla Mona
C. c. cornuta	Endemic	Islandwide
Cyclura ricordii	Endemic	Uncertain (see discussion in Schwartz, 1980b)
	Family Polychrotidae	
Anolis aliniger	Endemic	North Island invader of South Island
Anolis altavelensis	Endemic	South Island, Isla Alto Velo
Anolis alumina	Endemic	South Island
Anolis armouri	Endemic	South Island
Anolis bahorucoensis	Endemic	South Island
A. b. bahorucoensis		South Island
A. b. southerlandi		South Island
Anolis baleatus	Endemic	North Island
A. b. altager		North Island
A. b. caeruleolatus		North Island
A. b. fraudator		North Island
A. b. lineatacervix		North Island - Isla Saona
A. b. litorisilva		North Island
A. b. multistruppus		North Island
A. b. samanae		North Island
A. b. scelestus		North Island
A. b. sublimis		North Island
Anolis barahonae	Endemic	South Island
A. b. barahonae		South Island
A. b. albocellatus		South Island
A. b. ininquinatus		South Island
A. b. mulitus		South Island
Anolis barbouri	Endemic	South Island invader of North Island
Anolis brevirostris	Endemic	N. Island invader of S. .Island
A. b. brevirostris		Valle de Neiba / Plaine de Cul-de-Sac

Table 4.2 cont.

A. b. deserticola		North Island
A. b. wetmorei		South Island
Anolis caudalis	Endemic	North Island invader of South Island, Île de la Gonâve
Anolis chlorocyanus	Endemic	North Island invader of South Island
A. c. chlorocyanus		North Island invader of South Island
A. c. cyanostictus		North Island
Anolis christophei	Endemic	North Island
Anolis coelestinus	Endemic	South Island
A. c. coelestinus		South Island
A. c. demissus		South Island, Île Grande Cayemite
A. c. pecuarius		South Island, Île-à-Vache
Anolis cristatellus	Introduced	Puerto Rico Bank
A. c. cristatellus		Puerto Rico
Anolis cybotes	Endemic	Islandwide
A. c. cybotes		Islandwide
A. c. doris		Île de la Gonâve
A. c. ravifaux		North Island - Isla Saona, Isla Catalinita (?)
Anolis darlingtoni	Endemic	South Island
Anolis distichus	West Indian	Islandwide; Bahamas (introduced elsewhere)
A. d. aurifer	Endemic	South Island
A. d. dominicensis	Endemic	North and South islands
A. d. favillarum	Endemic	South Island
A. d. ignigularis	Endemic	North Island (introduced in Florida)
A. d. juliae	Endemic	South Island - Île-à-Vache
A. d. patruelis	Endemic	South Island - Île Grande Cayemite
A. d. properus	Endemic	North Island
A. d. ravitergum	Endemic	North Island
A. d. sejunctus	Endemic	North Island - Isla Saona
A. d. suppar	Endemic	South Island
A. d. tostus	Endemic	North Island - Isla Catalina

Table 4.2 cont.

A. d. vinosus	Endemic	South Island
Anolis dolichocephalus	Endemic	South Island
A. d. dolichocephalus		South Island
A. d. portusalus		South Island
A. d. sarmenticola		South Island
Anolis etheridgei	Endemic	North Island
Anolis eugenegrahami	Endemic	North Island
Anolis fowleri	Endemic	North Island
Anolis haetianus	Endemic	South Island
Anolis hendersoni	Endemic	South Island
A. h. hendersoni		South Island
A. h. ravidormitans		South Island
Anolis insolitus	Endemic	North Island
Anolis koopmani	Endemic	South Island
Anolis longitibialis	Endemic	South Island
A. l. longitibialis		South Island - Isla Beata
A. l. specuum		South Island
Anolis marcanoi	Endemic	North Island
Anolis marron	Endemic	South Island
Anolis monticola	Endemic	South Island
A. m. monticola		South Island
A. m. quadrisartus		South Island
Anolis olssoni	Endemic	North Island invader of South Island
A. o. olssoni		North Island
A. o. alienus		North Island
A. o. dominigensis		North Island
A. o. extentus		Islandwide
A. o. ferrugicauda		North Island
A. o. insulanus		Île de la Gonâve
A. o. montivagus		North Island
A. o. palloris		South Island
Anolis placidus	Endemic	North Island
Anolis porcatus	Introduced	Cuba
Anolis ricordii	Endemic	Islandwide
A. r. ricordii		North and South islands
A. r. leberi		South Island
A. r. subsolanus		South Island
A. r. viculus		South Island

Table 4.2 cont.

Anolis rimarum	Endemic	North Island
Anolis rupinae	Endemic	South Island
Anolis semilineatus	Endemic	Islandwide
Anolis sheplani	Endemic	South Island
Anolis shrevei	Endemic	North Island
Anolis singularis	Endemic	South Island invader of North Island (?), Île de la Gonâve
Anolis strahmi	Endemic	South Island
A. s. strahmi		South Island
A. s. abditus		South Island
Anolis websteri	Endemic	North Island
Anolis whitemani	Endemic	North Island
A. w. whitemani		North Island, Valle de Neiba/Plaine de Cul-de-Sac
A. w. breslini		North Island
A. w. lapidosus		North Island
	Family Scincidae	
Mabuya lineolata	Endemic	North Island
Mabuya bistriata (?)	Native (?)	Northern South America
M. b. sloanei	Native (?)	North Island (one record); Puerto Rico Bank, Turks and Caicos islands, Jamaica
	Family Teiidae	
Ameiva chrysolaema	Endemic	North Island invader of South Island
A. c. chrysolaema		Islandwide
A. c. abbotti		South Island - Isla Beata
A. c. alacris		North Island
A. c. boekeri		North Island
A. c. defensor		North Island
A. c. evulsa		South Island - Grosse Caye
A. c. ficta		South Island
A. c. jacta		North Island
A. c. parvoris		North Island
A. c. procax		North Island
A. c. quadrijugis		South Island

Table 4.2 cont.

A. c. regularis		North Island
A. c. richardthomasi		North Island - Isla Saona
A. c. secessa		Île de la Gonâve
A. c. umbratilis		Valle de Neiba
A. c. woodi		North Island - Île de la Tortue
Ameiva leberi	Endemic	South Island
Ameiva lineolata	Endemic	North Island invader of South Island
A. l. lineolata		Islandwide
A. l. beatensis		South Island - Isla Beata
A. l. meracula		North Island
A. l. perplicata		North Island
A. l. privigna		South Island
A. l. semota		North Island - Isla Catalina
Ameiva taeniura	Endemic	Islandwide; Navassa Island
A. t. taeniura		South Island
A. t. aequorea		South Island - Île-à-Vache
A. t. azuae		North Island
A. t. barbouri		North Island, Île de la Gonâve
A. t. ignobilis		North Island
A. t. meyerabichi		North Island
A. t. pentamerinthus		South Island – Île Grande Cayemite
A. t. regnatrix		South Island
A. t. rosamondae		North Island - Isla Saona
A. t. tofacea		North Island
A. t. vafra		North Island
A. t. varica		South Island
A. t. vulcanalis		North and South islands
	Family Tropiduridae	
Leiocephalus barahonensis	Endemic	South Island
L. b. barahonensis		South Island
L. b. aureus		South Island
L. b. beatanus		South Island - Isla Beata
L. b. oxygaster		South Island

Table 4.2 cont.

Leiocephalus endomychus	Endemic	North Island
Leiocephalus lunatus	Endemic	North Island
Leiocephalus melanochlorus	Endemic	South Island
L. m. melanochlorus		South Island
L. m. hypsistus		South Island
Leiocephalus personatus	Endemic	North Island invader of South Island
L. p. personatus		South Island
L. p. actites		North Island
L. p. agraulus		North Island
L. p. budeni		North Island
L. p. elattoprosopon		South Island
L. p. mentalis		North Island
L. p. poikilometes		North Island
L. p. pyrrholaemus		North Island
L. p. scalaris		North Island
L. p. socoensis		North Island
L. p. tarachodes		North Island
L. p. trujilloensis		North Island
Leiocephalus pratensis	Endemic	North Island
L. p. pratensis		North Island
L. p. chimarus		North Island - Île-à-Cabrit
Leiocephalus rhutidira	Endemic	North Island
Leiocephalus schreibersii	Endemic	North Island invader of South Island
L. s. schreibersii		Islandwide
L. s. nesomorus		North Island - Île de la Tortue
Leiocephalus semilineatus	Endemic	North Island invader of South Island
Leiocephalus vinculum	Endemic	Islandwide (?)
L. v. vinculum		Île de la Gonâve
L. v. altavelensis		South Island - Isla Alto Velo
	SUBORDER SERPENTES	
	Family Boidae	
Epicrates fordii	Endemic	North Island invader of South Island

Table 4.2 cont.

E. f. fordii		North Island, Valle de Neiba/ Plaine de Cul-de-Sac, Île de la Gonâve
E. f. agametus		North Island
E. f. manototus		North Island - Île-à-Cabrit
Epicrates gracilis	Endemic	Islandwide
E. g. gracilis		North Island
E. g. hapalus		South Island
Epicrates striatus	West Indian	Islandwide; Bahamas
E. s. striatus	Endemic	North and South Islands, Île de la Gonâve
E. s. exagistus	Endemic	South Island
E. s. warreni	Endemic	North Island - Île de la Tortue
	Family Colubridae	
Alsophis anomalus	Endemic	Islandwide
Alsophis melanichnus	Endemic	Islandwide
Antillophis parvifrons	Endemic (?)	Islandwide; Bahamas (status unknown)
A. p. parvifrons		South Island
A. p. alleni		Île de la Gonâve, Île Petite Gonâve
A. p. lincolni		South Island
A. p. niger		North Island
A. p. paraniger		North Island
A. p. protenus		North and South islands
A. p. rosamondae		South Island - Île-à-Vache
A. p. stygius		North Island - Isla Saona
A. p. tortuganus		North Island - Île de la Tortue
Darlingtonia haetiana	Endemic	South Island
D. h. haetiana		South Island
D. h. perfector		South Island
D. h. vaticinata		South Island
Hypsirhynchus ferox	Endemic	Islandwide
H. f. ferox		North and South islands
H. f. exedrus		North Island - Isla Saona

Table 4.2 cont.

H. f. paracrousis		Île de la Gonâve
H. f. scalaris		South Island
Ialtris agyrtes	Endemic	Islandwide
Ialtris dorsalis	Endemic	Islandwide
Ialtris parishi	Endemic	Islandwide
Uromacer catesbyi	Endemic	Islandwide
U. c. catesbyi		South Island
U. c. cereolineatus		South Island - Île Grande Cayemite, Île Petite Cayemite (?)
U. c. frondicolor		Île de la Gonâve
U. c. hariolatus		North Island
U. c. incháusteguii		North Island - Isla Saona
U. c. insulaevaccarum		South Island - Île-à-Vache
U. c. pampineus		North Island
Uromacer frenatus	Endemic	South Island invader of North Island, Île de la Gonâve
U. f. frenatus		South Island
U. f. chlorauges		South Island invader of North Island
U. f. dorsalis		Île de la Gonâve
U. f. wetmorei		South Island - Isla Beata
Uromacer oxyrhynchus	Endemic	North Island invader of South Island
	Family Leptotyphlopidae	
Leptotyphlops asbolepis	Endemic	North Island
Leptotyphlops calypso	Endemic	North Island
Leptotyphlops leptipilepta	Endemic	South Island
Leptotyphlops pyrites	Endemic	South Island
	Family Tropidophiidae	
Tropidophis haetianus	West Indian	Islandwide; Cuba, Jamaica
T. h. haetianus	Endemic (?)	North Island invader of South Island, Île de la Gonâve; Cuba
T. h. hemerus	Endemic	North Island
T. h. tiburonensis	Endemic	South Island
	Family Typhlopidae	
Typhlops capitulatus	Endemic	South Island, Plaine de Cul-de-Sac

Table 4.2 cont.

Typhlops gonavensis	Endemic	Île de la Gonâve
Typhlops hectus	Endemic	South Island invader of North Island
Typhlops pusillus	Endemic	North Island invader of South Island, Île de la Gonâve
Typhlops schwartzi	Endemic	North Island
Typhlops sulcatus	Endemic	South Island invader of North Island, Île de la Gonâve
Typhlops syntherus	Endemic	South Island
Typhlops tetrathyreus	Endemic	Uncertain, Plaine de Cul-de-Sac
Typhlops titanops	Endemic	South Island
	ORDER CROCODYLIA	
	Family Crocodylidae	
Crocodylus acutus	Native	Islandwide (historically); Neotropics

[a]The species list is taken from Powell et al. (1996), except where noted. Origins of endemic species are from Schwartz (1980b), except for taxa subsequently described or when the taxonomy has changed. Non-Hispaniolan distributions are from Schwartz and Henderson (1991). Endemism of subspecies is noted only for non-Hispaniolan endemic species.

[b]Based on allopatry, clear and consistent meristic and morphological differences (Thomas, 1965a), and comments in Thomas (1965c) and Powell (1992b, 1993), we recognize the main island populations as a species distinct from that of Île de la Gonâve.

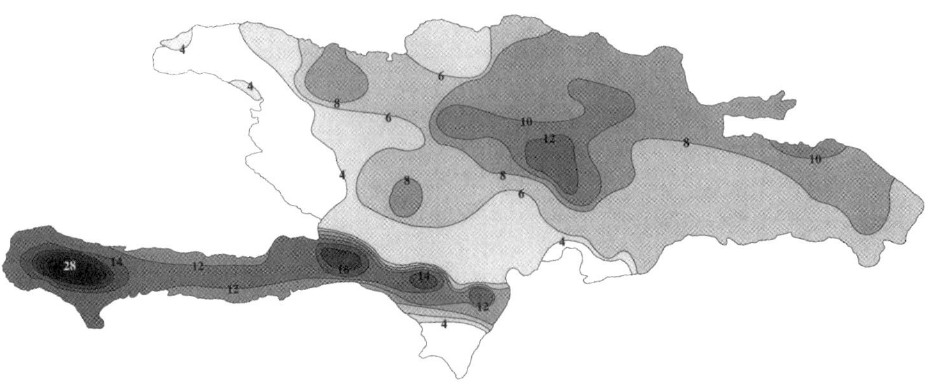

Figure 4.2. Amphibian species density on Hispaniola. Species density is the number of overlapping species distributions. This entity is approximately equal to the number of sympatric species but often is greater than the number of syntopic species. Contours with intervals of two species were determined by outlining and superimposing ranges based on records in Schwartz and Henderson (1991) with modifications by Hedges (1999). This figure is modified from Fig. 12 in Hedges (1999).

However, another consideration relative to the large number of reptilian (and a few amphibian) subspecies was raised by Powell (1993). Reptiles tend to have more morphological characters (e.g., scales) capable of being examined for variation (molecular data suggest that amphibians may actually demonstrate as much geographic variability as reptiles; S. B. Hedges, personal communication).

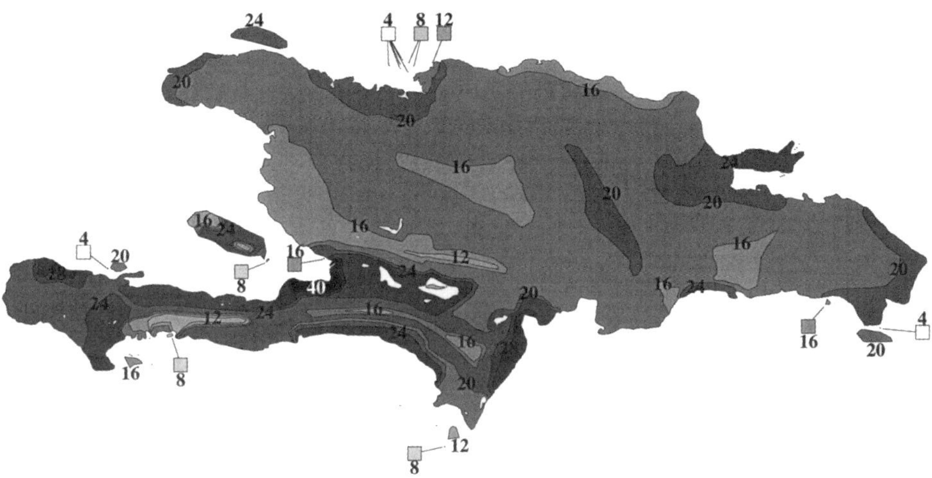

Figure 4.3. Squamate reptilian species density on Hispaniola. Species density is the number of overlapping species distributions. This entity is approximately equal to the number of sympatric species but often is greater than the number of syntopic species. Numbers are conservative estimates because species for which few records exist were not assumed to be present in all intervening areas of apparently suitable habitat. Contours with intervals of four species were determined by outlining and superimposing ranges based on records in Schwartz and Henderson (1991).

Based on an application of the evolutionary species concept (Wiley, 1978; Frost and Hillis, 1990), Powell (1993) identified subspecies with ranges that are apparently allopatric with those of conspecifics and formal diagnoses of morphological distinctiveness had been published (e.g., *Ameiva chrysolaema* in Fig. 4.4). Largely without reexamination and further study of each situation, he recognized 43 Hispaniolan species which might actually represent complexes containing as many as 92 specific taxa. Five of the identified subspecies have been elevated to full species status: *Leiocephalus endomychus* by Pregill (1992), *Bufo* (*Peltophryne*) *fractus*, *Aristelliger expectatus*, and *Hemidactylus haitianus* by Powell *et al.* (1996a), and *Amphisbaena hyporissor* (this chapter).

Figure 4.4. Ranges of representative species on Hispaniola. *Ameiva chrysolaema* (light gray) has a broad distribution on both North and South paleoislands. This species was considered a North Island invader of the South Island by Schwartz (1980b). *Sphaerodactylus samanensis* (medium gray) is a North Island endemic (Schwartz, 1980b) with a limited range restricted to the Los Haitises region along the southern shore of the Bahía de Samaná. *Eleutherodactylus chlorophenax* (black) is a South Island endemic (Schwartz, 1980b) with a limited range restricted to the western portion of the Massif de la Hotte. Range outlines are modified from Schell et al. (1993), Schwartz and Henderson (1991), and Cunningham *et al.* (1998), respectively.

If only a fraction of the additional taxonomic questions survive the scrutiny of detailed morphological and biochemical evaluations, the number of subspecies elevated to full species status will increase substantially the species richness of Hispaniola – with considerable implications for the development of conservation programs on the island.

The following is a summary of the distribution and presumed endemism of currently recognized taxa of Hispaniolan amphibians and reptiles. Most of the attributions regarding paleoisland origins are from Schwartz (1980b) and those based on subsequently described taxa are based largely on distribution. However, S. B. Hedges (personal communication) pointed out the difficulty in making such decisions based solely on distributional data. Obviously, the best information would come from phylogenetic studies, but sufficiently detailed analyses are lacking for many Hispaniolan forms. In addition, a valid alternative to "invasion" hypotheses in some instances (e.g., *Celestus agasepsoides* or *Typhlops pusillus*) is vicariance (see Crother and Guyer, 1996a,b): A continuous range in the Pleistocene, for example, may have been split by a change in the intervening habitat (e.g., drying or inundation of the Valle de Neiba/Plaine du Cul de Sac). As a result, the assignment of biogeographic histories presented here should be treated as preliminary until more detailed phylogenies become available for a greater number of the Hispaniolan taxa. Distributional data included in Schwartz and Henderson (1991) are not cited repeatedly; only subsequently published references are listed.

Amphibians

Of the 65 recognized species, one is an introduced ranid, four are bufonids, four are hylids, and the remainder are leptodactylids (Table 4.1). *Rana catesbeiana* has been introduced, probably intended as a food source, from North America. This species is widely distributed across the island wherever suitable ponds or slow-moving streams are available for breeding.

Of the four bufonids, one (*Bufo marinus*) was introduced to the island and has spread widely in lowland areas, especially in agricultural regions, but also in the arid Valle de Neiba/Plaine de Cul-de-Sac. The other three species are endemic to Hispaniola and all specifically to the North Island. *Bufo fluviaticus* has a restricted range in the Valle de Cibao, and *B. fractus* has a similarly restricted range in the eastern Dominican lowlands. Both areas are moderately to strongly mesic. *Bufo guentheri*, in contrast, is widely distributed across the North Island and has invaded the South Island via the Valle de Neiba/Plaine de Cul-de-Sac. Unlike the other two endemic species, *B. guentheri* is largely xerophilic. Hedges (1996a) recognized a single West Indian lineage (the *Bufo peltocephalus* group) of native bufonids, centered in Cuba, to which all endemic Hispaniolan species are assigned.

All of the four hylid species are Hispaniolan endemics and have islandwide distributions (i.e., North or South island origins cannot be determined). *Osteopilus dominicensis* is found in a variety of habitats, apparently not even requiring an arboraceous area (although these frogs are usually arboreal). Very adaptable and obviously capable of effectively utilizing severely altered habitats, females often deposit ova in artificial pools and fountains and adults are not infrequently edificarian in urban areas and resorts. The three species of *Hyla* are almost always associated with mesic broadleaf forests, although *H. pulchrilineata* is known to occupy forest remnants and "situations much disturbed by man" (Schwartz and Henderson, 1991). *Hyla vasta* is most common in gallery forests near streams, and *H. heilprini* nearly always is associated with rapidly flowing or torrential waters. Hedges (1996a) recognized two West Indian lineages of native hylids (*Hyla heilprini* and *Osteopilus*, to which he assigned the other Hispaniolan species currently placed in the genus *Hyla*).

The lone Hispaniolan leptodactylid in the genus *Leptodactylus, L. dominicensis*, is a North Island endemic restricted to the southern shore of the Bahía de Samaná. Hedges (1996a) placed this species with the closely related Puerto Rican Bank endemic, *L. albilabris*, into a single lineage.

All other Hispaniolan leptodactylids are assigned to the genus *Eleutherodactylus*. Of the 55 species, only three have islandwide distr i-butions, 17 are North Island endemics (none of which have invaded the South Island), and 36 are endemic to the South Island. Of the latter, three have invaded the North Island and all three have endemic North Island subspecies plus one subspecies restricted to the Valle de Neiba/Plaine de Cul-de-Sac.

When all Hispaniolan amphibians are considered, two species (*Bufo marinus* and *Rana catesbeiana*) were introduced and 63 (96.9%) are endemic (Hedges, 1997). Of the endemic forms, seven (11.1%) have islandwide distributions and their origins cannot be attributed to either of the paleoislands, 20 (31.7%) are North Island endemics, and 36 (57.1%) are South Island endemics. Only one North Island form has invaded the South Island, whereas three South Island forms have established populations on the North Island. Of the 20 subspecies, all of which are *Eleutherodactylus*, 10 are in three species with South Island origins, six in two species with North Island origins, and four in a single species with an islandwide distribution.

When distribution is related to physiographic provinces, high levels of endemism are also evident. By correlating the biogeographic regions of Henderson and Schwartz (1984a) with distributions of native amphibians, a picture of regional endemism develops (Table 4.3). Of the 62

Table 4.3. Distributions of Native Hispaniolan Amphibians Versus the Biogeographic Regions Used in Henderson and Schwartz (1984a).[a]

Species	Biogeographic Region															
	MH	MS	SB	PB	CS	GG	SN	PC	PN	CC	VC	NH	ER	LA	MG	TP
Bufo fluviaticus											X					
Bufo fractus												X				
Bufo guentheri					X		X			X						
Hyla heilprini	X	X					X			X		X		X		
Hyla pulchrilineata	X	X				X			X	X	X					
Hyla vasta	X	X	X							X		X				
Osteopilus dominensis	X	X	X	X	X	X	X	X		X	X	X	X	X	X	X
Eleutherodactylus abbotti	X	X	X					X	X	X	X	X	X	X	X	X
Eleutherodactylus alcoae			X	X												
Eleutherodactylus amadeus	X															
Eleutherodactylus apostates	X															
Eleutherodactylus armstrongi		X	X													
Eleutherodactylus audanti	X	X	X				X			X						
Eleutherodactylus auriculatoides							X									
Eleutherodactylus bakeri	X															
Eleutherodactylus brevirostris	X															
Eleutherodactylus caribe	X															
Eleutherodactylus chlorophenax	X															
Eleutherodactylus corona	X															
Eleutherodactylus counouspeus	X															
Eleutherodactylus darlingtoni		X														
Eleutherodactylus dolomedes	X															
Eleutherodactylus eunaster	X															
Eleutherodactylus flavescens												X	X			
Eleutherodactylus fowleri		X														
Eleutherodactylus furcyensis		X														
Eleutherodactylus glandulifer	X															
Eleutherodactylus glanduliferoides		X														
Eleutherodactylus glaphycompus	X															
Eleutherodactylus grahami									X							
Eleutherodactylus haitianus										X						
Eleutherodactylus heminota		X	X													

Table 4.3 cont.

Species	Biogeographic Region															
	MH	MS	SB	PB	CS	GG	SN	PC	PN	CC	VC	NH	ER	LA	MG	TP
Eleutherodactylus hypostenor	X	X	X													
Eleutherodactylus inoptatus	X	X	X				X			X	X	X	X			X
Eleutherodactylus jugans		X														
Eleutherodactylus lamprotes	X															
Eleutherodactylus leoncei		X														
Eleutherodactylus lucioi									X							
Eleutherodactylus minutus										X						
Eleutherodactylus montanus										X						
Eleutherodactylus sp. "N"							X									
Eleutherodactylus nortoni		X														
Eleutherodactylus oxyrhynchus	X	X														
Eleutherodactylus parabates							X									
Eleutherodactylus parapelates	X															
Eleutherodactylus patriciae										X						
Eleutherodactylus paulsoni	X															X
Eleutherodactylus pictissimus	X	X	X		X									X		X
Eleutherodactylus pituinus										X						
Eleutherodactylus poolei										X						
Eleutherodactylus probolaeus													X			
Eleutherodactylus rhodesi								X								
Eleutherodactylus rufifemoralis			X													
Eleutherodactylus ruthae	X	X	X				X			X		X	X			
Eleutherodactylus schmidti										X						
Eleutherodactylus sciagraphus	X															
Eleutherodactylus semipalmatus	X	X														
Eleutherodactylus thorectes	X															
Eleutherodactylus ventrilineatus	X															
Eleutherodactylus weinlandi							X			X		X	X			
Eleutherodactylus wetmorei	X	X								X						
Leptodactylus dominicensis												X				

[a]These regions are less detailed than those discussed in the text and illustrated in Figure 1 (see comments in Henderson and Schwartz, 1984a:3). However, the use of these regions, with subdivided areas combined and no consideration of satellite islands (see text), provides an adequate appreciation of regional endemism among Haitian and Dominican frogs. *Eleutherodactylus warreni* is omitted from this table because it is endemic and restricted to Île de la Tortue. Key to abbreviations: MH = Massif de la Hotte, MS = Massif de la Selle, SB = Sierra de Baoruco, PB = Península de Barahona, CS = Valle de Neiba/Plaine de Cul-de-Sac, GG = Golfe de la Gonâve, SN = Sierra de Neiba and Montagnes du Trou-d'Eau, PC = Plateau Central, PN = Presqu'île du Nord-Ouest, CC = Cordillera Central and Massif du Nord, VC = Valle de Cibao, NH = Plaine du Nord (north Haitian littoral plain extending eastward to include much of the northern and eastern República Dominicana), ER = eastern República Dominicana (largely mesic coastal plain), LA = Llanos de Azua, MG = Sierra Martín García, TP = Tiburon Peninsula.

species considered (see caption to Table 4.3), 42 (67.7%) are restricted to a single one of the 16 regions considered. Another seven (11.3%) are found only in two regions, and in all instances the regions are adjacent and similar in habitat (e.g., the Massif de la Hotte and Massif de la Selle). Six (9.7%) additional species are limited to three to five regions, all of which are either contiguous and similar in habitat, consist of upland areas and adjacent lowlands, and/or represent very similar situations (e.g., uplands) separated by apparently uninhabitable intervening areas of distinctly different habitats. Only seven species (11.3%) are found in as many as six regions and in areas with varying ecological conditions, and only two (*Osteopilus dominicensis* and *Eleutherodactylus abbotti*) are found in as many as three-fourths (≥12) of the regions considered. The hylid frogs were by far the most geographically diverse group, with no species found in fewer than five regions, and *O. dominicensis* in all but one.

Few amphibians have established populations on satellite islands. *Eleutherodactylus alcoae* occurs on Isla Beata, *E. inoptatus* has been found on Île de la Tortue, and *E. pictissimus* is known from Île-à-Vache. *Eleutherodactylus warreni* is endemic and restricted to Île de la Tortue and is the only satellite island endemic amphibian of Hispaniola. *Osteopilus dominicensis*, the only species found on more than one satellite, possibly a tribute to its ecological versatility, has populations on Île de la Gonâve, Île-à-Vache, Île Grande Cayemite, Île de la Tortue, and Isla Saona. No amphibians are known to occur on Navassa Island.

Reptiles

Of the 146 recognized species, one is a crocodilian, two are emydid turtles, and the remainder are squamates, with all three traditionally recognized suborders represented (Table 4.2). *Crocodylus acutus* has a broad historical distribution which includes southern Florida, the western Greater Antilles (Cuba, Hispaniola, and Jamaica), and the Neotropical mainland south along both coasts of southern México and Central and South America to northern Perú on the Pacific side and to Venezuela on the Caribbean (Atlantic) side (Groombridge, 1987). On Hispaniola, native populations once were found along many coastal areas as well as in a number of inland streams and lakes (Thorbjarnarson, 1988). Current populations appear to be restricted to saline Etang Saumatre, hypersaline Lago Enriquillo, and a few geographically proximate freshwater streams and lakes in the Valle de Neiba/Plaine de Cul-de-Sac (Schubert and

Santana, 1996), although evidence supporting the extirpation of crocodiles from the Montecristi region is inconclusive and further (nocturnal) surveys will be required to confirm this conclusion.

Hispaniola is the only West Indian island on which two species of native emydid turtles are known to occur (Seidel and Incháustegui Miranda, 1984). Both species are native to Hispaniola, and *Trachemys decorata* is endemic. Most populations are found in the Valle de Neiba/Plaine de Cul-de-Sac, but others are known from the Vallée de l'Artibonite on the North Island and from the western Tiburon Peninsula of the South Island. Schwartz (1980b) discussed the difficulty of tracing the origin of this species to either the North or South paleoisland. *Trachemys stejnegeri*, on the other hand, is endemic to the Bahamas, Puerto Rico, and Hispaniola (Seidel, 1996), with the populations on the latter constituting a subspecies (*T. s. vicina*) widespread on and endemic to the North Island. Hedges (1996a) placed all West Indian species of *Trachemys* in a single lineage, presumably of North American origin.

Amphisbaenian squamates are represented on Hispaniola by five species (Table 4.2), all of which are endemic. The only North Island endemic (*Amphisbaena manni*) has invaded the South Island, but only one (*A. innocens*) of the three South Island endemics contains North Island populations. The only currently recognized subspecies are of *A. hyporissor*, and the ranges of both are limited to the South Island. The fifth species (*A. gonavensis*) is restricted to Île de la Gonâve. In addition to the latter, *A. caudalis* is found only on satellite islands, Île Grande Cayemite and Presqu'île de Baradères off the northern coast of the Tiburon Peninsula (Haiti). Other insular populations include those of *A. hyporissor leberi* (Isla Beata), *A. innocens* (Île Grande Cayemite) and *A. manni* (Île de la Tortue); none are endemic to these islands. Hedges (1996a) placed all West Indian species in the genus *Amphisbaena* (including Cuban species formerly in *Cadea*) in a single lineage.

Lacertilian squamates constitute the most abundant and diverse of the herpetofaunal elements on Hispaniola. Seven families are represented: anguids by 12 species and 22 subspecies in one genus, gekkonids by 38 species and 41 subspecies in five genera, iguanids (sensu strictu) by two species and one subspecies in one genus, polychrotids by 42 species and 66 subspecies in one genus, scincids by two species and one subspecies in one genus, teiids by four species and 35 subspecies in one one, and tropidurids by 10 species and 24 subspecies in one genus.

All 12 species of Anguidae are placed in the genus *Celestus* and all are Hispaniolan endemics. Three are islandwide forms, six are North Island endemics, and three appear to have originated on the South Island. Two (*C. costatus* and *C. stenurus*) of the three islandwide species contain both North and South Island endemic subspecies. Of the North Island species, 1one (*C. curtissi*) is represented on the South Island by a South Island endemic subspecies. Of the South Island endemic species, one (*C. agasepsoides*) has invaded the North Island. All three subspecies of any South Island endemic, all in *C. haitianus*, are restricted to the South Island. Four subspecies of *C. costatus* are endemic to satellite islands (Île-à-Vache, Île de la Tortue, Isla Catalinita, Isla Saona), and another is endemic to Navassa Island (see below). All were listed by Powell (1993) as possible candidates for full-species recognition. Another population is known from the Cayos Siete Hermanos (Burns *et al.*, 1992). *Celestus curtissi* contains populations on Île de la Gonâve, Île de la Tortue, Isla Catalina, and Isla Saona. None represent endemic subspecies, although the populations on the two latter islands are unassigned subspecifically. Satellite island populations of *C. stenurus* are known to occur on Île-à-Vache, Île Grande Cayemite, Île-à-Cabrit, and the Cayos Siete Hermanos. Although the population on Île Grande Cayemite is unassigned, none appear to represent endemic subspecies. Hedges (1996a) assigned all Hispaniolan species in *Celestus*, including those formerly placed in *Diploglossus*, *Sauresia*, and *Wetmorena*, to a single West Indian lineage. In contrast, Savage and Lips (1993), resurrecting the classification which places species with claw sheaths in *Diploglossus* and those lacking sheaths in *Celestus*, retained *Sauresia* and *Wetmorena* and noted their affinity with Diploglossus. On this basis, Hispaniolan anguids represent two distinct lineages.

The 38 species and 41 subspecies of Hispaniolan gekkonids are placed in five genera. Both species of *Aristelliger* are islandwide endemics, and neither is polytypic. Populations of *A. expectatus* are known from Île de la Gonâve, Île Grande Cayemite, Île-à-Cabrit, Île de la Tortue, Isla Beata, Isla Alto Velo, and Cayo Pisaje. The only known satellite populations of *A. lar* are on Isla Catalina and the Cayos Siete Hermanos (but see Burns *et al.*, 1992 and Powell and Parmerlee, 1992). Both are in an Antillean *Aristelliger* lineage (Hedges, 1996a). *Gonatodes albugularis* is widely distributed through Central and northern South America and also is native to the western Antilles. All Hispaniolan populations, including those on Île de la Gonâve and Île-à-Cabrit, are assigned to the West Indian endemic subspecies, *G. a. notatus*, which also is found on Jamaica and the Cayman Islands. Hedges (1996a) recognized a single West Indian lineage. Two species in the genus *Hemidactylus* are found on Hispaniola, each representing separate West Indian lineages (Hedges, 1996a). *Hemidactylus haitianus* is a West Indian endemic also found on Cuba and Puerto Rico. The only known Hispaniolan satellite island population is on Isla Saona. *Hemidactylus mabouia* is widespread in the Neotropics, where it probably is native (but see Kluge, 1969 and Vanzolini, 1978). On Hispaniola, the species is known only from the area around Port-au-Prince and the extreme northeastern Dominican Republic. Because this species is known as an extremely successful colonizer (Meshaka *et al.*, 1994a,b,c), these populations are presumed to be native. *Phyllodactylus wirshingi* is a West Indian endemic found on Puerto Rico and Hispaniola. Populations on the latter constitute two North Island endemic subspecies, one of which (*P. w. hispaniolae*) has extended its range into the Valle de Neiba. Hedges (1996a) recognized this species as a distinct West Indian lineage.

The remaining 32 species and 38 subspecies of Hispaniolan gekkonids are all in the genus *Sphaerodactylus*. Of the 32 species, all are native and all but one are endemic. That species, *S. elegans* (known also from Cuba), is represented by an endemic subspecies (*S. e. punctatissimus*) which has an islandwide distributional pattern and also contains populations on Île de la Gonâve and Île Grande Cayemite. Twenty species are North Island endemics, four of which have invaded the South Island, and two of these contain South Island endemic subspecies. Eleven species are South Island endemics, of these, three (*S. copei*, *S. cryphius*, *S. omoglaux*) have invaded the Valle de Neiba/Plaine de Cul-de-Sac and one (*S. copei*) has extended its range onto the North Island and to Île de la Gonâve. *Sphaerodactylus altavelensis* has satellite island populations on Île-à-Cabrit and on Isla Alta Velo, the latter constituting an endemic (nominate) subspecies (although Powell *et al.*, 1989 tentatively attributed a specimen taken on the Península de Barahona to this subspecies). Populations of *S. copei* are on Île de la Gonâve, Île Grande Cayemite, and Île-à-Vache, with the first two recognized as satellite island endemic subspecies. *Sphaerodactylus difficilis euopter* is endemic to Île de la Tortue, but the population of *S. d. typhlopous* reported from the Cayos Siete Hermanos is not restricted to the islands. *Sphaerodactylus randi strahmi* has been found on, but is not restricted to Cayo Pisaje. Similarly, *S. savagei juanilloensis* is known from Isla Saona but also is found on the main island. However, the populations of *S. savagei* on Isla Catalinita is unassigned subspecifically. A population of *S. thompsoni* occurs on Isla Beata. All of the satellite island endemics were listed by Powell (1993) as deserving consideration as full species. Hedges (1996a) placed all West Indian *Sphaerodactylus* in a single lineage.

Both iguanid species are in the West Indian endemic genus *Cyclura*. *Cyclura ricordii* is restricted to Hispaniola, but its paleoisland affinities are uncertain (Schwartz, 1980b). *Cyclura cornuta* is an islandwide species on Hispaniola, where it is represented by the nominate subspecies. Endemic subspecies also occur on Isla Mona between Hispaniola and Puerto Rico and on Navassa Island. Powell (1993) suggested that these forms may be deserving of full-species status. These and all West Indian species of *Cyclura* represent a single lineage (Hedges, 1996a).

All 42 species and 66 subspecies of polychrotid lizards on Hispaniola are currently placed in the genus *Anolis*. Two species are introduced: *A. cristatellus* from the Puerto Rico Bank and *A. porcatus* from Cuba (Hispaniolan populations of both are assigned to nominate subspecies). The former is well established in and around La Romana and the latter in Santo Domingo. In both instances, their distributions end with the onset of largely native vegetation, in which they are replaced, respectively, by the ecologically similar endemics, *A. cybotes* and *A. chlorocyanus*. *Anolis distichus*, also known from the Bahamas and introduced into Florida, is native. With 12 endemic Hispaniolan subspecies on both the North and South islands, *A. distichus* is considered an islandwide form. Of the 39 endemic species, *Anolis cybotes*, *A. ricordii*, and *A. semilineatus* also are islandwide species. Seventeen species are recognized as North Island endemics. Of these, five (*A. aliniger*, *A. brevirostris*, *A. caudalis*, *A. chlorocyanus*, and *A. olssoni*) have invaded the South Island and another (*A. whitemani*) the Valle de Neiba/Plaine de Cul-de-Sac. Two of the South Island invaders (*A. brevirostris* and *A. olssoni*) contain subspecies endemic to that area. Of the 19 South Island endemics, only two (*A. barbouri* and *A. singularis*) have invaded the North Island and the latter is questionable; Schwartz and Henderson (1991) noted that North Island specimens of *A. singularis* from the Sierra de Martín García may represent a related form. *Anolis coelestinus* also is known from the Plaine de Cul-de-Sac. *Anolis altavelensis* is endemic to Isla Alto Velo. Satellite island endemic subspecies are known from Île de la Gonâve (*A. cybotes doris* and *A. olssoni insulanus*), Île Grand Cayemite [*A. coelestinus de-*

missus and *A. distichus patruelis* (which also may refer to the population on Île Petit Cayemite)], Île-à-Vache (*A. coelestinus pecuarius* and *A. distichus juliae*), Isla Beata (*A. l. longitibialis*), Isla Catalina (*A. distichus tostus*), and Isla Saona [*A. baleatus lineatacervix* and *A. cybotes ravifaux* (which also may refer to the population on Isla Catalinita), *A. distichus sejunctus*]. Non-endemic satellite island populations occur on Île de la Gonâve (*A. caudalis*, *A. chlorocyanus*, and *A. singularis*), Île de la Petit Gonâve (*A. caudalis*), Île de la Tortue (*A. chlorocyanus*, *A. cybotes*, *A. distichus*, and *A. semilineatus*), Île-à-Cabrit (*A. brevirostris*, *A. caudalis*, *A. cybotes*, *A. olssoni*, and *A. semilineatus*), Île Grande Cayemite (*A. cybotes* and *A. semilineatus*), Île-à-Vache (*A. cybotes*), Isla Catalina (*A. cybotes* and *A. semilineatus*), Isla Beata (*A. brevirostris*), Isla Saona (*A. chlorocyanus*), and the Cayos Siete Hermanos (*A. distichus*, *A. whitemani*) — although Schwartz had planned to recognize the population of *A. cybotes* on Isla Catalina at the subspecific level (R. W. Henderson, personal communication). Powell (1993) listed 16 satellite island endemics and other diagnosed allopatric subspecies as possibly deserving full-species status. Hedges (1996a) placed all West Indian species of *Anolis* (including those until recently assigned to *Chamaeleolis* and *Chamaelinorops*) in a single lineage.

Both Hispaniolan skinks are in the genus *Mabuya*. *Mabuya lineolata* is endemic and restricted to the North Island. This species constitutes a single West Indian lineage (Hedges, 1996a). The status of the other familial representative is much less certain. *Mabuya bistriata* is found widely throughout the West Indies, with the Hispaniolan population from near Santo Domingo assigned to *M. b. sloanei*, presumably the same subspecies found on Jamaica, the Turks and Caicos Islands, and the Puerto Rico Bank. The broad distribution of this lizard argues for consideration of the Hispaniolan specimens as being native, but a distribution centered on a cosmopolitan port suggests an introduction. Hedges (1996a) considered all of the West Indian populations of this lizard to be part of another single lineage.

All four species of teiids are in the genus *Ameiva* , and all are endemic to Hispaniola. *Ameiva taeniura* is an islandwide form, with endemic subspecies on both the North and South islands. Satellite island endemic subspecies occur on Île de la Gonâve, Île Grande Cayemite, Île-à-Vache, and Isla Saona (although that on Île de la Gonâve has also been recorded from the adjacent North Island). Additional insular populations are on Île Petite Cayemite and Isla Catalina. *Ameiva chrysolaema* and *A.*

lineolata are North Island endemics that have invaded the South Island. Both contain satellite island endemic subspecies: *A. chrysolaema* on Île de la Gonâve, Île de la Tortue, Grosse Caye, Isla Beata, and Isla Saona; and *A. lineolata* on Isla Beata and Isla Catalina. Other satellite island populations are known from Île-à-Cabrit (both species) and from Isla Catalina and the Cayos Siete Hermanos (*A. chrysolaema*). *Ameiva leberi* is a South Island endemic without any satellite island populations. Powell (1993) listed 21 satellite island endemics and other diagnosed allopatric subspecies as possibly deserving full-species status. All West Indian species of *Ameiva*, except populations of *A. ameiva* in the western Caribbean and on the southern Lesser Antilles, are assigned to a single lineage (Hedges, 1996a).

Five families of ophidian squamates are represented on Hispaniola: boids by three species and eight subspecies in one genus, colubrids by 11 species and 27 subspecies in six genera, leptotyphlopids by four species in one genus, tropidophiids by a single genus with one species and three subspecies, and typhlopids by nine species in one genus. No species of snakes have been introduced.

All three native species of Boidae are in the genus *Epicrates*. *Epicrates fordii* is a North Island endemic which has invaded the South Island through the Valle de Neiba/Plaine de Cul-de-Sac. Satellite island populations occur on Île de la Gonâve, Isla Saona, and Île-à-Cabrit, with the latter considered a subspecies endemic to this tiny coastal island. *Epicrates gracilis* and *E. striatus* are both islandwide species, with subspecies endemic to both paleoislands. The former is restricted to Hispaniola, but *E. striatus* also is native to the Bahamas, where an additional five subspecies occur. Hispaniolan satellite island populations of *E. striatus* are known from Isla Saona, Île de la Gonâve, Île Grande Cayemite, Île-à-Vache, and Île de la Tortue, but only the last is considered an endemic subspecies (listed by Powell, 1993 as possibly deserving of full-species status). Hedges (1996a) followed Kluge (1989) in concluding that all West Indian species of *Epicrates* form a monophyletic group.

The 11 species and 27 subspecies of Hispaniolan colubrids currently are placed in six genera. Both species of *Alsophis* are islandwide endemics. The few scattered localities for both species (only two for *A. melanichnus*) make any further estimation of origin impossible. Satellite island populations of *A. anomalus* are known from Île de la Tortue and Isla Beata. *Antillophis parvifrons* is another islandwide Hispaniolan species, although this snake also has been recorded from Little Inagua Island in the Bahamas (the status of this population is unknown; Schwartz and Henderson, 1991). Of the nine endemic subspecies, four are associated solely with satellite islands: *A. p. alleni* on Île de la Gonâve and Île Petit Gonâve, *A. p. rosamondae* on Île-à-Vache, *A. p. stygius* on Isla Saona, and *A. p. tortuganus* on Île de la Tortue. Powell (1993) considered these to be candidates for recognition as full species. Other satellite island populations are known from Île Grande Cayemite and Grosse Caye (*A. p. parvifrons*), Isla Beata (*A. p. lincolni*), and Isla Catalina (*A. p. paraniger*). South Island endemic *Darlingtonia haetiana* contains three su b-species, all restricted to the uplands of the three sequential ranges of the South Island. *Hypsirhynchus ferox* is another islandwide form, with subspecies on both paleoislands and two composed of satellite island populations: *H. f. exedrus* on Isla Saona and *H. f. paracrousis* on Île de la Gonâve. All three species of *Ialtris* also are considered islandwide forms, with the dearth of locality records for *I. agyrtes* (N = 3) and *I. parishi* (N = 2) precluding a more accurate picture of their origins. Both *I. dorsalis* and *I. parishi* are known from Île de la Tortue, and *I. dorsalis* also is known to occur on Île de la Gonâve and Île-à-Vache. The three species of *Uromacer* paint a more disparate picture. *Uromacer catesbyi* is an islandwide form with seven subspecies, five of which are restricted to satellite islands: *U. c. cereolineatus* to Île Grand Cayemite and presumably Île Petit Cayemite, *U. c. frondicolor* to Île de la Gonâve, *U. c. Incháusteguii* to Isla Saona, *U. c. insulaevaccarum* to Île-à-Vache, and *U. c. scandax* to Île de la Tortue. All may represent full species (Powell, 1993). *Uromacer frenatus* is a South Island endemic which has invaded the North Island (Smith and Powell, 1991). Endemic satellite populations are known from Île de la Gonâve (*U. f. dorsalis*) and Isla Beata (*U. f. wetmorei*). The monotypic species, *U. oxyrhynchus*, was considered by Schwartz (1980b) to be a North Island endemic which has invaded the South Island. Satellite island populations are known from Île de la Tortue, Isla Catalina, and Isla Saona. Powell (1993) included all satellite island endemic subspecies in his list of taxa deserving full-species status. Hedges (1996a) placed all the Hispaniolan colubrids into a single West Indian alsophine lineage.

The four species of *Leptotyphlops* are all endemic to Hispaniola, two (*L. asbolepis* and *L. calypso*) to the North Island and two (*L. leptipilepta* and *L. pyrites*) to the South Island. Hedges (1996a) followed Thomas *et al.* (1985) in placing all four Hispaniolan species with *L. bilineata* from the Lesser Antilles into a single West Indian radiation.

The lone Hispaniolan tropidophiid, *Tropidophis haetianus*, also is known from Cuba (site of a single record of the otherwise Hispaniolan nominate subspecies) and Jamaica, with three endemic subspecies. On Hispaniola, the species is an islandwide form containing subspecies endemic to both paleoislands. Satellite island populations are known from Île de la Gonâve and Île de la Tortue. All West Indian tropidophiids were placed in a single lineage by Hedges (1996a).

All nine species in the genus *Typhlops* are endemic to Hispaniola. Two (*T. pusillus* and *T. schwartzi*) are considered to be North Island endemics, and one of these (*T. pusillus*) has invaded the South Island and also is found on Île de la Gonâve. Five species are South Island endemics; of these, two (*T. syntherus* and *T. titanops*) are restricted to the South Island, two (*T. hectus* and *T. sulcatus*) have invaded the North Island, with *T. sulcatus* also known from Île de la Gonâve and from Navassa Island. The fifth (*T. capitulatus*) also has been recorded from the Plaine de Cul-de-Sac. *Typhlops gonavensis* is endemic to Île de la Gonâve, and the status of *T. tetrathyreus*, the range of which is largely restricted to the Plaine de Cul-de-Sac, is uncertain. In addition to those forms on Île de la Gonâve, satellite island populations are known from Île Grande Cayemite (*T. hectus*, *T. pusillus*, and *T. sulcatus*), Île de la Tortue (*T. pusillus*), Isla Alto Velo (*T. sulcatus*), and Isla Saona (*T. pusillus*). Hedges (1996a), citing Thomas (1989), placed all West Indian species of *Typhlops* in a single lineage.

When all 146 species of Hispaniolan reptiles are considered, only two species (*Anolis cristatellus* and *A. porcatus*) clearly were introduced [the status of *Mabuya bistriata* and *Hemidactylus mabouia* is unknown, but we will follow Schwartz and Henderson (1991) in treating the Hispaniolan populations of these species as native]. Of the native species, 14 have native populations elsewhere (other than Navassa Island) and, of these, eight are represented on Hispaniola by endemic subspecies. Conservatively treating only these eight and those restricted to Hispaniola as endemics, 94.5% of Hispaniolan reptiles are endemic. Of the endemic forms, 24 (17.4%) have islandwide distributions, 62 (44.9%) are endemic to the North Island, 48 (34.8%) are endemic to the South Island, two (1.4%) are endemic and restricted to Île de la Gonâve, and the status of two (1.4%) is uncertain. Of the North Island species, 19 (30.6%) have invaded the South Island and another the Valle de Neiba/Plaine de Cul-

Table 4.4 Distributions of native Hispaniolan squamate reptiles versus the biogeographic regions used in Henderson and Schwartz (1984a).[a]

Species	Biogeographic Region																													
	MH	MS	SB	PB	CS	GG	SN	PC	PN	CC	VC	NH	ER	LA	MG	TP	IT	IG	PG	GC	IP	IV	GR	IC	SH	CA	IS	CT	IB	AV
Amphisbaena caudalis																X				X										
Amphisbaena gonavensis																		X	X											
Amphisbaena innocens	X	X	X		X		X									X														
Amphisbaena hyporissor				X																									X	
Amphisbaena manni			X		X		X	X	X	X	X	X	X	X	X		X													
Celestus agasepsoides				X											X															
Celestus anelpistus												X																		
Celestus carraui											X	X																		
Celestus costatus	X	X	X	X			X	X	X	X	X	X	X			X	X					X			X	X	X			
Celestus curtissi			X	X	X	X	X		X				X				X	X									X	X		
Celestus darlingtoni										X																				
Celestus haetianus		X	X																											
Celestus macrotus		X																												
Celestus marcanoi										X																				
Celestus sepsoides	X	X	X							X		X	X			X		X		X										
Celestus stenurus	X				X	X	X	X	X	X	X	X	X	X	X	X				X		X		X	X					
Celestus warreni								X	X								X													
Aristelliger expectatus				X	X	X		X	X		X			X		X	X	X		X				X					X	X
Aristelliger lar				X	X							X	X	X		X									X			X		
Gonatodes albogularis					X	X										X		X												
Hemidactylus haitianus				X	X			X		X	X	X	X	X		X											X			
Hemidactylus mabouia				X								X																		
Phyllodactylus wirshingi					X				X					X																
Sphaerodactylus altavelensis		X		X	X	X		X	X					X		X														X
Sphaerodactylus armstrongi		X	X													X														

Table 4.4 cont.

	M	M	S	P	C	G	S	P	P	C	V	N	E	L	M	T	I	I	P	G	I	I	G	I	S	C	I	C	I	A
	H	S	B	B	S	G	N	C	N	C	C	H	R	A	G	P	T	G	G	C	P	V	R	C	H	A	S	T	B	V
Sphaerodactylus asterulus									X																					
Sphaerodactylus callocricus												X																		
Sphaerodactylus cinereus		X			X	X		X																						
Sphaerodactylus clenchi												X																		
Sphaerodactylus cochranae												X																		
Sphaerodactylus copei					X	X										X		X		X		X								
Sphaerodactylus cryphius					X																									
Sphaerodactylus darlingtoni							X					X	X		X															
Sphaerodactylus difficilis			X					X	X		X	X	X	X			X								X					
Sphaerodactylus elasmorhynchus	X																													
Sphaerodactylus elegans	X				X	X	X	X	X	X		X				X		X		X										
Sphaerodactylus epiurus												X																		
Sphaerodactylus ladae															X															
Sphaerodactylus lazelli															X															
Sphaerodactylus leucaster														X																
Sphaerodactylus nycteropus																X														
Sphaerodactylus ocoae												X		X																
Sphaerodactylus omoglaux					X																									
Sphaerodactylus perissodactylius															X															
Sphaerodactylus plummeri				X																										
Sphaerodactylus randi				X																										
Sphaerodactylus rhabdotus					X																									
Sphaerodactylus samanensis												X																		
Sphaerodactylus savagei													X													X	X			
Sphaerodactylus shrevei									X																					

Table 4.4 cont.

	MH	MS	SB	PB	CS	GG	SN	PC	PN	CC	VC	NH	ER	LA	MG	TP	IT	IG	PG	GC	IP	IV	GR	IC	SH	CA	IS	CT	IB	AV
Sphaerodactylus sommeri									X																					
Sphaerodactylus streptophorus		X		X												X														
Sphaerodactylus thompsoni				X																									X	
Sphaerodactylus williamsi									X																					
Sphaerodactylus zygaena																X														
Cyclura cornuta				X	X						X			X			X	X	X	X	X						X		X	
Cyclura ricordii				X	X																									
Anolis aliniger		X					X			X		X																		
Anolis altavelensis																														X
Anolis alumina			X	X												X														
Anolis armouri	X	X	X																											
Anolis bahorucoensis		X	X																											
Anolis baleatus										X	X	X	X		X												X			
Anolis barahonae			X	X																										
Anolis barbouri	X	X	X				X			X																				
Anolis brevirostris				X	X									X		X							X						X	
Anolis caudalis						X										X		X	X				X							
Anolis chlorocyanus		X			X	X	X	X	X	X	X	X	X	X			X	X									X			
Anolis christophei										X		X																		
Anolis coelestinus	X	X	X	X	X															X		X								
Anolis cybotes	X	X	X	X	X	X	X	X	X	X	X	X	X	X	X	X	X	X	X	X		X		X	X	X				
Anolis darlingtoni	X																													
Anolis distichus	X	X	X		X	X	X	X	X	X	X	X	X	X		X	X			X	X	X			X		X	X		
Anolis dolichocephalus	X															X														
Anolis etheridgei										X																				
Anolis eugenegrahami								X																						
Anolis fowleri										X																				
Anolis haetianus																X														
Anolis hendersoni	X	X														X														
Anolis insolitus										X																				
Anolis koopmani	X																													
Anolis longitibialis				X																									X	

Table 4.4 cont.

	MH	MS	SB	PB	CS	GG	SN	PC	PN	CC	VC	NH	ER	LA	MG	TP	IT	IG	PG	GC	IP	IV	GR	IC	SH	CA	IS	CT	IB	AV
Anolis marcanoi												X		X																
Anolis marron																X														
Anolis monticola	X																													
Anolis olssoni				X	X	X	X	X	X	X	X	X	X	X				X						X						
Anolis placidus							X																							
Anolis ricordii	X	X			X		X	X	X	X	X																			
Anolis rimarum										X																				
Anolis rupinae	X																													
Anolis semilineatus	X	X	X		X		X	X		X	X	X	X	X		X	X			X				X				X		
Anolis sheplani			X																											
Anolis shrevei										X																				
Anolis singularis	X	X	X												X			X												
Anolis strahmi			X	X																										
Anolis websteri						X		X	X																					
Anolis whitemani					X				X		X			X											X					
Mabuya lineolata									X		X		X																	
Mabuya bistriata													X																	
Ameiva chrysolaema				X	X	X		X	X		X	X	X	X		X	X	X					X	X	X		X	X	X	
Ameiva leberi				X																										
Ameiva lineolata				X	X	X		X	X		X													X				X	X	
Ameiva taeniura				X	X	X	X			X		X	X	X		X		X		X	X	X	X				X	X		
Leiocephalus barahonensis				X	X											X													X	
Leiocephalus endomychus								X																						
Leiocephalus lunatus													X														X	X		
Leiocephalus melanochlorus																X														
Leiocephalus personatus						X	X			X	X	X	X			X									X					
Leiocephalus pratensis								X																X						
Leiocephalus rhutidira									X																					
Leiocephalus schreibersii					X	X		X	X		X			X		X	X								X					
Leiocephalus semilineatus					X									X																
Leiocephalus vinculum																		X												X

Table 4.4 cont.

	M	M	S	P	C	G	S	P	P	C	V	N	E	L	M	T	I	I	P	G	I	I	G	I	S	C	I	C	I	A
	H	S	B	B	S	G	N	C	N	C	C	H	R	A	G	P	T	G	G	C	P	V	R	C	H	A	S	T	B	V
Epicrates fordii					X	X			X			X		X				X						X			X	X		
Epicrates gracilis			X		X							X				X														
Epicrates striatus	X	X	X	X	X		X	X	X	X	X	X	X			X	X	X				X					X			
Alsophis anomalus					X				X		X					X	X												X	
Alsophis melanichnus												X				X														
Antillophis parvifrons	X	X	X	X	X	X	X	X	X	X	X	X	X	X	X	X	X	X	X	X		X	X				X	X	X	
Darlingtonia haetiana	X	X	X																											
Hypsirhynchus ferox				X	X			X	X	X	X		X	X		X		X									X			
Ialtris agyrtes			X	X											X															
Ialtris dorsalis	X	X	X		X							X				X	X	X				X								
Ialtris parishi																X	X													
Uromacer catesbyi	X	X	X	X	X	X	X	X	X	X	X	X	X	X		X	X	X		X	X	X					X	X		
Uromacer frenatus				X	X										X	X		X		X		X					X			
Uromacer oxyrhynchus				X	X	X			X	X	X	X	X	X	X	X	X										X	X		
Leptotyphlops asbolepis															X															
Leptotyphlops calypso												X																		
Leptotyphlops leptipilepta		X																												
Leptotyphlops pyrites				X	X																									
Tropidophis haetianus	X	X	X	X	X	X	X		X	X	X	X	X	X		X	X	X												
Typhlops capitulatus					X											X														
Typhlops gonavensis																		X												
Typhlops hectus	X		X		X		X	X		X										X										
Typhlops pusillus		X	X		X	X	X	X	X	X	X	X	X	X	X	X	X	X		X							X			
Typhlops schwartzi												X	X	X																
Typhlops sulcatus				X	X	X										X		X		X										X
Typhlops syntherus				X																										

Table 4.4 cont.

Typhlops tetrathyreus	X			X		X
Typhlops titanops	X	X				

[a] These regions are less detailed than those discussed in the text and illustrated in Figure 1 (see comments in Henderson and Schwartz, 1984a:3). However, the use of these regions, with subdivided areas combined, provides an adequate appreciation of regional endemism among Haitian and Dominican amphisbaenians, lizards, and snakes. Key to abbreviations: MH = Massif de la Hotte, MS = Massif de la Selle, SB = Sierra de Baoruco, PB = Península de Barahona, CS = Valle de Neiba/Plaine de Cul-de-Sac, GG = Golfe de la Gonâve, SN = Sierra de Neiba and Montagnes du Trou-d'Eau, PC = Plateau Central, PN = Presqu'île du Nord-Ouest, CC = Cordillera Central and Massif du Nord, VC = Valle de Cibao, NH = Plaine du Nord (north Haitian littoral plain extending eastward to include much of the northern and eastern República Dominicana), ER = eastern República Dominicana (largely mesic coastal plain), LA = Llanos de Azua, MG = Sierra Martín García, TP = Tiburon Peninsula, IT = Île de la Tortue, IG = Île de la Gonâve, PG = Île de la Petit Gonâve, GC = Île Grande Cayemite, IP = Île Petit Cayemite, IV = Île-à-Vache, GR = Grosse Caye, IC = Île-à-Cabrit, SH = Cayos Siete Hermanos, CA = Isla Catalinita, IS = Isla Saona, CT = Isla Catalina, IB = Isla Beata, AV = Isla Alto Velo.

de-Sac, whereas only eight (16.6%) of the South Island endemics have invaded the North Island and another three are established widely in the Valle de Neiba/Plaine de Cul-de-Sac. Of the 228 endemic subspecies, 79 are in 15 polytypic species classified as islandwide forms, 95 are in 20 polytypic North Island endemic species, and 54 are in 18 South Island species.

In relating reptilian distributions to the biogeographic provinces, including satellite islands, defined in Henderson and Schwartz (1984a) (Table 4.4), regional endemism is less evident than that in amphibians (i.e., more reptilian than amphibian species are found in more regions or, conversely, fewer are restricted to one or very few regions). Only 141 native species were considered; the aquatic emydid turtles and one crocodilian were omitted, because any correlation with physiographic region would merely represent a coincidental juxtaposition of lakes, rivers, or coastal areas with the defined terrestrial regions. Of the species meeting the criteria for inclusion, 51 (36.2%) are restricted to only one of the 16 main island and 14 satellite island regions included in the analysis. Another 23 (16.3%) are found in only two, usually proximate and ecologically similar regions, and 25 more (17.7%) are found in three to five regions, which, as for amphibians, either are contiguous or similar in habitat. Forty-two (29.8%) species are found in six or more regions, but only two (*Anolis cybotes* and *Antillophis parvifrons*) are found in more than three-fourths (≥23) of the regions considered. However, if the satellite islands are disregarded, ten species are found in at least three-fourths ($\geq$12) of the main-island regions, with specimens of both *Anolis cybotes* and *Antillophis parvifrons* having been recorded from all 16. All familial groups except two contained at least one species found in as many as nine regions. *Mabuya lineolata* (Scincidae) was found in three regions (*M. bistriata* in only one) and only one species of *Leptotyphlops* was found in more than one region (*L. pyrites* in two).

Navassa Island

The known herpetofauna of Navassa Island (Table 4.5) consists of eight squamate species in two suborders, seven families, and eight genera. However, four have not been collected recently and may be extinct. Six other species have been recorded from the island but were regarded by Thomas (1966a) as not having been originally collected there (i.e., erroneous collection data had been recorded). *Aristelliger cochranae* and *Celestus costatus* have distinct Hispaniolan affinities. The former is closely related to and until recently was thought to be conspecific with Hispaniolan *A. expectatus*. The latter is a Navassan endemic subspecies (*C. c. badius*) of a widely distributed Hispaniolan form. *Sphaerodactylus becki* and *Anolis longiceps* are both endemic to Navassa Island. The former appears to be most closely related to Hispaniolan *S. copei*, and was, in fact, placed in the *S. copei* species group by Hass (1996). *Anolis longiceps* appears to be most closely related to *A. maynardi* of the Cayman Islands, and Schmidt (1921b) stated that *A. longiceps* is "apparently related to the Cuban *Anolis porcatus* through *Anolis maynardi*." Of the species for which recent specimens are not available, the iguanid *Cyclura cornuta* was represented on Navassa by an endemic subspecies [*C. c. onchiopsis* (= *C. c. nigerrima*)]. *Leiocephalus eremitus* has been confused with *L. melanochlorus* (Schmidt, 1921b), but Thomas (1966a) seemed confident that a distinct Navassan form once existed. *Typhlops sulcatus* is endemic to Hispaniola, whereas Navassan *Tropidophis melanurus* apparently represents an endemic subspecies (*T. m. bucculentus*) of a snake species endemic to Cuba. Thus, of the eight species rather reliably attributed to Navassa Island, six have Hispaniolan and two have Cuban affinities, although one of the latter may have reached Navassa via the Cayman Islands.

Historical Perspectives

Although pre-Columbian native Americans undoubtedly exploited some members of the herpetofauna (at least iguanas and turtles) as food (see Powell and Henderson, 1996b), the first historical references to any Hispaniolan amphibian or reptiles are those of Christopher Columbus (cited in Major, 1870; Morison, 1963), who made frequent references to iguanas and turtles in his writings. Also, Gonzalo de Oviedo (cited in Stoudemire, 1959), an early European chronicler, described an encounter with a very large (undoubtedly exaggerated) snake "at the foote of the Mountains called Pedernales." However, a formal taxonomy of Hispaniolan species was many years in coming, and realized only with the publication of *The Herpetology of Hispaniola* by Doris M. Cochran in 1941. Arguably the most comprehensive herpetofaunal account in the entire West Indian region, this book provided a foundation to which every subsequent authority has first turned when addressing questions pertaining to the amphibians and reptiles of the island.

Amphibians

Taxonomic authorities for Hispaniolan amphibians are listed in Table 4.6. The widespread toad, *Bufo marinus*, introduced on Hispaniola, was described by Linnaeus (1758), albeit not from a West Indian specimen. The next Hispaniolan species to be described, *Rana catesbeiana* by George Shaw (1802), also is introduced and was described from non-West Indian (North American) material. During the first half of the nineteenth century, European scientists, soliciting materials from any source, placed a large number of West Indian species into a systematic order (Powell and Henderson, 1996b). However, only two Hispaniolan endemics were included: *Osteopilus dominicensis*, described by Johann J. von Tschudi (1838), and *Eleutherodactylus oxyrhynchus*, described by Constant Duméril and Gabriel Bibron (1841) in their massive *Érpétologie Générale*. All other species and all subspecies occurring on the island have been described subsequently by North Americans. Edward Drinker Cope (1869, 1871) described two endemic hylids (*Hyla pulchrilineata* and *H. vasta*) before the turn of the century, but 59 of the 65 Hispaniolan species have been described since 1900. Thomas Barbour (1914, 1942) described three species of *Eleutherodactylus*, and another was described by Karl P. Schmidt (1919). However, Hispaniolan anuran

Table 4.5. Reptiles of Navassa Island (from Thomas, 1966b).

Species	Status	Faunal Association
CLASS REPTILIA ORDER SQUAMATA		
SUBORDER LACERTILIA		
Family Anguidae		
Celestus costatus	Native	Hispaniola
C. c. badius	Endemic	
Family Gekkonidae		
Aristelliger cochranae	Endemic	Hispaniola (*A. expectatus*)
Sphaerodactylus becki	Endemic	Hispaniola (*S. copei* group)
Family Iguanidae		
Cyclura cornuta *	Native	Hispaniola
C. c. nigerrima *	Endemic	
Family Polychrotidae		
Anolis longiceps	Endemic	Cuba (*A. porcatus*) via Cayman Islands (*A. maynardi*)
Family Tropiduridae		
Leiocephalus eremitus *	Endemic	Hispaniola (*L. melanochlorus*)
SUBORDER SERPENTES		
Family Tropidophiidae		
Tropidophis melanurus *	Native	Cuba
T. m. bucculentus *	Endemic	
Family Typhlopidae		
Typhlops sulcatus *	Native	Hispaniola

*Not collected recently; these species may be extinct or the Navassan populations may be extirpated.

taxonomy in the first half of the twentieth century was dominated by Cochran, G. Kingsley Noble, and Benjamin Shreve. Shreve named four species of *Eleutherodactylus* in a 1936 paper and three more with Ernest E. Williams in 1963. Cochran (1923a,b, 1932d, 1934b, 1935, 1937, 1938a, 1939, 1941) named 11 species and one currently recognized subspecies, including the sole representative of the genus *Leptodactylus* and an endemic toad. Noble, alone (1923a) and with William G. Hassler (1933), named eight species of *Eleutherodactylus*.

However, an appreciation for the tremendous diversity of Hispaniolan amphibians was left for the second half the century and was largely due to the prodigious efforts of Albert Schwartz. In a series of papers, Schwartz (1964a, 1964/1965, 1965a, c,d, 1966a, 1968c, 1971b, 1972, 1973c, 1976a,b, 1977a, 1979b, 1980a) named 21 species and 12 subspecies of *Eleutherodactylus* (including *E. neodreptus*, recently placed in the synonymy of *E. audanti*; Hedges, 1996d) and two species of endemic toads. In doing so, he was solely responsible for over one third of the known species and over 85% of the subspecies on the island. A pure taxonomist, Schwartz nevertheless sought to know the animals he studied in the field as well as in the more traditional context of museum specimens. As a result, his taxa often are identified by habitat use as well as by morphological and meristic characters. Although some workers have criticized the vast number of subspecies described by Schwartz, even the most ardent critics cannot deny that our understanding of geographic variation in the taxa addressed by him has been immeasurably enhanced.

Table 4.6. Taxonomic Authorities for Hispaniolan Amphibians.[a]

Authority	Number of taxa described	Percentages
Barbour (1914–1942)	3	4.6 (0) 3.8
Cochran (1923–1941)	11 (1)	16.9(7.1) 15.2
Cope (1869–1871)	2	3.1 (0) 2.5
Duméril and Bibron	1	1.5 (0) 1.3
Hedges (1988)	1	1.5 (0) 1.3
Hedges and Thomas (1987–1992)	4	6.2 (0) 5.1
Hedges, Thomas, and Franz (1987)	1	1.5 (0) 1.3
Linnaeus (1758)	1	1.5 (0) 1.3
Lynn (1958)	0 (1)	0 (7.1) 1.3
Noble (1923)	6	9.2 (0) 7.6
Noble and Hassler (1933)	2	3.1 (0) 2.5
Schmidt (1919)	1	1.5 (0) 1.3
Schwartz (1964–1980)	23 (12)	35.4 (85.7) 44.3
Shaw (1802)	1	1.5 (0) 1.3
Shreve (1936)	4	6.2 (0) 5.1
Shreve and Williams (1963)	3	4.6 (0) 3.8
Tschudi (1838)	1	1.5 (0) 1.3

[a] The range of publication dates is given in parentheses after each name. The number of currently recognized species described is followed by the number of subspecies currently accepted as valid (nominate subspecies are not included). Percentages are for species (subspecies) and for all currently recognized taxa at these levels.

William G. Lynn named another subspecies in 1958, but nearly all of the recent taxonomic work by scientists other than Schwartz has and is being done by S. Blair Hedges and colleagues. Hedges alone (1988), in three papers with Richard Thomas (1987, 1992a,b), and with Thomas and Richard Franz (1987) has named six species of *Eleutherodactylus*, and another (*E.* sp. "N" in Table 4.1) is being described at this time; based on an extrapolation of a graph illustrating the rate of species discoveries in the West Indies (Hedges, 1996a) and new discoveries not yet reported (S. B. Hedges, personal communication), more taxa apparently remain to be discovered.

In addition to the introduced *Bufo marinus*, Hispaniola supports three endemic toads. *Bufo guentheri* was described by Cochran in 1941, and *B. fluviaticus* and *B. fracta* by Schwartz in 1972 (the latter as a subspecies of *B. guentheri*). Charles S. Sonnini and Pierre-André Latreille (1801) and François-Marie Daudin (1802b) had described *B. gutturosus* and *B. strumosus*, respectively, but Cochran (1941) quoted a manuscript note by Leonhard Stejneger which stated that their inadequate descriptions were sufficient to preclude application of those names to any Hispaniolan (or West Indian) species. Nevertheless, both were subsequently and inappropriately used by Gravenhorst (1829) and Günther (1858 [1859]) in published references to Hispaniolan specimens. Shreve and Williams (1963) assigned *B. guentheri* to the "Caribbean section" of New World bufonids, and Pregill (1981a) resurrected the genus *Peltophryne* to accommodate these species. The latter generally was accepted, although Schwartz and Henderson (1988) emended the name to *Peltaphryne* (see comment in Powell, 1992a). Based on molecular evidence, Hedges *et al.* (1992b) questioned the validity of *Peltophryne*, and placed the Hispaniolan species (and other Antillean congeners) back into the genus *Bufo*. The latter action was formally acknowledged in Powell *et al.* (1996a).

Four hylid frogs, all endemic, occur on Hispaniola. *Hypsiboas* (= *Osteopilus*) *dominicensis* was described by Tschudi (1838) and *Hyla pulchrilineata* and *H. vasta* by Cope (1869, 1871, respectively) during the nineteenth century. *Hyla heilprini* was described by Noble (1923a). Dunn (1926a) proposed that the Hispaniolan and Jamaican species represented independent radiations, but that they were "allied." Trueb and Tyler (1973), using morphological evidence, concluded that at least six independent invasions of the West Indies occurred, all from South America. Hedges *et al.* (1992b) and Hedges (1996a), based on molecular evidence, supported as "most plausible" the South American origin of these frogs, but differ from earlier workers by suggesting that all West Indian hylids (except *H. heilprini*) represent a monophyletic group, that is to take the generic name *Osteopilus*). In contrast, Anderson (1996) used karyological data to strongly support the monophyly of *Osteopilus*, *sensu strictu*.

All other Hispaniolan anurans are in the family Leptodactylidae, and all but one are in the genus *Eleutherodactylus*. Heyer (1978) synonymized *Leptodactylus dominicensis*, originally described by Cochran (1923b), with *L. albilabris*, a closely related frog endemic to the Puerto Rico Bank. However, Schwartz and Henderson (1991) and Powell *et al.* (1996a) continued to recognize as full species these apparently closely related forms.

Tschudi (1838) named *Cornufer unicolor*, but this name has subsequently been placed in the synonymy of *Platymantis*, an Australasian ranid. Zweifel (1967) showed that Tschudi's lectotype was not a ranid and suggested it was an older name for *Eleutherodactylus inoptatus*, which had been described subsequently by Barbour (1914) as *Leptodactylus inoptatus*. This taxon was reassigned to *Eleutherodactylus* by Schmidt (1921a). To prevent confusion, Zweifel sought to suppress Tschudi's names, and the Commission of Zoological Nomenclature (Anonymous, 1978) removed *Cornufer* and *unicolor* from further consideration for eleutherodactyline frogs.

The first name coined for and still assigned to a Hispaniolan *Eleutherodactylus* is attributable to Duméril and Bibron (1841), who described *Hylodes oxyrhynchus*. *Eleutherodactylus ventrilineatus* was originally assigned to *Leptodactylus* by Shreve (1936). Substitute names were assigned by Barbour (1942; *E. haitianus* for *E. intermedius* Cochran 1941, not *E. intermedius* Barbour and Shreve 1937) and by Cochran (1937; *E. jugans* for *Leptodactylus* [= *Eleutherodactylus*] *darlingtoni* Cochran 1935:372, not *E. darlingtoni* Cochran 1935:368). In general and despite the confusion noted previously, the taxonomy of Hispaniolan forms in this speciose genus has been remarkably stable. Disagreements have centered largely around relationships within the genus and among subgenera.

Dunn (1926a) developed a taxonomic framework for the genus *Eleutherodactylus*. Variously refined in the intervening years, that initial plan was expressed more fully by Schwartz (e.g., 1965a) and Joglar (1989) on the basis of morphological data and largely supported by electrophoretic data presented in Hedges (1989a). Three of five currently recognized subgenera are represented in the West Indies, and all include Hispaniolan forms (Hedges, 1989a). The subgenus *Euhyas* is endemic to the western Greater Antilles (Cuba, Jamaica, and Hispaniola) and Isla Mona, although species have been introduced into México and the United States. All six species assigned to *Pelorius* are restricted to Hispaniola. The subgenus *Eleutherodactylus* occurs throughout the West Indies (except Jamaica) and on the mainland, although Hedges (1996a) cited evidence that the West Indian species form a monophyletic group. While recommending the retention of the currently assigned names until more information is available, Lynch (1996), based on cranial morphology, disagreed with Hedges (1989a) and Joglar (1989) in assignment of species to groups and in the suggested monophyly of *Pelorius*.

Reptiles

Taxonomic authorities for Hispaniolan reptiles are listed in Table 4.7. The earliest formal description of a Hispaniolan species was of *Lacerta* (= *Cyclura*) *cornuta* by Pierre-Joseph Bonnaterre in 1789. In retrospect, this was the first of a number of Hispaniolan species described during an era in which museum specialists, primarily European, solicited specimens from every continent in order to impose a taxonomic order on the reptiles of the world. A list of subsequent authorities reads almost like a "who's who in nineteenth century herpetology." François-Marie Daudin (1803) described *Scincus sloanii* (= *Mabuya bistriata sloanei*). Georges Cuvier (1807) described *Crocodilus* [sic] *acutus* from "St.-Dominique." Johann von Spix described *Scincus* (= *Mabuya*) *bistriatus* in 1825. *Hemidactylus mabouia* was described from a Lesser Antillean specimen by Alexandre Moreau de Jonnès in 1818. In 1830, Johann Wagler described *Sphaerodactylus cinereus*. William S. MacLeay (1834) described *Sphæriodactylus* [sic] *elegans*, albeit from a Cuban specimen. Constant Duméril and Gabriel Bibron (1836, 1837, 1839) and Duméril *et al.* (1854) included in their *Érpétologie Générale* seven species and one subspecies still recognized today. *Dendrophis* (= *Uromacer*) *catesbyi* and *Pristinotus* (= *Leiocephalus*) *schreibersii* were added to the growing list of Hispaniolan reptiles by Hermann Schlegel (1837) and Johann L. C. Gravenhorst (1837 [1838]), respectively. John Edward Gray (1840, 1852) described *Anolis porcatus* (from Cuba) and *Sauresia sepsoides*. Johann Gustav Fischer (1856, 1888) described *Homalochilus* (= *Epicrates*) *striatus*, *Chilabothrus* (= *Epicrates*) *gracilis*,

Table 4.7. Taxonomic Authorities for Hispaniolan Reptiles.[a]

Authority	Number of taxa described	Percentages
Arnold (1980)	2 (1)	1.4 (0.6) 0.9
Barbour (1914-1925)	3 (1)	2.1 (0.6) 1.2
Barbour and Carr (1940)	1	0.7 (0) 0.3
Bocourt (1870)	1	0.7 (0) 0.3
Bonnaterre (1789)	1 ([b])	0.7 (0) 0.3
Cochran (1923-1939)	15 (12)	10.3 (6.6) 8.3
Cope (1861-1879)	16 (3)	11.0 (1.7) 5.8
Cullom Schwartz (1980)	0 (2)	0 (1.1) 0.6
Cuvier (1807)	1	0.7 (0) 0.3
Daudin (1803)	0 (1)	0 (0.6) 0.3
Duméril and Bibron (1836-1854)	7 (1)	4.8 (0.6) 2.4
Dunn (1920)	1 (4)	0.7 2.2) 1.5
Fischer (1856-1888)	2 (1)	1.4 (0.6) 0.9
Gali and Schwartz (1982)	0 (1)	0 (0.6) 0.3
Gali *et al.* (1988)	0 (1)	0 (0.6) 0.3
Gans and Alexander (1962)	1	0.7 (0) 0.3
Garman (1887–1888)	1 (1)	0.7 (0.6) 0.6
Graham (1981)	1	0.7 (0) 0.3
Grant (1951)	1	0.7 (0) 0.3
Gravenhorst (1837)	1	0.7 (0) 0.3
Gray (1840-1852)	2	1.4 (0) 0.6
Günther (1858-1865)	4	2.7 (0) 1.2
Hedges and Thomas (1989)	1	0.7 (0) 0.3
Hertz (1976)	1	0.7 (0) 0.3
Incháustegui *et al.* (1985)	1	0.7 (0) 0.3
Jan (1867)	0 (1)	0 (0.6) 0.3
Kerster and Smith (1955)	1	0.7 (0) 0.3
Lazell (1961)	1	0.7 (0) 0.3
MacLeay (1834)	1	0.7 (0) 0.3
Meerwarth (1901)	1	0.7 (0) 0.3
Mertens (1938-1950)	1(5)	0.7 (2.8) 1.8
Moreau de Jonnès (1818)	1	0.7 (0) 0.3
Noble (1923)	1(3)	0.7 (1.7) 1.2
Noble and Hassler (1933)	5 (1)	3.4 (0.6) 1.8
Peters (1863)	1	0.7 (0) 0.3
Rand (1961)	1	0.7 (0) 0.3
Reinhardt and Lütken (1863)	0 (2)	0 (1.1) 0.6
Richmond (1964)	2	1.4 (0) 0.6

Table 4.7 cont.

Ruibal (1946)	1	0.7 (0) 0.3
Schlegel (1837)	1	0.7 (0) 0.3
Schmidt (1919-1928)	4	2.7 (0) 1.2
Schwartz (1964-1983)	10 (106)	6.8 (58.6) 35.6
Schwartz and Franz (1976)	1	0.7 (0) 0.3
Schwartz and Graham (1980)	1	0.7 (0) 0.3
Schwartz *et al.* (1979)	1	0.7 (0) 0.3
Schwartz and Henderson (1982)	0 (1)	0 (0.6) 0.3
Schwartz and Incháustegui (1976)	1	0.7 (0) 0.3
Schwartz and Jacobs (1989)	0 (1)	0 (0.6) 0.3
Schwartz and Klinikowski (1966)	1 (9)	0.7 (5.0) 3.1
Schwartz and Rossman (1976)	1	0.7 (0) 0.3
Schwartz and Thomas (1965-1977)	2 (3)	1.4 (1.7) 1.5
Sheplan and Schwartz (1974)	0 (4)	9 (2.2) 1.2
Shreve (1936-1968)	6 (4)	4.1 (2.2) 3.1
Spix (1825)	1	0.7 (0) 0.3
Steindachner (1867)	1	0.7 (0) 0.3
Thomas (1962-1989)	10 (2)	6.8 (1.1) 3.7
Thomas and Hedges (1988-1993)	5	3.4 (0) 1.5
Thomas *et al.* (1985)	3	2.1 (0) 0.9
Thomas and Schwartz (1965–1983)	5 (8)	3.4 (4.4) 4.0
Wagler (1830)	1	0.7 (0) 0.3
Weinland (1862)	1	0.7 (0) 0.3
Williams (1960-1975)	7 (1)	4.8 (0.6) 2.4
Williams and Rand (1969)	1	0.7 (0) 0.3
Williams and Webster (1974)	1	0.7 (0) 0.3

[a] The range of publication dates is given in parentheses after each name. The number of currently recognized species described is followed by the number of subspecies currently accepted as valid (nominate subspecies are not included). Percentages are for species (subspecies) and for all currently recognized taxa at these levels.

[b] Hispaniolan endemic nominate subspecies.

and *Ameiva regularis* (= *A. chrysolaema regularis*). Four additional reptilian species [*Pelophilus* (= *Epicrates*) *fordii*, *Hypsirhynchus ferox*, *Phylodryas* (= *Ialtris*) dorsalis, *Ahaetulla* (= *Uromacer*) *frenata*] were described by Albert Günther (1858, 1861, 1865). David Weinland (1862) described *Amphisbaena innocens*, and Johannes Reinhardt and Christian Lütken (1862) named two taxa currently recognized as subspecies; included among the latter was the exceedingly abundant *Anolis (distichus) dominicensis*, which Powell (1993) suggested was specifically distinct from Bahamian *A. distichus*. Wilhelm Peters (1863) described *Alsophis anomalus*. In 1867, Giorgio Jan and Ferdinando Sordelli described *Dromicus protenus* (= *Antillophis parvifrons protenus*) and Franz Steindachner *Sphaerodactylus copei*. Marie-Firmin Bocourt (1870) described *Anolis brevirostris*, and the end of the period of European dominance came with the description of *Hemidactylus haitianus* (as *H. brookii haitianus*) by Hermann Meerwarth in 1901.

In the latter half of the nineteenth century, a transition from European to North American scientists, and the North Americans would continue to dominate the taxonomy of Hispaniolan reptiles through the present day. Edward Drinker Cope (1861, 1861 [1862], 1862a,b, 1864, 1868, 1879) described 19 taxa, 16 currently recognized as species and three more as subspecies. Among Cope's species are the nearly ubiquitous green anoles, *Anolis chlorocyanus* and *A. coelestinus*, the widespread grass anole, *A. semilineatus*, as well as the exceedingly common *A. cybotes*, *A. distichus*, *Celestus stenurus*, *Leiocephalus personatus*, and *Dromicus* (= *Antillophis*) *parvifrons*. Samuel Garman (1887a,b) described *Sphaerodactylus picturatus* (= *S. copei picturatus*) and *Anolis haetianus*. As for amphibians, the pace accelerated after the turn of the century, and many of the same persons were responsible. Thomas Barbour (1914, 1925) named five species (one, *Anolis doris*, now considered a subspecies of *A. cybotes*), and with Archie Carr (1940) described [*Pseudemys* (= *Trachemys*) *decorata*]. Emmett R. Dunn (1920a,b) named a lizard and four snakes (all of the latter now are considered subspecies of *Antillophis parvifrons* or *Uromacer frenatus*). G. Kingsley Noble, alone (1923b) and with William G. Hassler (1933), described six species and four currently recognized subspecies. However, dominating this period in Hispaniolan taxonomy was Doris M. Cochran. In a series of papers (1923c, 1927, 1928b,c, 1931a,b, 1932a,b,c, 1933, 1934a,b, 1935, 1939), she named 27 currently recognized taxa (15 species and 12 subspecies), and represented among them were lizards, snakes, and an amphisbaenian. Acknowledging the importance of Cochran's contributions, Rodolfo Ruibal (1946) described *Sphaerodactylus cochranae*. Also prominent during this time, and an exception to the general dominance by North Americans of 20th century Hispaniolan taxonomy, was Robert Mertens, who described in three papers (1938, 1939, 1950) one currently recognized species (*Anolis aliniger*) and five additional taxa currently considered subspecies.

However, and again as in amphibians, an acknowledgment of the tremendous diversity among Hispaniolan reptiles came only in the second half of this century – and, again, was largely due to the efforts of Albert Schwartz. In an impressively long list of papers from 1964 to 1989, alone (1964b, 1965b,e, 1966b, 1967a,b, 1968a,b,d, 1969a,b, 1970a,b,c,d, 1971c,d,e, 1973a,b,d, 1974a,b, 1975a,b,c, 1976c,d, 1977b,c, 1978b,c, 1979a,c,d,e,f, 1980c, 1981, 1983) or with one or more of many collaborators (Cullom and Schwartz, 1980; Gali and Schwartz, 1982; Gali *et al.*, 1988; Incháustegui *et al.*, 1985; Schwartz and Franz, 1976; Schwartz and Graham, 1980; Schwartz *et al.*, 1979; Schwartz and Henderson, 1982; Schwartz and Incháustegui, 1976; Schwartz and Jacobs, 1989; Schwartz and Klinikowski, 1966; Schwartz and Rossman, 1976; Schwartz and Thomas, 1964 [1965], 1965, 1977; Sheplan and Schwartz, 1974; Thomas and Schwartz, 1965, 1967, 1977, 1983a,b), he named a total of 24 species and 136 subspecies, fully 16.4% of the species and an astounding 75.1% of the reptilian subspecies on Hispaniola. Although difficult to ascertain solely from reading and using the keys included in guides such as Henderson and Schwartz (1984a) and Henderson *et al.* (1984), Schwartz brought to his taxonomy a sense of an organism's total gestalt, along with its ecology and the often very detailed characters identified in his keys (Thomas, 1996). Herpetologists who have not worked on Hispaniola simply cannot appreciate the clarity of the differences among geographic races often separated more distinctly by habitat than by distance.

Still, Schwartz was not alone in elucidating the growing knowledge of reptilian diversity on Hispaniola. Although better known for his work elsewhere in the West Indies, Chapman Grant (1951) described *Celestus curtissi*. Harold Kerster and Hobart Smith (1955) described *Phyllodactylus wirshingi* from Puerto Rican material. James "Skip" Lazell, usually associated with the Lesser Antilles, named *Sphaerodactylus shrevei* in 1961. Carl Gans and A. A. Alexander (1962) described *Amphisbaena gonavensis* in their analysis of Hispaniolan amphisbaenians. Two species of *Typhlops* were described by Neil Richmond (1964) in a paper that began the process of elucidating the relationships among these scolecophidians. In 1976, Paul Hertz, better known for his studies of saurian thermal biology, named *Anolis alumina*. Sixto Incháustegui, a Dominican naturalist, with Schwartz (1976) and with Schwartz and Robert W.

Henderson (1985) identified *Diploglossus* (= *Celestus*) *marcanoi* and *D*. (= *Celestus*) *carraui*, respectively. Douglas Arnold (1980) identified two species and a subspecies in his analysis of the *Anolis brevirostris* complex.

However, despite of these many, frequently prominent contributors, a few persons (in addition to Schwartz) stand out. Ernest E. Williams (1960, 1962a,b, 1963a,b, 1965a,b, 1975) named seven species and one subspecies of *Anolis*, plus another species (*A. rupinae*) with Thomas Preston Webster in 1974 and still another (*A. insolitus*) with A. Stanley Rand in 1969. Rand, who with Williams applied the ecomorph concept to West Indian anoles (see Chapter 7), also described *A. koopmani* in 1961. Richard Thomas, alone (1965a,b,c, 1966b, 1968, 1971, 1974a, 1982, 1989), with Schwartz (Schwartz and Thomas, 1964 [1965], 1965, 1977; Thomas and Schwartz, 1965, 1967, 1977, 1983a,b), more recently with S. Blair Hedges (Hedges and Thomas, 1989a; Thomas and Hedges, 1988, 1989, 1992, 1993), and with Roy W. McDiarmid and Fred G. Thompson (1985) described a total of 26 species and 13 subspecies of Hispaniolan reptiles. Thus Thomas is second only to Schwartz in the total number of taxa described and, when only species are considered, he is responsible for more than any other single person. Hedges, also, is worthy of more than casual mention. With Thomas, he has identified six reptilian species, but of particular note are his recent synoptic work on origins and relationships of the herpetofauna and his explorations into many nearly inaccessible areas searching for additional taxa that may still lurk there.

The two species of *Trachemys*, described in this century by Barbour and Carr (1940, *T. decorata*) and Schmidt (1928, *T. stejnegeri*), have been subjected to various generic reassignments as a result of our frequently changing, but growing understanding of relationships among the American emydid turtles. Originally described as *Pseudemys*, McDowell (1964) placed both forms (along with species currently in *Chrysemys*, *Pseudemys*, and *Trachemys*) in the genus *Chrysemys*, *sensu lato*, but recognized at a subgeneric level an affinity between North American *T. scripta* and the West Indian forms. Seidel and Incháustegui Miranda (1984) elevated McDowell's subgenera to full generic status. In addition, considerable confusion has existed regarding the specific and subspecific assignments of the Hispaniolan forms. Mertens and Wermuth (1955) and Wermuth and Mertens (1961, 1977) had considered both taxa to be subspecies of *T. terrapen*; Pritchard (1979) had indicated that only one form, *T. decorata*, occurred on Hispaniola; and Williams (1956) and Schwartz and Thomas (1975) had considered *T. s. vicina* a subspecies of Cuban *T. decussata*. Seidel and Incháustegui Miranda (1984) established the current usage, which has been followed by all subsequent workers.

In their initial descriptions, all Hispaniolan amphisbaenians were placed in the genus *Amphisbaena*, and for all but two, *A. gonavensis* and *A. hyporissor*, the original assignment to species is still recognized. *Amphisbaena gonavensis* first was described by Gans and Alexander (1962) as a subspecies of *A. innocens*, and *A. hyporissor* by Thomas (1965a) as a main island subspecies of *A. gonavensis*. In addition, Gans and Alexander (1962) considered *A. caudalis* a subspecies of *A. innocens*, but Schwartz and Thomas (1975) noted that the two forms had been collected in sympatry on Île Grande Cayemite. Only minor confusion was caused by Schmidt (1928), who inadvertently created a *nomen nudum* by using *Amphisbaena weinlandi* for *A. innocens* Weinland.

Historically, Hispaniolan anguids have been assigned to the genera *Celestus*, *Diploglossus*, *Panolopus*, *Sauresia*, and *Wetmorena*. *Panolopus costatus* was described by Cope (1861 [1862]), but more recent confusion has centered on the differences between *Celestus* and *Diploglossus*, among which many of the species have been repeatedly and variously assigned. Following Strahm and Schwartz (1977), Henderson and Schwartz (1984a) and Henderson *et al.* (1984) distinguished these genera by the presence (*Diploglossus*) or absence of a "well-developed osteodermal radix." Savage and Lips (1993) rejected assignments to genera on the basis of this character because the presumed differences represent ontogenetic stages. Instead, they used the presence (*Diploglossus*) or absence (*Celestus*) of claw sheaths to distinguish these forms and, in doing so, placed all Hispaniolan forms formerly referred to *Diploglossus* in *Celestus*.

Sauresia sepsoides was described by Gray (1852) as a new genus of "Scincidae." Weinland (1862) described the same form as *Embryopus habichii* from Haitian material. In both instances, these lizards were distinguished from other known taxa in having only four toes. The monotypic genus *Wetmorena* was established by Cochran (1927) for *W. haetiana*, and was segregated from other West Indian anguids by the lack of external ear openings. As species in both genera have claw sheaths, Savage and Lips (1993) suggested that they are allied with *Diploglossus* and resulted from speciation within that group. Based on albumin ID and DNA sequence data [Hedges *et al.*, 1992b; Hass, Maxson, and Hedges, manuscript in preparation (cited in Hedges, 1996a)], both Hispaniolan endemic genera were placed in a monophyletic radiation of Antillean *Celestus*. These changes first appeared formally in Powell *et al.* (1996a).

The gekkonid genus *Aristelliger* was erected for *A. lar* by Cope (1861 [1862]). Grant (1931a) described *A. cochranae* from Navassa Island, and Cochran (1933) described *A. expectatus* from Haiti. Mertens (1939), without comment, sunk the latter to a subspecies of the former, and this was followed subsequently until *A. expectatus* was elevated back to full-species status by Powell *et al.* (1996a). Noble and Klingel (1932) erected the genus *Aristelligella* for small members of the genus *Aristelliger* with two or more asymmetrical terminal scansors (Bauer and Russell, 1993). Although not explicitly stated, that description implicated *A. barbouri* as the type species, although two species (the other being *A. cochranae*) were assigned to the new genus. Hecht (1951) and others have used *Aristelligella* as a subgenus, but others [R. I. Crombie, personal communication (cited in Bauer and Russell, 1993)] regard it as valid at the generic level.

Gymnodactylus (= *Gonatodes*) *notatus* was described by Reinhardt and Lütken (1862) from Haitian material. Barbour (1937), in his third list of West Indian amphibians and reptiles, placed the species in the genus *Gonatodes*, previously applied to *G. albigularis* [sic] by Fitzinger (1843). Vanzolini and Williams (1962) sunk *notatus* to a subspecies of *G. albogularis*.

The two Hispaniolan representatives of the genus *Hemidactylus* have obvious African affinities (Kluge, 1969). *Hemidactylus mabouia*, with a nearly pan-Neotropical distribution, was described from Lesser Antillean material (Duméril and Duméril, 1851; Stejneger, 1904; Smith and Taylor, 1950) as *Gecko mabouia* by Moreau de Jonnès in 1818, but still is considered conspecific with populations in Africa. *Hemidactylus haitianus* was described as *H. brookii haitianus* by Meerwarth (1901). The *H. brookii* complex [R. I. Crombie, personal communication (cited in Powell *et al.*, 1996a)] has a broad distribution in the Eastern Hemisphere, across equatorial Africa (including the Cape Verde Islands), probably several Indian Ocean islands (the taxonomic status of these populations is unclear), the Indian subcontinent, and through the Malay–Indonesian Archipelago. The Antillean populations were elevated to full-species status by Powell *et al.* (1996a).

Geckos in the genus *Phyllodactylus* originally were described by Gray (1830) from "tropical South America?" (Gray, 1845), but the type species, *P. pulcher*, is restricted to Barbados (Parker, 1935). These lizards are characterized by leaf-like terminal subdigital lamellae and have a broad, but spotty distribution throughout the Neotropics (Dixon, 1973a). The presence of this species in the Greater Antilles was established with the description of. *P wirshingi* from Isla Caja de Muertos, Puerto Rico, by Kerster and Smith in 1955. The first published acknowledgment that geckos in the genus *Phyllodactylus* occurred on Hispaniola was in a remark in Schwartz and Thomas (1975), in which they noted that "These lizards are similar to *Ph. wirshingi*, but we merely point out that the genus occurs on Hispaniola." Additional populations had been discovered by the time Schwartz (1979f) described the two endemic Hispaniolan subspecies, *P. w. hispaniolae* and *P. w. sommeri*.

The genus *Sphaerodactylus* was named by Wagler (1830), and all 32 Hispaniolan species were properly assigned at the time of their original descriptions. Morphological studies resulted in the recognition of several species groups (Thomas and Schwartz, 1966, 1983a; Shreve, 1968; Thomas, 1975; Schwartz and Graham, 1980; Schwartz and Garrido, 1981a, 1985; Schwartz, 1983), but these studies provided few insights into relationships among the groups. Hass (1991), based on starch-gel electrophoretic data and immunological comparisons of serum albumin, suggested that West Indian forms are monophyletic (*S. sputator* section of Hass, 1996) and that Greater Antillean *Sphaerodactylus* were partitioned into two main groups, the *cinereus* and *argus* series, both including Hispaniolan species. Using DNA sequence data, Hass (1996) modified her previous classification; Hispaniolan species were largely unaffected, except that *S. elasmorhynchus* was not assigned to a subsection.

The iguanid (*sensu strictu*) genus *Cyclura* is represented on Hispaniola by two species. *Cyclura cornuta* was the first Hispaniolan reptile to be described, as *Lacerta cornuta* by Bonnaterre (1789). Lacépède (1789) soon thereafter placed this species in the genus *Iguana*, but Wagler (1830) erected the monotypic genus *Metopoceros* for *cornuta*. Fitzinger (1843), however, treated *Metopoceros* as a subgenus of *Hypsilophus*. These iguanas were placed in the genus *Cyclura* , erected by Harlan (1824 [1825]) for *C. carinata* from the Bahamas, by Cope (1886).

Aloponotus ricordii was named by Duméril and Bibron (1837). Fitzinger (1843) recognized *Aloponotus* as a subgenus of *Hypsilophus*, but Boulenger (1885) placed *Aloponotus ricordii* Duméril and Bibron (1837), *Hypsilophus* (*Aloponotus*) *ricordii* Fitzinger (1843), and *Aloponotus ricordi* Gray (1845) in the synonomy of *Metopoceros cor-*

nutus (Bonnaterre, 1789); in the process misspelling each name. This species was placed in the genus *Cyclura* by Cochran (1924).

The names *Cyclura onchiopsis* and *C. nigerrima*, both described from Navassa Island by Cope (1885, 1886), have been used interchangeably; the confusion apparently resulting from the fact that the holotype of *C. nigerrima* (USNM 9974) bears a lower number than any syntype of *C. onchiopsis* (USNM 9977), although the latter name was used in an earlier publication. *Cyclura stejnegeri* was described by Barbour and Noble (1916) from Isla Mona in the Mona Channel between Hispaniola and Puerto Rico. Barbour (1937) relegated *C. stejnegeri* to subspecific status within *C. cornuta*, and Schwartz and Thomas (1975) did the same for *C. onchiopsis*. A number of efforts to elevate either or both to full species have not met with wide acceptance (see remark in Schwartz and Henderson, 1988:122). As a result, Hispaniolan populations continue to be recognized as constituting a nominate subspecies.

All 42 species of Hispaniolan anoles currently are placed in the genus *Anolis*. However, this consistency has not always characterized this group. Cope erected a number of genera and subgenera for various West Indian forms, and three Hispaniolan species were included: *Eupristus baleatus* (1864), *Anolis* (*Ctenocercus*) *coelestinus* (1862b), and *Anolis* (*Anolis*) *cybotes* (1862b). Cochran (1934b) erected the genus *Audantia* for *A. armouri*, presumably on the basis of a distinct "transverse gular fold." She subsequently described *Audantia shrevei* in 1939. Etheridge (1960), by inference, placed these species in the genus *Anolis*. Later, Williams (1963b) treated *A. armouri* as a subspecies of *A. cybotes*, but he elevated it back to a full species in 1976. In 1935, Cochran erected the genus *Xiphocercus* for *X. darlingtoni*. Williams (1962b) placed this taxon in *Anolis*, creating a nomenclatural conflict with *A. darlingtoni* Cochran (1939), for which Williams provided the substitute name, *A. etheridgei*. However, the most dramatic effort to generically partition the anoline lizards, by Guyer and Savage (1986), was based largely on morphology. Schwartz and Henderson (1988) applied these generic assignments to West Indian forms. As a result, Hispaniolan species were placed in the genera *Ctenonotus* (*altavelensis*, *armouri*, *brevirostris*, *caudalis*, *cristatellus*, *cybotes*, *distichus*, *eugenegrahami*, *haetianus*, *longitibialis*, *marcanoi*, *marron*, *shrevei*, *whitemani*) and *Semiurus* (*baleatus*, *barahonae*, *ricordii*), but none were assigned to the genera *Dactyloa* or *Norops* (see comments below regarding *Chamaelinorops*). Cannatella and de Queiroz (1988) and Williams (1989a) raised substantive questions regarding the validity of these genera, which despite their frequent use by workers on the Neotropical mainland (see note added in proof in Vitt and Zani, 1996b), have not subsequently been applied widely to West Indian forms. Hedges (1996a) suggested that classifications such as those of Williams (1976) and Burnell and Hedges (1990), in which informal categories like series and groups were used, are preferable, as they allow for disagreement and realignment without affecting formal classification.

The morphologically (Forsgaard, 1983) and ecologically (Flores *et al.*, 1994) distinctive *Anolis barbouri* was placed in the monotypic genus *Chamaelinorops* by Schmidt (1919). Although presumably from Navassa Island, Thomas (1966a) restricted the type locality to the Massif de la Hotte in Haiti. In the interim, *C. wetmorei* had been described by Cochran (1928a) from Hispaniola. Hass *et al.* (1993), based largely on DNA sequence data, synonymized *Chamaelinorops* (and Cuban *Chamaeleolis*) within *Anolis*. The lack of agreement over taxonomy reflects the dichotomy of opinions regarding the origin of the species. Etheridge (1960) and Wyles and Gorman (1980a) suggested that this lizard is very derived, having arisen within *Anolis*, but Williams (1977) and Case and Williams (1987) argued that the species is very primitive, maybe the most primitive of all extant anoles.

General relationships among West Indian anoline lizards, postulated by Etheridge (1960) and Williams (1976), largely were based on osteology and karyotypes. Subsequent work, based on albumin ID data (e.g., Wyles and Gorman, 1980; Shochat and Dessauer, 1981; Hass *et al.*, 1993), allozyme data (Burnell and Hedges, 1990), and DNA sequence data (Hass *et al.*, 1993), has refined our understanding of those relationships. Many of the series and species groups recognized by Williams (1976) were supported, although his major alpha/beta dichotomy was not.

The following Hispaniolan species originally were described as subspecies: *Anolis aliniger* (*A. chloro-cyanus aliniger* Mertens, 1939), *A. altavelensis* (*A. dominicensis altavelensis* Noble and Hassler, 1933), *A. barahonae* (*A. ricordii barahonae* Williams, 1962a), *A. caudalis* (*A. dominicensis caudalis* Cochran, 1932b), and *A. dolichocephalus* (*A. hendersoni dolichocephalus* Williams, 1963a). *Anolis bahorucoensis* has been sunk to subspecies (*A. hendersoni baharucoensis* [sic] Williams, 1963a) and subsequently elevated back to a full species. *Anolis cybotes doris* (Barbour, 1925) and *A. distichus dominicensis* (Reinhardt and Lütken, 1862) originally were described as species, but have been relegated subsequently to subspecific status. Most described species which later were synonymized can be attributed to scattered collecting activities during a time when an understanding of the diversity and biogeography of Hispaniolan forms was woefully incomplete; included in this group are *A. laeviceps* (= *A. chlorocyanus*) (Lichtenstein, 1856), *A. latirostris* (= *A. coelestinus*) from Navassa Island (Schmidt, 1919) (also see below), and *A. biauritus* (= *A. distichus*) (Meerwarth, 1901). However, the situation involving *A. cochranae* (= *A. semilineatus*), described by Williams and Rand (1961), occurred more recently. Additional confusion has prevailed regarding the status of closely related forms, notably among the distichoids (*A. distichus* and the *A. brevirostris* complex) (Arnold, 1980; Williams and Case, 1986) and cybotoids (Williams, 1963b; Case and Williams, 1988; also see remarks in Schwartz and Henderson, 1988).

Mabuya lineolata was described as an endemic Hispaniolan species by Noble and Hassler (1933). Unfortunately, the taxonomic status of *Mabuya bistriata*, known only from the vicinity of Santo Domingo, is considerably less clear-cut. Until Powell *et al.* (1996a), this species was listed among the Hispaniolan fauna as *M. mabouya*. Schwartz and Henderson (1988) presented a rather detailed synonymy, but also said that "the taxonomy of Antillean *Mabuya* is not so simple as current nomenclature indicates; however, the study of this group is complicated by the extinction or virtual extinction of a number of island populations." They then referred the reader to "more complete synonymies" in Dunn (1935) and Peters and Donoso-Barros (1970). Although the latter has been updated (Peters and Donoso-Barros, 1986), no clarification of the status of West Indian populations was forthcoming. Avila-Pires (1995) addressed the status of Amazonia *Mabuya* in what is now considered the *M. mabouya* complex, with revisions suggesting that the West Indian forms are, in fact, *M. bistriata*. Although this nomenclature was used by Hedges (1996a), Murphy (1996), and Powell *et al.* (1996a), the latter stated that "the assignment of Antillean populations remains unresolved." The Hispaniolan material currently is assigned to *M. bistriata sloanei*, originally described as *Scincus sloanii* by Daudin (1803).

The polytypic species *Ameiva chrysolaema* was described by Cope (1868). *Amiva* [sic] *vittipunctata* (Cope, 1871) and *Cnemidophorus affinis* (Fischer, 1883) were subsequently placed in the synonymy of the nominate subspecies. *Ameiva abbotti* was described by Noble (1923b) as a form endemic to Isla Beata, and *Ameiva regularis* by Fischer (1888) from northern Haiti; both are now relegated to subspecific status. Additional subspecies were described by Cochran (1934b, *A. c. woodi*), Mertens (1938, *A. c. boekeri*), Schwartz (1968a, *A. c. quadrijugis*), and Schwartz (1973a, *A. c. evulsa*), but the definitive analysis of this species is attributed to Schwartz and Klinikowski (1966). They named nine currently recognized subspecies, clarified the status of those forms previously described, and also named *A. leberi* as a patternless subspecies of *A. chrysolaema*. Schwartz and Thomas (1975) elevated the latter, stating "that *A. ch. ficta* and *A. leberi* are broadly syntopic without intergradation for about 30 kilometers along the Pedernales–Oviedo road, and *A. leberi* occurs at the type-locality of *A. ch. ficta*." However, Sproston *et al.* (1998) found no ecological segregation between the two forms and suggested that the status of *A. leberi* be reevaluated.

Ameiva lineolata was described by Duméril and Bibron (1939). Noble (1923b) named *A. beatensis* from Isla Beata, but Mertens (1939) relegated it to a subspecies of *A. lineolata*. Four additional subspecies were defined by Schwartz (1965e). A similar history applies to *A. taeniura*. Originally described by Cope (1862a), insular species subsequently were described on Île de la Gonâve (*A. barbouri* by Cochran, 1928a), Isla Saona (*A. rosamondae* by Cochran, 1934b), and Navassa Island (*A. navassae* by Schmidt, 1919). *Ameiva t. meyerabichi* was named by Mertens (1939) in a paper in which he also relegated *A. barbouri* and *A. rosamondae* to subspecific status. However, the definitive work on the species was by Schwartz (1967b), in which he described eight additional subspecies and sunk *A. navassae* into the synonymy of *A. taeniura*. Another insular subspecies, *A. t. pentamerinthus* from Île Grande Cayemite, was named by Schwartz (1968a).

Frost and Etheridge (1989) determined that lizards in the endemic West Indian genus *Leiocephalus* were in the same clade as South American *Tropidurus* and *Liolemus*. They referred to this clade as the family Tropiduridae, and placed *Leiocephalus* in the subfamily Leiocephalinae. Pregill (1992), in his analysis of the West Indian forms, followed this treatment. However, Hedges *et al.* (1992b) and C. A. Hass, *et al.* [manuscript in preparation (cited in Hedges, 1996a)] presented albumin ID and DNA sequence data that suggested a closer affinity between *Leiocephalus* and North American Crotaphytidae. Although unresolved (may the distinctiveness of these lizards warrant recognition at the familial level?), little doubt exists that Hispaniola, with ten extant species, is the center of their radiation, which currently encompasses the West Indies north of Puerto Rico [although material known only from fossils indicates that the range of the genus once included all of the major islands and banks of the West Indies, as least as far south as Martinique (Pregill *et al.*, 1988)].

The first Hispaniolan species to be described was *Pristonotus schreibersii* (Gravenhorst, 1837 [1838]), later assigned to *Steironotus* by Fitzinger (1843), and finally to *Leiocephalus* Gray, 1827 by Cope (1868). Cope also described *L. melanochlorus* (1862b), *Liocephalus* [sic] *macropus* (1862b), *Liocephalus* [sic] *trigeminatus* (= *L. personatus*) (1862b), and *Liocephalus* [sic] *eremitus* (1868), the last from Navassa Island. *Leiocephalus barahonensis* was described by Schmidt (1921a), treated as a subspecies of *L. personatus* by Mertens (1939) and Cochran (1941), and reelevated to full-species status by Schwartz (1967a). Noble and Hassler (1933) described *L. altavelensis* from Isla Alto Velo, the placement of which has subsequently caused some confusion. Schwartz (1967a) suggested that the population might represent (1) an aberrant form of *L. barahonensis*, (2) a distinct species, or (3) a subspecies of *L. vinculum* – despite its peculiar distribution far removed from the nearest putative *L. vinculum* on Île de la Gonâve. He least liked the first and eventually settled for the third option, although Pregill (1992) presented a case for the first (currently used) and Powell (1993) also suggested that full-species status might be most accurate. Cochran (1928b) erected the genus *Hispaniolus* for *H. pratensis*. She distinguished *Hispaniolus* from *Leiocephalus* on the basis of the former lacking dorsal or caudal crests. Etheridge (1966) placed *Hispaniolus* into the synonymy of *Leiocephalus*. Cochran (1932a) first used the combination *L. personatus*, and she also named *L. lunatus* (1934a), *L. semilineatus* (1941), and *L. vinculum* (1941) as subspecies of the former. All were elevated to full species by Schwartz (1967a), who also suggested that these species, with *L. barahonensis*, constituted the *L. personatus* group; Pregill (1992), however, found little evidence to support that characterization. *Leiocephalus endomychus*, originally described as a subspecies of *L. vinculum* by Schwartz (1967a), was elevated to a full species by Pregill (1992). Schwartz (1979c) described *L. rhutidira*.

The three Hispaniolan boids, all currently placed in the genus *Epicrates*, were originally described as *Pelophilus fordii* by Günther (1861), *Homalochilus striatus* by Fischer (1856), and *Chilabothrus gracilis* by Fischer (1888). In that same paper, Fischer (1888) also allocated the equally slender and largely arboreal *P. fordii* to *Chilabothrus*. Although Steindachner assigned *striatus* to *Epicrates* in 1863, the morphologically distinct *fordii* and *gracilis* were not reassigned until

Boulenger (1893) did so in his *Catalogue of the Snakes in the British Museum*. Cope (1862a) described *Homalochilus multisectus*, the type locality of which was restricted to the vicinity of Santo Domingo by Sheplan and Schwartz (1974); it currently is relegated to the synonymy of the nominate subspecies of *E. striatus* (Stull, 1935).

The sole Hispaniolan tropidophiid, *Tropidophis haetianus*, was described as *Ungualia* [sic] *haetiana* by Cope (1879), the confusion due to the similarity with the Neotropical mainland genus *Ungaliophis*, which is placed with *Trophidophis* in a distinct family. *Tropidophis conjunctus*, described by Fischer (1888), is relegated to the synonymy of the nominate subspecies, *T. h. haetianus*. Cochran (1924) treated *T. haetianus* as a subspecies of Cuban *T. maculatus*, a designation followed by Stull (1928) and others until Schwartz and Marsh (1960) justified treatment of these forms as a distinct species.

Among Hispaniolan colubrids, two species of racers have been described: *Alsophis melanichnus* by Cope (1862a) and *Zamenis* (= *Alsophis*) *anomalus* by Peters (1863). *Dromicus parvifrons* also was described by Cope (1862a), and was allocated to *Antillophis* by Maglio (1970). Subsequent to the original description, a number of species and subspecies were described and assigned variously to *Dromicus*, *Leimadophis*, and *Leptophis*. Jan and Sordelli (1867) described *Dromicus protenus*, and Fischer (1883) *Leptophis frenatus* (with an erroneous type locality listed as Sierra Leone). Both were placed within *Antillophis parvifrons* (as *A. p. protenus*) by Maglio (1970). Dunn (1920b) named *Leimadophis alleni* from Île de la Gonâve and *L. tortuganus* from Île de la Tortue; both forms were relegated to subspecific status by Maglio (1970). Currently recognized subspecies originally placed in genera other than *Antillophis* are: *Leimadophis parvifrons niger* (Dunn, 1920b), *Leimadophis parvifrons lincolni* (Cochran, 1931a), *Dromicus parvifrons rosamondae* (Cochran, 1934b), and *Dromicus parvifrons paraniger* and *Dromicus parvifrons stygius* (Thomas and Schwartz, 1965). The monotypic genus *Darlingtonia* was erected by Cochran (1935) for *D. haetiana*. Recently, Hedges and Garrido (1992b) suggested that this species may be reassigned to *Arrhyton*. In contrast, unpublished allozyme data (B. I. Crother, personal communication) indicate that including *Darlingtonia* in *Arrhyton* would render the genus polyphyletic; instead, these data suggest that *Darlingtonia* and Jamaican *Arrhyton* form a group independent of Cuban and Puerto Rican *Arrhyton*. The monotypic genus *Hypsirhynchus* was erected and *H. ferox* was described by Günther (1858). Cope

(1862a) described *H. scalaris* from Haitian material, but Schwartz (1971c) relegated it to subspecific status. Günther (1858) also described *Philodryas* (= *Ialtris*) *dorsalis*. Cope (1862a) erected the genus *Ialtris* for *I. vultuosa*, which subsequently was placed into the synonymy of *I. dorsalis*, although confusion over the generic name did not cease until well into the twentieth century (Powell and Henderson, 1994). Günther's (1862) *Dromicus mentalis* and Werner's (1909) *Dromicus w-nigrum* contributed to the confusion, as did Cope's (1863b) own unjustified emendation of *Ialtris* to *Jaltris* (Smith and Wallach, 1993). *Ialtris parishi* (Cochran, 1932c) and *I. agyrtes* (Schwartz and Rossman, 1976) were described as such. Similar confusion has existed regarding snakes in the endemic genus *Uromacer*. *Dendrophis catesbyi* was described by Schlegel (1837), but Duméril *et al.* (1854) placed it in the newly erected genus *Uromacer* (Duméril, 1853) along with the newly described *U. oxyrhynchus* (although the latter was erroneously assigned a type locality of Senegal). In 1861, Duméril erected the monotypic genus *Megalocercus* for Schlegel's *D. catesbyi*. Shortly thereafter, Günther (1865) described *Ahaetulla frenata* and Garman (1887b) *U. inornatus*. The former was reassigned to *Uromacer* and the latter sunk into the synonymy of *U. frenatus* by Boulenger (1894). Further complicating an understanding of these snakes were the descriptions of three satellite island forms as new species. Dunn (1920b) named *U. scandax* from Île de la Tortue (relegated to a subspecies of *U. catesbyi* by Mertens, 1939). Dunn (1920b) also described *U. dorsalis* from Île de la Gonâve, and Cochran (1931a) described *U. wetmorei* from Isla Beata. The former was sunk to subspecific status (within *U. frenatus*) by Schwartz (1979d) and the latter by Horn (1969). Interestingly, Schwartz (1979d) subsequently recognized the validity of *U. wetmorei*, but Schwartz and Thomas (1975) followed Horn, and Henderson and Schwartz (1984b) formally acknowledged that action. Lazell (1983) suggested that the Antillean xenodontine genera *Alsophis*, "*Dromicus*," *Arrhyton*, and *Liophis* were weakly differentiated and used the name *Liophis* for members of this complex; nevertheless, subsequent workers largely have followed Schwartz and Henderson (1988) in using the more "classical" names.

The history of scolecophidian taxonomy of Hispaniolan forms is simultaneously straightforward and relatively complex. The first leptotyphlopid discovered on the island was *Leptotyphlops pyrites*, described by Thomas (1965b). All of the other three known species were described by Thomas *et al.* (1985). In contrast, the situation regarding Hispaniolan typhlopid snakes is more circuitous, with the current state of knowledge incomplete until publication of the analysis by Thomas (1989). Much of the confusion occurred because the name *Typhlops lumbricalis* (but not *Anguis lumbricalis* of Linnaeus, 1758) was variously attributed to Hispaniolan material (Cochran, 1924, 1941; Richmond, 1964; Thomas, 1974a) and subsequently assigned to several species. Cope (1868) described *T. sulcatus*, originally from Navassa Island, and Barbour (1914) described *T. pusillus*, and only these names and *T. lumbricalis* were used until Richmond's (1964) study, in which he named *T. capitulatus*, *T. gonavensis*, and *T. haitiensis*. The latter was placed in the synonymy of *T. sulcatus* in a paper by Thomas (1965c), in which he also named *T. syntherus* and relegated *T. gonavensis* to a subspecies of *T. capitulatus*. *Typhlops hectus* was described by Thomas (1974a) for a part of the material which had been referred to as *T. lumbricalis*. In his doctoral dissertation, Thomas (1976) made reference to three new forms, all previously designated in the literature as *T. lumbricalis*, as *Typhlops* I. Subspecies A, *Typhlops* I. Subspecies B, and *Typhlops* II. With their formal descriptions by Thomas (1989) as *T. schwartzi*, *T. tetrathyreus*, and *T. titanops*, respectively, the currently accepted status of the nine Hispaniolan species finally was resolved.

The final component of the Hispaniolan herpetofauna and the lone-crocodilian representative on the island is *Crocodylus acutus*, described by Cuvier (1807) from "St.-Dominique."

Navassa Island

The three taxonomic authorities for the known Navassan reptiles are listed in Table 4.8. Thomas (1966a) summarized the state of the herpetofauna and provided a history of collecting activity on the island. Cope (1868), based on specimens collected by W. J. Rasin, described *Celestus badius*, *Liocephalus* [sic] *eremitus* , *Typhlops sulcatus*, and *Ungalia* (= *Tropidophis*) *bucculentus*. In 1885, he described *Cyclura onchiopsis*. Mertens (1939) relegated *C. badius* to a subspecies in *Celestus costatus*, an allocation with which Schwartz (1964b), in his study of Hispaniolan *Diploglossus costatus*, concurred. *Leiocephalus eremitus* was described from a single specimen in the Rasin collection. A later specimen, collected by R. H. Beck, was actually a specimen of *L. melanochlorus* (Thomas, 1966a), and was probably taken from the Les Cayes region of Haiti. It was the latter specimen on which comments in Etheridge (1966) were based. Thomas (1966a) examined the original holotype, commented on its unique features, and stated that: "The distinctness of this form plus the confidence generated by the W. J. Rasin specimens indicate that Navassa once possessed an endemic species of *Leiocephalus* which is now extinct." *Typhlops sulcatus* also was described from a single specimen. Richmond (1964) noted the existence of *T. sulcatus*, but did not compare it to *T. haitiensis*, which he named in the same paper. Thomas (1965c) first allocated a Hispaniolan specimen to *T. sulcatus*, and later (1966b) confirmed his suspicion that *T. haitiensis* belonged in the synonymy of *T. sulcatus*. Cope (1868) suggested an affinity between *Ungalia bucculentus* and *Tropidophis melanurus*, but Boulenger (1893) listed it in the synonymy of *T. maculata*, and Stull (1928) considered it a subspecies of *T. pardalis*. Bailey (1937) noted that Stejneger (1917) had supported Cope's original contention, as did he. Schwartz and Marsh (1960) concurred. Formal publication of the currently recognized trinomial appeared first in Thomas (1966a). Cope (1886) described *Cyclura nigerrima*, and the name often has been confused with *C. onchiopsis* (Cope, 1885), even after clarification in Schwartz and Thomas (1975). Regardless of name, the Navassan *Cyclura* was regarded as a distinct species by Barbour and Noble (1916), Schmidt (1919, 1921b), and Cochran (1941). However, Barbour (1937) treated it as a subspecies of *C. cornuta* and expressed doubts of its distinctiveness at even that level. Mertens (1939) agreed with the latter, but Cochran (1941) remarked on "striking differences." Thomas (1966a), without having examined specimens, "arbitrarily" used the trinomen, the convention largely followed since.

Table 4.8. Taxonomic Authorities for Native Navassan Reptiles (Thomas, 1966).[a]

Authority	Number of taxa described	Percentages
Cope (1868)	2 (3)	40.0 (100.0) 62.5
Grant (1931)	1	20.0 (0) 12.5
Schmidt (1921)	2	40.0 (0) 25.0

[a]The publication date is given in parentheses after each name. The number of currently recognized species described is followed by the number of subspecies currently accepted as valid (nominate subspecies are not included). Percentages are for species (subspecies) and for all currently recognized taxa at these levels.

R. J. Beck spent a week on Navassa in 1917 (Wetmore and Swales, 1931) and collected the greatest number of Navassan specimens currently in any collection. On the basis of this material, Schmidt (1919) described *Sphaerodactylus becki*, *Anolis longiceps*, *A. latirostris*, *Chamaelinorops* (= *Anolis*) *barbouri*, *Ameiva navassae*, and noted the "probable" presence on the island of *Anolis olssoni*. *Anolis longiceps* is a large anole in the *carolinensis* group, and its Cuban (via the Cayman Islands) affinity has been discussed previously. In contrast, *A. latirostris*, characterized by a broad spatulate snout, was probably taken in Haiti, where Beck spent about a month prior to his time on Navassa Island. Thomas (1966a) examined the specimen, determined it was damaged (crushing the snout), and placed *A. latirostris* in the synonymy of *A. coelestinus*. *Chamaelinorops* (= *Anolis*) *barbouri* is another example of a specimen the origin of which was erroneously recorded. Undoubtedly these lizards were taken in Haiti by Beck from mesic upland habitats which are absent on Navassa. The questions regarding *Ameiva navassae* are unresolved. Described from a single specimen, Thomas (1966a) assumed that the collection data were incorrect and placed this taxon in the synonymy of *A. taeniura*, with a type locality restricted to the western part of the Tiburon Peninsula. He noted, however, that Schwartz, in a personal communication, had indicated that this specimen was at or above the range in several counts for *A. taeniura* in the Les Cayes–Camp Perrin region. Subsequently, Schwartz (1967b) treated *navassae* as a subspecies of *A. taeniura* endemic to Navassa Island, and that allocation has been followed since. That Schmidt (1919) had some suspicions regarding the origin of Beck's specimens is obvious from his note that *Anolis olssoni* was "probably from Navassa Island." In addition to the question of its origin, the specimen turned out to be an example of *A. semilineatus* [Schmidt, 1921b; E. E. Williams, personal communication (cited in Thomas, 1966a)]. Schmidt (1921b) also noted the presence on Navassa of *Sphaerodactylus cinereus* and *A. distichus*. Like *A. semilineatus*, these undoubtedly were collected by Beck in Haiti (Thomas, 1966a).

Aristelliger cochranae was described by Grant (1931a), and shortly thereafter Cochran (1933) described *A. expectatus* from Hispaniola. Cochran (1941) discussed the distinctiveness of the two forms, and counts by Thomas (1966b) generally supported her contentions. Nevertheless, Thomas (1966b) chose to follow Mertens (1939) in considering these geckos conspecific. This was generally accepted until Powell *et al.* (1996a), based largely on published data, comments in Powell (1993), and prior use by Powell and Parmerlee (1993), once again elevated the Hispaniolan form to full-species status.

Fossil Record

In general, the West Indian fossil record is sparse. However, due to the fortunate series of "accidents" that led to the formation of amber deposits on Hispaniola, at least a few specimens are preserved in their entirety, safely encased in hardened resin. The only known frog is an *Eleutherodactylus* sp. from Upper Eocene to late Pleistocene amber (Poinar and Cannatella, 1987), although Poinar (1992) referred to six other Dominican frogs in amber, none have been "verified for authenticity or described." That tortoises, *Geochelone* sp., were found historically on Navassa Island and Hispaniola is supported by late Pleistocene to Holocene material reported by Auffenberg (1967) and Franz and Woods (1983), respectively. Etheridge (1965) described late Pleistocene material from the Dominican Republic as *Aristelliger lar*, *Celestus costatus*, *C. stenurus*, *Diploglossus* (= *Celestus*) sp., *Anolis chlorocyanus*, *A. cybotes*, *A. ricordii*, *Leiocephalus apertosulcus* (†), *L. personatus*, *Ameiva chrysolaema*, and *A. taeniura*. Etheridge (1965) referred to the material listed above as *Diploglossus* (= *Celestus*) sp. as *D. stenurus*, but based on its large body size, Schwartz (1970a) assumed the specimen was *D.* (= *Celestus*) *warreni*. However, the subsequent discovery of two additional large anguids on Hispaniola (Schwartz *et al.*, 1979; Incháustegui *et al.*, 1985) and the fact that Holocene dwarfism has occurred in some Antillean lizards (Pregill, 1986) renders identification of this material to species impossible. *Leiocephalus apertosulcus* was described from a right dentary, but numerous cranial and postcranial bones were also referred to this species, the extinction of which was presumed to be pre-Columbian. The holotype of *Anolis dominicanus* (†) (Rieppel, 1980) is preserved in early Miocene amber. Pregill (1984) described *Leiocephalus anonymous*, also from a right dentary, which was found by A. J. Poole around 1927 "in cave sediment probably no older than latest Pleistocene" (Pregill, 1992). Pregill (1984) referred more material to this species, and speculated that *L. anonymous* may have persisted into historical times. In the same year, Böhme (1984) described *Sphaerodactylus dommeli* from Oligocene amber.

Current Status of the Herpetofauna

Hispaniola has been subjected to the generally negative impact of non-native Americans for a longer period than any other area in the Western Hemisphere. Still, many species of Hispaniolan amphibians and especially reptiles appear to thrive despite (or maybe because of) human alterations to habitats. *Osteopilus dominicensis* is known to breed in ornamental fountains and even in heavily contaminated drainage ditches in the center of urban areas (R. Powell, personal observation). Schwartz and Henderson (1991) remarked on the fact that *Hyla pulchrilineata* frequently occurs "in situations much disturbed by man." *Eleutherodactylus weinlandi* is encountered commonly in resort areas along the southeastern Dominican coast, where it freely takes advantage of mesic situations created by the construction of ornamental pools and by the watering of lawns and golf courses (R. Powell, personal observation). Several species of *Eleutherodactylus* often are found frequenting the debris resulting from agricultural practices and human wastefulness (Schwartz and Henderson, 1991). Some anoles, notably *Anolis distichus* and members of the *A. brevirostris* complex, *A. chlorocyanus* and its South Island counterpart *A. coelestinus*, *A. cybotes*, and the grass anoles are nearly ubiquitous, coexist successfully with humans (see Chapter 7), and may, in some instances, actually function as human commensals — a role more commonly associated with some geckos. And a few geckos similarly do, in fact, thrive. *Hemidactylus haitianus* is almost as ubiquitous at night in urban areas as are some of the anoles by day. *Sphaerodactylus difficilis* frequently occurs in human habitations and often makes use of the debris that inevitably accompanies human activities (R. Powell, personal observation). Even the generally more secretive *Aristelliger expectatus* occasionally plays the role of human commensal (Lynxwiler *et al.*, 1991). A few anguids (e.g., *Celestus costatus* and *C. stenurus*) take advantage of human-made clearings and the resultant litter, and *Ameiva chrysolaema* and *A. taeniura* often are seen foraging in open areas of human design, even in streets and on construction sites in urban areas. When weeds and scrub were allowed to grow in a previously "manicured" park in Barahona, the density of the *A. chrysolaema* population declined (R. Powell, personal observation). Snakes, in general, fare less well in the company of *humans*, who often actively persecute them, but *Antillophis parvifrons* occurs frequently in urban areas (e.g., the main street in Sosua, downtown Barahona, and an intensely developed resort area near La Romana; R. Powell, personal observation). An adult *Hypsirhynchus ferox* was collected crossing the main highway along the coast in Barahona, *Uromacer catesbyi* is not infrequently encountered in the urban parks of Santo Domingo, and *Typhlops syntherus*, to the consternation of the local inhabitants, has been taken around an old foundation in a heavily cultivated garden along Highway 44 in Paraiso (Powell, personal observation).

However, many species fare less well. Powell and Henderson (1996a) noted (admittedly in discussing the entire West Indies, but the implications for Hispaniola are evident):

> Habitat destruction is rampant and forests are rapidly disappearing in order to produce charcoal and plant crops... The introduction of the mongoose is correlated with the decline or disappearance of several snake taxa and populations... The introduction of cats, dogs, and rats has had a catastrophic effect on some *Cyclura* populations. Human numbers are booming and development (largely for the tourist trade) continues unabated.

Obviously, the habitat specialists and species with restricted ranges are most vulnerable, although the most attention generally has been devoted to a few species frequently exploited by humans and formally considered endangered. The sea turtles, although not discussed elsewhere in this chapter, largely have been eliminated from Haitian beaches, primarily due to the commercial trade for "tortoise shell" but also for meat and eggs (Ottenwalder, 1996). Those on the Dominican side of the island have fared little better. Crocodiles (*C. acutus*) once were common along most coastal areas, but as a result of uncontrolled hunting for meat, hides, and parts presumably with aphrodisiac qualities, now appear to be restricted to the largely saline lakes in the Valle de Neiba/Plaine de Cul-de-Sac (Thorbjarnarson, 1988; Schubert and Santana, 1996). Both *Cyclura cornuta* and *C. ricordii* have been exploited for food, and populations densities are low in most areas. Similarly, populations of the two aquatic turtles in the genus *Trachemys* have been severely depleted (Schubert, 1993; J. Ottenwalder and Gomez, unpublished manuscript; J. A. Ottenwalder, personal observation), and the appearance of turtles and their eggs in markets continues to aggravate the situation. However, in addition to these well-documented examples, several other species may be in even worse condition.

The very large anguid, *Celestus anelpistus*, may be restricted to a habitat (broadleaf lowland forest) which is nearly gone. Workers clearing the area from which the type series was taken had not previously seen these large lizards (Schwartz *et al.*, 1979), suggesting their rarity, and none have been collected since (Henderson, 1988). The colubrid snake *Alsophis melanichnus* has not been collected for several decades and may be extinct (Schwartz and Henderson, 1991; Schubert, 1993; Powell and Henderson, 1998). Two species in the genus *Ialtris*, *I. agyrtes* and *I. parishi*, are at best very rare (males of *I. agyrtes* have never been collected). Although known locality records are scattered and this may be a reflection of secretive habits, this is unlikely, as both are large and probably are actively foraging and diurnally active (Schwartz and Henderson, 1991). Instead, the paucity of records undoubtedly speaks to the rarity of these species. Based on well-documented events on Hispaniola and elsewhere in the West Indies (Henderson and Sajdak, 1986; Henderson, 1992), the most probable cause is predation by the mongoose (*Herpestes javanicus*).

Despite recent concerns over declines in amphibian populations (Heyer *et al.*, 1994, and references cited therein), Hedges (1993) concluded that declaring any West Indian frog species extinct was premature [but see Joglar and Burrowes (1996) regarding the probable extinctions of Puerto Rican frogs]. Hedges (1993), however, stressed that habitat destruction (e.g., rampant deforestation) could have a catastrophic effect on the frog fauna once all forests had been destroyed. Both the Dominican Republic and Haiti are listed among the top ten tropical countries in terms of net annual deforestation, with rates of 2.5% and 3.9%, respectively (World Resources Institute, 1994). Whereas less than 1% of Haitian forests remain and many amphibians are known to occur only in moist upland forests, Hedges (1993) suggested that Haiti soon may become the world's "first major biodiversity disaster." On the other hand, Henderson and Powell (see Chapter 7) suggested the possibility that the frogs" of Haiti (and possibly all of Hispaniola) will endure as long as second growth forests, banana plantations, and similar degraded habitats persist." Their explanation is that the fauna may be more euryoecious than what is generally thought to be the case. S. B. Hedges (personal communication) noted in response that many species clearly are dependent on natural (primary) vegetation. Among these are "leaf litter" species (e.g., *Eleutherodactylus thorectes*, *E. ventrilineatus*), stream "specialists" (stream quality being a reflection of forest quality), and forms associated closely with orchids and/or bromeliads. Also, amphibians may "explode" following clear-cutting or slash-and-burn activities (J. M. Ruder, A. K. Howard, and J. D. Forester, unpublished data; S. B. Hedges, personal communication), giving the impression that they may thrive in heavily disturbed habitats, although this probably is correlated with a temporary explosion of invertebrates exploiting the overabundance of decaying vegetation. Inevitably, however, the "explosion" is followed by a "crash," after which only a few common species remain. Regardless, when habitat destruction is combined with the impact of introduced predators, such as the mongoose, black rat (*Rattus rattus*), feral cat (*Felis domesticus*), and even large, alien amphibians like *Rana catesbeiana* and *Bufo marinus* (Hedges and Thomas, 1991; Hedges, 1993, 1996a), the outlook for many Hispaniolan amphibians is grim — especially because these predators may operate in undisturbed habitats.

Both Haiti and the Dominican Republic have set aside protected areas, some of which contain the few remaining large stands of natural vegetation. Unfortunately, enforcement of the laws establishing and protecting these areas is spotty, especially in Haiti. Ottenwalder (1996) vividly documented the abject failure of Haitian authorities to enforce laws designed for the conservation of natural resources, and Hedges (1996a) recorded seeing "trees being cut and removed from one of the last remaining stands of natural forest in the country" – in the "center of Pic Macaya National Park." The situation is considerably better in the Dominican Republic (Ottenwalder, 1989), where the Departamento de Vida Silvestre and the Dirección Nacional de Parques (Hoppe, 1989; Valdez Sierra and Mateo Féliz, 1992) are actively engaged in surveying the natural resources of that country, educating of the populace, and conserving areas of particular importance. However, and despite laws making it illegal to cut trees, clear-cutting and burning are common, even in national parks.

The situation on Navassa Island is complicated by the number of species that have been erroneously reported to occur on the island. However, Thomas (1966a), based on his own work and the results of five trips by various persons during which reptiles were collected, indicated that four species, all of which have been collected recently, are extant, and four additional species, none of which have been collected recently, are presumed to be extinct. Such a high rate of extirpation may be the combined effect of human alteration of habitats and the very small size and relative isolation of the island, with the latter resulting in some parts of the fauna never having been well established and others particularly vulnerable to perturbations of their limited habitats. Of the four presumably extinct forms, *Cyclura cornuta onchiopsis*, like the large iguanas elsewhere in the West Indies, is very sensitive to human activities, and may have been driven to extinction (Thomas, 1966a). The reason for the extinction of *Leiocephalus eremitus* is more difficult to explain, because the lizards in this genus usually are sufficiently euryoecious to tolerate, if not thrive, under the onslaught of human habitat alterations – and this has raised some doubts regarding the validity of the taxon (but see Thomas, 1966a). Similarly difficult to explain is the status of *Typhlops sulcatus*. Although having a lifestyle that often precludes the ready collection of specimens, as a result of the number of trips and collectors that have visited Navassa over the years, surely additional material would have been discovered if the species was still extant. Again, however, confidence in the validity of the Rasin collection warrants recognition of *T. sulcatus* as a former member of the Navassan herpetofauna. Thomas (1966a) stated that it "is difficult to believe that [*Tropidophis melanurus*] *bucculentus* has been exterminated on Navassa; further collecting may yield more specimens of this form."

Hedges (1997), relative to the entire West Indies, recommended that (1) deforestation be controlled, (2) park boundaries and laws against the illegal felling of trees be enforced, (3) human populations be controlled, and (4) additional protected areas be set aside. In particular reference to Hispaniola, he suggested the enlargement of Parc National Pic Macaya and Park National Morne La Visite in Haiti. Two regions of comparable biotic value in the Dominican Republic have benefited from recent government action. The Sierra de Neiba National Park was established in 1995 (Decree 221-95), and highlands in the eastern Sierra de Baoruco (encompassing Loma Remigio, Loma Pie de Palo, and Loma Trocha de Pey) were protected as the Reserva Biológica Padre Miguel Domingo Fuertes (category 1B IUCN) in July 1996 (Decree 233, Article 11). Removal or control of introduced predators also must receive a high priority [see Vogel *et al.* (1996) and Tolson (1996) for discussion of efforts on Jamaica and Puerto Rican satellite islands, respectively]. Finally, Hedges and Thomas (1991) and Hedges (1997) presented a strong case for the promotion of scientific collection and continued exploration, primarily by relaxing the often restrictive requirements for obtaining collecting or export permits.

5

The Puerto Rico Area

Richard Thomas

Biology Department
PO Box 23360
University of Puerto Rico-Río Piedras
San Juan, Puerto Rico 00931

Introduction: The Region

The Puerto Rico Area includes approximately 200 islands and cays on three submarine banks: The Mona Bank, including Monito, the Desecheo Bank, The Puerto Rico Bank, and the St. Croix Bank (Fig. 5.1). The native herpetofauna of this archipelago comprises three genera of anurans, one emydid turtle, and 15 genera of squamates, with a total of approximately 80 species. Puerto Rico, with an area of about 8682 km^2, is the largest island. With Pleistocene lowering of sea level, the most recent episode of which ended during the interval from 15,000 to 8000 years ago, the Puerto Rico Bank constituted a single elongate island that extended from Puerto Rico to Anegada (Heatwole and MacKenzie, 1966) having an area of around 21,000 km^2. The Puerto Rico Bank includes Puerto Rico and over 80 associated cays, which might harbor populations of reptiles or amphibians. To the east of Puerto Rico are the Passage Islands of Culebra and Vieques (and 22 associated cays), and the 93 islands of the American and British Virgin Islands (MacLean *et al.*, 1977; Heatwole *et al.*, 1981; Lazell, 1983). St. Croix and its three associated cays lie off the Bank 58 km to the south of St. Thomas. To the west of the Bank lie the islands of Desecheo (21 km west of Puerto Rico), Mona (61 km west of Puerto Rico), and its satellite, Monito, which have herpetofaunas partly or wholly derived from the Puerto Rico Bank.

The Puerto Rico Bank has a plutonic core that is overlain in coastal plains of the north and south coasts of Puerto Rico by thick deposits of Miocene to Oligocene limestone, giving rise to various karst terrains, particularly prominent and extreme in the north where one finds haystack karst (Monroe, 1976). The southwestern corner of Puerto Rico is an accreted wedge of sea-floor containing ultramafic and serpentine rocks

(Case *et al.*, 1984; Burke, 1988). Except for the island of Anegada, a low-lying limestone platform at the eastern extreme of the Bank, along with most of the cays surrounding Puerto Rico, and a large part of Vieques, there are no extensive limestone exposures elsewhere on the Bank. Some of the Virgin Islands do have areas of limestone. Mona and Monito Islands, which lie off the bank to the west are elevated limestone platforms. Desecheo is igneous in composition.

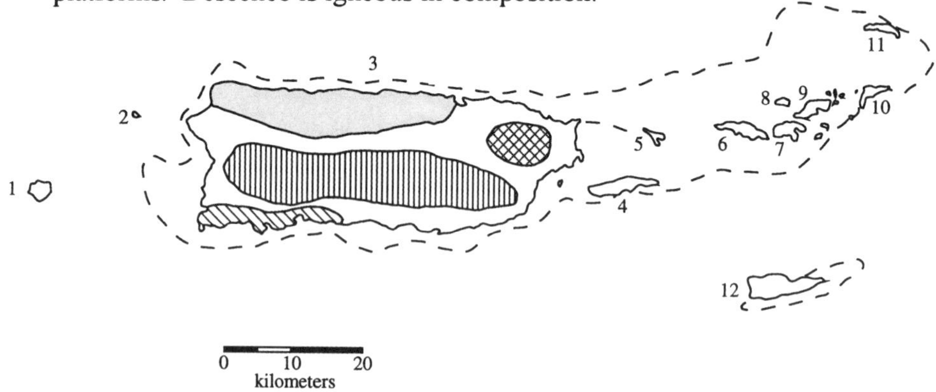

Figure 5.1. The islands of Puerto Rico area: 1, Mona; 2, Desecheo; 3, Puerto Rico; 4, Vieques; 5, Culebra; 6, St. Thomas; 7, St. John; 8, Jost Van Dyke; 9, Tortola; 10, Virgin Gorda; 11, Anegada; 12, St. Croix. The dotted line delimits the Puerto Rico Bank. Within Puerto Rico the approximate limits of major regions are indicated: northern karst area (shaded), Cordillera Central and associated highlands (vertical hatching), the Luquillo Massif (cross hatching), and the xeric extreme (diagonal hatching).

The geological subdivisions notwithstanding, the overriding determinant of distributions of amphibians and reptiles is probably moisture. Beyond Puerto Rico, the other islands of the area are more xerophytic: scrub to dry woods with the most mesic areas being found in ravines and on the highest parts of the islands. Ewel and Whitmore (1973), employing Holdridge's Life Zone system, note that on the Passage Islands and U.S. Virgin islands the areas are climatically capable of supporting Subtropical Moist Forest (= Life Zone) range from 72% on St. Thomas to 0% on Culebra. Tortola has a substantial amount of Subtropical Moist Forest habitat. St. Croix, although larger than any of the eastern bank islands, has only 16% in the Moist Tropical Forest Life Zone. Mona is reported by Ewel and Whitmore to be 100% Subtropical Dry Forest. These generally xeric conditions of the eastern islands extend to the eastern extreme of Puerto Rico, but the uplift of the Luquillo Mountains continuing into the Cordillera Central and the northern coastal plain are mesic, with the wettest areas being in the higher elevations of the Luquillo massif, where the annual rainfall may exceed 4000 mm/year. The south coast of Puerto Rico is in a rain shadow that increases in dryness toward the west with the areas of least rainfall receiving about one meter a year and producing the most extremely xerophytic conditions where this falls on a limestone substrate. Mona and Desecheo are xerophytic but not so extreme as the driest part of the southwestern coast of Puerto Rico. Many of the Puerto Rico Bank islands appear slightly tilted toward the south. This results in many islands having rocky or sandy "high-energy" shores on the north and "drowned" coastlines on the south with mangrove areas and mangrove cays along the south coasts of the larger islands. Along the east and south coasts of Puerto Rico are numerous cays and islets of various sizes and elevations. The south coast cays are more homogeneous: low, with extensive mangrove and with limestone, coral rubble, and sand substrates, although some have small rocky areas, and Caja de Muertos, the largest and farthest offshore, is rocky and elevated (90 m).

A Brief History of Herpetology in the Puerto Rico Area

Endemic species were first described from the Puerto Rico area by Duméril and Bibron in their Erpétologie générale (1844). Major additions to the herpetofauna were made by Stejneger (1904), Schmidt (1928), and Grant in the 1930s. Descriptions of new species have continued with diminishing frequency into recent years (Hedges and Thomas, 1991a). Thomas and Joglar (1996) discussed the history of the development of herpetology in the area.

The Puerto Rican Herpetofauna within the Greater Antillean Assemblages

The Puerto Rico Area shares with the rest of the Greater Antilles a fauna of widespread genera. Each major Greater Antillean island bank has an *Eleutherodactylus* fauna, an *Anolis* fauna, a *Sphaerodactylus* fauna, at least one anguid lizard (*Diploglossus* or *Celestus*), at least one large *Alsophis*, one or more small generalized colubrids, at least one *Typhlops*, and a freshwater emydid turtle of the genus *Trachemys* (two on Hispaniola). In addition, most islands have one or more amphisbaenids (lacking on Jamaica and the Bahamas), one or more hylid frogs (lacking on Puerto Rico), and one or more tropidophiid snakes (lacking on Puerto

Rico). However, for post-Pleistocene extinctions on Jamaica and Puerto Rico, the iguanid genus *Leiocephalus* would also be a part of the widespread Greater Antillean fauna (Pregill, 1981b).

Puerto Rico, like Jamaica, has no endemic genera of amphibians and reptiles. In Cuba and Hispaniola a number of species in the past have been accorded generic status, because they depart strikingly in morphology from their nearest relatives. To avoid an epizootic of paraphyly, several of these genera have been sunk into larger parent genera, e. g., *Chamaelinorops* and *Chamaeleolis* into *Anolis* (Hass *et al.*, 1993), the anguids *Sauresia* and *Wetmorena* into *Celestus*, *Cadea* into *Amphisbaena* (Hedges, 1996b). Despite the nomenclatural changes, the fact remains that those "genera" represent significant departures in morphology from their congeners. The Puerto Rico Area lacks any such strikingly divergent species, the most divergent certainly being the live-bearing *Eleutherodactylus jasperi*, whose uniqueness lies in its mode of reproduction. In external morphology, it is a fairly standard arboreal member of its genus.

Patterns of Distribution: The Puerto Rico Bank

Of the 80 or so species in the entire area, 57 occur on Puerto Rico, and, of those, 37 are confined to that island and its immediate offshore cays. As one progresses east of Puerto Rico into the Passage Islands and the Virgin Islands, there is a marked drop-off in species, no doubt due largely to the reduction in area of the eastern islands but also due to the absence of high-elevation, moist habitat.

Distribution within Puerto Rico

Within Puerto Rico several patterns of distribution are found among the amphibians and reptiles. First, island wide species are found at almost all elevations: *Leptodactylus albilabris*, *Eleutherodactylus antillensis*, *Eleutherodactylus coqui*, and *Anolis cristatellus*. It is unclear to what extent *L. albilabris*, *E. antillensis*, and *A. cristatellus* occur at very high elevations naturally; they prefer open habitats and may have extended their elevational ranges along roads and clearings. *L. albilabris* is arguably the most ecologically tolerant of all Puerto Rican amphibians and reptiles; it occurs from the most xeric conditions to wet montane forest, albeit in disturbed situations, and is the first frog in the assembly sequence to be found on small islands. Second, there are widespread, generally low to moderate elevation species (*Eleutherodactylus antillensis*, *E. cochranae*, *Leptodactylus albilabris*, *Anolis cristatellus*, *A. pulchellus*,

Epicrates inornatus, *Alsophis portoricensis*, *Arrhyton exiguum*, *Sphaerodactylus macrolepis*, etc.). Third, there are widespread mostly upland mesophilic species (*Eleutherodactylus brittoni*, *E. eneidae*, *E. wightmanae*, *Anolis cuvieri* , *A. gundlachi*, *A. krugi*, *A. occultus*, *Diploglossus plei*, *Sphaerodactylus klauberi*, *Typhlops rostellatus*). Fourth, there are seven species restricted to the xeric southern coastal area (*Ameiva wetmorei*, *Anolis cooki*, *Anolis poncensis*, *Phyllodactylus wirshingi*, *S. roosevelti*, *Amphisbaena xera*, *Typhlops granti*). Beyond that, there is a miscellany of restricted distributions, some whose areas of restriction are ecologically or physiographically definable (*Eleutherodactylus unicolor* restricted to the high elevations of the Luquillo Mountains, *E. jasperi* restricted to the Sierra de Cayey, *Amphisbaena schmidti*, restricted to the northern karst area, *E. cooki* restricted to the Sierra de Panduras). and some which are not: *Amphisbaena bakeri*, restricted to western, mostly interior Puerto Rico in both karst and non-karst areas; *Sphaerodactylus gaigeae* and *Typhlops hypomethes* with U-shaped distributions around the eastern end of the island; *Sphaerodactylus nicholsi*, occupying the southwestern coastal region with an outlier population on the north coast; *Sphaerodactylus townsendi*, the sister species of the former, ranging from the southeast to the east and onto the eastern cays and Vieques. Over much of the region the distribution of species is more determined by moisture and elevation (or an interaction of the two). The ranges of species liking open, lowland habitat were presumably more restricted in pre-Columbian times, when Puerto Rico was more heavily forested, and those preferring closed canopy forest were presumably more extensively distributed. In the 1940s Puerto Rico reached a maximum of deforestation due to cultivation and pasturage, such that only 6% of the area was forested (including shade crops such as coffee). With economic changes and the dwindling of agriculture, there has been a recrudescence of forest so that 35% of the land was in forest by 1987 (Birdsey and Weaver, 1987). This certainly must fall far short of the percentage of aboriginal forest. Hedges (1999) analyzed the distributions of amphibians in the Puerto Rico Area and found that the species density is greatest in the Luquillo Mountains (El Yunque), which although not the highest peaks in the island they are the wettest. The occurrence of the highest densities around high mountain peaks is a recurrent pattern (Hedges, 1999).

The largest of the eastern Puerto Rico Bank islands is St. Thomas with an area of 77 km^2. The large islands of the eastern part of the bank, (Culebra, Vieques, St. Thomas, St. John, Tortola, Virgin Gorda, and Anegada) all have approximately 14-19 species. Most of the species of these islands are bank wide, occurring also on Puerto Rico, but there is a small set of eastern bank endemics: *E. schwartzi*, *Anolis ernestwilliamsi*, *Sphaerodactylus parthenopion*, *Amphisbaena fenestrata*, *Typhlops cat a-pontus*, and *Typhlops richardi*. The giant *Anolis roosevelti*, described from Culebra in 1931 but not found alive since (Grant, 1931b), probably also belongs in this category. *Eleutherodactylus lentus* of the eastern Bank islands also occurs off the Bank on St. Croix.

Aside from the eastern bank endemics, the Passage Islands and the Virgin Islands for the most part share a group of widespread sun-tolerant species that are found in the lowlands of Puerto Rico: *Eleutherodactylus antillensis*, *Eleutherodactylus cochranae* , *Leptodactylus albilabris*, *Ameiva exsul*, *Anolis cristatellus* , *Anolis pulchellus*, *Anolis stratulus* , *Sphaerodactylus macrolepis* , *Alsophis portoricensis* , and *Arrhyton e x-iguum*. The foregoing species are the common, conspicuous part of the shared herpetofauna. The skink, *Mabuya mabouya*, although not common in Puerto Rico, is also part of this shared lowland assemblage, as is the toad *Bufo lemur*, which occurs on Puerto Rico and Virgin Gorda; it was no doubt continuously distributed in the past. The ecological vicariants, *Typhlops platycephalus* , *hypomethes*, *richardi*, and *catpontus* are also part of this bank-wide assemblage of species. *Sphaerodactylus parthenopion*, known only from the Virgin Islands of Tortola and Virgin Gorda, has its closest relative in *S. nicholsi* of southwestern Puerto Rico. Several species are endemic (or possibly relict endemics) to one or more of these eastern islands. *Anolis roosevelti* and *Amphisbaena fenestrata* are eastern bank species without known or obvious sister species on Puerto Rico, although one presumes that their closest relatives are (or were) found on Puerto Rico. The emydid turtle, *Trachemys stejnegeri* extends east of Puerto Rico to Vieques and possibly Culebra (Seidel, 1988a).

Lazell (1983, Table 1) presented a listing of the British Virgin Islands, their areas, elevations and species of amphibians and reptiles. The table demonstrates, in effect, the assembly rules for the herpetofaunas of these islands. The sequence is basically hierarchical with the species occupying the very small islands occurring also on the larger islands, but there is some irregularity in what species are found on the larger islands. The species that will occupy all islands from the smallest to the largest is *Anolis cristatellus* (or its daucusite equivalent, *A. ernestwilliamsi*); next is *Sphaerodactylus macrolepis*. and next *Ameiva exsul*, followed sequentially by *Anolis stratulus*, *Mabuya mabouya*, and so on. The smallest of the British Virgin Islands to be occupied by a frog (*Leptodactylus albilabris*) is Jost Van Dyke the arwa of which is 840 ha. The assembly sequence is basically similar for the La Cordillera Cays, the Passage Islands, and the U. S. Virgin Islands (Heatwole *et al.*, 1981; R. Thomas, unpublished data), except that west of Culebra and its cays, *Sphaerodactylus townsendi* replaces *S. macrolepis* as the small cay species, *S. macrolepis* being found only on high cays (Palominos, Piñeros). In the La Cordillera cays *Leptodactylus albilabris* occupies Cayo Icacos (Heatwole *et al.*, 1981), which is smaller at 139 ha than Jost Van Dyke. On the southern coastal cays of Puerto Rico, a similar sequence of occupancy occurs, but it appears that *Ameiva exsul* will occupy some low cays with mostly halophytic undergrowth that *Sphaerodactylus townsendi* does not. An indication of the rigidity with which the assembly rules hold sway, at least in the smaller range of island size, is the failure of anoles introduced onto a small cay by Levins and Heatwole (1973). They introduced, at different times, numbers of *Anolis stratulus* and *Anolis pulchellus* onto a small, vegetated sand cay (Cayo Palominitos) off eastern Puerto Rico, and both introductions failed. Both of these species are members of the widespread, heliothermic lizards occupying many large and small islands of the bank. This cay, although variable in size, is below the size at which the assembly sequence would predict occupancy by these species. *Anolis cristatellus* and *Ameiva exsul* occupy the island naturally.

The islands of the Puerto Rico area occupying banks separate from the Puerto Rico Bank tend to have high endemicity and a greater proportion of species from beyond the Area than do islands on the Bank, although their peripheral positions near Hispaniola (Mona) or the Lesser Antilles (St. Croix) are probably crucial; Desecheo, closer to Puerto Rico has no Hispaniola-derived species. Mona has species related to Puerto Rico Bank species (six of nine), but *Cyclura cornuta* is Hispaniolan in affinity, as, very likely, is *Eleutherodactylus monensis*, which resembles *E. probolaeus*, found in similar habitats on the eastern tip of Hispaniola (R. Thomas, personal observation). Three of Mona's nine amphibians and reptiles are considered endemic species. Two of Monito's three reptiles (*Anolis* cf. *monensis* and *Mabuya mabouya*) are closest to species found on Mona, but *Sphaerodactylus micropithecus* is morphologically peculiar and of uncertain affinities. St. Croix, lying off the bank 58 km to the south of St. Thomas is much larger than any of the eastern Bank islands (230 km^2) but has a smaller herpetofauna (13 species) than any of these islands. However, it has a more highly endemic fauna with four endemic species (*Ameiva polops*, *Anolis acutus*, *Sphaerodactylus beattyi*, and *Alsophis sancticrucis*).

It is of great interest to know how the fauna of the Puerto Rico area is related to other parts of the Antillean Region: What ancient connections may be indicated by patterns of phylogenetic relationship, or what pattern of overwater dispersal may likewise have formed linkages of relationship. Some of the Puerto Rico Area species occur beyond the archipelago or show obvious affinities with species that do: The freshwater turtle, *Trachemys stejnegeri* occurs on the Puerto Rico Bank, Hispaniola, and Great Inagua Island in the southern Bahamas (Seidel, 1988a). A population on Marie Galante in the Guadeloupean Archipelago is apparently introduced (Seidel and Adkins, 1987; Seidel, 1988a). *Hemidactylus brooki* (Hispaniola, Cuba), *Phyllodactylus wirshingi* (Hispaniola), *Sphaerodactylus macrolepis* (the Anguilla Bank), *Mabuya mabouya* (widespread in the Antilles extending to South America); the subspecies to which the Puerto Rico area populations pertain, *M. m. sloanei*, is Greater Antillean in distribution, with *M. m. mabouya* occurring in the Lesser Antilles. However, the systematics of this "species" is probably not well represented by the nomenclature. The Mona Island rock iguana is closely related to the Hispaniolan *Cyclura cornuta*, and is considered by most to be subspecifically related to that form (Schwartz and Henderson, 1991). *Sphaerodactylus macrolepis* ranges throughout the Puerto Rico Bank but occurs off the bank on St. Croix and on the Anguilla Bank, where the populations are recognized as *S. macrolepis parvus*. A *Typhlops* occurring in the Turks and Caicos Islands is very close to *T. platycephalus* (Thomas, 1976); similarly *Anolis scriptus* of the southern Bahamas, including the Turks and Caicos islands is remarkably close to *Anolis cristatellus* in morphology (R. Thomas, unpublished data) and in genetic distance (Gorman *et al.*, 1980a). The study by Gorman *et al.* (1980a) also showed that *A. desechensis* of Desecheo Island is only weakly, if at all, differentiated from *A. cristatellus*, a fact also remarked upon by Lazell (1983). The frog *Leptodactylus albilabris* is very close to the Hispaniolan *L. dominicensis* (Schwartz and Thomas, 1975).

A number of more explicit attempts than the preceding kinds of data have been made to assess phylogenetic relationships of taxa having species in the Puerto Rico Area. The ever popular anoles have received the lion's share of the attention. Etheridge's (1960) morphologically based study was the first but it did not propose a total evolutionary tree, although it did recognize taxonomic groupings. Williams' (1972) analysis of ecomorph evolution presumed that the Puerto Rican anoles constitute a single radiation. Shochat and Dessauer (1981) used immunological distance to assess evolutionary relationships of the anoles. Gorman *et al.* (1980a, 1983) specifically focused on Puerto Rican anoles. Guyer and Savage (1986) combined molecular and morphological data sets to obtain a phylogenetic tree of the anoles, which indicated that the Puerto Rican species have primarily Hispaniolan affinities. Burnell and Hedges (1990) analyzed West Indian *Anolis* phylogeny using electrophoretic data, and Hass *et al.* (1993) derived phylogenies based on sequence divergence of serum albumin and DNA sequence data. These molecular-based studies indicate relationships of the Puerto Rico Area anoles with both the Greater Antilles (Jamaica, Cuba, Hispaniola) and the Lesser Antilles. Roughgarden (1995) combined data from a number of sources and concluded that the Puerto Rican anoles have a sister-group relationship with the *bimaculatus* group anoles from the Lesser Antilles north of Martinique. Unfortunately, his tree was made only for anoles of Puerto Rico and the Lesser Antilles.

Joglar (1989) using morphological data and Hedges (1989a) using allozyme data derived phylogenies for the West Indian *Eleutherodactylus*. Joglar's study resolved only major groups of West Indian species, which are all widely distributed; Hedges' results, on the other hand, indicated membership of the Puerto Rican species in a clade with the Lesser Antillean species.

Tolson (1987) analyzed cloacal gland secretion of *Epicrates* to derive a phylogeny, and Kluge (1988) extended this work adding morphological data to obtain a phylogeny of *Epicrates*, which indicated that *E. inornatus* form a monophyletic group with the Jamaican *E. subflavus*, and that *E. monensis* is part of a clade whose other members are in Hispaniola.

Hass (1991, 1996) used allozyme, microcomplement fixation, and mitochondrial DNA data to derive phylogenies of *Sphaerodactylus*. In both of these papers she showed that the Puerto Rican *Sphaerodactylus* are part of the *argus* series, which includes taxa from Puerto Rico (four), Hispaniola (four), Jamaica (two), and Cuba (one). Although they did not try to derive a phylogeny of the species, Murphy *et al.* (1984) examined the relationship of two species of Puerto Rican *Sphaerodactylus* using allozyme data, showing that *S. townsendi* and *S. nicholsi* are separate species, as their morphology suggests. The genetic distance of those two from *S. roosevelti* is within the range of closely related species (Gorman *et al.*, 1980a).

Thomas (1989) proposed a phylogeny of *Typhlops* based on morphological data, and Hedges and Thomas (1991) looked at the relationships of the Puerto Rican *Typhlops* using allozyme data. Thomas' results indicated that most of the Puerto Rican species belong in with a group having both Hispaniolan and Lesser Antillean members. *Typhlops rostellatus* belongs to another Hispaniolan clade and has presumably invaded Puerto Rico more recently than the other Puerto Rican species, which seem to be part of a radiation within Puerto Rico. The allozyme data confirm that *T. rostellatus* is well separate from the *platycephalus-richardi-hypomethes* complex.

It seems highly likely that the Puerto Rico Bank herpetofauna has phyletic relationships both to the west with Hispaniola, Jamaica, and the Bahamas, and to the southeast with the Lesser Antilles.

The Incursive and Anthropochore Herpetofauna

One of the problems in considering any fauna is which of its components are present by virtue of natural dispersal phenomena (whether vicariant or waif) and which have been brought into the area by humans, perhaps transported from areas far beyond reach of natural dispersal and having no import to the deciphering of biotic history that gave rise to the fauna. In general, we recognize that species found in the region and nowhere else are almost certainly native species. On Puerto Rico no one doubts that *Eleutherodactylus portoricensis* is a native species, and no one doubts that *Hyla cinerea* is an introduced one. Species present problems when their plausible histories are natural dispersal or anthropochory. Most of the introduced species on Puerto Rico present no question of their endemicity: *Bufo marinus*, a South American species, was introduced in the 1920s to combat insect pests in the cane fields (Rivero, 1978). Puerto Rico is the only Greater Antillean island to lack native hylid frogs, but three, *Hyla cinerea* (North America), *Osteopilus septentrionalis* (Cuba), and *Scinax rubra* (South America), have been introduced (Schwartz and Henderson, 1991; Thomas and Joglar, 1996). *Osteopilus septentrionalis* was also introduced on St. Thomas and St. Croix, as was *Eleutherodactylus coqui* (Schwartz and Henderson, 1991). The North American *Rana catesbeiana* was introduced as a potential food source (not realized) during the early part of the century (Rivero, 1978); it is widespread and common on Puerto Rico. Likewise the crocodilian, *Caiman crocodilus*, established on the north coast of Puerto Rico and present on Vieques (whether breeding or not is not known), was certainly introduced by man, probably as the result of the pet trade, sometime in the 1950s or 1960s (personal observation). The Cuban Rock Iguana, *Cyclura nubila*, established on one small island, Isla Magueyes in the bay of Parguera, is known to have been established from captive animals (Rivero, 1978).

Perhaps the first species to be suspected of being introduced are the two species of *Hemidactylus*: *H. brooki* and *H. mabouia*, which, because populations of both species exist in west Africa, were assumed by many to have been brought over during the slave trade (see Kluge, 1969 for a review). However, Kluge (1969) demonstrated the likelihood that the New World populations of *H. brooki* and *H. mabouia* are native to the New World, having become established by natural means. As late as the

1930s, when Grant (1932) wrote about *Hemidactylus* only *H. brooki* was known from Puerto Rico (Grant had a substantial collection from widespread localities), and Schmidt in the 1920s (Schmidt, 1926, 1928, had not collected them). *Hemidactylus mabouia* was known only from Vieques and St. Thomas. The situation is quite different today; *H. mabouia* is widespread on Puerto Rico, in some places enormously abundant (personal observation). It is now known from 34 islands of the Puerto Rico area (Lazell, 1983; Heatwole *et al.* 1981; R. Thomas, unpublished data). In my experience on the cays around Puerto Rico, Vieques, and Culebra it is present only on those cays that have had human habitations and has so far been found on six out of 61 cays investigated. Puerto Rico is at the northern extreme of the distribution of this species, and it is likely that *H. mabouia* is extending its range, in part due to human agency. *Hemidactylus brooki*, on the other hand, appears not to be a small island species; it is known from a few of the offshore islands of Cuba and Hispaniola (Schwartz and Henderson, 1991).

Similar problems are presented by the tortoise *Geochelone carbonaria* and the lizards *Thecadactylus rapicaudus* and *Iguana iguana*. All three of these reptiles occur from South America through the Lesser Antilles and into the Puerto Rico region (Schwartz and Henderson, 1991). Were they were brought by Amerindian migrants, by Europeans, or by natural dispersal? It is, of course possible that a mixture of the two processes has occurred. It is pretty certain that *Iguana iguana* has become established on the Puerto Rican main island in the last 30 years or so, presumably from the pet trade. *Iguana iguana* has been in the Virgin Islands for a longer time, however, and Lazell (1973, 1983; personal communication) is convinced that it is native there and in the Lesser Antilles. *Geochelone carbonaria* is also probably native to some areas of its occurrence in the Antilles (Lazell, 1993), but almost certainly it has been introduced in some areas (see Censky, 1988 for a discussion). There are records of this species from Puerto Rico, but it is not clear that there have ever been breeding populations (Censky, 1988).

6

The Lesser Antillean Fauna

Ellen J. Censky [1] *and Hinrich Kaiser* [2]

[1] *Connecticut State Museum of Natural Hisotry, University of Connecticut, Storrs, Connecticut 06269*

[2] *Department of Biology, La Sierra University, Riverside, California 92515*

Introduction

The Lesser Antilles extend from South America northward in a long arc of more than a dozen major islands and hundreds of small islets (Fig. 6.1). The islands form a discrete group separated from the Virgin Islands to the north by the Anegada Passage and from Trinidad and Tobago to the south by the Tobago Trough. They originated as a volcanic arc during the Late Cretaceous (Perfit and Williams, 1989) and were uplifted into their present location on the eastern edge of the Caribbean Plate in the Early Oligocene (Malfait and Dinkelman, 1972). These islands are clustered on undersea banks, each of which was emergent as one large island until the post-Pleistocene rise in sea levels. As sea levels rose, large islands fragmented into many smaller ones. The present-day islands vary in size from Guadeloupe, the largest (1706 km^2), to the tiniest of cays (Willet, 1987). Furthermore, profound differences in elevation exist, ranging from 1397 m on Martinique to only about 39 m on Barbuda and Anguilla.

Two separate arcs exist, an outer, older limestone arc and an inner, younger volcanic arc. The inner one still has active volcanoes and extends from Grenada through the western half of Guadeloupe (Basse Terre) on to Saba and is composed of igneous, extrusive rock. The outer limestone arc extends from Marie Galante and eastern Guadeloupe (Grande Terre) northward to Anguilla and Sombrero. Its igneous layer has eroded away, and it is capped with oceanic limestone. The limestone arc tends to be lower in elevation (highest elevation: 484 m on St. Martin)

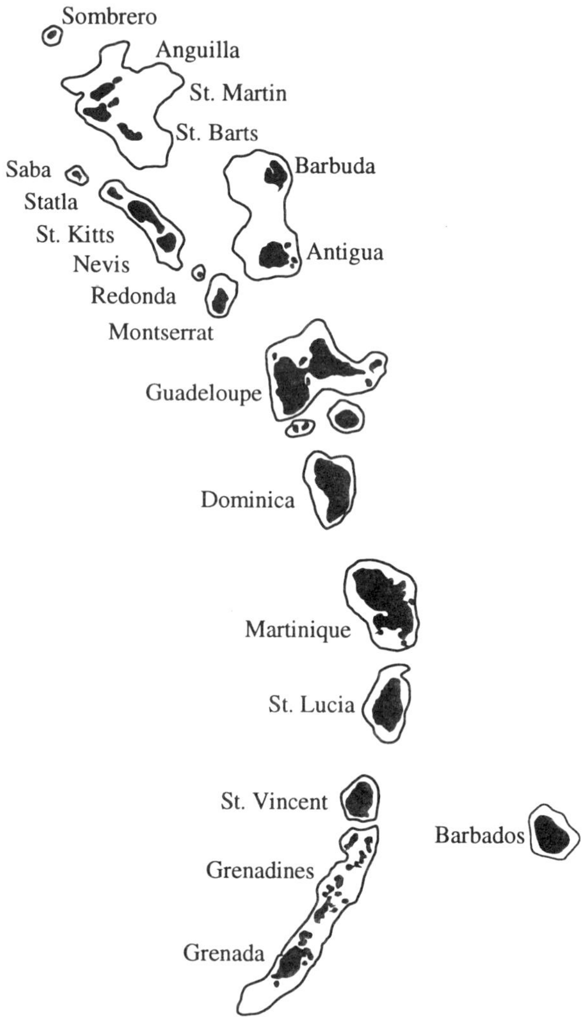

Figure 6.1. The islands of the Lesser Antilles. The map depicts the relative locations of the major islands. The maps of individual islands are contoured, with each contour line representing 1000 ft.

Table 6.1. Endemism in the Lesser Antilles

	Lesser Antillean Endemism		
Island	Bank endemic (%)	Lesser Antillean endemic (%)	Number of Species[a]
Anguilla	47	67	7/3/5
St. Martin/Maarten	38	61	5/3/5
St. Barthelemy	56	70	5/2/3
Barbuda	38	75	3/3/2
Antigua	38	66	5/3/4
Saba	14	57	1/3/3
St. Eustasius	10	73	1/7/3
St. Christophers (Kitts)	9	67	1/7/4
Nevis	8	75	1/8/3
Redonda	50		2/0/2
Monserrat	21	57	3/5/6
Guadeloupe	25	55	5/6/9
Dominica	22	78	4/10/4
Martinique	28	61	5/6/7
St. Lucia	29	62	6/7/8
St. Vincent	18	65	3/8/6
Grenadines	27	60	4/5/6
Grenada	30	45	6/3/11
Barbados	40	70	4/3/3

[a] Number of Bank endemics/number of Lesser Antillean endemics excluding bank endemics/number of non-Antillean species.
Thin lines separate the Banks of islands.

than the volcanic arc (highest elevation: 1397 m on Martinique). The amount of rainfall that each island receives depends, in part, on its elevation. The low, limestone islands receive less annual rainfall (mean = 1106 mm) than the higher volcanic islands [mean = 4193 mm; Eastern Caribbean Natural Area Management Program (ECNAMP), 1980]. According to the characterization of biomes established by Holdridge (1967), all the limestone caribees are classified as Tropical Dry Forest. Biomes on islands of the volcanic arc, on the other hand, range from Tropical Dry Forest in the low lying areas to Tropical Moist Forest at higher elevations. Guadeloupe (Basse Terre), Dominica, Martinique, and St. Vincent also have areas of Subtropical Rain Forest.

The number of amphibians and reptiles in the Lesser Antilles is related to island size, diversity of habitat, and distance from the mainland or another large island. The literature lists 95 species of amphibians and reptiles in the Lesser Antilles, six introduced, five known only from the original series of specimens, two known only from the fossil record, and two vagrants. The large islands from Guadeloupe south all have relatively high diversity, whereas the islands north of Guadeloupe, which all are smaller in size, have lower diversity (Table 6.1). Endemism in the Lesser Antilles is high and can be examined at two levels, bank endemism and Lesser Antillean endemism. Bank endemics are those species occurring on only one bank of islands (e.g., Anguilla Bank = Anguilla, St. Martin, St. Barthélemy; Dominica Bank = Dominica). Lesser Antillean endemics are those species which occur on more than one bank but are restricted to the Lesser Antilles. Bank endemism ranges from a low of 9 or 10% of species on St. Kitts Bank to a high of 38-56% of species on the Anguilla Bank (Table 6.1). Lesser Antillean endemism ranges from a low of 45% of the species on Grenada to 78% of the species on Dominica (Table 6.1). Because Grenada lies closest to the mainland, it is not surprising that it shares a greater proportion of its fauna with South America. Equally, Dominica has a high level of endemism because it is a large, tropical island with many habitat types and is relatively distant from the mainland. What is surprising is that the St. Kitts Bank, while having a majority of Lesser Antillean endemics (67-75%), has a very low level of bank endemism (8-10%); recent extinctions have probably contributed to this (e.g., *Cyclura* and *Leiocephalus*) (Table 6.1). Herein, we outline and discuss the diversity and taxonomic history of the Lesser Antillean herpetofauna. In terms of both biogeography and conservation, this fauna is very important because of the high levels of endemism and because it is a faunal link between the diverse herpetofaunas of South America and the Greater Antilles.

Table 6.2. List of Species by Island.

Species	Island																		
	AN	SM	SB	BA	AT	SA	SE	SC	NE	RE	MO	GU	DO	MA	SL	SV	GS	GR	B
Bufo marinus					I			I	I		I	I		I	I	I		I	I
Colostethus chalcopis														*					
Eleutherodactylus amplinympha													*						
E. barlagnei												*							
E. euphronides																		*	
E. johnstonei	I	*		*	*	*	*	*	*		*	I	I	*	*	*	I	*	*
E. martinisensis			I		?							*	*	*					
E. pinchoni												*							
E. shrevei																*			
Leptodactylus fallax											(*)		*						
L. validus																*	*	*	
Osteopilus septentrionalis		I																	
Scinax rubra															I				
Geochelone carbonaria	*	*	*	*	*		D	D			*		D		D	D	*	*	
G. sombrerensis	+																		
Pelusios subniger												I							
Trachemys stejnegeri[1]												I							
T. scripta												I							
Ameiva ameiva[1]																(*)	*	*	
A. atrata										*									
A. cineracea												(*)							
A. corax	**																		
A. corvina	**																		
A. erythrocephala							*	*	*										
A. fuscata													*						
A. griswoldi				*	*														
A. major														(*)					
A. plei[2]	*	*	*																
A. pluvianotata[2]										*	*								
Anolis aeneus																	*	*	
A. bimaculatus[2]				*	*		*	*	*										
A. extremus															I				*
A. ferreus												**							
A. gingivinus	*	*	*																
A. griseus																*			
A. lividus											*								
A. luciae															*				
A. marmoratus[11]												*							
A. nubilis										*									
A. oculatus[4]													*						
A. richardi																	*	*	
A. roquet[6]														*					
A. sabanus						*													
A. trinitatis																*			
A. wattsi[4]	*	*		*	*		*	*	*						I				
Bachia heteropus[1]																	*	*	
Cnemidophorus vanzoi															**				
Diploglossus montisserrati											(*)								
Gecko gecko														I					
Gonatodes albogularis[1]														(*)					

Table 6.2 cont.

Gymnophthalmus pleei[3]													*	*				
G. speciosus																	*	
G. underwoodi				I?							I?	*			*	*		*
Hemidactylus mabouia	*	*			*	*	*	*	*	*	*	*	*	*	*	*	*	*	*
H. palaichthus														*				
Iguana delicatissima	*	(*)	*		*		*		(*)			*	*	*					
I. iguana						*					*	*			*	*	*	*	
Kentropyx borckiana																			*
Leiocephalus cuneus	+			+	+														
L. herminieri													(*)						
Mabuya mabouya[1]	*	*	*							*	*	*	*	(*)	(*)	(*)	*	*	*
Phyllodactylus pulcher																			*
Sphaerodactylus elegantulus				*	*														
S. fantasticus[9]											*	*	I?						
S. kirbyi																	*		
S. macrolepis[1]	*	*	*																
S. microlepis[2]													?		*				
S. sabanus						*	*	*	*										
S. sputator	*	*	*				*	*	*										
S. vincenti[9]													*	*	*	*			
Thecadactylus rapicauda	*	*	*	*	*	*	*	*	*		*	*	*	*	*	*	*	*	
Alsophis antigua					**														
A. antillensis[6]											*	*	*						
A. rijersmai	*	(*)	*																
A. rufiventris						*	*	(*)	(*)										
Boa constrictor[2]													*		*				
Bothrops caribbaea															*				
B. lanceolata														*					
Chironius vincenti																*			
Clelia clelia[2]																		*	
C. errabunda													*		*				
Corallus enydris[1]																*	*	*	
Liophis cursor														(*)					
L. juliae[3]												*	*						
L. melanotus																		(*)	
L. ornatus															**				
L. perfuscus																			*
Leptotyphlops bilineata												?		*	*				*
L. tenella					?														
Mastigodryas bruesi																*	*	*	
Pseudoboa neuwiedi																		(*)	
Ramphotyphlops braminus	I																		
Typhlops dominicanus[2]												*	*						
T. monastus[2]				*	*			*	*		*								
T. tasymicris																		*	
Caiman crocodylus[1]																	V		
Crocodylus intermedius																		V	
TOTAL SPECIES	14	13	10	8	12	7	11	12	12	4	14	20	18	18	21	17	15	20	10

Note: ? = doubltful occurrence, (*) = endangered or extirpated, + = fossil, ** = satellite island only, I = Introduced population, D = *denticulata or tuberculata*, V = Vagrant. A superscript after the species name indicates the number of subspecies in the Lesser Antilles. AN, Anguilla; SM, St. Martin/Maarten; SB, St. Barthelemy; BA, Barbuda; AT, Antigua; SA, Saba; SE, St. Eustatius; SC, St. Christophers (Kitts); NE, Nevis; RE, Redonda; MO, Monserrat; GU, Guadeloupe; DO, Dominica; MA, Martinique; SL, St. Lucia; SV, St. Vincent; GS, Grenadines; GR, Grenada; B, Barbados.

Diversity and Taxonomic History

Frogs

There are now 13 frog species extant in the Lesser Antilles (Table 6.2), and although this is not a high level of diversity by West Indian standards (Schwartz and Henderson, 1991), the biological and historical origins are remarkable (Kaiser, 1995). Three species have been introduced (*Bufo marinus*, *Osteopilus septentrionalis*, *Scinax rubra*), whereas three regionally endemic species (*Eleutherodactylus johnstonei*, *E. martinicensis*, *Leptodactylus validus*) have been introduced on various islands within the Lesser Antilles, and, in the case of *E. johnstonei*, to places well beyond (Table 6.2; Kaiser, 1997; Kaiser and Hardy, 1994; Kaiser and Henderson, 1994). The remaining seven species are single-island endemics.

French explorers and missionaries traveling in the late eighteenth century were among the first to return from the West Indies with anuran material, among it the first documented specimens from the Lesser Antilles. Resident scientists at the Museu National d'Histoire Naturelle in Paris, foremost among them F. Daudin, A. –M. –C. Duméril, A. H. Duméril, and G. Bibron, made their workplace a focus of amphibian research and began to place systematic order upon the received specimens, soliciting assistance from colleagues throughout Europe. One of the results of these collaborations was the first description of a Lesser Antillean frog, namely that of *Hylodes martinicensis*, by J. J. Tschudi (1838). Given the technical limitations of the time, most Lesser Antillean specimens were placed into already existing taxa, and it has taken to this day to recognize the true diversity of the anuran fauna.

Tschudi's (1838) description of *H. martinicensis* carries with it an interesting anecdote that seems symptomatic for some of the nomenclatural hiccups in the nineteenth Century (see Lizards). During a visit to Paris, Tschudi examined a variety of specimens from the New World for his project on "Batrachians" (1838), including those cataloged by Bibron. A jar containing six small brown frogs purportedly from Martinique (but see Schwartz, 1967c) and shelved preliminarily under the name "*Hyla martinicensis*" was ready to be cataloged and described as "*Eleutherodactylus martinicensis*" by Duméril and Bibron. The specimens were of great interest to Tschudi since they were among the first ones from that island and certainly an important addition to his study of frog diversity. In his 1838 text, Tschudi used the name "*Hylodes*

Martinisensis" for these specimens, thereby providing a name priority to which Duméril and Bibron deferred in their classic herpetology text (Duméril and Bibron, 1841). However, unbeknownst to Tschudi, Bibron had changed the label of the jar to *Eleutherodactylus martinicensis* since these frogs were obviously not members of *Hyla*, indicating that this was to be the name used in a future volume of *Érpétologie Générale*. To avoid a later discrepancy between their book series and the specimen labels during relabeling of jars, they mentioned that name in their detailed description of the specimens. The name on the jar in Paris is still *Eleutherodactylus martinicensis*. A nomenclatorial clarification of the name *Eleutherodactylus* was attempted by Myers (1962), but given the number of taxonomic names that now hinge on this taxonomic eccentricity (512 at last published count; Lynch and Duellman, 1997) a final decision has not emerged.

The only other Lesser Antillean frog described in the nineteenth Century was *Leptodactylus validus* (Garman, 1887c) from St. Vincent. Placed in synonymy with *L. wagneri* (Heyer, 1970), it remained in the obscurity of classification until Heyer's (1994) revision of the *L. wagneri* complex. The description of an endemic *Leptodactylus* from Dominica (Müller, 1926) added an interesting data point to biogeographic studies because this member of the *pentadactylus* group (Heyer, 1979) is geographically distant from the other species in that group.

After the original description of *Eleutherodactylus* frogs from the Lesser Antilles (Tschudi, 1838; Duméril and Bibron, 1841), descriptive systematic and taxonomic work resumed with the work of Thomas Barbour (1914). His monograph on the West Indian herpetofauna included the description of *Eleutherodactylus johnstonei*, and although in subsequent publications (Barbour, 1930a, 1935, 1937) the same author cast implicit doubts on the existence of the taxon (Kaiser and Hardy, 1994), its distinction has now been ascertained (Kaiser, 1992). A surge of interest in *Eleutherodactylus* occurred between 1958 and 1969, beginning with the description of *Hyla barbudensis* from a fossil ilium from Barbuda (Auffenberg, 1958). This species was placed in *Eleutherodactylus* by Lynch (1966), and synonymized with *E. johnstonei* by Pregill *et al.* (1988). The collections and work by the late Albert Schwartz (Schwartz, 1967c, 1969) became the baseline for contemporary study of *Eleutherodactylus*. *Eleutherodactylus barlagnei* was described by Lynch (1965) from material collected by J. D. Lazell and P. Barlagne. In his description, Lynch could find few external differences between the new species and *E. martinicensis*, whereas Schwartz (1967c:51) called it "[c]ertainly the most striking and distinctive of the Lesser Antillean *Eleutherodactylus*." Schwartz attributed Lynch's difficulty in differentiating the new species to the state of the type series. He provided a very detailed description. In the same paper, Schwartz (1967c) described *E. pinchoni*, a miniaturized frog sympatric with *E. martinicensis* and *E. barlagnei* on the Basse Terre portion of Guadeloupe.

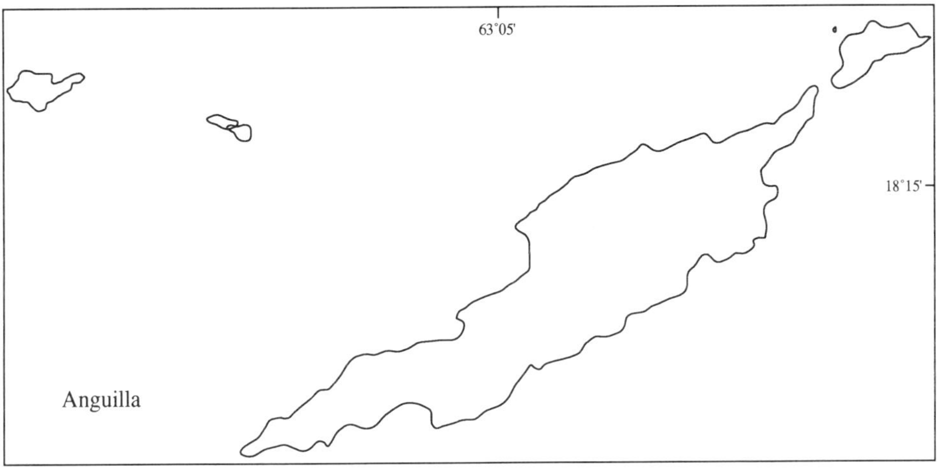

A recent surge in research activity has resulted in the discovery of two new species, *Eleutherodactylus amplinympha* on Dominica (Kaiser *et al.*, 1994a) and *Colostethus chalcopis* on Martinique (Kaiser *et al.*, 1994b). It also led to the removal of *E. euphronides* and *E. shrevei* from the synonymy of *E. urichi* (Kaiser *et al.*, 1994c) and of *Leptodactylus validus* from *L. wagneri* (Heyer, 1994). The dendrobatid *C. chalcopis* is a remarkable addition to the West Indian herpetofauna because it is the only dendrobatid endemic on an oceanic island. Its tadpole is atypical for the genus (Kaiser and Altig, 1994) and a systematic study of the relationship of this taxon with mainland congeners is now being completed by the junior author and colleagues. The presence of *E. amplinympha* in high forests of Dominica marks the continuation of an interesting trend among Lesser Antillean frogs. With the exception of St.

Lucia, all major islands in the volcanic arc (Guadeloupe, Dominica, Martinique, St. Vincent, Grenada) harbor at least one single-island endemic frog species, and most have a single-island endemic *Eleutherodactylus*. Looking at the distribution of taxa and simply filling the "endemism gaps," the more widely distributed *E. martinicensis* may have originated on Martinique and the very effective colonizer *E. johnstonei* on St. Lucia.

Lizards

Compared with the relatively straightforward taxonomy of frogs, taxonomy of the lizards of these islands is less stable. For some groups, consensus is lacking at any taxonomic level above species [see Guyer and Savage (1986) and Williams (1989a) for *Anolis* and Frost and Etheridge (1989) and Lazell (1992) for Iguanidae]. The diversity of lizards on each island in the Lesser Antilles is not as great as that found in the Greater Antilles but greater than that of Lesser Antillean anurans. However, taking the Lesser Antilles as a unit, lizard diversity includes 12 genera and 42 species. In addition, one species has recently been introduced (*Gekko gecko*), whereas five others have not been collected since the original specimens were taken (+*Ameiva cineracea*, *Ameiva major*, *Gonatodes albogularis*, *Diploglossus montserrati*, and +*Leiocephalus herminieri*), and two species are known only from the fossil record (*Leiocephalus cuneus*, *Cyclura* sp.). Many of the genera are widespread (Fig. 6.2) in the Lesser Antilles (*Anolis*, *Ameiva*, *Iguana*, *Thecadactylus*, *Hemidactylus*, *Sphaerodactylus*, and *Mabuya*), whereas a few are restricted to one or a few islands (*Kentropyx*, *Gymnophthalmus*, *Bachia*, *Phyllodactylus*).

There are two genera of extant Iguania in the Lesser Antilles, *Anolis* (Polychridae) and *Iguana* (Iguanidae), and two that are extinct, *Leiocephalus* (Tropiduridae) and *Cyclura* (Iguanidae). Whereas there are only two species of *Iguana* in the Lesser Antilles, *Anolis* is represented by 16 species, most of which are bank endemics. By 1900, no fewer than 33 species of *Anolis* had been described by such noted men as Samuel Garman (1887b), Edward Drinker Cope (1864), John Edward Gray (1840), and A. -M. -C. Duméril and G. Bibron (1837). However, Boulenger (1887) synonymized 11 of Garman's (1887b) 13 new species into four species (*virgatus* = *gingivinus*; *asper*, *nubilus*, *speciosus*, *sabanus*, *lividus* = *leachii*; *gentilis*, *cinereus*, *extremus*, *vincenti*, *luciae* = *alligator*; and *griseus*, *trossulus* = *richardi*). One year later, Günther

(1888) split *Anolis leachii* and placed *lividus* and *sabanus* into *alliceus*. Barbour (1914) disputed both Boulenger's and Günther's taxonomy and elevated many of these variants to full species (*asper*, *speciosus*, *lividus*, *sabanus*, *nubilus*, *griseus*, *vincenti*, *luciae*, *extremus*, *trossulus*). Barbour (1915, 1923) also described four new species (*antigua*, *barbudensis*, *forresti*, *terraealtae*), and Fowler (1918) described one new species (*mayeri*).

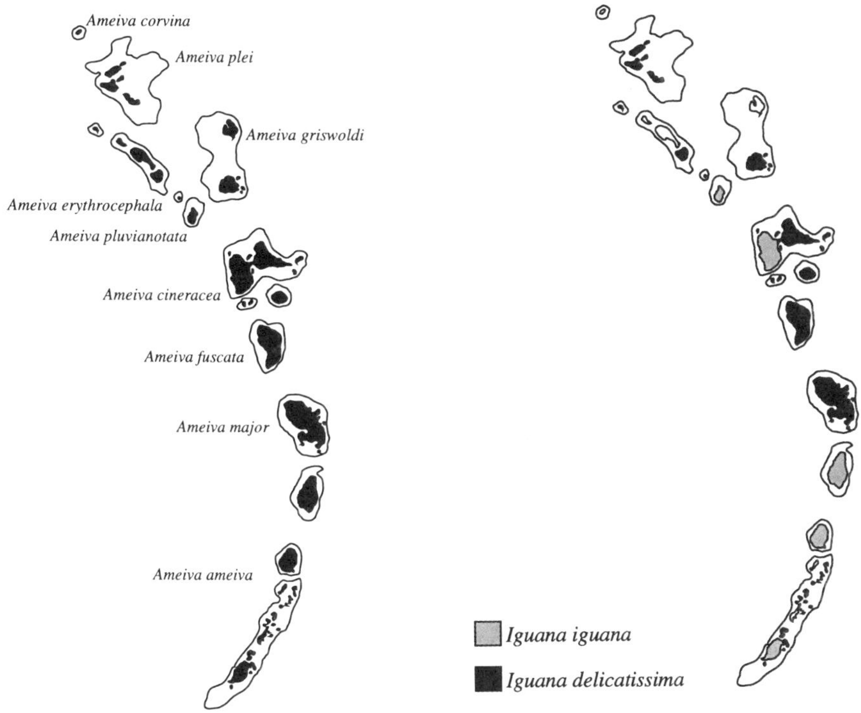

Figure 6.2. Examples of two types of widespread taxa. The two species of *Iguana* are distributed throughout the islands and do not exhibit single-island or island bank endemism. The genus *Ameiva* is also widely distributed in the Lesser Antilles, but is represented by nine endemic species, none widespread like the *Iguana* taxa.

With new material that included live "colorful" specimens, Underwood (1959b) revised the Lesser Antillean *Anolis*. He recognized three groups: the *wattsi* group from the Anguilla Bank, the Antigua Bank and the St. Kitts Bank; the *bimaculatus* group from the islands extending from Dominica northward; and the *roquet* group ranging from

Martinique southward. Once again, many of the full species were relegated to subspecific status or synonymized, reducing the number of species in the Lesser Antilles to nine: group: *bimaculatus* (six subspecies), *marmoratus*, *ferreus*, *speciosus* (two subspecies), *occulatus*, and one undescribed species from Desirade; *roquet* group: *roquet* (three subspecies), *trinitatus* (four subspecies), and *richardii* (two subspecies); *wattsi* group: *Anolis foresti* was placed in the synonymy of *wattsi*.

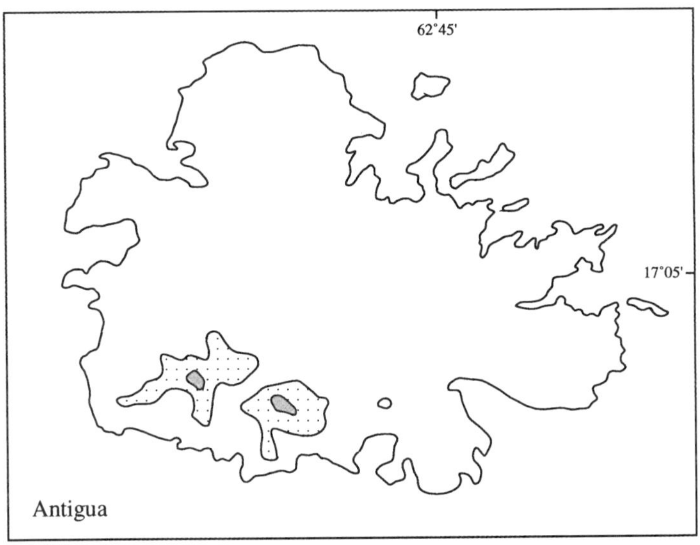

Although he doubted the conspecific relationship of *gingivinus* to *bimaculatus*, Williams (1962c) nonetheless reviewed the *bimaculatus*

group and retained all trinomials. In addition, he described a new species from Barbuda, *Anolis alter* , and placed it in the *wattsi* group. Lazell (1962, 1964a) subdivided *oculatus* into four subspecies and *marmoratus* into 12 subspecies, including six forms that Underwood (1959b) listed as members of different species.

Lazell (1972) retained only two taxa unchanged nomenclaturally from Underwood (1959b), *bimaculatus leachii* and *oculatus*. Many of Underwood's subspecies were once again elevated to full species (*gingivinus*, *lividus*, *sabanus*, *luciae*, *richardi*, *griseus*), some of the species that Underwood had synonymized were taken out of synonymy (*extremus*, *aeneus*, *nubilus*), and some subspecies were placed as subspecies of other species. Lazell also synonymized *Anolis alter* with *wattsi*. He followed Etheridge's (1959) arrangement of species groups which recognized only two in the Lesser Antilles, *bimaculatus* with Greater Antillean affinities and *roquet* with South American affinities. Lazell's work substantiated the conclusions on the *roquet* species group by Gorman and colleagues (Gorman and Dessauer 1965, 1966; Gorman and Atkins 1968a,b) which examined karyology and blood proteins. Work by Burnell and Hedges (1990) has verified these conclusions. Starch gel electrophoretic studies (Gorman and Kim, 1976) have shown that the genetic divergence of the northern Lesser Antillean *bimaculatus* group is less than that of the southern roquet group, suggesting that *bimaculatus* is evolutionarily younger.

Much of the confusion in Lesser Antillean *Anolis* systematics lies in the fact that these species are extremely variable in morphological characters and coloration. Williams (1959) stated "it is probable that the rather subtle scale characters that distinguish them would have been dismissed...had these species not first been recognized by color or behavior or both." Today, 16 species and 27 subspecies are recognized in the Lesser Antilles, though it should be noted that Malhotra and Thorpe (1991) challenged the validity of subspecies, specifically *oculatus* subspecies. Their research demonstrated that the pattern of variation in characters systems was incongruent, indicating an ecogenetic rather than phylogenetic origin. Each bank of islands has its own endemic *Anolis* except the St. Kitts and Antigua Banks which share two species (*bimaculatus* and *wattsi*; *wattsi* also occurs on the Anguilla Bank), and the Grenada Bank which shares its two species with Trinidad and South America (*richardi*, *aeneus*; Table 6.2). In addition, St. Vincent (and Young Island; Lazell and Sinclair, 1990a) has a population of a species known from Trinidad (*trinitatus*) and two species of *Anolis* (*extremus* and *wattsi*) have been introduced to St. Lucia. Late quaternary fossils of *Anolis* have been found on the islands of Anguilla, Antigua, Barbuda, Guadeloupe, Barbados, St. Eustatius, and St. Kitts. *Anolis* from the latter two islands were from cultural remains (Pregill *et al.*, 1994).

Both species of *Iguana* occur in the Lesser Antilles. *Iguana delicatissima* is a Lesser Antillean endemic, occurring on a number of islands from Martinique northward (Table 6.2). *Iguana iguana* has a more widespread distribution occurring from northern Mexico through Central America to the Tropic of Capricorn in South America and into the West Indies. Barring a few exceptions in the Lesser Antilles, *Iguana iguana* occurs on all islands where *I. delicatissima* does not occur (Table 6.2). Those exceptions are Barbados, Barbuda and Redonda where no *Iguana* have been recorded, and Guadeluope where both species occur.

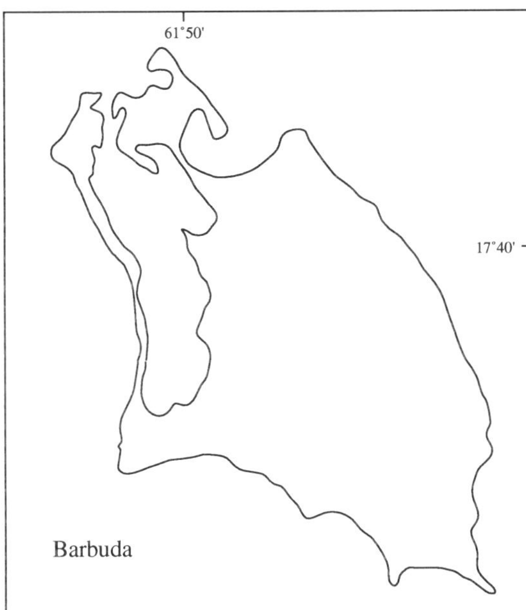

Lacerta iguana (= *Iguana*) was one of the first species described from the Lesser Antilles (Linnaeus, 1758). Only ten years later *Iguana delicatissima* was described, along with *Iguana tuberculata* (Laurenti, 1768). Boulenger (1885) recognized Laurenti's two species of *Iguana*, but also placed *rhinolopha* Weigmann as a subspecies of *tuberculata* and gave its distribution as Central America and the West

Indies. Dunn (1934) synonymized *tuberculata* with *Iguana iguana*, and supported Boulenger's assignment of *rhinolopha*. Lazell (1973) retained two species in his review of Lesser Antillean *Iguana*, but did not recognize *rhinolopha* as a subspecies.

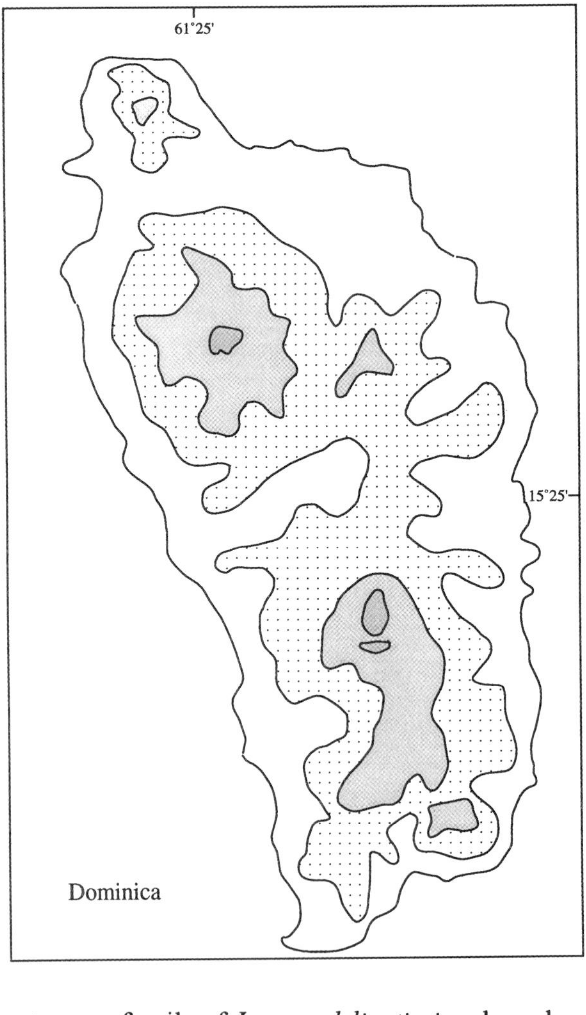

Late quaternary fossils of *Iguana delicatissima* have been discovered in cultural remains on St. Eustatius, St. Kitts and Antigua. Noncultural remains of *Iguana iguana* have been found on Barbados and ? *Cyclura* noncultural remains have been found on Barbuda (Pregill *et al.*, 1994).

Because it is likely that Amerindians transported *Iguana*s between islands, the present distribution of these two species is probably not an accurate reflection of their prehistoric range (Steadman *et al.*, 1984).

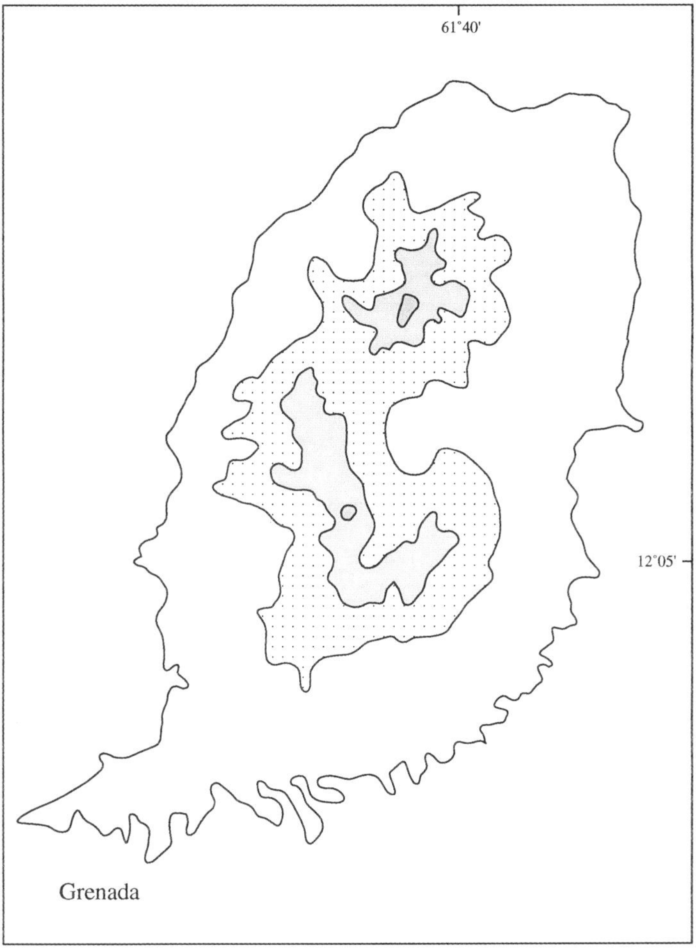

The only tropidurid recorded in the Lesser Antilles is *Leiocephalus*. Duméril and Bibron (1837) described *Leiocephalus herminieri* (three specimens) and Boulenger (1885) added a fourth specimen to the series. Although no specimens have been taken since the original series, Pregill (1992) reported a fifth specimen, collected at the same time, in Rijksmuseum van Natuurlijke Historie (RMNH 2888). There has always been doubt about the type locality, which has been listed as Martinique. Pregill (1992) stated that "there is nothing that assures us that *Leiocephalus herminieri* ...is endemic to that island."

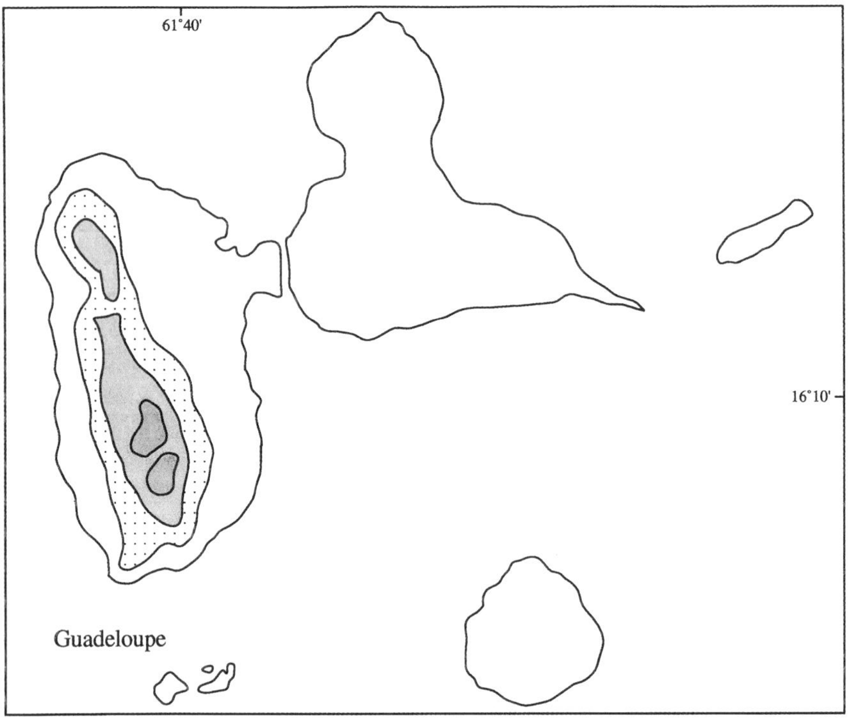

The other species of *Leiocephalus* from the Lesser Antilles is the extinct *L. cuneus*, which is known from fossils taken on Barbuda and Antigua and possibly Anguilla and Guadeloupe (*Leiocephalus* cf. *cuneus*; Pregill, 1992). The occurrence of *Leiocephalus* specimens (fossil and otherwise) on a number of banks in the Lesser Antilles led Pregill (1992) to state that *Leiocephalus* probably was distributed throughout a good portion of the Lesser Antilles.

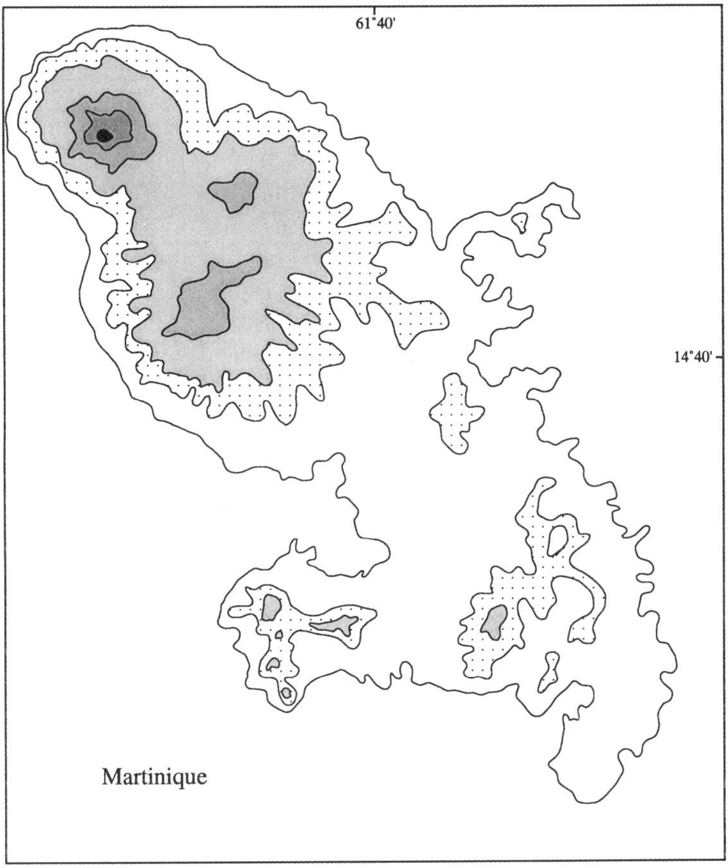

There are three genera of medium-sized teiids in the Lesser Antilles. *Ameiva* is widespread and diverse, occurring throughout the Lesser Antilles. The other two (*Kentropyx*, *Cnemidophorus*) are restricted to a single island each. As with *Anolis*, early workers on *Ameiva* described a different species from each of the islands in the Lesser Antilles (Garman, 1887d; Gray, 1838a,b; Cope, 1869), and of the 19 described species, only

11 appear valid today. Barbour and Noble (1915) were the first to do a revision of the genus *Ameiva* which also included the Lesser Antillean species. For the most part, they retained one species per island, though they did report *Ameiva plei* from two islands, St. Martin and St. Barthélemy, and *Ameiva aquilina* from St. Vincent, Grenada and possibly the Grenadines. They also synonymized *scutata* and *analifera* with *plei* and described a new species (*cineracea*) from Guadeloupe. Subsequent to this review, Barbour (1916) described a new species from Antigua (*griswoldi*) and Schmidt (1920) described a new species from Nevis (*nevisana*). This scheme of Lesser Antillean *Ameiva* systematics stood until Baskin and Williams (1966) undertook a reevaluation of the group. They recognized one species for each bank of islands in the Lesser Antilles, except for *Ameiva ameiva* which occurs on two banks, St. Vincent and Grenada. As a result, they synonymized a number of named forms. They also named a new species, *Ameiva vanzoi* from Maria Islands, off St. Lucia. This was later placed in the genus *Cnemidophorus* based on the tongue structure (Presch, 1971). This is the only species of *Cnemidophorus* known from the Lesser Antilles and is currently listed as an endangered species (USFWS 50 CFR 17.11).

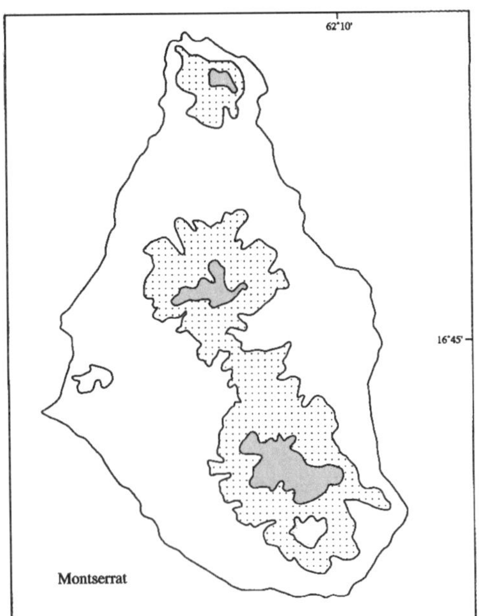

Tuck and Hardy (1973) reexamined the type of *Ameiva surinamensis tobaganus* Cope and concluded that it could not have come from Tobago, but probably was from St. Vincent or Grenada. Therefore, *Ameiva ameiva aquilina* became a junior synonym of *A. a. tobagana*. In 1992, Censky and Paulson revised the Anguilla Bank *Ameiva* (*Ameiva plei*), and described a second species, *Ameiva corax*, from the tiny island of Little Scrub, off northern Anguilla. This species is one of three melanistic species of *Ameiva* found in the Lesser Antilles. The other two are *corvina* on Sombrero and *pluvianotata atrata* on Redonda. All three species are found on small, barren islands with dark substrate. It has been suggested these species exhibit cryptic coloration due to selective predation (Censky and Paulson, 1992). Today, there are 9 extant species of *Ameiva* in the Lesser Antilles (Table 6.2). One species, *major*, is known only from the type series and a second species, *cineracea*, is known from the type series and the fossil record. Both are apparently extinct. *Ameiva ameiva tobagana*, while apparently common in the Grenadines, is considered extinct on St. Vincent (Corke, 1992). Noncultural fossil remains of *Ameiva* have been found on Anguilla, Barbuda, Antigua, and Guadeloupe, whereas cultural fossil remains have been found on St. Kitts, Statia, and Antigua (Pregill *et al.*, 1994). All remains are late Quaternary.

The other medium-sized teiid found in the Lesser Antilles is *Kentropyx*. *Kentropyx* is a South American lizard with only one species reaching the Lesser Antilles (*Kentropyx borckiana*). Cope (1861) was the first to report *Kentropyx* from Barbados. He identified the species as *Centropyx intermedius* Gray. Later, Peters (1869) noted that a species had been described by Merrem (1809) and named "Die Borckische Eidechse", but Peters (1869) was the first to use the binomial *C. Borckiana*. Garman (1887d) described a species, *Centropyx copii*, from Barbados which was later synonymyzed with *borckiana* (Gallagher and Dixon, 1980). This species was thought to be extinct (Westermann, 1953); however, Underwood (1962) disputed this based on visual records.

There are two genera of microteiids in the Lesser Antilles, *Gymnophthalmus* and *Bachia*. Both occur in the southern Lesser Antilles with *Gymnophthalmus* occurring on all major islands from Guadeloupe southward, whereas *Bachia* occurs only on Grenada and some of the Grenadines. Bocourt (1881) described two species of *Gymnophthalmus* from the Lesser Antilles, *pleei* from Martinique and *luetkenii* from St.

Lucia. Later collectors (Garman, 1887d; Boulenger, 1891), failed to find *G. luetkenii* on St. Lucia, but did find *G. pleei*. As a result, Barbour (1914) stated that the locality for the type of *luetkenii* was undoubtedly incorrect. With fresh materials collected by Schwartz and colleagues, Thomas (1965d) noted some distinctions between island populations and described three subspecies of *G. pleei*: the nominate subspecies from Martinique, *luetkenii* from St. Lucia, and *nesydrion* from Maria Islands off St. Lucia.

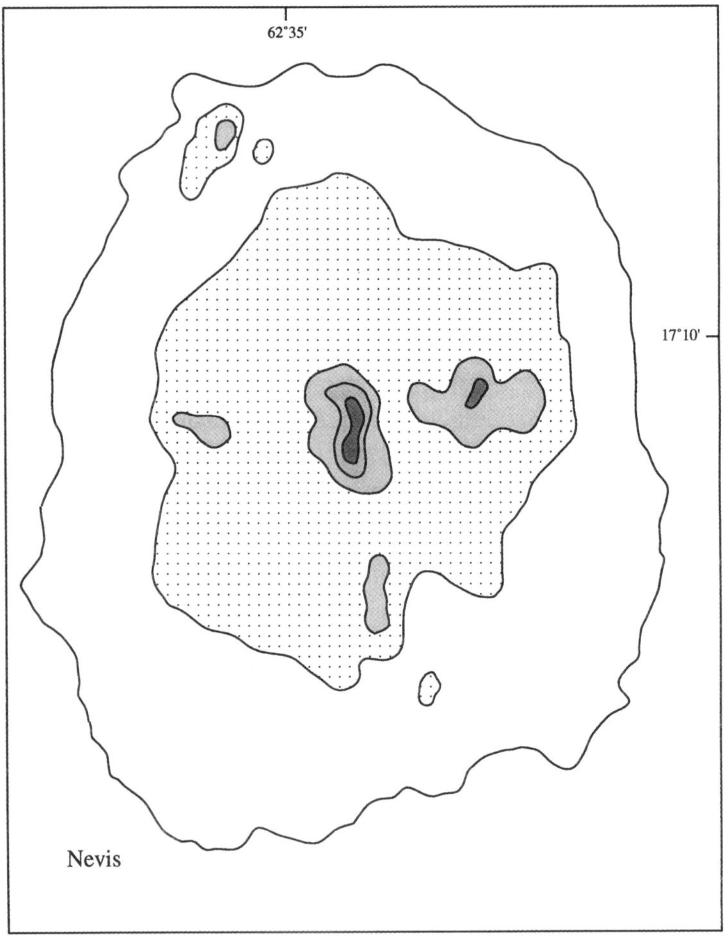

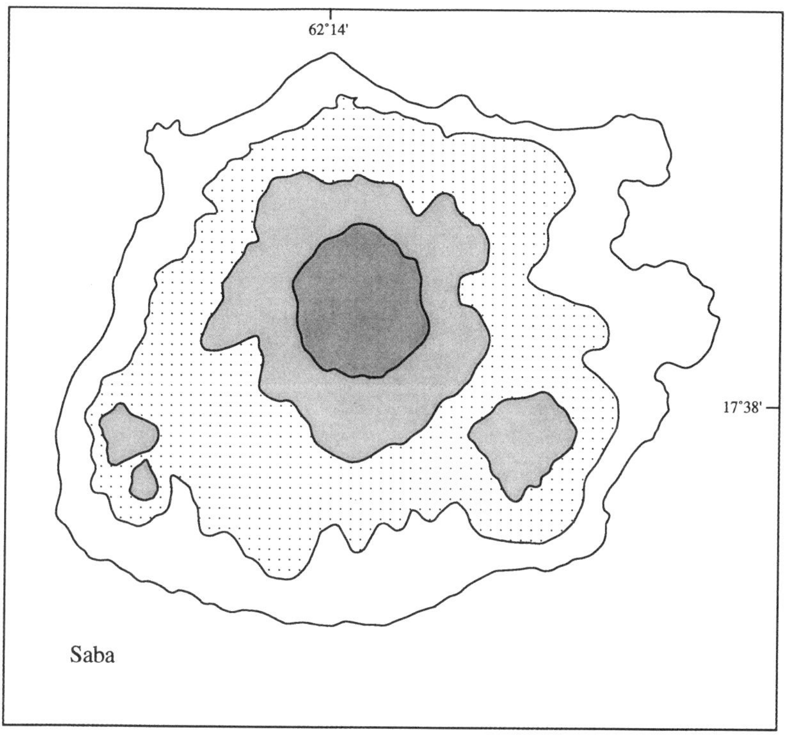
62°14'
17°38'
Saba

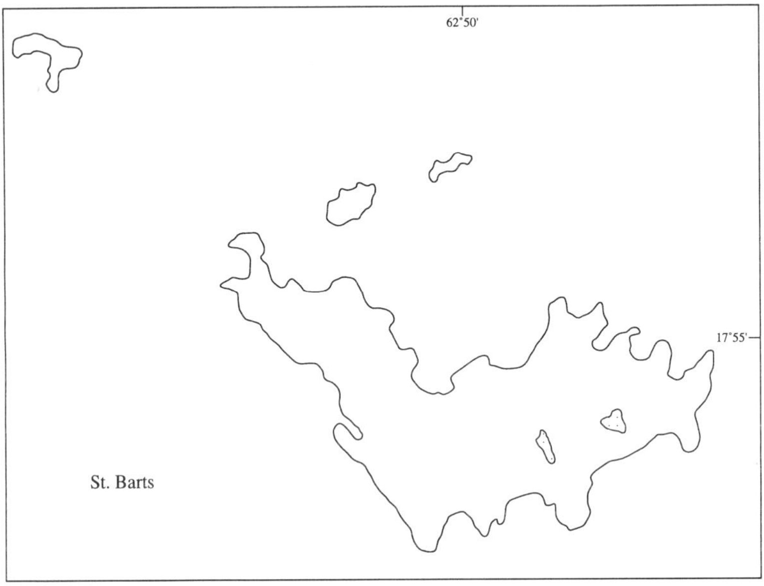

Gymnophthalmus underwoodi was described by Grant (1958) from Barbados. Thomas (1965d) found very little morphological differentiation between specimens from Barbados and Trinidad and the adjacent mainland, but chose to retain the species status of *underwoodi*, and Hoogmoed (1973) concurred by assigning Suriname specimens to the species *underwoodi*. Recently, *G. underwoodi* has been found on St. Vincent, Guadeloupe, and Dominica (Brooks, 1983; tentatively identified as *pleei*, but redetermined by Vanzolini, 1990), Bequia Island in the Grenadines (Lazell and Sinclair, 1990b), and Barbuda (Censky and Lindsay, 1997). While Schwartz and Henderson (1988) considered the Guadeloupe and St. Vincent populations as probable introductions, Corke (1990) disputed this because of the presence of a well-established population on an uninhabited islet off St. Vincent, which he considered an unlikely port for introduction. Thomas (1965d) considered this species to be parthenogenic because no males had, as yet, been found. Interestingly, Brooks (1983) reported one reproductively mature male from Dominica. A South American species of *Gymnophthalmus*, *G. speciosus*, has recently been reported from the southern Lesser Antillean island of Grenada (Vanzolini, 1990). This species occurs from northern South America to Central America.

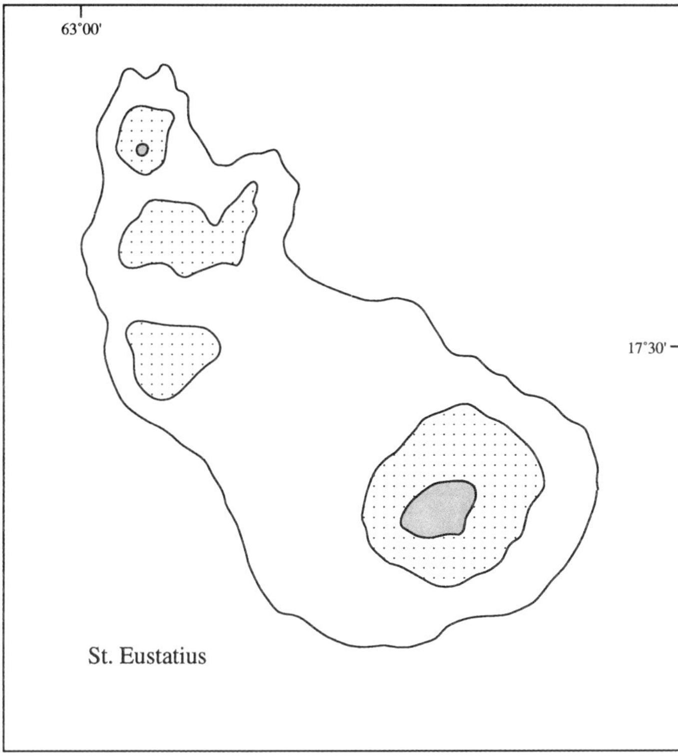

Garman (1887d) was the first to report *Bachia* (= *Scolecosaurus*) from the Lesser Antillean island of Grenada, and he referred his single specimen to *Scolecosaurus cuvieri*. Barbour (1914), with additional specimens from Grenada, noted their distinctiveness and described them as *Scolecosa[u]rus alleni*. Later, Barbour (1933) described a subspecies of *alleni* from the Grenadines, *Scolecosaurus alleni parviceps*. Vanzolini (1961) considered the distinctions between *Scolecosaurus* and *Bachia* to be inconsistent and placed *Scolecosaurus* in the synonymy of *Bachia*. In a review of southeastern Caribbean microteiids, Thomas (1965d) placed *Bachia alleni parviceps* in synonymy with *B. alleni alleni*. Dixon (1973b) then revised the genus *Bachia* placing *alleni*, along with 5 South American species, in synonymy with *heteropa*, based on the presence of lateral hexagonal scales, 4 digits on forelimbs, and the presence of interparietal and frontonasals. The specimens from Grenada, Grenadines, and Tobago were placed in the subspecies *Bachia heteropa alleni*. No fossil remains of microteiids have been found.

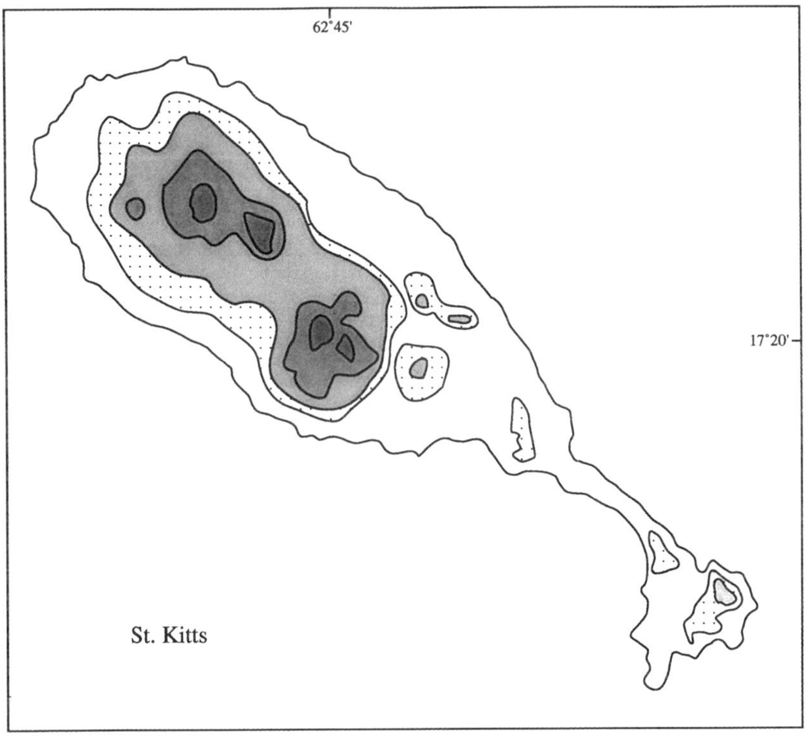

There are four extant genera of geckos in the Lesser Antilles (*Hemidactylus*, *Thecadactylus*, *Sphaerodactylus*, *Phyllodactylus*), one recently introduced species (*Gecko*), and one extinct species (*Gonatodes*). Two species of *Hemidactylus* are found in the Lesser Antilles, the widespread (*H. mabouia*) and the less widely distributed *H. palaichthus* from the southern Lesser Antillean islands. *Hemidactylus mabouia* is found on all major islands, with the exception of St. Barthélemy and Barbuda; a specimen from Redonda was photographed by K. Lindsay and the photo is deposited at Carnegie Museum. Though Du Tertre first reported "*Hemidactylus*" from the Lesser Antilles in 1654 (Kluge, 1969), the species was not described until 1818 by Moreau de Jonnés as *Gecko mabouia*. However, Duméril and Bibron (1836) were the first to use the name *Hemidactylus* for this species. Moreau de Jonnés noted that this species was distributed widely in the Greater Antilles,

Lesser Antilles, and South America. Stejneger (1904) later restricted the type locality to St. Vincent. Kluge (1969), in a review of the New World *mabouia-brooki* complex, reexamined the type and though it was clear to him that the type specimen of *H. mabouia* was actually a *brooki*, he chose, for nomenclatural stability, to retain the name *mabouia* for the Lesser Antillean radiation. This species is of African origin, and it has been inferred that its presence in the New World is due to transport aboard slaving ships (Grant, 1959). However, Kluge (1969) believed its presence more likely to be due to natural dispersal across an ocean barrier, with subsequent further colonization. He concluded this because of the significant differences between Old World and New World populations and the nonrandom differences in New World populations.

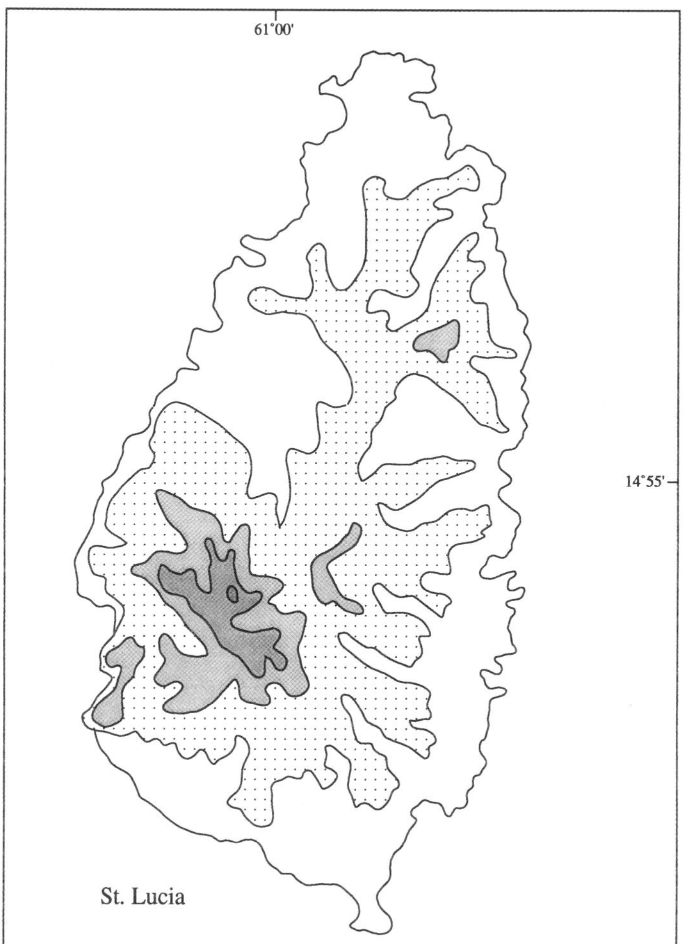

The second species of *Hemidactylus* which occurs in the Lesser Antilles was described as *palaichthus* by Kluge in 1969. The species was known only from Maria Major Island off of St. Lucia until Corke (1987) reported it from the main island of St. Lucia and additional islets. This species is also found on Trinidad and in north central South America. It is an endemic New World species (Kluge, 1969). Noncultural, late Quaternary fossils of *Hemidactylus* cf *mabouia* have been found on Guadeloupe (Pregill *et al.*, 1994).

Thecadactylus rapicauda is the only species assigned to this genus. It occurs throughout the Lesser Antilles, as well as in South America, and in Central America north to Mexico. The only Lesser Antillean islands that it has not been recorded from are Redonda and Barbados (Garman, 1887a; Barbour, 1914; Schwartz and Thomas, 1975). Non-cultural, late Quaternary fossil remains of *Thecadactylus* have been found on Anguilla, Barbuda, Antigua, and Guadeloupe (Pregill *et al.*, 1994).

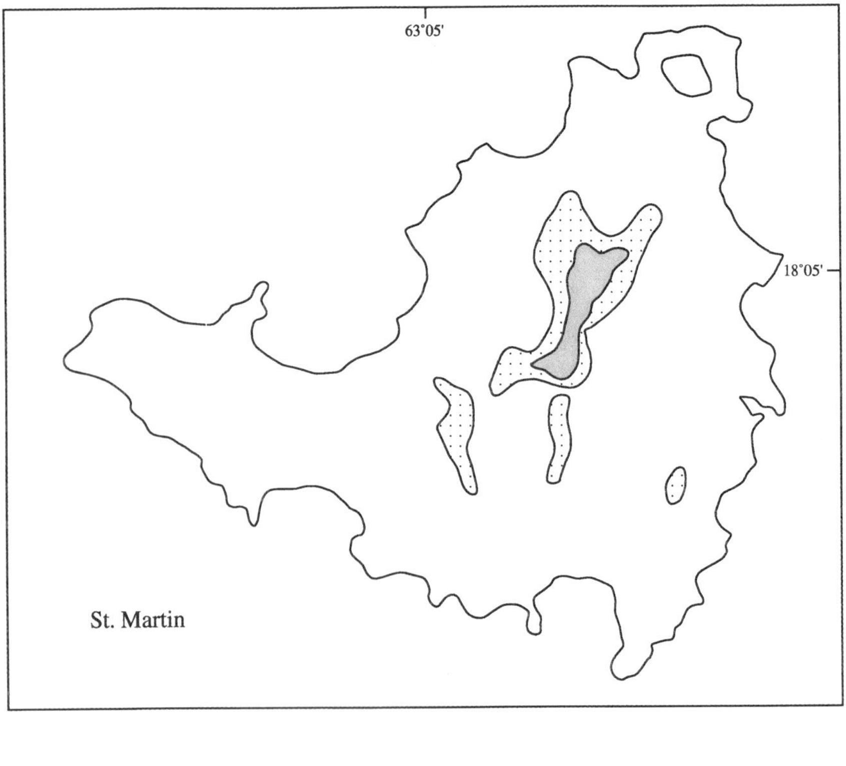

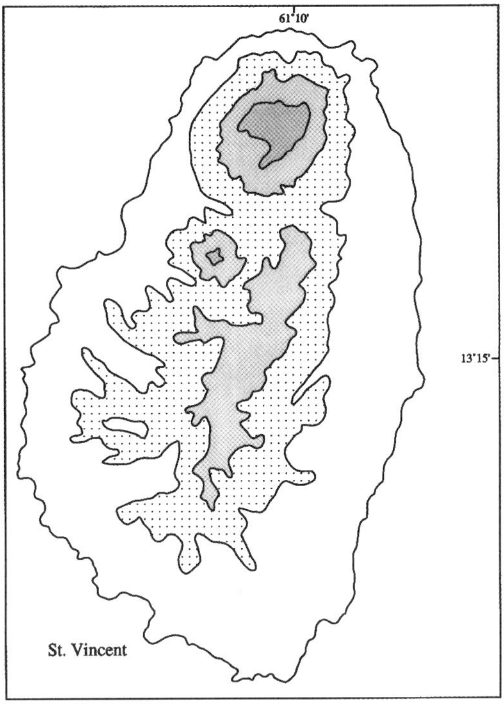

Sphaerodactylus is the most speciose genus of geckos in the Lesser Antilles. Eight species and 20 subspecies have been reported thus far (Schwartz and Thomas, 1975; Schwartz and Henderson, 1988, Lazell, 1994). As with *Anolis* and *Ameiva*, the historical tendency to name one species per island has resulted in numerous names, many of which have been relegated to synonymy. Sparrman (1784) was the first to describe this lizard in the Caribbean, with his description of *Lacerta sputator* (= *Sphaerodactylus sputator*) from St. Eustatius. Barbour (1923) considered *sputator* a dichromatic species; however, King (1962a), showed that the purported dichromatism was actually due to the fact that two species, *sputator* and *sabanus* were represented in the syntypes. *Sphaerodactylus sabanus* was described from a series of specimens taken on the island of Saba (Cochran, 1938b). Garman (1887a) described *pictus* from St. Kitts, which King (1962a) placed in the synonymy of *sputator*. Today, *sabanus* occurs on the St. Kitts Bank and Saba, while *sputator* is known from the Anguilla Bank and the St. Kitts Bank. Another species found on the Anguilla Bank is *Sphaerodactylus macrolepis* . Nine subspecies of *macrolepis* have been described from the greater Puerto Rico region (Thomas and Schwartz, 1966), but only one subspecies (*macrolepis parvus*) occurs in the Lesser Antilles (King, 1962a).

In 1836, Duméril and Bibron described *Sphaerodactylus fantasticus* from Martinique. However, Barbour (1915) questioned whether the specimens described actually came from Martinique because it was one of those received from Plée whose localities have been questioned (Censky and Paulson, 1992; Pregill, 1992). Barbour restricted the range of *fantasticus* to Guadeloupe based on scale characters and described the species from Martinique as *Sphaerodactylus festus* (Barbour, 1915). King (1962a) agreed with Barbour's locality designation for *fantasticus* and described a new subspecies from Montserrat, *Sphaerodactylus fantasticus ligniservulus*. Thomas (1964) reviewed the group and described six more subspecies, confined to the Guadeloupe complex of islands, and one subspecies on Dominica, though Evans (1989) stated that the subspecies *fuga* was introduced. Boulenger (1885) described a specimen from Antigua as *fantasticus*, but it was later put in synonymy with *Sphaerodactylus elegantulus*. *Sphaerodactylus fantasticus* is known to occur on Montserrat, Dominica, and Guadeloupe and its associated islands and cays.

Sphaerodactylus elegantulus was first described in 1917 by Barbour from one specimen collected on Antigua. Parker (1933), with little reservation, placed a female specimen from St. Lucia in this species, noting that there was little doubt about the locality, and that *elegantulus* must either have recently been introduced onto St. Lucia or it naturally cooccurred with *microlepis*. Because no *S. elegantulus* had been reported from St. Lucia since that time, King (1962a) suggested that it was more reasonable to assume that it belonged to the species *S. vincenti* because of the proximity of St. Lucia to Dominica, and because of the superficial similarity between *elegantulus* and *vincenti*.

Sphaerodactylus vincenti was described by Boulenger (1891) from several specimens from St. Vincent. In his revision of the Lesser Antillean *Sphaerodactylus*, King (1962a) placed *festus* as a subspecies of *vincenti* due to the overlap of many characters. For the same reason, he placed *monilifer* from Dominica, a species originally part of a series listed by Günther (1888) as *S. copii* and later described by Barbour (1921) as *monilifer*, in synonymy with *vincenti festus*. In a review of *S. vincenti*, Schwartz (1964) named an additional five subspecies from

Martinique. He also named a new subspecies of *vincenti* from St. Lucia and resurrected *monilifer* as a subspecies of *vincenti* on Dominica.

Sphaerodactylus microlepis was described by Reinhardt and Lütken (1862) and the type locality was listed as St. Croix. However, Barbour (1921) indicated that the type locality was in error, stating that the species was from St. Lucia. Boulenger (1891) added Dominica as a locality for *microlepis*, and King (1962a) followed suit noting that "Martinique lies between the islands of St. Lucia and Dominica and that future collecting may reveal this species on that island." However, in 1965, Schwartz considered the species to occur on only St. Lucia, stating that there was only one record of one specimen for Dominica and until further specimens were found, it should not be considered to occur on this island. In the same paper, Schwartz described a subspecies *S. microlepis thomasi* from the small Maria Islands, off the southern coast of St. Lucia. Interestingly, the first specimens of *Sphaerodactylus* taken from the Grenadines (Bequia) were as recent as 1989 (Lazell and Sinclair, 1990c) and were described as *Sphaerodactylus kirbyi* (Lazell, 1994).

Sphaerodactylus is widespread in the Lesser Antilles (Table 6.2), with at least one species occurring on each bank of islands except Barbados (a photograph of a *Sphaerodactylus* sp. from Redonda is deposited in the Carnegie Museum and a series of specimens were recently collected; H. Kaiser, personal observation). The only *Sphaerodactylus* that is endemic to a single bank of islands is *S. elegantulus* on the Antigua Bank. *Sphaerodactylus microlepis* may also be a single-bank endemic, occurring on St. Lucia and satellites, with a disputed distribution on Dominica. *Sphaerodactylus sputator* and *S. sabanus* occur on two banks each (Anguilla and St. Kitts, and Saba and St. Kitts, respectively). *Sphaerodactylus vincenti* and *S. fantasticus* occur on a number of banks. The only fossil remains of *Sphaerodactylus* in the Lesser Antilles were found on Anguilla (Pregill *et al.*, 1994). They have not been identified to species.

The relationships among Lesser Antillean *Sphaerodactylus* are based primarily on similarity and proximity; entry into the Lesser Antilles is presumed to have been from South America via stepping stones to the Greater Antilles (King, 1962b). Hass's (1991) molecular study of West Indian *Sphaerodactylus* included a few of the species found in the Lesser Antilles. Her data suggested a single mainland dispersal to Hispaniola, then outward migration from there, including two separate events into the Lesser Antilles. Her data supported a monophyletic West Indian group. Though she included only three Lesser Antillean species (*sputator*, *sabanus*, and *fantasticus*), she was able to demonstrate a relationship between *sputator* and Greater Antillean species. *Sphaerodactylus fantasticus* and *sabanus* were most divergent and she suggested that these two species along with *elegantulus*, *microlepis* and *vincenti* form a group. Hass' recent study (1996) determined that *sputator* was as different from the Greater Antillean group as are *sabanus* and *fantasticus*. She concluded that all Lesser Antillean species form a group, but cautioned that further study is needed.

There is only one species of *Phyllodactylus* extant in the Lesser Antilles, although several names appear in the literature. Gray described *Phyllodactylus pulcher* from a specimen of unknown origin (= Tropical America; Gray, 1845). Later, Cope (1862a) described a species from Barbados as *Phyllodactylus spatulatus*. Parker (1935), upon examination of type material of both species, failed to find any differences between them and synonymized *spatulatus* with *pulcher*. Dixon (1962) confirmed this decision. Dixon also described a new species of *Phyllodactylus* from Grenada, naming it *Phyllodactylus underwoodi*. Later, Dixon and Huey (1970) synonymized *underwoodi* with *P. ventralis*, stating that locality data of the label on the type was in error [Grenada = New Grenada (Colombia)]. *Phyllodactylus pulcher* occurs only on Barbados.

Gonatodes albogularis was described by Duméril and Bibron in 1836 from a specimen presumably collected by Plée in Martinique. It is the only record of the genus in the Lesser Antilles and, as such, has been doubted (Barbour and Ramsden, 1919). Vanzolini and Williams (1962) stated that collecting had not been intensive enough to claim this unequivocally. To date, no *Gonatodes* have been collected in the Lesser Antilles.

Gecko gecko was introduced on to the island of Martinique about 20 years ago. Henderson *et al.* (1993) report that it is limited to areas of human dwellings.

Anguid lizards are represented in the Lesser Antilles by only one specimen of a single species, *Diploglossus montisserrati* from Montserrat (Underwood, 1964). Although many of the islands of the Lesser Antilles have experienced extensive collecting over the past three decades, no further specimens have been found.

There is only one representative of the family Scincidae in the Lesser Antilles, *Mabuya mabouya* (= "*sloanei* ", R. Crombie, personal communication), though many more species have been described. Between 1831 and 1887, six species of *Mabuya* were described from the Lesser Antilles, each from a different island or islands [Gray (1831); *aeneus*; Duméril and Bibron (1839); *mabouia*; Cope (1862a); *lanceolata*; Bocourt (1879); *metallica*; Garman (1887d); *dominicana* and *luciae*]. These were all synonymized with *M. mabouya* by Dunn (1936). In this same paper he also reported the species for the first time from St. Martin, Redonda, and Marie Galante, Montserrat. It has since been reported from Anguilla and St. Barthélemy (Schwartz and Thomas, 1975). Barbour (1930b), however, listed the *Mabuya* on Martinique, St. Lucia, St. Vincent, and Barbados as extinct and linked their extinction to the introduction of the mongoose. Corke (1992) stated that they were extinct on Martinique and that he failed to find them on St. Lucia and St. Vincent, suggesting that they are either very rare or extinct. However, there are recent reports of the species from Barbados (Fraser *et al.*, 1990). There have been no fossil remains of *Mabuya* found to date in the Lesser Antilles.

Snakes

The taxonomy of Lesser Antillean snakes has had a very interesting history that is intimately related to human settlement of the islands and the importation of exotic animals. Slave laborers working in sugarcane fields were inexperienced with the fauna on their new island homes, and they would usually kill and mangle a snake on sight for fear of their lives. Bad experiences in mainland colonies made amateur naturalists, such as missionaries, equally wary. Thus, scientific collecting of snakes was not of sufficient volume until the early 1960s (Lazell, 1964b), and a consolidation of the taxonomy was achieved only after that time. Considering the sizes of islands in the Lesser Antilles, however, there exists a fairly high level of snake diversity and endemism. The total of 32 species contains 13 subspecies, and their description spans our entire taxonomic history, beginning with Carolus Linnaeus.

Three taxa (*Boa constrictor*, *B. c. orophias*, and *Corallus hortulanus*) were described by Linnaeus (1758), and these are the only members of the family Boidae found in the Lesser Antilles. Lesser Antillean boids are limited in distribution. Whereas *Boa* occurs on Dominica (subspecies *nebulosa*) and St. Lucia (subspecies *orophias*), *Corallus hortulanus cooki* is limited to St. Vincent, Grenada, and some of the Grenadines. In a review of Lesser Antillean *Boa*, Lazell (1964b) clarified some of the taxonomy and provided a biogeographical background. He considered the Dominican form *nebulosa* the terminus of a "stepped-cline series," which, by his definition of a subspecies, is a sharply defined peculiar geographic variant that is diagnostically homogeneous (Lazell, 1964b). The subspecies *B. c. nebulosus* was described by Lazell (1964b) in the genus *Constrictor* but was transferred to *Boa* by Peters and Orejas-Miranda (1970). *Corallus cooki* was described by Gray (1842), but later reduced to the rank of subspecies of *Corallus enydris* by Forcart (1951) who also synonymized Barbour's (1914) *Boa grenadensis* with *C. e. cooki*. McDiarmid *et al.* (1996) addressed the confusion surrounding the proper scientific name for this species and placed *B. enydris* in synonymy of *Corallus hortulanus*. As with *B. c. nebulosus*, there are indications that the island populations of *C. hortulanus* are morphologically distinguishable (R. W. Henderson, personal communication). *Boa constrictor* was recovered from the Indian Creek archaeological site on Antigua (Pregill *et al.*, 1988).

Lazell (1964b) removed both Caribbean members of *Bothrops* from the synonymy of *B. atrox* and reinstated them as full species, *B. lanceolata* restricted to Martinique, and *B. caribbaea* confined to St. Lucia. As with his investigations of *Boa*, he capitalized on his own recent collections and the availability of increased numbers of specimens from what might best be considered the "*B. atrox* complex" of South and Central America (Villa, 1984). Garman's (1887e) taxon *Trigonocephalus* is a junior synonym of *Bothrops*, and Lazell (1964b) designated a lectotype to stabilize the taxonomy. Lazell (1964b) also designated Martinique as the type locality for Lacépèdes' (1789) species *Coluber lanceolatus* and validated the name of the taxon as given by Hoge (1952). Interestingly, the distributions of *B. caribbaea* and *Boa constrictor orophias* are largely coastal and overlap completely, with *B. c. orophias* ranging to slightly higher elevations. Such a distribution may be indicative of relatively recent colonization combined with a slow spread through the inhabited areas that provide sufficient rodent abundance.

Lesser Antillean snakes of the genus *Typhlops* are small, secretive burrowers and consequently relatively rare in museum collections. They have therefore posed a taxonomic challenge that has only recently been addressed in the research of Thomas (1966c, 1974b, 1989) and Richmond (1966). *Typhlops dominicana* was apparently described by Stejneger (1904) on the basis of Boulenger's (1893) description of some

T. richardi (Schwartz and Thomas, 1975). Its nominate subspecies is limited to Dominica (Schwartz and Henderson, 1991). Richmond (1966) described *T. guadeloupensis* as an endemic species from Guadeloupe, but after being listed by Schwartz and Thomas (1975), this form was reduced to subspecies rank within *T. dominicana* (Thomas, 1976). Subsequent check lists use the subspecific form without further comment (Hedges and Thomas, 1989a; Schwartz and Henderson, 1991; Schwartz *et al.*, 1978). Thomas (1966c) described *T. monastus* from Montserrat as well as the subspecies *T. m. geotomus* (occurring on Antigua, Barbuda, Great Bird, Nevis, St. Kitts), differentiating the two taxa by middorsal scale counts. He considered these forms most closely related to the Jamaican *T. jamaicensis* (Thomas, 1989). *Typhlops tasymicris* is restricted to Grenada, and thought to be a relic of waif dispersal from northern South America (Thomas, 1989). Given the difficulty of obtaining series of specimens sufficient for detailed comparisons, a further refinement of the taxonomy of Lesser Antillean *Typhlops* cannot be excluded, especially in view of the as yet uncertain relationships of some key taxa (Thomas, 1989).

Another genus in the family Typhlopidae was recently recorded from the Lesser Antilles. Censky and Hodge (1997) reported *Ramphotyphlops braminus* from the island of Anguilla. Two specimens were collected and were undoubtedly introductions.

Two members of the family Leptotyphlopidae have been recorded from the Lesser Antilles, both in the genus *Leptotyphlops*. The range of the more widely distributed *L. bilineata* was at one point thought to include Barbados, Guadeloupe, Martinique, and St. Lucia (e.g., Barbour, 1937). Its presence on Guadeloupe, as indicated by Duméril and Bibron (1844), is doubtful since there are no recent specimens from that island (Schwartz and Thomas, 1975). Its name also represents a taxonomic curiosity: named *Typhlops bilineatus* and figured in Schlegel (1844), it was described by Duméril and Bibron (1844) and placed into the genus *Leptotyphlops* by Barbour (1914). The occurrence of the second species, *L. tenella*, on Antigua is highly questionable and based only on a single record. Whereas Schwartz and Thomas (1975) and Schwartz and Henderson (1991) list the species for Antigua, Schwartz and Henderson (1985) omitted it from their checklist.

The remaining Lesser Antillean snakes are members of the family Colubridae. Of the four extant species of *Alsophis*, *A. antillensis* has the widest range. Although Schlegel (1837) gave the range for his taxon

Psammophis antillensis simply as "Antilles", its range was restricted to Guadeloupe by Brongersma (1937). The nominate subspecies is now known to occur on Guadeloupe and Marie-Galante (Schwartz and Thomas, 1975), with the subspecies *danforthi*, *manselli*, *sanctonum*, and *sibonius* endemic to Terre-de-Bas, Montserrat, Terre-de-Haut, and Dominica, respectively. Schwartz (1966c) revised the taxonomy of *A. antillensis* by transferring both Parker's (1933) *A. leucomelas manselli* and Cope's (1879) *A. sibonius* to subspecific status under *A. antillensis*. Lazell (1967) placed Barbour's (1915) taxon *sanctonum* into the synonymy of *A. antillensis* at the rank of subspecies. Two other species of *Alsophis* have regionally restricted distributions. *Alsophis rijersmai* is native to the Anguilla Bank (Anguilla, St. Martin, St. Barthélemy), and *A. rufiventris* occurs on Saba and the St. Kitts Bank (Schwartz and Henderson, 1991). *Alsophis rijersmai* was only recently seen on mongoose-infested St. Maarten after a 40-year hiatus (Powell *et al.*, 1992). The former includes in its synonymy Garman's (1887e) *cinereus* (Schwartz and Thomas, 1975). The Antiguan *Alsophis* has had a difficult nomenclatorial history which is still unresolved. Schwartz and Henderson (1991) revived Parker's (1933) name *antiguae* as the specific name for Antiguan *Alsophis* and provided some largely historical evidence (see also Pregill *et al.*, 1988). Parker (1933) described *A. leucomelas antiguae* and subsequently elevated it to full species (Parker, 1936). With no discussion, Schwartz (1966c) reduced it in rank to subspecies. Schwartz *et al.* (1978) remarked that an earlier taxon than Parker's (1933), *punctifer*, described by Martin (1838), might be the appropriate subspecific name for the Antiguan taxon, and that further study was necessary to investigate this question. Henderson (1990) placed the only extant population of *A. antiguae*, on Great Bird Island, into its own subspecies, *A. a. sajdaki*.

The genus *Liophis* is represented by five species and displays an extreme degree of single-island endemism. Each described species occurs basically on a single island, and the distribution of the genus in the Lesser Antilles is nearly continuous along the island chain, reaching from Grenada in the south to Guadeloupe in the north. The detailed work by Dixon (1980, 1981) provided the basis for the transfer of Lesser Antillean *Dromicus* to *Liophis*, reviving Cope's (1862b) genus. *Liophis cursor* is a very rare species that used to occur abundantly on Martinique before the introduction of the mongoose (Lazell, 1967; Henderson and Sajdak, 1986). It was previously thought to be extinct and is now possibly restricted to a few dozen specimens on Rocher de Diamant, a small satellite island off Martinique (Lazell, 1967). Cope's (1862b) *L. putnami* is a junior synonym of this species. To the north, *L. juliae* occurs on Dominica in its nominate subspecific form, with *L. j. copeae* and *L. j. mariae* native to Guadeloupe and Marie-Galante, respectively. Parker (1933) synonymized Barbour's (1914) *Leimadophis mariae* with *L. mariae*. Three species of *Liophis* occur on southern Lesser Antillean islands. The situation of *L. ornatus* on St. Lucia is similar to that of *L. cursor* on Martinique: the species is now only extant on one small, mongoose free satellite island, Maria Major (Corke, 1987). *Liophis melanotus* exists on Grenada in the Lesser Antilles, but is more widely distributed on the continent (e.g. Trinidad, Tobago, northern South America; Schwartz and Henderson, 1991). The Barbadian populations of *Liophis*, *L. perfuscus*, have a very restricted distribution in the hilly, extreme east central portion of the island which is relatively mongoose free and not affected by sugar cane monoculture.

Four other snakes occur in the Lesser Antilles, and all are scarce in collections as well as in the wild. *Chironius vincenti* is endemic to St. Vincent, where it had been considered extinct until 1988 (Henderson *et al.*, 1988). Taxonomically, Schwartz and Thomas (1975) combined Boulenger's (1891) *Herpetodryas carinatus* var. *vincenti* to the current name to distinguish the St. Vincent and mainland populations. Although the name *C. carinatus* has also been applied to these snakes, only a questionable record from Guadeloupe (Boulenger, 1894) actually places *C. carinatus*, a species widespread on the mainland, on a Lesser Antillean island. The widespread snake *Clelia clelia* occurs in its nominate subspecific form throughout much of the neotropics and until recently was regarded as part of the fauna on Dominica and St. Lucia. Greer (1965) taxonomically flagged the populations on Grenada by placing them into the subspecies *groomei*. However, this subspecific designation was not recognized by Peters and Orejas-Miranda (1970), and has not been further validated. The populations on Dominica and St. Lucia were described by Underwood (1993) as a new species, *Clelia errabunda*. Garman collected three specimens of *Pseudoboa neuwiedi* from Grenada in 1883. It has not been taken since and has been considered extirpated (Underwood, 1962). Lastly, *Mastigodryas bruesi* is a native to the southern Lesser Antilles that was described by Barbour (1914) in the genus *Alsophis*. Stuart (1941) placed Barbour's taxon into the genus *Mastigodryas*.

Turtles

There are four species of turtles on the Lesser Antillean islands, though three (*Pelusios subniger*, *Trachemys scripta*, *Trachemys stejnegeri*) and possibly the fourth (*Geochelone carbonaria*) have been introduced. *Pelusios subniger* is an African species that was introduced onto Guadeloupe where it is moderately common (Schwartz and Thomas, 1975). *Trachemys stejnegeri stejnegeri* occurs throughout low elevations in Puerto Rico and it has been introduced to the island of Marie Galante, off of Guadeloupe (Seidel, 1988a; Schwartz and Thomas, 1975). Likewise, *Trachemys scripta* has been introduced onto Guadeloupe (Schwartz and Henderson, 1991).

Populations of *Geochelone* are known from every major island and many smaller islands in the Lesser Antilles, except Barbados and Sombrero where other extinct tortoises have been reported (Censky, 1988). Most of these populations have been identified as *Geochelone carbonaria*, and those that have been reported as *G. denticulata* were listed before a distinction was made between the two species and may well be *carbonaria*. There have been four hypotheses proposed for *Geochelone* distribution in the Caribbean: (1) natural dispersal, (2) introduction by prehistoric indians, (3) introduction by early European settlers, and (4) recent introduction as escaped pets. None of these can be irrefutably supported by the distributional data, and the current distribution may be the result of a combination of dispersal modes (Censky, 1988). The extinct tortoise, *Geochelone sombrerensis*, was described from remains found on Sombrero island in the northern Lesser Antilles (Leidy, 1868). Fossil bones of *Geochelone* sp. were found on Barbados (Ray, 1964) and *Testudines* sp. from cultural remains on Barbuda (Pregill *et al.* 1994).

Crocodiles

Crocodiles are not native to the Lesser Antilles. Two species have been recorded in the southern Lesser Antilles. *Caiman crocodylus*, a Central and South American species, is apparently an occasional vagrant to the Grenadines (Schwartz and Henderson, 1991). In addition, in September 1910, a *Crocodylus intermedius* washed ashore in Grenada (Underwood, 1962).

Biogeography

Although there is considerable diversity and high endemicity in a variety of genera (e.g., *Anolis*, *Ameiva*, *Eleutherodactylus*, and *Liophis*), a fairly clear picture of the biogeographic origins for some of the Lesser Antillean herpetofauna can be constructed. In general terms, there seem to be two faunal stop gaps, one in the north between the Anguilla Bank an the Virgin Islands and the other in the south between St. Lucia and St. Vincent. The southern gap was already considered important by Lescure (1987), when he observed a progressive reduction of species diversity for amphibians and reptiles between South America, Trinidad, Tobago, and Grenada and St. Vincent in the southern Lesser Antilles. However, as gross diversity decreases when moving north into the island chain, single-island endemism increases, with some of its highest levels reached in the center of the island chain (Table 6.1), at the greatest distance from the diverse herpetofaunas in the Greater Antilles or South America.

All conceivable variations on distributional motif occur for amphibians and reptiles in the Lesser Antilles. Some taxa are endemic only on tiny specks of land (e.g., *Anolis nubilus* or *Ameiva atrata* on Redonda, *Ameiva corax* on Little Scrub Island, *Alsophis antiguae sajdaki* on Great Bird Island), some are native to small islands (e.g., *Anolis sabanus* on Saba, *Diploglossus montisserrati* on Montserrat), some to larger islands (e.g., *Eleutherodactylus euphronides* on Grenada, *Liophis perfuscus* on Barbados), others are Bank endemics (e.g. *Alsophis rijersmai* on the Anguilla Bank, *Ameiva erythrocephala* on the St. Kitts Bank) or regional endemics (both *Eleutherodactylus martinicensis* and *Gymnophthalmus pleei* occur in the central Lesser Antilles), and only *Eleutherodactylus johnstonei* and an *Iguana* species occur on all major islands. Multiple waif dispersal, with subsequent in situ speciation under a variety of selective pressures and extinctions on different islands, is the most encompassing and most easily invoked mechanism to populate the islands with the species we observe today. However, in some cases (e.g., *Eleutherodactylus*, *Iguana*) human-mitigated introductions, either cultural or accidental, are very important to consider because of the nonrandom distribution of some taxa (Pregill *et al.*, 1994; see Kaiser, 1992, for *Eleutherodactylus*).

As a major faunal division in the southern Lesser Antilles, the St. Vincent passage is of unrivaled importance. There are 11 species (one frog, two crocodilians, four lizards, four snakes) that exist in northern

South America and on Grenada and/or St. Vincent whose distribution terminus is the St. Vincent passage. This is the major influx from a more highly diverse fauna. In contrast, at the northern extreme of the Lesser Antilles only *Sphaerodactylus macrolepis* occurs in both the Greater and Lesser Antilles. However, the assumption that the herpetofauna of the Lesser Antilles has its ancestry in both the Greater Antilles and South America is borne out by studies of specific relationships and the Quaternary fossil record (e.g., Hass, 1991; Kaiser *et al.*, 1994b). Given that the occasional and random dispersal event from island to island within the Lesser Antilles still occurs, either by natural means or by human introduction, we can assume that the present-day distribution of taxa is only a vignette in time, and that species ranges in this unique island system may be in continuous flux.

Conservation

The small sizes and limited resources of the islands of the Lesser Antilles, in relation to the human populations, have created serious problems of resource depletion and exploitation. Many of the islands have had a long history of environmental degradation due to numerous factors. Chief among them was the alteration of habitat which began shortly after Europeans arrived with tobacco and sugar-cane. Recently, the tourism industry has taken its toll on a number of islands, with hotel complexes constructed at the expense of natural habitat. While some species appear to do well in altered habitats (*Anolis*, *Hemidactylus*, *Eleutherodactylus*), other species do not (*Iguana*). Seventy-nine populations and/or species of vertebrates no longer occur in the Lesser Antilles since the late Quaternary, and most have been extirpated since European/African settlement (G. Pregill, personal communication).

Another factor that has had an impact on the herpetofauna of the islands is the introduction of exotic animals, particularly mongoose and goat. Barbour (1930) suggested that the mongoose was responsible for the extirpation of a number of populations of reptiles. However, Baskin and Williams (1966) disputed this, stating that there was no simple relationship between reptile population declines and the presence of mongoose. Corke (1992) found that ground lizards and skinks are abundant *only* on mongoose-free islands in the windwards. He did not unequivocally argue that the mongoose is the culprit but stated that more research into anthropogenic causes needs to be done. However, the mongoose is probably responsible for the extirpation of a number of populations of *Alsophis* (Sajdak and Henderson, 1991). Direct killing of snakes has led to the decline of other species, especially on islands where venomous species occur (Sajdak and Henderson, 1991; Corke, 1992). Goats and other herbivores have had a decimating impact on at least one *Iguana* population (Anguilla, E. Censky, personal observation), while other *Iguana* populations have been impacted by hunting (St. Martin, Nevis, St. Lucia, St. Vincent; Corke, 1992; M. Day, personal communication). *Iguana delicatissima* on Dominica, however, remains healthy even though hunting occurs (M. Day, personal communication).

The loss of biodiversity has not gone unnoticed by many governments in the Lesser Antilles. In the early 1980s, a cooperative effort between the Caribbean Conservation Association and the University of Michigan Wildlife Management Center established ENCAMP. ENCAMP produced a series of technical reports with general guidelines for sustainable management of natural resources. Since then, the Island Resources Foundation, with support from the United Nations Development Program and in collaboration with a number of non-government organizations, has produced in-depth summaries and specific guidelines (Country Environmental Profile) for many of these islands. These reports are based on the information available, scant as it is. A general lack of information on population biology and ecology impedes the efforts put forth by these governments toward recovery and management programs. It has become clear that general systematic, ecology, and survey work (especially on the cays) is needed on many of these islands.

7

West Indian Herpetoecology

Robert W. Henderson[1] *and Robert Powell*[2]

[1] *Section of Vertebrate Zoology, Milwaukee Public Museum, 800 W. Wells Street, Milwaukee, Wisconsin 53233-1478*

[2] *Department of Natural Sciences, Avila College, 11901 Wornall Road, Kansas City, Missouri 64145*

The islands, especially the shores, are teeming with lizards of every color, of every variety of marking, and of all sizes.

Ober, 1880

Another individual of this species [Leiocephalus carinatus] *seemed to exist in part on food scraps dropped by diners at a hotel in South Andros; this lizard would lurk behind a large curtain, darting out to the appropriate table and back as the occasion arose.*

Schoener *et al.*, 1982

Introduction

Aside from the blatantly obvious physical attributes of West Indian islands, the amphibians and reptiles of the region have proven to be ideal tools for biologists with an interest in ecology. Among the alluring qualities of the herpetofauna are its intriguing distribution, wonderful diversity, and the frequently high population densities. New species of West Indian amphibians and reptiles are still being described on a routine basis (especially from Cuba and Hispaniola), the evolutionary species concept (*sensu* Frost and Hillis, 1990) as applied to the West Indian herpetofauna (e.g., Powell, 1993) has provocative implications (e.g., as many as 25 species of *Ameiva* may occur on Hispaniola instead of the currently recognized four species), and West Indian biogeography remains controversial (e.g., Williams, 1989b; Hedges *et al.*, 1992b; Crother and Guyer, 1996). However, research publications that focus on aspects of the ecology of the West Indian herpetofauna are now much more prevalent than taxonomic and systematic publications. For example, Powell and Henderson (1996b) determined that, through the 1960s, taxonomic publications far outnumbered nonsystematic publications in West Indian herpetology. The 1970s was a transitional decade, and during the 1980s the emphasis had shifted dramatically to nontaxonomic subjects; this trend has continued into the 1990s. The approaches to collecting ecological data and the kinds of ecological data collected also have changed dramatically, from largely anecdotal accounts with varying degrees of reliability to quantification of structural aspects of the habitat and problem-oriented research addressing a wonderfully diverse array of topics ranging from habitat selection (e.g., Chandler and Tolson, 1990), foraging behavior (e.g., Lewis, 1989), dietary analyses (e.g., Floyd and Jenssen, 1983), population densities (e.g., Schoener and Schoener, 1980), and social behavior (e.g., Stamps, 1976) to the ecological context of parasite communities in *Anolis* (Dobson *et al.*, 1992), ultraviolet vision and anole communication (Fleishman *et al.*, 1993), and genetic programming in behavioral ecology (Koza *et al.*, 1992).

Here we briefly review topics that have been important in West Indian herpetoecology (species diversity and distribution), citing relevant literature as sources of additional, more detailed information. Additionally, we have concentrated on several topics (population densities, trophic relations, and ecological versatility) that until now have been addressed only casually (or in reference to only one or two species) and that we view as focal to the ecology of West Indian amphibians and reptiles. Throughout, we have tried to make comparisons between the herpetofaunas of the West Indies and the Neotropical mainland. The taxonomy used herein follows the recent checklist of West Indian amphibians and reptiles by Powell *et al.* (1996a).

History

Species by species summaries of the current state of our ecological knowledge of the West Indian herpetofauna appear in Schwartz and Henderson (1991). They estimated that ecological knowledge was limited to about 5% of the herpetofauna. In retrospect, this estimate is low; about 12% of the herpetofauna has received active attention (i.e., has met one or more of the following criteria: animals were marked and released, portions of the habitat were quantified, thermal responses to the habitat were determined, population densities were determined, prolonged observations of social behavior were made, and/or activity ranges were determined; it precludes fortuitous observations associated with collecting, stomach content analyses devoid of fieldwork, and/or observations made of strictly captive animals). Not surprisingly (because of their often high densities, diurnal activity, and ubiquity), more lizard species have received attention (about 19% of the lizard fauna) than either frogs (ca. 4%) or snakes (ca. 5%), and 72% of the lizards that met the criteria were *Anolis*.

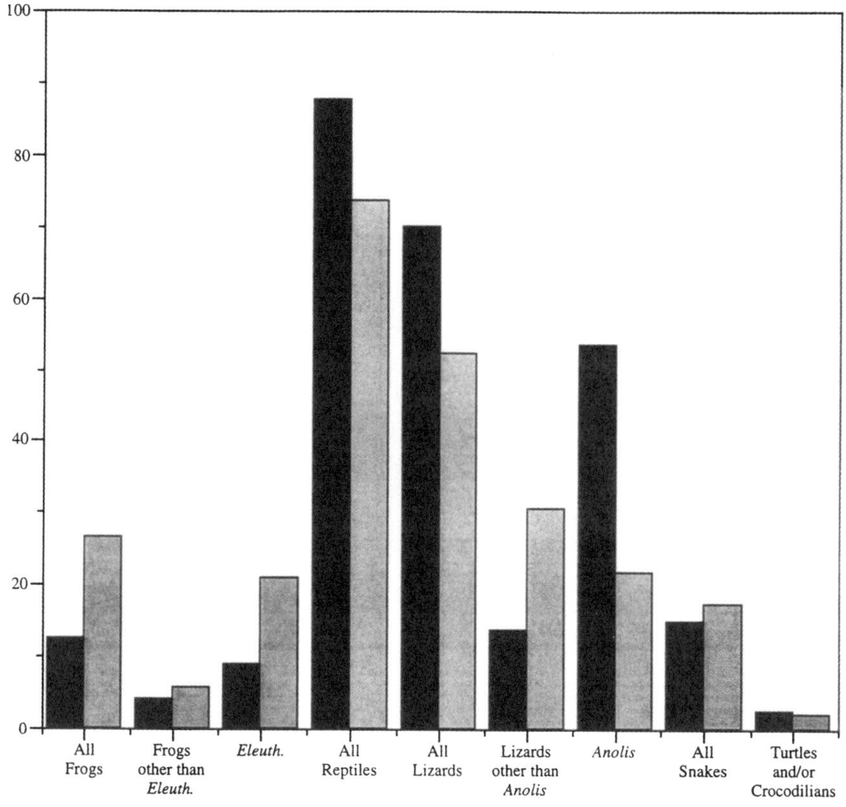

Figure 7.1. The relationship between taxonomic group and the amount of published research in West Indian herpetoecology. Solid bars indicate precentage of publications and gray bars indicate proportion of the herpetofauna the taxonomic group represents. For example, there have been a disproportionate number of publications concerned with *Anolis*, whereas lizards other than *Anolis* have been relatively neglected. *Eleuth., Eleutherodactylus.*

An exhaustive literature analysis is beyond the scope of this chapter. Natural history summaries and species by species reviews of the literature are available in Schwartz and Henderson (1991). Of the ecologically oriented papers published between 1960 and 1990 and cited by Schwartz and Henderson (1991) in either their Bibliography or Literature Cited sections, 12.6% dealt with frogs and 87.4% with reptiles. Of the frog papers, 68.4% focused on *Eleutherodactylus*. The papers on reptiles included a very high percentage that addressed lizard ecology (79.9%) compared to 17% for snakes and 3% for turtles and crocodilians combined. Figure 7.1 summarizes the taxonomic distribution of the published papers. The number of papers that used anoles as their subject species is highly disproportionate to the rest of the herpetofauna but hardly surprising in light of their ubiquity, density, and visibility. They comprised 64.4% of all papers on reptiles and 53.3% of all papers on frogs or reptiles. Papers with frogs other than *Eleutherodactylus* or with snakes, crocodilians, and/or turtles as their focal organisms have appeared in about the same proportion as reflected by their diversity within the West Indian herpetofauna.

Early Contributions

Although more comprehensive histories of West Indian herpetology appear elsewhere (Powell and Henderson, 1996b; see Chapter 1), here we want to broaden the ecological scope presented therein and to specifically put the study of ecological topics into an historical context. By doing so, we can gain a vantage point for the developmental chronology of an ecological perspective toward the West Indian herpetofauna and see how several key individuals have influenced the advancement of Antillean herpetoecology.

In reviewing the earliest literature that might contain potentially useful observations regarding the West Indian herpetofauna, we determined that the best to be hoped for was accurate anecdotal observations [and we, like Fitch (1987), are advocates of anecdotal observations, excluding them from criteria set earlier only because they do not indicate a concentrated effort to do research on a particular species or problem]. Most early contributions originated primarily from descriptions of habitats and, less frequently, from accounts of the herpetofauna. The letters and diaries of Christopher Columbus, for example, provided descriptions of, in many instances, a West Indies that no longer exists. In 1493, on Hispaniola, he wrote of "trees of a thousand kinds and tall, and they seem to touch the sky" (Major, 1870), describing forests that were long ago sacrificed for ship timbers.

By the early to mid-nineteenth century, natural history exploration in the neotropics produced a number of detailed accounts in the "A Naturalist in ..." genre (e.g., Gosse, 1851; Ober, 1880; Bell,

1889). Certainly the single best example for the West Indies is Gosse's (1851) *A Naturalist's Sojourn in Jamaica.* Gosse's most detailed and appealing accounts deal with Jamaica's rich lizard fauna:

> One feature with which a stranger [to Jamaica] cannot fail to be struck on his arrival in the island, and which is essentially tropical, is the abundance of the Lizards that everywhere meet his eye. As soon as he ever sets foot on the beach...he sees on every hand the beautifully-coloured and meek-faced Ground Lizard (*Ameiva dorsalis*) scratching like a bird among the sand.
>
> If he looks into the outbuildings of the estates, the mill-house, or the boiling house, or the cattle sheds, a singular croaking sound above his head causes him to look up...It is the gecko, or Croaking Lizard [*Aristelliger praesignis*], a nocturnal animal in its chief activity, but alwas to be seen in these places, or in hollow trees, even by day.
>
> "The stranger walks into the dwelling-house. Lizards, lizards, still meet his eye. The little anoles ([*A. grahami, A. opalinus*] &c) are chasing each other in and out between the jalousies, now stopping to protrude from the throat a broad disk of brilliant colour, crimson or orange, like the petal of a flower, then withdrawing it, and again displaying it in coquettish play...Another is running up and down on the plastered wall, catching the ants as they roam in black lines over its whited surface; and another leaps from the top of some piece of furniture upon the back of the visitor's chair, and scampers nimbly along the collar of his coat.
>
> He lifts the window sash; and instantly there run out on the sill two or three minute Lizards of a new kind, allied to the Gecko, the common Pallette-tip (*Spahaeriodactylus* [sic] *argus*).
>
> In the woods he would meet with other kinds. On the trunks of the trees he might frequently see the Venus [*Anolis garmani*], as it is provincially called; a Lizard much like the Anoles of the houses, of a rich grass-green colour, with orange throat disk, but much larger and fiercer: or in the eastern parts of the island the great Iguana [*Cyclura collei*], with its dorsal crest like the teeth of a saw running all down its back.

Gosse (1851) devoted an impressive number of pages to describing *Anolis* behavior (especially of *A. grahami* and *A. opalinus*: "They are particularly numerous in the lieux."), habitat, and distribution; little did he know how important these ubiquitous lizards would become in the study of many aspects of vertebrate ecology and evolution.

Other observations provide an historical context; for example, his premongoose observations on the abundance of the probably now-extinct racer *Alsophis ater*. Gosse described the "Gray Snake," *Natrix capistrata* (= *A. ater*), and gave maximum snout-vent length (SVL) as 16 in.; he described the diet as including *Anolis* and gekkonids; he observed that "though not rare, it can scarcely be called abundant...I have found it scattered in distant and varying localities, in the lowlands and on the mountain-tops." In 1872, the mongoose was introduced to Jamaica (Espeut, 1882); after 1930, specimens of *A. ater* were no longer added to museum collections (Henderson, 1992); by 1940, Grant (cited in Lynn and Grant, 1940) reported that he had not met "anyone who had seen a live specimen in years;" and, as of this writing (April 1996), despite intensive herpetological fieldwork in

Jamaica, no specimen of *A. ater* has been cataloged into a collection in over 60 years.

Toward the end of the nineteenth century, Gundlach (1894) published his *Erpetologia Cubana* which presented interesting and useful anecdotes regarding distribution, abundance, and behavior of some members of the Cuban herpetofauna. Garman, who initiated the long tradition of West Indian fieldwork by herpetologists at Harvard's Museum of Comparative Zoology, produced a series of papers on West Indian reptiles and included field observations in his accounts (e.g., Garman, 1887e).

The Twentieth Century

In the first half of the Twentieth Century, summary papers on various West Indian herpetofaunas were published (e.g., Stejneger, 1904; Barbour, 1914; Barbour and Ramsden, 1919; Schmidt, 1928; Lynn and Grant, 1940; Mertens, 1939, 1940; Cochran, 1941). Each contained interesting observations on aspects of natural history, but many of the observations were not firsthand. Barbour, especially, relied on secondhand accounts or made offhand judgments on the abundance of many species and subsequently has been chastised by several authors (e.g., Curtis, 1946; Lazell, 1972). Notable exceptions to the anecdotal "ecology" of the first six decades of the twentieth century are the publications of Wolcott (1923) on the diets of Puerto Rican lizards, Noble (e.g., 1933) on lizard mating behavior, and Evans (1938) on territoriality in Cuban *Anolis*.

Not until the 1960s did the approach to studying the ecology of West Indian reptiles take several innovative turns. We believe that three catalysts were primarily responsible: (1) Ecology in general was coming into its own as a scientific discipline and no reason existed to preclude the West Indies as a geographic arena, especially considering their proximity to the United States; (2) Albert Schwartz had just started his remarkable inventory of the West Indian herpetofauna and was clearly demonstrating the remarkable species diversity of the area; and (3) Ernest Williams began his multidisciplinary analyses of anole biology.

Indeed, that the animals involved in these ecological innovations were Gosse's "exceedingly interesting" anoles seems altogether appropriate. Preliminarily, Oliver (1948) characterized habitat preferences in four species of *Anolis* on Bimini by perch height. Subsequently, Ruibal (1961) used body temperature and "thermal habitats" to describe ecological differences in five species of Cuban anoles. In 1961, Collette investigated relationships between aspects of ecology and morphology in several species of Cuban *Anolis*, and he presented a stylized habitat diagram indicating niches of five species. Rand (1962) used perch height to characterize microhabitat in three species of Hispaniolan *Anolis* under captive conditions. Rand (1964b) used morphology, climatic habitat (including body temperatures), and structural habitat (adding the critical dimension of perch diameter to the by now standard perch height dimension) to define the ecological distribution of *Anolis* in Puerto Rico, and he used the term "structural habitat" to define the perch characteristics. Rand (1967a) refined his methods and did a similar analysis of anoles in the Kingston (Jamaica) area, and Schoener (1968) and Schoener and Gorman (1968) took the analyses a logical step farther by adding an examination and quantification of diets. However, the common thread connecting Ruibal, Collette, Rand, Schoener, and Gorman was Ernest E. Williams of Harvard University: all came under his influence early in their research careers.

The Past 30 Years

Ecological research on the West Indian herpetofauna during the past 30 years has been largely problem oriented. Few autecological studies have been produced, but excellent exceptions to this trend include detailed accounts for *Cyclura carinata* (Iverson, 1979), *Anolis lineatopus* (Rand, 1967b), and a Haitian population of *Crocodylus acutus* (Thorbjarnarson, 1988). In addition, via series of papers, several species have been studied more extensively and intensively than others. One frog, *Eleutherodactylus coqui* on Puerto Rico, has been more thoroughly investigated than any other, primarily through the efforts of Margaret M. Stewart and her co-workers (e.g., Stewart, 1985; Stewart and Rand, 1991; Stewart and Woolbright, 1996; Taigen *et al.*, 1984; Townsend and Stewart, 1986; Woolbright, 1985; Woolbright and Stewart, 1987). Similarly, Judy Stamps chose *Anolis aeneus* on Grenada for intensive socioecological research (e.g., Stamps, 1976, 1983, 1987, 1988; Stamps and Tanaka, 1981), and Jonathan Roughgarden and colleagues (summarized in Roughgarden, 1995) addressed the ecology of Lesser Antillean *Anolis*, and various authors have contributed to an understanding of the ecology of *Anolis sagrei*. Much of the content of these papers has been summarized in Schwartz and Henderson (1991). More recently, Ellen Censky (1995a,b, 1996) has conducted detailed field studies with *Ameiva plei*, and Allen Lewis and coworkers have examined intriguing aspects of the natural history of *Ameiva exsul* (Lewis, 1986, 1989; Lewis and Saliva, 1987; Lewis *et al.*, 1996). No snake species to date has received comparable attention, but Tolson's work (1988, 1996; Chandler and Tolson, 1990) on *Epicrates monensis*, and Henderson's work on *Corallus* in the West Indies (1993a; Henderson and Winstel, 1995; Henderson and Henderson, 1995; Henderson *et al.*, 1996) and field and laboratory research with *Alsophis portoricensis* (e.g., Thomas and Prieto-Hernández, 1985; Rodriguez-Robles and Leal, 1993; J. C. Gillingham, unpublished data), come closest.

The concept of the structural habitat has continued to be an important one in West Indian herpetology. It has provided a convenient means of defining and quantifying strategic habitat variables in arboreal (or scansorial) snakes and lizards, especially in situations in which congenerics occur syntopically. Studies of structural aspects of anole habitats have subsequently proliferated (e.g., Schoener, 1968; Schoener and Gorman, 1968; Rand and Williams, 1969; Schoener and Schoener, 1971a,b; Roughgarden *et al.*, 1981), and the concept has recently been applied to West Indian snakes by Henderson *et al.* (1982) and Henderson and Winstel (1995). Chandler and Tolson (1990) have examined this concept in the context of predator (*E. monensis*) and prey (*Anolis cristatellus*) relations. An important concern in the study of structural habitats, and one that is usually overlooked, is that of habitat availability (Schoener, 1975); species may differ in structural habitat only because the habitats in which they live differ in what is available.

Rand noted that anoles with similar structural habitats also shared morphological similarities; he suggested that these similarities could be the result of being closely related or the result of convergence due to environmental similarities (= ecomorph). Williams (1972, 1983) refined and expanded the ecomorph concept. Ecomorphs are species of different phyletic origin but with similar morphological adaptations to similar niches. Moermond (1979) carried the ecomorph concept farther, including orientation and movements of anoles within a structural habitat matrix. He found that the perch matrix affects the rapidity of movements on surfaces and the frequency of jumps and distances jumped between surfaces; also, species can be separated into jumpers, runners, and crawlers based on movement type and limb and body proportions. Furthermore, Moermond (1979) suggested that availability of structural habitats may help explain the presence or absence of various ecomorphs, and that the local spatial distribution of these microhabitats influences the within-habitat distribution of each species. These data have been independently confirmed in applications to the anoles of other Greater Antillean communities (e.g., Estrada and Rodriguez, 1984; Losos, 1990a,b; Powell and

Russell, 1992) and on the Neotropical mainland (e.g., Pounds, 1988). Losos (1990a,b,c) used the ecomorph concept and found evidence to support the hypothesis that morphology, performance capability, ecology, and behavior have evolved synchronously. In discussing the application of the ecomorph concept and ecomorphology to Neotropical snake assemblages, Cadle and Greene (1993) stressed the need for maintaining an historical perspective when analyzing ecomorphs and ecomorphology. The ecomorph concept has been subsequently applied to West Indian species of *Eleutherodactylus* (Hedges, 1989a) and *Sphaerodactylus* (Thomas *et al.*, 1992) but not nearly as intensively or thoroughly as for *Anolis*. These remain fruitful areas for future research.

Like the structural habitat concept, studies of other aspects of West Indian herpetoecology have undergone a gradual refinement. Research on thermal biology focused on *Anolis* and, again, began with Ruibal (1961), Rand (1967b), and Schoener and Gorman (1968). However, new students, again with ties to Harvard and E. E. Williams (Huey and Webster, 1975, 1976; Hertz, 1981, 1983; Hertz *et al.*, 1993), established ambitious research protocols. Similarly, analyses of diets in frogs (Woolbright and Stewart, 1987; Duer *et al.*, 1992), anoles (e.g., Floyd and Jenssen, 1983; Schoener and Toft, 1983; Pacala and Roughgarden, 1985; Schoener and Spiller, 1987; Bullock *et al.*, 1993; Cullen and Powell, 1994), other lizards (Powell *et al.*, 1990a; Cunningham *et al.*, 1993), snakes (Henderson *et al.*, 1987a, 1988a; Henderson and Crother, 1989), and mixed reptile communities (White *et al.*, 1992) examined seasonal effects, sexual differences, and/or variation in diets by habitat differences.

Some Aspects of the Ecology of the West Indian Herpetofauna

Several aspects of the ecology of the West Indian herpetofauna that we believe have made it especially attractive include: (1) species diversity, (2) distributions, (3) population densities, (4) trophic relations, and (5) versatility. In brief discussions on each of these five topics, we have provided data, compared these with herpetofaunas on the Neotropical mainland, and offered explanations as to why things are as they are in the West Indies. Our treatment is not exhaustive, but the literature cited provides an overview and a starting point for those interested in pursuing particular facets of our discussion.

Origin and Diversity

West Indian islands collectively support one of the richest terrestrial vertebrate faunas in the Western Hemisphere. Considering that the mammalian fauna is relatively small [145 species (80% endemic) historically and about 63 extant species; Hedges, 1996a] and the avian fauna is modest (about 425 species of which only 35% are endemic; Hedges, 1996a), the herpetofauna obviously represents, in respect to number of species, the dominant vertebrate group in the West Indies [622 total described species (169 amphibians and 453 reptiles) of which 572 (92%) are endemic; Hedges, 1996a,b]. Among amphibians, 99% are endemic to the West Indies, and 93% of reptiles are endemic to the area. Seventy-seven independent lineages were identified by Hedges (1996b) and, where a source area within the New World could be determined (42 of 53 lineages), 79% show a South American origin [with smaller contributions from Central America (8/53, 15%) and North America (3/53, 6%)]. Although *Eleutherodactylus* originated by either vicariance or dispersal, and the xantusiid lizard *Cricosaura typica* may have originated by vicariance, "all other lineages appear to have arrived by dispersal during the Cenozoic, and all but nine lineages in the last half of the Cenozoic (30-0 mya)" (Hedges, 1996; but see Page and Lydeard, 1994). Three genera dominate in terms of numbers of species: *Eleutherodactylus* (139 species), *Sphaerodactylus* (79), and *Anolis* (138). Combined, they represent about 57% of the entire herpetofauna. The next most diverse genera are *Typhlops* (represented by 23 species), *Leiocephalus* (23 species), *Celestus* (19 species), *Ameiva* (19 species), *Tropidophis* (13 species), *Amphisbaena* (16 species), *Bufo* (12 species), *Arrhyton* (12 species), and *Alsophis* (11 species). All other genera are represented by fewer than 10 species. Roughgarden (1995) offered an explanation for the reptile-dominated faunas of the Caribbean (which contrast markedly with the avian-dominated faunas of Pacific islands). He believes that

> The Caribbean islands are old enough to have accumulated a reptile fauna before birds and mammals diversified. Because the Caribbean Islands remained at warm latitudes, and because the islands are small enough to lead to relatively little predation on lizards, the islands today may show communities that resemble those of the Cretaceous.

Table 7.1. The Number of Genera and Species of Native Frogs, Lizards, and Snakes Occurring at West Indian and Mainland Localities of Varying Area and the Densities of Frogs + Lizards + Snakes at These Localities.

	Genera/Species			
Location	Frogs	Lizards	Snakes	Species/km^2
West Indies (ca 214,305/ km^2	6/166	22/321	23/105	0.003
Hispaniola (73,147 km^2)	5/64	11/107	10/28	0.003
Dominica (790 km^2)	2/3	6/7	4/4	0.018
Guatemala and Belize (131,852 km^2)[1]	7/77	27/75	56/119	0.002
Costa Rica (50894 km^2)[2]	21/120	30/68	59/128	0.006
La Selva (15 km^2,Costa Rica)[3]	14/49	15/27	37/47	8.2
Barro Colorado Is. (15 km^2, Panama Canal Zone)[3]	13/49	14/26	32/47	8.1
Santa Cecilia (3 km^2, Amazonian Ecuador)[3]	30/86	21/30	34/53	56.3
Manu (10 km^2, Cocha Cashu, Amazonian Perú)[3]	22/75	19/16	31/31	7.2

[1]From Campbell and Vannini (1989).
[2]From Savage and Villa (1986).
[3]From Duellman, (1990).

Drawing comparisons of species diversity between the West Indies and the Neotropical mainland is difficult since so many variables can affect the numbers (e.g., area, topography, latitude, vegetation). The larger the areas compared, the more similar the numbers. For example, Guatemala + Belize have 0.002 species/km^2, and Hispaniola has 0.003 species/km^2; Costa Rica has 0.006 species/km^2 (Table 7.1). As the areas become smaller, however, fewer opportunities for comparison exist because specific mainland sites have received intensive study and, with the exception of El Verde, Puerto Rico (Reagan and Waide, 1996), no West Indian site has been studied in the manner of, for example, Santa Cecilia (Duellman, 1978) or La Selva (Guyer, 1990). Nevertheless, at specific sites in the West Indies, considerable evidence indicates species-rich herpetocommunities. For example, up to 24 species of anurans, including 20 species of *Eleutherodactylus*, occur together in the Castillon area (ca. 1.0 km^2) on the north slope of the distal La Hotte range in Haiti, and 16 anuran species cooccur at Soroa, Pinar del Río, Cuba (Hedges, 1996c). At La Palma in the Cordillera Central of the Dominican Republic, seven species of *Anolis* coexist in a very limited area (Rand and Williams, 1969). Similarly, in the heavily disturbed courtyard of a hotel in the coastal town of Barahona (Barahona Province, Dominican Republic), Powell and colleagues (unpublished data) encountered 13 lizard species, including five species of *Anolis* (an additional six species, including two *Anolis*, could be added if the area within 5 km of Barahona was included).

In contrast, Duellman (1990), for example, reported 86 species of anurans (including 15 species of *Eleutherodactylus*), 31 lizards and amphisbaenians (but only 6 species of *Anolis*), and 53 snakes at Santa Cecilia (Upper Amazon of Ecuador); Donnelly and Guyer (1994) encountered 23 frog species at a single pond at La Selva (Costa Rica), and eight species of *Anolis* also occur there (Duellman, 1990). Surprisingly, only ten species of *Anolis* are recorded from the entire region of Amazonian Brazil (Avila-Pires, 1995).

Inventories of herpetofaunas in the West Indies indicate that, despite their richness, they often lack the diversity found on the Neotropical mainland (especially in rain forest habitat). However, these communities more than compensate for this lack of comparative species diversity with spectacular numbers of individual animals.

Distribution

Although many factors may be involved (e.g., physiography and proximity to other land masses), larger West Indian islands harbor more species of amphibians and reptiles than do smaller islands (e.g., Darlington, 1957; MacArthur and Wilson, 1967). However, few exposed land masses in the West Indies which sustain more than herb-stage vegetation (Lazell, 1972) are devoid of, at least, reptiles, and West Indian islands harbor from one to nearly 200 described species of frogs and reptiles. In examining the distributions of frogs, lizards, and snakes, we have determined that nearly 40% have multiple-island distributions (MIDs; although perhaps occupying the same island bank), but the differences between frogs and reptiles is considerable and undoubtedly reflects the superior overwater colonizing ability of reptiles. Only 15.8% of the frog species have MIDs, whereas 47.2% of the lizards and 53.6% of the snakes have MIDs. Williams (1969) suggested certain characteristics of successful colonizers. Although anoles were the focus of his discussion, the criteria apply to other taxa as well: (1) inhabitants of open forest or savanna, ecotonal; (2) not a deep shade, rain forest, or montane taxon; and (3) an ecological generalist (not a specialist). More reptilian than anuran species meet these criteria.

In another analysis of factors that affect distributions, Schoener and Schoener (1983a) conducted experiments on 521 small islands in the Bahamas and concluded that the occurrence of lizard species (especially *Anolis*) is related to: (1) island area; (2) structural vegetation diversity; (3) abundance of vegetation and, in some instances, altitude (although the latter may merely reflect a correlation between plant diversity and increasing elevation); (4) distance to the nearest main island (more than distance to the nearest island); and (5) competitive interactions. Unpublished work based on data preliminarily presented in Burns *et al.* (1992) indicates that similar factors apply to the distribution of lizards on the Cayos Siete Hermanos, a small archipelago off the north coast of Hispaniola.

Population Densities

Although the West Indies (especially the Greater Antilles) harbor a rich anuran fauna (primarily *Eleutherodactylus*, which comprises nearly 80% of the known frog species), it is a reptile-dominated vertebrate fauna (>70% of known species), and lizards especially predominate. Not only do they attain high densities [Andrews (1979) suggested that food might be the only restraint on density] but also snakes and other predators prey more frequently on lizards than on invertebrates, frogs, birds, and mammals. Table 7.2 summarizes current knowledge of the densities of West Indian frogs, lizards, and snakes. In comparison to areas outside the West Indies, population densities determined for West Indian species (especially lizards) are singularly amazing.

Table 7.2. Population Densities of West Indian Frogs, Lizards, and Snakes[a].

Species	Island	Density (No./ha.)	Reference
Anura			
Leptodactylidae			
Eleutherodactylus antillensis	Puerto Rico	400	Stewart and Woolbright (1996)
E. coqui	Puerto Rico	100-23,000	Stewart and Pough, (1983); Stewart and Woolbright (1996)
E. hedricki	Puerto Rico	200	Stewart and Woolbright (1996)
E. portoricensis	Puerto Rico	800	Stewart and Woolbright (1996)
E. richmondi	Puerto Rico	100	Stewart and Woolbright (1996)
E. wightmanae	Puerto Rico	400	Stewart and Woolbright (1996)
E. cundalli, E. gossei, E. johnstonei, E. planirostris	Jamaica	Four species combined: 4,635	Stewart and Martin (1980)
Sauria			
Gekkonidae			
Sphaerodactylus clenchi	Hispaniola	4000	Cheng (1983)
S. macrolepis	Guana, B.V.I	1600-2600	Lazell (1991)
S. vincenti	Martinique	8200	Leclair and Provencher (1988)
Polychrotidae			
Anolis aeneus	Grenada	130-1080	Roughgarden *et al.* (1983)
A. acutus	St. Croix	2000-5600	Ruibal and Philibosian (1974)
A. angusticeps	S. Bimini	2500	Schoener and Schoener (1980)
A. bimaculatus	St. Eustatius	2220-3000	Roughgarden (1995)
A. brevirostris	Hispaniola	800	Moster *et al.* (1992)

Table 7.2 cont.

A. cristatellus	Guana, B.V.I.	4000-7200	Lazell (1991)
A. cuvieri	Puerto Rico	1.33	Dial *et al.* (1994)
A. cybotes	Hispaniola	367	Fobes *et al.* (1992)
	Hispaniola	136-144	Schell *et al.* (1993)
A. distichus	Abaco	9700	Schoener and Schoener (1980)
A. evermanni	Puerto Rico	2308	Dial *et al.* (1994)
A. gingivinus	St. Martin	760-12,980	Roughgarden *et al.* (1983)
A. griseus	St. Vincent	5500	Roughgarden *et al.* (1983)
A. gundlachi	Puerto Rico	1976	Reagan (1992)
A. oculatus	Dominica	2148	Bullock and Evans (1990)
A. olssoni	Hispaniola	608-650	Smith *et al.* (1992)
A. pulchellus	Puerto Rico	Up to 20,000	Gorman and Harwood (1977)
	Guana, B.V.I.	1000-1800	Lazell (1991)
A. richardi	Grenada	1630-1800	Roughgarden *et al.* (1983)
A. sagrei	Abaco	9700	Schoener and Schoener (1980)
A. stratulus	Puerto Rico	25,870	Reagan (1992)
	Puerto Rico	2339	Dial *et al.* (1994)
	Guana, B.V.I.	3000-5400	Lazell (1991)
A. trinitatus	St. Vincent	5000	Roughgarden *et al.* (1983)
A. wattsi	Barbuda	3510	Roughgarden *et al.* (1983)
	St. Maarten	5010	
	St. Christopher	4970	
	St. Eustatius	4780-9850	Roughgarden (1985)
Tropiduridae			
Leiocephalus carinatus	Bahamas	700	Schoener *et al.* (1982)
L. schreibersii	Hispaniola	143	Schreiber *et al.* (1993)
Iguanidae			
Cyclura carinata	Pine Cay, Caicos Island	90.3 (juv.) >31.1 (adults)	Iverson (1979)

Table 7.2 cont.

C. pinguis	Anegada, B.V.I.	2.03	Carey (1975)
Scincidae			
Mabuya mabouya	Dominica	751	Bullock and Evans (1990)
Teiidae			
Ameiva chrysolaema	Hispaniola	144	Schell *et al.* (1993)
A. fuscata	Dominica	379	Bullock and Evans (1990)
Serpentes			
Typhlopidae			
Typhlops richardi	Guana, B.V.I.	Ca. 580	Lazell (1991)
Boidae			
Corallus grenadensis	Grenada	19-61	Henderson and Winstel (1992) and R. Henderson, Unpublished data
Epicrates monensis	Cayo Diablo	>100	Tolson (1988)

[a]All densities have been extrapolated to number/ha. In some cases, only the maximum density is presented for a given study, even when there was variation in densities from locality to locality.

Frogs

Relatively few data are available regarding frog densities. The highest recorded density for mainland *Eleutherodactylus* is 4586/ha for *E. stejnegarianus* in a Costa Rican cloud forest at 1200 m; *E. bransfordi* occurred at a density of 635/ha in wet forest at 100 m in Costa Rica (Scott, 1976). These data are closer to those for West Indian frog densities than are mainland densities for lizards, but still pale in comparison to >20,000 *E. coqui*/ha in Puerto Rico (Stewart and Pough, 1983; Stewart and Woolbright, 1996).

Lizards

Based on data for areas extralimital to the West Indies, the highest densities recorded are for a gekkonid (*Gonatodes antillensis*) on Bonaire (4200/ha; Bennet and Gorman, 1979), skinks (*Mabuya* spp.) in the Seychelles (up to 3600/ha; Cheke, 1984), and a temperate skink (1500/ha; Turner, 1960). An anole (*Anolis bonairensis*) on Bonaire reached 1318/ha (Bennett and Gorman, 1979). Species of *Phelsuma*, diurnal geckos that inhabit anoline niches, attain densities of 200-300/ha on the Seychelles Islands (Cheke, 1984). On Guam, Rodda *et al.* (1995) determined lizard densities of up to 20,000/ha, but this included several species and most were introduced. With one exception, all of these records are from tropical islands. More significantly, on the Neotropical mainland, anole densities are low in comparison to those in the West Indies (Table 7.3). Andrews (1979) found anoles to be two or three times more abundant at a site on Dominica than at a site in Costa Rica and also that the most common species in the West Indies reached higher densities than the most common species on the Neotropical mainland. Andrews offered possible reasons for the disparity in mainland versus island population densities (discussed later). Lieberman (1986), in comparing forest leaf litter herpetofauna versus cacao litter herpetofauna at La Selva, Costa

Table 7.3. Population Densities of *Eleutherodactylus*, *Anolis*, and *Ameiva* on the Neotropical Mainland

Species	Location	Density (No./ha)	Reference
Eleutherodactylus			
bransfordi	Costa Rica	635	Scott (1976)
fitzingeri	Panama	2060	Heatwole and Sexton (1966)
stejnegarianus	Costa Rica	4586	Scott (1976)
Anolis			
fuscoauratus	Peru	18.1	Duellman (1987)
humilis	Costa Rica	91	Scott (1976)
limifrons	Panama	1886	Andrews (1979)
nebulosus	Mexico	900	Jenssen (1970)
polylepis	Costa Rica	224	Scott (1976)
Ameiva amieva	Peru	4.6	Duellman (1987)
quadrilineata	Costa Rica	29.7-63.8	Hirth (1963)

Rica, found that the forest site contained greater species diversity but that the cacao site harbored greater densities of frogs and lizards.

Duellman (1987) presented biomass data for a lizard community at a rain forest locality in the Upper Amazon Basin of Perú. Twenty-one species accounted for 1040 g/ha. The only comparable data are those presented by Bullock and Evans (1990) for three widespread lizard species (*Anolis oculatus*, *Mabuya mabouya*, and *Ameiva fuscata*) on Dominica, where the maximum biomass occurred in coastal woodlands and was recorded as 44,700 g/ha. In lowland rain forest habitat on Dominica, the highest recorded biomass was 950 g/ha for *A. oculatus* alone; in comparison, in Amazonian Perú, *A. fuscoauratus* accounted for only 20.3 g/ha. Similarly, *Ameiva ameiva* accounted for 193.4 g/ha in Peruvian rain forest (in clearings) (Duellman, 1987), but *A. fuscata* accounted for up to 25,070 g/ha in coastal woodland habitat on Dominica (Bullock and Evans, 1990).

Snakes

Few data are available for population densities of snakes in the West Indies or on the Neotropical mainland. Scott (1976) calculated a density of 22 snakes/ha of all leaf-litter species combined in a Costa Rican wet forest, and Henderson (1974), based on mark and recapture study at mangrove edge in Belize, determined that *Oxybelis aeneus*, a slender, arboreal lizard eater, occurred at densities of 28-35/ha. At best we can provide a density index for the sake of comparison, but because the indices are based on samples in different kinds of habitat, comparisons must be viewed cautiously. Densities of vertebrate-eating snakes in the West Indies appear to be high when compared to continental data (Turner, 1977; Lillywhite and Henderson, 1993), although high snake abundance on islands outside of the West Indies has also been reported (e.g., Klauber, 1956:529). Density data are available for two arboreal boids in the West Indies: *Epicrates monensis*, a small (ca. 1.0 m SVL) saurophagous species, occurs at densities of >100/ha on a satellite island off Puerto Rico (Tolson, 1988), and *Corallus grenadensis*, a lizard- and mammal-eating species occurred at densities of 19-61/ha in agriculturally disturbed habitats on Grenada (Henderson and Winstel, 1992; R.W. Henderson, unpublished data).

Murphy *et al.* (1994) calculated rates of encounters with snakes in several tropical forest communities: two snake communities in Southeast Asia (Thailand and Borneo) yielded about 1 snake/day; two sites on the Neotropical mainland (Guyana, Ecuador) yielded 0.39 and 0.38 snakes/day; and average daily snake encounters at a site on the continental island of Trinidad was 2.3. Although comparable data for a West Indian site are lacking, encountering 5-10 snakes/hr in some habitats (not rain forest) is not unusual (but representing only one or two species), snake "markets" (i.e., where money is offered to people living in a particular area in exchange for snakes, and dozens of people may join the hunt) have often yielded over 100 vertebrate-eating snakes within a 24hr period, and 955 specimens of the secretive arthropod-eating *Typhlops pusillus* were accumulated in approxomately 36 hr (Schwartz and Henderson, 1991). Henderson and Henderson (1995) encountered up to 4.5 *C. grenadensis*/hr in xeric scrub on Grenada.

The population densities of amphibians and reptiles in the West Indies exhibit a predictable pattern. Arthropod-eating taxa for which we have some index of density (frogs, lizards, and typhlopid snakes) often occur at high densities (sometimes phenomenally so); frog and lizard predators occur at lower densities, and mammal predators occur at the lowest densities. This pattern is not unique to the West Indies, and herpetofaunas on the Neotropical mainland exhibit similar relationships between diet and relative abundance (e.g., Duellman, 1978). The available evidence indicates that, in general, population densities of amphibians and reptiles in the West Indies, in a variety of habitats but including rain forest, are often many times higher than those on the Neotropical mainland. Assuming density compensation (MacArthur *et al.*, 1972; Case, 1975; Case *et al.*, 1979; Williamson, 1981), higher population densities are anticipated in the West Indies relative to the mainland, but the numbers are nevertheless extraordinary even relative to other islands.

Trophic relationships of the herpetofauna

Excluding turtles and crocodilians (a total of 12 species), the West Indian herpetofauna is composed trophically of 11 herbivores (iguanid lizards; 1.8% of 622 species), 74 snakes that eat vertebrates (11.9%), and 537 species of frogs, lizards, amphisbaenians, and snakes that prey predominantly on arthropods (86.3%) (Table 7.4).

Trophic Generalists and Specialists

Although placing a species at one end or the other of a broad spectrum ranging from "specialist" to "generalist" is tempting, in reality most species fall somewhere in between. Such labeling of West Indian amphibians and reptiles is difficult for a suite of reasons: (1) Dietary analyses have usually been done without corresponding data on the relative abundance of prey species in the habitat; (2) we do not

Table 7.4. Diets of West Indian Frogs and Squamate Reptiles[a]

Genus	No. of Species	Distribution	Diet
Frogs			
Bufonidae			
Bufo	12	G, L	A
Dendrobatidae			
Colostethus	1	L	A
Hylidae			
Hyla	3	G	A
Osteopilus	9	B, G	A, (F)
Leptodactylidae			
Eleutherodactylus	139	B, G, L	A
Leptodactylus	5	G, L	A
Lizards			
Anguidae			
Celestus	19	G	A
Diploglossus	3	G, L	A, (L)
Gymnopthalmidae			
Bachia	1	L	A
Gymnopthalmus	2	L	A
Polychrotidae			
Anolis	138	B, G, L	(P)[b]
Iguanidae			
Ctenosaura	1	W	P,A, (L)
Cyclura	8	B, G	P,A, (L)
Iguana	2	G, L	P
Tropiduridae			
Leiocephalus	23	B, G	A, (L), (P)
Gekkonidae			
Aristelliger	6	B, G	A
Gonatodes	1	G	A
Hemidactylus	4	B, G, L	A
Phyllodactylus	2	G, L	A
Sphaerodactylus	79	B, G, L	A
Tarentola	2	B, G	A
Thecadactylus	1	G, L	A
Scincidae			
Mabuya	2	G, L	A
Teiidae			
Ameiva	19	B, G, L, W	A, (L)

Table 7.4 cont.

Cnemidophorus	2	L, W	A
Kentropyx	1	L	A
Tretioscincus	1	W	A
Xantusiidae			
Cricosaura	1	G	A
Snakes			
Boidae			
Boa	1	L	M
Corallus	1	L	L, M, (B)
Epicrates	9	B, G	L, M, (B)
Colubridae			
Alsophis	11	B, G, L	F, L, M, (B)
Antillophis	2	G	F, L
Arrhyton	12	G	F, L
Chironius	1	L	F
Clelia	2	L	S(?), M
Coniophanes	1		
Darlingtonia	1	G	F
Hypsirhynchus	1	G	L
Ialtris	3	G	L, M, S
Liophis	4	L	F, L
Mastigodryas	1	L	F, L
Nerodia	1	G	Fi
Tretanorhinus	1	G	Fi
Uromacer	3	G	F, L
Leptotyphlopidae			
Leptotyphlops	8	B, G, L	A
Tropidophiidae			
Tropidophis	13	B, G	F, L, M, (B)
Typhlopidae			
Typhlops	23	B, G, L	A
Elapidae			
Micrurus	1	W	L, S
Viperidae			
Bothrops	2	L	M

[a]Distribution: B = Bahamas, G = Greater Antilles (including Virgin Islands and Cayman Islands), L = Lesser Antilles, W = western Caribbean. Diet: A = arthropods, B = birds, F = frogs, Fi = fish, L = lizards, M = mammals, P = plants, S = snakes (letters designating prey that appear in parentheses indicate prey catagories of lesser importance). Genera that are not native to the region have been omitted.

[b]Giant species (>100 mm SVL) only (Bowersox et al., 1994b)

know specifically what most West Indian frogs and reptiles eat and although extrapolating what we know about Species A to Species B is tempting, it is also dangerous; (3) diet for a given species may not be the same from one site to the next, and a species that is a habitat "generalist" will exhibit trophic differences over its range; (4) community studies are virtually absent for the West Indian herpetofauna, and we do not know how, for example, syntopic leaf litter frogs and lizards partition trophic resources; and (5) the terms (specialist versus generalist) used can be misleading because there are different degrees of specialization. For example, the Hispaniolan endemic colubrid snake genus *Uromacer* has three recognized species: *U. catesbyi*, an active forager, *U. frenatus*, an ambush forager, and *U. oxyrhynchus*, another ambush forager. *Uromacer catesbyi* undergoes an ontogenetic shift in diet from primarily lizards (*Anolis*) to primarily hylid frogs (*Osteopilus dominicensis*); *U. frenatus* undergoes an ontogenetic shift from primarily scansorial lizards (*Anolis*) to a high percentage of ground-dwelling lizards (*Ameiva*); *U. oxyrhynchus* eats primarily *Anolis* throughout its life. Of the three species, *U. catesbyi* exhibits the most taxonomically dramatic diet shift (lizards to frogs), *U. frenatus* eats only lizards (a specialist?), but a variety of lizards (a lizard generalist?), and *U. oxyrhynchus* is an *Anolis* specialist. But even this oversimplifies the situation. For example, *U. frenatus* occurs in xeric areas where *Ameiva* occurs at high densities and numbers of *Anolis* are reduced, so its diet might reflect prey availability more than anything else (Henderson *et al.*, 1987c, 1988a). Lieberman (1986:56), in describing a Costa Rican leaf litter herpetofauna, concluded that "all species ate the arthropods in proportions significantly different from those found in the can traps...In a strict sense, none of the herpetofauna species are absolute dietary generalists. Each species has its own position along a continuum from absolute specialist to absolute generalist."

Frogs

Few diet analyses have been conducted for West Indian anurans, and the species most often studied to date have been leptodactylids. The available information, including purely anecdotal data, represent only 12 species (8.7% of the West Indian *Eleutherodactylus*). Jones' (1982) Puerto Rican study indicated that species of *Eleutherodactylus* are opportunistic, taking arthropod prey (including a high percentage of ants) in about the same proportions that they occur in the habitat. However, a disproportionate number of large prey items were taken. Similar types of studies on the Neotropical mainland indicate that

Table 7.5. Foraging and Diet Charactersitics of Some West Indian Frogs and Lizards and Their Relationship to Island Size[a].

Genus/ Island area (km^2)	Foraging Mode	Adaptive Zone	Most Frequently Consumed Prey	Reference
Frogs				
Bufonidae				
Bufo	S	G		
111,463			Hymenoptera (Formicidae)	Sampedro Marin and Fundora (1982)
Hylidae				
Osteopilus	S	S		
111,463			Isopoda, [Orthoptera]	Meshaka (1996)
76,193			Coleoptera	Duer et a.l. (1992)
207			Diptera [Orthoptera]	Meshaka (1996)
185			Orthoptera	Meshaka (1996)
Leptodactylidae				
Eleutherodactylus	S	LG		
76,193			Hymenoptera (Formicidae), [Coleoptera], [Orthoptera]	Howard *et al.* (unpublished manuscript)
76,193			Hymenoptera (Formicidae)	Howard *et al.* (unpublished manuscript)
11,526			Hymenoptera (Formicidae)	Schwartz and Henderson (1991)
11,526			Hymenoptera (Formicidae)	Schwartz and Henderson (1991)
11,526			Hymenoptera (Formicidae)	Schwartz and Henderson (1991)
11,526			Hymenoptera (Formicidae)	Schwartz and Henderson (1991)
11,526			Hymenoptera (Formicidae)	Schwartz and Henderson (1991)
8,865			Homoptera	Jones (1982)
49			Hymenoptera (Formicidae)	Jones (1982)
49			Areanae	Jones (1982)

Table 7.5 cont.

Leptodactylus	S	G		
751			Orthoptera	Brooks (1982)
Lizards				
Gekkonidae				
Aristelliger	S	A		
<0.15			Coleoptera [Orthoptera]	Burns *et al.* (1992)
Sphaerodactylus	A	G		
76,193			Isoptera	Cunningham *et al.* (1993)
76,193			Coleoptera	Cunningham *et al.* (1993)
8,865			Acari, [Araneida], [Collembola], [Isopoda]	Thomas and Kessler (1996)
8,865			Collembola	Thomas and Kessler (1996)
Polychrotidae				
Anolis ("typical")	S	G, A		
<0.15			Hymenoptera (Formicidae)	Burns *et al.* (1992)
<0.15			Hymenoptera (Formicidae)	Burns *et al.* (1992)
ca. 10			Hymenoptera (Formicidae)	Schoener (1968)
ca. 10			Homoptera [Diptera]	Schoener (1968)
ca. 10			Diptera	Schoener (1968)
ca. 10			Diptera, [Hymenoptera (Formicidae)]	Schoener (1968)
20			Hymenoptera (Formicidae)	Roughgarden (1995)
20			Lepidoptera larva [Hymenoptera (Formicidae)]	Roughgarden (1995)
311			Hymenoptera (Formicidae)	Schoener and Gorman (1968)
751			Hymenoptera (Formicidae)	Andrews (1979)

Table 7.5 cont.

1,102			Hymenoptera (Formicidae)	Schoener and Gorman (1968)
11,424			Hymenoptera (Formicidae)	Rand (1976b)
76,193			Hymenoptera (Formicidae)	Fobes *et al.* (1992)
76,193			Hymenoptera (Formicidae)	Lenart *et al.* (1994)
76,193			Hymenoptera (Formicidae)	Lenart *et al.* (1997)
Anolis ("giant")	A, S	A		
76,193			Lepidoptera (larvae)	Bowersox *et al.* (1994b)
311			Hymenoptera (Formicidae)	Schoener and Gorman (1968)
Tropiduridae				
Leiocephalus	S	L, G		
<0.15			Hymenoptera (Formicidae)	Burns *et al.* (1992)
<0.15			Hymenoptera (Formicidae)	Burns *et al.* (1992)
76			Hemiptera, [Isopoda]	Schoener *et al.* (1982)
1,544			Lepidoptera (larvae), [Plants], [Coleoptera]	Schoener *et al.* (1982)
76,193			Hymenoptera (Formicidae)	Schreiber *et al.* (1993)
76,193			Hymenoptera (Formicidae), [Coleoptera]	Lahey *et al.* (unpublished manuscript)
Anguidae				
Celestus	A	G		
76,193			Coleoptera, [Orthoptera]	White *et al.* (1992)
76,193			Dictyoptera, [Lepidoptera]	Cisek *et al.* (1990)
76,193			Diplopoda	Inchaústegui *et al.* (1985)
Diploglossus	A	G		

Table 7.5 cont.

8,865			[Dermaptera]	Thomas and Kessler (1996)
Teiidae				
Ameiva	A	G		
<0.15			Dictyoptera	Burns *et al.* (1992)
91			Coleoptera	Censky (1996)
751			Oligochaeta, [Coleoptera]	Bullock and Jury (1990)
8,865			Gastropoda	Lewis (1989)
76,193			Dictyoptera, [Diptera]	Powell *et al.* (1989)
76,193			Isoptera, [Coleoptera larvae and adults]	Sproston *et al.* (unpublished manuscript)
76,193			Isoptera, [Coleoptera]	Sproston *et al.* (unpublished manuscript)v
76,193			Isoptera, [Orthoptera], [Arachnida]	Sproston *et al.* (unpublished manuscript)

[a]Under "Foraging Mode" A = active and S = sit-and-wait; under "Adaptive Zone" A = arboreal/scansorial, G = ground-dwelling, and LG = largely ground-dwelling. Usually only the orders that contribute the most prey items are listed. Prey orders in brackets indicate that they occurred nearly as frequently as the order listed first.

considerable interspecific variation exists in *Eleutherodactylus* diets (Lieberman, 1986). The data addressing opportunism are equivocal. According to Toft (1980), a disproportionate number of large prey items were taken by Panamanian species, which are characterized as being in a "non-ant specialist" guild. In contrast, Lieberman's (1986) Costa Rican study indicated that species of *Eleutherodactylus* took a wide range of prey sizes, consumed ants in proportion to their occurrence in the leaf litter, and should be considered generalists.

Andrews (1979), in discussing diet differences between mainland and West Indian anoles, suggested that ants in the diet of West Indian species of *Anolis* was an indication that island anoles were food limited relative to their mainland counterparts (but see below). This hypothesis could also be applied to *Eleutherodactylus*, but an alternative hypothesis is that West Indian frogs and reptiles are, in general, eurytrophic, their trophic plasticity contributes to their high densities, and they are eating those prey items (e.g., ants) which they most frequently encounter. In a Puerto Rican study of *Eleutherodactylus coqui*, Woolbright and Stewart (1987) found that prey size was correlated with frog size (large frogs ate one or two large prey items, whereas smaller frogs ate several small items), and that foraging success was reduced during the dry season. In contrast, Lieberman (1986) found that arthropod abundance peaked at the end of the dry season at La Selva, Costa Rica (forest and cacao), suggesting that foraging success for leaf litter herpetofauna (including several species of *Eleutherodactylus*) would also peak.

In contrast to *Eleutherodactylus*, members of the endemic hylid genus *Osteopilus* (*dominicensis* and *septentrionalis*) prey often on coleopterans and orthopterans and these orders contributed substantially to the total volume of prey (Table 7.5); ants are virtually absent from their diets (Duer *et al.*, 1992; Meshaka, 1996). Sit-and-wait foragers in the bufonid genus *Bufo* [formerly *Peltophryne*; *guentheri* (Parsons, 1995) *lemur* (Schmidt, 1928), *longinasus* (Valdés de la Osa and Ruiz García, 1980), *peltocephalus* (Sampedro Marín and Torres Fundora, 1982)] also include a large number of ants in their diets (up to 84% by frequency of occurrence). Similarly, on the mainland, the diet of *Bufo haematiticus* at La Selva (Costa Rica) included a high proportion of ants (88%) in its diet, as did *Dendrobates pumilio* (63%) and *Gastrophryne pictiventris* (85%).

Lizards

Not surprisingly, most of what we know about the diet of West Indian lizards is based on Anolis. Andrews (1979) compared sites on the Central American mainland and on Dominica. She found that: Small items, particularly ants, are relatively more abundant in island than mainland habitats and more abundant in the diets of island than mainland anoles as well. The large numbers of prey in the stomachs of island anoles is a necessary consequence of small average prey sizes. Thus, a comparison of diets suggest that while mainland anoles have the "option" of feeding on large, high-yield prey items, island anoles do not.

She concluded that indirect evidence indicated "that anoles on islands are food limited relative to anoles on the mainland." Contrary to Andrews' conclusions, work by Guyer (1988a,b) on the Neotropical mainland has shown that food does indeed limit anole (*A. humilis*) population density: the numbers of adult males and females were observed to increase in the presence of supplemental food.

Subsequent work has illustrated further the importance of ants in the diets of West Indian anoles (e.g., Floyd and Jenssen, 1983; Burns et al., 1992; Table 7.5). Within the anoline fauna, exceptions to this rule are the so-called giant species (SVL > 100 mm), which prey most often on lepidopteran larvae, coleopterans, orthopterans, and fruit (Andrews, 1979; Bowersox et al., 1994b). We are uncertain why ants represent an inferior food source (Andrews, 1979). Although we know little or nothing specific about most Neotropical frog and lizard diets (mainland and insular faunas), except that most feed predominantly on arthropods, we do know that ants are the most ubiquitous and abundant (numbers and biomass) arthropods in the Neotropics (Wilson, 1991). If most West Indian species are opportunistic predators, predation on the most frequently encountered prey species is expected. Since frog and lizard species in the West Indies are capable of consuming up to 50,000-340,000 prey items/ha/day (Reagan, 1996), an abundant, presumably energetically low-cost food source seems a necessary prerequisite for maintenance of the highest frog and lizard population densities in the world.

When the diets of nonanoline, arthropod-eating lizards are examined (Table 7.5), a different picture emerges. Ants, although often included, do not comprise a significant proportion of the diets. Predation on ants may be a function of foraging strategy. Anoles are typically sit-and-wait foragers (a known exception being the active forager *A. valencienni*; Hicks and Trivers, 1983), as are the other conspicuous ant predators in the West Indies (members of the tropidurid genus *Leiocephalus*). Those species that do not include ants in their diets, or at least do not include them in large numbers, are active foragers, highly arboreal, or are largely nocturnal. Vitt and de Carvalho (1995) found that in the lavrado area of northern Brazil, lizards exhibit a similar, but not unexpected, dichotomy: sit-and-wait species (an *Anolis* and a tropidurid) feed on mobile invertebrates (including many ants), whereas active foragers (teiids) take prey that is predominantly inactive, clustered, and which includes relatively few ants. Since this is a reflection of lizard foraging mode and degree of prey mobility, that the same pattern prevails in the West Indies is not surprising.

Floyd and Jenssen (1983) suggested that "high availability and ease of capture are primary factors influencing the lizards' [*Anolis opalinus* on Jamaica] choice of food." Work by Schoener (1968) and Schoener et al. (1982) has indicated that island size (food per unit area is more abundant on larger islands and there is a possible absence of large insects on small islands) and the presence or absence of congenerics can influence dietary composition and breadth. In closing, we contend that by eating the most abundant food (usually ants), lizard foraging "territories" require less space, thereby increasing the potential number of lizards that can occupy a particular area or volume of habitat. The opportunistic foraging of many West Indian frogs, lizards, and snakes is probably in large part responsible for the high population densities achieved by many species.

Snakes

Of the 74 known vertebrate-eating snake species in the West Indies, only two (*Chironius vincenti* and *Darlingtonia haetiana*) are frog specialists (Henderson and Schwartz, 1986; Henderson and Haas, 1993), three prey exclusively on lizards (but will accept other vertebrates in captivity), about 60 take frogs and lizards, none are bird specialists although large boids (e.g., *Epicrates striatus*, *Corallus* spp.) and colubrids (e.g., *Alsophis cantherigerus*) do occasionally include them in their diets, and eight or nine prey frequently (but not exclusively) on mammals. Figure 7.2 presents a partial food web focusing on Cuban populations of *Alsophis cantherigerus* and illustrates trophic relationships between members of a herpetofauna. But even with the few components included, the web in Figure 7.2 is oversimplified; *Osteopilus* will eat conspecifics and *Anolis* (Meshaka, 1996), and *Anolis* will eat other *Anolis*, etc.

Among colubrid snakes, Henderson and Crother (1989) found that of 707 prey items, 74.9% were lizards, 20.8% frogs, 1.7% snakes, 1.6% mammals, 0.4% fishes, and 0.3% birds; combined, frogs and lizards accounted for 95.7% of the prey of colubrids compared to only 1.9% for birds and mammals. Invertebrates were virtually absent from the diets of West Indian colubrids, whereas invertebrate predation is common in some (but not all; Vitt and Vangilder, 1983) communities on the Neotropical mainland. Predation on invertebrates by New World colubrids is phylogenetically restricted and species or groups that prey largely or exclusively on invertebrates (e.g., *Atractus*, *Dipsas*, *Geophis*) do not occur in the West Indies (Cadle and Greene, 1993).

At the generic level, *Anolis* comprised 56.8% of the prey items, *Eleutherodactylus* 13.5%, *Ameiva* 9.5%, *Osteopilus* 7.2%, and *Leiocephalus* 4.4%. Among lizards, 75.8% were anoles (and anoles comprise ca. 41% of the lizard species in the West Indies), and among the frogs, 64.6% were *Eleutherodactylus* (which comprise about 79% of the frog species). Although these data would be more meaningful if we had corresponding data on encounter rates for anoles and other

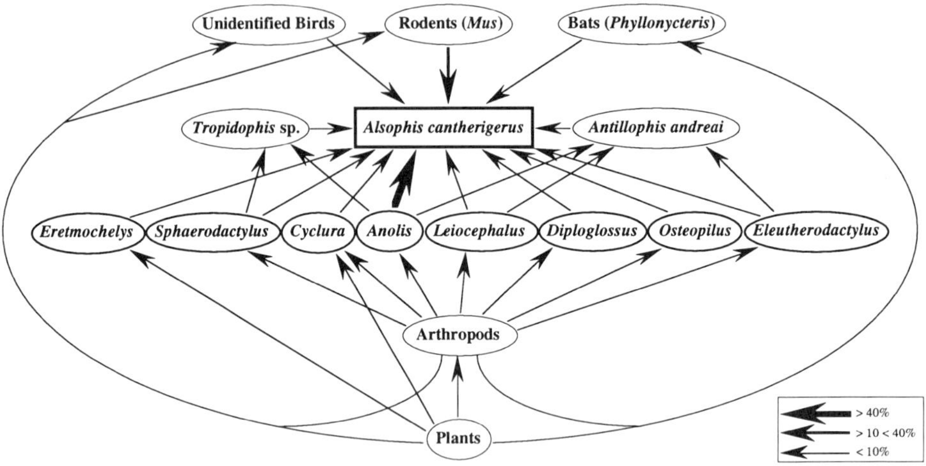

Figure 7.2. A graphic representation of trophic relationships with Cuban populations of the colubrid snake *Alsophis cantherigerus* as the focal species. Width of the arrows indicates percentage each prey group contributed to the diet of *A. cantherigerus*. No attempt was made to quantify the contributions of plants and arthropods to any other group (i.e., arrow thickness has no significance). The diet of *A. cantherigerus* was based on the examination of stomach contents of preserved specimens (Henderson and Sajdak, 1996).

lizards in the West Indies, considering that the colubrid fauna is composed of diurnal foragers and that anoles are diurnal, ubiquitous (occur from ground level to the crowns of trees in a wide range of habitats), and attain extremely high population densities, the high rate of predation on anoles is hardly a surprise. Those snake species that prey on lizards other than *Anolis* tend to be either ground-dwelling (e.g., *Hypsirhynchus ferox*) or ambush foragers (e.g., *H. ferox*, *Uromacer frenatus*), and all exhibit an ontogenetic shift in diet from anoles to nonanoline lizards.

Predation on *Anolis* and, to a considerably lesser extent, *Eleutherodactylus*, was widespread across geographic boundaries in the West Indies; all major islands and island groups supported colubrid snake faunas that preyed predominantly on anoles. Figures 7.3 and 7.4 illustrate the importance of anoles in the diets of West Indian colubrids despite different foraging modes and adaptive zones (Fig. 7.3) and regardless of island size (Fig. 7.4). The latter figure also illustrates how geography affects trophic niche breadths. Similarly, Rodríguez-Robles and Greene (1996) determined that in the Greater Antilles small (<800 mm SVL) and medium-sized (> 800 but < 1200

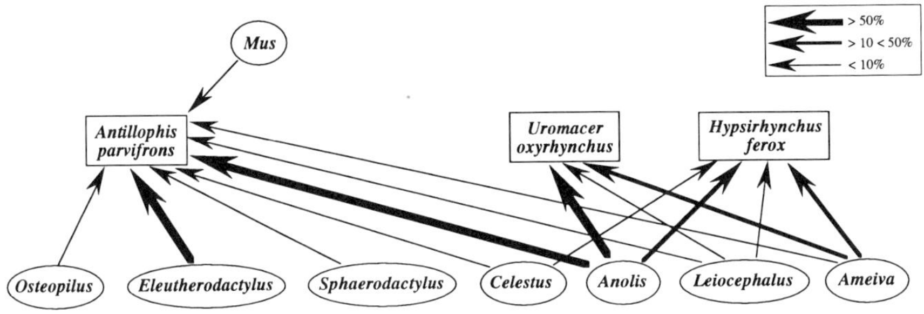

Figure 7.3. Regardless of foraging mode or adaptive zone, anoles are the primary prey of colubrid snakes in the West Indies. *Antillophis parvifrons* is a ground-dwelling active forager (data from Henderson et al., 1987a), *Uromacer oxyrhynchus* is an arboreal sit-and-wait strategist (data from Henderson et al., 1987b), and *Hypsirhynchus ferox* is a sit-and-wait ground-dweller (data from Henderson, 1984). All three species are diurnal, all are endemic to Hispaniola, and all occur sympatrically (and probably syntopically). Arrow width indicates percentage members of each genus contribute to the diet of the three snake species based on examination of stomach contents of preserved specimens.

mm SVL) snakes prey mainly or entirely on frogs and especially lizards, and that most large (> 1200 mm SVL) species eat primarily mammals. Among colubrids on the four largest islands in the Greater Antilles, species richness is positively correlated with prey species richness, and no island supports more than one large species that preys predominantly on mammals (i.e., *Epicrates* spp.).

Henderson and Crother (1989) noted,

> On the neotropical mainland, anoles are not nearly as conspicuous, do not occur at such high densities, do not comprise such a major portion of the lizard fauna, and do not contribute as much to the diets of most snake species as they do in the Antilles.

Using data gleaned from Duellman (1978) and Dixon and Soini (1986), of 158 prey items of colubrid snakes, 48.1% were frogs, 21.5% were lizards, and 13.3% were birds and mammals combined; only 3.3% of the prey items were anoles and they comprised only 14.7% of the lizard prey (and 18-21% of the lizard species in the faunas). Similarly, Greene (1988) and Braker and Greene (1994) have noted the high incidence of frogs in the diets of snakes in Costa Rica ("about half of the species feed largely or entirely on frogs" at La Selva; Braker and Greene, 1994). Henderson (1993a,b) described the high incidence of lizards (almost exclusively *Anolis*) in the diets of

West Indian species of the arboreal boid genus Corallus, compared to the Neotropical mainland where lizards are nearly absent from tree

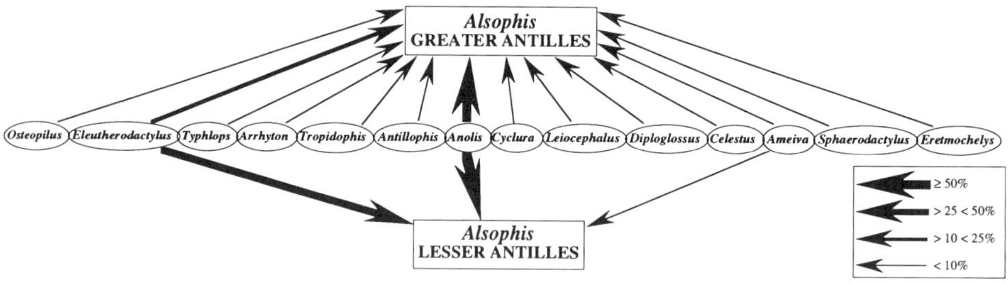

Figure 7.4. Although all West Indian species in the colubrid snake genus *Alsophis* prey heavily on *Anolis*, species occuring in the prey-rich Greater Antilles take a wide variety of frogs and reptiles. Lesser Antillean *Alsophis* are more limited in exploitable frogs and reptiles. Width of the arrow indicates percentage contribution members of each prey genus made to either Greater or Lesser Antillean *Alsophis* diets (based on examination of stomach contents of preserved specimens; Henderson and Sajdak, 1996).

boa diets. Phylogenetic analyses by Rodríguez-Robles and Greene (1996) suggested that size reduction occurred three times within *Epicrates*, once in mainland *E. cenchria* and twice within the Antillean clade. Their hypothesis "that small size evolved in Caribbean *Epicrates* in response to a shift in adult diet to smaller lizard prey" was supported by their analysis.

Trophically, anoles are the most widely exploited vertebrates in the West Indies. Fitch and Henderson (1987) noted that, for *Anolis bahorucoensis*, "every individual...is constantly in danger, from several directions simultaneously, and can survive only by instantaneous escape responses that demand priority over every other kind of activity." The same may well apply to most other species of *Anolis* in the West Indies [with the possible exception of the giant (>100 mm SVL) species], and Henderson and Crother (1989) suggested that members of no other vertebrate-eating snake fauna of similar diversity prey to such a large extent on members of a single genus of prey organisms anywhere else in the world.

Overall, we know more about the diets of a higher percentage of colubrid snakes in the West Indies than either frogs or lizards. Although several species might be labeled specialists (*Darlingtonia haetiana*, *Uromacer oxyrhynchus*), we feel that most are best described as opportunistic. Anoles account for more than 50% (by frequency of occurrence) of the prey items in the diets of most West Indian colubrids (Henderson *et al.*, 1988a; Henderson and Crother, 1989; Henderson and Sajdak, 1996) and they also make substantial contributions to the diets of boids and tropidophiids (Henderson *et al.*, 1987a; Henderson and Crother 1989; Henderson, 1993a). Colubrids that prey primarily or exclusively on invertebrates (the "Goo-eaters" of Cadle and Greene, 1993) are absent from the West Indies, as are all trophically "specialized" mainland snake genera. But despite the high incidence of anoles in the diets of West Indian vertebrate-eating snakes, we do not consider these snakes anole specialists. Species of *Anolis* are, more often than not, the dominant vertebrate species at a given site in the West Indies; this is not true on the mainland (e.g., Duellman, 1987). In 1991, at a disturbed site at Barahona, R. Powell (unpublished data) found about 135 *Anolis* to 19 *Ameiva* and 18 *Leiocephalus*. On Dominica, Bullock and Evans found the maximum density of *Anolis oculatus* at a coastal woodland site to be 25 times greater than maximum density for *Ameiva fuscata*. Conversely, the greatest difference wherein *Ameiva* was the dominant species was 2.1 times greater than for *Anolis*. At a xeric site on Hispaniola where the ground-dwelling, *Ameiva*- and *Leiocephalus*-eating colubrid *Hypsirhynchus ferox* occurred, Powell *et al.* (1996b) encountered 287 *Ameiva*+*Leiocephalus* compared to 107 *Anolis* (2.6 times more primarily ground-dwelling lizards than scansorial anoles). We believe, then, that West Indian vertebrate-eating snakes are exploiting those prey most frequently encountered, based on their adaptive zone (e.g., arboreal or ground-dwelling), foraging mode, and habitat.

Predators of Frogs and Reptiles

Based on published accounts and personal observations, we tabulated a list of over 100 predators of frogs and reptiles in the West Indies, and it is surely far from complete. Nearly 80% of these predators are other frogs and reptiles (Table 7.6). Virtually all the predation on frogs and reptiles by frogs and lizards is infrequent and opportunistic. Species of West Indian frogs and most reptiles that are primarily arthropod predators, but will prey occasionally on vertebrates, include: *Leptodactylus fallax* (Brooks, 1982; Lescure, 1979), *Osteopilus dominicensis* (Duer *et al.*, 1992), *O. septentrionalis* (Meshaka, 1996), *Hemidactylus haetianus* (Bowersox *et al.*, 1994a), *Anolis chlorocyanus* (Bowersox *et al.*, 1994a), *A. cybotes* (Fitch and Henderson, 1987; Lenart *et al.*, 1996), *A. vermiculatus* (González-

Table 7.6. Known Predators (invertebrates, anurans, etc.) of West Indian Frogs and Reptiles (listed by genus) Determined by Stomach Content Analysis or Field Observation.

	Predators						
Genus	Inv.	Anu.	Liz.	Sna.	Cro.	Bir.	Mam.
Anura							
Bufo	X			X			X
Eleutherodactylus	X	X	X	X		X	X
Hyla				X			
Osteopilus		X	X			X	
Sauria							
Hemidactylus				X			
Sphaerodactylus	X		X	X		X	
Anolis	X	X	X	X		X	X
Leiocephalus			X	X			
Cyclura			X	X		X	X
Iguana				X		X	X
Mabuya				X		X	
Celestus				X			
Diploglossus				X			
Ameiva				X		X	X
Testudines							
Trachemys					X		
Amphisbaenia							
Amphisbaena	X			X			
Serpentes							
Typhlops				X		X	
Boa		X					
Corallus							X
Tropidophis				X		X	
Alsophis				X		X	X
Antillophis				X		X	
Arrhyton				X			
Ialtris						X	
Liophis						X	
Uromacer						X	

Inv., Invertebrates; Anu., anurans; Liz., lizards; Sna., snakes; Cro., crcodilians; Bir., birds; Mam.,

Bermuda and Rodríguez-Schettino, 1982), *Leiocephalus barahonensis* (Lahey *et al.*, 1998). *L. schreibersii* (Jenssen *et al.*, 1989), *Cyclura carinata* (Iverson, 1979), *Ameiva chrysolaema* (Sproston *et al.*, 1998), and small *Crocodylus acutus* (Thorbjarnarson, 1988). Snakes and birds represent the most significant predators of frogs and reptiles, although predation by birds may vary from island to island (Wunderle, 1981; Adolph and Roughgarden, 1983, McLaughlin and Roughgarden, 1989). In any case, the list of bird species is certainly inadequate in providing an accurate picture of avian predation pressure. More complete analyses of trophic relationships in West Indian faunas (not restricted to the herpetofauna) appear in Goldwasser and Roughgarden (1993), Roughgarden (1995), Reagan (1996), Reagan *et al.* (1996), and Stewart and Woolbright (1996). All, however, reinforce the importance of anoles in West Indian food webs and analysis of a rain forest community on Puerto Rico indicated that frogs and lizards dominate nocturnal (*Eleutherodactylus*) and diurnal (*Anolis*) webs (Reagan, 1996; Reagan *et al.*, 1996).

Apparently, it is a widely held belief that West Indian anoles receive little predation pressure relative to mainland lizards (McLaughlin and Roughgarden, 1989; Roughgarden, 1995), and Andrews (1979), in discussing the high densities of anoles in the West Indies compared to the Neotropical mainland, commented that island anoles have fewer predators to contend with, noting that fewer snake species occurred at island sites. Certainly this is true, but this does not mean that fewer snakes are present. Almost all West Indian snakes, with the exception of scolecophidians, prey on anoles at some time in

Table 7.6 cont.

mammals.

Note: Preadators are listed by species as follows: **Invertebrates**: ant (*Crematogaster brevispinosa*), spiders (*Olios antiguensis*, *Stasnia portoricensis*, *Oligoctenus ottleyi*, and *Avicularia laeta*), amblypygid (*Phrynus longipes*), scorpion (*Tityus obtusus*), centipede (*Scolopendra alternans*), crab (*Epilobocera situatifrons*); **Anurans**: *Eleutherodactylus coqui*, *E. wightmanae*, *Leptodactylus fallax*, *Osteopilus dominicensis*, *O. septentrionalis*; **Lizards**: *Anolis coelestinus*, *A. cristatellus*, *A. cybotes*, *A equestris*, *A. grahami*, *A. gundlachi*, *A. lineatopus*, *A. richardi*, *A stratulus*, *A. vermiculatus*, *Leiocephalus carinatus*, *Cyclura carinata*, *Ameiva exsul*, *A. plei*, *A. taeniura*, *Sphaerodactylus klauberi*; **Snakes**: virtually all boids, tropidophiids, and colubrids (about 70 species); **Crocodilians**: *Crocodylus acutus*, *C. rhombifera*; **Birds**: *Buteo jamaicensis*, *B. platypterus*, *B. ridgewayi*, *Falco sparverius*, *Tyto alba*, *Otus nudipes*, *Egretta thula*, *Florida caerulea*, *Bubulcus ibis*, *Coccyzus minor*, *Saurothera merlini*, *S. vieilloti*, *Crotophagous ani*, *Tyrannus dominicensis*, *Melanerpes portoricensis*, *Myiarchus tyrannulus*, *Mimus gilvus*, *Todus mexicanus*, *Turdus nudigenis*, *T. plumbeus*, *Margarops fuscatus*, *Ferminia cerverai*, *Quiscalus lugubris*; **Mammal**: *Didelphis marsupialis*, *Rattus rattus*, *Felis catus*, *Canis familiaris*, *Herpestes javanicus*.

their life history, and many prey predominantly or exclusively on anoles throughout their lives. In addition, some snakes, especially small colubrids and boids, may occur in very high densities in the West Indies and be the major predators of frogs and lizards. Similarly, although islands harbor fewer potential prey species for snakes, it does not mean that they harbor fewer potential prey items (Henderson and Bourgeois, 1993; Henderson and Sajdak, 1996).

Versatility

A herpetofauna derived largely from successful colonizers might be expected to exhibit the colonizing characteristics of its ancestors (Williams, 1969; see also Distribution). Because successful colonizers must be ecological generalists capable of surviving, if not thriving, in varied habitats, the question arises whether or not West Indian amphibians and reptiles include more species exploiting a broad spectrum of ecological variables than members of the mainland Neotropical herpetofauna.

Such data are lacking for most of the West Indies and island-mainland comparisons are difficult to make (e.g., assessment of vegetative zones or communities on islands and on the mainland may not have been comparably evaluated). Nevertheless, comparisons of distributions of Greater Antillean amphibians and reptiles with those of the mainland may be instructive. Unfortunately, the only comparable information exists solely for one major taxonomic group (snakes) in only one area (Honduras) (Wilson and Meyer, 1985). When we compared the distributions of Honduran snakes relative to physiographic regions and Holdridge's life zones with those of Hispaniola (Tables 7.7 and 7.8), however, no obvious pattern emerged.

The two snake faunas differ substantially. That of Honduras consists of seven families and 95 species (two were omitted from the following comparisons, see caption to Table 7.7), only two of which are scolecophidians. In contrast, the snakes of Hispaniola represent only five families (elapids and viperids are absent) and 28 species, but 13 are scolecophidian arthropod eaters. Nevertheless, only 12 Honduran snakes (13% of the snake fauna) occupy two-thirds (five or all six) of the physiographic regions identified by Wilson and Meyer (1985) and only four (4%) are found in all regions. On Hispaniola, with its varied topography and 18 identified physiographic regions (see caption to Table 7.7), six species (21%) occupy two-thirds or more (>12) of the regions considered and one species (4%) is found in

Table 7.7. Numbers and Percentages (in parentheses) of Honduran and Hispaniolan Snake Species Relative to the Number of Physiographic Regions in Which They Occur[a].

	Number of Physiographic Regions								
Location	1	2	3	4	5	6	7	8	9
Honduras	25	24	18	14	8	4	–	–	–
	(27)	(26)	(19)	(15)	(9)	(4)			
Hispaniola	7	5	3	0	2	1	1	1	1
	(25)	(18)	(11)		(7)	(4)	(4)	(4)	(4)

Location	10	11	12	13	14	15	16	17	18
Honduras	–	–	–	–	–	–	–	–	–
Hispaniola	0	1	0	0	1	1	3	0	1
		(4)			(4)	(4)	(11)		(4)

[a]The Honduran species *Micrurus ruatanus* and *Pelamis platurus* are excluded from consideration because the distribution of the former is restricted to a satellite island and the latter is marine. The six physiographic provinces of Honduras are those identified by Wilson and Meyer (1985). The 18 Hispaniolan regions are those of Henderson and Schwartz (1984) with subdivided areas combined and all satellite islands except Ile de la Gonâve excluded.

all 18. However, at the other extreme, 49 Honduran snakes (53%) are found in one-third or fewer (2 or less) regions and 25 (27%) are found in only one, whereas 18 Hispaniolan species (64%) are limited to one-third or fewer (6 or less) regions and seven species (25%) are restricted to one region. When distribution is examined relative to vegetative life zones (Table 7.8), differences are even more equivocal. Only five Honduran snakes (5%) occur in more than two-thirds (>6) of the eight life zones identified by Wilson and Meyer (1985) and none is found in all eight. Forty-two species (45%) are found in less than one-third (<2) of the life zones, with 24 (26%) restricted to one, but another 28 species (30%) are found in three life zones. On Hispaniola, only one species (4%) lives in more than two-thirds (7 or more) of nine life zones and 18 (64%) occupy one-third (3) or fewer life zones, with seven species (25%) restricted to only one zone.

Although these data suggest that a relatively few more Hispaniolan snakes occupy a greater variety of physiographic regions than do Honduran snakes, this does not hold true when considering distributions relative to vegetative life zones, and the numbers of specialists (those species occupying few physiographic provinces or

Table 7.8. Numbers and Percentages (in parentheses) of Honduran, Cuban, and Hispaniolan Snake Species Relative to the Number of Physiographic Regions in Which They Occur[a].

	Number of Physiographic Regions								
Location	1	2	3	4	5	6	7	8	9
Honduras	24	18	28	15	3	4	1	0	–
	(26)	(19)	(30)	(16)	(3)	(4)	(1)		
Cuba	5	5	3	1	0	1	1	3	4
	(22)	(22)	(13)	(4)		(4)	(4)	(13)	(17)
Hispaniola	7	4	7	4	3	2	1	0	0
	(25)	(14)	(25)	(14)	(11)	(7)	(4)		

[a]The Honduran species *Micrurus ruatanus* and *Pelamis platurus* are excluded from consideration because the distribution of the former is restricted to a satellite island and the latter is marine. The eight vegetational formations of Honduras are those identified by Wilson and Meyer (1985). The "ecogeographic" distribution of Cuban snakes is from Rodriguez (1993). The nine Hispaniolan vegetational zones are taken from maps of Haiti and the Dominican Republic produced by the Organización de Estados Americanos.

vegetative life zones) are comparable. Nor does the vastly greater number of scolecophidians on Hispaniola substantially alter the results; the 13 species include some with restricted and some with more varied distributions in about equal numbers to those of macrostomatan snakes (although no leptotyphlopid or typhlopid snakes are included among the most versatile species).

Using data from Rodríguez-Schettino (1993; see also Rodríguez-Schettino, 1986 and Estrada, 1993b]) on the distribution of snakes in Cuba relative to "ecogeographic" zones, somewhat different results emerge (Table 7.8). We used only data from nine zones on "mainland" Cuba, disregarding snake distributions from Isla de la Juventud and offshore archipelagos. The percentages of snakes occupying one or two zones are comparable to those of Honduras and Hispaniola, but 17% of the snake fauna (4 species) occupied all nine zones in Cuba, whereas no snakes did this on Hispaniola or in Honduras. Thirteen Cuban species (56.5%) occur in one-third of the zones, and nine species (39%) occur in two-thirds. The most widespread Cuban species fell into one of two categories: (1) species belonging to genera represented by only one species (*Epicrates*, *Alsophis*, *Antillophis*, *Tretanorhinus*), or (2) species belonging to genera represented by more than one species wherein one species was widespread and congenerics were geographically (and ecologically) restricted (*Typhlops*, *Tropidophis*, *Arrhyton*). The results of our mainland-island comparisons for the sake of determining versatility of a fauna remain equivocal, but we suggest that further investigation of this topic would be fruitful.

In addition to the possibility that a large number of ecological generalists among West Indian species might be a reflection of their ancestral status as good colonizers, the versatility of these amphibians and reptiles is an important and intriguing consideration in lieu of historical and recent responses of the herpetofauna to human influences. To fully understand the ecology of many organisms, and this may be especially true of vertebrates on islands, the impact of human history must be taken into account (Pregill, 1986; Pregill *et al.*, 1994). As on the Yucatán Peninsula (Lee, 1996), the historical introduction of predators and competitors and habitat modifications in the West Indies (Henderson *et al.*, 1996) have ramifications for species abundance and distribution and may result in extirpation or extinction (Iverson, 1978; Henderson, 1992; Hedges, 1993; Kaiser and Henderson, 1994).

Henderson (1992) noted that the herpetofauna of the Lesser Antilles is especially versatile and appears to be tolerant of habitat degradation (notable exceptions include forest-adapted *Eleutherodactylus*, *Diploglossus montiserrati*, and *Chironius vincenti*), and to date no species has been lost to habitat destruction. Similarly, Hedges (1993) concluded that it was premature to declare any frog species in the West Indies extinct, although several species have not been collected in recent years (but see Joglar and Burrowes, 1996). Hedges (1993) also stressed, however, that habitat destruction (e.g., rampant deforestation) could have catastrophic effects on the frog fauna once all forests had been destroyed (e.g., forest cover in Haiti is now at <1% of historical levels). Based on the knowledge that (1) 99% of the original forest in Haiti has been eliminated and (2) apparently no Haitian frogs have become extinct, can we conclude that elimination of the last 1% will not have a catastrophic effect? Perhaps the anuran fauna of Haiti (and possibly all of Hispaniola) will endure as long as second growth forests, banana plantations, and similar degraded habitats persist because the fauna is euryoecious.

Perhaps better examples could be drawn from studies of West Indian lizards, and biologists working in the West Indies have been often struck by the exceedingly opportunistic foraging of lizard species. In the Bahamas, Schoener *et al.* (1982) observed *Leiocephalus* (1) feeding on bread, peanut butter, jelly, crackers, and cheese; (2) lapping water from ice cubes and quickly learning to successfully solicit those ice cubes; and (3) feeding on food scraps dropped on the floor by diners at a hotel. One of us (R. Henderson), while eating canned beans and tuna on Great Bird Island (off Antigua), had *Anolis bimaculatus* and *Ameiva griswoldi* crawl into the cans to eat. Wijffels (1997) suggested that good food availability and absence of predators (and therefore a longer life span) contributed to the large size attained by *Anolis bimaculatus* in and around human dwellings on St. Eustatius. Artificial lighting has resulted in a number of traditionally diurnal anoles occupying nocturnal niches (Schwartz and Henderson, 1991; Powell and Henderson, 1992; Bowersox *et al.*, 1994a) more commonly associated with gekkonids (usually *Hemidactylus* spp.). The possible impact on gecko populations has not been assessed, but recent (May 1994) observations of *Anolis cybotes* and *Sphaerodactylus difficilis* in nocturnal syntopy in Cabarete, Dominican Republic, suggest that, because *A. cybotes* is an opportunistic predator of *S. difficilis* (Schwartz and Henderson, 1991; Lenart *et al.*, 1996), this temporal niche expansion may provide increased opportunities for saurophagy by *A. cybotes*.

Some species appear to have benefited from the presence and activity of humans. *Eleutherodactylus johnstonei* is widely distributed throughout the Lesser Antilles and many locales outside of the region (Kaiser and Hardy, 1994). The species obviously has taken advantage of its ecological versatility and opportunities for human-mediated transport to establish populations on previously frog-free islands and even in the presence of native congeners (Kaiser and Henderson, 1994), and "as habitat degradation progresses rapidly on most Caribbean islands, the advance of *E. johnstonei* will likely proceed in step, and at the expense of local endemics."

Another example is that of heliophilic anoles, which often are very abundant in human-modified habitats. *Anolis porcatus* is rarely seen in Cuban forests, except high in the canopy and in occasional tree gaps, and *A. sagrei* is never seen in densely forested areas (J. B. Losos, in litt.). Both species, however, are ubiquitous in open, disturbed habitats. Also, consider Hispaniolan anoles. A few species are nearly everywhere (e.g., *A. chlorocyanus*/*coelestinus*, *A. cybotes*, *A. distichus*/*brevirostris*, and the grass anoles) and these forms generally thrive in altered environments [although *A. chlorocyanus* has largely been displaced by introduced Cuban *A. porcatus* in urban Santo Domingo (Powell *et al.*, 1990b) and *A. cybotes* by the Puerto Rican bank generalist, *A. cristatellus*, in the less-than-pristine environs of La Romana (Fitch *et al.*, 1989)].

Henderson and Winstel (1995) and Henderson *et al.* (1996) documented exploitation of human-modified habitats by the arboreal boa *Corallus grenadensis* in an area of mixed agriculture and uncultivated woodlands on Grenada. Not surprisingly, habitat devoid of trees or with only solitary trees (i.e., devoid of contiguous tree crown habitat) precluded the presence of *C. grenadensis*. Small tree boas (<500 mm SVL) were encountered most often in uncultivated woods (78.1%), and medium-sized tree boas (500-1000 mm SVL) occurred in uncultivated woods (64.3%) more often than in cultivated trees (35.7%), but used the latter more often than small snakes (21.9%). Large tree boas (>1000 mm SVL) were more often encountered in mango trees (48.2%) and other cultivated trees (22.2%) than in uncultivated areas (29.6%). The cultivated trees used by the tree boas (e.g., mango, breadfruit, nutmeg, cacao) did not occur on the island until, in some cases, 150-200 years before present. Similarly, the rodent species (*Oryzomys* sp.; L. Lippold, personal communication) presumably exploited by adult *C. grenadensis* on pre-Columbian Grenada is now extinct; the rodents (*Mus*, *Rattus*) that adult *C. grenadensis* now exploit as food did not occur on the island until about 1500.

However, unlike the highly visible examples presented above, a preponderance of West Indian amphibians and reptiles rarely occur in severely altered situations. These animals apparently are, at least in most cases, descendants of effective colonizers (Hedges, 1996; but see also Crother and Guyer, 1996), but many species have evolved in situ, presumably through natural selection, into ecological specialists. To find them one must look in specific habitats (e.g., *Anolis barbouri* in montane ravines with abundant leaf litter and well shaded by intact forest canopy; Flores *et al.*, 1994). Because individual habitats are frequently restricted in area, the distributions and population sizes of these species are generally small.

Summary

The following is a hypothetical scenario that offers speculation for why things are as they are in the West Indies. The West Indian herpetofauna is largely derived from the herpetofauna of the South American mainland, and most lineages appear to have reached the West Indies by overwater dispersal. Potentially, many of the colonizers were animals of coastal habitats, forest edge, and other open situations, making them more tolerant of high ambient temperatures and sun-drenched habitats than, for example, forest-dwelling species. This tolerance would, presumably, increase their chances of surviving a prolonged overwater journey during which shade would be minimal.

The West Indian herpetofauna is depressed in the number of families and genera present but rich in species due, in large part, to the spectacular radiations of three genera (*Eleutherodactylus*, *Sphaerodactylus*, *Anolis*). West Indian frogs, lizards, and snakes tend to occur in extremely high densities compared to mainland species. Even in rain forest habitat, anoles, for example, occur at densities hundreds of times higher than those of mainland populations, and their biomass may approach the biomass of entire lizard communities on the mainland. In short, lizard communities on the Neotropical mainland consist of more *species of lizards*, but West Indian communities are composed of *more lizards*. In turn, the West Indian snake fauna preys largely on frogs and lizards, a food source that is ubiquitous and abundant. West Indian frogs and reptiles are, in general, habitat and trophic opportunists. Exceptions, of course, exist, especially in the rich faunas of the Greater Antilles where in situ speciation has produced habitat and trophic specialists. On the smaller islands of the Lesser Antilles, ecological generalization and versatility are even more widespread.

Future Emphasis

The characteristics of the West Indian herpetofauna that, in part, have accounted for the exciting ecological research that has already been done (inter and intraisland diversity, the adaptive radiations of several genera, the extremely localized distributions of some species and the broad, multi-island distributions of others, and the rarity of some species and the ubiquity and phenomenally high densities of others) remain lures for future research.

Predicting the future is a risky venture; still, the belief that ecological investigations will become more important in the twenty-first century is easy to support. Island-bound human populations continue to grow and their need for food, living space, amusement, and resources capable of being exploited in search of wealth will inevitably increase the pace at which the habitats of West Indian animals are altered. As a result, ecological studies will be necessary as a springboard not only for posing and addressing intellectual questions, but for the very pragmatic consideration of issues pertaining to the conservation of the region's flora and fauna. With this in mind, biologists must be willing to examine interactions in severely altered habitats (Powell *et al.*, 1996b), contrary to our inclination to focus on the remaining pristine areas. Also, scientists must collaborate actively with persons in those regional and national agencies whose responsibilities include conservation, agriculture, tourism, and urban development. Although many amphibian and reptilian species are ecologically versatile, others are not. Some anurans have very specialized habitat requirements (Hedges, 1993; Kaiser and Henderson, 1994; Joglar and Burrowes, 1996) and some endemic anurans apparently are experiencing difficulty withstanding the dual onslaught of habitat destruction and the introduction of *Bufo marinus* (e.g., Kaiser and Henderson, 1994). Turtles, both freshwater and marine species, are being heavily exploited for food. Many lizard species appear to coexist successfully with humans, and a few even thrive as human commensals, but exploitation of *Cyclura* for food and of many species for the pet trade has had a measurable effect. At least the larger species of snakes are vulnerable to continued human efforts at extermination and a number have been brought to the edge of extinction by the mongoose (Henderson, 1992; Tolson and Henderson, 1993).

In light of these realities and the resultant universal need to consider applications to conservation (Greene, 1994), we have chosen not to focus on the applied but rather on the more classical types of questions that have been overlooked to date or to which past or present work is leading. In addition to issues raised specifically by conservation biology and the need to continue ongoing investigations by many biologists, we believe there are three major areas in need of attention; all are conventional, all have been neglected, and each will ultimately contribute to our ability to conserve individual species, communities, and habitats.

Community Studies

The study of herpetological communities in the West Indies has been largely confined to congenerics and almost exclusively centered on *Anolis* (e.g., Rand, 1967a; Rand and Williams, 1969; Moermond, 1979; Schoener and Schoener, 1980; Roughgarden *et al.*, 1983, Losos, 1992) and to a lesser degree *Eleutherodactylus* (Stewart, 1979; Jones, 1982). On the Neotropical mainland, analyses of taxonomically diverse frog (e.g., Toft, 1980; Lieberman, 1986), lizard (e.g., Rand and Humphrey, 1968; Duellman, 1987; Vitt and de Carvalho, 1995), and snake (e.g., Vitt and Van Gilder, 1983; Cadle and Greene, 1993) faunas, or of an entire herpetofauna (e.g., Duellman, 1978, 1990; Duellman and Mendelson, 1995), are more common. Similar studies are virtually lacking for the West Indies, although the recent work of

Bullock and Evans (1990) and Thomas and Kessler (1996) addresses taxonomically diverse communities. But these, like the recent syntheses of Reagan (1996), and Reagan *et al.* (1996), have focused on communities in moist montane forests, whereas similarly diverse communities in deserts, dry forests, and chaparral have been largely neglected (Powell *et al.*, 1996b). Studies by Censky (1996), Reagan (1996), and Tolson (1996) include references to competition, predation, and habitat destruction, respectively, involving sympatric species other than amphibians and reptiles. Because similar situations exist elsewhere, future community studies should not be limited to the herpetofauna. Rather, we should seek to build on the growing body of trophic data in order to elucidate more fully the entire gamut of relationships in which West Indian frogs and reptiles play such prominent roles. Regardless of the taxonomic breadth of community studies in the West Indies, geographically they constitute a fertile arena for inter- and intraisland comparisons.

Autecologies

With the exception of those noted earlier, autecological studies are sorely lacking for the West Indian herpetofauna. As Vitt and Zani (1996) noted in an article on a Neotropical mainland lizard:

> A vast majority of insights leading to major discoveries have been based on patterns revealed by analyses of large data sets taken directly from the natural history literature...but we still lack data for most species. Logistics of field research in tropical countries has contributed partially to the general lack of data for many [species in many] areas but lack of appreciation for the value of individual species studies...has undoubtedly contributed indirectly by making it difficult to acquire research support.

Even the basic natural history of frogs, with the notable exception of *Eleutherodactylus coqui*, has been largely ignored. Some species of *Bufo* and *Osteopilus* are very common, and species of *Eleutherodactylus* are frequently ubiquitous. Considering the current global interest in amphibian populations, island populations of frogs should be more attractive than ever as subjects of intensive ecological pursuits.

Despite intensive efforts, many aspects of sea turtle ecology remain a mystery, and the few remaining viable populations of freshwater turtles and crocodilians deserve attention. Along with anurans in general, detailed investigations of nonanoline lizard ecology are virtually absent from the literature. Interest in *Cyclura* has remained fairly constant, but they are large and spectacular in appearance. The diminutive gekkonids in the genus *Sphaerodactylus* offer opportunities at the other size extreme. They often occur in high densities (Table 7.2), little is known about social interactions, and nothing is known about movement ecology or habitat utilization. Considering the diversity and distribution of *Sphaerodactylus*, our lack of knowledge is surprising. Likewise, the large gekkonids of the genus *Aristelliger*, a magnificent *Ameiva* fauna, and a host of intriguing anguids in an amazing array of sizes beg for attention.

Considering the disproportionate amount of attention anoles have received in the past 30 years, it is surprising that an autecological approach to studying anoline ecology has been ignored. One spectacular segment of the anoline fauna has been neglected completely: the giant species, and these lizards (e.g., *A. barahonae*, *A. smallwoodi*) are large enough to monitor with radio telemetry (Losos *et al.*, 1990). With the exception of the boines, the snake fauna has, been largely ignored ecologically. The dwarf boas (*Tropidophis*) and many colubrids (e.g., *Alsophis*, *Antillophis*, *Uromacer*) are common enough for mark and recapture methods or large enough for the use of radio telemetry.

Amphibians and Reptiles in Modified Habitats

Few, if any, ecologically oriented studies of amphibians and reptiles in the West Indies have been conducted in habitats that have not been affected, to one degree or another, by humans. The long history of humans in the West Indies provides a fertile ground for examining their role in geographic distribution and ecology. Lee (1996), for example, has done extensive analyses of the role of the Mayan culture in influencing the present-day herpetofauna (distribution and ecology) of the Yucatán Peninsula of México. In the West Indies, Kaiser (1992) has masterfully documented how trade-mediated human introductions may have been the single most important factor in creating the current distribution of *Eleutherodactylus* in the Lesser Antilles. Henderson *et al.* (1996) described and documented post-Columbian changes in the ecology of the arboreal boid *Corallus grenadensis* in Grenada; this species has had to adapt to new habitats, new prey species, and new predators during the past 500 years. Similarly, Henderson *et al.* (1987a) provided a scenario which described a shift in prey species by the boid *Epicrates striatus* on Hispaniola from now-extinct rodents and insectivores to European-introduced *Mus* and *Rattus*; the shift probably became complete in the early twentieth century.

Opportunities abound for investigations of human-disturbed habitats (e.g., Meshaka, 1996), because few, if any, species of frogs and reptiles in the West Indies have not been forced to survive in human-modified habitats. For example, at one site in Barahona, Dominican Republic, at least 13 lizard species coexist in a severely altered area of <1400 m^2 (Fobes *et al.*, 1992; Schell *et al.*, 1993b; Schreiber *et al.*, 1993). Similarly, Olson (1995) observed amphibians and reptiles to be diverse and common in Haitian courtyards. Studies of anoles using fence posts and coconut palms, geckos and frogs exploiting the prey-attracting properties of electric lights and the cover of human dwellings, or *Alsophis* using sheets of discarded plywood for cover are all common examples of amphibians and reptiles exploiting human modifications of habitat. Some are subtle, others obvious, but all are examples of portions of the herpetofauna adapting to human disturbances. The West Indian herpetofauna, among other insular and mainland faunas, is in a state of flux. The introduction of predators (*Herpestes javanicus*: Baskin and Williams, 1966; Henderson, 1992; cats and dogs: Iverson, 1978; *Rattus rattus*: Tolson, 1988), prey species (*R. rattus*, *Mus musculus*: Henderson *et al.*, 1996), and competitors (e.g., *Eleutherodactylus johnstonei*: Kaiser and Henderson, 1994; various *Anolis*: Fitch *et al.*, 1989; Losos *et al.*, 1993) contribute to modifications in the present-day distributions of many species (Case and Bolger, 1991; Case *et al.*, 1992; Losos *et al.*, 1993). Determining how the herpetofauna is responding to these modifications, be they subtle or flagrant, should be a top priority.

8

Evolutionary Relationships

Brian I. Crother
Department of Biology
Southeastern Louisiana University
Hammond, Louisiana 70402

Nothing in evolution makes sense except when seen in the light of phylogeny.

–Jay M. Savage, 1997 Society of Systematic Biologists Presidential Address

Introduction

The West Indies hosts at least 585 species of reptiles and amphibians in some 70 genera representing Anura (157 species, 11 genera), Testudinata (8 species and 4 genera; excluding sea turtles), Sauria (317 species and 29 genera), Serpentes (99 species and 24 genera), and Crocodylia (4 species and 2 genera) (Schwartz and Henderson, 1991). Given these large numbers of taxa for a quite definable biogeographic unit (see Chapter 10), the task of estimating the phylogenetic relationships within these groups is and has been no simple task. However, the continuation of the task has become more important than ever with the current and looming extinction crises and the consequent necessity to understand global biodiversity. To identify crisis areas for conservation, or just simply the "best" (= most diverse?) areas for preservation, methods have been developed that employ phylogenies as the baseline data (e.g., Crozier, 1992, 1997; Faith 1992, 1994; Forey *et al.*, 1994; Vane-Wright *et al.*, 1991). Thus, at what one might characterize as one of the cutting edges of applied biology is phylogeny. Although the application of phylogeny to conservation may seem surprising, it is not surprising how central phylogeny has become in comparative studies of arguably all biological disciplines. Its importance can perhaps be measured by the number of texts that have been recently published on the comparative method in biology: e.g., Brooks and McLennan (1991), Eggleton and Vane-Wright (1994), Harvey and Pagel (1991), and Stearns (1992). In addition to conservation and comparative biology in the broad sense (whether it be organismal or molecular), phylogeny has also become the cornerstone for the study of historical biogeography, regardless of the method (Crother and Guyer, 1996). Finally, phylogenies, especially when constructed from different data sets, can be used to assess the accuracy of phylogenetic methods themselves.

Based on the necessity of phylogeny in biology, both applied and basic, it has become important to address phylogenetic questions of all groups. This, with the aid of modern methods, has led to the production of phylogenies at rates never seen before, which means phylogenies are being reconstructed that are in some cases less than optimal and thus have led to and will lead to controversy and debate over their veracity (accuracy). This chapter will attempt to review all the phylogenetic (not taxonomic) work that has been conducted with Caribbean herpetofaunal groups and in the process discuss some of the debates that have already arisen in our attempts to produce best estimates of phylogeny. [Hedges (1996) made general comments on the relationships of most West Indian taxa, but because those comments were not accompanied by data I have eschewed including his notions in the following review.] In this review, I will also assess the phylogenies for general robustness, which admittedly will be biased in favor of conscientious cladistic analyses and hypotheses based on multiple data sets. I will also point out taxonomic areas in need of work and possible directions of future work.

Anura

The great majority of the West Indian anurans are endemics, and it is these groups that will be focused on in the succeeding discussion. Thus for example, for taxa such as *Rana grylio*, *Hyla crucifer*, and *Gastrophryne carolinensis* there will not be a discussion of phylogeny.

Bufonidae

There are two genera of bufonids found in the West Indies, one which is endemic. The genus *Bufo* is represented by the widespread and introduced *B. marinus* and as such will not be given further attention. The endemic genus, as concluded by Pregill (1981a) is *Peltophryne* and it is composed of 10 species (following Schwartz and Henderson, 1991).

The relationships within the genus are not well understood and no cladograms have been presented. Ruibal (1959) and Schwartz (1972) seemed to suggest that *P. peltacephala*, *gundlachi*, and *cataulaciceps* formed one group and that *empusa*, *fluviatica*, *guentheri*, *lemur*, and *taladai* formed a second group with *longinasa* apparently not related to either group. Tihen (1962) considered the Caribbean taxa possibly related to a group consisting of North and South American species, but this was broken up by Martin (1972). Cei *et al.* (1972) hypothesized that *peltacephala* shared a relationship with the *Bufo granulosa* group of South America. Finally, Pregill (1981a) discussed the cranial morphology and its evolutionary implications for the West Indian taxa. Unfortunately, but correctly so, Pregill (1981a) did not construct a phylogeny from his data because of outgroup difficulties. However, he did make some general phylogenetic statements about possible relationships. Pregill (1981a) suggested that *fluviatica* and *longinasa* were the most primitive taxa; *lemur* and *guentheri* were probably sister taxa; *taldai*, *gundlachi*, *cataulaciceps*, and *empusa* may form a monophyletic group; and *peltacephala* was unique and not yet placeable.

The intergeneric phylogeny, as the intrageneric phylogeny, has not been estimated in any detail. In a phylogenetic analysis focusing on the position of *Frostius*, Cannatella (1986) noted that *Peltophryne* (as well as *Crepidophryne* and *Rhamphophryne*) was easily separated from all other neotropical bufonids but did not include it in the analyses and resultant cladograms. However, Graybeal and Cannatella (1995) later noted that the monophyly of *Peltophryne* was questionable.

In summary, the genus *Peltophryne* remains unaddressed phylogenetically at both the interspecific and intergeneric levels; i.e., baseline phylogenetic hypotheses have yet to be constructed for this group. Also, other than Pregill's (1981a) work, the previous comments on interspecific relationships have been "only conjectural treatment" (Pregill, 1981a:273).

Hylidae

The level of phylogenetic knowledge for the hylid radiations in the West Indies is similar to that of the bufonids; no cladistic hypotheses have been put forth even though the taxa have received attention, albeit none as a group since Trueb and Tyler (1974). Dunn (1926a) considered the Jamaican hylids, *Calypthahyla crucialis*, *Hyla marianae*, *H. wilderi*, and *Osteopilus brunneus* to form a monophyletic unit derived from Hispaniola. Dunn (1926a) also considered the members of the Hispaniolan group, *Hyla heilprini*, *H. pulchrilineata*, *H. vasta*, and *Osteopilus dominicensis* to be each others' closest relatives. And finally, Dunn (1926a) considered the entire Antillean radiation (thus including the Cuban *septentrionalis*) to be derived from a single invasion event, therefore implying their monophyly. Noble (1927) and Shreve and Williams (1963) discussed other possibilities but essentially concluded to follow Dunn (1926a). Duellman (1970), in his monograph on Middle American hylid frogs, disputed both of Dunn's (1926a) island groups. He cautiously accepted Trueb's (1970) first discussion on the casque-headed frogs, in which she suggested a *septentrionalis* group composed of *septentrionalis*, *dominicensis*, *vasta*, *brunneus*, and *lichenata*. Acceptance of this group breaks up both of Dunn's (1926a) groups. Duellman (1970) went further with the Hispaniolan hylids and indicated that *heilprini* and *pulchrilineata* were unique lineages and that neither were related to the *septentrionalis* group of Trueb (1970) or the other *Hyla* of Jamaica. Schwartz and Fowler (1973) reviewed the ideas of Dunn (1926a), Shreve and Williams (1963) and Trueb (1970) and decided that they had little argument against Trueb's *septentrionalis* group. Schwartz and Fowler (1973) also noted, in similar fashion to Duellman (1970), that the relationships among the small hylids of Jamaica and Hispaniola were unclear and entertained the idea put forth by Dunn (1926a) that they were neotenic forms of the larger hylids in the *septentrionalis* group.

The most recent work on the entire Antillean radiation of hylids is that of Trueb and Tyler (1974). They noted that there were three recent introductions (*Ololygon rubra* from South America to St. Lucia, *Hyla squirella* from North America to Grand Bahama, and *H. cinerea* from North America to Puerto Rico) and that their affinities are not with any of the nine endemic taxa. Trueb and Tyler (1974) rejected the monophyletic and neotenic hypotheses of Dunn (1926a), and based on morphological differences suggested that the West Indian hylids were composed of six distinct lineages, each related to mainland components and not to each other. These lineages are *marianae-wilderi*, *heilprini*, *pulchrilineata*, *septentrionalis-dominicensis-brunneus*, *vasta*, and *lichenata*. The hypothesized mainland relationships of *heilprini* possibly were ("highly reminiscent of", Trueb and Tyler, 1974:32) with the *albomarginata* group, the *septentrionalis* group with *Osteocephalus*, *vasta* with the *boans* group, and *lichenata* perhaps with *Osteocephalus* or *Trachycephalus*. The other taxa could not be placed.

The possible disparate relationships among the West Indian hylids is suggestive of a complex history for the group and would thus make estimation of their phylogenetic relationships difficult. However, based on the information of Trueb and Tyler (1974), any future cladistic study on the West Indian hylids must not assume monophyly of the group and must include numerous mainland taxa in order to clarify the extent of the nonmonophyletic nature of these hylids.

Leptodactylidae

The genus *Leptodactylus* is represented in the West Indies, according to Schwartz and Henderson (1991), by five species, three of which are endemic (*albilabris*, *dominicensis*, *fallax*) and two that are South American (*insularum*, *wagneri*). However, the recognition of these taxa was disputed by Heyer (1978, 1979). Heyer (1978) presented evidence which indicated that *albilabris* and *dominicensis* were conspecifics and consequently he synonymized *dominicensis*. Heyer (1970) figured a phylogeny of the *melanotus* group that included *wagneri*, but later (1994) he considered West Indian *wagneri* to be possibly confounded with other populations and still considers its status to be unresolved. Curiously, Schwartz and Henderson did not mention Heyer's work, nor did they discuss reasons for the acceptance of *dominicensis*. As a general statement about *Leptodactylus* relationships, they apparently do not form a monophyletic group, with each taxon possibly being associated with a different mainland lineage. *Leptodactylus fallax* is a member of the *pentadactylus* species group and is hypothesized to belong to a clade of unspecified relationships which include *flavopictus*, *knudseni*, *labyrinthicus*, and *pentadactylus*. *Leptodactylus albilabris* (*sensu* Heyer) is a member of the *fuscus* species group and is hypothesized to belong to a clade of unspecified relationships which include *bufonius*, *fuscus*, *gracilis*, *labialis*, *latinasus*, *mystaceus*, and *mystacinus*. Bogart (1974) presented karyotype data for *dominicensis* (= *albilabris*) that did not dispute its membership in the *fuscus* group. Explicit phylogenetic hypotheses are yet to be constructed for either of these presumptive clades.

Unlike *Leptodactylus*, the genus *Eleutherodactylus* (over 130 species) has had several explicit phylogenetic hypotheses put forth based on morphological (Joglar, 1989; Kaiser *et al.*, 1994b,c; Lynch, 1996), allozyme (Hedges, 1989a,b, 1991; Kaiser *et al.*, 1994d, 1996; Smith *et al.*, 1981), immunological data (Hass and Hedges, 1991), and multiple data sets (Kaiser, 1996). Hedges (1989a) and Joglar (1989) attempted to examine all the species, but only Hedges presented hypotheses for most of the taxa (Fig. 8.1). Joglar (1989) presented summarized phylogenies. Perhaps the best way to assess the state of West Indian *Eleutherodactylus* phylogeny would be to search for congruence among the various data sets. Starting at the most inclusive level, the phylogenies of Hedges (1989a) and Joglar (1989) are difficult to reconcile because of the summarized nature of Joglar's phylogeny, which was based on a cladistic analysis of morphological characters. However, Joglar suggested that the *auriculatus* group, *unistrigatus* group, and the *recordii* group [these groups are somewhat similar in content to Hedge's (1989a) groups] form parts of a polytomy which also included *Syrrhopus* and *Tomodactylus*. Sister to this polytomy was the *inoptatus* group.

At best, Joglar's hypothesis is consistent with one of Hedges distance-based (Cavalli-Sforza distance) trees. Hedges presented two trees based on allozymes with all of the taxa he studied (84 species), one constructed with UPGMA (unweighted pair-group method of averages) and the other with distance Wagner. As mentioned previously, the UPGMA tree was consistent with Joglar's hypothesis in that the *inoptatus* group was outside the other groups. However, the two hypotheses of Hedges (1989a) were not congruent with each other and in fact were quite different in the relationships of the groups. The UPGMA tree suggested (((I,II)III)IV), whereas the distance Wagner tree showed ((I,IV)III)(II,III). Of these four major groups of Hedges, none were monophyletic, but in the distance Wagner tree his group I (composed of Jamaican taxa) did fall out in a single clade.

Hedges (1989a,b) presented several different hypotheses on the relationships of the Jamaican group. In a paper exclusively on Jamaican *Eleutherodactylus*, Hedges (1989b) constructed two distance-based phylogenies (UPGMA and distance Wagner with Cavalli-Sforza distance) and a character-based phylogeny (FREQPARS). None of these hypotheses are congruent with each other nor with the trees in the larger work (1989a).

Hass and Hedges (1991) constructed a phylogeny of five *Eleutherodactylus* taxa (of different subgenera) from immunological distance data clustered with distance Wagner. Comparison of these relationships ((*gossei*, *nubicola*)*planirostris*)(*montanus*, *inoptatus*) to the allozyme relationships, ((((*gossei*, *nubicola*)*inoptatus*)*montanus*)*planirostris*) and (((*gossei*, *planirostris*)(*nubicola*,*montanus*))*inoptatus*), reveals the absence of congruence between the immunological data and the allozyme data.

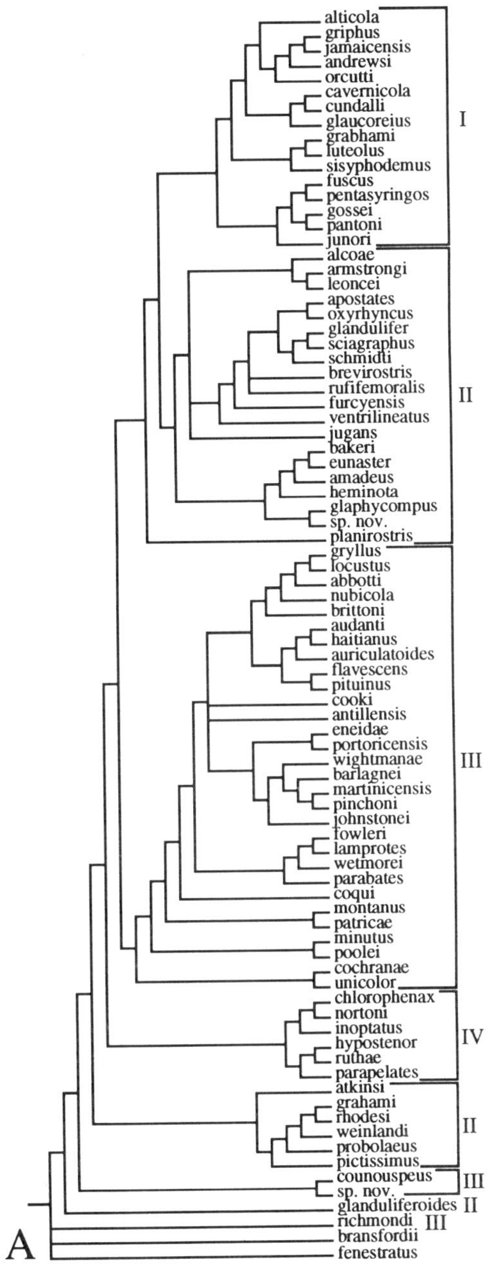

Figure 8.1. Phylogeny estimates of West Indian *Eleutherodactylus* based on allozyme data. (A) UPGMA tree. (B) Distance Wagner tree. Branch lengths do not correspond to distance estimates (redrawn from Hedges, 1989).

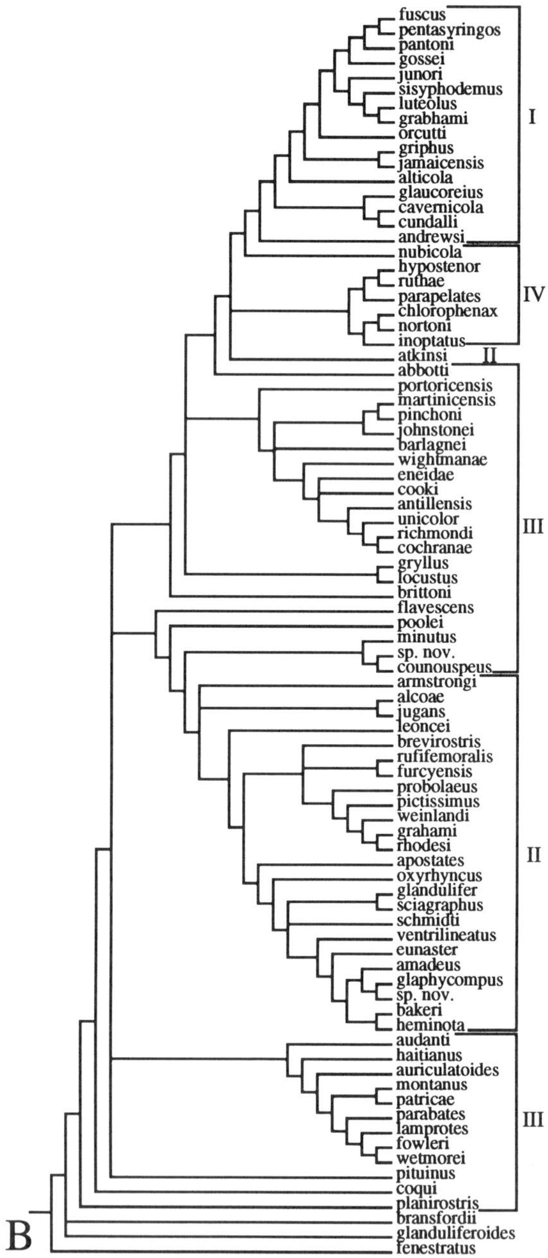
fuscus
pentasyringos
pantoni
gossei
junori
sisyphodemus
luteolus
grabhami
orcutti
griphus
jamaicensis
alticola
glaucoreius
cavernicola
cundalli
andrewsi
nubicola
hypostenor
ruthae
parapelates
chlorophenax
nortoni
inoptatus
atkinsi
abbotti
portoricensis
martinicensis
pinchoni
johnstonei
barlagnei
wightmanae
eneidae
cooki
antillensis
unicolor
richmondi
cochranae
gryllus
locustus
brittoni
flavescens
poolei
minutus
sp. nov.
counouspeus
armstrongi
alcoae
jugans
leoncei
brevirostris
rufifemoralis
furcyensis
probolaeus
pictissimus
weinlandi
grahami
rhodesi
apostates
oxyrhyncus
glandulifer
sciagraphus
schmidti
ventrilineatus
eunaster
amadeus
glaphycompus
sp. nov.
bakeri
heminota
audanti
haitianus
auriculatoides
montanus
patricae
parabates
lamprotes
fowleri
wetmorei
pituinus
coqui
planirostris
bransfordii
glanduliferoides
fenestratus
I
IV
II
III
II
III
B

Smith *et al.* (1981) published an allozyme-based distance phylogeny of part of the Puerto Rican *Eleutherodactylus* species assemblage. In comparison with the results of Hedges (1989a) there is great disparity. Hedges, in his complete trees, depicts the Puerto Rican bank species as nonmonophyletic, which is an assumption made by Smith *et al.* However, in one of Hedges' (1989a:356) character-based trees, the Puerto Rican taxa are included in the same clade (unresolved relationships) but joined with the Lesser Antillean taxa.

Joglar (1983) also published an allozyme study of the Puerto Rican *Eleutherodactylus*, but his study had a more complete sampling than Smith *et al.* (1981). However, Joglar's study shared other limitations with the Smith *et al.* study: distance-based phenograms and assumed monophyly. Joglar presented two hypotheses from two different distances, both clustered with UPGMA. The two phenograms shared some similarities in sister taxa and basal lineages but otherwise the two trees differed in structure. In some aspects the Joglar and Smith *et al.* trees were consistent, but the consistency differed between the two Joglar trees. As far as comparison with the Hedges (1989a) hypothesis is concerned, the remarks are essentially the same as for the comparison between the Smith *et al.* (1981) and the Hedges tree: congruence is lacking. The species pairs that are resolved in Hedges [(*portoricensis* , *enieidae*) and (*gryllus*, *locustus*)] were not found by Joglar. Again, as in the Smith *et al.* study, perhaps the assumption of monophyly confounded the results, or perhaps as noted by Page and Lydeard (1994), the data derived from sequential electrophoresis (the method employed by Hedges, 1989a) is intractable for phylogenetic analysis because the relationships between the numerous character states are uncertain.

To complicate matters, Lynch (1996) reexamined the phylogeny of the subgenus *Pelorius* which was erected by Hedges (1989a) and as - signed six Hispaniolan species. Hedges found two three-taxon sister clades, one resolved and the other not. Lynch (1996), employing outgroup comparisons, could not find resolution in either clade and concluded that support for the two clades was at best weak or simply absent.

Kaiser (1996) summarized multiple (five) data sets into a consensus hypothesis for the Lesser Antillean (LA) *Eleutherodactylus*. He co n-cluded that the LA assemblage was not monophyletic, with the northern clade probably related to Greater Antillean forms and the southern clade probably related to South American taxa. Again, the search for congruence between Kaiser's conclusions and Hedges' work failed. Hedges

(1989a) presented four different hypotheses for his LA representatives. The two distance-based hypotheses (UPGMA and distance Wagner) were incongruent (Fig. 8.1) with each other and with Kaiser (1996). Hedges (1989a) also presented two character-based hypotheses, one which placed the LA taxa in an unresolved clade with the Puerto Rican taxa and the other which separated the LA taxa but left them unresolved.

As noted by Hedges (1989a), the groups he recovered are largely similar to the groups recognized by Dunn (1926a) and Schwartz (e.g., 1965c, 1969c, 1976b,). However, because of the persistent incongruities among the modern hypotheses (as noted previously) questions arise as to why resolution has remained elusive. The obvious answers are inconsistent approaches to phylogeny reconstruction coupled with data perhaps inadequate for resolving the problem. Also, compounding the problem is simply the size of the group. This is an enormous group of over 130 taxa and thus represents a Ruthian task. Possibly the best way out of the current morass of West Indian *Eleutherodactylus* phylogeny would be through the analysis of additional data sets and a subsequent total-evidence approach. Then perhaps a stronger phylogenetic signal will emerge to provide a robust hypothesis of West Indian *Eleutherodactylus* evolution.

Sauria

Amphisbaenidae

This morphologically peculiar and phylogenetically enigmatic squamate group (e.g., Gans, 1967, 1978; Rage, 1982; Gauthier *et al.*, 1988; Presch, 1988; Schwenk, 1988) is represented in the West Indies by two genera, *Amphisbaenea* and *Cadea*, and 14 species. Equally as intriguing as the morphology and evolution is their biogeographic distribution in the West Indies (the global distribution is also an eye-catcher for the biogeographer): the Greater Antilles sans Jamaica. Although the group is seemingly tantalizing for systematists and certainly has not been ignored (e.g., Vanzolini, 1951; Zug and Schwartz, 1958; Gans and Alexander, 1962; Thomas, 1965a, 1966d), no explicit phylogenies have been produced for the West Indian radiation.

Vanzolini (1951) presented a systematic rearrangement but did not discuss relationships. Zug and Schwartz (1958) examined only *Cadea*, and Thomas (1965a, 1966d) mostly described new diversity. Only Gans and Alexander (1962), in their monograph on West Indian *Amphisbanea* and *Cadea*, even remotely considered relationships. At the outset they expressed caution at the evaluation of phylogeny and decided that any explicit statements would be made at a later date. Gans and Alexander did note that the taxa showed high similarity in external characters and that the inclusion of internal characters did little to further sort out phylogeny. A single suggestion of relationships was made based on body annuli. Three groups were identified: *innocens - cubana cubana*, *cubana barbouri - manni - caeca*, and *bakeri - fenestrata*. However, they stressed that no overall pattern emerged other than that *manni* appeared distinct from the others.

The apparent inability of morphology to elucidate relationships even within the West Indies assemblage might suggest an ancient and homoplastic group. The ancient aspect is almost certainly the case, given that its current global distribution may have been affected by the breakup of Pangea (Gans, 1990). Perhaps sequences from conservative DNA regions are required to address the problem, and apparently at this time molecular data are being collected (C. Gans, personal communication).

Anguidae

This family is represented in the West Indies by five genera and 23 species. Four of the genera and 22 of the species belong to the subfamily Diploglossinae and all these species are endemic. The fifth genus (and 23rd species) is *Ophisaurus ventralis*, which was introduced to Grand Cayman Island, where it is possibly now established (Schwartz and Henderson, 1991). The other genera in the West Indies are *Celestus*, *Diploglossus*, *Sauresia*, and *Wetmorena*, with the latter two genera endemic to Hispaniola. The definition, and thus the contents of these genera remain in flux, and perhaps the main reason is because of the absence of any explicit and robust phylogenetic hypotheses.

Ideas of relationship and content have been suggested and discussed and have largely been centered around scuttelation, sheathed or exposed claws, and osteoderm patterns (Boulenger, 1885; Dunn, 1939; Grant, 1940b; Savage and Lips, 1993; Schwartz, 1970; Strahm and Schwartz, 1977; Underwood, 1959a). The key character for delimiting the taxa was sheathed or exposed claws until Strahm and Schwartz (1977) recognized that the claws were not retractile. The character Strahm and Schwartz (1977) relied on was osteoderm pattern, and based on this character they reorganized *Celestus* and *Diploglossus*. Savage and Lips (1993) reemphasized the importance of the sheathed or exposed claw character (and deemphasized the osteoderm character following Wilson *et al.*, 1986) and once again reorganized the genera along traditional lines. Savage and Lips (1993) attempted a phylogenetic analysis using eight external morphological traits but achieved only an unresolved polychotomy. Thus, phylogenetic support for their classification remained elusive. However, Savage and Lips (1993) presented a biogeographic argument to support their classification over that of Strahm and Schwartz (1977). Regardless, the phylogeny of the Caribbean anguid fauna remains unresolved.

Gekkonidae

The gekkonid assemblage is rather interesting in that although there are seven genera (*Aristelliger*, *Gonatodes*, *Hemidactylus*, *Phyllodactylus*, *Sphaerodactylus*, *Tarentola*, *Thecadactylus*) found in the Caribbean region, none are endemic. *Sphaerodactylus* is by far the largest radiation, but *Aristelliger* is the best candidate for a true West Indian genus, with five of the six species found exclusively in the West Indies and the other taxon located on western Caribbean islands and Belize. Overall, this genus has received little attention and so the interspecific phylogeny remains purely speculative. Schwartz and Crombie (1975) proposed that *Aristelliger* was composed of two distinct groups, one containing *cochranae* and *barbouri* [this followed Hecht (1951), who actually placed these taxa in the subgenus *Aristelligella*] and the other composed of the other four taxa (*georgeensis*, *hechti*, *lar*, *praesignis*). Within this latter group, Schwartz and Crombie (1975) went only so far as to say that *hechti* is probably closer to *lar* and *praesignis* than to *georgeensis*. Interestingly, they cautioned that additional species may be hidden under these four names and further speculation on phylogeny seemed premature. It is surprising that recent attempts have not been made to estimate the phylogeny of this genus; it seems perfect for an allozyme data set.

The phylogenetic position of the genus *Aristelliger* was figured by Russell (1979) based on aspects of foot structure. Russell placed *Aristelliger* as the sister to a clade composed of *Gekko* and *Chondrodactylus*, two Old World taxa.

As currently understood, the genus *Gonatodes* is represented in the Caribbean islands of Cuba, Jamaica, Hispaniola, and Grand Cayman by a single species, *albogularis* (but note that subspecific differentiation is recognized; Vanzolini and Williams, 1962). To my knowledge, no explicit interspecific phylogenetic hypotheses have been published for the genus. The position of the genus has been speculated upon many times

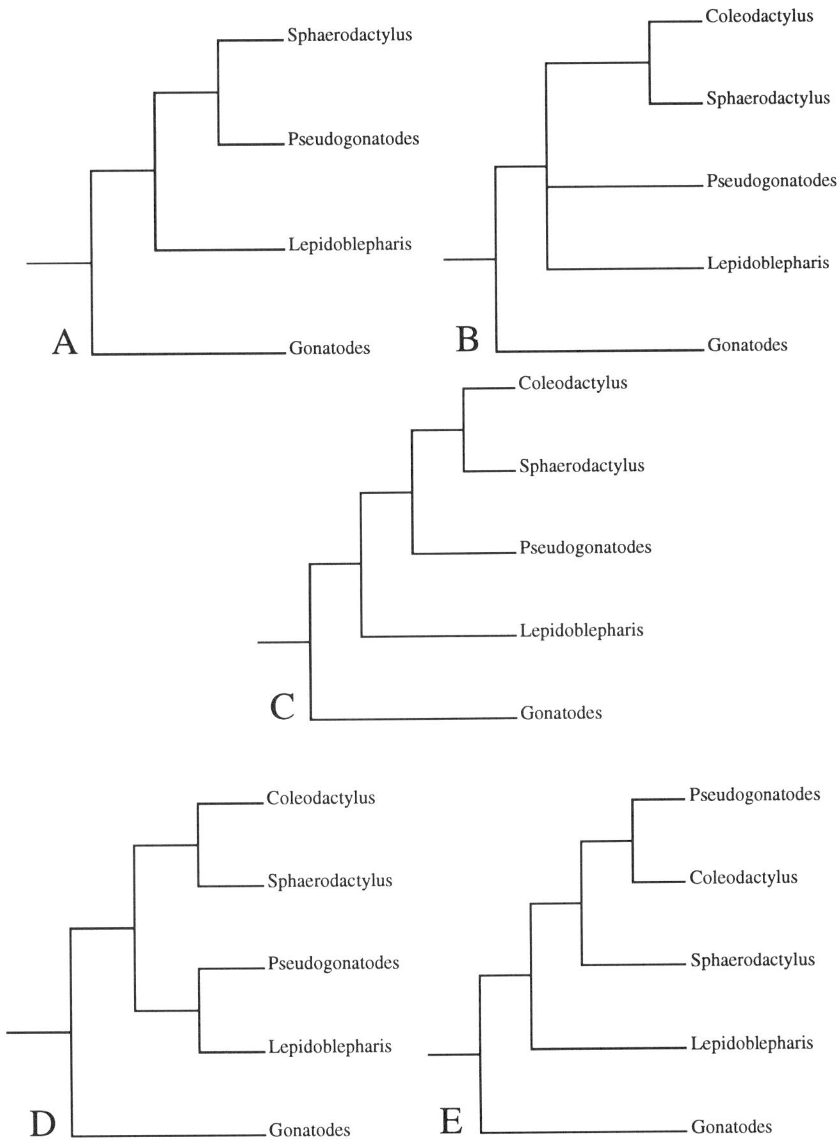

Figure 8.2. Various phylogeny estimates of gecko genera *Gonatodes* and *Sphaerodactylus*. (A) Noble, 1921, (B) Park, 1926, (C) Vanzolini, 1968, (D) Russell, 1972, (E) Kluge, 1995 (redrawn from Kluge, 1995).

(Noble, 1921; Parker, 1926b; Vanzolini, 1968; Russell, 1972); in addition, a rigorous cladistic analysis has also been conducted (Kluge, 1995). Remarkably, these five hypotheses are congruent on the placement of *Gonatodes*: basal to the other sphaerodactyl genera (Fig. 8.2).

The apparently amazing dispersal capabilities of gekkonid lizards are exemplified in the genus *Hemidactylus*. Five species are recognized in the Caribbean region: two are clearly human introductions (*garnotii*, *turcicus*) and so will not be considered further, two others are most probably natural trans-Atlantic migrants (*brooki*, *mabouia*), and the fifth is a Lesser Antillean-northeastern South American endemic. Kluge (1969) argued strongly within a phylogenetic context that the widespread *brookii* and *mabouia* were conspecifics with their African populations. He also convincingly argued that they must have arrived to the New World via trans-Atlantic rafting and not by human introduction. Kluge (1969), (when computer-assisted phylogeny estimation was still primordial) constructed a Wagner diagram and found *mabouia* and *brookii* as sister taxa but also depicted *brookii* as paraphyletic with regard to the New World endemic *palaichthus*. Based on Kluge's hypothesis, I suspect biochemical data may be useful here in teasing out patterns of differentiation. At the generic level, Russell (1979) suggested that *Hemidactylus* was the sister of part of *Cyrtodactylus*.

Another broad-ranging taxon with representatives in the Caribbean is *Phyllodactylus*. Two species are recognized, *wirshingi* on Hispaniola and Puerto Rico and *pulcher* on Barbados. Schwartz (1979f) argued that *wirshingi* was a natural resident and had undergone subsequent diversification, at least at the subspecific level. The relationships of these taxa have not been explicitly examined, although Dixon (1962) speculated that *wirshingi* was closely related to *martini* of the Dutch Leeward Islands. Russell (1979) included *Phyllodactylus* and placed it as the sister to the clade containing *Aristelliger*.

The latest list (Powell *et al.*, 1996a) of Caribbean *Sphaerodactylus* totaled 78 species (another dozen or so are found in Central and South America), which represents the third largest herpetofaunal radiation in the region [behind the obvious *Eleutherodactylus* and *Anolis* (as conceptualized by Williams, 1976, and not Guyer and Savage, 1986; but see the following discussion of polychrotid lizards)]. Given the size of the radiation it is no surprise that many ideas have been put forth on the relationships of various parts of the radiation and that really only one worker (Hass, 1991, 1996) has attempted to resolve the phylogeny of the entire West Indian radiation.

Three different hypotheses have been presented about the relationships of the genus: (*Pseudogonatodes*, *Sphaerodactylus*) (Noble, 1921); (*Coleodactylus*, *Sphaerodactylus*) (Parker, 1926b; Vanzolini, 1968; Russel, 1972); and ((*Coleodactylus*, *Pseudogonatodes*) *Sphaerodactylus*) (Kluge, 1995). Kluge's hypothesis represents the only detailed cladistic analysis and so at this time should be considered the best estimate of *Sphaerodactylus* relationships at the generic level.

The intrageneric relationships of *Sphaerodactylus* are a bit more complicated. As with other groups, various speculative comments have been made about the relationships of parts of the genus. For example, King (1962:48) in the following quote suggested the Lesser Antillean forms were not monophyletic: "The species in the Lesser Antilles are closely related to and possibly conspecific with species widespread in the Greater Antilles (*S. macrolepis* with *S. difficilis*, *S, grandisquamis*, and *S. notatus*) and on the mainland (*S. molei* with *S. lineolatus*)." Schwartz and Thomas (1983) addressed King's notions in a study of the *difficilis* complex. Based on six morphological characters, they figured a dendrogram (as noted in Chapter 1, the only tree to be found in a Schwartz taxonomy paper) of the complex and found little to support King's relationships but did support King's ideas of nonmonophyletic island radiations. Also, Schwartz and Garrido (1985) indicated that the Cuban assemblage of *Sphaerodactylus* was derived from multiple areas, and thus was also not monophyletic. Such narratives need not be further examined because in the case of *Sphaerodactylus*, modern attempts at estimating the phylogeny have been put forth (Hass, 1991, 1996).

Hass (1991) employed allozyme data derived from sequential electrophoresis (Singh *et al.*, 1976) to "determine the major groups of West Indian *Sphaerodactylus* and not the relationships within these groups." Hass included 48 species in this study and presented a character-based bootstrap proportion consensus tree and a distance Wagner tree (she also presented a cladogram showing the character distributions but because this tree was only 1 of 1000, the relationships are not discussed). A strict consensus tree of the parsimony analysis was not shown (presumably because of lack of resolution: Hass mentioned that only four species pairs were resolved). The character-based bootstrap tree was mostly unresolved and did not have a high proportion for the support of West Indian monophyly. The distance tree was fully resolved (as it must be) but out of the 45 nodes, only two had bootstrap proportions less than 90%, 13 had proportions less than 50%, and four had proportions of zero. Therefore, the robustness of this tree, although fully resolved, must be questioned.

Page and Lydeard (1994) questioned the phylogeny and in a parsimony reanalysis they essentially obtained the same largely unresolved tree that Hass (1991) presented. Page and Lydeard suggested the problem lay in the data derived from sequential electrophoresis. Hass reported 152 alleles from 15 loci and Page and Lydeard noted that the relationship between the character states could not be known and that as the number of states increase, the informativeness of those characters decreases

Hass (1996) added a mitochondrial DNA (mtDNA) data set (16S gene, 330 bases) to the fray with a reduced number of taxa (28). Phylogenetic hypotheses were derived from neighbor-joining (distance) and maximum parsimony (character) analyses. Twenty-one most parsimonious trees were found and although there were no measures to interpret the robustness of the parsimony tree or the fit of the data, the majority-rule consensus tree had only four nodes not found in 100% of the 21 trees. The neighbor-joining and parsimony trees were largely consistent, but as might be expected they were not fully congruent (Fig. 8.3). Most of the incongruence was in the internal nodes, although some of the smaller terminal groups also were rearranged. Even so, Hass (1996) seems to have at least corroborated some of the traditional groupings within *Sphaerodactylus*, such as the *argus* series, *ramsdeni* group, *nigropunctatus* group, and part of the *difficilis* complex. She also found (with three exemplar species), contra to previous assertions, a monophyletic LA clade.

Resolution in terms of relationships of these groups must await further analyses and overcome a possibly detrimental sampling problem. An immediate possibility in terms of analysis would be to combine the allozyme data with the mtDNA for the 20 overlapping taxa. (On the other hand, this may be inappropriate given the independent history of the mtDNA. Because of this, the mtDNA is probably more appropriately employed as a consilient data set.) The covariation in the mtDNA data may help organize the multitude of alleles in the allozyme data set. The sampling problem refers back to work done by Harris and Kluge (1984) on Central American *Sphaerodactylus*. Their cladistic study could not conclude that the mainland species were monophyletic. If this is correct, then the only possible relatives would be in the West Indies. Thus, the implicit assumption employed in Hass' studies that the West Indian *Sphaerodactylus* are monophyletic may be incorrect. This may be the reason why the relationships among the groups appear unstable.

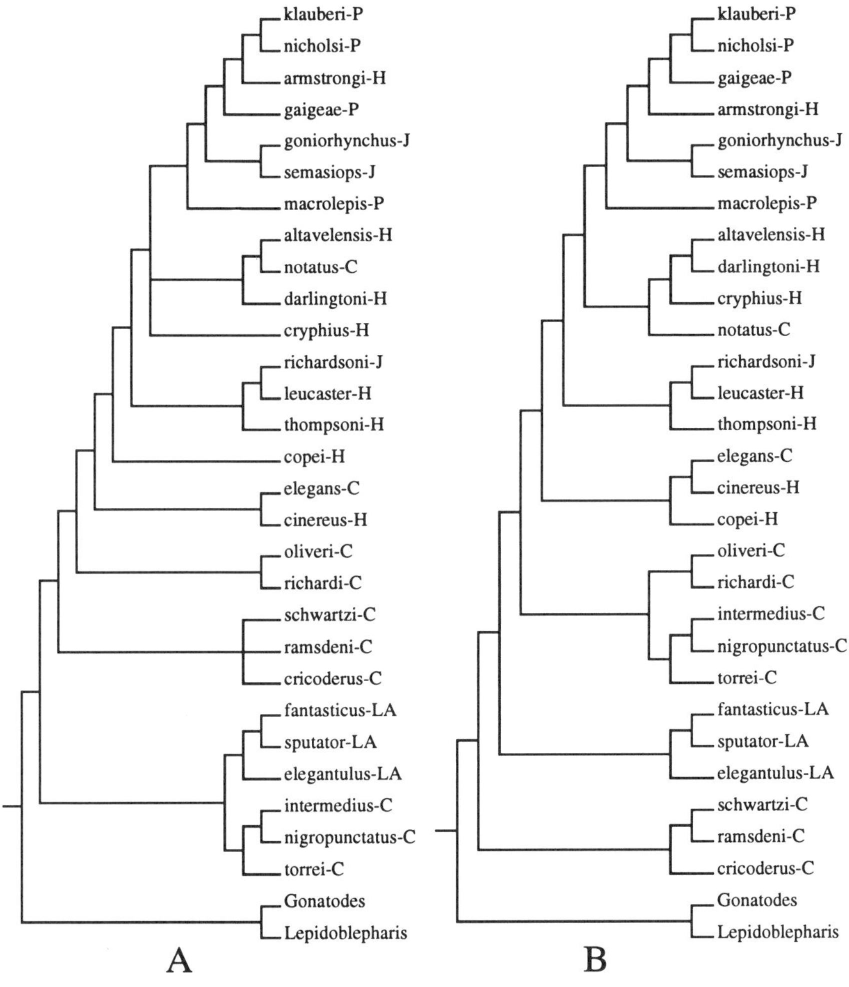

Figure 8.3. Phylogeny estimate of *Sphaerodactylus* based on mitochondrial 16S DNA sequence. (A) Majority-rule consensus of 21 most parsimonious trees. (B) Neighbor-joining tree (redrawn from Hass, 1996).

Essentially no phylogenetic work has been conducted with *Thecadactylus* (*rapicauda*; the genus is monotypic). By implication of Kluge's (1967) classification of *Thecadactylus* into the Gekkoninae, the genus is related to other West Indian genera such as *Aristelliger* and *Tarentola*. Kluge (1967) inferred, based on the absence of postanal slits, *Thecadactylus* probably shared a close history with *Aristelliger* and *Lygodactylus*. Hoogmoed (1973:64) on the other hand considered Kluge's (1967) hypothesis "a little far fetched" and opined that the loss of postanal sacs, which Kluge considered a synapomorphy, was independent.

Tarentola was revised by Joger (1984) and he placed the single New World species into its own subgenus, *Neotarentola*. The phylogenetic relationships of this taxon were not discussed

Iguanidae

The cladistic revolution in classification, that being the storage of information of monophyly and sister relationships instead of grades, brought about great changes in large groups that were subjected to cladistic analysis and subsequently re-classified. The Iguanidae of 10 years ago bears little resemblance to the current family in terms of its contents due to the efforts of Frost and Etheridge (1989). Following Frost and Etheridge, the Caribbean has only two representative genera, *Iguana* and *Cyclura*. The former contains two species (*iguana* and *delicatissima*) and the latter contains eight extant species (*ricordi*, *carinata*, *nubila*, *cychlura*, *cornuta*, *collei*, *rileyi*, *pinguis*). Two taxa known only from skeletal material (*mattea* Miller, 1918; *portoricensis* Barbour, 1919) were sunk into synonymy with *pinguis* by Pregill (1981b).

Barbour and Noble (1916) and Schwartz and Carey (1977) considered *Cyclura* to be the sister to *Ctenosaura*. Avery and Tanner (1971) proposed that *Cyclura* was the sister to a clade composed of *Ctenosaura* and *Sauromalus*. Kevin de Queiroz (1987) demonstrated that *Iguana* and *Cyclura* were each others closest relatives and a subsequent analysis (Norell and de Queiroz, 1991) with fossil taxa found *Cyclura* to be the sister of an *Iguana-Pumilia* (extinct) clade. The Norell and de Queiroz (1991) analysis fully resolved the Iguanidae phylogeny, which had previously been considered as part of an unresolved trichotomy with *Ctenosaura* and an *Amblyrhynchus-Conolophus-Sauromalus* clade (de Queiroz, 1987; Etheridge and de Queiroz, 1988; Frost and Etheridge, 1989). The most recent estimate of iguanid phylogeny (Sites *et al.*, 1996) combined the morphological data with sequences from two regions of mtDNA. Sites *et al.* performed a rigorous phylogenetic analysis which yielded a strikingly different hypothesis from that of Norell and deQueiroz (1991). Sites *et al.* found *Cyclura* to be the sister to a large clade that included *Ctenosaura*, *Amblyrhynchus*, *Conolophus*, *Iguana* and *Sauromalus* (Fig. 8.4). In this same study, *Iguana* and *Sauromalus* were supported as sister taxa. Given the nature of the Sites *et al.* (1996) study, their hypothesis must be considered the current best estimate of the placement of *Cyclura* and *Iguana*, although one must keep in mind that the novel hypothesis may be a result of independent evolution of the mtDNA. Interestingly, *Cyclura* appears to have diverged relatively early in the history of the iguanas, perhaps hinting that the divergence was associated with the earliest proto-Antilles vicariance events.

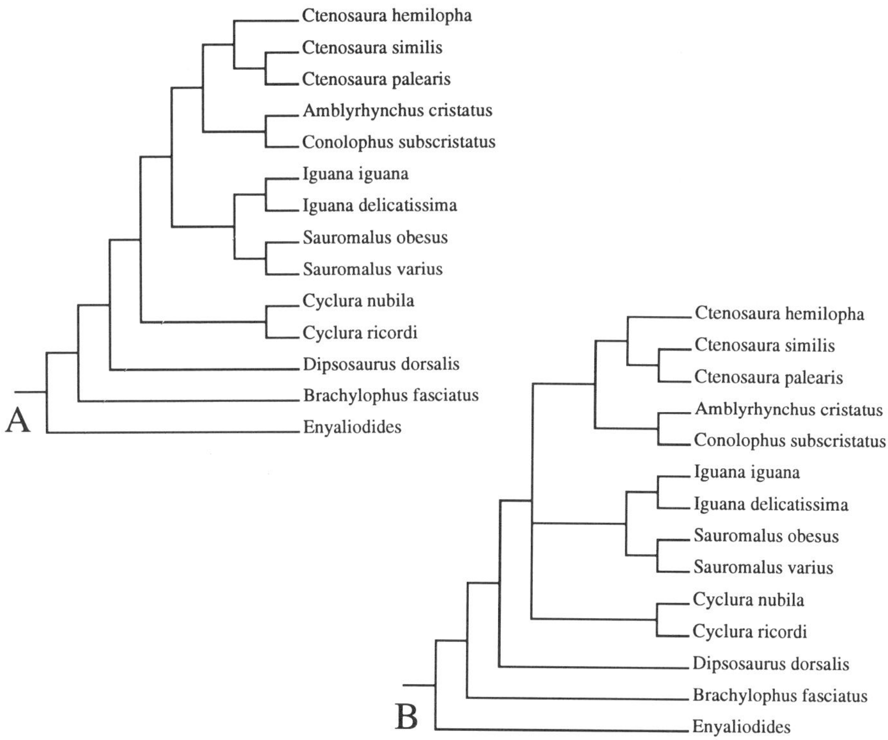

Figure 8.4. Phylogeny estimate of iguanid taxa based on mitochondrial DNA. (A) Maximum-parsimony tree with characters of three or more homoplasies deleted, (B) Consensus tree of the two most parsimonious trees derived from consistent characters only (redrawn from Sites *et al.*, 1995).

Given that *Iguana* contains only two species, which was verified by Lazell (1973), it is safe to say that the phylogeny is worked out. Perhaps the nagging question about *I. iguana* in the Caribbean is whether or not they are distinct from the mainland or are recent imports through Carib Indian movements. Lazell (1973) concluded that the Caribbean populations were distinct and as such populated the islands via natural means and not human transport. A recent report by Censky *et al.* (1998) of *I. iguana* dispersal in the LA corroborated Lazell's hypothesis.

The intrageneric phylogeny of *Cyclura* has been discussed and estimated, with the first key study addressing the question being that of Schwartz and Carey (1977). Based on the examination of 39 external morphological characters, they proposed a hypothesis of evolutionary history. They suggested that two groups were present, one composed of *ricordi* and *carinata* and the other composed of the remaining taxa (the fossil taxa were not included in the study). In the latter group, they believed that *pinguis* was outside of the other five taxa, and the relationships of these five could not be resolved with respect to each other. Schwartz and Carey (1977: 90) made a peculiar observation and conclusion about a subspecies of *rileyi* and the species *cychlura*,

> It well may be that *C. r. cristata* should be regarded as either a distinct species, or that it should be combined nomeclatorially with *C. cychlura*. Neither course recommends itself; the first beclouds the affinities, the second violates our concepts of *cychlura* both in juvenile pattern and in head scuttelation.

This problem is instructive in that it suggests homoplasy may be a factor in resolving this group, and that if *rileyi* is in fact a paraphyletic species, fundamental systematic problems remain in our understanding of *Cyclura*.

Hollingsworth (1998) inferred the phylogeny of *Cyclura* while attempting to understand the history of *Sauromalus*. Hollingsworth examined morphology and included polymorphic characters and coded them for analysis two ways. One approach followed Campbell and Frost (1993, polymorphisms were assigned discrete states) and the other coding scheme followed Wiens (1995) in which the frequencies of the polymorphisms were coded into bins. Regardless of the potential problems associated with frequency data of any kind (e.g. Crother, 1990; Murphy, 1993), Hollingsworth (1998) corroborated Schwartz and Carey's (1977) hypotheses with the exception that Hollingsworth found *pinguis* as the sister to all the *Cyclura* (((*rileyi* , *collei*)*nubila*)*cornuta*, *cychlura*))(*ricordii*, *carinata*)).

At the time of this writing, there is an ongoing study using mtDNA sequence to address the phylogeny of *Cyclura* (C. Malone, personal communication).

Polychrotidae

The representative taxa of this iguanian family include the anole genera *Chamaelinorops* (monotypic), *Chamaeleolis* (three species), and *Anolis* [140 species; whether *Anolis* is composed of multiple genera sensu Guyer and Savage (1986) remains in debate; see below for further discussion]. It is clear that these three genera plus *Phenacosaurus* form a monophyletic group (Etheridge, 1960; Williams, 1977; Peterson, 1983a; Guyer and Savage, 1986; Frost and Etheridge, 1989), and this "anole" group seems clearly related to *Polychrus*. In contrast, the hypothesized relationships among these genera have varied. The phylograms of Etheridge (1960, redrawn here as Fig. 8.5) indicate a paraphyletic *Anolis*, with the other genera internested. My interpretation of Etheridge differs from the tree of Paull *et al.* (1975; also a redrawing of Etheridge) even though they also depict a paraphyletic *Anolis* (Fig. 8.5). Williams (1977) presented another hypothesis and he drew a monophyletic *Anolis* with *Phenacosaurus* as the sister ((((alpha, beta) *Phenacosaurus*) *Chamaeleolis*) *Chamaelinorops*) (Fig. 8.5). Wyles and Gorman (1980) presented immunological distance data that they interpreted as suggesting *Chamaelinorops* was more closely related to other Hispaniolan *Anolis* than were other *Anolis* (thus implying a paraphyletic *Anolis*). Case and Williams (1987) employed morphology and allozymes to test the Wyles and Gorman (1980) hypothesis and successfully rejected the paraphyly notion as described by Wyles and Gorman (1980). Guyer and Savage (1986), in their provocative reanalysis of anole phylogeny, also found a monophyletic *Anolis* (sensu Etheridge, 1960) relative to the other anole genera. Hass *et al.* (1993) brought mtDNA and immunological data (ID) to the question of *Chamaeleolis*, *Chamaelinorops*, and *Anolis* relationships (they did not include *Phenacosaurus*). Based on their ID and mtDNA trees they concluded that *Anolis* was paraphyletic with regard to the other two genera. An interesting observation on the Hass *et al.* (1993) study was the choice of outgroups. The distantly related tropidurid genus *Leiocephalus* was used as the outgroup for the ID analysis. In the mtDNA analysis, *Leiocephalus*, *Polychrus*, and *Eumeces* were used and tellingly,

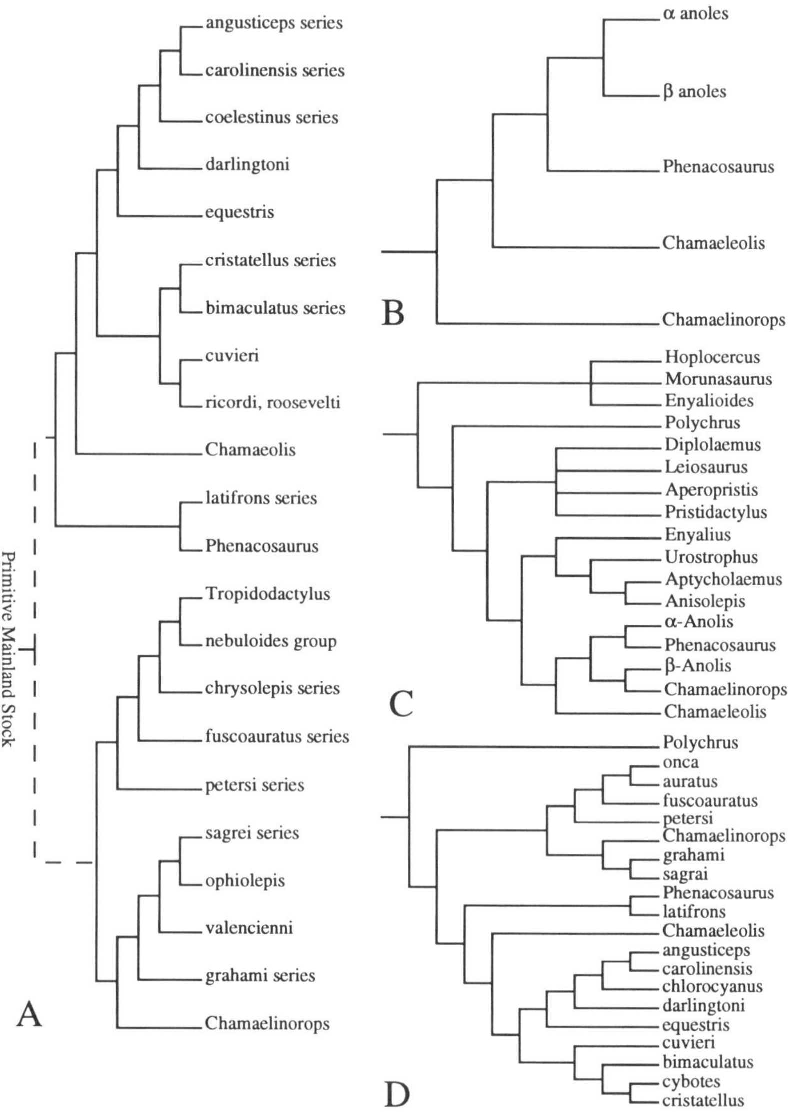

Figure 8.5. Phylogeny estimates of higher level anoline relationships. These are essentially interpretations of Etheridge, 1960. (A) Interpretation herein. (B) Hypothesis of Williams, 1977. (C) Interpretation of Etheridge by Paull *et al.*, 1975. (D) Interpretation of Etheridge by Guyer and Savage, 1986.

Leiocephalus, and not *Polychrus*, fell out as the sister to the anoles. Morphologically, evidence is lacking for a *Leiocephalus* - anole relationship (e.g., Etheridge, 1960, 1966; Frost and Etheridge, 1989; Pregill, 1992), whereas evidence for a closer *Polychrus* - anole relationship is not lacking (e.g., Etheridge, 1960; Peterson, 1983a,b; Guyer and Savage, 1986; Frost and Etheridge, 1989). As far as the relatively simple question of intergeneric relationships is concerned, there currently is no resolution.

Resolving the relationships within *Anolis* (*sensu* Etheridge) is a far more daunting task than estimating the phylogeny of anole genera. It is daunting for the simple reason that there are so many taxa. Savage and Guyer (1989) listed 286 species total, and of these 140 are found in the Caribbean islands and the Bahamas. What is so difficult about 140 taxa is that the approximate number of rooted trees that one must whittle down to a single phylogeny is about 3×10^{279}, which is far greater than the number of particles in the known universe. Given that, is it any wonder that although this group has received much attention over the past couple decades, little has been agreed on by the various workers?

Before I discuss the various interspecific studies, I want to give attention to the study by Guyer and Savage (1986). Guyer and Savage synthesized the osteological data of Etheridge (1960), the multitude of immunological distance studies, and the available karyotypic data to construct a consensus tree of anole relationships. Based on tnis synthesis they subsequently re-classfied *Anolis* and broke it into several genera: *Anolis*, *Ctenonotus*, *Dactyloa*, *Norops*, and *Semiurus*. The erection of these genera was based on the discovery of five monophyletic groups in their final tree, which in itself is not such a controversial method. Although Schwartz and Henderson (1988) immediately embraced the Guyer and Savage classification, a subsequent volume (Schwartz and Henderson, 1991) did not, and two pointed rebuttals (Cannatella and de Queiroz, 1988; Williams, 1989a) to Guyer and Savage followed. Guyer and Savage (1992) contested the arguments against their original study, added more data, and conducted further analyses. Interestingly, the 1992 work by Guyer and Savage is almost never cited and the reasons given for the consistent dismissal of their phylogeny and classification are based on the Cannatella and de Queiroz (1988) and Williams (1989a) rebuttals. Have the Guyer and Savage (1986, 1992) studies and subsequent classification been given a fair scientific treatment? I recommend that all current and aspiring anolologists reread these four papers and pay special attention to the Guyer and Savage 1992 paper in which they argued point by point their work, included suggestions of Cannatella and de Queiroz and Williams, add more data to the analysis, and yet resolve essentially the same set of relationships (Fig. 8.6). I find these to be compelling reasons why the Guyer and Savage studies should not be overlooked and should be subjected to the same test of rejection as any other hypothesis. Whether or not their hypothesis is rejected, it is certain that Guyer and Savage stirred up much activity toward resolving the immense problem of anole phylogeny and classification.

Below the genus level, hypotheses of anole relationships are many and are largely incongruent. Although there are earlier systematic studies on anoles, the dissertation of Etheridge (1960) is effectively the baseline for all subsequent anole phylogeny hypotheses. Etheridge proposed two main divisions of anoles, the Alpha Section and the Beta Section. Both sections have representatives in the West Indies, dominated in species numbers by the Alpha Section. Etheridge figured phylogenies for both sections but mostly at the level of series (Fig. 8.5). In general Etheridge considered the sections monophyletic with respect to each other and that they were derived from a mainland ancestor(s). The Beta Section was confined to Jamaican and Cuban forms and the Alpha section was found elsewhere throughout the Caribbean region.

Lazell (1966) examined the relationships of the Jamaican *Anolis reconditus* and figured a theoretical schema of four taxa which he considered a monophyletic group. Lazell disagreed with Etheridge (1960) by not considering *garmani* part of this group. Lazell (1972) later turned his attention to the Lesser Antillean anoles. Here, he presented two diagrammatic schemes which depicted the hypothesized temporal sequence of colonization events coupled with relationships. These figures are difficult to interpret, and Lazell readily admits his diagrams are not objective renderings. He stated (1972:98),

> To reconstruct the evolution of the groups on an objective foundation is impossible. I shall reconstruct it, then, on a subjective basis. The result cannot be shown to be either right or wrong; one need not believe it or disbelieve it; there is no need to like it or dislike it, there is no necessity to even look at it. In what follows, I shall make abstract representations of real objects and arrange them in patterns that appeal to me. The couching of phrases in speculative terms now ends, but I practice no deception. Even as a mobile constructed out of odd bits of hospital apparatus is not medicine, so this is not science. In all respects, it is the purest of art.

One certainly cannot object to or criticize such honesty. Thankfully, however, the epistemological, theoretical, and empirical foundations of modern systematics are solid, and as such it can be argued that phylogenetics is science, not art, and thus the hypotheses are correctly viewed as ideas open to refutation.

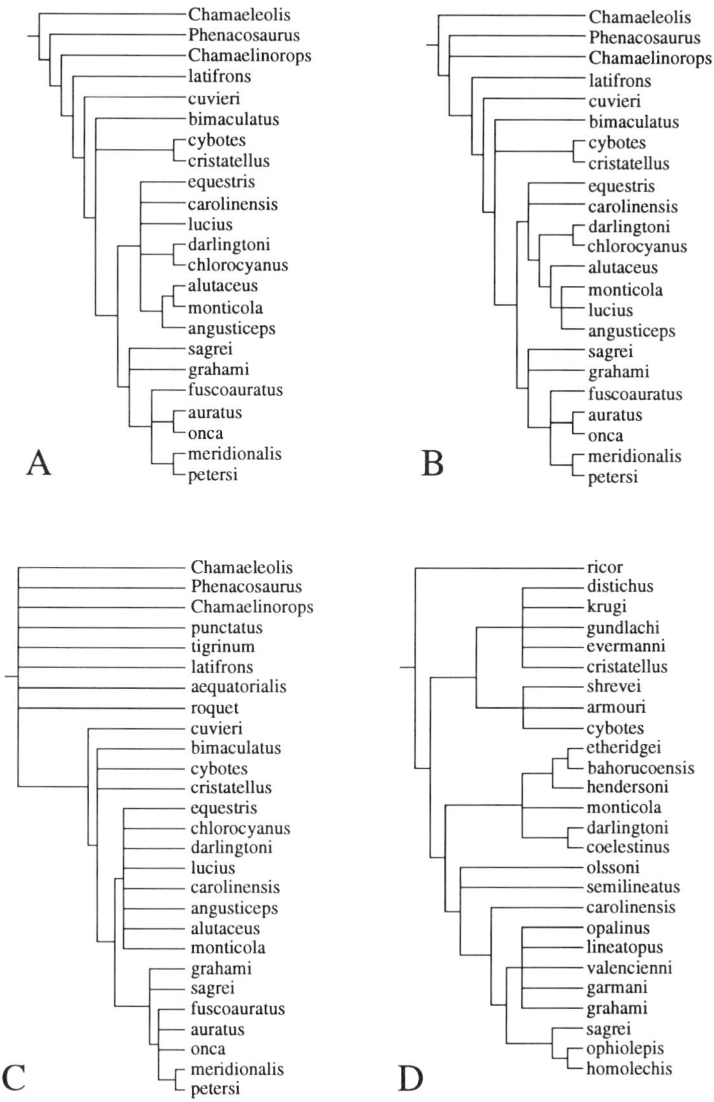

Figure 8.6. Phylogeny estimates of anoline lizards derived from the parsimony analysis of osteological data. (A) Estimate based on modified data set from Guyer and Savage (1986); successive approximation weighted analysis. (B) Estimate based on modified data set from Cannatella and de Quieroz (1989); successive approximation weighted analysis. (C) Estimate based on modified data set from Williams (1989). Unweighted analysis. (D) Estimate from combined osteology, allozyme, and karyotype data recovered with successive approximation weighted analysis (redrawn from Guyer and Savage, 1992).

Gorman and associates were the first to consistently put forth phylogenetic hypotheses of anoles based on objective criteria, and they remained essentially the lone group to do so for almost 20 years. Albeit, it can easily be argued that the first phylogenies were every bit as subjective as Lazell's.

Gorman and Atkins (1969) figured a proposed phylogeny of *bimaculatus*, *acutus*, and Puerto Rican species groups based on karyotypes and lactate dehydrogenase mobility. Although the *roquet* group was also discussed, a phylogeny was not forthcoming until 1974. Yang *et al.* (1974) produced a genetic distance-based dendrogram which was interpreted as an estimate of phylogeny. The phylogeny was at odds with the Gorman and Atkins (1969) tree and thus forced a reevaluation of their colonization scenario. For heuristic purposes, I recoded the Yang *et al.* allozyme data set with the locus as the character and the alleles as the states. Where polymorphisms occurred, frequencies >0.90 were considered fixed and frequencies <0.90 were coded as uncertain. Based on Gorman and Atkins (1969), *Anolis luciae* was used to root the tree alone and together with *blanquillanus* and *bonairensis*. A branch and bound search on PAUP 3.1.1 (Swofford, 1993) recovered 56 most parsimonious trees at 26 steps with high descriptive statistics (CI = 0.938, RI = 0.923, RC = 0.888). However, the strict consensus tree yielded a complete polytomy of the ingroup taxa except for *blanquillanus* and *bonairensis*, which formed a separate clade.

In 1976 Gorman and Kim took the same approach as Yang *et al.* (1974) to estimate the relationships of the other Lesser Antillean anoles: the *bimaculatus* group. A single fully resolved dendrogram was obtained. Following the protocol I described previously, I also reanalyzed these data. For this analysis I employed the *wattsi* complex and *acutus*, as discussed by Gorman and Kim (1976), as outgroup taxa. The branch and bound search was stopped at 4200 most parsimonious trees, each at 21 steps (CI = 0.905, RI = 0.889, RC = 0.804). The strict consensus tree corroborated some of the clades found by Gorman and Kim. The biggest surprise was that *acutus* (considered outside of the other groups) was internested within the *bimaculatus* group as the sister to *pogus*, results which are apparently at odds with chromosomal and morphological evidence. When *acutus* was used as the sole outgroup, the monophyly of the *wattsi* series could not be substantiated. These results are at odds with widely held views of anole relationships (e.g., Williams, 1976; Roughgarden *et al.*, 1987; Roughgarden, 1995). However, Roughgarden *et al.*

(1987) and Roughgarden (1995) also disagree with Gorman and Kim (1976) and the reanalysis herein in the placement of *leachi* and *bimaculatus* (Fig. 8.7). The relationships of the eastern Caribbean anoles remain unresolved, probably for the reasons discussed later.

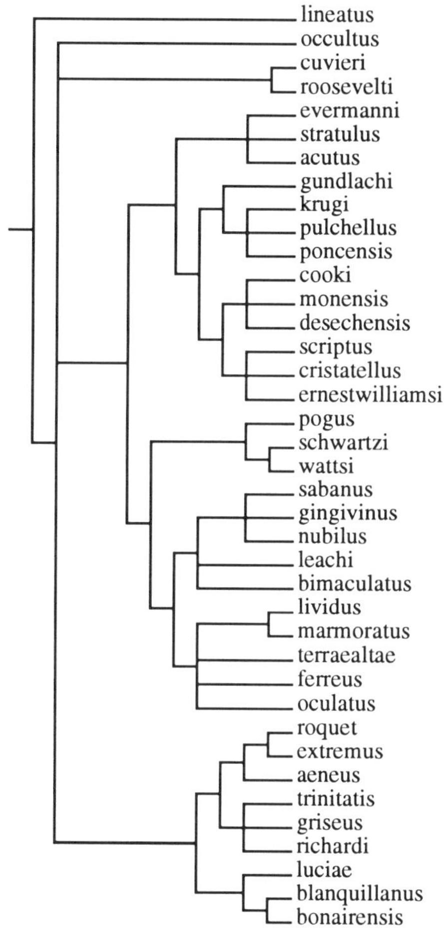

Figure 8.7. Phylogeny estimate of eastern Caribbean anoles based on squamation, karyotypes, allozymes, and immunological data. This is essentially a hand drawn consensus tree of the aforementioned data. The relationships in the *bimaculatus* group are the same as in Roughgarden (1987) and Losos (1992) (redrawn from Roughgarden, 1995).

The phylogenetic relationships of Puerto Rican anoles have been figured several times (Williams, 1972; Gorman *et al.*, 1980a, 1983; Burnell and Hedges, 1990; Roughgarden, 1995). Williams (1972) synthesized osteological, allozyme, and karyotypic data to arrive at his phylogeny. As an interesting aside compare the following quote from Williams (1972:74) about his feelings on his proposed phylogeny with the previous quote of Lazell (1972): "The several kinds of evidence are in very good agreement and the dendrogram can be taken to be a highly probable representation of real relationships."

Gorman *et al.* (1980a), in an incomplete sampling of the Puerto Rican taxa found *monensis* and *cooki* to be sisters and *cristatellus* paraphyletic with regard to *scriptus*. Later, Gorman *et al.* (1983) employed allozymes to examine the relationships of Puerto Rican bank anoles and anoles from elsewhere. They conducted two studies, one with all 10 Puerto Rican species and five others, and a second with eight Puerto Rican species and three others. Both a phenogram (UPGMA) and a cladogram (Wagner) were constructed for each study. The results were incongruent among and between studies and methods. Based on these results and karyotype data, Gorman *et al.* (1983) also presented a summary cladogram. None of the Gorman *et al.* (1983) trees were congruent with Williams' (1972) hypotheses, the main reason probably being the non-monophyletic nature of the Puerto Rican assemblage. The various analyses in Burnell and Hedges (1990) and the schema of Roughgarden (1995) further suggest that the Puerto Rican anoles are not monophyletic. Given that, attempts at understanding the phylogenetic relationships of Puerto Rican anoles without broad taxon sampling are/were doomed to inaccuracy.

Like Puerto Rico, Jamaica has an enticingly small number of anole species and as such has been subjected to phylogenetic analysis under the assumption that the assemblage is monophyletic (with the exception of *sagrei*). Underwood and Williams (1959) evaluated the Jamaican anoles and offered a systematic arrangement for them. Hedges and Burnell (1990) drew a phylogeny of Jamaican anoles as monophyletic, which they considered an interpretation of Underwood and Williams (1959). Interestingly, Underwood and Williams (1959) did not propose a monophyletic radiation of Jamaican anoles, and in fact were convinced otherwise (Underwood and Williams, 1959:9): "These are grouped only as we see them in the island. Without a doubt the Jamaican groups are only parts of more widely distributed groups." Lazell (1966) could be interpreted in the same way: if *garmani* and *valencienni* were not related to the other Jamaican taxa, did their relationships belong elsewhere?

In the two distance analyses of their allozyme data of 49 West Indian species, Burnell and Hedges (1990) found the Jamaican anoles monophyletic, including *sagrei*, but in the parsimony analysis they formed two groups, one related to Hispaniolan forms and the other related to members of the *carolinensis* series. Examination of the bootstrap values shown on these trees may hint at the robustness of the data and thus of the trees. The best of the three trees had only 3/44 (6.8%) nodes with a bootstrap proportion >80%. The monophyly of Jamaican anoles should be questioned.

Hedges and Burnell (1990) argued that they demonstrated the monophyly of Jamaican anoles by finding them monophyletic relative to five other species (and with *sagrei* as the only beta anole). I argue that monophyly relative to five other taxa when another 200+ species exist is not a clear demonstration. Although I think Hedges and Burnell may be correct because of historical biogeographic considerations, their demonstration is not convincing.

The anole assemblages of Cuba and Hispaniola are large in number (54 and 42 species, respectively) and decidedly nonmonophyletic. As such, no detailed phylogenetic hypotheses of all the taxa have been proposed. However, Poe (1998) included 37 species in a study of twig anole (Williams, 1983) relationships. His combined data set approach rejected twig anole monophyly and corroborated the hypotheses that the Cuban and Hispaniolan radiations were non-monophyletic. Regardless, estimation of the phylogenies of the radiations will only be forthcoming in large, inclusive studies which attempt to understand anoles as a whole.

Overall, the systematics of the West Indian anoles (as well as the rest of the anoles) remain in flux, although this was largely not the case after the publication of Williams' (1976,a,b) two influential papers on the classification of anoles. He used a number of formal and informal categories to imply relationship (monophyly?), and these implications have been widely followed since [although Williams (1989a) has argued otherwise]. Guyer and Savage (1986) figured the implied relationships of Etheridge (1960) and/or Williams (1976a,b) (Fig. 8.5).

Although the Etheridge/Williams scheme has been widely used, it has also been challenged. The assumptions of monophyly and relationship have been questioned numerous times from multiple data sources; for example, Gorman *et al.* (1980), Wyles and Gorman (1980), Shochat and Dessauer (1981), Guyer and Savage (1986, 1992), Burnell and Hedges (1990), and Hass *et al.* (1993). Gorman *et al.* (1980b) used albumin immunological data and found incongruence between their study and their interpretation of Williams (1976a,b) for eastern Caribbean anoles. Wyles and Gorman (1980) used immunology again and could not support the classic relationships based on osteology. Shochat and Dessauer (1981) followed along the same path and noted discrepancies. The studies by Guyer and Savage (1986, 1992) were discussed previously and suffice to say they challenged the classic classification based on multiple lines of evidence. Burnell and Hedges (1990) employed data derived from sequential electrophoresis, found their results incongruent with the Etheridge/Williams scheme, and reclassified the West Indian anoles following the category conventions of Etheridge and Williams. Hass *et al.* (1993) examined albumin immunology and 16S mtDNA sequence and also found relationships at odds with Etheridge and Williams.

Although the previous studies all reject the Etheridge/Williams ideas of anole relationships, they are not all in agreement among themselves (see the accompanying figures). Several probable reasons exist for the incongruence: sample, methodology, and data. Exemplar sampling or incomplete samples, while perhaps sometimes necessary, can result in misleading phylogenies and certainly incongruence when different studies include different samples. Inconsistent methodological approaches can lead to perceived incongruences, when in fact the data indeed covary on the same history (Kluge, 1991). Different data sets, although they theoretically covary on the same history, can also result in different hypotheses (e.g., gene trees versus species trees). Of course, the combination of these three problems probably acts synergistically to exacerbate the incongruency. With anole phylogeny, I suspect this is what has occurred.

Where does that leave anole systematics? At the time of this writing a number of labs are bringing DNA sequences, more complete taxon sampling, and I predict consistent phylogenetic approaches to test the controversial ideas of not only Guyer and Savage but also Etheridge, Williams, and all the rest who have sought to understand anole phylogeny.

Scincidae

The Scincidae is taxonomically depauperate in the Caribbean, with only a single genus and two species present. The genus *Mabuya* is widespread, with a range that includes the Greater and Lesser Antilles (exclusive of Cuba), northern South America into Panama, Africa, Madagascar, and southern Asia. In the region of interest here, one of the two species is endemic to Hispaniola (*lineolata*) and the other (*bistriata*) encompasses the remainder of the West Indian distribution of the genus (Schwartz and Henderson, 1991).

The relationships of New World *Mabuya* remain enigmatic. Dunn (1935) suggested that their affinities were with African forms but that relationships between *mabouya* and *lineolata* were obscured by the fact that *lineolata* was unique and highly differentiated from its geographic counterpart. Since Dunn's comments, no studies have attempted to better understand the phylogeny of *Mabuya*. Perhaps the closest to addressing *Mabuya* phylogeny was Greer's (1986) work on the monophyly of the Lygosominae. Clearly, detailed phylogenetic studies of *Mabuya* are needed and would help clarify the relationships of the Caribbean forms.

Teiidae

This New World lizard family is represented in the West Indies by four genera: *Ameiva*, *Bachia*, *Cnemidophorus*, and *Gymnopthalmus*. As currently understood, there are 20 species of *Ameiva*, 1 species of *Bachia*, 2 species of *Cnemidophorus* , and 2 species of *Gymnopthalmus*. Amazingly (or perhaps not amazing at all), for such a widespread and important radiation there have been no modern, rigorous phylogenetic studies of these taxa. Interestingly though, there have been two schemes of phylogeny drawn, one on *Bachia* and relatives (Dixon, 1973) and the other on *Ameiva* (Barbour and Noble, 1915). Otherwise, any notions of relationship for these are only narrative (one might consider the phylograms as drawn narratives) and some require interpretation.

Barbour and Noble (1915) presented their concept of the relationships of *Ameiva* in diagrammatic fashion (Fig. 8.8). In general, they recognized three distinct groups of *Ameiva*: (1) an *A. ameiva* group that contained the South America, Lesser Antillean, Puerto Rican and Hispaniolan species (*vittipunctata* = *chrysolaema*), (2) an *A. undulata* group with relatives in Cuba, Jamaica, and Hispaniola, and (3) a Bahamian - Puerto Rican bank group.

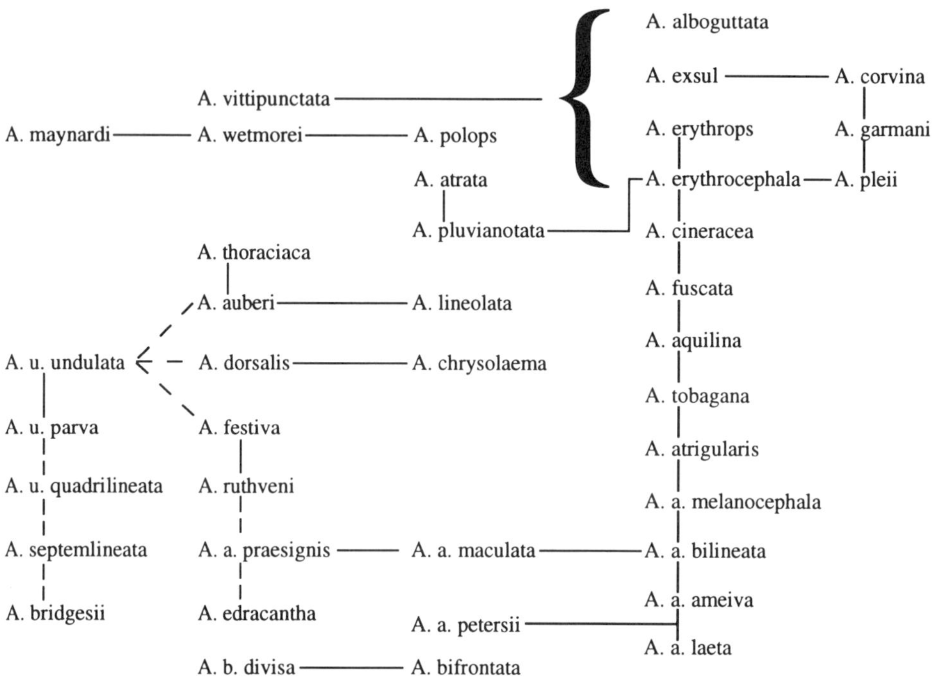

Figure 8.8. Phylogeny estimate of *Ameiva* based on morphology. Solid lines represent close relationship, dashed lines indicate the taxa are related but with unknown taxa intervening (Interpreted and redrawn from Barbour and Noble, 1915).

Cochran (1941) did not agree with Barbour and Noble and in her revisions she repeatedly broke up Barbour and Noble's second and third groups. For example, Cochran suggested that *wetmorei* and *lineolata* were related and that *polops* and *taeniura* (considered *lineolata* by Barbour and Noble, 1915) were closest relatives, and she synonymized a South American member (*vittipunctata*) with a Central American member (*chrysolaema*).

Baskins and Williams (1966) evaluated the LA *Ameiva* and drew some tentative conclusions about relationships. They considered the LA *Ameiva* to be composed of three lineages, a northern group (*erythrocephala*, *griswoldi*, *pleei*, *pluvionotata*, *corvina*, and *fuscata*), a southern group containing *cineracea* and *major*, and a third group composed of only *ameiva*. Within the northern group, Baskins and Williams only speculated that *erythrocephala* and *griswoldi* were sisters relative to the rest that had uncertain affinities. By default, the two extinct members of the southern group were considered to have sister-taxa status and *ameiva* was related to mainland taxa. They also speculated that if the Lesser Antillean taxa had any relationship to Greater Antillean taxa, it might be with *chrysolaema* and/or *exsul*.

Heatwole and Torres (1967) examined the Puerto Rican Bank taxa and concluded that *wetmorei*, *exsul*, and *polops* were not related but derived from independent stocks. They did not speculate as to what these stocks may have been.

Schwartz (1970) proposed a close relationship between *auberi* and *dorsalis*, which is at odds with the schema of Barbour and Noble (1915). He further suggested that *chrysolaema* might possibly be part of that group. He also accepted Cochran's (1941) idea that *taeniura* and *polops* were closely related but then added that *maynardi* and *wetmorei* also were relatives of *taeniura-polops*. No mention was made of *lineolata*, which Cochran (1941) thought was related to *wetmorei*. Schwartz (1970) also argued that the entire West Indian *Ameiva* radiation was derived from a South American ancestor.

Dixon (1973) reviewed the relationships of *Bachia* and related genera and presented in a phylogram his concept of phylogeny. The lone West Indian form, *B. heteropus* (found on the Grenada Bank and northeast South America), was depicted as being the sister to *pallidiceps*, which is a northern South American taxon that ranges into Panama.

The species of *Cnemidophorus* are confined to tiny islands in the western Caribbean and the LA. *Cnemidophorus lemniscatus* is widespread on the Central and South American mainland and is also found on two small islands off Nicaragua. The other species, *vanzoi*, is restricted to an islet of St. Lucia. Initially, *C. vanzoi* was described as an *Ameiva* (Baskins and Williams, 1966), but Presch (1971) demonstrated that it belonged to *Cnemidophorus* based on distinctive tongue morphology. As an *Ameiva*, the distribution made some sense, as a *Cnemidophorus*, it appears highly unusual. Presch (1971) considered *vanzoi* to be a member of the *lemniscatus* group.

Thomas (1965d) reviewed the *Gymnopthalmus* of the southeastern Caribbean and concluded that *G. pleei* (found on the central LA islands of Guadeloupe, Dominica, Martinique, and St. Lucia) was related to *G. lineatus* (Curaçao and mainland). The other species, *G. underwoodi*

(Guadeloupe, St. Vincent, Barbados) could not be placed but Thomas (1965d) suggested its affinities lay with mainland forms.

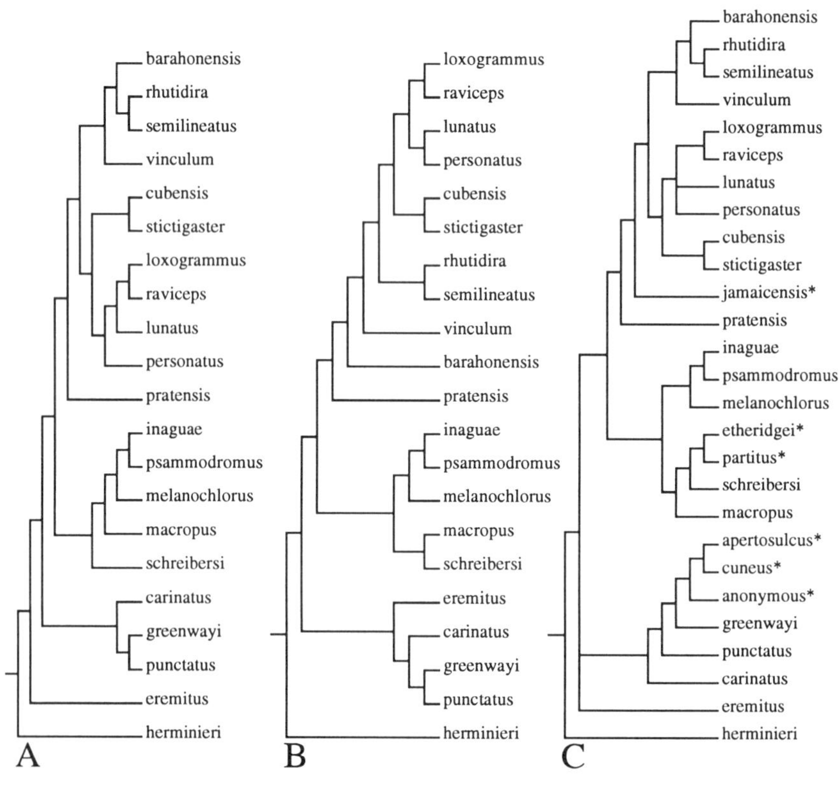

Figure 8.9. Phylogeny estimates of *Leiocephalus* based on parsimony analyses of morphological data. (A) Tree derived from an analysis with the software HENNIG86. (B) Tree derived from an analysis of the same data as A but with the software PAUP. (C) Strict and Adams consensus tree of parsimony analysis which included fossil taxa (denoted by aterisks) (redrawn from Pregill, 1992).

Tropiduridae

A single endemic genus (*Leiocephalus*, as diagnosed by Etheridge, 1966) representative of the Tropiduridae is found in the Caribbean islands. Frost and Etheridge (1989) proposed that *Leiocephalus* was the sister taxon to a South American clade composed of the *Stenocercus* group, *Tropidurus* group, and *Uranoscodon*. Pregill (1992) supported the Frost and Etheridge (1989) hypothesis with additional morphological characters.

The genus *Leiocephalus* is fairly diverse in numbers of species, and they are arguably the most distinctive terrestrial lizards on the islands. There are either 21 or 22 extant species (depending on whether you refer to Pregill, 1992, or Schwartz and Henderson, 1991) and all possess a remarkably upwards curled tail [thus the "curly-tailed lizard" moniker; Pregill (1992) colorfully described the tail as resembling a "coiled watch spring"]. In addition to the extant species, there are two recently extinct and another six which are probably late Holocene in age but no older than late Pleistocene (Pregill, 1992).

Schwartz was by the far the most prolific worker on *Leiocephalus* , publishing no less than 16 papers on variation and taxonomy in the genus (see Pregill, 1992, for reference list). However, the only intrageneric phylogeny proposed was by Pregill (1992). The phylogeny was morphology based, included the fossil taxa, and was analyzed within a cladistic framework. The evidence was well documented and the phylogenies fairly well supported. As such, the conclusions of Pregill (1992) should be considered good estimates of *Leiocephalus* phylogeny. The trees without fossil taxa differ somewhat from the trees with fossil taxa, so both are depicted herein (Fig. 8.9). The phylogenies suggest *Leiocephalus* had a complex evolutionary history, i.e., the island radiations were not found to be monophyletic and may reflect the complex geological history of the Caribbean region or indicate a taxon of great dispersal abilities.

Xantusiidae

A single member of this enigmatic family is found in the Caribbean: the Cuban endemic *Cricosaura typica*. Superficially, with the family containing only three (or possibly four) genera, the phylogenetic problem would appear to be simple. This, however, has not been the case, with the frustration aptly described by Bezy (1972:1): "Not only have xantusiid lizards been troublesome to students of 'higher classification,' but those unfortunate taxonomists who have been lured into extensive studies of the systematics of the family have suffered greater torments." Over 25 years later Bezy's comment remains true.

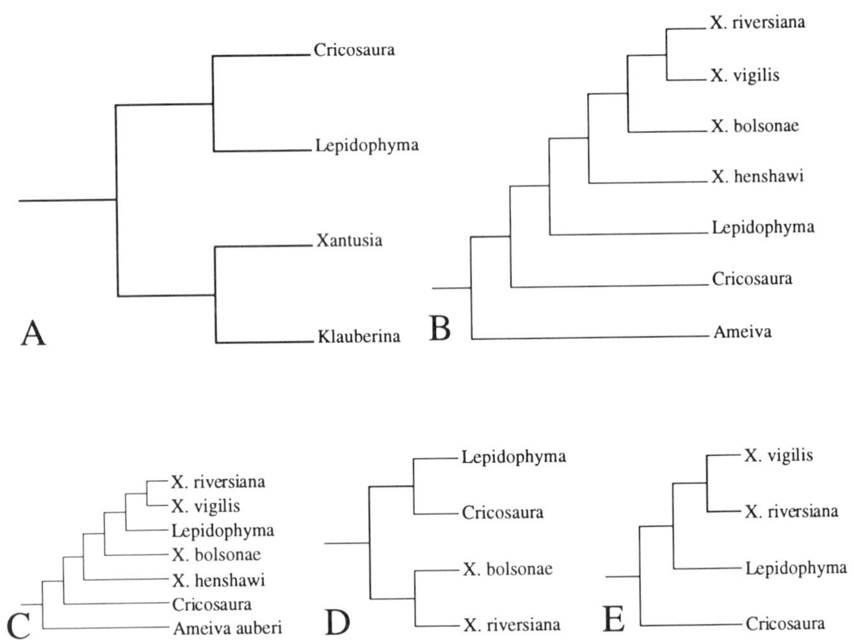

Figure 8.10. Phylogeny estimates of *Cricosaura.* (A) Estimate derived from the phylogenetic analysis of morphological characters (redrawn from Crother *et al.*, 1986). (B) Estimate derived from analysis of mtDNA sequence (redrawn from Hedges *et al.*, 1991). Neighbor-joining tree from combined sequences. Branch lengths do not correspond to distance estimates. (C-E) Phylogeny estimates of *Cricosaura* relative to other xantusiids based on reanalyses of mitochondrial DNA used in Hedges (1991) and with added morphological data (redrawn from Crother and Presch, 1993). All trees are based on parsimony analyses. (C) cytochrome b data alone, unweighted, ambiguities removed, (D) combined molecular and morphological data unweighted with base ambiguities, (E) combined data weighted with base ambiguities.

Savage (1955, 1963, 1964) can be considered the first worker to discuss the evolutionary relationships of *Cricosaura* to the other xantusiid genera. Savage (1955, 1964) examined external and osteological morphological characters and suggested that *Klauberina* (= *Xantusia riversiana*) was the ancestor of the other genera but that *Cricosaura* was its closest ally, although distantly related. So distant in fact, that Savage (1963) placed *Cricosaura* in its own subfamily. Interestingly, in interpretations of Savage's relationship hypotheses, neither Schatzinger (1980) nor Crother *et al.* (1986) depicted a *Cricosaura–Klauberina* clade (Fig. 8.10). Schatzinger (1980), in a study of *Palaeoxantusia*, diagrammed *Cricosaura* as the sister to the rest of the genera and noted that his conclusions were similar to those arrived at by Savage. Crother *et al.* (1986) in a cladistic reanalysis of Savage's morphological data, depicted two different Savage hypotheses, one phenetic and the other evolutionary. The phenetic tree was the same as Schatzinger's diagram, and the evolutionary tree placed *Lepidophyma* as the sister of *Cricosaura.* I am hard pressed to explain the absence of the *Cricosaura– Klauberina* hypothesis in Crother *et al.* (1986). Regardless, Crother *et al.* found *Cricosaura* to be the sister taxon of *Lepidophyma.*

Bezy (1972) presented karyotypic data on all the xantusiids except for *Cricosaura*, but Hass and Hedges (1992) later published the karyotype of *Cricosaura.* Hass and Hedges concluded that karyotype data were consistent with a hypothesis they (Hedges *et al.*, 1991) proposed based on mitochondrial DNA. This hypothesis stated that *Cricosaura* was the sister to the other genera. However, based on Bezy's (1972) conclusions concerning the evolution of diploid number and number of acrocentric chromosomes, the assertion of Hass and Hedges (1992) is incorrect. Bezy (1972) suggested that diploid number and acrocentric counts have evolved from high numbers to low numbers. As such, Bezy considered *Xantusia* primitive (2N = 40, 11 pairs acrocentrics), *Lepidophyma* derived (2N = 38, 9-10 pairs), and, to follow Bezy's logic, *Cricosaura* the most derived (2N = 24, 6 pairs). This scenario suggests a *Cricosaura-Lepidophyma* relationship, which is at odds with Hass and Hedges (1992) and Hedges *et al.* (1991) but in concordance with Crother *et al.* (1986).

The Hedges *et al.* (1991) study has been the only one to employ DNA sequence to the question of xantusiid relationships, and perhaps as a result of this the authors (Hedges *et al.*, 1991:768) make the claim that their study "provides the first robust estimate of intergeneric relationships in this family" As mentioned previously, they concluded that *Cricosaura* was the sister to the other xantusiid genera (Fig. 8.10). Their conclusions however were challenged by Crother and Presch (1993), who reanalyzed the data and demonstrated that in fact a robust estimate was not achieved by Hedges *et al.* (1991). Instead, the results of the reanalyses yielded all manner of relationships (Fig. 8.10), strongly indicating that the final word on xantusiid relationships is yet to come (Crother and Presch, 1994; Hedges and Bezy, 1993, 1994).

Yet another data set has been examined in xantusiid lizards: scale microstructure. Peterson and Bezy (1985) found *Cricosaura* as the sister to *Xantusia*, but in a later study (Bezy and Peterson, 1988), they retreated from their hypothesis and concluded that scale microstructure failed to resolve the relationships among the xantusiid genera.

What can be concluded about the state of phylogenetic knowledge for *Cricosaura*? Although much effort has been expended on reco n-structing the phylogeny of xantusiids, it seems clear that further work remains.

Serpentes

The snake fauna in the West Indies is represented by seven families, 23 genera and 108 species [Powell *et al.*, (1996a) listed 107 species of snakes; I include *Micrurus ruatanus*, an endemic in the western Caribbean]. The Colubridae make up the largest fraction of the assemblage with 17 genera (six endemic) and 48 species (43 endemic). Although the remainder of the families (Boidae, Elapidae, Leptotyphlopidae, Tropidopheidae, Typhlopidae, and Viperidae) contain no endemic genera, all but five of the species are endemic to the West Indies/Caribbean islands. Of the 11 nonendemic species, only two (*Elaphe guttata* and *Diadophis punctatus*) are considered to be human introductions (as such, these two taxa will not be discussed further).

Boidae

Three genera of boids (*Boa*, *Corallus*, and *Epicrates*) are found in the Caribbean islands and none are endemic. *Boa* and *Corallus* are represented by their widespread mainland species, *constrictor* and *hortu l-lanus*, respectively. *Epicrates*, on the other hand, is represented by 10 endemic species, found throughout the Greater Antilles and the Bahamas.

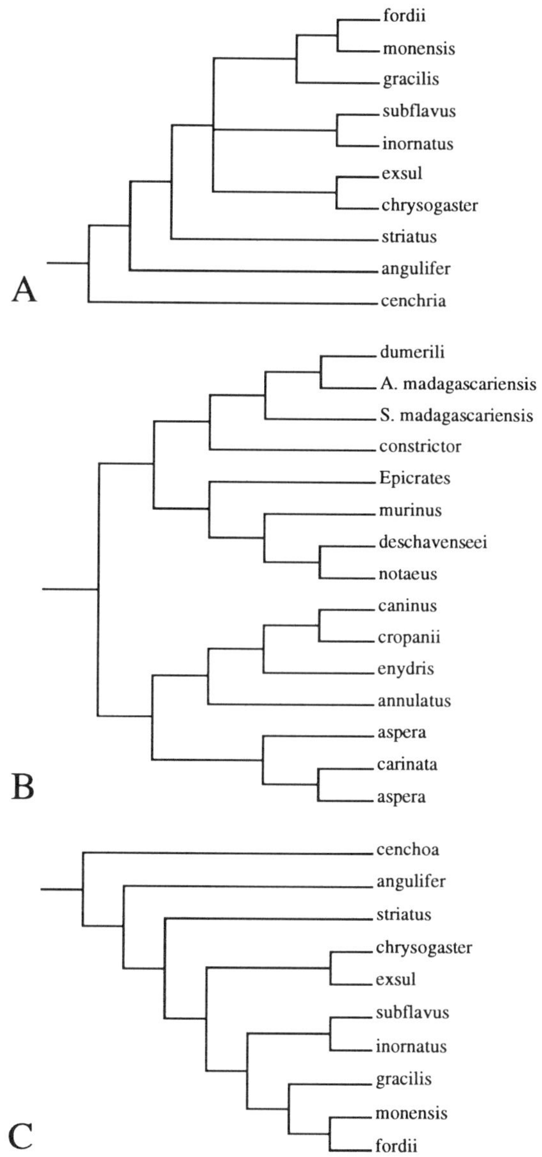

Figure 8.11. Phylogeny estimates of boine snakes. (A) Estimate of the interspecific relationships of the genus *Epicrates* derived from a parsimony analysis of morphological and biochemical data (Kluge, 1989) (A, B, redrawn from Kluge, 1991). (B) Most parsimonious estimate of boine phylogeny based on morphological data (Kluge, 1991). (C) Phylogeny estimate of *Epicrates* based on a parsimony analysis of morphology and lipid data (redrawn from Tolson, 1987).

Differentiation of *Boa constrictor* apparently has not occurred at the specific level in the LA but some variation from the mainland form is present, which led Lazell (1964) to assign subspecific epithets to the island forms. Lazell (1964) considered the island *Boa* to be the end of a step cline. Until recently, *Boa* was monotypic, containing only *constrictor*. Kluge (1991) synonymized *Acrantophis* and *Sanzinia* with *Boa* based on his phylogenetic analysis of boine snakes. According to Kluge (1991) the sister group to *B. constrictor* contains the Madagascaran species *manditra*, *dumerili*, and *madagascariensis* (Fig. 8.11). Heise *et al.* (1995) included *B. constrictor* in their mtDNA based phylogeny of snakes and most peculiarly found *Boa* as the sister to colubroids instead of to *Python*, a result not supported by either Kluge (1991) or Cundall *et al.* (1993). The reasons for this peculiar arrangement may be in the mtDNA data set. The mtDNA may not be conserved enough for the question of higher snake phylogeny, the estimate may be befuddled by ancestral polymorphism problems, and it is possible that the phylogeny reflects a gene tree instead of a species tree. (Obviously it is a gene tree. The implication is that the gene evolution did not track the species evolution.) These matters are for discussion elsewhere.

The phylogenetics of the species *Corallus hortullanus* (= *enydris*, McDiarmid *et al.*, 1996) has been addressed at the intrageneric (Kluge, 1991) and intraspecific (Henderson and Hedges, 1995) levels. Henderson and Hedges examined single individuals from diverse geographic localities, including St. Vincent, Grenada, and Trinidad. They used a 307-base pair stretch of cytochrome *b* in the mtDNA to address the monophyly of *hortullanus* and to discover the possible origin of the West Indian populations. Monophyly was supported but the intraspecific results were intriguing in that Panama clustered with Trinidad, whereas the West Indian individuals were sisters, in a clade with Guyana, Brazil, and Peru. At the interspecific level, Kluge found *hortullanus* nested within the other two species of *Corallus* and *Xenoboa* (((*caninus*, *cropanii*) *hortullanus*) *annulatus*) (Fig. 8.11) and given these results he synonymized *Xenoboa* with *Corallus*.

Sheplan and Schwartz (1974) were the first workers to examine the systematics of all the species of Antillean *Epicrates* in a single work. At the time of their work, the systematics of *Epicrates* was, as they stated, "in a chaotic state." Based on the external characteristics of these snakes, they attempted to add order to the chaos. In retrospect, given the recent studies of Tolson (1987) and Kluge (1989), Sheplan and Schwartz (1974) provided strong predictions about *Epicrates* phylogeny. They suggested two groups, one comprising the smaller-size species and the other comprising the larger species. The small species are apparently monophyletic (with the possible exception of *exsul*), but the larger species are not (Kluge, 1989; Tolson, 1987). Sheplan and Schwartz suggested that *angulifer* was basal, that *inornatus* and *subflavus* were sisters, and that *fordii* and *monensis* were also sister taxa. All these points were reaffirmed by Kluge (1989) and Tolson (1987). The only major difference between Sheplan and Schwartz (1974) and Kluge (1989) and Tolson (1987) was the placement of *exsul*. Kluge and Tolson both relied on morphological, osteological, and skin and scent lipids for their phylogenies (Kluge added 44 more morphological characters to Tolson's original data set). In Kluge's study, when the morphological data were analyzed separately *exsul* was placed in the smaller-sized clade, *a la* Sheplan and Schwartz. Tolson's morphological data alone never supported that arrangement. Tolson and Kluge's studies did differ and their results differed only in the level of resolution (Fig. 8.11). Tolson used *Corallus* as the outgroup, whereas Kluge used a hypothetical ancestor composed of both *Corallus* and *Eunectes*, and of course the two studies differed in that Kluge added 44 morphological characters. Regardless, their total evidence phylogenies were 100% consistent with each other, departing only in the relative positions of *chrysogaster-exsul* and *inornatus -subflavus* clades. Kluge's hypothesis was simply unresolved at that point.

At the intergeneric level several hypotheses have been figured for the relationships of *Epicrates*. Kluge (1991) provided interpretations of narratives and redrawings of these relationships, including Schwaner and Dessauer (1981), Underwood (1976), and McDowell (1979). These phylogenies and Kluge's (1991) were all generally similar in depicting the possibility of an *Epicrates-Corallus* relationship.

Colubridae

As mentioned previously, the colubrid snakes are represented in the Caribbean/West Indies by 48 species, 43 of which are endemic. Also mentioned previously, two of these (*Diadophis punctata* and *Elaphe guttata*) are considered human introductions and will not be considered further.

Only two species of colubrine snakes are found in the Antilles and both are endemics. *Chironius vincenti* and *Mastigodryas bruesi* are both found on St. Vincent but *C. vincenti* ranges further south throughout the

Grenadines and Grenada. These genera of colubrines have been included in but one phylogenetic study, an extraordinarily ambitious allozyme project conducted to address the phylogeny of snakes (Dowling *et al.*, 1996). Without using *vincenti* or *bruesi*, but using other representatives of these genera, Dowling *et al.* (1996) placed *Chironius* and *Mastigodryas* in a large clade composed of a somewhat eclectic collection of colubrids (Fig. 8.12). Do these genera really share a close relationship with *Dasypeltis*?

Dowling *et al.* (1996) used slowly evolving loci as data as discovered through sequential electrophoresis. Typically, allozyme data reach their phylogenetic utility limits at the genus level (intrasubfamilial level?; Buth, 1984). Even with the sequential protocol, clearly the data were strained to construct a snake phylogeny with such ancient origins. This was evidenced by the numerous notations by the authors concerning misplaced taxa. It can be safely said that further work needs to be done on the phylogeny of these taxa.

A single natricine species, *Nerodia clarkii*, is distributed across the northern coast of Cuba in the Antilles, but the extent of the distribution is throughout the coast of the southeastern United States from Texas to Florida. Lawson (1987) conducted the only explicit phylogenetic analysis of *Nerodia* taxa. He used allozymes and performed both discrete (with parsimony) and distance-based analyses. The parsimony tree suggested a *clarkii-sipedon* clade, two of the distance based estimates indicate a *clarkii-fasciata* clade, and the third distance tree shows *clarkii* as the sister to a large clade that contains four other species of *Nerodia* (not including *sipedon* or *fasciata*) and *Thamnophis couchii*. The lack of congruence is a problem, but Kluge (1991) correctly pointed out that incongruence derived from comparison of phylogenies based on differing methodologies and philosophies only serves to obscure the evidence present in the data. Because of my allegiance to the cladistic parsimony paradigm, I suggest that *clarkii-sipedon* might be the best estimate of relationship only tentatively, however, given recent improvements in our understanding of coding allozyme data. Unfortunately, Lawson's (1987) data could not be recoded and reanalyzed because the outgroup data were not included in the publication.

Figure 8.12. Phylogeny estimate of snakes based on a UPGMA analysis of allozyme data (condensed and redrawn from Dowling *et al.*, 1996).

The remaining 43 species belong to the still enigmatic xenodontine-dipsadine group. Several of these taxa have been variously considered part of a single radiation (*Alsophis*, *Antillophis*, *Arrhyton*, *Darlingtonia*, *Hypsirhynchus*, and *Uromacer*) or several unrelated monophyletic groups. Because of this history, these taxa will be considered together. The other genera, *Clelia*, *Coniophanes*, *Liophis*, *Pseudoboa*, and *Tretanorhinus*, have not been considered associated with the former taxa or each other and thus will be reviewed separately.

Two species of *Clelia*, *errabunda* and *clelia*, are recognized on the islands and are both from the southern LA (Underwood, 1993). In a narrative, Underwood (1993) suggested that the previous two *Clelia* species were in the same group as *scytalina* and *equatoriana*. However, Dessauer *et al.* (1987), in a phenogram based on immunological distance data, found *scytalina* and *clelia* to be unrelated, with *Clelia* in fact paraphyletic with respect to *Pseudoboa* and *Oxyrhopus* (which in turn is paraphyletic with respect to *Clelia* and *Pseudoboa*). *Pseudoboa neuwiedii* was suggested to be the sister to *P. nigra* (Dessauer *et al.*, 1987), and in Cadle (1984a) *Clelia* and *Pseudoboa* are shown as sister taxa. Jenner and Dowling (1985) also show *Clelia* and *Pseudoboa* as sister taxa. Dowling *et al.* (1996), in the same allozyme paper mentioned previously, included a single species of *Clelia* (*rustica*) and it fell out as the sister to *Liophis viridis* (Fig. 8.12).

The western Caribbean island of Isla San Andrés hosts the endemic *Coniophanes andresensis*. No phylogenies have been estimated which include this species but the genus has been explicitly included in phylogenetic analyses (Cadle, 1984a,b, 1988; Crother, in review; Jenner and Dowling, 1985), and Myers and Campbell (1981) discussed the possible relationships of *Coniophanes*. Crother (in review) tried two methods of coding allozyme data and found the position of *Coniophanes* to be unstable. With the locus as the character and polymorphisms coded as uncertain, the genus fell out in a clade containing *Geophis*, *Imantodes*, *Conophis*, *Crisantophis*, and *Enulius* (Fig. 8.13). With the data coded using the presence and absence of alleles (alleles as characters), *Coniophanes* was placed as a basal lineage. Jenner and Dowling (1985) presented a hypothesis that depicted *Coniophanes* as the sister to *Leptodeira*, but the tree included only eight xenodontine genera and was largely an attempt to fit morphological data with immunological data. Myers and Campbell (1981) proposed in a "theory of relationships" a sister relationship between *Rhadinaea* and *Coniophanes*, and Cadle

(1984b), based on immunological distances, partly supported Myers and Campbell (1981) and suggested that *Coniophanes* was related either to some part of *Rhadinaea* or to *Urotheca* (= *Pliocercus*). In a phylogeny of exemplar taxa, Cadle (1984a, 1988) found *Coniophanes* as either sister to a *Sibon-Geophis* clade or to a *Leptodeira-Eridiphas* clade. Finally, in a study on *Tretanorhinus*, Pinou and Dowling (1994) placed *Coniophanes* with *Sibon*.

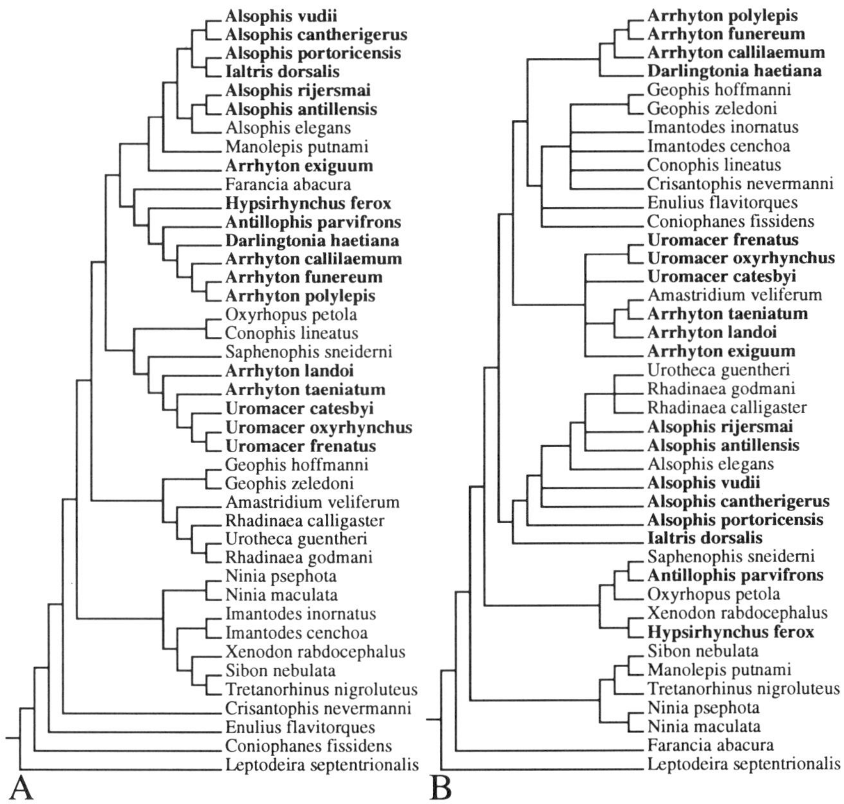

Figure 8.13. Phylogeny estimate of xenodontine snakes based on allozyme data. Bold type denote West Indian species. (A) Strict consensus tree of parsimony analysis with the alleles considered characters. (B) Strict consensus tree of parsimony analysis with uncertain polymorphisms removed (redrawn from Crother, in review).

Tretanorhinus is the only other genus with species on both the islands and the mainland. *Tretanorhinus variablilis* is found on Cuba and the Cayman Islands. Three species are found on the mainland, from southern Mexico into northern South America (Colombia and northwestern Ecuador). There have been no interspecific phylogenetic studies conducted on this genus, and, in fact, the mainland and island forms have not been compared in the same paper to date. Dunn (1939) worked on the mainland forms and Wood (1939) revised the island forms. In neither study did the authors speculate on the overall interspecific relationships of all the species (Dunn did mention that *taeniatus* and *moquardi* were related). Pinou and Dowling (1994) demonstrated that *Tretanorhinus* (using only the Cuban *variabilis*) was not a natricine but clearly related to dipsadines. Their phylogeny shows *Tretanorhinus* relative to *Leptodeira*, *Sibon*, and *Coniophanes*. Dowling *et al.* (1996), in their attempt to sort out snake phylogeny with allozymes, recovered a clade that contained *Pseustes*, *Leptophis*, and *Tretanorhinus*. The placement of *Tretanorhinus* in this most peculiar clade was by the authors admission, a misplacement. Crother (in review), like Dowling *et al.* (1996), used *nigroluteus* to represent the genus and found it to share a relationship with *Sibon* and with dipsadines in general, a hypothesis similar to that of Pinou and Dowling (1994).

The genus *Liophis* is represented in the Caribbean islands by seven species: *perfuscus* on Barbados, *juliae* on Dominica, Guadeloupe and Marie-Galante, *ornatus* on St. Lucia, *cursor* on Martinique, *melanotus* on Grenada (and Trinidad, Tobago, and northern South America), and *triscalis* on Curaçao [*triscalis* is included in this discussion because of Dixon's (1981) comments]. Maglio (1970), based on morphological data, inferred a phylogeny (Fig. 8.14) of the Caribbean *Liophis* (considered *Dromicus* by Maglio), with the exclusion of *triscalis* because of its distribution in the Dutch Leeward Islands (outside the scope of Maglio's work). The species *melanotus* was placed as the sister to the rest of the endemic Antillean forms. Dixon (1981) took exception with Maglio's concept of relationships among the Caribbean *Liophis*. First, Dixon did not consider *melanotus* closely related to the eastern Caribbean species, but did believe that *triscalis* should be part of the Antillean group. He also concluded that those five Caribbean species (not including *melanotus*) were not closely related to any mainland forms of *Liophis*. That is the current status of our knowledge on the phylogeny of the Caribbean *Liophis*.

The remaining genera of snakes (*Alsophis*, *Antillophis*, *Arrhyton*, *Darlingtonia*, *Hypsirhynchus*, *Ialtris*, and *Uromacer*) found in the West Indies have been variously considered either as multiple lineages (three or four) or as a monophyletic group. As such, they will initially be discussed together with regard to the question of monophyly. Dunn (1932) did not present an explicit phylogeny but argued that this assemblage of snakes was not monophyletic. Maglio (1970) presented the first explicit phylogenetic hypotheses for these taxa and he agreed with Dunn (1932) on one point: the assemblage was not monophyletic. Maglio (1970) depicted three groups: (1) the *cantherigerus* group (*Alsophis*, *Hypsirhynchus*, *Uromacer*), (2) the *andreae* group (*Antillophis*), and (3) the *funereum* group (*Arrhyton*, *Darlingtonia*) (Fig. 8.14). Cadle (1984a) collected immunological data for a subset of taxa from Maglio's *cantherigerus* group and, based on the small amount of differentiation exhibited, suggested that if his sample was typical of the West Indian radiation, then that radiation must be "rather compact" (Cadle, 1984a:16; I interpret this as meaning monophyletic, especially given his 1985 biogeographic scenario). In phylogenies of exemplars, Cadle placed *Alsophis* with *Hydrodynastes* and with *Clelia-Helicops*. Crother (in review), like Maglio (1970), recovered multiple groups of the West Indian taxa (Fig. 8.13). The mainland members of the groups varied depending on the allozyme coding method. As far as the West Indian members were concerned, *Arrhyton exiguum*, *Antillophis parvifrons*, and *Hypsirhynchus* were the only unstable taxa. Otherwise, one group contained *Alsophis* and *Ialtris* (a genus Maglio could not place), a second group contained the Jamaican *Arrhyton* (*funereum*, *callilaemum*, *polylepis*) and *Darlingtonia*, and a third group included *Uromacer* and the Cuban *Arrhyton* (*taeniatum*, *landoi*). Dowling *et al.* (1996) also found the Antillean xenodontines to be nonmonophyletic, with taxa scattered throughout a large xenodontine clade and *Arrhyton funereus* [sic] as part of the unresolved basal lineage between the scolecophidians and the rest of the snakes. It may be safe to assume that based on the phylogenetic evidence compiled to date, the monophyly of these taxa is seriously in doubt, although the relationships among these groups remain in question.

Concordance of interspecific hypotheses of phylogeny are also lacking. In the Antilles the genus *Alsophis* is currently composed of 11 species, several of which are extremely rare or possibly extinct. As with most of the xenodontine genera in the Antilles, Maglio (1970) was the first to estimate phylogenetic relationships for *Alsophis*. Maglio suggested a paraphyletic genus, inserting *Hypsirhynchus* and *Uromacer* as a clade related to *ater* (Fig. 8.14). *Alsophis vudii* and *cantherigerus* formed a clade that was the sister to the previously mentioned *ater* clade. *A. portoricensis* was nested within a clade of LA species and the Hispaniolan taxa were split: *melanichnus* with the LA clade and *anomalus* with the other clade. The proposed sister group to Maglio's paraphyletic *Alsophis* contained *Philodryas* and *Conophis*.

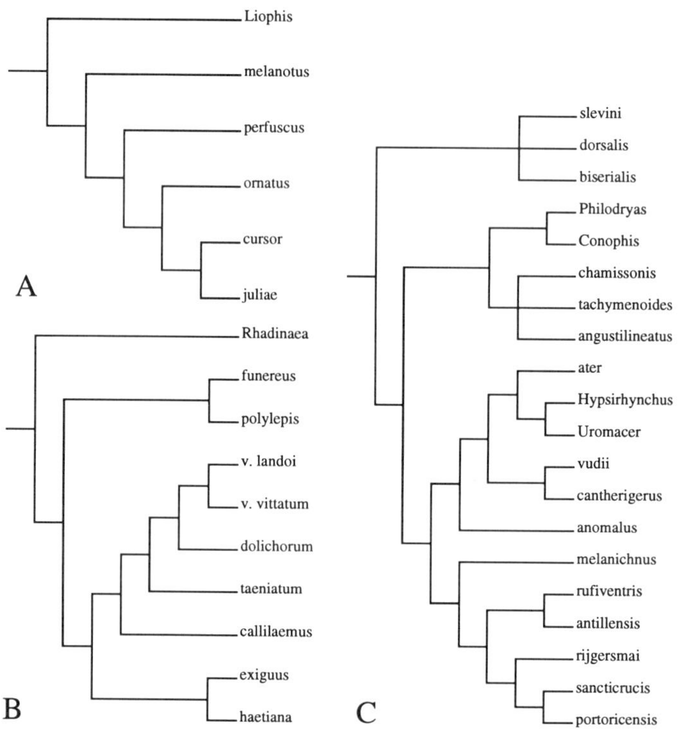

Figure 8.14. Phylogeny estimate of West Indian xenodontines based on morphology. (A) Estimate of *Liophis*. (B) Estimate of *Arrhyton* and *Darlingtonia*. (C) Estimate of *Alsophis, Uromacer*, and *Hypsirhynchus*. Specific epithets all refer to *Alsophis* (redrawn from Maglio, 1970).

Cadle (1984a) presented immunological data from *cantherigerus* , *vudii*, *rufiventris*, and *Hypsirhynchus* and concluded that his findings were totally consistent with Maglio's (1970) phylogeny. The species *cantherigerus* and *vudii* were essentially indistinguishable (sister taxa?), with *Hypsirhynchus* nearer to them than *rufiventris*, a LA *Alsophis*. The immunological data did not support a *Philodryas* or *Conophis* relationship with the Antillean *Alsophis*. However, Jenner and Dowling (1985) depicted a *Philodryas* - *Alsophis* relationship in their tree. Cadle (1988), in a depiction of some broad colubroid relationships, placed *Alsophis* in a clade with *Clelia* and *Helicops*.

The allozyme data of Crother (in review) also suggested a paraphyletic *Alsophis*, but with respect to different taxa. The preferred hypothesis (note that both hypotheses place *Ialtris* and *A. elegans* with the Antillean *Alsophis*) in Crother (in review) depicts two clades, one composed of the Greater Antillean and Bahamian species (*vudii*, *cantherigerus*, *portoricensis*) and *Ialtris dorsalis* and the other clade composed of the Lesser Antillean species *rijersmai* and *antillensis* plus the South American *Alsophis elegans* (Fig. 8.13). These results support the *vudii-cantherigerus* sister relationship (also supported in Crother and Hillis, 1995) and the close relationship among the LA taxa. However, in no cases were *Uromacer* or *Hypsirhynchus* found nested within *Alsophis*. The relationship of *Ialtris* with *Alsophis* was mentioned as a possibility by Maglio (1970) but only because he otherwise could not ascertain the place of *Ialtris*. Schwartz and Rossman (1976) strongly questioned Maglio's comment but held off making a decision until further data could be collected. Based on the allozyme data, Crother (in review) suggested synonymizing *Ialtris* with *Alsophis* to make the latter genus monophyletic. However, the Dowling *et al.* (1996) study indicated a very different picture of *Alsophis* and *Ialtris*. First, the two *Alsophis* representatives (*cantherigerus* and *portoricensis*) were not found to be related and *Ialtris* was even more distant, internested in fact in a clade composed of *Uromacer* and *Hypsirhynchus*, which will be discussed later.

The genus *Antillophis* has two species, *andreae* on Cuba and *parvifrons* on Hispaniola. Because the interspecific relationships represent a trivial problem, the focus will be on the possible sister taxa to the genus. Maglio (1970) suggested that *Antillophis* was derived from a mainland stock, specifically something close to *Saphenophis boursieri*, and that both these groups might be derived from mainland *Alsophis*. Myers (1973) took exception to this hypothesis and did not find evidence to suggest that *boursieri* and *Antillophis* were closely related (certainly not to the point of being placed in the same genus, as mentioned by Maglio). Crother (in review) included *Antillophis parvifrons* and *Saphenophis sneiderni* in his study. Crother's final conclusion was that *Antillophis* could not be reliably placed because of its instability on the trees. However, interestingly, in the locus-as-character-based phylogeny, *Antillophis* fell out as the sister to *Saphenophis*!

The Dowling *et al.* (1996) work also included *A. parvifrons* and in their tree *Antillophis* fell out in a clade containing *Alsophis*, *Arrhyton*, and the southern African lycodontine *Lamprophis* (Fig. 8.12).

The West Indian endemic genus *Arrhyton* currently has 12 species, but Crother (in review) has suggested a change that reduces the number to nine. Because of the apparent similarity and closeness of relationship with *Darlingtonia*, these taxa will be considered together. Maglio (1970) depicted a paraphyletic *Arrhyton* (Fig. 8.14), which included the genus *Darlingtonia*. His phylogenetic concept regarded *funereum* and *polylepis* as a basal sister group; *exiguum* and *Darlingtonia* as sisters, and the other Jamaican form, *callilaemum*, as the sister to a clade composed of the Cuban species. Schwartz and Garrido (1981b) thought the phylogeny was reasonable, but countered it with an alternative. They agreed with the Cuban monophyly (and broke the taxa into three groups), but thought *funereum*, *polylepis*, *callilaemum*, and *exiguum* also belonged in a single group (implied monophyly?). However, Schwartz and Garrido (1981b) and later Hedges and Garrido (1992b) concluded that more data were needed to get a better understanding of *Arrhyton* phylogeny.

Dowling *et al.* (1996) included five species of *Arrhyton* and *Darlingtonia* and their phylogeny supported none of the previous ideas except possibly that *exiguum* perhaps does not belong to *Arrhyton* (Schwartz and Garrido, 1981b). Otherwise, the Cuban forms were not monophyletic (the Jamaican *callilaemum* fell out as the sister to *landoi*), *Darlingtonia* was not related to *Arrhyton*, and *funereum* was not related to other *Arrhyton*. Crother (in review) found *Arrhyton* not to be a cohesive group but rather more along the lines of Schwartz and Garrido (1981b) than Dowling *et al.* (1996). The best supported clade indicates a sister relationship between *Darlingtonia* and the Jamaican *Arrhyton*, with *Darlingtonia* the sister to a clade that shows *funereum* and *polylepis* as sister taxa relative to *callilaemum* (Fig. 8.13). Buden (1966) first suggested the *funereum*-*polylepis* relationship. Crother and Hillis (1995) supported the monophyly of Jamaican *Arrhyton*, but the relationships of *callilaemum* and *funereum* were reversed. The robustness of the *Darlingtonia*–Jamaican *Arrhyton* clade led Crother (in review) to subsume the Jamaican taxa into the genus *Darlingtonia*. The two Cuban taxa included in Crother (in review) fell out in the same clade in both trees but never as sister taxa and always with *Uromacer* . The Puerto Rican *exiguum* was unstable (Fig. 8.13) and thus its relationships could not be placed.

Maglio (1970) considered the monotypic *Hypsirhynchus* as the closest relative to *Uromacer* and this clade was nested within the genus *Alsophis* (see discussion on *Alsophis*). As with most of the West Indian snakes, little has been done regarding the phylogenetic position of *Hypsirhynchus ferox*. In Cadle's (1984a) work he somewhat confirmed Maglio's notion that *Hypsirhynchus* belonged within *Alsophis* when he reported that *ferox* was more similar to some species of *Alsophis* than other *Alsophis*. Dowling *et al.* (1996) also supported the *Hypsirhynchus* – *Uromacer* relationship, but Crother (in review) did not. In the Crother study, *Hypsirhynchus* proved to be unstable, from being variously placed in a South American clade paired with *Xenodon* to being in a clade with *Darlingtonia*, *Antillophis*, and *Farancia*.

The final colubrid snake genus to be considered is *Uromacer*. This genus is endemic to Hispaniola, and the three species possess a distinctive treesnake morph. Horn (1969) was the first to speculate on the interspecific relationships of *Uromacer*. My interpretation of his arguments is that *catesbyi* is the sister to a clade of the more derived *frenatus-oxyrhynchus*. Maglio (1970) amazingly did not discuss the relationships within *Uromacer* but did emphatically point out the monophyly of the genus. Schwartz (1970) seemed to agree with Horn's (1969) idea of relationship, stating that there was a certain division between the short-snouted forms (*catesbyi*) and the long-snouted forms (*frenatus* , *oxyrhynchus*). Dowling *et al.* (1996), however, suggested a nonmonophyletic *Uromacer*, with one population of *catesbyi* and *frenatus* related to *Hypsirhynchus* and another population of *catesbyi* related to *Ialtris*. Crother (in review) found more traditional relationships within *Uromacer*, one in which ((*frenatus*, *oxyrhynchus*) *catesbyi*) was found, which is congruent with the early ideas, and the other tree simply has *catesbyi* as unresolved but in the same clade as *frenatus–oxyrhynchus*.

So where are we with regard to West Indian colubrid snake phylogeny? It is clear that we know little about the relationships of most of these taxa, but at least most taxa have been examined within some sort of a phylogenetic framework. However, most of the approaches are not rigorous discrete character-based parsimony, and no other sophisticated approaches have been applied because largely the data have not gone beyond allozymes (with the exception of Crother and Hillis, 1995). Much work is yet to be done before the West Indian colubrid assemblage, especially the xenodontines, is considered resolved.

Elapidae

Two species of coralsnakes are found on islands in the western Caribbean: *Micrurus nigrocinctus* and *M. ruatanus*. *Micrurus nigrocinctus* is found on the Greater Corn Islands and Corn Island, east of Nicaragua (this range of this species also includes all of Mesoamerica and northern Colombia). *Micrurus ruatanus* is endemic to Isla de Roatán, which lies off the northern coast of Honduras. Only a single attempt has been made to estimate the intrageneric phylogenetic relationships of *Micrurus* (Slowinski, 1995). Slowinski used allozymes and morphology in a combined analysis within a parsimony framework and found *M. ruatanus* and *M. nigrocinctus* to be in the same unresolved clade as *diastema*, *distans*, and a sister group composed of *fulvius* and *alleni*. The clade was supported by a single character which was typical of the result. It seems that coralsnake evolution has been conserved, making shared derived characters difficult to find. Slowinski (1995) did express caution about his phylogeny because many species of coralsnakes were not included in his study.

Leptotyphlopidae

There are eight species of *Leptotyphlops* recorded from the Caribbean region. Of these, *goudotii* and *tenella* are clearly mainland forms. *Leptotyphlops goudotii* is Mesoamerican and *tenella* is South American and is known from only a single record in Antigua, which may be in error (Thomas, 1965b). A third taxon, *columbi*, is found only in the Bahamas and is not apparently related to any of the other Caribbean species. The genus is widely distributed in both the Old and New Worlds, but Thomas (1965b) considered the West Indian radiation (at least the Hispaniolan radiation) monophyletic. However, in a later paper, Thomas *et al.* (1985) were uncertain about the monophyletic status and considered the possibility that the island taxa may represent relictual distributions.

There are no phylogenies for the West Indian species. Thomas *et al.* (1985) discussed in detail all the Hispaniolan taxa and included a key to all the Caribbean forms. They presented a narrative phylogeny for the Hispaniolan taxa which suggested a *calypso–asbolepis–leptepilepta* clade that was separate from *pyrites* and *bilineatus*. With the data available in Tables 1 and 2 of Thomas *et al.* (1985), I performed a parsimony analysis to estimate the phylogeny of the *bilineatus* group. All the characters from their Table 1 were included but only eye size, color, and rows fused in reduction were used from their Table 2 (the others were uninformative or uncodable). The data were analyzed as unordered and an all zero outgroup was included. An exhaustive search using PAUP 3.1 (Swofford, 1993) yielded four most parsimonious trees with a CI = 1.00. The strict consensus tree showed a ((*pyrites*, *bilineatus*)*calypso*, *asbolepis*, *leptepilepta*) set of relationships. The power of this hypothesis is questionable, given the assumption of monophyly and the absence of true outgroup characters.

Tropidopheidae

This family is represented in the West Indies by the genus *Tropidophis*. There are 16 species in this genus; all but three are found in the West Indies. There have been no phylogenies proposed for the species in the genus, although Stull (1928), Schwartz and Marsh (1960), and Schwartz and Garrido (1975) have made comments about possible species groups in their taxonomic revisions of aspects of *Tropidophis*. Bogert (1968) agreed that *Tropidophis* probably comprised a natural group but was critical of the discussions by Stull (1928) and Schwartz and Marsh (1960). In a recent review of West Indian boas, Tolson and Henderson (1993; they included *Tropidophis* although the genus is not in the Boidae), in their chapter on phylogeny, noted the previous revisers but did not substantially build on those earlier ideas. The genus itself has recently received phylogenetic attention by Kluge (1991), Cundall *et al.* (1993), Heise *et al.* (1995), and Dowling *et al.* (1996) in the context of overall snake phylogeny.

Kluge (1991) used morphology and estimated *Tropidophis* as the sister to *Trachyboa* in a clade with other members of the family. The family in turn was the sister to a clade composed of the bolyeroids, *Acrochordus*, and the colubroids. Cundall *et al.* (1993) also utilized morphological data within a parsimony framework and, in their total evidence estimate, found the family to be the sister to the *Acrochordus*–colubroid clade. The difference between the previous two hypotheses in the placement of *Tropidophis* is simply the relative positions of the bolyeroids and tropidophids. The most recent hypothesis (Heise *et al.*, 1995) is based on mtDNA sequence from the 12S and 16S regions. The mtDNA supported, albeit weakly based on the confidence probabilities at the nodes, *Tropi-*

dophis as the sister to a clade of *Loxocemus–Python*. These relationships were quite different from the former two hypotheses and other narrative ideas about snake phylogeny. The differences may be due to the methodological approaches. Kluge (1991) and Cundall *et al.* (1993) worked within a cladistic parsimony framework, whereas Heise *et al.* (1995) employed a neighbor-joining approach with the Jukes-Cantor model. An additional reason why the morphology may differ from the mtDNA is because the Heise *et al.* phylogeny is a gene tree which does not necessarily reflect the species tree.

The phylogenetic estimate by Dowling *et al.* (1996) depicted a most unusual set of relationships for *Tropidophis canus* and *haetianus*. First, *Tropidophis* was not monophyletic. Second, *T. canus* was internested in a clade that contained *Atractaspis* and *Heterodon*, among others (Fig. 8.12). Third, *T. haetianus* was placed unresolved between the Scolecophidia and the rest of the snakes. Again, the allozyme-based higher level snake phylogeny of Dowling *et al.* presents disconcerting sets of relationships.

Typhlopidae

According to Powell *et al.* (1996a) there are 24 species of *Typhlops* in the Antilles. There have been three attempts at understanding the evolutionary relationships of these taxa. Two studies (Hedges, 1989c; Hedges and Thomas, 1991) used allozyme data in a phylogenetic context to search for cryptic species in Jamaica and Puerto Rico, respectively. Unfortunately, the phylogenetic hypotheses were characterless distance-based estimates which obscure any character support that may or may not be present in the data. Hedges (1989c) concluded that no cryptic species existed on Jamaica. Interestingly, however, an examination of the data shows two populations that each have unique alleles present for two loci. These data may actually suggest the presence of two diagnosable populations on Jamaica. A similar problem exists in the Hedges and Thomas (1991) assessment of Puerto Rican *Typhlops*. In this study, they report three distinct clades among the *richardi* populations, two of which they grant specific status. However, an examination of the allozyme data resulted in zero diagnosable characters for either clade. This would have been obvious in a character-based analysis [which they say they did (Hedges and Thomas, 1991:449) but I could not find]. My own phylogenetic reanalysis of these data did not support either *platycephalus* or *hypomethes*. Dowling *et al.* (1996) found *richardi*, *platycephalus*, and *hy-*

pomethes to form a trichotomy, which was sister to *jamaicensis*. Clearly, the validity of these taxa should be questioned. Thomas (1989) provided a "cladogram of relationships" for all the Antillean *Typhlops* (Fig. 8.15). His hypothesis was based on morphological data and apparently constructed by hand (quite a feat considering there are well over 10^{21} possible trees). As a heuristic exercise, I (and Oscar Flores-Villela) organized Thomas' data into a matrix of 22 characters and estimated the phylogeny using PAUP 3.1 (Swofford, 1993; with the standard options of a heuristic search). We examined only the large clade and used *caecatus* as the outgroup.

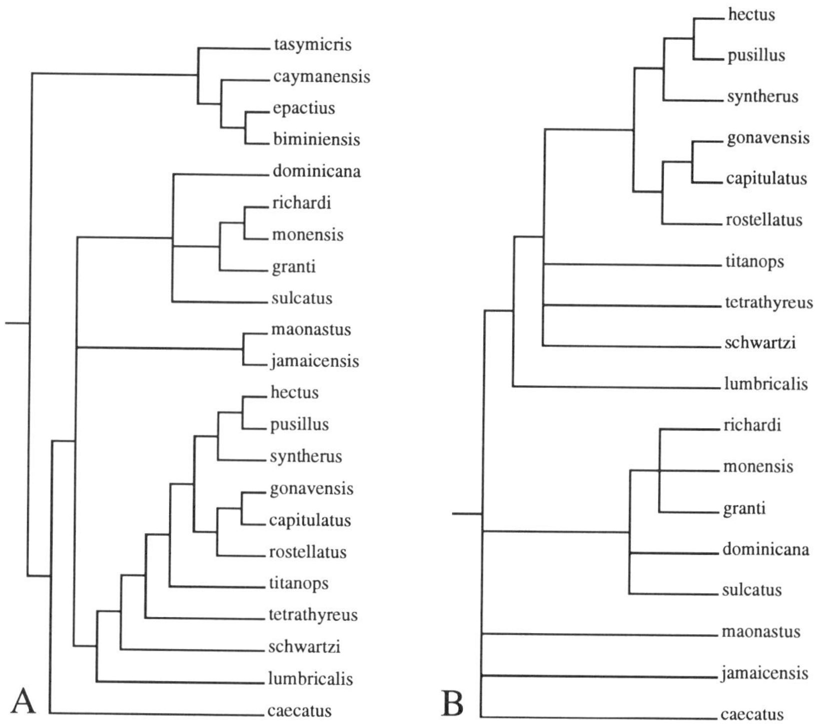

Figure 8.15. Phylogeny estimates of West Indian *Typhlops*. (A) Hand-calculated, morphology based hypothesis (redrawn from Thomas, 1989). (B) Parsimony reanalysis of the same data. See text for analysis details.

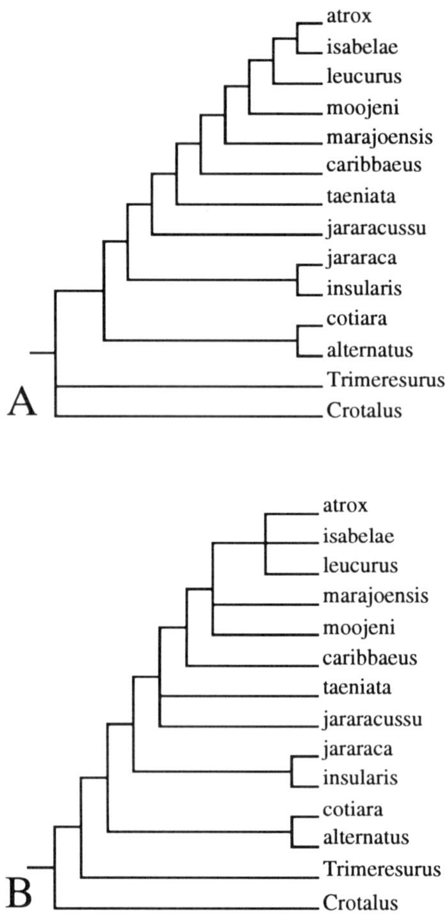

Figure 8.16. Phylogeny estimates of *Bothrops* and *Bothriopsis* based on mitochondrial DNA cyto - chrome b sequence. (A) 75% majority-rule tree from 11 most parsimonious trees discovered in a parsimony analysis. (B) Semistrict consensus tree of the parsimony consensus tree, a maximum likelihood tree, and two Fitch-Margoliash trees (one with and one without the molecular clock assumption) (redrawn from Salomao *et al.*, 1997).

The analysis resulted in a single most parsimonious tree (24 steps; CI = 0.826; RI = 0.947) that was mostly congruent with Thomas' hypothesis

(Fig. 8.15). The only difference is that the PAUP tree is less resolved (probably because we could not translate all of Thomas' characters into a matrix) and *monastus* and *jamaicensis* are not sister taxa but are placed basally with unresolved relationships. It can be concluded that Thomas' data are fairly robust, but again, more data are needed to solidify Thomas' conclusions.

Viperidae

Two species of *Bothrops*, *caribbaea* and *lanceolatus*, are currently recognized and are restricted to St. Lucia and Martinique, respectively. Historically, these two taxa have been considered as either *B. atrox* or related to *B. atrox* and the recent work supports these concepts. Amaral (1925, 1954) considered *lanceolatus* as an island form of *B. atrox* and sunk the species twice, the second action after Hoge (1952) reelevated the name. Lazell (1964) noted the difficulties in diagnosing the species from the *atrox* complex and admitted that no unique characters existed for either species. Regardless, he found various combinations of characters that allowed for his reelevating the names. Gosner (1987) reexamined a large series of both Antillean species with mainland *atrox/asper* and also concluded that *caribbaea* and *lanceolatus* were probably just *atrox/asper* variants based on the finding that none of the color patterns of the Antillean forms were unique. S. Werman (personal communication, unpublished data) found the island forms to be virtually identical with the *atrox* complex.

The only phylogenies for these taxa are based on 565 (Salomao *et al.*, 1997) and 520 nucleotides (Wüster *et al.*, 1997) of cytochrome *b*. Both studies found *caribbaea* [*lanceolatus* was not included, although Wüster (personal communication) said some DNA sequence was available and that based on that evidence the position of *lanceolatus* remained enigmatic] the sister taxon to the *atrox* complex, but the authors noted that no conclusions could be made about the reality of the species in the *atrox* complex (Fig. 8.16). Curiously, the implication was that *caribbaea* is real. Unfortunately, no characters were shown diagnosing any lineages on the trees, so evaluation of the clades was impossible. In addition, there is no way to ascertain whether or not the sequences used in the two studies are mutually exclusive or from the same region.

The nucleotide data support the notion that the island forms are at least closely related to, if not actually part of, the *atrox* complex. However, clearly a more complete study needs to be undertaken.

Testudines

There are four extant genera of turtles (excluding seaturtles) in the West Indies, and of these only one has endemic species. The taxa *Geochelone carbonaria* and *Pelusios subniger* are considered to be introductions (Pritchard and Trebbau, 1984; Censky, 1988), and *Kinosternon scorpiodes* is found only in the extreme western Caribbean off the shore of Nicaragua where its conspecifics are found throughout Central America into Texas and possibly into South America. Thus, only the relationships of the fourth genus, *Trachemys*, will be considered.

As currently recognized (the history of the contents of this genus has been somewhat tortured, Seidel, 1988a), four species (*terrapen, decorata, decussata, stejnegeri*) are contained in *Trachemys* and the phylogeny of this group appears for the most part to be well supported. Biochemical data (Seidel and Atkins, 1987) and morphological data (Seidel, 1988a) were analyzed as a combined data set within a cladistic framework. The resultant hypothesis (Fig. 8.17) suggested a paraphyletic West Indian *Trachemys* with respect to mainland Neotropical *T. scripta* (Seidel, 1988a).

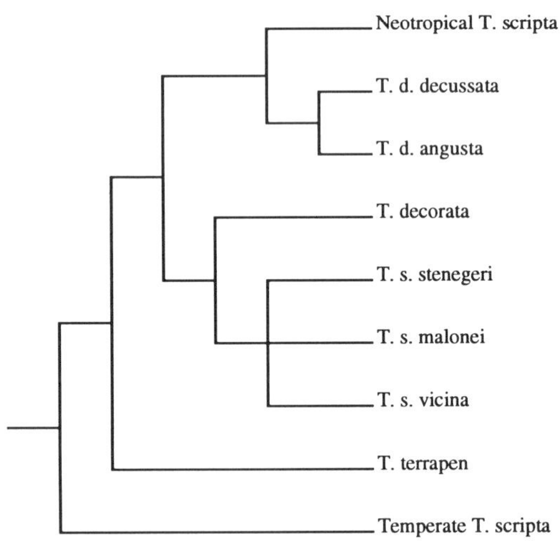

Figure 8.17. Phylogeny estimate of *Trachemys* based on a cladistic analysis of morphology and allozymes (redrawn from Seidel, 1996).

With *Pseudemys* as the root, mainland *T. scripta* fell out as the sister to the W.I. *Trachemys* with the other *scripta*: ((((*scripta*, *decussata*)(*stejnegeri*, *decorata*))*terrapen*)*scripta*). The obvious problem is the status of the taxon "*scripta*" because apparently the label is hiding additional diversity.

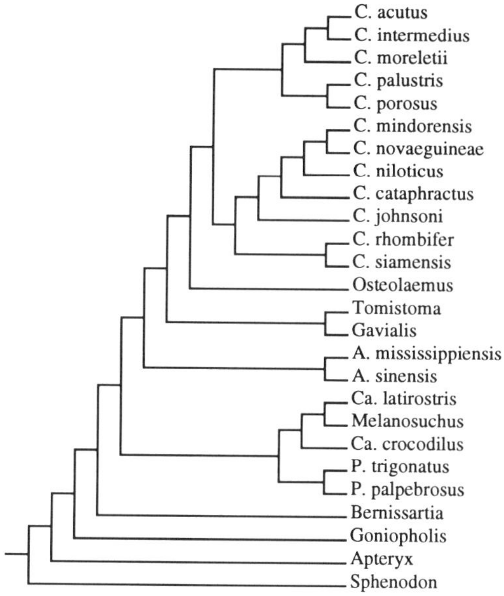

Figure 8.18. Phylogeny estimate of crocodilians based on a parsimony analysis of combined data from rDNA RFLP, mtDNA RFLP, 12S sequence, and morphology. The estimate is the single most parsimonious tree (redrawn from Poe, 1996).

Crocodylia

Four crocodilians are found in the West Indies: *Caiman crocodilus*, *Crocodylus acutus*, *C. intermedius*, and *C. rhombifer*. Of these, one is introduced (Puerto Rico; *Caiman*), two have been recorded as rare vagrants in the Grenadines (*Caiman* and *Crocodylus intermedius*), one is widespread throughout the Greater Antilles and mainland shores of Central America, southeastern North America and northern South America (*C. acutus*), and the fourth is endemic to Cuba (*C. rhombifer*). Gill (1968), in a zoogeographic work, suggested that *rhombifer* and *moreletii* were "certainly closely related" and that any apparent relationship between *acutus* and *intermedius* was only superficial and historical.

Karyotypes were employed by Cohen and Gans (1970) and in a quite convoluted scenario their data indicated an *acutus-johnsoni* group which possibly gave rise to *rhombifer* which was grouped by itself. Also, *rhombifer* may have been derived from *palustris* or *palustris* may have been derived from *rhombifer*. According to their schema, the *acutus* group may have been derived from either *porosus* or a group containing *intermedius*, *niloticus*, and *novaeguineae*. The schema presented by Cohen and Gans was based on certain assumptions about chromosomal change and thus represents what they called the "probable path of change."

Ecological data in the form of nesting habits were brought to bear on the phylogeny of crocodilians by Greer (1970). His phylogeny proposed a large clade which included *acutus*, *intermedius*, *rhombifer*, *niloticus*, *palustris*, *johnsoni*, and *siamensis*. His subdivision of the clade showed an *acutus- intermedius-rhombifer* group. Later, Campbell (1972) suggested that nesting habits do not reflect phylogenetic history but instead reflect ecological similarities.

Brooks (1981) conducted a cladistic analysis on morphological data of crocodile species-specific digenean parasites and built a crocodile phylogeny based on the results. In this classic work utilizing the nonrandom association of hosts and parasites he found an *acutus-rhombifer* clade sister to a *moreletii-intermedius* clade. This study was expanded (Brooks and O'Grady, 1989) to include data from nematode parasites and biogeographic information. In every outcome, the *acutus-rhombifer* sister relationship was supported but the sister relationship to that clade was unresolved.

Densmore (1983) used albumin immunodiffusion data clustered by UPGMA and found *acutus* sister to a clade containing *palustris*, *porosus*, *cataphractus*, and *johnsoni*. *Crocodilus rhombifer* was not included in the study. In the same paper, Densmore presented a hypothesis based on the allozyme presence/absence of alleles data and represented by an unrooted Wagner tree. In this case he found *acutus* and *rhombifer* in a clade with *siamensis* and *porosus* . Densmore and White (1991) followed up these data with restriction length fragment data from whole mitochondrial digests and nuclear rDNA digests and analyzed them using phenetic (UPGMA) and compatibility or clique analyses. The compatibility tree of the rDNA data exhibited an *acutus-intermedius* clade and a *moreletii-rhombifer* clade. The relationships of these two clades were unresolved because they were part of a seven branch polytomy. Two different compatibility analyses of the mtDNA data yielded two hypotheses presented as an unrooted networks. One hypothesis indicates an *acutus-moreletii* clade and includes *rhombifer* in a large group composed of *niloticus* , *novaeguineae*, *mindorensis*, and *palustris*. The other compatibility network suggested that *acutus* was unique, perhaps sister to *moreletii*, and that *rhombifer* was again part of a larger group but with a different membership: *palustris* , *niloticus*, and *siamensis*. Additionally, Densmore and White combined the data into a phenetic analysis using UPGMA as the clustering algorithm. This result suggested an *acutus-moreletii* clade with *rhombifer* the sister taxon and *intermedius* the sister to the clade composed of *acutus-moreletii-rhombifer*.

Poe (1996) presented a careful combined (total-evidence) analysis of all the previous character sets. He recovered a single most parsimonious tree (Fig. 8.18) from all the discrete characters. *Crocodilus acutus* was found to be the sister of *intermedius*, and *rhombifer* fell out as the sister to *C. siamensis*, but both of these clades had decay indices of only 1. Whereas the *C. acutus - C. intermedius* clade is reasonable, the *rhombifer - siamensis* relationship stretches the biogeographic imagination. Regardless, I consider the hypothesis of Poe to be the best estimate of crocodilian phylogeny.

Conclusions

A review of phylogeny is, in a way, simple relative to a review of ecology, for example. This is because the boundaries of the goal are defined: find all the figured phylogenies. Although I set out to do just that and have accumulated a large number of phylogenetic hypotheses for the West Indian herpetofauna, I undoubtedly missed some. Even if we assume this as true, the missing phylogenies would not alter my conclusions (nor alter the conclusions of the reader). The main point of summation derived from the review is that only a minuscule fraction of the West Indian herpetofauna is well-known phylogenetically. Few relationships have been estimated with multiple data sets and rigorous phylogenetic methodology, with the result being stability among those relationships (= highly corroborated hypotheses). For example, the taxon *Cricosaura*, although the subject of multiple data sets and phylogenetic analyses, has been unstable in its phylogenetic position and so is not considered to be "known" phylogenetically. I would argue that perhaps *Epicrates*, crocodilians, and *Trachemys* (even with the sticky *scripta* problem) could be considered to have well corroborated phylogenies.

Unfortunately, this pessimistic view of our state of the knowledge of West Indian herpetofaunal phylogeny is not helpful to the nonsystematist who needs a phylogeny to conduct a comparative study. One suggestion would be to use each of the available phylogenetic hypotheses which would lead to multiple testable hypotheses from the comparative study. It is perhaps more work, but certainly would be enlightening and might in fact lead to an empirical reason to choose one of the competing phylogenetic hypotheses over the others (such a study could be reasonably pared down if only philosophically justified approaches to phylogenetic inference are examined). The argument is not tautological but is analogous to using phylogenetic trees to choose among competing (or present novel) geological hypotheses. Another approach, which I think is the more justifiable of the two, would be to include the ecological, behavioral, etc., data with the original data used to estimate the cladogram and then reanalyze the data together. If care is taken in coding the ecological characters in question, concerns of tautology, non-independence, and bias become moot (Luckow and Bruneau, 1997).

The optimistic view is that phylogenetic herpetology in the West Indies is wide open (although I suspect there are those who would argue otherwise) and waiting to be done. To this end, a number of labs have taken up the challenge and are currently working on West Indian problems. The assessment of accuracy of phylogenies is a difficult proposition. I would argue impossible because the *true* phylogeny is unknowable. Instead, we are left with hypotheses that are resiliant to falsification. Multiple data sets are a necessity to establishing highly corroborated phylogenies. A priori, no single data set is superior to another (although for methods this is a different matter), and as such both molecular and morphological based phylogenies need to be estimated. Finally, these multiple data sets should be combined to give the overall best estimate of character covariation (congruence) and thus phylogenetic relationships.

Appendix.

Studies Which Include Figured Estimates of Phylogeny for West Indian Species.

Species	Reference
Anura	
Leptodactylidae	
Eleutherodactylus	Hass and Hedges 1991
	Hedges 1989 a,b, 1991
	Joglar 1983, 1989
	Kaiser *et al.* 1994 a,b,c, 1996
	Lynch 1996
	Smith *et al.* 1981
Leptodactylus	Heyer 1970
Sauria	
Gekkonidae	
Aristelliger	Russell 1979
Gonatodes	Kluge 1995
Hemidactylus	Kluge 1969
	Russell 1979
Phyllodactylus	Russell 1979
Sphaerodactylus	Schwartz and Thomas 1983
	Hass 1991, 1996
	Kluge 1995
	Page and Lydeard 1994
Iguanidae	
Cyclura and *Iguana*	de Queiroz 1987
	Etheridge and de Queiroz 1988
	Frost and Etheridge 1989
	Hollingsworth 1998
	Norell and de Queiroz 1991
	Sites *et al.* 1996

Polychrotidae	
Anolines	Burnell and Hedges 1990
	Cannatella and de Queiroz 1989
	Case and Williams 1987
	Etheridge 1960
	Frost and Etheridge 1989
	Gorman and Atkins 1969
	Gorman and Kim 1976
	Gorman *et al.* 1980, 1983
	Guyer and Savage 1986, 1992
	Hass *et al.* 1993
	Hedges and Burnell 1990
	Lazell 1966, 1972
	Paull *et al.* 1975
	Peterson 1983
	Poe 1998
	Roughgarden 1995
	Shochat and Dessauer 1981
	Williams 1972, 1977
	Wyles and Gorman 1980
	Yang *et al.* 1974
Teiidae	
Ameiva	Barbour and Noble 1915
Bachia	Dixon 1973
Tropiduridae	
Leiocephalus	Frost and Etheridge 1989
	Pregill 1992
Xantusiidae	
Cricosaura	Crother *et al.* 1986
	Crother and Presch 1993
	Hedges *et al.* 1991
	Peterson and Bezy 1985
	Schatzinger 1980
Serpentes	
Boidae	
Boa	Cadle 1988
	Heise *et al.* 1995
	Underwood 1976
Corallus	Henderson and Hedges 1995
	Kluge 1991
	Underwood 1976

Epicrates	Kluge 1989, 1991
	Tolson 1987
	Underwood 1976
Colubridae	
Alsophis	Cadle 1984, 1988
	Crother in review
	Crother and Hillis 1995
	Dowling *et al.* 1996
	Jenner and Dowling 1985
	Maglio 1970
Antillophis	Crother in review
	Crother and Hillis 1995
	Dowling *et al.* 1996
	Maglio 1970
Arrhyton	Crother in review
	Crother and Hillis 1995
	Dowling *et al.* 1996
	Maglio 1970
Chironius	Dowling *et al.* 1996
Clelia	Cadle 1984, 1988
	Dessauer *et al.* 1987
	Dowling *et al.* 1996
	Jenner and Dowling 1985
Coniophanes	Cadle 1984, 1988
	Crother in review
	Jenner and Dowling 1985
	Myers and Campbell 1981
	Pinou and Dowling 1994
Darlingtonia	Crother in review
	Dowling *et al.* 1996
	Maglio 1970
Hypsirhynchus	Cadle 1984
	Crother in review
	Crother and Hillis 1995
	Dowling *et al.* 1996
	Maglio 1970
Ialtris	Crother in review
	Dowling *et al.* 1996
Liophis	Dowling *et al.* 1996
	Maglio 1970
Mastigodryas	Dowling *et al.* 1996

Nerodia	Lawson 1987
Pseudoboa	Cadle 1984
	Dessauer *et al.* 1987
	Dowling *et al.* 1996
	Jenner and Dowling 1985
Tretanorhinus	Crother in review
	Dowling *et al.* 1996
	Pinou and Dowling 1994
Uromacer	Crother in review
	Crother and Hillis 1995
	Dowling *et al.* 1996
	Maglio 1970
Elapidae	
Micrurus	Slowinski 1995
Tropidopheidae	
Tropidophis	Cundall *et al.* 1993
	Dowling *et al.* 1996
	Heise *et al.* 1995
	Kluge 1991
Typhlopidae	
Typhlops	Dowling *et al.* 1996
	Hedges 1989
	Hedges and Thomas 1991
	Thomas 1989
Viperidae	
Bothrops	Salamao *et al.* 1997
Testudinata	
Emydidae	
Trachemys	Seidel 1987, 1988, 1996
Crocodylia	
Crocodylidae	
Crocodilus	Brooks 1981
	Brooks and O'Grady 1989
	Cohen and Gans 1970
	Densmore 1983
	Densmore and White 1991
	Greer 1970
	Poe 1996

9

Ecological and Historical Biogeography of the Caribbean

Gregory K. Pregill [1] *and Brian I. Crother* [2]
[1]*Department of Biology, The University of San Diego, 5998 Alcalá Park, San Diego, California 92110-2492*
[2]*Department of Biology, Southeastern Louisiana University, Hammond, Louisiana 70402*

Introduction

The modern era of biogeography began about 30 years ago consequent to four unrelated but more or less synchronous events: the formalization of plate tectonics and a mobile geology of the earth's crust; advances in radiometric dating techniques; advances in phylogeny reconstruction; and publication of MacArthur and Wilson's (1967) Theory of Island Biogeography. At this point, biogeography went in two directions. One was ecological. Insularity became the focus of study on community evolution and the processes of species interaction. Ecological biogeography thus probes such issues as equilibrium dynamics, taxon cycles, and competition in their capacity to effect species assortment.

The other direction in biogeography was the traditional search for the geographic origin of clades. Historical biogeography depends on phylogeny, levels of endemism, fluctuations in sea level, and past geomorphologies such as connections between a continent and an island, or one island with another. The two approaches to biogeography largely operate independently of one another, although they both attempt to explain the same thing: the geographic distribution of life. Historical biogeography finds no advantage in field experiments and models that strive to reveal ecological relationships. This is because the results obtained usually apply to a narrow set of circumstances from an instant in geological time. Nonetheless, the results of field manipulations and their models accumulate, and the general patterns that emerge from them make ecological biogeography a successful enterprise. However, ecological biogeography lacks a sturdy historical framework, and often it is difficult to identify which characteristics of a population are opportunistic responses to the environment of the time, and which are the products of history. Historical and ecological biogeography are interwoven, but especially for vertebrates this relationship is not well understood.

The Caribbean biota has been a spectacular inspiration to both approaches. In this chapter we review the herpetological components of each. Of the two, the historical approach is pursued by most herpetologists who study the West Indies. However, the amphibians and reptiles themselves, particularly lizards, are favored subjects of ecologists. Although we make no attempt to construct a unifying set of historical and ecological objectives, we do elaborate on why events in the Caribbean during the past 10,000 years are significant to both.

Historical Biogeography

Thoughts on the origin of the West Indian biota can easily be categorized into two periods: the previcariance era and the vicariance era. The temporal and philosophical boundary between these two periods is a fairly clean paradigm shift in general explanations about distribution. Although herpetologists were mostly minor players in the previcariance era, they took on substantially larger roles afterwards. Williams' (1989b) thorough review of the ideas and participants in Caribbean biogeography consolidated the history for subsequent discussions (e.g., Crother and Guyer, 1996). It is not our intention to cover all of the same ground here; however, some overlap with Williams is inescapable.

Previcariance Era

The central debate during the pre-vicariance era was whether land vertebrates dispersed to the islands over water, or directly by way of land bridges. Although both were dispersalist arguments, they differed in that overwater processes were, of necessity, random or chance events; land bridges implied concordant movement. The debate was constrained by the presumed stability of continents, which demanded the active movement (dispersal) of organisms across the landscape. Additionally, there was a prevalent assumption, steeped in religious views of creation, that taxa arose from centers of origin, further requiring dispersal explanations to account for distributions.

The entrenched foundation of the pre-vicariance attitude left little room for options, and for their part few herpetologists considered possibilities other than overwater dispersal. Indeed, if the continents were fixed then there was no acceptable alternative. However, the land bridge hypothesis had one of its greatest champions in the Harvard herpetologist Thomas Barbour (e.g., 1914, 1915). Barbour argued against the prevalent theory of flotsam and jetsam dispersal as espoused by Matthew (1915) and others. For example, Barbour and Noble (1915:418) noted,

> This revision, which is a detailed study of a single genus of strictly terrestrial teid (sic) lizards, shows clearly the close relationship and origin from a common stock of many of the Antillean forms. The data derived from this study seem to argue strongly against the flotsam and jetsam theory.

Approximately 20 years later, Barbour (1937) continued advocating non-overwater dispersal, although in words echoed four decades later by the vicariance proponent Gareth Nelson, he allowed that some random dispersal must have occurred: Barbour (1937:79) stated, "There has been undoubtedly some dispersal by flotsam and jetsam" and Nelson (1974:556) noted, "If so, many episodes of dispersal - which no doubt have occurred." But Barbour (1937:80) went on to say,

> However, to conclude that all of the animals of an island, such as any of the Greater Antilles, have been derived by any or all of these causes [various random events] is to support an explanation which is to my way of thinking infinitely less probable than to postulate extensive changes in land form in a region where so much tectionic (sic) movement is evident on every hand. That the separation of the Greater Antilles took place a long time ago is certain, for *Cadea* was not derived from *Amphisbaena* nor *Cricocaura* (sic) differentiated from its Xantusiid forebears except in a very long time.

The confidence Barbour displayed in his 1937 paper was partly attained from Schuchert's (1935) *Historical Geology of the Antillean-Caribbean Region* published two years earlier. Schuchert described the origin of the Greater Antilles in a manner not unlike modern tectonic models, and he concluded that land bridges were the best explanation to account for the West Indian biota.

Barbour's position on the origin of the West Indian biota is remarkable for three reasons: (1) until the vicariance revolution he essentially stood alone against random, overwater dispersal; (2) his reasoning presaged vicariance but without mechanism or method; and (3) he argued mostly from his knowledge of reptile and amphibian distributions and systematics (at the time most Caribbean zoogeography derived from the study of mammals). As strong as Barbour's arguments were, however, he had little support for them and could offer no geological mechanism to make land bridges credible. The overwater dispersal school was dogma, as expressed, for example, by Darlington (1938:297),

> Both of the two independent lines of valid evidence indicate, then, that the fauna of the Greater Antilles has been derived from the mainland across water...so far as I can see, no other hypothesis will fit the facts.

and finished by Simpson (1956:7):

> When I first drafted the present paper it seemed useless to reopen a question that had been so thoroughly discussed and had, I felt, been settled.

In the late previcariance era, the tempest over the Antillean biota as a whole spilled onto discussions about the distribution of species among the islands themselves. Various overwater dispersal scenarios proposed complex retrodictions of movement, with taxa making there way from one island to another and sometimes back again. Two examples from herpetology are Gorman and Atkins (1969) and Maglio (1970) (Fig. 9.1). Many of these scenarios were creative efforts at drawing lines and arrows, but their plausibility was difficult if not impossible to test. Ultimately, this problem of testability led to general dissatisfaction with random dispersal as an a priori explanation for Antillean land vertebrates.

Vicariance Era

The shift to a vicariance rationale in biogeography occurred fairly rapidly. In the 1970's, confluence of the independent but essentially simultaneous revolutions in geology and systematics were applied to the then obscure ideas of Leon Croizat's (1958, 1964) panbiogeography (Croizat *et al.*, 1974). Mechanisms of seafloor spreading and plate tectonics replaced the stable geology of the earths crust with drifting continents and the opening of ocean basins (e.g., Isacks *et al.*, 1968). The first tectonic model for the Caribbean plate soon followed (Malfait and Dinkelman, 1972). At about the same time, the english translation of Willi Hennig's *Phylogenetic Systematics* in 1966 proclaimed a fresh methodology for phylogenetic reconstruction. Its halting but ultimate embrace by systematists is chronicled in a lively book by Hull (1988).

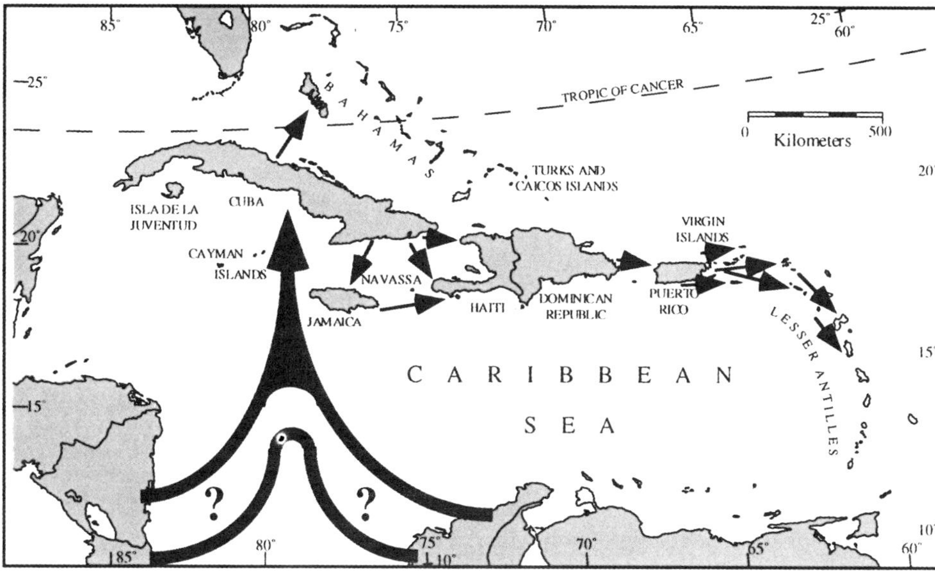

A

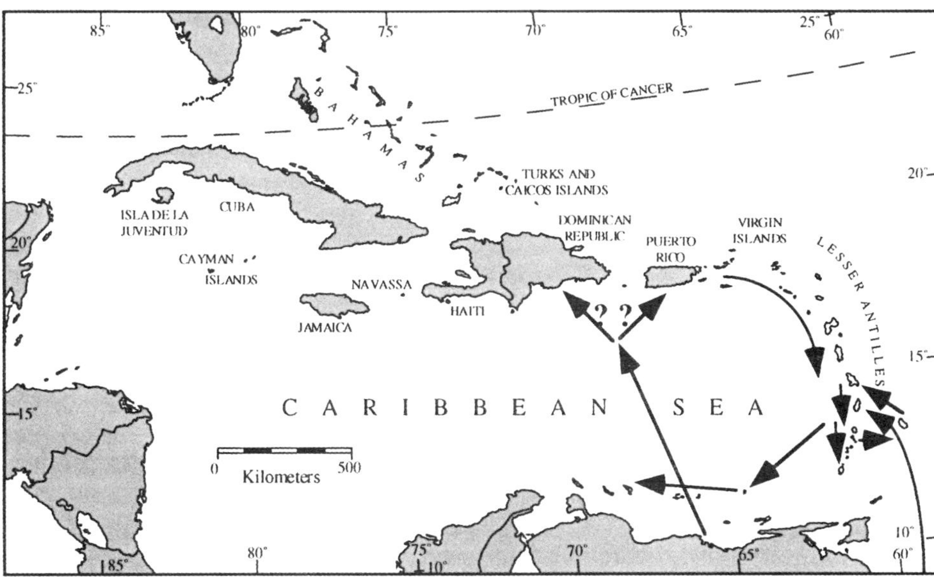

B

Figure 9.1. Example dispersal scenarios for West Indian reptiles. (A) Example for xenodontine snakes (redrawn from Maglio, 1970). (B) Example for anoline lizards (redrawn from Gorman and Atkins, 1969).

Croizat devised a simple scheme for plotting the range of species that allowed him to visualize the concordant distribution of diverse taxa. He argued that these "tracks", as he called them, must result from processes that affected entire biotas. Distributions outside the generalized tracks could be explained by specific instances of dispersal.

The vicariance school of historical biogeography began to materialize when Gareth Nelson (1969, 1973, 1974) brought attention to Croizat's ideas. Nelson combined new knowledge on geographic histories based on tectonics, with Hennigian phylogenetics (cladistics). The result was vicariance biogeography. The premise had elegant simplicity: if areas had historical relationships, and if taxa had historical relationships, then when area and taxonomic relationships were congruent the history of the areas explained the distributional history of the taxa. Thus, geographic history is the independent variable and biology the dependent variable. The method's power is its testability, so long as both geographic history and phylogeny are recoverable.

Among the first to apply vicariance methods to real problems was Donn Rosen (1976, 1978, 1985), who in doing so substantially altered the forum for Antillean biogeography. Rosen (1976) argued, based on then current tectonic models (mostly Malfait and Dinkelman, 1972), that a proto-Antilles archipelago was situated between North and South America (SA) from the late Cretaceous to the early Paleocene. This archipelago derived its fauna from the continents to the north and south. As the proto-Antilles moved eastward relative to North and South America in the early Cenozoic, it carried the antecedents of the Antillean biota with it.

Following the publication of Rosen's model, interest in Caribbean biogeography was renewed. Perhaps because amphibians and reptiles are the dominant terrestrial vertebrates (diversity and numbers) in the Antilles, herpetologists have been active contributors. Pregill (1981b) argued that the geophysical evidence for a proto-Antilles was equivocal, and that the fossil record (continental) suggested that many genera of West Indian vertebrates appeared subsequent to the formation of the Greater Antilles. Hedges (1982) rebutted Pregill's (1981b) contention that geologic evidence was insufficient, but did not comment on the fossil record or the composition of the modern fauna, which continued to be points of disagreement (e.g., Williams, 1989b; Crother and Guyer, 1996).

In 1982 Jay Savage published his "Enigma" paper. Savage emphasized the uniqueness of the Mesomerican (MA) herpetofauna and its relevance to historical biogeographic questions in the region. He modified Rosen's (1976) original model and elaborated on the complex, interrelated histories of the Caribbean and Mesoamerica. He also concluded that there must have been a late Cretaceous-early Paleocene land connection between MA and SA. Savage (1966, 1982) is not alone in arguing for a distinctive MA biota. Cadle (1985), Darwin and Welden (1992), and Crother *et al.* (1992) provided additional support. The reality of a MA-SA connection, although implied in some form by most tectonic models, remains unproven geologically (Crother and Guyer, 1996; Perfit and Williams, 1989). Biologically, however, a MS-SA connection is practically demanded.

Williams (1989b) addressed Savage's (1982) paper in detail. In particular, he disagreed that there was a late Cretaceous-early Paleocene connection between MA and SA, and he also maintained that the Tertiary fossil record was important despite its absence in the region. In tropical continental latitudes, and on most oceanic islands, the vertebrate fossil record is largely silent about events of the Cenozoic before the Quaternary. Tropical soils and climate are not conducive to the long-term preservation of bone, and all but the largest islands lack sufficient deposits in which bones of terrestrial vertebrates have their best chance of being preserved over millions of years. Historical biogeography of the Caribbean is thus disadvantaged by the lack of hard evidence from early periods of the Cenozoic. At issue is whether this gap can be filled in reasonably with assumptions. Minimizing assumptions on the one hand means that negative evidence (the lack of appropriate fossils), however lamentable, is irrelevant to the analysis. On the other hand, a comparatively rich fossil record from temperate North and South America might suggest minimum ages of circum-Caribbean sister taxa. For instance, if passerine birds are unknown before the Miocene and colubrid snakes before the Oligocene, then their appearance in the Caribbean would have postdated the rafting of the proto-Antilles (or whatever the geographic configuration of the islands was in the Paleocene). Still, the reality of such a history can never be demonstrated by corroborating evidence; it is speculation that can only be falsified, in this case by the discovery of older fossils.

In the past 15 or so years numerous herpetofaunal studies have addressed Caribbean biogeography, but there has been no solid consensus to emerge from the effort (e.g., Burnell and Hedges, 1990; Cadle, 1985; Crother, 1989; Crother *et al.*, 1986; Guyer and Savage, 1986; Hass,

1991; Hass and Hedges, 1991; Hass *et al.*, 1993; Hedges, 1989a; Hedges and Burnell, 1990; Hedges *et al.*, 1991; Jenner, 1981; Joglar, 1989; Kluge, 1988, 1995; Page and Lydeard, 1994; Pregill, 1981a, 1992; Roughgarden, 1995; Savage and Lips, 1993; Schwartz and Garrido, 1985; Seidel, 1988a, 1996; Sites *et al.*, 1996; Thomas, 1989; Tolson, 1987; Williams, 1989a). The most recent variables of the vicariance debate were added by Hedges *et al.* (1992b; also in Hedges 1996a). They reported pairwise comparisons of immunological distance (ID) data for selected West Indian and mainland taxa. Based on a molecular clock of those data, they concluded that the taxon divergences were too recent to have been associated with a late Cretaceous vicariance event. Accordingly, they argued for overwater dispersal of the herpetofauna. In rebuttal, Crother and Guyer (1996) combined much of the available data (including nonherpetofaunal phylogenies) into a single parsimony analysis and obtained a consistent result that supported a vicariance hypothesis. But there was more at stake. Hedges *et al.* (1992b) suggested that giant tsunamis – generated by a late Cretaceous bolide impact on the north coast of the Yucatan – scoured the Antilles of all but the most tenacious weeds and arthropods. Hence, land vertebrates must have recolonized later in the Cenozoic.

Page and Lydeard (1994), Crother and Guyer (1996), and Guyer and Crother (1996) argued against the molecular clock/bolide hypothesis [see Hedges *et al.* (1994) and Hedges (1996c,e) for counter arguments]. Page and Lydeard (1994) restated the importance of phylogenetic analysis in recovering biogeographic history. Crother and Guyer (1996) revealed problems with the tsunami hypothesis, the molecular clock calibrations, and the pairwise comparisons. Other studies (e.g., Keller *et al.*, 1997; Stinnesbeck *et al.*, 1997) found little evidence of impact-generated, short-term (over hours or days) tsunami deposits. Even if an impact did occur, physical models dispute the possibility of waves reaching the dimensions (4 km tall) required to inundate the entire West Indies (Gault and Sonett, 1982; see Crother and Guyer, 1996).

The molecular clock is a tantalizing and potentially powerful tool. Unfortunately, it is beset with problems, and is sloppy with respect to calibration. Consider the demonstration by Hillis *et al.* (1996), who idealized a molecular clock with 95% confidence limits and only stochastic variation as a source of error. At an arbitrarily assigned rate of one substitution per 1 million years (my), taxa separated by 15 my could have between 8 and 22 substitutions. However, an age of 5 my could also have

8 substitutions, such that neither age (5 or 15 my) can be rejected. Hillis *et al.* (1996:538) also examined ID data (like that used by Hedges *et al.*, 1992) from birds, and concluded

> "*The calibration of albumin divergence based on immunological comparisons among birds shows that the confidence limits of new predicted values of time may be so large as to not exclude any reasonable possibility*" (emphasis ours).

They illustrated their point (Hillis *et al.*, 1996:538, fig. 14). For an estimated divergence time of approximately 80 my, the ID values within the 95% confidence intervals range from 20 to 90. For a divergence time of 20 my, the range of ID values within the 95% confidence intervals is 0 - 45. Thus, for ID values between 20 and 45, neither 20 nor 80 my ages can be rejected at 95% confidence.

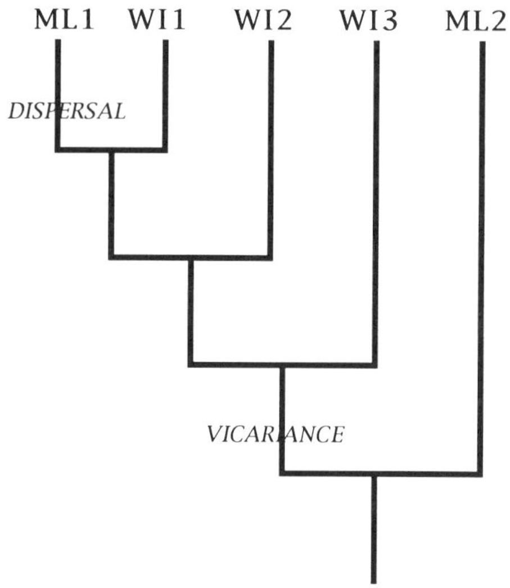

Figure 9.2. The importance of phylogeny for choosing taxa for pairwise comparisons when employing a molecular clock. ML = mainland, WI = West Indies. This "known" hypothetical phylogeny depicts a clade whose origins are in the mainland (ML2), populated the West Indies through vicariance, and subsequently dispersed back to the mainland (ML1). See text for further discussion.

The clock problems notwithstanding, Hedges *et al.* (1992) used nonindependent pairwise comparisons in their analysis. Crother and Guyer (1996) exacerbated the problem by mimicking an expanded data set. This technique fails to account for branch lengths (which imply rates of change), assuming that a phylogeny is correct (Hillis *et al.*, 1996). Branch length estimation itself is a source of error, but that aside, not taking it into account in pairwise comparisons can, by hiding varying rates of change, grossly underestimate calculation of confidence limits (Hillis *et al.*, 1996:534, fig. 9).

Additionally, the choice of taxa in pairwise comparison is critical for molecular clock applications to biogeography. In this sense (and for the moment ignoring branch lengths), a robust estimate of phylogeny is required. Assume a clade has several mainland (ML) and West Indian (WI) taxa. Further assume that the ancestors are ML, that subsequent vicariance led to the WI radiation, and that recent dispersal back to the ML occurred. From Fig. 9.2, if ML1 and any of the WI taxa are compared pairwise, an underestimate of divergence time will result. The closest estimate would come only from comparing ML2 and WI3. The only way to know this a priori is through the phylogeny.

Historical biogeography continues to struggle with the exigency of data while it frets over methods and principles. Caribbean biogeography has produced data of lasting value, and important conceptual advances. It has also encouraged remarkably imaginative speculation. We maintain, as have others, that phylogeny remains the basis for tackling the problems.

Ecological Biogeography

By whatever means an island accumulates a fauna – in units (vicariance), incrementally (dispersal) or by a combination of the two – the structural inertia of that fauna has biogeographic implications of its own. Ecological biogeography is the modern synthesis of those implications codified by MacArthur and Wilson (1963, 1967), and consecrated by Lack (1976). Its objective is to seek explanations about the composition of insular floras and faunas that are independent of history and phylogeny. Lack viewed island faunas as stable entities, a result of adaptation along continuous but nonoverlapping resource utilization by individual species. MacArthur and Wilson proposed that the identity of species would change through immigration and extinction, and that the number of species present on an island was dependent on area and isolation. In the Caribbean, ecological biogeography found its cause célèbre in *Anolis* lizards. Following Etheridge's (1960) pioneer study of their relationships, anoles rapidly ascended to the rank of "model organism." The catapult was Ernest Williams' (1969, 1972; Rand, 1969; Rand and Williams, 1969) concept of the ecomorph, which characterized the repetitive adaptations of anoles to similar habitats on different islands. Subsequently, *Anolis* became the fruit fly of evolutionary ecology (e.g., Roughgarden, 1995). Although the literature is vast and varied (Schwartz and Henderson, 1991), our purpose here is simply to assess how these ecological studies have enriched Caribbean biogeography with respect to land vertebrates, and reptiles and amphibians in particular. We do not consider all ramifications, only the basic tenets of the MacArthur and Wilson model. We are especially interested in the reality of its application in the face of empirical evidence, both historical and contemporary. This line of inquiry is admittedly something of a strawman, a case of comparing apples with oranges. The intent of ecological biogeography is to imagine what could happen rather than to discover what did. On the other hand, we believe that history is inseparable from biogeography. In particular, events of the Holocene left indelible marks on Caribbean island biotas. The most trenchant of these was the arrival of humans.

MacArthur and Wilson and Lack were not the first to consider island biogeography in an ecological context. Island size, distance, and other potential determinants of species number had long been recognized by biogeographers (Gleason, 1922). In the West Indies, Darlington (1957) suggested that the number of amphibian and reptile species tended to increase by a factor of two for each 10-fold increase in island area. Levins and Heatwole (1963) presented statistical formalization of amphibian and reptile distribution throughout the Puerto Rican Bank based on factors of dispersal and species survival (see also Heatwole and MacKenzie, 1966). MacArthur and Wilson wanted to bring area and distance together in a unifying theme. For any island, they reasoned, there should exist an equilibrium number of species (S), based on an island's size and distance from the source pool. The concept of equilibrium supposes that the number of species on an island is constantly subject to increase by newly arriving colonists, but is balanced by concomitant extinction of the residents. Rates of immigration and extinction are specific to the island and taxon. For vagile taxa such as birds and bats, S is achieved more quickly than for less mobile groups such as lizards and frogs. An island also must remain ecologically constant over a sufficient, but unspecified, period of time for the dynamics of immigration and extinction to attain a stable intersect.

MacArthur and Wilson's model was greeted with unrestrained enthusiasm by some biogeographers, with ridicule by others, and with skeptical curiosity by most. It is intuitively obvious that larger islands would have more species than smaller ones and distant islands fewer species than closer islands of equivalent size. That species number (S) is not attributable to anything more than chance alone was a frequent criticism. In one of the most stinging critiques, Sauer (1969:590) wrote:

> In short, the equilibrium model and its derivatives suffer from extreme oversimplification by treating islands as functional units with no attention to internal habitat diversity and by treating species as interchangeable with no allowance for genetic or geographical diversity. This is not even good as a first approximation, because it filters out the interpretable signal instead of the random noise. A model that visualizes various sizes of assemblages of characterless species on various sizes of featureless plains is essentially absurd, since it excludes the very basis of genesis and continued co-existence of multiple species.

Taking the model at face value, few would disagree with Sauer (1969). However, it has kept breathing for over 30 years, sustained not by the reality of an equilibrium number of species per se, but by pursuit of the ecological reasons that explain nonequilibrium (Case and Cody, 1987). Hence, for amphibians and reptiles we would need to know, or at least have a sense of, (1) the capability of an animal to disperse, at what distances, and at what frequency; and (2) the factors that determine successful dispersal– that is, an individual(s) actually becoming established on a new island once it arrived.

Dispersal Capability

Dispersal capability is nearly impossible to determine even generally, and not at all with much precision; very few studies have examined this problem. Amphibians and reptiles have moved between Caribbean islands, no doubt, but there are only limited means for them to do so. Rafts and flotsam are the most obvious vessels available to transport them and their eggs (e.g., Censky *et al.*, 1998; King, 1962a). For amphibians, there are no viable alternatives except by becoming airborne during hurricanes. Lizards and snakes can float, and depending on their tolerance for saltwater, passive drifting on surface currents could make nearby islands accessible (Schoener and Schoener, 1983b). Some anoles will voluntarily leap into the sea if stranded on wave-swept rocks, and can survive for at least 24 hr before making landfall (Schoener and Schoener, 1984).

Still, a rationale for *continuous* immigration does not hold up for most, if not all, nonvolant terrestrial vertebrates. The frequency of over water dispersal is remote and erratic even for supposedly capable lizards such as anoles. Were dispersal to occur successfully with some regularity, the diversity of West Indian *Anolis* species probably would be lower than it is currently. In the same way, the Lesser Antilles should have more than the present maximum of two species on any island.

Species-Area Relationships

Assuming that a lizard does disperse to a new island, the chances for successful colonization will depend on its health, sex, and the opportunities that await it. Ecological biogeography assesses these chances by extrapolating from the structure of existing communities. The most germane of the "assembly rules" that determine community structure are species per area, habitat complexity, and species interactions (e.g., Diamond, 1975; Haefner, 1988). Tests of the species-area relationship alone have been conducted for numerous taxa and islands. Most often the test is in the manner: $S = c\,A^z$, where area, A, is the independent variable, c is the y-intercept of the regression line, and z is a value that gives the slope of the regression line fitted to log-transformed data. During the 1970s many studies of insular plants and animals (although practically none on amphibians or reptiles) derived various values for z in hopes of uncovering a meaningful coefficient. The statistical interpretations were explored in detail, as were the assumptions of a log-normal distribution and abundance of species. At best, any unique theoretical basis for the S/A relationship is inconclusive beyond the fact that habitat diversity and area per se typically show a positive relationship to species number (Connor and McCoy, 1979). In a recent example, the incidence of most plant species on small islands surrounding Andros and the Exuma Cays in the Bahamas was positively related to some measure of area and negatively related to distance (Morrison, 1997). Cays with diverse vegetation are significantly higher and closer to mainland islands then cays naturally devoid. In another study, Dodd and Franz (1996) found that 42% of the distributional variation in the herpetofauna of the Exuma Cays Land and Sea Park was explained by island area. However, there was no correlation for lizards alone on the core islands nor with distance between satellite and core islands. They also observed that cats and rats negatively affected local populations and probably their data as well.

The slope (z) and intercept (c) are essentially devoid of biological meaning for the simple reason that the number of species occurring on an island is a function of multiple causes, not simply area alone. No manner of statistical tinkering will consistently reflect the possibilities (Connor and McCoy, 1979). For example, in the British Virgin Islands Lazell (1983) tallied the number of lizard species present on 43 cays ranging in size from 0.1 to 5444 ha. Only two of the cays fit the species-area curve; the rest had "too many" species. Moreover, cays of 10 ha or less should, according to the math, have zero species. Using recomputed values for z, 26% of the cays had too few species, 32% had too many, and the minimum area required to support one species was reduced to 50 m^2. In terms of percentage error (the actual number of species too few or too many), a better fit was obtained by incorporating elevation into area values, but 33 of 43 islands still did not conform to predictions (Lazell, 1983). In some cases, "too few" species may be an artifact of insufficient collecting, and the island in question might, with additional field work, conform to the equation. Alternatively, the island could simply be so inhospitable that no lizard would survive on it, or perhaps the species never dispersed there in the first place. By contrast, "too many" species requires complex explanations and a bending of the assembly rules. Anoles are capable of occupying almost any piece of land above high tide that supports "more than herb-stage vegetation" (Lazell, 1983:115). Such is nearly true as well for other West Indian lizards such as *Sphaerodactylus*. Hence, species with low vagility but high adaptability inevitably will defy the species-area equation but in opposite directions so that too few or too many species, respectively, will result.

Species Turnover

The equilibrium model supposes that immigration and extinction are continuous and linked. The outcome is a turnover in species composition. Most reputed examples of species turnover on West Indian islands come from short-term studies on birds and Lesser Antillean anoles. Turnover results from the extinction of a resident species as a consequence of an invader becoming established. Turnover, then, is driven by species interaction, especially competition. On smaller islands with fewer species, rates of turnover should be relatively high (e.g., Schoener and Schoener, 1983b; but see Simberloff and Boeklin, 1991). On larger, species-rich islands invader success should be lower because diverse, strongly interacting communities manifest emergent properties that make them resistant to newcomers (Case, 1991). Lack (1976), by contrast, viewed interaction as important, but in such a way that turnover eventually approaches zero. Short term interactions should give way to stability in the long term because species evolve niche specializations based on opportunity and ecological resources.

Species turnover has its origin in the "taxon-cycle" first conceptualized by Wilson's (1961) studies on Melanesian ants. The concept was applied early on to West Indian birds by assuming that a species progresses through stages of distribution beginning with expansion in range, differentiation, contraction, and finally extinction (Ricklefs, 1970; Ricklefs and Cox, 1972, 1978). That these stages are successional is at best dubious. They are merely a set of distributional patterns that almost any vertebrate in any archipelago will fall into, and there are several possible explanations to account for each without ever going through such a "cycle" (Pregill and Olson, 1981).

For reptiles, a taxon cycle was proposed for two species of *Sphaerodactylus* lizards on St. Croix (MacLean and Holt, 1979). Distributional information gathered there in the 1930s was compared with the results of field work 40 years later. During that interval the range of the endemic *S. beattyi* had contracted, whereas that of an invading species, *S. macrolepis*, had expanded. The two lizards are similar in size and habitat preferences, but their distributions are sharply defined on the island. They segregate in part by different physiological tolerances for water loss; *S. beattyi* is better adapted to xeric microhabitats (MacLean and Holt, 1979). If the range of either species would be greater on the island in the absence of the other, then competitive interaction is probably operating. However, this is not a taxon cycle in the sense of Wilson (1961) or Ricklefs and Cox (1972, 1978), which would require that the species actually passed through the stages from expansion to extinction.

In the West Indies the taxon cycle achieved notoriety in studies of Lesser Antillean anoles. The distribution of these lizards is characterized by the repetition of size morphs on either one- (intermediate body size) or two-species (large and small body size) islands. A variant of the taxon cycle explains this pattern in the form of an invasion/coevolution model (Pacala and Roughgarden, 1982; Roughgarden and Pacala, 1989; Roughgarden, 1995). A larger, invading species outcompetes a smaller resident species and eventually pushes it to extinction. The invader then stabilizes at the intermediate solitary size, whence, potentially, the cycle repeats itself. This model has received considerable scrutiny, and other authors have argued for character displacement or size assortment within the constraints of phylogeny (Losos, 1992b; Miles and Dunham, 1996; Giannasi *et al.*, 1997). For example, two species in sympatry may evolve in different directions to avoid competition, viz size displacement. Conversely, perhaps only species of different size could coexist on the island in the first place and therefore must have evolved allopatrically.

We add that mean body size of insular lizards varies over time in response to changing circumstances of habitat quality and predation that are independent of interactions with other species of lizards. There is ample fossil evidence, for example, demonstrating a reduction in body size during the Holocene for species in the Caribbean and elsewhere. This reduction correlates with human settlement and associated declines in environmental quality, and introductions of nonnative predators (Pregill, 1986; Pregill and Dye, 1989). Proof of a taxon cycle must come from data showing that a species actually went through the stages, whether the stages are defined as changing distribution patterns or as evolutionary changes in size. Given the time involved, a fossil record presumably would be required, and thus far the record reveals no such pattern.

Holocene Extinctions

We noted earlier that historical evidence has minimal relevance in ecological biogeography, which is why the Holocene fossil record is an under-utilized resource in such studies (but see Case *et al.*, 1992). Events of the past 10,000 years, however, have altered the ecological complexion of most all islands through extinction and landscape transformations. As a result, whatever natural processes of species composition remain are not easily identified. The West Indian Holocene record, although far from complete, now lists hundreds of vertebrate populations and species that disappeared locally, regionally, or completely. About 80% of the West Indian land mammal fauna no longer exists (Morgan and Woods, 1986; MacPhee and Marx, 1997), and many of the 20 orders and 64 families of resident birds also include extinct representatives (Olson and Hilgartner, 1982; Cuello, 1988). Large tortoises, freshwater turtles, snakes, and lizards vanished as well (e.g., Olson *et al.*, 1990). This comparatively recent loss of diversity means that present ecological associations do not reflect a full roster of participating species. It means that caution is called for when interpreting current distributions, especially those that appear relictual, or are incongruous with respect to a broader pattern. In many instances current distributions have resulted from recent extinctions on intervening islands. Extinction is of course predicted by the equilibrium hypotheses as a consequence of new immigrants colonizing an island. The difference, however, is that none of the Holocene extinctions is attributable to species turnover. For example, *Cyclura pinguis* is known historically only from tiny Anegada Island, but until the late Holocene this large rock iguana ranged throughout much of the Puerto Rican Bank (Pregill, 1981b). On a broader front, at least some of the current Greater Antillean vertebrate fauna actually extended well into the Lesser Antilles during most of the Holocene. Paleofaunas from the Leeward Islands of Anguilla, Antigua, and Barbuda record several extinct vertebrates (species and genera) that today are confined to the Greater Antilles. Among these is the Puerto Rican Parrot *Amazona vittata* that today survives in small numbers in Puerto Rico's Luquillo forest and is regarded as endemic to that island. Other examples include a thick-knee (*Burhinus* sp.), a crow (*Corvus* sp.), the bats *Mormoops blainvillei*, *Macrotus waterhousii*, and the extinct *Phyllonycteris major*, and lizards of the genera *Cyclura* and *Leiocephalus* (Steadman *et al.*, 1984a; Pregill *et al.*, 1988, 1994).

Much of the evidence for a faunal collapse during the Holocene occurs near the time of first human contact. We do not know, however, if the Holocene extinctions of West Indian vertebrates were relatively abrupt for some or most taxa, were gradual over thousands of years for others, or the extent to which humans were the proximate cause in either case (Steadman *et al.*, 1984a; Morgan and Woods, 1986; MacPhee *et al.*, 1989; Pregill *et al.*, 1994). Nevertheless, that the losses post-date human arrival is, by itself, strong evidence of direct or indirect human involvement (Martin and Steadman, 1998) There are, unfortunately, very few well-dated, fossil-bearing strata available to stabilize the chronology. Some extinctions may be related to changes in climate, vegetation, land area, sea level or other non-anthropogenic effects associated with the glacial/interglacial transition 10,000-12,000 years ago, or with other natural phenomena later in the Holocene. The late Pleistocene climate of the West Indies appears to have been characterized by aridity that gave way to more mesic conditions at the beginning of the Holocene. The fossil record includes a high proportion of xeric-adapted species that became extinct at that time, especially in the Greater Antilles (Pregill and

Olson, 1981). As in other parts of the globe, there may have been mid-Holocene climatic fluctuations of sufficient magnitude that the environmental changes brought on by people were exacerbated. For example, intense late Holocene aridity is indicated in Madagascar that acted in concert with human hunting and fire regimes. The combined effects were disastrous for native vertebrates, especially birds and mammals (Burney, 1993). Sediment cores from the Persian Gulf also indicate two to three centuries of severe drought in this region about 2000-3000 years ago (Kerr, 1998).

The West Indian Holocene record derives from both paleontological (non-cultural) and archaeological (cultural) contexts. Non-cultural fossils are almost exclusively the accumulated remains of owl pellets deposited in caves and sinkholes. Bone records from archaeological sites derive from the "kitchen middens" of Amerindians. Compared with the Greater Antilles, there are far fewer non-cultural sites in the Lesser Antilles, although archaeological sites abound (Pregill *et al.*, 1994). The most desirable vertebrate record would be one from a site spanning three key horizons: (1) Prehuman – deposits in place prior to the arrival of people to the Antilles; (2) Prehistoric (pre-Columbian) – deposits dating from the appearance of indigenous people up to the time that Europeans and Africans arrived, about 500 years ago; and (3) Historic (post-Columbian) – deposits that date from the time of European/African contact. For any island, calibrating the prehuman/human interface is critical in order to establish a minimum age for anthropogenic effect. Next, it is important to ascertain which species persisted into the Historic period (and there were some) but disappeared soon after Europeans began unloading their ships. Besides their technology, Europeans brought exotic predators and competitors both domestic (goat, pig, dog, cat) and commensal (black rat). Historic settlement may have accelerated extirpation that began in the Prehistoric period, or in some cases may have initiated it.

Dated, bone-bearing strata directly associated with first human contact are rare in the Antilles. However, the chronology of human arrival itself is retrievable from archaeological sites that demonstrate an established cultural presence. Still, earliest cultural evidence may lag by several or more centuries the initial arrival of small human cohorts. Even these small, early groups could have modified the landscape through repeated burning. For example, in Puerto Rico sediment core analysis from Lake Tortuguero on the north central coast revealed a spike in charcoal stratigraphy indicative of human presence at ca. 5500 years before present (ybp), approximately two millennia earlier than the oldest archaeological dates from that island (Burney *et al.*, 1994). This puts the occupation of Puerto Rico in synchrony with Hispaniola and Cuba, which have archaeological sites dated to around 6000 ybp (Moore, 1991). This is a minimum age and people may well have arrived in the Greater Antilles (except Jamaica) as early as 7000 ybp (Rouse, 1989).

In the Lesser Antilles, the colonization by Amerindian peoples from coastal regions of northeast South America occurred by 2000 BC (or perhaps even earlier) in the period known as the Archaic Age of that region (Keegan and Diamond, 1987; Rouse, 1992). Lesser Antillean Archaic Age sites are usually coastal and often near mangrove stands (Davis, 1982). These people were neither agrarian nor potters. Faunal remains from their occupation sites are poorly studied, and indicate only that the first Amerindians primarily exploited marine and estuarine animals, especially mollusks (Pregill *et al.*, 1994). Their impact on the terrestrial vertebrate fauna is not known. By about 500-400 BC, another northward migration began, but of ceramic-producing people that for the next thousand years developed agricultural practices, and exploited terrestrial as well as marine resources. They also transported exotic species from South America (e.g., the agouti *Dasyprocta* sp.), and shuttled native ones (e.g., *I. iguana*, some parrots, *Amazona*) among the islands (Steadman *et al.*, 1984b). Within about 200 years following European settlement (late 1700s) most aboriginal peoples in the West Indies had been extirpated through warfare, disease, enslavement, and interbreeding, although remnant groups have persisted on a few islands even to the present.

In both the Greater and Lesser Antilles there were several millennia of human-engendered changes to the landscape and fauna before biologists arrived to collect plants and animals, and to study ecosystems. Native vertebrates on smaller islands were especially vulnerable from the beginning. In the Lesser Antilles, the Holocene fossil record shows that the current vertebrate fauna is not nearly as rich as it was several thousand years ago. From Sombrero to Grenada human activities of recent millennia resulted in the loss of indigenous rice rats, birds, lizards, snakes, and tortoises. At least 79 indigenous populations or species of vertebrates became extinct on islands from which fossil specimens of them are known (Pregill *et al.*, 1994). The extinct and extirpated herpetofauna, for instance, is represented by 18 species from 10 islands and in multiple trophic levels (Table 9.1). Because of poor chronologies, it is

Table 9.1. Extinct and Extirpated Amphibians and Reptiles of the Lesser Antilles[a].

Species (location)	Extinct, historic record	Extinct, no historic record	Extinct, known only from fossils
Leptodactylus fallax (St. Kitts, Guadeloupe, St. Lucia)	X		
Geochelone sp. (Barbados)			X
Geochelone sombrerensis (Sombrero)			X
Anolis bimaculatus giant morph[b] (Antigua, Barbuda)			X
Anolis wattsi (Anguilla)	X		
cf. *Cyclura* sp. (Barbuda)			X
Iguana delicatissima (St. Kitts)		X	
Iguana iguana (Barbados)		X	
Iguana sp. (Marie Galante)		X	
Leiocephalus cuneus (Antigua, Barbuda)			X
Leiocephalus cf. cuneus (Anguilla, Guadeloupe)			X
*Leiocephalus herminieri** (Martinique)	X		
Ameiva cineracea (Guadeloupe)	X		
*Ameiva major** (Martinique)	X		
Boa constrictor (Antigua)		X	
Boidae sp. (Antigua)			X
Alsophis sp. (Barbuda)		X	
cf. *Clelia clelia* (Barbuda)		X	

*Taxa with an asterisk have no fossil record and have become extinct historically (from Pregill *et al.*, 1994).

[b] Pregill *et al.* (1988).

difficult to estimate which and how many species persisted into the Historic Age on a given island. Only two of the 15 extinct or extirpated species of reptiles with a fossil record are known historically by specimens. Six others are extant elsewhere in the Lesser Antilles, and seven are known by fossils only.

Holocene faunal losses are not peculiar to the West Indies. Worldwide, islands experienced various degrees of extinction of native species coincident with first human contact (Burney, 1993). Pacific island ecosystems from Hawaii to New Zealand were profoundly altered by the people who first settled them, sometimes within a few generations (Kirch and Hunt, 1997). Overall, the pattern for oceanic islands follows that of the peopling of the New World and Australia, which suffered substantial losses in their megafaunas at and during the time humans colonized them. There is not universal agreement about the causes of the continental extinctions (Martin and Klein,1984) or that in North America native peoples who arrived there around 12,000 years ago (Reitz *et al.*, 1996) had anything to do with them. The point, however, is that the European's "New World" of 500 years ago, both continental and insular, had already been modified by people to varying degrees.

For equilibrium biogeography there is an important message: the distribution and abundance of plants and animals in the West Indies (and on other islands) cannot be viewed principally as the outcome of natural ecological processes. Contemporary faunas and populations are merely a fragmented subset of what existed a few thousand or even hundreds of years ago. Habitat utilization and trophic structure have been rearranged, and if this is not taken into account attempts to model island biotas realistically will fall short. Anoles, for example, may be more abundant today because of their ability to exploit patchy, discontinuous habitats resulting from human land use (Rand and Williams, 1969). MacArthur, Wilson, and Lack are deservedly among the greatest names in twentieth-century biogeography. At the time of their syntheses on island biotas, there was no knowledge of the Holocene and its consequences. Today, without applying this knowledge, ecological biogeography founders in its attempt to elucidate species origins and distributional patterns. In a practical sense, it will remain stuck in the subjunctive so long as history is ignored.

Coda

When we back away from Caribbean biogeography the most solvent contributions in view are the careful documentation of species on each and every island, now and in the past. When practiced with sound taxonomy, survey and inventory represent scholarship of lasting value. Albert Schwartz appreciated this better than anyone. In the West Indies new records of taxa are still being made, albeit at a slower pace than several decades ago (see Chapter 1). Still, surprising discoveries await, especially paleontological ones. Contemporary herpetologists and paleontologists are in a foot-race to find them before they are lost to ever-expanding human land use. The effort, expense, and skill required often go unrecognized while disproportionate celebrity is extended to speculative ventures in ecological theory. Without detailed knowledge of distributions and variation, biogeography is stuck at ground zero. Put another way, the Schwartz and Thomas (1976) checklist is the most valuable contribution to the biogeography of West Indian amphibians and reptiles in the past 50 years.

10

The West Indies and Middle America: Contrasting Origins and Diversity

William E. Duellman
Natural History Museum and Department of Systematics and Ecology
The University of Kansas, Lawrence, Kansas 66045

Introduction

The geography of North and South America is unique in having two large continents attached to one another on one side by a narrow connection and partially bridged on the other side by a series of islands. Historically, the two continents never have been closer, because they were widely separated since the breakup of Pangaea and have drifted closer throughout the millions of years since the fragmentation of Gondwanaland. Geological evidence, albeit not always consistent, suggests that what is now Central America was an archipelago in the late Cretaceous and subsequently some of these islands moved eastward to form what is now the Greater Antilles. By the Miocene another chain of volcanic islands emerged between nuclear Central America and South America; continued orogeny and erosion eventually formed a long, narrow peninsula which in the late Pliocene connected with South America. Also during the Cretaceous, another chain of islands, the Lesser Antilles, was inserted between the Greater Antilles and South America. These historical events and their biogeographic significance have been discussed at length by various authors in two volumes–Stehli and Webb (1985) and Woods (1989).

Although the herpetofaunas of Middle America and the West Indies have been studied intensively for more than half a century, no critical comparisons of the two faunas have been made. This is especially compelling because both regions received many lineages of their biotas from the same source areas–North America and South America–yet their compositions are extremely different. Thus, the biogeography of these two areas provides a fascinating challenge for retrodiction. This challenge is especially intriguing because no single set of hierarchical explanations seems to apply to the West Indian herpetofauna, a problem addressed eloquently by Williams (1989b).

Materials and Methods

For purposes of this analysis, Middle America is defined as the Central American countries and southeastern Mexico east of the Isthmus of Tehuantepec; therefore, the rest of Mexico is part of North America. The West Indies exclude the Netherlands Antilles (Aruba, Bonaire, and Curaçao) and the continental islands of Trinidad and Tobago. Areas of regions were obtained from various atlases, and that for Mexico east of the Isthmus of Tehuantepec was determined by use of a Micro-Plan II image analysis system (Laboratory Computer Systems, Inc., Cambridge, Ma.). Composition and taxonomy of the West Indian herpetofauna was updated from Schwartz and Henderson (1991). For Middle America, these data were summarized from Villa *et al.* (1988).

Origins of the Herpetofaunas

Matthewsian biogeographers (e.g., Dunn, 1931; Darlington, 1938) believed in the fixity of continental platforms and interpreted biogeographic patterns entirely by the mechanism of dispersal. With respect to Middle America and the West Indies, this complacency was shattered by Rosen's (1976) application of revolutionary geological interpretations of plate tectonics and elaborated upon by Rosen's (1978) broader ramifications on historical biogeography. Rosen's biogeographic scenario was based primarily on Malfait and Dinkelman's (1972) plate tectonic model that postulated a series of islands, the proto-Antilles, forming in the region that is now lower Central America and drifting eastward to become the Greater Antilles; subsequently, another series of islands formed, expanded, and interconnected to form the present-day lower Central America, while the basement of most of the Lesser Antilles was exposed in the upper Cretaceous, and the islands were formed principally by volcanism in the Tertiary (Speed *et al.*, 1979). A major fault zone lies between Dominica and Martinique; this Dominican Fault and associated geological evidence strongly suggests independent histories of the northern and southern Lesser Antilles (Bouysse, 1984; Speed, 1985). The geological evidence has been expanded and reinterpreted in a variety of ways (e.g., Donnelly, 1989; Perfit and Williams, 1989; Pindell and Barrett, 1990; Pitman *et al.*, 1993); according to Pittman *et al.*, the complexity of the plates and their associated tectonic events obviates a consensus of interpretation.

This geological evidence necessitated re-evaluation of biogeographic data. Conflicting hypotheses of dispersal versus vicariance have been proposed for both West Indian and Middle American phyletic lineages of the biota. The vicariance model involves the separation (vicariance event) of a continuous distribution of a taxon by the formation of a barrier, whereas the dispersal model involves the dispersal of organisms across a pre-existing barrier. The implications relative to the proposed proto-Antilles and the lower Central American archipelago are obvious. Savage (1982) championed the vicariance model for the Middle American and West Indian herpetofaunas, whereas others, notably Pregill (1981c), Williams (1989b), and Hedges (1996b) supported the dispersal model for the West Indian herpetofauna. Williams (1989b:22), a staunch dispersalist, argued

> Between these two competing historical hypotheses [dispersal and vicariance] I do not see any a priori grounds for choice. Drifting islands do not make vicariance per se more plausible or dispersal less so. In fact, in the whole history of the West Indies, as now understood, I do not see how dispersal on any present evidence can be ruled out. A moving Antillean complex alters distances between islands and mainlands and among the islands. It makes dispersal easier or more difficult. It says, however, nothing per se about faunas or how they got where they are. Nearness assists dispersal, distance makes dispersal difficult and very improbable. Nearness is not enough for vicariance, and distance makes it, by definition, impossible. To falsify dispersal in the Caribbean even in the earliest Tertiary will require evidence that we do not now have.

More recent and an entirely different kind of geological evidence now pervades the controversy. A presumed huge bolide impact off the coast of Yucatan is estimated to have created a huge crater (300 km across) at the Cretaceous/Tertiary boundary (Hildebrand and Boynton, 1990; Sharpton *et al.*, 1993). Hedges *et al.* (1992b) speculated that the proto-Antilles were the closest islands to the impact site and likely could have been devastated by the effects of this impact, especially giant tsunamis–waves estimated to have been four or five km in height at the point of impact (Hildebrand and Boynton, 1990). In this scenario, the biota on these Noah's arks (sensu McKenna, 1973) might have been obliterated, or the diversity might have been greatly reduced. However, because the impact site was in shallow water (<200 m), the wave height would have been no more than 200 m at the point of impact and greatly diminishing peripherally (Ward *et al.*, 1985). Thus, Crother and Guyer (1996) argued that the bolide impact would not have obliterated the biotas on islands having major physical relief and thereby discarded the hypothesis of Hedges *et al.* (1992b) that the biota of the Greater Antilles was mostly eliminated at Cretaceous/Tertiary boundary.

The major problem in dealing with the biogeography of the herpetofaunas of Middle America and the West Indies is a lack of evidence defining monophyletic lineages and the absence of phylogenetic analyses of taxa within those lineages. The two largest genera, *Anolis* and *Eleutherodactylus,* inhabiting both regions have been subjected to at least partial analyses. Guyer and Savage (1987) subjected data on *Anolis* to phylogenetic analyses resulting in the recognition of five genera of species formerly placed in *Anolis,* and they proposed a biogeographic hypothesis based on vicariance. Their methodology was evaluated critically by Cannatella and de Quieroz (1989), who concluded that Guyer and Savage's "proposed taxonomy is neither demanded by, nor consistent with, the results of phylogenetic analysis." Williams (1989a) rigorously criticized Guyer and Savage's work on the basis of "serious errors and confusions" and recommended that neither the databases nor the taxonomy be accepted without reexamination of the primary evidence (but see rebuttal by Guyer and Savage, 1992). Roughgarden (1990, 1995) reviewed the biogeography of *Anolis* in Puerto Rico and the Lesser Antilles; he concluded that the *Anolis occulatus, cuvieri,* and *cristatellus* groups (with the exception of the Bahaman *A. scriptus* in the latter group) are restricted to islands on the Puerto Rican bank, whereas the *Anolis bimaculatus* group is restricted to the Lesser Antilles north of the Dominica Channel and the *Anolis roqueti* group inhabits the Lesser Antilles to the south of that channel.

Investigations by Lynch (1986), Hedges (1989a), and Joglar (1989) have resulted in the recognition of five subgenera of *Eleutherodactylus.* Some major discrepancies exist between Hedges' (1989a) results based on immunological distances and Joglar's (1989) results based on morphology. The subgenera *Euhyas* and *Pelorius* are endemic to the West Indies, and the phylogenetically composite subgenus *Eleutherodactylus* occurs in the West Indies and Middle America and is widespread in South America. The subgenus *Syrrhophus* is principally Mexican, whereas the subgenus *Craugastor* is widespread in Middle America and also extends into northwestern South America. Of these supposed five lineages, only *Craugastor* is demonstrably monophyletic (Lynch and Duellman, 1997).

Only a few other genera have been subjected to significant phylogenetic analyses. According to Pregill (1981a), the West Indian toads are a monophyletic group (*Peltophryne*), but this lineage is derived from within the South American *Bufo granulosus* group (Hedges, 1996b). Hylid frogs are highly diverse in Middle America (*Agalychnis, Anotheca,*

Duellmanohyla, Plectrohyla, Ptychohyla, and *Triprion* endemic or nearly so) and far less diverse in the West Indies (five species of *Hyla* and the endemic genera *Calyptahyla* and *Osteopilus,* the latter with three species) (Trueb and Tyler, 1974). However, Hedges (1996b) remarked that molecular data indicate that, with the exception of *Hyla heilprini,* the West Indian hylids form a monophyletic group. The relationships of most of the lineages of hylids within Central America remain obscure, but a phylogeny of *Plectrohyla* is available (Duellman and Campbell, 1992; Wilson *et al.*, 1994), as is one for the genera *Duellmanohyla* and *Ptychohyla* (Campbell and Smith, 1992). The *Eleutherodactylus rostralis* group of *Eleutherodactylus* was subjected to a phylogenetic analysis by Savage (1987). Although detailed phylogenetic analyses of the highly diverse Middle American salamander fauna have not been published, many studies (e.g., Wake, 1987; Wake and Lynch, 1976; Wake and Elias, 1983; Wake and Johnson, 1989) of the bolitoglossines reveal relationships with North America and separate radiations in the Mexican highlands, nuclear Central America, and lower Central America.

Among lizards, the gekkonid genus *Sphaerodactylus* occurs on the mainland in Middle and South America but is especially diverse in the West Indies. With two possible exceptions, the mainland species represent a monophyletic lineage (Harris and Kluge, 1984), and Hass (1991) postulated that the West Indian species are a monophyletic lineage. The iguanian lizard genus *Leiocephalus* is demonstrably monophyletic (Pregill, 1992). However, its relationships remain controversial. Although placed in its own subfamily as a sister group to the South American Tropidurinae by Frost and Etheridge (1989), Hedges *et al.* (1992b) suggested that *Leiocephalus* is more closely related to North American crotaphytids and phrynosomatids. On the other hand, recent evidence places *Leiocephalus* as the sister group to Madagascaran oplurines (D. R. Frost, personal communication). Xantusiid lizards are distributed in southwestern North America and Middle America with *Cricosaura typica* endemic to Cuba. Analyses by Crother *et al.* (1986) placed *Cricosaura* as the sister group to the Middle American *Lepidophyma,* but Hedges and Bezy (1994) concluded that *Cricosaura* is the basal lineage in the family. The systematic relationships among West Indian and mainland species of *Celestus* and *Diploglossus* are in such disarray that generic assignment is controversial for some species, but *Sauresia* and *Wetmorena* are part of a radiation of West Indian *Celestus* (Campbell and Camarillo, 1994; Hedges *et al.*, 1992b; Hedges, 1996b). The mainland gerrhonotine anguids were analyzed by Good (1988), but his results were challenged by Campbell and Frost (1993) in their analysis of *Abronia.*

Among the snakes in the West Indies, the speciose genera *Epicrates* and *Tropidophis* are represented on the South American mainland by only one and three species, respectively; the West Indian lineages are monophyletic (Kluge, 1989; Hedges *et al.*, 1992b). Within the morass of the Colubridae, six "alsophine" (Xenodontinae) genera (*Antillophis, Arrhyton, Darlingtonia, Hypsirhynchus, Ialtris,* and *Uromacer*) are endemic to the Greater Antilles, and *Alsophis* occurs in the Lesser Antilles and South America. Although there is immunological evidence for relationships of the West Indian genera with South American xenodontines (Cadle, 1984a, 1985), the monophyly of the alsophines remains questionable. Cadle (1984a,b,c, 1985) identified two major clades of xenodontine colubrids—a South American clade and a Middle American clade. Of the six West Indian endemic *Leptotyphlops,* all but *L. columbi* in the Bahamas represent a single radiation (Thomas *et al.*, 1985), but their relation with mainland American taxa is unknown. The majority of New World *Typhlops* occur in the West Indies; Thomas (1989) identified two groups, one with African and the other with New World species, but Hedges *et al.* (1992) indicated that the West Indian taxa form a monophyletic group. Mainland taxa are even less well-known; the only substantive phylogenetic analyses of Middle American genera are those of the colubrid *Adelphicos* by Campbell and Ford (1982) and Campbell and Brodie (1988), and of the viperid *Bothriechis* by Crother *et al.* (1992).

Savage (1982) made a strong argument for the recognition of a distinctive Mesoamerican herpetofauna, the earliest origins of which were from the ancestral tropical American herpetofauna that was fragmented by the inundation of the land connection between Central and South America in the early Tertiary; this biota was enhanced by subsequent isolation of stocks of northern affinities. Thus, ancestral stocks of South American and North American origin differentiated *in situ* through vicariance, and these vicariants speciated throughout most of the Tertiary. The result was a distinctive Mesoamerican herpetofauna amounting to more than 70 genera and hundreds of species, including essentially all bolitoglossine salamanders; frog genera such as *Atelophryniscus, Plectrohyla,* and *Ptychohyla*; lizard genera such as *Corytophanes, Laemanctus,* and *Lepidophyma*; snake genera such as *Conophis, Drymobius,* and *Scaphiodontophis*; and turtle genera such as *Claudius* and *Staurotypus.* The distinctiveness of the Mesoamerican biota was obscured somewhat by the biotic interchange across the Panamanian isthmus beginning in the Late Pliocene, when South American elements dispersed into Central America; these invaders include centrolenid, dendrobatid, and hemiphractine frogs, gymnophthalmid lizards, and diverse kinds of snakes (e.g., *Boa, Bothrops, Epicrates, Liophis,* and *Xenodon*). Likewise, a few

North American groups dispersed into Central America; these included *Rana, Sceloporus, Coluber, Kinosternon,* and natricine snakes. Some of these North American groups (e.g., *Rana, Sceloporus,* and *Kinosternon*) had undergone major speciation in the southwestern parts of their ranges, principally on the Mexican Plateau. Likewise, many Mesoamerican groups dispersed into South America, including *Bolitoglossa, Oedipina, Agalychnis, Smilisca, Basiliscus, Coniophanes,* and colubrine snakes.

Whereas the evolution and biogeography of the Mesoamerican herpetofauna is reasonably well substantiated (albeit there are some controversial groups), Savage's (1982) attempt to explain the origin of the Greater Antillean herpetofauna by a vicariance model involving vicariance of the archaic Mesoamerican elements on the proto-Antilles suffers from more controversial evidence. Hedges (1996b) estimated divergence times from albumin immunological data (see Chapter 9) of most lineages of the West Indian herpetofauna and concluded that most lineages came from South American sources (supported by patterns of ocean currents). Furthermore, he postulated that stocks dispersed into the West Indies throughout the Cenozoic. According to Hedges (1996a), the *Euhyas* stock of *Eleutherodactylus* and the xantusiid lizard *Cricosaura typica* are the only examples of Middle America/Greater Antillean vicariance that possibly were involved with the proto-Antilles. However, Crother and Guyer (1996) recalibrated Hedges' "clocks" and revised divergence dates based on geological evidence; they concluded that with the exception of lineages endemic to Jamaica, more than 60% of the lineages could have been shared by the mainland and the Greater Antilles.

Savage (1982) and Hedges (1996b) agreed that the herpetofauna of the more recent Lesser Antilles was derived from South America by dispersal; in fact, many of the species inhabiting these islands are distributed on the South American mainland. However, there is geological evidence that there was continuity between northern South America and the southern Antillean plate in the Cretaceous (Speed, 1985); there is a distinct difference between the herpetofaunas in the northern and southern Lesser Antilles (Roughgarden, 1995). Thus, some of the lineages in the Lesser Antilles may be the result of vicariance.

Most of the West Indian herpetofauna seems to have been derived from South America, although many lineages (e.g., *Eleutherodactylus unistrigatus* group, *Sphaerodactylus, Gonatodes, Mabuya, Ameiva*) are diverse in South America, Middle America, and the West Indies. Only two genera, *Celestus* and *Tretanorhinus,* are shared exclusively by the Greater Antilles and Middle America. If the Greater Antillean herpetofauna was derived by vicariance from the Mesoamerican herpetofauna in the early Tertiary, one might expect more lineages or pairs of lineages to be common to the two regions. Whereas the North American contribution to the Middle American herpetofauna was great and included early stocks that differentiated in Middle America (e.g., bolitoglossine salamanders, sceloporine lizards, *Eumeces, Thamnophis,* and *Crotalus*) and later dispersalists, North America made an insignificant contribution to the herpetofauna of the West Indies—only turtles of the genus *Trachemys* and the natricine snake *Nerodia clarki.*

Diversity of the Herpetofaunas

To anyone contemplating the composition of the herpetofaunas of the West Indies and Middle America, immediately obvious are the absence of caecilians and especially salamanders in the West Indies and the diversity and abundance of some genera, especially *Eleutherodactylus, Anolis,* and *Sphaerodactylus,* in the West Indies. As emphasized by Williams (1989b), the herpetofauna of the West Indies is depauperate in the number of genera (lineages) represented. However, contrary to a commonly held misconception, the islands support a herpetofauna rich in species. The native, non-marine herpetofauna of Middle America consists of 793 species in an area of 692,886 km^2, whereas that of the West Indies consists of 613 species within a land area of 232,392 km^2. Because of the much greater area of Middle America, the density of species per area (1.14 per square km^4) is less than one-half that in the West Indies (2.64 per square km^4).

Within Middle America, the largest number of species occur in Guatemala and Costa Rica (with Panama a close third), whereas in the West Indies, the largest numbers are in Hispaniola and Cuba (Table 10.1). The herpetofaunal diversity in Guatemala (366 species with a density of 3.4 per square km^4) and Costa Rica (361 species with a density of 7.1 per square km^4) is associated with the highlands in both countries, whereas that in Panama (350 species with a density of 4.5 per square km^4) is enhanced by the continuation of the Cordillera de Talamanca from Costa Rica into Panama, as well as numerous South American taxa (e.g., *Pleurodema, Pipa, Enyalioides,* and *Atractus*) that enter eastern Panama. In contrast, the primarily lowland countries, Belize and El Salvador, have the smallest herpetofaunas–131 species with a density of 5.7 per square km^4 and 120 species with a density of 5.6 per square km^4, respectively. Densities for other Middle American units are as follows: southeastern Mexico, 1.3; Honduras, 2.1; and Nicaragua, 1.6 species per square km^4. The low species density in southeastern Mexico reflects the large area of the Yucatan Peninsula, in which relatively few

Table 10.1. Comparative Composition of the Native, Non-Marine Herpetofaunas in Middle America and the West Indies. Areas are in square kilometers; numbers are species.

Geographic Unit	Area	Anura	Amphibians	Reptiles	Total
SE Mexico	239,555	73	94	216	310
Belize	22,965	27	29	102	131
Guatemala	108,889	95	133	233	366
Honduras	112,088	57	74	162	236
El Salvador	21,393	25	31	89	120
Nicaragua	130,000	52	58	152	210
Costa Rica	50,900	120	149	212	361
Panama	77,096	127	156	194	350
Middle America	692,886	246	337	456	793
Cuba	114,524	50	50	101	151
Hispaniola	76,484	63	63	138	201
Jamaica	10,962	23	23	31	54
Puerto Rico	8,897	20	20	47	67
Bahama Bank	14,368	2	2	41	43
Lesser Antilles	7,157	10	10	84	94
West Indies	232,392	167	167	446	613

species occur; also, the low density in Nicaragua is a result of the absence of extensive highland regions.

Among the Greater Antilles, Hispaniola has the largest herpetofauna (201 species) and a higher density of species per area (2.63 per square km^4) than Cuba with 151 species and a density of 1.3 per square km^4. However, Puerto Rico has 67 species with a density of 7.5 per square km^4, and Jamaica has 54 species with a density of 4.9 per square km^4. The large number of species (94) and high density (13.1 per square km^4) in the Lesser Antilles has two causal factors: (1) the high degree of endemism (especially of *Anolis* and *Sphaerodactylus*) on particular islands and (2) the large number of South American taxa (e.g., *Bachia, Kentropyx, Thecadactylus, Corallus,* and *Bothrops*) present on the is - lands. In contrast, the Bahama Bank has only 43 species with a density of 2.9 per square km^4. The low density is owing to the fact that only two amphibians are present, and many of the species of reptiles are widespread on the islands.

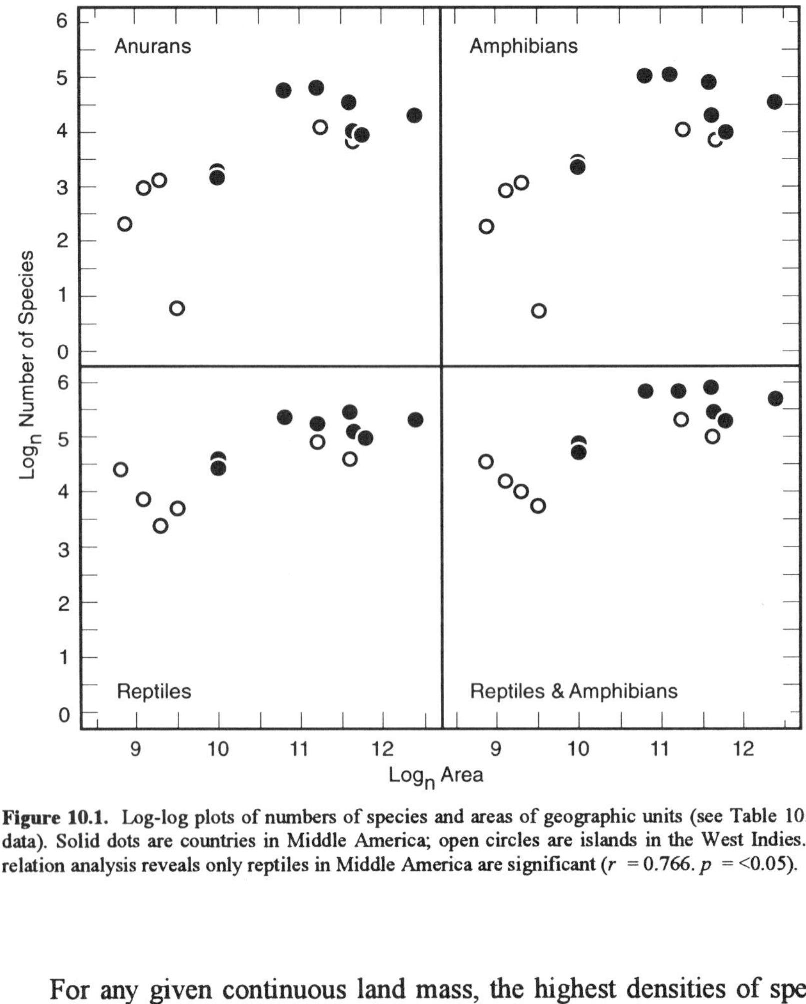

Figure 10.1. Log-log plots of numbers of species and areas of geographic units (see Table 10.1 for data). Solid dots are countries in Middle America; open circles are islands in the West Indies. Correlation analysis reveals only reptiles in Middle America are significant (r = 0.766. p = <0.05).

For any given continuous land mass, the highest densities of species are in Puerto Rico (7.5 per square km^4) and Costa Rica (7.1 per square km^4), but the actual numbers of species are vastly different–67 in Puerto Rico and 361 in Costa Rica. In comparing West Indian islands with continental areas of like size, there is a great difference in species richness. For example, Cuba (114,524 km^2) has 151 species, whereas Honduras (112,088 km^2) has 236 species. Likewise, Hispaniola (76,484 km^2) has 201 species, whereas Panama (77,096 km^2) has 350 species.

These differences would not be nearly so great if the lowlands on the mainland were excluded and the uplands were treated as islands. Even so, there is remarkable diversity in the uplands of Middle America; the highlands of nuclear Central America contain 126 species of amphibians, whereas there are 138 species in the highlands of Costa Rica and western Panama (J. A. Campbell, personal communication).

The theory of island biogeography contends that larger islands harbor more species (MacArthur and Wilson, 1963). However, this does not hold for the Greater Antillean herpetofauna. Cuba is about twice the size of Hispaniola but has only 75% as many species as Hispaniola. Jamaica is slightly larger than Puerto Rico but has only 80% as many species. A correlation analysis of numbers of species of anurans, amphibians, reptiles, and amphibians and reptiles in areas within these geographic units in the West Indies and Middle America in Table 10.1 revealed only one significant correlation ($r = 0.766$, $P < 0.05$)–area vs. number of species of reptiles in Middle America (Fig. 10.1).

As expected, because of insularity, endemism is higher in the West Indies than in Middle America. In the West Indies, 98% of the amphibians and 93% of the reptiles are endemic, whereas in Middle America, 73% of the amphibians and 65% of the reptiles are endemic. In Middle America, endemism is highest in the highlands of nuclear Central America (Chiapas, Mexico, eastward to western Honduras) and in the Central Highlands and Cordillera de Talamanca in Costa Rica and Panama.

The faunal unbalance between Middle America and the West Indies is indicative of independent histories of their biotas. Many significant elements in the Middle American herpetofauna are absent in the West Indies, including caecilians, salamanders, centrolenid, microhylid, ranid, and most genera of hylid frogs. Although the West Indian reptile fauna is diverse, it is missing most genera of iguanid and teiid lizards and colubrid snakes that occur in Middle America.

On the other hand, some genera are far more diverse in the West Indies than they are in Middle America. For example, amphisbaenians are represented by 14 species in the West Indies but only three species in Middle America and *Typhlops* by 23 in the West Indies and only three in Middle America. By far the most striking examples of speciation in the West Indies are frogs of the genus *Eleutherodactylus* and lizards of the genera *Anolis* and *Sphaerodactylus* (Fig. 10.2). Of the Middle American geographic units, Costa Rica has the highest densities of all three genera: *Eleutherodactylus*, 0.61 per square km^4, *Anolis,* 0.51 per square km^4, and *Sphaerodactylus,* 0.10 per square km^4. Hispaniola has the highest numbers of these species among single islands in the West Indies:

Eleutherodactylus, 0.73 per square km^4, *Anolis,* 0.54 per square km^4, and *Sphaerodactylus,* 0.45 per square km^4. The densities in Hispaniola are not that much higher than in Costa Rica.

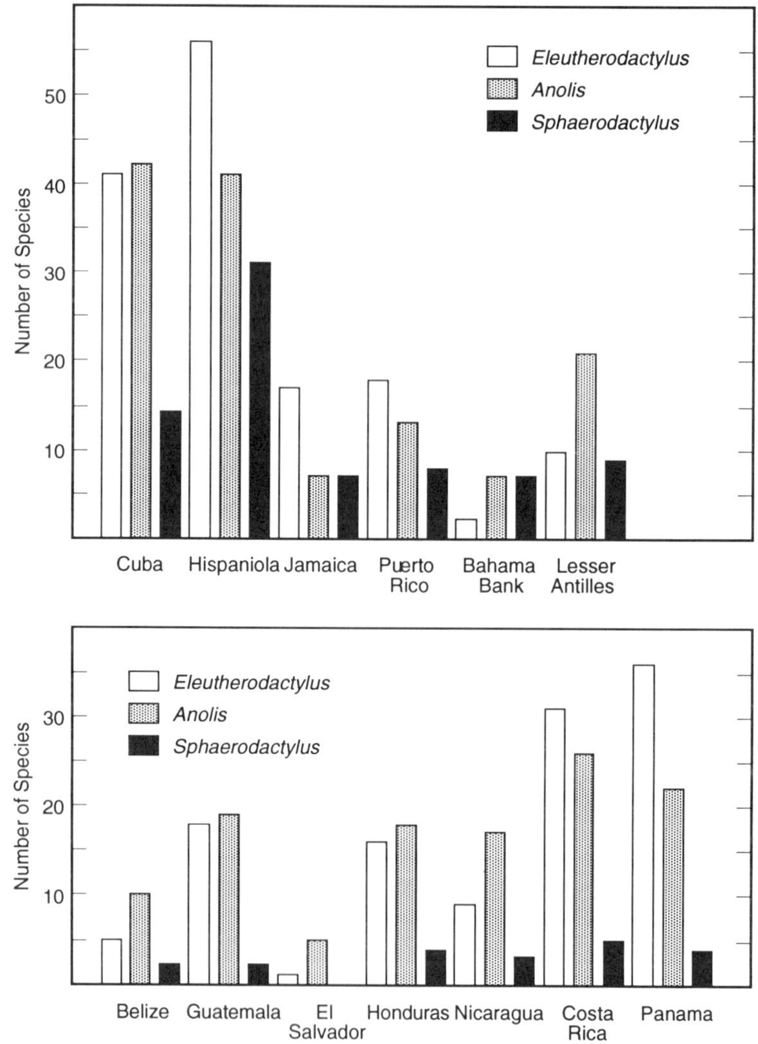

Figure 10.2. Comparative numbers of species in three genera in Middle American countries and on islands in the West Indies.

On the other hand, for the much smaller island of Puerto Rico the densities are as follows: *Eleutherodactylus*, 2.02 per square km^4 , *Anolis,* 1.46 per square km^4, and *Sphaerodactylus,* 0.80 per square km^4. The highest densities–*Eleutherodactylus* (1.40 per square km^4), *Anolis* (2.93 per square km^4), and *Sphaerodactylus* (1.26 per square km^4)–for a geographic unit are in the Lesser Antilles, where many species are endemic to one of many islands.

Discussion

The patterns of distribution and speciation in the West Indian herpetofauna involve far fewer lineages than those in Middle America. Has the absence of certain lineages "permitted" the great diversity of some lineages (e.g., *Eleutherodactylus, Anolis,* and *Sphaerodactylus*) in the West Indies? Unfortunately, few comparative ecological data are available. Most ecological studies in the West Indies have involved acquisition of data to demonstrate potential interspecific competition (e.g., Schoener, 1968; Schoener and Gorman, 1968), although Williams (1969), Schoener and Schoener (1971a,b), and Roughgarden (1995) presented data on habitat utilization by *Anolis.* Most studies on Middle American anoles have dealt with single species (e.g., Andrews, 1971, on *A. polylepis;* Fitch 1972, on *A. tropidolepis.*

On the other hand, in the West Indies there are no studies comparable to the thorough investigations of communities, such as at Barro Colorado Island, Panama (Rand and Myers, 1990) and La Selva, Costa Rica (Guyer, 1990), which were compared by Duellman (1990). Because the total numbers of species of amphibians and reptiles at these two sites (133 and 134 species, respectively) approach the total for the entire island of Cuba (150), it seems likely that communities at given sites in the West Indies will be significantly smaller than on the mainland of Middle America or South America.

Knowledge of the Middle American and West Indian herpetofaunas is approaching puberty. It will reach maturity only when the historical geology is clarified precisely, when the fauna is known taxonomically, when biogeographic patterns have been established on the basis of congruent cladograms of unrelated taxa, and when comparable ecological studies reveal the composition, patterns of resource utilization, and reproductive biology within communities. I hope that the foregoing essay will provide inspiration and initiative for workers to undertake these tasks so that sometime in the not too distant future a more precise comparison of two fascinating herpetofaunas will be possible.

References

Adams, W. E., K. W. Pickersgill, and G. Underwood. 1957. The development of the carotid arch in *Sphaerodactylus argus* Gosse, with some remarks on "Inselbildungen" in the aortic arches. J. Morphol. 121: 399-423.

Adler, K. 1989. Herpetologists of the Past, pp. 5-125 in Contributions to the History of Herpetology, edited by K. Adler. Society for the Study of Amphibians and Reptiles Contributions in Herpetology, Ithaca, NY.

Adolph, S. C., and J. Roughgarden. 1983. Foraging by passerine birds and *Anolis* lizards on St. Eustatius (Neth. Antilles): implications for interclass competition and predation. Oecologia 56: 313-317.

Ahrenfeldt, R. H. 1953. Two British anatomical studies on American reptiles (1650-1750) – I. Hans Sloane: comparative anatomy of the American crocodile. Herpetologica 9: 79-86.

Ahrenfeldt, R. H. 1954. Identification of the Amphibia and Reptilia recorded in Jamaica by Hans Sloane (1688-89). Copeia 1954: 105-111.

Amaral, A. 1925. On the differentiation of the species *Bothrops atrox* (Linné, 1758), *B. jararaca* (Wied, 1824), and *B. jararacussu* Lacerda, 1884. Contrib. Harvard Inst. Trop. Biol. Med. 2: 22-43.

Amaral, A. 1954. Contribuiçao ao conhecimento dos ofidios neotropicos XXXV. A propósito de revalidaçao de *Coluber lanceolatus* Lacépède, 1789. Mem. Inst. Butantan 26: 207-214.

Anderson, K. 1996. A karyological perspective on the monophyly of the hylid genus *Osteopilus*, pp. 157-168 in Contributions to West Indian Herpetology: A Tribute to Albert Schwartz, edited by R. Powell, and R. W. Henderson. Society for the Study of Amphibians and Reptiles Contributions in Herpetology, Ithaca, NY.

Andrews, E. A. 1892. Notes on the fauna of Jamaica. Johns Hopkins Univ. Circ. 97: 72-77.

Andrews, R. 1971. Structural habitat and time budget of a tropical *Anolis* lizard. Ecology 52: 262-270.

Andrews, R. 1976. Growth rate in island and mainland anoline lizards. Copeia 1976: 477-482.

Andrews, R. 1979. Evolution of life histories: a comparison of *Anolis* lizards from matched island and mainland habitats. Breviora 454: 1-51.

Anonymous. 1943. Notes from members. Nat. Hist. Notes Nat. Hist. Soc. Jamaica 1: 7.

Anonymous. 1950. Corrections and comments. Nat. Hist. Notes Nat. Hist. Soc. Jamaica 4: 66.

Anonymous. 1967. An annotated checklist of the Amphibia of Jamaica. Caribbean Biol. Center Newsletter 2: 15-16.

Anonymous. 1978. Opinion 1104/relative precedence of *Cornufer* Tschudi, 1838, and *Platymantis* Günther, 1858 (Amphibia Salientia). Bull. Zool. Nomencl. 34: 222-233.

Anonymous. 1989. Biggest little islands in the hemisphere. The Nature Conservancy 39: 32.

Arnold, D. L. 1980. Geographic variation in *Anolis brevirostris* (Sauria: Iguanidae) in Hispaniola. Breviora 461: 1-31.

Asprey, G. F., and R. G. Robbins. 1953. The vegetation of Jamaica. Ecol. Monogr. 23: 359-412.

Auffenberg, W. 1958. A small fossil herpetofauna from Barbuda, Leeward Islands, with the description of a new species of *Hyla*. Q. J. Florida Acad. Sci. 21: 248-254.

Auffenberg, W. 1967. Notes on West Indian tortoises. Herpetologica 31: 34-44.

Avila-Pires, T. C. S. 1995. Lizards of Brazilian Amazonia (Reptilia: Squamata). Zool. Verhand. 299.

Bailey, J. R. 1937. A review of some recent *Tropidophis* material. Proc. New England Zoöl. Club 16: 41-52.

Barbour, T. 1910. Notes on the herpetology of Jamaica. Bull. Museum Comp. Zool 52: 273-301.

Barbour, T. 1914. A contribution to the zoogeography of the West Indies, with especial reference to amphibians and reptiles. Mem. Museum of Comp. Zool. 44: 205-359.

Barbour, T. 1915. Recent notes regarding West Indian reptiles and amphibians. Proc. Biol. Soc. Washington 27: 71-78.

Barbour, T. 1916. Additional notes on West Indian reptiles and amphibians. Proc. Biol. Soc. Washington 29: 215-220.

Barbour, T. 1917. A new Antillean *Sphaerodactylus*. Proc. Biol. Soc. Washington 30: 163-164.

Barbour, T. 1921. *Sphaerodactylus*. Mem. Museum Comp. Zool. 47: 217-278.

Barbour, T. 1922. An annotated list of the Amphibia and Reptilia of Jamaica, pp. 666-669 in Handbook of Jamaica for 1922.

Barbour, T. 1923. West Indian investigations of 1922. Occ. Pap. Museum Zool. Univ. Michigan 132: 1-9.

Barbour, T. 1925. New Neotropical lizards. Proc. Biol. Soc. Washington 38: 101-102.

Barbour, T. 1930a. A list of Antillean reptiles and amphibians. Zoologica (New York) 11: 61-116.

Barbour, T. 1930b. Some faunistic changes in the Lesser Antilles. Proc. New England Zool. Club 11: 73-85.

Barbour, T. 1933. Notes on *Scolecosaurus*. Copeia 1933: 74-77.

Barbour, T. 1935. A second list of Antillean reptiles and amphibians. Zoologica 19: 75-141.

Barbour, T. 1937. Third list of Antillean reptiles and amphibians. Bull. Museum Comp. Zool. 82: 73-166.

Barbour, T. 1942. Two preoccupied names. Copeia 1942: 179.

Barbour, T., and A. F. Carr. 1940. Antillean terrapins. Mem. Museum Comp. Zool. 54: 381-415.

Barbour, T., and G. K. Noble. 1915. A revision of the lizards of the genus *Ameiva*. Bull. Museum Comp. Zool. 59: 417-479.

Barbour, T., and G. K. Noble. 1916. A revision of the lizards of the genus *Cyclura*. Bull. Museum Comp. Zool. 60: 139-164.

Barbour, T., and G. K. Noble. 1920. Some amphibians from northwestern Peru, with revision of the genera *Phyllobates* and *Telmatobius*. Bull. Museum Comp. Zool. 63: 304-427.

Barbour, T., and C. T. Ramsden. 1919. The herpetology of Cuba. Mem. Museum Comp. Zool. 47: 71-213.

Barbour, T., and B. J. Shreve. 1937. Novitates cubanae. Bull. Museum Comp. Zool. 80: 377-387.

Baskin, J. N., and E. E. Williams. 1966. The Lesser Antillean *Ameiva* (Sauria, Teiidae). Re-evaluation, zoogeography, and effects of predation. Studies on the Fauna of Curacao and other Caribbean Islands 23: 144-177.

Bauer, A. M., and A. P. Russell. 1993. *Aristelliger*. Cat. Am. Amphib. Rept. 565: 1-565.4.

Bell, H. J. 1889. Obeah; Witchcraft in the West Indies. Sampson Low, Marston, Searle and Rivington.

Bennett, A. F., and G. C. Gorman. 1979. Population density and energetics of lizards on a tropical island. Oecologia 42: 339-358.

Bezy, R. L. 1972. Karyotypic variation and evolution of the lizards in the family Xantusiidae. Contr. Sci. Nat. Hist. Museum Los Angeles Co. 227: 1-29.

Bezy, R. L., and J. A. Peterson. 1988. The microstructure of scale surfaces in the xantusiid lizard genus *Lepidophyma*. Herpetologica 44: 281-289.

Birdsey, R. A., and P. L. Weaver. 1987. Forest area trends in Puerto Rico. U.S. Forest Service Res. Note 331: 1-13.

Blake, J. A. 1986. Complex chromosomal variation in natural populations of the Jamaican lizard *Anolis grahami*. Genetica 69: 3-17.

Bocourt, M.-F. 1870. Description de quelques sauriens nouvoux originaires de l'Amérique méridionale. Nouv. Arch. Museum Hist. Nat. Paris 6: 11-18.

Bocourt, M.-F. 1870-1900. Estudes sur les reptiles et les batraciens. Miss. Sci. Mex. et Am. cent. 3: 1-1012.

Bogart, J. P. 1974. A karyosystematic study of frogs in the genus *Leptodactylus* (Anura: Leptodactylidae). Copeia 1974: 728-737.

Bogart, J. P. 1981. Chromosome studies in *Sminthillus* from Cuba and *Eleutherodactylus* from Cuba and Puerto Rico (Anura: Leptodactylidae). Life Sci. Contrib., Royal Ontario Museum 129: 1-22.

Bogart, J. P., and S. B. Hedges. 1995. Rapid chromosome evolution in Jamaican frogs of the genus *Eleutherodactylus* (Leptodactylidae). J. Zool. London 235: 9-31.

Bogert, C. M. 1968. The variations and affinities of the dwarf boas of the genus *Ungaliophis*. Novitates 2340: 1-26.

Böhme, W. 1984. Erstfund eines fossilen Kugelfingergeckos (Sauria: Gekkonidae) aus dominikanischem Bernstein (Oligozän von Hispaniola, Antillen). Salamandra 20: 212-220.

Bonnaterre, P.-J. 1789. Erpétologie, in Encyclopédie Méthodique. Tableau Encyclopédique et Méthodique de Trois Règnes de la Nature. Panckoucke, Paris.

Boulenger, G. A. 1882. Catalogue of the Batrachia Gradientia s. Caudata and Batrachia Apoda in the Collection of the British Museum, Trustees of the British Museum, London.

Boulenger, G. A. 1883-1896. Catalogue of the Snakes in the British Museum (Natural History). Trustees of the British Museum, London.

Boulenger, G. A. 1885. Catalogue of the Specimens of Lizards in the Collection of the British Museum (Natural History). Trustees of the British Museum, London.

Boulenger, G. A. 1887. Reptilia and Batrachia. Zool. Rec. 24: 1-34.

Boulenger, G. A. 1889. Catalogue of the Chelonians, Rhychocephalians, and Crocodiles in the British Museum (Natural History). Trustees of the British Museum, London.

Boulenger, G. A. 1891. On reptiles, batrachians, and fishes from the lesser West Indies. Proc. Zool. Soc. London 1891: 351-357.

Boulenger, G. A. 1893. Catalogue of the snakes of the British Museum (Natural History). Volume I. Trustees of the British Museum, London.

Boulenger, G. A. 1894. Catalogue of the snakes of the British Museum (Natural History). Volume II. Trustees of the British Museum, London.

Boulenger, G. A. 1906. Reptiles and Batrachians, pp. 517-531 in The History of the Collections contained in the Natural History Departments of the British Museum. Trustees of the British Museum, London.

Bourgeois, R. W. 1995. Herpetological publications of Albert Schwartz, pp. 21-27 in Contributions to West Indian Herpetology: A Tribute to Albert Schwartz, edited by R. Powell, and R. W. Henderson. Society for the Study of Amphibians and Reptiles Contributions in Herpetology, Ithaca, NY.

Bouysee, P. 1984. The Lesser Antilles island arc: structure and geodynamic evolution. Initial Reports, Deep Sea Drilling Project, 78A: 83-103. Government Printing Office, Washington DC., U.S.

Bowersox, S. R., S. Calderón, G. Cisper, R. S. Garcia, C. Huntington A. Lathrop, L. Lenart, J. S. Parmerlee, Jr., R. Powell, A. Queral, D. D. Smith, S. P. Sowell, and K. C. Zippel. 1994a. Miscellaneous natural history notes on amphibians and reptiles from the Dominican Republic. Bull. Chicago Herpetol. Soc. 29: 54-55.

Bowersox, S. R., S. Calderón, R. Powell, J. J. Parmerlee, D. D. Smith, and A. Lathrop. 1994b. Nahrung eines Riesenanolis, *Anolis barahonae*, von Hispaniola, mit einer Zusammenfassung des Nahrungsspektrums westindischer Riesenanolis-Arten. Salamandra 30: 155-160.

Bowin, C. 1975. The geology of Hispaniola, pp. 501-522 in Ocean Basins and Margins. Vol. 3, Gulf Coast, Mexico, and the Caribbean, edited by A. E. M. Nairn, and F. G. Stehli. Plenum, New York.

Braker, H. E., and H. W. Greene. 1994. Population biology: life histories, abundance, demography, and predator-prey interactions, pp. 244-255 in La Selva: Ecology and Natural History of a Neotropical Rain Forest, edited by L. A. McDade, S. Bawa, H. A. Hespenheide and G. S. Hartshorn. Univ. Chicago Press, Chicago.

Brongersma, L. D. 1937. The types of *Psammophis antillensis* Schlegel. Zool. Meded. 20: 1-5.

Brooks, D. R. 1981. Hennig's parasitological method: a proposed solution. Syst. Zool. 30: 229-249.

Brooks, D. R., and D. A. McLennan. 1991. Phylogeny, Ecology and Behavior: a Research Program. Univ. of Chicago Press, Chicago.

Brooks, D. R. and R. T. O'Grady. 1989. Crocodilians and their helminth parasites: macroevolutionary considerations. Am. Zool. 29:873-883.

Brooks, E. S. J. 1954. Sir Hans Sloane. The Great Collector and His Circle. Batworth Press, London.

Brooks, G. R., Jr. 1982. An analysis of prey consumed by the anuran, *Leptodactylus fallax*, from Dominica, West Indies. Biotropica 24: 209-214.

Brooks, G. R. 1983. *Gymnophthalmus pleei* Bocourt: an addition to the lizard fauna of Dominica, West Indies. Herpetol. Review 14: 31-32.

Browne, P. 1789. The Civil and Natural History of Jamaica with Complete Linnaean Indices, London.

Buden, D. W. 1966. An evaluation of Jamaican *Dromicus* (Serpentes, Colubridae) with the description of a new species. Breviora 238: 1-10.

Buide, M. S. 1967. Lista de los anfidios y reptiles de cuba. Torreia 1:60.

Bullock, D. J., and P. G. H. Evans. 1990. The distribution, density and biomass of terrestrial reptiles in Dominica, West Indies. J. Zool. London 222: 421-443.

Bullock, D. J., and H. M. Jury. 1990. A note on the feeding habits of *Ameiva fuscata* from Dominica, Lesser Antilles. Herpetol. J. 1: 532-534.

Bullock, D. J., H. M. Jury, and P. G. H. Evans. 1993. Foraging ecology in the lizard *Anolis oculatus* (Iguanidae) from Dominica, West Indies. J. Zool. London 230: 19-30.

Burke, K. 1988. Tectonic evolution of the Caribbean. Annu. Rev. Earth Planet. Sci. 16: 201-230.

Burnell, K. L., and S. B. Hedges. 1990. Relationships of West Indian *Anolis* (Sauria: Iguanidae): An approach using slow-evolving protein loci. Carib. J. Sci. 26: 7-30.

Burney, D. A. 1993. Recent animal extinctions: recipes for disaster. Am. Sci. 81: 530-541.

Burney, D. A., L. P. Burney, and R. D. E. MacPhee. 1994. Holocene charcoal stratigraphy from Laguna Tortuguero, Puerto Rico, and the timing of human arrival on the island. J. Archaeo. Sci. 21:273-281.

Burns, J. K., C. A. Cunningham, R. A. Dupuis, M. N. Trask, J. S. Tulloch, R. Powell, J. S. Parmerlee, Jr., K. L. Kopecky, and M. L. Jolley. 1992. Lizards of the Cayos Siete Hermanos, Dominican Republic, Hispaniola. Bull. Chicago Herpetol. Soc. 27: 225-232.

Buskirk, R. E. 1985. Zoogeographic patterns and tectonic history of Jamaica and the northern Caribbean. J. Biogeogr. 12: 445-461.

Buth, D. G. 1984. The application of electrophoretic data in systematic studies. Annu. Rev. Ecol. Syst. 15: 501-522.

Cadle, J. E. 1984a. Molecular systematics of neotropical xenodontine snakes: I. South American xenodontines. Herpetologica 40: 8-20.

Cadle, J. E. 1984b. Molecular systematics of neotropical xenodontine snakes: II. Central American xenodontines. Herpetologica 40: 21-30.

Cadle, J. E. 1984c. Molecular systematics of neotropical xenodontine snakes. III. Overview of xenodontine phylogeny and history of New World snakes. Copeia 1984: 641-652.

Cadle, J. E. 1985. The neotropical colubrid snake fauna (Serpentes: Colubridae): lineage components and biogeography. Syst. Zool. 34: 1-20.

Cadle, J. E. 1988. Phylogenetic relationships among advanced snakes: A molecular perspective. Univ. California Publ. Zool. 119: i-x, 1-77.

Cadle, J. E., and H. W. Greene. 1993. Phylogenetic patterns, biogeography, and ecological structure of neotropical snake assemblages, pp. 281-293 in Species Diversity in Ecological Communities: Historical and Geographical Perspectives, edited by R. E. Ricklefs, and D. Schleter. Univ. Chicago Press, Chicago.

Campbell, H. W. 1972. Ecological or phylogenetic interpretations of crocodilian nesting habits. Nature 238: 404-405.

Campbell, J. A. 1994. A new lizard of the genus *Diploglossus* (Anguidae: Diploglossinae) from Mexico, with a review of Mexican and northern Central American species. Herpetologica 50: 193-209.

Campbell, J. A., and J. E. D. Brodie. 1988. A new colubrid snake of the genus *Adelphicos* from Guatemala. Herpetologica 44: 416-422.

Campbell, J. A., and L. S. Ford. 1982. Phylogenetic relationships of the colubrid snakes of the genus *Adelphicos* in the highlands of Middle America. Occas. Pap. Museum Nat. Hist. Univ. Kansas 100: 1-22.

Campbell, J. A., and D. R. Frost. 1993. Anguid lizards of the genus *Abronia*: revisionary notes, descriptions of four new species, and phylogenetic analysis, and key. Bull. Am. Museum Nat. Hist. 216: 1-121.

Campbell, J. A., and E. N. Smith. 1992. A new frog of the genus *Ptychohyla* (Hylidae) from the Sierra de Santa Cruz, Guatemala, and description of a new genus of Middle American stream-breeding treefrogs. Herpetologica 48: 153-167.

Campbell, J. A., and J. P. Vannini. 1989. Distribution of amphibians and reptiles in Guatemala and Belize. Proc. West. Found. Vert. Zool. 4: 1-21.

Cannatella, D. C. 1986. A new genus of Bufonid (Anura) from South America, and phylogenetic relationships of the neotropical genera. Herpetologica 42: 197-205.

Cannatella, D. C., and K. de Queiroz. 1988. Phylogenetic systematics of the anoles: is a new taxonomy warranted? Syst. Zool. 38: 57-69.

Capote, R. P., R. E. Ricardoi, and A. V. González. 1989. 1. Vegetación Actual, 1:1,000,000. X. Flora and Vegetación. Nuevo Atlas Nacional de Cuba: Academia de Ciencias de Cuba.

Carey, W. M. 1975. The rock iguana, *Cyclura pinguis*, on Anegada, British Virgin Islands, with notes on *Cyclura ricordi* and *Cyclura cornuta* of Hispaniola. Bull. Florida State Museum, Biol. Sci. 19: 189-233.

Case, J. E., T. L. Holcombe, and R. G. Martin. 1984. Map of geological provinces in the Caribbean region, pp. 1-30 in The Caribbean-South American Plate Boundary and Regional Tectonics, edited by W. E. Bonini, R. B. Hargrave, and R. Shagam. Geol. Soc. Am. Mem., Boulder, CO.

Case, S. M., and E. E. Williams. 1984. A study of a contact zone in the *Anolis distichus* complex in the central Dominican Republic. Herpetologica 40: 118-137.

Case, S. M., and E. E. Williams. 1987. The cybotoid anoles and *Chamaelinorops*: evidence of mosaic evolution. Zool. J. Linnaean Soc. 91: 325-341.

Case, T. J. 1975. Species numbers, density compensation, and colonizing ability of lizards on islands in the Gulf of California. Ecology 56: 3-18.

Case, T. J. 1991. Invasion resistance, species build-up and community collapse in metapopulation models with interspecies competition. Biol. J. Linnean Soc. 42: 239-266.

Case, T. J., and D. T. Bolger. 1991. The role of introduced species in shaping the distribution and abundance of island reptiles. Evol. Ecol. 5: 272-290.

Case, T. J., D. T. Bolger, and A. D. Richman. 1992. Reptilian extinctions: the last ten thousand years, pp. 91-125 in Conservation Biology: the Theory and Practice of Nature Conservation, Preservation, and Management, edited by P. L. Fiedler, and S. K. Jain. Chapman and Hall, New York.

Case, T. J., and M. L. Cody. 1987. Testing theories of island biogeography. Am. Sci. 75: 402-410.

Case, T. J., M. E. Gilpin, and J. M. Diamond. 1979. Overexploitation, interference competition, and excess density compensation in insular faunas. Am. Nat. 113: 843-854.

Cei, J. M., V. Erspamer, and M. Roseghini. 1972. Biogenic amines, pp. 233-242 in Evolution in the genus *Bufo*, edited by W. F. Blair. Univ. Texas Press, Austin, TX.

Censky, E. J. 1988. *Geochelone carbonaria* (Reptilia: Testudines) in the West Indies. Florida Sci. 50: 108-114.

Censky, E. J. 1995a. Mating strategy and reproductive success in a teiid lizard, *Ameiva plei*. Behaviour 132: 529-557.

Censky, E. J. 1995b. Reproduction in two Lesser Antillean populations of *Ameiva plei* (Teiidae). J. Herpetol. 29.

Censky, E. J. 1996. The evolution of sexual size dimorphism in the teiid lizard *Ameiva plei*: A test of alternative hypotheses, pp. 277-289 in Contributions to West Indian Herpetology: A Tribute to Albert Schwartz, edited by R. Powell, and R. W. Henderson. Society for the Study of Amphibians and Reptiles Contributions in Herpetology, Ithaca, NY.

Censky, E. J., and K. Hodge. 1997. Geographic distribution: *Ramphotyphlops braminus*. Herpetol. Rev. 28: 210.

Censky, E. J., K. Hodge, and J. Dudley. 1998. Over-water dispersal of lizards due to hurricanes. Nature 395: 556.

Censky, E. J., and K. Lindsay. 1997. Geographic distribution: *Gymnopthalmus underwoodi*. Herpetol. Re. 28: 210.

Censky, E. J., and D. R. Paulson. 1992. Revision of the *Ameiva* (Reptilia: Teiidae) of the Anguilla Bank, West Indies. Ann. Carnegie Museum 61: 177-195.

Chandler, C. R., and P. J. Tolson. 1990. Habitat use by a boid snake, *Epicrates monensis*, and its anoline prey, *Anolis cristatellus*. J. Herpetol. 24: 151-157.

Cheke, A. S. 1984. Lizards of the Seychelles, pp. 331-360 in Biogeography and Ecology of the Seychelles Islands, edited by D. R. Stoddart. W. Junk Publ., The Hague.

Cheng, H. Y. 1983. Predation, tail regeneration and reproduction in two sympatric geckos (*Sphaerodactylus*) in Hispaniola. 26th Annu. Meetg Soc. Stud. Amphib. Rept.: 52.

Cisek, J. M., C. A. Cunningham, R. A. Dupuis, R. P. Seibolt, and R. Powell. 1990. *Wetmorena haetiana* (NCN). Food Habits. Herpetol. Rev. 21: 62.

Cochran, D. M. 1923a. A new species of *Eleutherodactylus* from the Dominican Repuplic [sic]. Proc. Biol. Soc. Washington 36: 93-94.

Cochran, D. M. 1923b. A new frog of the genus *Leptodactylus*. J. Washington Acad. Sci. 13: 184-185.

Cochran, D. M. 1923c. A new *Anolis* from Haiti. J. Washington Acad. Sci. 13: 225-226.

Cochran, D. M. 1924. Notes on the herpetological collections made by Dr. W. L. Abbott on the island of Haiti. Proc. U.S. Natl. Museum 66: 1-15.

Cochran, D. M. 1927. A new genus of anguid lizards from Haiti. Proc. Biol. Soc. Washington 40: 91-92.

Cochran, D. M. 1928a. A new species of *Chamaelinorops* from Haiti. Proc. Biol. Soc. Washington 41: 45-47.

Cochran, D. M. 1928b. A new genus and species of lizard, *Hispaniolus pratensis*, from the Haitian Republic. Proc. Biol. Soc. Washington 41: 49-51.

Cochran, D. M. 1928c. The herpetological collections made in Haiti and its adjoining islands by Walter J. Eyerdam. Proc. Biol. Soc. Washington 41: 53-59.

Cochran, D. M. 1931a. New reptiles from Beata Island, Dominican Republic. Proc. Biol. Soc. Washington 44: 89-92.

Cochran, D. M. 1931b. A new lizard from Haiti, *Sphaerodactylus stejnegeri*. Copeia 1931: 89-91.

Cochran, D. M. 1932a. Two new subspecies of lizards of the genus *Leiocephalus* from Hispaniola. Proc. Biol. Soc. Washington 45: 177-182.

Cochran, D. M. 1932b. Two new lizards from Hispaniola. Proc. Biol. Soc. Washington 45: 183-187.

Cochran, D. M. 1932c. A new snake, *Ialtris parishi*, from the Republic of Haiti. Proc. Biol. Soc. Washington 45: 189-190.

Cochran, D. M. 1932d. A new frog, *Eleutherodactylus wetmorei*, from the Republic of Haiti. Proc. Biol. Soc. Washington 45: 191-193.

Cochran, D. M. 1933. A new gecko from Haiti, *Aristelliger expectatus*. Proc. Biol. Soc. Washington 46: 33-35.

Cochran, D. M. 1934a. A new lizard, *Leiocephalus personatus lunatus*, from the Dominican Republic. Occas. Pap. Boston Soc. Nat. Hist. 8: 153-156.

Cochran, D. M. 1934b. Herpetological collections made in Hispaniola by the Utowana Expedition, 1934. Occas. Pap. Boston Soc. Nat. Hist. 8: 163-188.

Cochran, D. M. 1935. New reptiles and amphibians collected in Haiti by P. J. Darlington. Proc. Boston Soc. Nat. Hist. 40: 367-375.

Cochran, D. M. 1937. A necessary change in an amphibian name. J. Washington Acad. Sci. 27: 312.

Cochran, D. M. 1938a. A new species of frog from Haiti. Proc. Biol. Soc. Washington 51: 93-94.

Cochran, D. M. 1938b. Reptiles and amphibians from the Lesser Antilles collected by Dr. S. T. Danforth. Proc. Biol. Soc. Washington 51: 147-156.

Cochran, D. M. 1939. Diagnoses of three new lizards and a frog from the Dominican Republic. Proc. New England Zool. Club 18: 1-3.

Cochran, D. M. 1941. The herpetology of Hispaniola. Bull. U.S. Nat. Museum 177: vii + 398.

Cochran, D. M. 1961. Type specimens of reptiles and amphibians in the United States National Museum. U.S. Nat. Museum Bull. 220: xv + 291.

Cockerell, T. D. A. 1894. A little known Jamaican naturalist, Dr. Anthony Robinson. Am. Nat. 28: 775-790.

Cocteau, J. T., and G. Bibron. 1843. Reptiles. Hist. Fis. Pol. Nat. Cuba 4: 1-143.

Cohen, M. M., and C. Gans. 1970. The chromosomes of the order Crocodilia. Cytogenetics 9: 81-105.

Collette, B. B. 1961. Correlations between ecology and morphology in anoline lizards from Havana, Cuba and southern Florida. Bull. Museum Comp. Zool. 125: 137-162.

Connor, E. F., and E. D. McCoy. 1979. The statistics and biology of the species-area relationship. Am. Nat. 113: 791-833.

Cope, E. D. 1861. Notes and descriptions of anoles. Proc. Acad. Nat. Sci. Philadelphia 13: 208-215.

Cope, E. D. 1861 (1862). On the genera *Panolopus*, *Centropyx*, *Aristelliger* and *Sphaerodactylus*. Proc. Acad. Nat. Sci. Philadelphia 13: 494-500.

Cope, E. D. 1862a. Synopsis of the species of *Holcosus* and *Ameiva*, with diagnoses of new West Indian and South American Colubridae. Proc. Acad. Nat. Sci. Philadelphia 14: 60-82.

Cope, E. D. 1862b. Contributions to Neotropical saurology. Proc. Acad. Nat. Sci. Philadelphia 14: 176-188.

Cope, E. D. 1863a. On *Trachycephalus*, *Scaphiopus* and other American Batrachia. Proc. Acad. Nat. Sci. Philadelphia 15: 43-54.

Cope, E. D. 1863b. In Errata and addenda. Proc. Acad. Nat. Sci. Philadelphia 14: 594.

Cope, E. D. 1864. Contributions to the herpetology of tropical America. Proc. Acad. Nat. Sci. Philadelphia 16: 166-181.

Cope, E. D. 1868. An examination of the Reptilia and Batrachia obtained by the Orton Expedition to Equador and the Upper Amazon, with notes on other species. Proc. Acad. Nat. Sci. Philadelphia 20: 96-140.

Cope, E. D. 1869. Seventh contribution to the herpetology of tropical America. Proc. Am. Philos. Soc. 11: 147-169.

Cope, E. D. 1871. Ninth contribution to the herpetology of tropical America. Proc. Acad. Nat. Sci. Philadelphia 23: 200-224.

Cope, E. D. 1879. Eleventh contribution to the herpetology of tropical America. Proc. Am. Philos. Soc. 18: 261-277.

Cope, E. D. 1885. The large iguanas of the Greater Antilles. Am. Nat. 19: 1005-1006.

Cope, E. D. 1886. On the species of Iguaninae. Proc. Am. Philos. Soc. 23: 261-271.

Cope, E. D. 1895. The Batrachia and Reptilia of the Univ. of Pennsylvania West Indian Expedition of 1890 and 1891. Proc. Acad. Nat. Sci. Philadelphia 46: 429-442.

Corke, D. 1987. Reptile conservation on the Maria Islands (St. Lucia, West Indies). Biol. Conserv. 40: 263-279.

Corke, D. 1992. The status and conservation needs of the terrestrial herpetofauna of the Windward Islands (West Indies). Biol. Conserv. 62: 47-58.

Cousens, P. N. 1956. Notes on the Jamaican and Cayman Island lizards of the genus *Celestus*. Breviora 56: 1-6.

Croizat, L. 1958. Panbiogeography. Published by the author. Caracas, Venezuela.

Croizat, L. 1964. Space, Time and Form: the biological synthesis. Published by the author, Caracas, Venezuela.

Croizat, L., G. Nelson, and D. E. Rosen. 1974. Centers of origin and related concepts. Syst. Zool. 23: 265-287.

Crombie, R. I. 1973. Comment on the proposed suppression of *Hyla crucialis* (Amphibia). Z.N.(S.) 1982. Bull. Zool. Nomencl. 30: 4-6.

Crombie, R. I. 1974. The ecology, behavior and systematics of Jamaican hylid frogs. Yearbook Am. Philos. Soc. 1973: 304-305.

Crombie, R. I. 1977. A new species of frog of the genus *Eleutherodactylus* (Amphibia: Leptodactylidae) from the Cockpit Country of Jamaica. Proc. Biol. Soc. Washington 90: 194-204.

Crombie, R. I. 1985. Another new forest-dwelling frog (Leptodactylidae: *Eleutherodactylus*) from the Cockpit Country of Jamaica. Trans. San Diego Soc. Nat. Hist. 21: 145-153.

Crombie, R. I., D. W. Steadman, and J. C. Barber. 1984. A preliminary survey of the vertebrates of Cabarita Island, St. Mary Parish, Jamaica. Atoll. Res. Bull. 280: 1-12.

Crother, B. I. 1990. Is 'some better than none' or do allele frequencies contain phylogenetically useful information. Cladistics 6(2): 277-281.

Crother, B. I. 1999. Phylogenetic relationships among West Indian xenodontine snakes (Serpentes; Colubridae) with comments on the phylogeny of some mainland xenodontines. Contemporary Herpetology.

Crother, B. I., J. A. Campbell, and D. M. Hillis. 1992. Phylogeny and historical biogeography of the palm-pitvipers, genus *Bothriechis*: biochemical and morphological evidence, pp. 1-20 in Biology of the Pitvipers, edited by J. A. Campbell, and J. E. D. Brodie. Selva, Tyler. TX.

Crother, B. I., and C. Guyer. 1996. Caribbean historical biogeography: was the dispersal-vicariance debate eliminated by an extraterrestrial bolide? Herpetologica 52: 440-465.

Crother, B. I., and D. M. Hillis. 1995. Nuclear ribosomal DNA restriction sites, phylogenetic information, and the phylogeny of some xenodontine (Colubridae) snakes. J. Herpetol. 29:316-320.

Crother, B. I., M. M. Miyamoto, and W. F. Presch. 1986. Phylogeny and biogeography of the lizard family Xantusiidae. Syst. Zool. 35: 37-45.

Crother, B. I., and W. F. Presch. 1992. The phylogeny of xantusiid lizards: The concern for analysis in the search for a best estimate of phylogeny. Mol. Phylogenet. Evol. 1: 289-294.

Crother, B. I., and W. Presch. 1994. Xantusiid lizards, concern for analysis, and the search for a best estimate of phylogeny: further comments. Mol. Phylo. Evol. (3)3: 272-275.

Crozier, R. H. 1992. Genetic diversity and the agony of choice. Biol. Conserv. 61: 11-15.

Crozier, R. H. 1997. Preserving the information content of species: genetic diversity, phylogeny, and conservation worth. Annu. Rev. Ecol. Syst. 28: 243-68.

Cuello, J. P. 1988. Lista de las aves fósiles de al región neotropical y de las islas Antillanas. Paula-Coutianas Porto alegre. 2: 3-79.

Cullen, D. J., and R. Powell. 1994. A comparison of food habits of a montane and lowland population of *Anolis distichus* (Lacertilia: Polychrotidae) from the Dominican Republic. Bull. Maryland Herpetol. Soc. 30: 62-66.

Cullom, S. J., and A. Schwartz. 1980. Variation in the Hispaniolan giant anole *Anolis barahonae* with the description of two new subspecies. Herpetologica 36: 93-98.

Cundall, D., V. Wallach, and D. A. Rossman. 1993. The systematic relationships of the snake genus *Anomochilus*. Zool. J. Linnean Soc. 109: 275-299.

Cunningham, C. A., R. Powell, and S. B. Hedges. 1998. *Eleutherodactylus chlorophenax*. Cat. Am. Amphib. Rept. 648: 1-2.

Cunningham, C. A., J. S. Tulloch, and R. Powell. 1993. Food habits of four Hispaniolan species of *Sphaerodactylus* (Lacertilia: Gekkonidae). Herpetol. Nat. Hist. 1: 91-93.

Curtis, A. 1947. Prevalence of snakes in Haiti. Herpetologica 3: 224.

Cuvier, G. 1807. Sur les différentes espèces de crocodiles vivans et sur leurs caractères distinctifs. Annu. Natl. Museum Hist. Nat. Paris 10: 8-66.

Darlington, P. J. 1938. The origin of the fauna of the Greater Antilles, with discussion of dispersal of animals over water and through the air. Q. Rev. Biol. 13: 274-300.

Darlington, P. J. 1957. Zoogeography: The Geographical Distribution of Animals. J. Wiley & Sons, Inc., New York.

Darwin, S. P., and A. L. Welden (Editors). 1992. Biogeography of Mesoamerica, Special Publ. No. 1. Tulane Studies Zool. Bot., New Orleans, LA.

Daudin, F.-M. 1802a. Histoire Naturelle, Générale et Particuliere, des Reptiles. L'Imprimerie de F. Dufart, Paris.

Daudin, F.-M. 1802b. Histoire Naturelle des Rainettes, des Grenouilles et des Crapauds. Lerrault, Paris.

Daudin, F.-M. 1803. Histoire Naturelle des Reptiles. Lerrault, Paris.

Davis, D. D. 1982. Archaic settlement and resource exploitation in the Lesser Antilles: preliminary information from Antigua. Carib. J. Sci. 17: 107-122.
de Beer, G. R. 1953. Sir Hans Sloane and the British Museum. Oxford Univ. Press, New York.

de Queiroz, K. 1987. Phylogenetic systematics of iguanine lizards: a comparative osteological study. Univ. California Publ. Zool. 118: 1-203.

Densmore, L. D. 1983. Biochemical and immunological systematics of the order Crocodilia. pp397-465 in Evolutionary Biology, Vol. 16, edited by M. K. Hecht, B. Wallace, and G. H. Prance. Plenum, New York.

Densmore, L. D., and R. D. Owen. 1989. Molecular systematics of the order Crocodilia. Am. Zool. 29: 831-841.

Densmore, L. D., and P. S. White. 1991. The systematics and evolution of the Crocodilia as suggested by restriction endonuclease analysis of mitochondrial and nuclear ribosomal DNA. Copeia 1991: 602-615.

Dessauer, H. C., J. E. Cadle, and R. Lawson. 1987. Patterns of snake evolution suggested by their proteins. Fieldiana Zool. 1376: 1-34.

Dial, R., J. Roughgarden, and S. C. Tobin. 1994. Notes on the absolute abundance of canopy anoles, *Anolis cuvieri*, *A. stratulus*, and *A. evermanni* (Lacertilia: Polychridae) in the Luquillo Forest, Puerto Rico. Carib. J. Sci. 30: 278-279.

Diamond, J. 1975. Assembly of species communities, pp. 342-444 in Ecology and Evolution of Communities, edited by M. L. Cody and J. Diamond. Harvard Univ. Press, Cambridge.

Diaz, L. M., A. R. Estrada, and L. V. Moreno. 1996. A new species of *Anolis* (Sauria: Iguanidae) from the Sierra de Trinidad, Sancti Spíritu, Cuba. Carib. J. Sci. 32(1): 54-58.

Dickerson, M. C. 1916. Description of a new amphisbaenian collected by the late Charles S. Mead in 1911, on the Isle of Pines, Cuba. Bull. Am. Museum Nat. Hist. 35: 659-662.

Diesel, R., G. Baurle, and P. Vogel. 1995. Cave breeding and froglet transport: a novel pattern of anuran brood care in the Jamaican frog, *Eleutherodactylus cundalli*. Copeia 2: 354-360.

Dixon, J. R. 1962. The leaf-toed geckos, genus *Phyllodactylus*, of northeastern South America. Southwest. Nat. 7: 211-226.

Dixon, J. R. 1973a. *Phyllodactylus*. Cat. Am. Amphib. Rept. 141: 1-2.

Dixon, J. R. 1973b. A systematic review of the teiid lizards, genus *Bachia*, with remarks on *Heterodactylus* and *Anotosaura*. Univ. Kansas Museum Nat. Hist. Misc. Publ. 57: 1-47.

Dixon, J. R. 1980. The neotropical colubrid snake genus *Liophis*. The generic concept. Milwaukee Publ. Museum Contr. Biol. Geol. 31: 1-40.

Dixon, J. R. 1981. The neotropical colubrid snake genus *Liophis*: The eastern Caribbean complex. Copeia 1981: 296-304.

Dixon, J. R., and R. B. Huey. 1970. Systematics of the lizards of the gekkonid genus *Phyllodactylus* of mainland South America. Los Angeles County Museum Contrib. Sci. 192: 1-78.

Dixon, J. R., and P. Soini. 1986. The Reptiles of the Upper Amazon Basin, Iquitos Region, Peru. Milwaukee Public Museum, Milwaukee, WI.

Dobson, A. P., S. V. Pacala, J. D. Roughgarden, E. R. Carper, and E. A. Harris. 1992. The parasites of *Anolis* lizards in the northwestern Lesser Antilles. I. Patterns of distribution and abundance. Oecologia 91: 110-117.

Dodd, C. K., and R. Franz. 1996. Species richness and biogeography of the herpetofauna in the Exuma Cays Land and Sea Park, Bahamas, pp. 175-194 in Contributions to West Indian Herpetology: A Tribute to Albert Schwartz, edited by R. Powell, and R. W. Henderson. Society for the Study of Amphibians and Reptiles Contributions in Herpetology, Ithaca, NY.

Donnelly, M. A., and C. Guyer. 1994. Patterns of reproduction and habitat use in an assemblage of neotropical frogs. Oecologia 98(3-4): 291-302.

Donnelly, T. W. 1989. History of marine barriers and terrestrial connections: Caribbean paleogeographic inference from pelagic sediment analysis, pp. 103-118 in Biogeography of the West Indies: Past, Present, and Future, edited by C. A. Woods. Sandhill Crane Press, Gainesville.

Dowling, H. G., R. Highton, G. C. Maha, and L. R. Maxson. 1983. Biochemical evaluation of colubrid snake phylogeny. J. Zool. London 201: 309-329.

Dowling, H. G., C. A. Hass, S. B. Hedges, and R. Highton. 1996. Snake relationships revealed by slow-evolving proteins: a preliminary survey. J. Zool. London 240: 1-28.

Downer, A., and R. Sutton. 1990. Birds of Jamaica - A Photographic Field Guide. Cambridge Univ. Press., Cambridge, UK.

Drewry, G. E., and K. Jones. 1976. A new ovoviviparous frog, *Eleutherodactylus jasperi* (Amphibia, Anura, Leptodactylidae) from Puerto Rico. J. Herpetol. 10: 161-165.

Duellman, W. E. 1970. The hylid frogs of Middle America. Monogr. Museum Nat. Hist. Univ. Kansas 1: 1-748.

Duellman, W. E. 1978. The biology of an equatorial herpetofauna in Amazonian Ecuador. Misc. Publ. Museum Nat. Hist., Univ. Kansas 65: 1-352.

Duellman, W. E. 1987. Lizards in an Amazonian rain forest community: resource utilization and abundance. Nat. Geog. Res. 3: 489-500.

Duellman, W. E. 1989. Tropical herpetofauna communities: patterns of community structure in neotropical rainforests, pp. 61-88 in Vertebrates in Complex Tropical Systems, edited by M. L. Harmelin-Vivien, and F. Bourlière. Springer-Verlag, New York.

Duellman, W. E. 1990. Herpetofaunas in neotropical rainforests: comparative composition, history, and resource use, pp. 455-505 in Four Neotropical Rainforests, edited by A. H. Gentry. Yale Univ. Press, New Haven, CT.

Duellman, W. E. 1993. Amphibian species of the world: additions and corrections. Univ. Kansas Museum Nat. Hist. Spec. Publ. 21: 1-372.

Duellman, W. E., and J. A. Campbell. 1992. Hylid frogs of the genus *Plectrohyla*: systematics and phylogenetic relationships. Misc. Publ. Museum Zool. Univ. Michigan 181: 1-32.

Duellman, W. E., and J. R. Mendelson, III. 1995. Amphibians and reptiles from northern Departamento Loreto, Peru: taxonomy and biogeography. Univ. Kansas Sci. Bull. 55: 329-376.

Duellman, W. E., and L. Trueb. 1986. Biology of Amphibians. McGraw-Hill, New York.

Duer, C. K., J. M. Cisek, and R. Powell. 1992. Food habits of *Osteopilus dominicensis* (Anura: Hylidae). Carib. J. Sci. 28: 226-228.

Duméril, A. M. C. 1853. Prodrome de la classification des reptiles ophidiens. Troisième ordre de la classe des reptiles. Les ophidiens. Mem. Acad. Sci. France 23: 399-535.

Duméril, A. M. C. 1861. Reptiles et poissons de l'Afrique occidentale. Arch. Museum d'Hist. Nat. 10: 137-240.

Duméril, A. M. C., and G. Bibron. 1834. Érpétologie Générale ou Histoire Naturelle Compléte des Reptiles. Imprimerie et Fonderiede Fain, Paris.

Duméril, A. M. C., and G. Bibron. 1836. Érpétologie Générale ou Histoire Naturelle Compléte des Reptiles. Vol. 3. Libraire Encyclopédique de Roret, Paris.

Duméril, A. M. C., and G. Bibron. 1837. Érpétologie Générale ou Histoire Naturelle Compléte des Reptiles. Vol. 4. Libraire Encyclopédique de Roret, Paris.

Duméril, A. M. C., and G. Bibron. 1839. Érpétologie Générale ou Histoire Naturelle Compléte des Reptiles. Vol. 5. Libraire Encyclopédique de Roret, Paris.

Duméril, A. M. C., and G. Bibron. 1841. Érpétologie Générale ou Histoire Naturelle Compléte des Reptiles. Vol. 8. Libraire Encyclopédique de Roret, Paris.

Duméril, A. M. C., and G. Bibron. 1844. Érpétologie Générale ou Histoire Naturelle Compléte des Reptiles. Libraire Encyclopédique de Roret, Paris.

Duméril, A. M. C., and A. H. A. Duméril. 1851. Catalogue Méthodique de la Collection des Reptiles du Muséum d'Histoire Naturelle du Paris. Gide et Baudry, Paris.

Duméril, A. M. C., G. Bibron, and A. H. A. Duméril. 1854. Érpétologie Générale ou Histoire Naturelle Complète des Reptiles. Vol. 7. Libraire Encyclopédique Roret, Paris.

Dunn, E. R. 1920a. A new lizard from Haiti. Proc. New England Zool. Club 7: 33-34.

Dunn, E. R. 1920b. On the Haitian snakes of the genera *Leimadophis* and *Uromacer*. Proc. New England Zool. Club 7: 37-44.

Dunn, E. R. 1925. A new tree-toad from Jamaica. Occ. Pap. Boston Soc. Nat. Hist. 5: 161-162.

Dunn, E. R. 1926a. The frogs of Jamaica. Proc. Boston Soc. Nat. Hist. 38: 111-130.

Dunn, E. R. 1926b. Additional frogs from Cuba. Occ. Pap. Boston Soc. Nat. Hist. 5(2):209-215.

Dunn, E. R. 1927. Key to Jamaican frogs and notes on their habits, pp. 532-538 in Handbook of Jamaica.

Dunn, E. R. 1928. Two new frogs from Jamaica. Proc. U.S. Natl. Museum 74: 1-2.

Dunn, E. R. 1929. On the trail of the snorer. Nature 13: 110-112.

Dunn, E. R. 1931. The herpetological fauna of the Americas. Copeia 1931: 106-119.

Dunn, E. R. 1932. The colubrid snakes of the Greater Antilles. Copeia 1932: 89-92.

Dunn, E. R. 1934. Notes on *Iguana*. Copeia 1934: 1-4.

Dunn, E. R. 1936. Notes on American Mabuyas. Proc. Acad. Natl. Sci. Philadelphia 87: 533-557.

Dunn, E. R. 1939. Mainland forms of the snake genus *Tretanorhinus*. Copeia 1939: 212-217.

Du Tertre, J. 1654. Histoire générale, des Isles de S. Christophe, de la Guadeloupe, de la Martinique, et autres dans l'Amérique, Paris.

ENCAMP. 1980. Survey of conservation priorities in the Lesser Antilles. Caribbean Conservation Association. School of Natural Resources, Univ. Michigan.

Eggleton, P., and D. Vane-Wright. 1994. Pattern and Process: Phylogenetic Approaches to Ecological Problems. Academic Press, New York.

Espeut, W. B. 1882. On the acclimitization of the Indian mungoos in Jamaica. Proc. Zool. Soc. London 1882: 712-714.

Estrada, A. R. 1992. Lista de anfibios y reptiles de Cayo Saetía, Costa N. de Mayari, Holguín, Cuba. Comun. Breves Zool. 14.

Estrada, A. R. 1993a. Herpetofauna del Archipiélago de los Canarreos, Cuba. Poeyana 431: 1-19.

Estrada, A. R. 1993b. Anfibios y reptiles de Cayo Coco, Archipiélago de Sabana-Camagüey, Cuba. Poeyana 432: 1-21.

Estrada, A. R. 1994. Herpetofauna de la cuenca Banao-Higuanojo, Sancti Spíritus, Cuba. Rev. Acad. Colombiana Cien. 73.

Estrada, A. R., and R. Alonso. 1997. Nueva especie del grupo *limbatus* (Leptodactylidae: *Eleutherodactylus*) de la región oriental de Cuba. Carib. J. Sci. 33: 41-44.

Estrada, A. R., and L. F. de Armas. 1998. Apuntes ecológicos sobre *Cricosaura typica*. Carib. J. Sci, in press.

Estrada, A. R., and O. H. Garrido. 1990. Nueva subespecie de *Anolis jubar* (Lacertilia, Iguanidae) para Cayo Coco y la Loma de Cunagua, Ciego de Avila, Cuba. Biologia IV(1): 71-79.

Estrada, A. R., and O. H. Garrido. 1991. Dos nuevas especies de *Anolis* (Lacertilia: Iguanidae) de la Región Oriental de Cuba. Carib. J. Sci. 27(3-4): 146-161.

Estrada, A. R., and S. B. Hedges. 1991. Nueva espeie de *Eleutherodactylus* (Anura: Leptodactylidae) de la Región Oriental de Cuba. Carib. J. Sci. 27(3-4): 139-145.

Estrada, A. R., and S. B. Hedges. 1996a. A new frog of the genus *Eleutherodactylus* from eastern Cuba (Anura: Leptodactylidae). Herpetologica 52(3): 435-439.

Estrada, A. R., and S. B. Hedges. 1996b. A new diminutive frog of the genus *Eleutherodactylus* from Cuba (Anura, Leptodactylidae). Copeia 1996 (4): 852-859.

Estrada, A. R., and S. B. Hedges. 1997a. Nueva especie de *Eleutherodactylus* (Anura: Leptodactylidae) del macizo Sagua-Baracoa, Cuba. Carib. J. Sci. 33(3-4): 222-226.

Estrada, A. R., and S. B. Hedges. 1997b. A new species of frog from Meseta de Cabo Cruz, eastern Cuba (Leptodactylidae: *Eleutherodactylus*). Carib. J. Sci. 33(3-4): 227-236.

Estrada, A. R., and S. B. Hedges. 1998. Sistemática de las rana ribereñas de Cuba (Leptodactylide: *Eleutherodactylus*) con la descripción de una nueva especie. Carib. J. Sci., in press.

Estrada, A. R., and J. Novo. 1985a. A claraciones sobre la distribución de *Eleutherodactylus pinarensis* Dunn (Amphibia: Anura: Leptodactylidae) en la occidente de Cuba. Poeyana 302: 1-6.

Estrada, A. R., and J. Novo. 1985b. Nueva especie de *Eleutherodactylus* del grupo *ricordi* (Anura:Leptodactylidae) del occidente de Cuba. Poeyana 303: 1-10.

Estrada, A. R., and A. S. Rodriguez. 1984. Análisis de la ecomorfología de 23 especies de lagartos cubanos del género *Anolis*. Ciencias Biológicos 12: 91-104.

Estrada, A. R., J. Novo, and L. V. Moreno. 1989. Redescripción de *Eleutherodactylus symingtoni* Schwartz (Anura:Leptodactylidae) del occidente de Cuba. Biología III(2): 155-165.

Etheridge, R. 1960. The relationships of the anoles (Reptilia: Sauria: Iguanidae): an interpretation based on skeletal morphology. PhD. Dissertation, Univ. of Michigan, Ann Arbor.

Etheridge, R. E. 1965. Fossil lizards from the Dominican Republic. Q. J. Florida Acad. Sci. 28: 83-105.

Etheridge, R. E. 1966. The systematic relationships of West Indian and South American lizards referred to the iguanid genus *Leiocephalus*. Copeia 1966: 79-91.

Etheridge, R. E. and K. de Queiroz. 1988. A phylogeny of iguanidae, pp. 283-368 in Phylogenetic Relationships of the Lizard Families: Essays Commemorating Charles L. Camp, edited by R. Estes and G. Pregill. Stanford Univ. Press, Stanford.

Evans, L. T. 1938. Cuban field studies on territoriality of the lizard, *Anolis sagrei*. Comp. Psychol. 25: 97-125.

Evans, P. 1989. Herpetofauna of the Commonwealth of Dominica (Windward I., Lesser Antilles), West Indies. British Herpetol. Soc. Bull. 28: 5-7.

Ewel, J. J., and J. L. Whitmore. 1973. The ecological life zones of Puerto Rico and the U.S. Virgin Islands. Forest Service Res. Paper 18: 1-72.

Faith, D. P. 1992. Conservation evaluation and phylogenetic diversity. Biol. Conserv. 61: 1-10.

Faith, D. P. 1994. Phylogenetic diversity: a general framework for the prediction of feature diversity. Philos. Trans. R. Soc. London Ser. B 345:45-48.

Fischer, J. G. 1856. Neue Schlangen des Hamburgischen Naturhistorischen Museums. Abh. Naturwiss. Ver. Hamburg 3: 79-116.

Fischer, J. G. 1883. Beschreibungen neuer Reptilien. Oster-Progr. Akad. Gymn. Hamburg 1883: 1-17.

Fischer, J. G. 1888. Ueber eine Kollection Reptilien und Amphibien von Hayti. Jahrb. Wiss. Anst. Hamburg 5: 23-45.

Fitch, H. S. 1972. Ecology of *Anolis tropidolepis* in Costa Rican cloud forest. Herpetologica 28: 10-21.

Fitch, H. S. 1987. The sin of anecdotal writing. Herpetol. Rev. 18: 68.

Fitch, H. S., and R. W. Henderson. 1987. Ecological and ethological parameters in *Anolis bahorucoensis*, a species having rudimentary development of the dewlap. Amphib.-Rept. 8: 69-80.

Fitch, H. S., R. W. Henderson, and H. Guarisco. 1989. Aspects of the ecology of an introduced anole: *Anolis cristatellus* in the Dominican Republic. Amphib.-Rept. 10: 307-320.

Fitzinger, L. 1843. Systema Reptilium. Fasciculus primus. Amblyglossae. Braumüller et Seidel, Vindobonae.

Fleishman, L. J., E. R. Loew, and M. Leal. 1993. Ultraviolet vision in lizards. Nature 365: 397.

Flores, G. 1984. Comparative osteology, relationships, and evolution in Jamaican frogs of the genus *Eleutherodactylus*, pp. 122. Harvard Univ., Cambridge, MA.

Flores, G., J. H. Lenzycki, and J. J. Palumbo. 1994. An ecological study of the endemic Hispaniolan anoline lizard, *Chamaelinorops barbouri* (Lacertilia: Iguanidae). Breviora 499: 1-23.

Floyd, H. B., and T. A. Jenssen. 1983. Food habits of the Jamaican lizard, *Anolis opalinus*: resource partitioning and seasonal effects examined. Copeia 1983: 319-331.

Fobes, T. M., R. Powell, J. J. S. Parmerlee, A. Lathrop, and D. D. Smith. 1992. Natural history of *Anolis cybotes* (Sauria: Polychridae) from an altered habitat in Barahona, Dominican Republic. Carib. J. Sci. 28: 200-207.

Forcart, L. 1951. Nomenclature remarks on some generic names of the snake family Boidae. Herpetologica 7: 197-199.

Forey, P. I., C. J. Humphries, and R. I. Vane-Wright. 1994. Systematics and Conservation Evaluation. Clarendon, Oxford, UK.

Forsgaard, K. 1983. The axial skeleton of *Chamaelinorops*, pp. 284-295 in Advances in Herpetology and Evolutionary Biology, edited by A. G. J. Rhodin, and K. Miyata. Museum Comp. Zool., Cambridge, MA.

Fowler, H. W. 1918. Some amphibians and reptiles from Porto Rico and the Virgin Islands. Publ. Carnegie Inst. Washington 252: 1-16.

Franz, R., and C. A. Woods. 1983. A fossil tortoise from Hispaniola. J. Herpetol. 17: 79-81.

Frazer, H., S. Carrington, A. Forde, and J. Gilmore. 1990. A-Z of Barbados Heritage. Heinemann, Jamaica.

Frost, D. R., and R. Etheridge. 1989. A phylogenetic analysis and taxonomy of iguanian lizards (Reptilia: Squamata). Univ. Kansas Museum Nat. Hist. Misc. Publ. 81: 1-65.

Frost, D. R., and D. M. Hillis. 1990. Species in concept and practice: Herpetological applications. Herpetologica 46: 87-104.

Gali, F., and A. Schwartz. 1982. A new subspecies of *Leiocephalus personatus* from the República Dominicana. J. Herpetol. 16: 177-179.

Gali, F., A. Schwartz, and A. Suarez. 1988. A new subspecies of *Leiocephalus personatus* (Sauria: Iguanidae) from Haiti. Proc. Biol. Soc. Washington 101: 1-3.

Gallagher, D. S., and J. R. Dixon. 1980. A new lizard (Sauria: Teiidae: *Kentropyx*) from Brazil. Copeia 1980: 616-620.

Gans, C. 1964. *Amphisbaena schmidti*, a third species of the genus from Puerto Rico (Amphibaenia: Reptilia). Breviora 198: 1-11.

Gans, C. 1967. The diphyletism of the amphisbaenia (Reptilia): a re-evaluation based upon chromosome counts. Copeia 1967: 485-487.

Gans, C. 1978. The characteristics and affinities of the Amphisbaenia. Trans. Zool. Soc. London 34:347-416.

Gans, C. 1990. Patterns in amphisbaenian biogeography: a preliminary analysis, pp. 133-143 in Vertebrates in the Tropics, edited by G. Peters, and R. Hutterer. Museum Alexander Koenig, Bonn.

Gans, C., and A. A. Alexander. 1962. Studies on amphisbaenids (Amphisbaenia, Reptilia). 2. On the amphisbaenids of the Antilles. Bull. Museum Comp. Zool. 128: 67-158.

Garman, S. 1887a. On West Indian Gekkonidæ and Anguidæ. Bull. Essex Inst. 19: 17-24.

Garman, S. 1887b. On West Indian reptiles. Iguanidae. Bull. Essex Inst. 19: 25-30.

Garman, S. 1887c. West Indian Batrachians in the Museum of Comparative Zoology. Bull. Essex Inst. 19: 13-16.

Garman, S. 1887d. On West Indian reptiles and batrachians in the Museum of Comparative Zoology at Cambridge, Mass. Bull. Essex Inst. 19: 1-16.

Garman, S. 1887e. On West Indian reptiles in the Museum of Comparative Zoology at Cambridge, Mass. Proc. Am. Philos. Soc. 24: 278-286.

Garrick, L. D., R. L. Sutton, and J. W. Lang. 1985. Observations on the largest Jamaican tree frog, *Calyptahyla crucialis*. Carib. J. Sci. 21: 159-162.

Garrido, O. H. 1972. *Anolis bremeri* Barbour (Lacertilia: Iguanidae) en el occidente de Cuba Isla de Pinos. Carib. J. Sci. 12(1-2): 59-77.

Garrido, O. H. 1973a. Nueva especie de *Leiocephalus* (Lacertilia, Iguanidae) para Cuba. Poeyana (116): 1-19.

Garrido, O. H. 1973b. Distribución y variación de *Anolis homolechis* Cope (Lacertilia: Iguanidae) en Cuba. Poeyana 120: 1-68.

Garrido, O. H. 1975a. Nuevos reptiles del archipiélago cubano. Poeyana 141: 1-58.

Garrido, O. H. 1975b. Distribución y variación de *Anolis argillaceus* Cope (Lacertilia: Iguanidae) en Cuba. Poeyana 142: 1-28.

Garrido, O. H. 1975c. Distribución y variación del complejo *Anolis cyanopleurus* (Lacertilia: Iguanidae) en Cuba. Poeyana 143: 1-60.

Garrido, O. H. 1975d. Variación de *Anolis argillaceus* Cope (Lacertilia: Iguanidae) en el occidente de Cuba y la Isla de Pinos. Poeyana 144: 1-18.

Garrido, O. H. 1981. Nueva subespecie de *Anolis equestris* (Sauria: Iguanidae) para Cuba, con comentarios sobre la distribución y afinidades de otras poblaciones del complejo. Poeyana 232: 1-15.

Garrido, O. H. 1982a. Descripción de una nueva especie cubana de *Chamaeleolis* (Lacertilia: Iguanidae), con notas sobre su comportamiento. Poeyana 236: 1-25.

Garrido, O. H. 1982b. Nueva especie de *Anolis* (Lacertilia: Iguanidae) para Cuba. Doñana Acta Vert. 9: 31-37.

Garrido, O. H. 1983. Nueva especie de *Anolis* de la Sierra del Turquino. Carib. J. Sci. 19(3-4): 71-76.

Garrido, O. H. 1985. Nueva subespecie de *Anolis isolepis* (Lacertilia: Iguanidae) para Cuba. Doñana Acta Vert. 12(1): 41-49.

Garrido, O. H. 1988. Nueva especie para la ciencia de *Anolis* (Lacertilia: Iguanidae) de Cuba perteneciente al complejo *argillaceus*. Doñana Acta Vert. 15(1): 45-57.

Garrido, O. H. 1990. Nueva especie de *Anolis* de la sección beta (Lacertilia: Iguanidae) para Cuba. Biología 4: 157-162.

Garrido, O. H., and S.B. Hedges. 1992. Three new grass anoles from Cuba (Squamata: Iguanidae). Carib. J. Sci. 28(1-2): 21-29.

Garrido, O. H., and M. L. Jaume. 1984. Catálogo descriptivo de los Anfibios y Reptiles de Cuba. Doñana Acta Vert. 11(2): 5-128.

Garrido, O. H., O. Pérez-Beato, and L. V. Moreno. 1991. Nueva especie de *Chamaeleolis* (Lacertilia: Iguanidae) para Cuba. Carib. J. Sci. 27(3-4): 162-168.

Garrido, O. H., and A. Schwartz. 1972. The Cuban *Anolis spectrum* complex (Sauria, Iguanidae). Proc. Biol. Soc. Washington 85(45): 509-522.

Gault, D. E., and C. P. Sonnet. 1982. Laboratory simulation of pelagic asteroidal impact: atmospheric injection, benthic topography, and the surface wave radiation field, pp. 69-92 in Geological Implications of Impacts of Large Asteroids and Comets on the Earth, edited by L. T. Silver, and P. H. Schultz. Geol. Soc. Am. Spec. Pap. 190. Boulder, CO.

Gauthier, J. A., R. E. Estes, and K. de Queiroz. 1988. A phylogenetic analysis of Lepidosauromorpha, pp. 15-98 in Phylogenetic Relationships of the Lizard Families: Essays Commemorating Charles L. Camp, edited by R. Estes and G. Pregill. Stanford Univ. Press, Stanford, CA.

Giannasi, N., R. S. Thorpe, and A. Malhorta. 1997. Introduction of *Anolis* species to the island of St. Lucia, West Indies: testing for hybrids using multivariate morphometrics. J. Herpetol. 31: 586-589.

Gleason, H. A. 1922. On the relation between species and area. Ecology 3: 158-162.

Goin, C. J. 1950. Color pattern inheritance in some frogs of the genus *Eleutherodactylus*. Bull. Chicago Acad. Sci. 9: 1-15.

Goin, C. J. 1953. Rediscovery of the frog *Litoria luteola* Gosse in Jamaica. Occ. Pap. Inst. Jamaica 7: 1-4.

Goin, C. J. 1954. Remarks on the evolution of color pattern in the *gossei* group of the frog genus *Eleutherodactylus*. Ann. Carnegie Museum 33: 185-195.

Goin, C. J. 1958. Further studies on color pattern inheritance in the frog, *Eleutherodactylus nubicola*. Yearbook Am. Philos. Soc.: 248-250.

Goin, C. J. 1959a. Notes on the maxillary dentition of some frogs of the genera *Eleutherodactylus* and *Leptodactylus*. Herpetologica 15: 134-136.

Goin, C. J. 1959b. A synonym and a homonym in the frog genus *Hyla*. Copeia 4: 340-341.

Goin, C. J. 1960. Pattern variation in the frog *Eleutherodactylus nubicola* Lynn. Bull. Florida State Museum 5: 243-258.

Goin, C. J., and B. W. Cooper. 1950. Notes on a collection of amphibians from Jamaica. Occ. Pap. Inst. Jamaica 4: 1-9.

Goldwasser, L., and J. Roughgarden. 1993. Construction and analysis of a large Caribbean food web. Ecology 74: 1216-1233.

González-Bermúdez, F., and L. Rodríguez-Schettino. 1982. Datos etoecológicos sobre *Anolis vermiculatus* (Sauria: Iguanidae). Poeyana 245: 1-18.

Good, D. A. 1988. Phylogenetic relationships among gerrhonotine lizards. An analysis of external morphology. Univ. California Publ. Zool. 121: 1-139.

Gorman, G. C., and L. Atkins. 1968a. Natural hybridization between two sibling species of *Anolis* lizards: chromosome cytology. Science 159: 1358-1360.

Gorman, G. C., and L. Atkins. 1968b. New karyotypic data for 16 species of *Anolis* (Sauria: Iguanidae) from Cuba, Jamaica, and the Cayman Islands. Herpetologica 24: 13-21.

Gorman, G. C., and L. Atkins. 1969. The zoogeography of Lesser Antillean *Anolis* lizards – an analysis based upon chromosomes and lactic dehydrogenase. Bull. Museum Comp. Zool. 138: 53-80.

Gorman, G. C., D. G. Buth, M. Soulé, and S. Y. Yang. 1980a. The relationships of the *Anolis cristatellus* species group: Electrophoretic analysis. J. Herpetol. 14: 269-278.

Gorman, G. C., D. Buth, M. Soulé, and S. Y. Yang. 1983. The relationships of the Puerto Rican *Anolis*: Electrophoretic and karyotypic studies, pp. 626-642 in Advances in Herpetology and Evolutionary Biology, edited by A. G. J. Rhodin, and K. Miyata. Museum Comp. Zool., Cambridge.

Gorman, G. C., D. G. Buth, and J. S. Wyles. 1980b. *Anolis* lizards of the Eastern Caribbean: A case study of evolution. III. A cladistic analysis of albumin immunological data, and the definition of species groups. Syst. Zool. 29: 143-158.

Gorman, G. C., and H. C. Dessauer. 1965. Hemoglobin and transferrin electrophoresis and relationships of island populations of *Anolis* lizards. Science 150: 1454-1455.

Gorman, G. C., and H. C. Dessauer. 1966. The relationships of *Anolis* of the *roquet* species group (Sauria: Iguanidae) - I. Electrophoretic comparison of blood proteins. Comp. Biochem. Physiol. 19: 845-853.

Gorman, G. C., and R. Harwood. 1977. Notes on population density, vagility, and activitiy patterns of the Puerto Rican grass lizard, *Anolis pulchellus* (Reptilia, Lacertiia, Iguanidae). J. Herpetol. 11: 363-368.

Gorman, G. C., and Y. J. Kim. 1976. *Anolis* lizards of the eastern Caribbean: a case study in evolution. II. Genetic relationships and genetic variation of the *bimaculatus* group. Syst. Zool. 25: 62-77.

Gosner, K. L. 1987. Observations on Lesser Antillean pit vipers. J. Herpetol. 21: 78-80.

Gosse, P. H. 1847. The Birds of Jamaica. John Van Voorst, London.

Gosse, P. H. 1848a. On the habits of *Ameiva dorsalis*. Proc. Zool. Soc. London 16: 24-27.

Gosse, P. H. 1848b. On the habits of *Cyclura lophoma*, an iguaniform lizard. Proc. Zool. Soc. London 16: 99-104.

Gosse, P. H. 1848c. On the habits of *Mabuya agilis*. Proc. Zool. Soc. London 16: 59-62.

Gosse, P. H. 1850. Description of a new genus and six new species of Saurian reptiles. Ann. Mag. Nat. Hist. 6: 344-448.

Gosse, P. H. 1851. A Naturalist's Sojourn in Jamaica. Longman, Brown, Green, & Longmans, London.

Grant, C. 1931a. A new species of *Aristelliger* from Navassa. J. Dept. Agric. Puerto Rico 4: 399-400.

Grant, C. 1931b. A new species and two new subspecies of the genus *Anolis*. J. Dept. Agric. Puerto Rico 15: 219-222.

Grant, C. 1931c. Reestablishment of a scincid lost since 1837. J. Dept. Agric. Porto Rico 15: 217-218.

Grant, C. 1932. The hemidactyls of the Porto Rico region. J. Dept. Agric. Puerto Rico 16: 51-57.

Grant, C. 1938. The Jamaican *Dromicus funereus* re-established. Copeia 1938: 83-86.

Grant, C. 1939a. Additional data on Jamaican snakes of the genus *Dromicus*. Copeia 1939: 105-106.

Grant, C. 1939b. Two new sphaerodactyls from Jamaica. Copeia 1939: 7-13.

Grant, C. 1940a. Notes on the reptiles and amphibians of Jamaica, with diagnoses of new species and subspecies. Jamaica To-Day 15: 151-157.

Grant, C. 1940b. The herpetology of Jamaica II. The reptiles. Bull. Inst. Jamaica Sci. Ser. 1: 61-148.

Grant, C. 1940c. The herpetology of the Cayman Islands. Bull. Inst. Jamaica Sci. Ser. 2: iv + 65.

Grant, C. 1944a. New sphaerodactyls from Cuba and the Isle of Pines. Herpetologica 2(6):118-125.

Grant, C. 1944b. Scale structure in Jamaican lizards of the genus *Celestus*. Copeia 1944: 109-111.

Grant, C. 1946. Note from Major Chapman Grant. Nat. Hist. Notes Nat. Hist. Soc. Jamaica 3: 78.

Grant, C. 1951. The specific characters of the Celesti, with description of a new species of *Celestus* (Sauria, Anguidae). Copeia 1951: 67-69.

Grant, C. 1958. A new *Gymnophthalmus* (Reptilia, Teiidae) from Barbados. Herpetologica 14: 227-228.

Grant, C. 1959. Herpetology of Barbados, B.W.I. Herpetologica 15: 97-101.

Grant, C., and C. R. D. Sola. 1934. Antillean tortoises and terrapins: distribution, status, and habits of *Testudo* and *Pseudemys*. Copeia 1934: 73-79.

Gravenhorst, J. L. C. 1829. Reptilia Musei Zoologici Vratislaviensis Recensita et Descriptiva. Fasciculus Primus continens Chelonids et Batrachia. Sumptibus Leopoldi Vossi, Lipsiae.

Gravenhorst, J. L. C. 1837/1838. Beiträge zur genaueren Kenntniss einiger Eidechsengattungen. Nova Acta Acad. Caes. Leop.-Carol. 18: 712-784.

Gray, J. E. 1827. A description of a new genus and some new species of saurian reptiles; with a revision of the species of chameleons. Philos. Mag. 2: 207-214.

Gray, J. E. 1830. Specilegia Zoologica, or Original Figures and Short Systematic Descriptions of New and Unfigured Animals. Part II. Trustees of the British Museum, London.

Gray, J. E. 1831. A synopsis species of the class Reptilia, pp. 1-110 in The Class Reptilia, Arranged by Baron C. Cuvier, with Specific Descriptions. The Animal Kingdom Arranged in Conformity with its Organization by the Baron Cuvier, with Additional Descriptions of all the Species Hitherto Named, and of Many Others. Edited by E. Griffiths and E. Pidgeon. Vol. 9., Whitaker, London.

Gray, J. E. 1838a. Catalog of the slender-tongued Saurians, with descriptions of many new genera and species. Annot. Nat. Hist. I: 274-283.

Gray, J. E. 1838b. Catalogue of the slender-tongued Saurians, with descriptions of many new genera and species. Annu. Mag. Nat. Hist. 2: 287-293.

Gray, J. E. 1840. Catalogue of the species of reptiles collected in Cuba by W. S. MacLeay, Esp. – with some notes on their habits extracted from his MS. Annu. Mag. Nat. Hist. 5: 108-115.

Gray, J. E. 1842. Synopsis of the species of prehensile-tailed snakes, or family Boidae. Zool. Misc. 41-46.

Gray, J. E. 1845. Catalogue of the Specimens of Lizards in the Collection of the British Museum. Newman, London.

Gray, J. E. 1852. Description of *Sauresia*, a new genus of Scincidae from St. Domingo. Annu. Mag. Nat. Hist. 10: 281-282.

Graybeal, A., and D. C. Cannatella. 1995. A new taxon of Bufonidae from Peru, with descriptions of two new species and a review of the phylogenetic status of supraspecific Bufonid taxa. Herpetologica 51: 105-131

Greene, H. W. 1988. Species richness in tropical predators, pp. 259-280 in Tropical Rainforests: Diversity and Conservation, edited by C. F. Almeda, and M. Pringle. California Acad. Sci. and Pacific Div. American Assoc. Adv. Sci., San Francisco.

Greene, H. W. 1994. Systematics and natural history, foundations for understanding and conserving biodiversity. Am. Zool. 34: 48-56.

Greer, A. E. 1965. A new subspecies of *Clelia clelia* (Serpentes: Colubridae) from the island of Grenada. Breviora 223: 1-6.

Greer, A. E. 1967. Notes on the mode of reproduction in anguid lizards. Herpetologica 23:94-99.

Greer, A. E. 1970. Evolutionary and systematic significance of crocodilian nesting habits. Nature 227: 523-524.

Griffiths, I. 1959. The phylogeny of *Sminthillus limbatus* and the status of the Brachycephalidae (Amphibia: Salientia). Proc. Zool. Soc., London 132: 457-487.

Groombridge, B. 1987. The distribution and status of world crocodilians, pp. 9-21 in Wildlife Management: Crocodiles and Alligators, edited by G. Webb, C. Manolis and P. Whitehead. Surrey Beatty & Sons, Chipping Norton, Australia.

Gundlach, J. 1867. Revista y catalogo de los reptiles cubanos. Rep. Fisico-Natural Isla Cuba 2: 102-119.

Gundlach, J. 1880. Contribución a la Erpetología Cubana. G. Montiel, La Habana.

Günther, A. C. L. G. 1858. Catalogue of Colubrine Snakes in the Collection of the British Museum. Trustees of the British Museum, London.

Günther, A. C. L. G. 1858 (1859). Catalogue of the Batrachia Salientia in the Collection of the British Museum. Trustees of the British Museum, London.

Günther, A. C. L. G. 1861. On a new species of the family Boidae. Proc. Zool. Soc. London 1861: 142.

Günther, A. C. L. G. 1862. On new species of snakes in the collection of the British Museum. Annu. Mag. Nat. Hist. 9: 124-132.

Günther, A. C. L. G. 1865. Fourth account of new species of snakes in the collection of the British Museum. Annu. Mag. Nat. Hist. 15: 89-98.

Günther, A. 1888. Notes on reptiles and frogs from Dominica, West Indies. Annu. Mag. Nat. Hist. 2: 362-366.

Günther, A. E. 1980. The Founders of Science at the British Museum, 1753-1900. Halesworth, Suffolk, UK.

Guyer, C. 1988a. Food supplementation in a tropical mainland anole, *Norops humilis*: demographic effects. Ecology 69: 350-361.

Guyer, C. 1988b. Food supplementation in a tropical mainland anole, *Norops humilis*: effects on individuals. Ecology 69: 362-369.

Guyer, C. 1990. The herpetofauna of La Selva, Costa Rica, pp. 371-385 in Four Neotropical Rainforests, edited by A. H. Gentry. Yale Univ. Press, New Haven, CT.

Guyer, C., and B. I. Crother. 1996. Additional comments on the origin of the West Indian herpetofauna. Herpetologica 52: 620-622.

Guyer, C., and J. M. Savage. 1986. Cladistic relationships among anoles (Sauria: Iguanidae). Syst. Zool. 35: 509-531.

Guyer, C., and J. M. Savage. 1992. Anole systematics revisited. Syst. Biol. 41: 89-120.

Haefner, J. W. 1988. Assembly rules for Greater Antillean *Anolis* lizards. Oecologia 74: 551-565.

Hager, J., and T. A. Zanoni. 1993. La vegetatión natural de la República Dominicana: una nueva clasificación. Moscosoa 7: 39-81.

Hallowell, E. 1857. Notes on the reptiles in the collection of the Academy of Natural Sciences of Philad'a. Proc. Acad. Nat. Sci. Philadelphia 8: 221-238.

Hardy, J.D. 1967. Geographic variation in the West Indian lizard, *Anolis angusticeps*, with the description of a new form, *Anolis angusticeps paternus*, subsp. nov., from the Isle of Pines, Cuba (Reptilia: Iguanidae). Carib. J. Sci. 6(1-2): 23-31.

Harlan, R. 1824/1825. Description of two species of Linnæan *Lacerta*, not before described, and construction of the new genus *Cyclura*. J. Acad. Nat. Sci. Philadelphia 4: 242-251.

Harlan, R. 1826a. Descriptions of several new species of batrachian reptiles, with observations on the larvae of frogs. Am. J. Sci. Arts 10: 53-65.

Harlan, R. 1826b. A new species of *Hyla*. Am. J. Sci. Arts 10: 64-65.

Harris, D. M., and A. G. Kluge. 1984. The *Sphaerodactylus* (Sauria: Gekkonidae) of Middle America. Occas. Pap. Museum Zool. Univ. Michigan 706: 1-59.

Harvey, P. H., and M. D. Pagel. 1991. The Comparative Method in Evolutionary Biology. Oxford Univ. Press, Oxford, UK.

Hass, C. A. 1991. Evolution and biogeography of West Indian *Sphaerodactylus* (Sauria: Gekkonidae): a molecular approach. J. Zool. (London) 225: 525-561.

Hass, C. A. 1996. Relationships among West Indian geckos of the genus *Sphaerodactylus*: a preliminary analysis of mitochondrial 16S ribosomal RNA sequences, pp. 175-194 in Contributions to West Indian Herpetology: A Tribute to Albert Schwartz, edited by R. Powell, and R. W. Henderson. Society for the Study of Amphibians and Reptiles Contributions in Herpetology, Ithaca, NY.

Hass, C. A., and S. B. Hedges. 1991. Albumin evolution in West Indian frogs of the genus *Eleutherodactylus* (Leptodactylidae): Caribbean biogeography and a calibration of the albumin immunological clock. J. Zool. London 225: 413-426.

Hass, C. A., and S. B. Hedges. 1992. Karyotype of the Cuban lizard *Cricosaura typica* and its implications for Xantusiid phylogeny. Copeia 1992: 563-565.

Hass, C. A., S. B. Hedges, and L. R. Maxson. 1992. The relationships of West Indian anguid lizards. 1992 ASIH/HL Joint Meetg. Progr.: 115. (abstract).

Hass, C. A., S. B. Hedges, and L. Maxon. 1993. Molecular insights into the relationships and biogeography of West Indian anoline lizards. Biochem. Syst. Ecol. 21: 97-114.

Heatwole, H., R. Levins, and M. D. Byer. 1981. Biogeography of the Puerto Rican Bank. Atoll. Res. Bull. 251: 1-55.

Heatwole, H., and F. MacKenzie. 1966. Herpetogeography of Puerto Rico. IV. Paleogeography, faunal similarity and endemism. Evolution 21: 429-438.

Heatwole, H., and O. J. Sexton. 1966. Herpetofaunal comparisons between two climatic zones in Panama. Am. Midl. Natur. 75: 45-60.

Heatwole, H., and F. Torres. 1967. Distribution and geographic variation of the ameivas of Puerto Rico and the Virgin Islands. Studies Fauna Curaçao Carib. Isl. 24: 64-111.

Hecht, M. K. 1951. Fossil lizards of the West Indian genus *Aristelliger* (Gekkonidae). Am. Museum Novit. 1538: 1-33.

Hecht, M. K. 1952. Natural selection in the lizard genus *Aristelliger*. Evolution 6: 112-124.

Hedges, S. B. 1982. Caribbean biogeography: implications from recent plate tectonic studies. Syst. Zool. 31: 518-522.

Hedges, S. B. 1987. Vocalization and habitat preference of the Jamaican tree-frog, *Hyla marianae* (Anura: Hylidae). Carib. J. Sci. 23: 380-384.

Hedges, S. B. 1988. A new diminutive frog from Hispaniola (Leptodactylidae: *Eleutherodactylus*). Copeia 1988: 636-641.

Hedges, S. B. 1989a. Evolution and biogeography of West Indian frogs of the genus *Eleutherodactylus*: slow-evolving loci and the major groups, pp. 305-370 in Biogeography of the West Indies: Past, Present, and Future, edited by C. A. Woods. Sandhill Crane Press, Gainesville. FL.

Hedges, S. B. 1989b. An island radiation: allozyme evolution in Jamaican frogs of the genus *Eleutherodactylus* (Anura, Leptodactylidae). Carib. J. Sci. 25: 123-147.

Hedges, S. B. 1989c. Geographic protein variation in the Jamaican blind snake, *Typhlops jamaicensis* (Serpentes: Typhlopidae). Carib. J. Sci. 25: 93-96.

Hedges, S. B. 1991. Electrophoretic and morphologic variation in *Eleutherodactylus glaphycompus* (Anura: Leptodactylidae) of Hispaniola. J. Herpetol. 25: 10-17.

Hedges, S. B. 1993. Global amphibian declines: a perspective from the Caribbean. Biodiver. Conserv. 2: 290-303.

Hedges, S. B. 1996a. Historical biogeography of West Indian vertebrates. Annu. Rev. Ecol. Syst. 27: 163-196.

Hedges, S. B. 1996b. The origin of West Indian amphibians and reptiles, pp. 95-128 in Contributions to West Indian Herpetology: A Tribute to Albert Schwartz, edited by R. Powell, and R. W. Henderson. Society for the Study of Amphibians and Reptiles Contributions in Herpetology, Ithaca, NY.

Hedges, S. B. 1996c. Vicariance and dispersal in Caribbean biogeography. Herpetologica 52: 466-473.

Hedges, S. B. 1996d. The Hispaniolan frog *Eleutherodactylus neodreptus* Schwartz (Anura: Leptodactylidae) is a synonym of *E. audanti* Cochran. Carib. J. Sci. 32: 248.

Hedges, S. B. 1996e. More on West Indian zoogeography. Herpetologica 52: 622-624.

Hedges, S. B. 1999. Distribution patterns of amphibians in the West Indies, pp. 000-000 in Regional Patterns of Amphibian Distribution: A Global Perspective, edited by W. E. Duellman. Johns Hopkins Univ. Press, Baltimore, Maryland.

Hedges, S. B. and R. L. Bezy. 1993. Phylogeny of xantusiid lizards: concern for data and analysis. Mol. Phylogen. Evol. 2:76-87.

Hedges, S. B., and R. L. Bezy. 1994. Xantusiid lizards and phylogenetic inference. Mol. Phylogen. Evol. 3: 275-278.

Hedges, S. B., R. L. Bezy, and L. R. Maxson. 1991. Phylogenetic relationships and biogeography of xantusiid lizards, inferred from mitochondrial DNA sequences. Mol. Biol. Evol. 8: 767-780.

Hedges, S. B., and K. L. Burnell. 1990. The Jamaican radiation of *Anolis* (Sauria: Iguanidae): an analysis of relationships and biogeography using sequential electrophoresis. Carib. J. Sci. 26: 31-44.

Hedges, S. B., A. R. Estrada, and R. Thomas. 1992a. Three new species of *Eleutherodactylus* from eastern Cuba, with notes on vocalizations of other species (Anura:Leptodactylidae). Herpetol. Monograph, 6:68-83.

Hedges, S. B. and O. H. Garrido. 1992a. A new species of *Tropidophis* from Cuba (Serpentes: Tropidophidae). Copeia 1992(3): 820-825.

Hedges, S. B., and O. H. Garrido. 1992b. Cuban snakes of the genus *Arrhyton*: two new species and a reconsideration of *A. redimitum* Cope. Herpetologica 48: 168-177.

Hedges, S. B., and O. H. Garrido. 1993. A new species of gecko (*Sphaerodactylus*) from Central Cuba. J. Herpetol. 27(3): 300-306.

Hedges, S. B., C. A. Hass, and T. K. Maugel. 1989. Physiological color change in snakes. J. Herpetol. 23: 450-455.

Hedges, S. B., L. González, and A. R. Estrada. 1995. Rediscovery of the Cuban frog *Eleutherodactylus cubanus* and *E. turquinensis* (Anura: Leptodactylidae). Carib. J. Sci. 31(3-4): 327-332.

Hedges, S. B., C. A. Hass, and L. R. Maxson. 1992b. Caribbean biogeography: molecular evidence for dispersal in West Indian terrestrial vertebrates. Proc. Natl. Acad. Sci. (USA) 89: 1909-1913.

Hedges, S. B., C. A. Hass, and L. R. Maxson. 1994. Reply: towards a biogeography of the Caribbean. Cladistics 10: 43-55.

Hedges, S. B., and R. Thomas. 1987. A new burrowing frog from Hispaniola with comments on the *inoptatus* group of the genus *Eleutherodactylus* (Anura: Leptodactylidae). Herpetologica 43: 269-279.

Hedges, S. B., and R. Thomas. 1989a. Supplement to West Indian amphibians and reptiles: a check-list. Milwaukee Publ. Museum Contr. Biol. Geol. 77: 1-11.

Hedges, S. B., and R. Thomas. 1989b. A new species of *Anolis* (Sauria: Iguanidae) from the Sierra de Neiba, Hispaniola. Herpetologica 45: 330-336.

Hedges, S. B., and R. Thomas. 1991a. The importance of systematic research in the conservation of amphibian and reptile populations, pp. 56-61 in Status y Distribución de los Reptiles y Anfibios de la Región de Puerto Rico, edited by J. A. Moreno. Dept. Recur. Nat. Puerto Rico.

Hedges, S. B., and R. Thomas. 1991b. Cryptic species of snakes (Typhlopidae: *Typhlops*) from the Puerto Rico Bank detected by protein electrophoresis. Herpetologica 47: 448-459.

Hedges, S. B., and R. Thomas. 1992a. A new marsh-dwelling species of *Eleutherodactylus* from Haiti (Anura: Leptodactylidae). J. Herpetol. 26: 191-195.

Hedges, S. B., and R. Thomas. 1992b. Two new species of *Eleutherodactylus* from remnant cloud forest in Haiti (Anura, Leptodactylidae). Herpetologica 48: 351-358.

Hedges, S. B., and R. Thomas. 1998. A new anguid lizard from Cuba. Copeia 1998: 97-103.

Hedges, S. B., R. Thomas, and R. Franz. 1987. A new species of *Eleutherodactylus* (Anura, Leptodactylidae) from the Massif de la Hotte, Haiti. Copeia 1987: 943-949.

Heise, P. J., L. R. Maxson, H. G. Dowling, and S. B. Hedges. 1995. Higher-level snake phylogeny inferred from mitochondrial DNA sequences of 12S rRNA and 16S rRNA genes. Mol. Biol. Evol. 12(2): 259-265.

Henderson, R. W. 1974. Aspects of the ecology of the neotropical vine snake, *Oxybelis aeneus* (Wagler). Herpetologica 30: 19-24.

Henderson, R. W. 1984. The diet of the Hispaniolan snake *Hypsirhynchus ferox* (Colubridae). Amphib.-Rept. 5: 367-371.

Henderson, R. W. 1988a. *Diploglossus anelpistus*. Cat. Am. Amphib. Rept. 424: 1.

Henderson, R. W. 1988b. *Diploglossus carraui*. Cat. Am. Amphib. Rept. 425: 1.

Henderson, R. W. 1990. A new subspecies of *Alsophis antiguae* (Parker) from Great Bird Island (Antigua), Lesser Antilles. Carib. J. Sci. 25: 119-122.

Henderson, R. W. 1992. Consequences of predator introductions and habitat destruction on amphibians and reptiles in the post-Columbus West Indies. Carib. J. Sci. 28: 1-10.

Henderson, R. W. 1993a. Foraging and diet in West Indian *Corallus enydris* (Serpentes: Boidae). J. Herpetol. 27: 24-28.

Henderson, R. W. 1993b. On the diets of some arboreal boids. Herpetol. Nat. Hist. 1: 91-96.

Henderson, R. W., M. H. Binder, and R. A. Sajdak. 1982. Ecological relationships of the tree snakes *Uromacer catesbyi* and *U. oxyrhynchus* (Colubridae) on Isla Saona, República Dominicana. Amphib.-Rept. 2: 153-163.

Henderson, R. W., and R. W. Bourgeois. 1993. Notes on the diets of West Indian *Liophis* (Serpentes: Colubridae). Carib. J. Sci. 29: 253-254.

Henderson, R. W., and B. I. Crother. 1989. Biogeographic patterns of predation in West Indian colubrid snakes, pp. 479-517 in Biogeography of the West Indies: Past, Present, and Future, edited by C. A. Woods. Sandhill Crane Press, Inc., Gainesville.

Henderson, R. W., B. I. Crother, T. A. Noeske-Hallin, A. Schwartz, and C. R. Dethloff. 1987a. The diet of the Hispaniolan snake *Antillophis parvifrons* (Colubridae). J. Herpetol. 21: 330-334.

Henderson, R. W., A. Delatte, and T. J. McCarthy. 1993. *Gekko gecko* (Sauria: Gekkonidae) established on Martinique, French West Indies. Carib. J. Sci. 29: 128-129.

Henderson, R. W., and G. T. Haas. 1993. Status of the West Indian snake *Chironius vincenti*. Oryx 27: 181-184.

Henderson, R. W., and S. B. Hedges. 1995. Origin of West Indian populations of the geographically widespread boa *Corallus enydris* inferred from mitochondrial DNA sequences. Mol. Phylo. Evol. 4: 88-92.

Henderson, R. W., and K. F. Henderson. 1995. Altitudinal variation in body temperature in foraging tree boas (*Corallus enydris*) on Grenada. Carib. J. Sci. 31: 73-76.

Henderson, R. W., T. A. Noeske-Hallin, B. I. Crother, and A. Schwartz. 1988a. The diets of Hispaniolan colubrid snakes II. Prey species, prey size, and phylogeny. Herpetologica 44: 55-70.

Henderson, R. W., T. A. Noeske-Hallin, J. A. Ottenwalder, and A. Schwartz. 1987b. On the diet of the boa *Epicrates striatus* on Hispaniola, with notes on *E. fordi* and *E. gracilis*. Amphib.-Rept. 8: 251-258.

Henderson, R. W., R. Powell, J. C. Daltry, and M. L. Day. 1996. *Alsophis antiguae*. Cat. Amer. Amphib. Rept. (632):1-3.

Henderson, R. W., and R. A. Sajdak. 1986. West Indian racers: a disappearing act or a second chance? Lore, Milwaukee Publ. Museum 36: 13-18.

Henderson, R. W., and R. A. Sajdak. 1996. Diets of West Indian racers (Colubridae: *Alsophis*): Composition and biogeographic implications, pp. 327-338 in Contributions to West Indian Herpetology: A Tribute to Albert Schwartz, edited by R. Powell, and R. W. Henderson. Society for the Study of Amphibians and Reptiles Contributions in Herpetology, Ithaca, NY.

Henderson, R. W., R. A. Sajdak, and R. M. Henderson. 1988b. The rediscovery of the West Indian colubrid snake *Chironius vincenti*. Amphib.-Rept. 9: 415-416.

Henderson, R. W., and A. Schwartz. 1984a. A guide to the identification of the amphibians and reptiles of Hispaniola. Milwaukee Public Museum Spec. Publ. Biol. Geol. 4: 1-70.

Henderson, R. W., and A. Schwartz. 1984b. *Uromacer frenatus*. Cat. Am. Amphib. Rept. 357.1-357.2.

Henderson, R. W., and A. Schwartz. 1986. The diet of the Hispaniolan colubrid snake, *Darlingtonia haetiana*. Copeia 1986: 529-531.

Henderson, R. W., A. Schwartz, and S. J. Incháustegui. 1984. Guía para la identificación de los anfibios y reptiles de la Hispaniola. Museum Nac. Hist. Nat. Ser. Monogr. 1: 1-128.

Henderson, R. W., A. Schwartz, and T. A. Noeske-Hallin. 1987c. Food habits of three colubrid tree snakes (genus *Uromacer*) on Hispaniola. Herpetologica 43: 235-242.

Henderson, R. W., and R. A. Winstel. 1992. Activity patterns, temperature relationships, and habitat utilization in *Corallus enydris* (Serpentes: Boidae) on Grenada. Carib. J. Sci. 28: 229-232.

Henderson, R. W., and R. A. Winstel. 1995. Aspects of habitat selection by an arboreal boa (*Corallus enydris*) in an area of mixed agriculture on Grenada. J. Herpetol. 29: 272-275.

Henderson, R. W., R. A. Winstel, and J. Friesch. 1996. *Corallus hortulanus* (Serpentes: Boidae) in the post-Columbian West Indies: New habitats, new prey species, and new predators, pp. 417-423 in Contributions to West Indian Herpetology: A Tribute to Albert Schwartz, edited by R. Powell, and R. W. Henderson. Society for the Study of Amphibians and Reptiles Contributions in Herpetology, Ithaca, NY.

Hennig, W. 1966. Phylogenetic Systematics. Univ. of Illinois Press, Urbana, IL.

Hertz, P. E. 1976. *Anolis alumina*, new species of grass anole from the Barahona Peninsula of Hispaniola. Breviora 437: 1-19.

Hertz, P. E. 1981. Adaptation to altitude in two West Indian anoles (Reptilia: Iguanidae): field thermal biology and physiological ecology. J. Zool. London 195: 25-37.

Hertz, P. E. 1983. Eurythermy and niche breadth in West Indian *Anolis* lizards: a reappraisal, pp. 472-483 in Advances in Herpetology and Evolutionary Biology: Essays in Honor of Ernest E. Williams, edited by A. G. J. Rhodin, and K. I. Miyata. Museum Comp. Zool., Cambridge, MA.

Hertz, P. E., R. B. Huey, and R. D. Stevenson. 1993. Evaluating temperature regulation by field-active ectotherms: the fallacy of the inappropriate question. Am. Nat. 142: 796-818.

Heyer, W. R. 1969. The adaptive ecology of the species groups of the genus *Leptodactylus* (Amphibia, Leptodactylidae). Evolution 23: 421-428.

Heyer, W. R. 1970. Studies on frogs of the genus *Leptodactylus* (Amphibia: Leptodactylidae). VI. Biosystematics of the *melanotus* group. Contrib. Sci. Nat. Hist. Museum Los Angeles County. 191: 1-48.

Heyer, W. R. 1974. Systematics of the *marmoratus* group of the frog genus *Leptodactylus* (Amphibia, Leptodactylidae). Los Angeles County Museum Nat. Hist. Contrib. Sci. 253: 1-50.

Heyer, W. R. 1978. Systematics of the *fuscus* group of the frog genus *Leptodactylus* (Amphibia, Leptodactylidae). Smithsonian Contrib. Zool. 301: 1-43.

Heyer, W. R. 1979. Systematics of the *pentadactylus* species group of the frog genus *Leptodactylus* (Amphibia: Leptodactylidae). Smithsonian Contrib. Zool. 301: 1-43.

Heyer, W. R. 1994. Variation within the *Leptodactylus podicipinus-wagneri* complex of frogs (Amphibia: Leptodactylidae). Smithsonian Contrib. Zool. 546: 1-124.

Heyer, W. R., M. A. Donnelly, R. W. McDiarmid, L.-A. C. Hayek, and M. S. Foster. 1994. Measuring and Monitoring Biological Diversity: Standard Methods for Amphibians, Smithsonian Institution Press, Washington, DC.

Hicks, R. 1973. New studies on a montane lizard of Jamaica, *Anolis reconditus*. Breviora 404: 1-23.

Hicks, R. A., and R. L. Trivers. 1983. The social behavior of *Anolis valencienni*, pp. 570-595 in Advances in Herpetology and Evolutionary Biology: Essays in Honor of Ernest E. Williams, edited by A. G. J. Rhodin, and K. I. Miyata. Museum Comp. Zool. Harvard Univ., Cambridge, MA.

Hildebrand, A. R., and W. V. Boynton. 1990. Proximal Cretaceous-Tertiary boundary impact deposits in the Caribbean. Science 248: 843-847.

Hillis, D. M., B. K. Mable, and C. Moritz. 1996. Applications of molecular systematics: the state of the field and a look to the future, pp. 515-543 in Molecular Systematics, edited by D. M. Hillis, C. Moritz, and B. K. Mable. Sinauer Associates, Sunderland, MA.

Hirth, H. F. 1963. The ecology of two lizards on a tropical beach. Ecol. Monogr. 33: 83-112.

Hoge, A. R. 1952. Notas erpetolgicas. Revalida o de *Bothrops lanceolata* (Lace). Mem. Inst. Butantan 24: 231-236.

Holdridge, L. R. 1967. Life zone ecology. Tropical Sci. Center, San Jose, Costa Rica.

Hollingsworth, B. D. 1998. The systematics of chuckwallas (Sauraomalus) with a phylogentic analysis of other iguanid lizards. Herp. Monog. 12: 38-191.

Hoogmoed, M. S. 1973. Notes on the herpetofauna of Surinam IV. Junk, The Hague.

Hoppe, J. 1989. Los Parques Nacionales de la República Dominican. Ed. Corripio, Santo Domingo.

Horn, H. S. 1969. Polymorphism and evolution of the Hispaniolan snake genus *Uromacer* (Colubridae). Breviora 324: 1-23.

Howard, A. K., J. D. Forester, J. M. Ruder, R. Powell, and J. J. S. Parmerlee. MS. The diets of two sympatric frogs: *Eleutherodactylus abbotti* and *E. armstrongi* (Leptodactylidae) from the Sierra de Baoruco, Hispaniola. Unpublished manuscript.

Huebeck, C., and P. Mann. 1991. Structural geology and Cenozoic tectonic history of the southeastern termination of the Cordillera Central, Dominican Republic. Geol. Soc. Am. Spec. Pap. 262: 315-336.

Huey, R. B., and T. P. Webster. 1975. Thermal biology of a solitary lizard: *Anolis marmoratus* of Guadeloupe, Lesser Antilles. Ecology 56: 445-452.

Huey, R. B., and T. P. Webster. 1976. Thermal biology of *Anolis* lizards in a complex fauna: the *cristatellus* group on Puerto Rico. Ecology 57: 985-994.

Hull, D. L. 1988. Science as a Process. Univ. of Chicago Press, Chicago, IL.

Humbolt, A. 1856. The island of Cuba. Derby & Jackson, New York.

Inchaústegui, S. J., A. Schwartz, and R. W. Henderson. 1985. Hispaniolan giant *Diploglossus* (Sauria: Anguidae): description of a new species and notes on the ecology of *D. warreni*. Amphib.-Rept. 6: 195-201.

Institute, W. R. 1994. World Resources 1994-95. Oxford Univ. Press, New York.

Isacks, B., J. Oliver, and L. R. Sykes. 1968. Seismology and the new global tectonics. J. Geophys. Res. 73: 5855-5899.

Iverson, J. B. 1978. The impact of feral cats and dogs on poulations of the West Indian rock iguana, *Cyclura carinata*. Biol. Conserv. 14: 63-73.

Iverson, J. B. 1979. Behavior and ecology of the rock iguana *Cyclura carinata*. Bull. Florida State Museum, Biol. Sci. 24: 175-358.

Jan, G., and F. Sordelli. 1867. Iconographie Générale des Ophidiens. Livr. 25. Published by authors, Milan.

Jeffrey-Smith, M. 1946. The whistling frog increases. Nat. Hist. Notes Nat. Hist. Soc. Jamaica 3: 26.

Jenner, J. V. 1981. A zoogeographic study and the taxonomy of the xenodontine colubrid snakes. Ph.D. Dissertation. New York Univ.

Jenner, J. V., and H. G. Dowling. 1985. Taxonomy of American xenodontine snakes: the tribe Pseudoboini. Herpetologica 41: 161-172.

Jenssen, T. A. 1970. The ethoecology of *Anolis nebulosus* (Sauria, Iguanidae). J. Herpetol. 4: 1-38.

Jenssen, T. A. 1973. Shift in structural habitat of *Anolis opalinus* due to congeneric competition. Ecology 54: 863-869.

Jenssen, T. A. 1977. Morphological, behavioral and electrophoretic evidence of hybridization between the lizards, *Anolis grahami* and *Anolis lineatopus neckeri*, on Jamaica. Copeia 1977: 270-276.

Jenssen, T. A. 1979. Display modifiers of *Anolis opalinus* (Lacertilia: Iguanidae). Herpetologica 35: 21-30.

Jenssen, T. A., and R. M. Andrews. 1984. Seasonal growth rates in the Jamaican lizard, *Anolis opalinus*. J. Herpetol. 18: 338-341.

Jenssen, T. A., and S. C. Nunez. 1994. Male and female reproductive cycles of the Jamaican lizard, *Anolis opalinus*. Copeia 1994: 767-780.

Jenssen, T. A., D. L. Marcellini, K. A. Buhlmann, and P. H. Goforth. 1989. Differential infanticide in adult curly-tailed lizards, *Leiocephalus schreibersi*. Anim. Behav. 38: 1054-1061.

Joger, U. 1984. Taxonomische Revision der Gattung *Tarentola* (Reptilia: Gekkonidae). Bonn Zool. Beitr. 35: 129-174.

Joglar, R. L. 1983. Estudio fenetico del genero *Eleutherodactylus* en Puerto Rico. Carib. J. Sci. 19: 33-40.

Joglar, R. L. 1989. Phylogenetic relationships of the West Indian frogs of the genus *Eleutherodactylus*: a morphological analysis, pp. 371-404 in Biogeography of the West Indies: Past, Present, and Future, edited by C. A. Woods. Sandhill Crane Press, Gainesville, FL.

Joglar, R. L., and P. A. Burrowes. 1996. Declining amphibian populations in Puerto Rico, pp. 371-380 in Contributions to West Indian Herpetology: A Tribute to Albert Schwartz, edited by R. Powell, and R. W. Henderson. Society for the Study of Amphibians and Reptiles Contributions in Herpetology, Ithaca, NY.

Jones, K. L. 1982. Prey patterns and trophic niche overlap in four species of Caribbean frogs, pp. 49-55 in Herpetological Communities, edited by J. N. Jr. Scott. Research Report 13, United States Fish and Wildlife Service.

Kaiser, H. 1992. The trade-mediated introduction of *Eleutherodactylus martinicensis* (Anura: Leptodactylidae) on St. Barthélémy, French Antilles, and its implications for Lesser Antillean Biogeography. J. Herpetol. 26: 264-273.

Kaiser, H. 1996. Systematics and biogeography of Eastern Caribbean *Eleutherodactylus* (Anura: Leptodactylidae): consensus from a multidisciplinary approach, pp. 129-140 in Contributions to West Indian Herpetology: A Tribute to Albert Schwartz, edited by R. Powell, and R. W. Henderson. Society for the Study of Amphibians and Reptiles Contributions in Herpetology, Ithaca, NY.

Kaiser, H. 1997. Origins and introductions of the Caribbean frog, *Eleutherodactylus johnstonei* (Leptodactylidae): management and conservation concerns. Biodivers. Conserv. in press.

Kaiser, H., and R. Altig. 1994. The atypical tadpole of the dendrobatid frog, *Colostethus chalcopis*, from Martinique, French Antilles. J. Herpetol. 28: 374-378.

Kaiser, H., L. A. Coloma, and H. M. Gray. 1994b. A new species of *Colostethus* (Anura: Dendrobatidae) from Martinique, French Antilles. Herpetologica 50: 23-32.

Kaiser, H., D. M. Green, and M. Schmid. 1994a. Systematics and biogeography of Eastern Caribbean *Eleutherodactylus* (Anura: Leptodactylidae) with the description of a new species from Dominica. Can. J. Zool. 72: 2217-2237.

Kaiser, H., and J. J. D. Hardy. 1994. *Eleutherodactylus martinicensis*. Cat. Am. Amphib. Rept. 582: 1-4.

Kaiser, H., J. J. D. Hardy, and D. M. Green. 1994c. Taxonomic status of Caribbean and South American frogs currently ascribed to *Eleutherodactylus urichi* (Anura: Leptodactylidae). Copeia 1994: 780-796.

Kaiser, H., and R. W. Henderson. 1994. The conservation status of Lesser Antillean frogs. Herpetol. Nat. Hist. 2: 41-56.

Kaiser, H., and H. Kaiser. 1994. *Eleutherodactylus johnstonei*. Cat. Am. Amphib. Rept. 581: 1-5.

Kaiser, H., T. F. Sharbel, and D. M. Green. 1994d. Systematics and biogeography of eastern Caribbean *Eleutherodactylus* (Anura: Leptodactylidae): evidence from allozymes. Amphib.-Rept. 15: 375-394.

Keegan, W. F., and J. M. Diamond. 1987. Colonization by humans: a biogeographical perspective. Adv. Archeol. Method Theory. 10: 49-92.

Keller, G., J. G. Lopez-Oliva, W. Stinnesbeck, and T. Adatte. 1997. Age, stratigraphy, and deposition of near K/T siliciclastic deposits in Mexico: Relation to bolide impact? GSA Bull. 109: 410-428.

Kerr, R. A. 1998. Research News: Sea-floor dust shows drought felled Akkadian Empire. Science 279: 325-326.

Kerster, H. W., and H. M. Smith. 1955. The identity of the Puerto Rican species of *Phyllodactylus* (Reptilia: Squamata). Herpetologica 11: 229-232.

Khudoley, K. M., and A. A. Myerhoff. 1971. Paleogeography and geological history of the Greater Antilles. Geol. Soc. of Am. Memoirs 129: 1-199.

King, F. W. 1962a. Systematics of Lesser Antillean lizards of the genus *Sphaerodactylus*. Bull. Florida State Museum 7: 1-52.

King, F. W. 1962b. The occurrence of rafts for dispersal of land animals into the West Indies. Q. J. Florida Acad. Sci. 25: 46-52.

Kirch, P. V., and T. L. Hunt (Editors). 1997. Historical Ecology in the Pacific Islands. Prehistoric Environmental and Landscape Change. Yale Univ. Press, New Haven, CT.

Klauber, L. M. 1956. Rattlesnakes: Their Habits, Life Histories, and Influence on Mankind. Univ. California Press, Berkeley.

Kluge, A. G. 1964. A revision of the South American gekkonid lizard genus *Homonota* Gray. Novitates 2193: 1-41.

Kluge, A. G. 1969. The evolution and geographical origin of the New World *Hemidactylus mabouia-brookii* complex (Gekkonidae, Sauria). Misc. Publ. Museum Zool. Univ. Michigan 138: 1-78.

Kluge, A. G. 1988. A concern for the evidence and a phylogenetic hypothesis of cladistic relationships among *Epicrates* (Boidae, Serpentes). Syst. Zool. 38: 7-25.

Kluge, A. G. 1991. Boine snake phylogeny and research cycles. Misc. Publ. Museum Zool. Univ. Michigan 178: i-iv + 1-58.

Kluge, A. G. 1995. Cladistic relationships of sphaerodactyl lizards. Novitates 3139: 1-23.

Koza, J. R., J. Roughgarden, and J. P. Rice. 1992. Evolution of food-foraging strategies for the Caribbean *Anolis* lizard using genetic programming. Adapt. Behav. 1: 171-199.

Lacépède, B.-G. - É. 1789. Histoire Naturelle des Quadrupèds Ovipares et des Serpens. Hôtel de Thou, Paris.

Lack, D. 1976. Island Biology Illustrated by the Land Birds of Jamaica. Studies in Ecology Vol. 3. Univ. of California Press, Berkeley.

Laessle, A. M. 1961. A micro-limnological study of Jamaican bromeliads. Ecology 42: 499-517.

Lahey, G. J., R. A. Sosa, S. M. Micco, E. J. Censky, and R. Powell. MS. On the natural history of *Leiocephalus barahonensis* (Lacertilia: Tropiduridae) from the Barahona Peninsula, Hispaniola.

Lambert, M. R. K. 1986. Herpetology in Jamaica with general reference to conservation. British Herpetol. Soc. Bull. 18.

Landwer, A. J., G. W. Ferguson, R. Herber, and M. Brewer. 1995. Habitat use of introduced and native anoles (Iguanidae: *Anolis*) along the northern coast of Jamaica. Texas J. Sci. 47: 45-52.

Lannoo, M. J., D. S. Townsend, and R. J. Wassersug. 1987. Larval life in the leaves: arboreal tadpole types, with special attention to the morphology, ecology and behavior of the oophagous *Osteopilus brunneus* (Hylidae) larvae. Fieldiana Zool. 38: iv + 31.

Laurentí, J. N. 1768. Specimen Medicum, Exhibens Synopsin [sic] Reptilium Emendatam cum Experimentis circa Venena, Univ. of Vienna, Vienna.

Lawson, R. 1987. Molecular studies of Thamnophiine snakes: 1. The phylogeny of the genus *Nerodia*. J. Herpetol. 21: 140-157.

Lazell, J. D., Jr. 1961. A new species of *Sphaerodactylus* from northern Haiti. Breviora 139: 1-5.

Lazell, J. D., Jr. 1962. The anoles (Sauria: Iguanidae) of the eastern Caribbean, Part V. Geographic differentiation in *Anolis oculatus* on Dominica. Bull. Museum Comp. Zool. 127: 466-475.

Lazell, J. D., Jr. 1964a. The anoles (Sauria: Iguanidae) of the Guadeloupéen archipelago. Bull. Museum Comp. Zool. 131: 359-401.

Lazell, J. D., Jr. 1964b. The Lesser Antillean representatives of *Bothrops* and *Constrictor*. Bull. Museum Comp. Zool. 132: 245-273.

Lazell, J. D., Jr. 1966. Contributions to the herpetology of Jamaica – Studies on *Anolis reconditus* Underwood and Williams. Bull. Inst. Jamaica Sci. Ser. 18: 5-15.

Lazell, J. D., Jr. 1967. Wiederentdeckung von zwei angeblich ausgestorbenen Schlangenarten der westindischen Inseln. Salamandra 3: 91-97.

Lazell, J. D., Jr. 1972. The anoles (Sauria, Iguanidae) of the Lesser Antilles. Bull. Museum Comp. Zool. 143: 1-115.

Lazell, J. D., Jr. 1973. The lizard genus *Iguana* in the Lesser Antilles. Bull. Museum Comp. Zool. 145: 1-28.

Lazell, J. D., Jr. 1983. Biogeography of the herpetofauna of the British Virgin Islands, with description of a new anole (Sauria: Iguanidae), pp. 99-117 in Advances in Herpetology and Evolutionary Biology. Essays in Honor of Ernest E. Williams, edited by A. G. J. Rhodin, and K. Miyata. Museum Comp. Zool., Cambridge. MA.

Lazell, J. D., Jr. 1991. The herpetofauna of Guana island: diversity, adundance, rarity, and conservation, pp. 28-33 in Status y Distribución de los Reptiles y Anfibios de la Region de Puerto Rico, edited by J. A. Moreno. Depto. Rec. Natur., Puerto Rico.

Lazell, J. D., Jr. 1992. The family Iguanidae: disagreement with Frost and Etheridge (1989). Herpetol. Rev. 23: 109-112.

Lazell, J. D., Jr. 1993. Tortoise, cf. *Geochelone carbonaria*, from the Pleistocene of Anguilla, northern Lesser Antilles. J. Herpetol. 27: 485-486.

Lazell, J. D., Jr. 1994. A new *Sphaerodactylus* (Sauria: Gekkonidae) from Bequia, Grenada Bank, Lesser Antilles. Breviora 496: 1-20.

Lazell, J. D., Jr. 1996. Careening Island and the Goat Islands: evidence for the arid-insular invasion wave theory of dichopatric speciation in Jamaica, pp. 195-205 in Contributions to West Indian Herpetology: a Tribute to Albert Schwartz, edited by R. Powell, and R. W. Henderson. Society for the Study of Amphibians and Reptiles Contributions in Herpetology, Ithaca, NY.

Lazell, J. D., Jr., and T. Sinclair. 1990a. *Anolis trinitatus*. Herpetol. Rev. 21: 96.

Lazell, J. D., Jr., and T. Sinclair. 1990b. Geographic Distribution: *Gymnophthalmus underwoodi*. Herpetol. Rev. 21: 96.

Lazell, J. D., Jr., and T. Sinclair. 1990c. Geographic Distribution: *Sphaerodactylus* cf. *vincenti*. Herpetol. Rev. 21: 97.

Leclair, R., Jr., and G. Provencher. 1988. Demographic traits of a small sphaerodactyline lizard from Martinique, West Indies, pp. 127-128 in Herpetologists' League-ASIH Annu. Meeting. (abstract)

Lee, J. C. 1996. Pre-columbian human influences on the biogeography of amphibians and reptiles in the Yucatan Peninsula, pp. 435-436 in Contributions to West Indian Herpetology: A Tribute to Albert Schwartz, edited by R. Powell, and R. W. Henderson. SSAR Contrib. Herpetol., Ithaca.

Leidy, J. 1868. Notes on some vertebrate remains from the West Indian islands. Proc. Acad. Natl. Sci. Philadelphia 20: 178-180.

Lenart, L. A., R. Powell, J. J. S. Parmerlee, D. D. Smith, and A. Lathrop. 1994. The diet and a gastric parasite of *Anolis armouri*, a cybotoid anole from montane pine forest in southern Hispaniola. Herpetol. Nat. Hist. 2: 97-100.

Lenart, L. A., R. Powell, J. J. S. Parmerlee, A. Lathrop, and D. D. Smith. 1997. Anoline diversity in three differentially altered habitats in the Sierra de Baoruco, República Dominicana, Hispaniola. Biotropica 29: 117-123.

Lescure, J. 1979. Étude taxonomique et eco-ethologique d'un amphibien des Petites Antilles: *Leptodactylus fallax* Müller, 1926 (Leptodactylidae). Bull. Museum Natur. Hist. Nat. Paris 4e sér., 1, sect A: 757-774.

Lescure, J. 1987. Le peuplement en reptiles et amphibiens des Petites Antilles. Bull. Soc. Zool. France 112: 327-342.

Levins, R., and H. Heatwole. 1963. On the distribution of organisms on islands. Carib. J. Sci. 3:173-177.

Levins, R., and H. Heatwole. 1973. Biogeography of the Puerto Rican Bank: Introduction of species onto Palominitos Island. Ecology 54: 1056-1064.

Lewis, A. R. 1986. Body size and growth in two populations of the Puerto Rican ground lizard (Teiidae). J. Herpetol. 20: 190-195.

Lewis, A. R. 1989. Diet selection and depression of prey abundance by an intensively foraging lizard. J. Herpetol. 23: 164-170.

Lewis, A. R., and J. F. Saliva. 1987. Effects of sex and size on home range, dominance, and activity budgets in *Ameiva exsul* (Lacertilia: Teiidae). Herpetologica 43: 374-383.

Lewis, A. R., G. Tirado, A. Aponte, J. Blanco, and J. Sepulveda. 1996. Social control of male reproductive success in *Ameiva exsul*, pp. 445-449 in Contributions to West Indian Herpetology: A Tribute to Albert Schwartz, edited by R. Powell, and R. W. Henderson. Society for the Study of Amphibians and Reptiles Contributions in Herpetology, Ithaca, NY.

Lewis, C. B. 1941. The frogs of the Clydesdale region. Nat. Hist. Notes Nat. Hist. Soc. Jamaica 1: 10-12.

Lewis, C. B. 1943. Frog eggs. Nat. Hist. Notes Nat. Hist. Soc. Jamaica 2: 35.

Lewis, C. B. 1950. Gosse's illustration of Jamaican reptiles and amphibia. Nat. Hist. Notes Nat. Hist. Soc. Jamaica 4: 103.

Lewis, J. F. 1980. Cenozoic tectonic evolution and sedimentation in Hispaniola, pp. 65-73 in Transactions of the Ninth Caribbean Geologic Conference. Amigo Hogar, Santo Domingo.

Lewis, J. F., and G. Draper. 1990. Geology and tectonic evolution of the northern Caribbean margin, pp. 77-140 in The Geology of North America. Vol. H. The Caribbean Region, edited by G. Dengo, and J. E. Case. Geological Society America, Boulder, CO.

Licht, P., and G. C. Gorman. 1970. Reproductive and fat cycles in Caribbean *Anolis* lizards. Univ. California Pub. Zool. 95: 1-52.

Lichtenstein, H. 1856. Nomenclatur Reptilium et Amphibiorum Musei Zoologici Berolinensis. Königl. Akad. Wiss., Berlin.

Lieberman, S. S. 1986. Ecology of the leaf litter herpetofauna of a Neotropical rain forest: La Selva Costa Rica. Acta Zool. Mexicana 15: 1-72.

Lillywhite, H. B., and R. W. Henderson. 1993. Behavioral and functional ecology of arboreal snakes, pp. 1-48 in Snakes: Ecology and Behavior, edited by R. A. Seigel, and J. T. Collins. McGraw-Hill, NY.

Linnaeus, C. 1758. Systema Naturae per Regna Tria Naturae, Secundum Classes, Ordines, Genera, Species, cum Characteribus, Differentiis, Synonymis, Locis. L. Salvii, Stockholm.

Linnaeus, C. 1939. Systema Naturae. Regnum Animale. Trustees of the British Museum (Natural History), London. [Reprint].

Losos, J. B. 1990a. Ecomorphology, performance capability, and scaling of West Indian *Anolis* lizards: an evolutionary analysis. Ecol. Mongr. 60: 369-388.

Losos, J. B. 1990b. Concordant evolution of locomotor behaviour, display rate and morphology in *Anolis* lizards. Anim. Behav. 39: 879-890.

Losos, J. B. 1990c. The evolution of form and function: morphology and locomotor performance in West Indian *Anolis* lizards. Evolution 44: 1189-1203.

Losos, J. B. 1992a. The evolution of convergent structure in Caribbean *Anolis* communities. Syst. Biol. 41: 403-420.

Losos, J. B. 1992b. A critical comparison of the taxon-cycle and character-displacement models for size evolution of *Anolis* lizards in the Lesser Antilles. Copeia 1992: 279-288.

Losos, J. B. 1994. Integrative approaches to evolutionary ecology: *Anolis* lizards as model systems. Annu. Rev. Ecol. Syst. 25: 467-493.

Losos, J. B., M. R. Gannon, W. J. Pfeiffer, and R. B. Waide. 1990. Notes on the ecology and behavior of *Anolis cuvieri* (Lacertilia: Iguanidae) in Puerto Rico. Carib. J. Sci. 26: 65-66.

Losos, J. B., J. C. Marks, and T. W. Schoener. 1993. Habitat use and ecological interactions of an introduced and native species of *Anolis* lizard on Grand Cayman, with a review of the outcomes of anole introductions. Oecologia 95: 525-532.

Losos, J. B., and B. Sinervo. 1989. The effects of morphology and perch diameter on sprint performance of *Anolis* lizards. J. Exp. Biol. 145: 23-30.

Luckow, M. and A. Bruneau. 1997. Circularity and independence in phylogenetic tests of ecological hypotheses. Cladistics 13: 145-151.

Lynch, J. D. 1965. A new species of *Eleutherodactylus* from Guadeloupe, West Indies. Breviora 220: 1-7.

Lynch, J. D. 1966. The status of the tree frog, *Hyla barbudensis* Auffenberg, from Barbuda, British West Indies. Copeia 1966: 524-530.

Lynch, J. D. 1971. Evolutionary relationships, osteology, and zoo-geography of leptodactyloid frogs. Misc. Publ. Museum Nat. Hist. Univ. Kansas 53: 1-238.

Lynch, J. D. 1986. The definition of the Middle American clade of *Eleutherodactylus* based on jaw musculature (Amphibia: Leptodactylidae). Herpetologica 42: 248-258.

Lynch, J. 1991. Three replacement names for preoccupied names in the genus *Eleutherodactylus* (Amphibia: Leptodactylidae). Copeia 1991(4): 1138-1139.

Lynch, J. D. 1996. The relationships of the Hispaniolan frogs of the subgenus *Pelorius* (*Eleutherodactylus*: Leptodactylidae), pp. 141-155 in Contributions to West Indian Herpetology: A Tribute to Albert Schwartz, edited by R. Powell, and R. W. Henderson. Society for the Study of Amphibians and Reptiles Contributions in Herpetology, Ithaca, NY.

Lynch, J., and W. E. Duellman. 1997. Frogs of the genus *Eleutherodactylus* (Leptodatylidae) in Western Ecuador: systematics, ecology, biogeography. Univ. Kansas Nat. Hist. Museum Spec. Publ. 23: 1-236.

Lynn, W. G. 1936. A study of the thyroid in embryos of *Eleutherodactylus nubicola*. Anat. Rec. 64: 525-535.

Lynn, W. G. 1937. Two new frogs from Jamaica. Herpetologica 1: 88-90.

Lynn, W. G. 1940. The herpetology of Jamaica I. Amphibians. Bull. Inst. Jamaica Sci. Ser. 1: 1-60.

Lynn, W. G. 1942. The embryology of *Eleutherodactylus nubicola*, an anuran which has no tadpole stage. Carnegie Inst. Washington Publ. 541: 27-62.

Lynn, W. G. 1944. Duplicitas anterior in the toad, *Eleutherodactylus alticola*. Anat. Rec. 89: 345-351.

Lynn, W. G. 1954. Description of a new frog of the genus *Eleutherodactylus* from Jamaica. J. Washington Acad. Sci. 44: 400-402.

Lynn, W. G. 1958. Some amphibians from Haiti and a new subspecies of *Eleutherodactylus schmidti*. Herpetologica 14: 153-157.

Lynn, W. G., and J. N. Dent. 1942. A new frog of the genus *Eleutherodactylus* from Jamaica. Herpetologica 2: 71-74.

Lynn, W. G., and J. N. Dent. 1943. Notes on Jamaican amphibians. Copeia 1943: 234-242.

Lynn, W. G., and C. Grant. 1940. The herpetology of Jamaica. Bull. Inst. Jamaica Sci. Ser. 1: 1-148.

Lynn, W. G., and A. M. Peadon. 1955. The role of the thyroid gland in direct development in the anuran, *Eleutherodactylus martinicensis*. Growth 19: 263-286.

Lynn, W. G., and G. A. Walsh. 1957. The morphology of the thyroid gland in the Lacertilia. Herpetologica 13: 157-172.

Lynxwiler, J. R., R. Powell, J. J. S. Parmerlee, A. Lathrop, and D. D. Smith. 1991. Notes on the natural history of *Hemidactylus brookii haitianus* (Sauria: Gekkonidae) from the Dominican Republic. Dactylus 1: 1-9.

MacArthur, R. H., J. M. Diamond, and J. R. Karr. 1972. Density compensation in island faunas. Ecology 53: 330-342.

MacArthur, R. H., and E. O. Wilson. 1963. An equilibrium theory of insular zoogeography. Evolution 17: 373-387.

MacArthur, R. H., and E. O. Wilson. 1967. The Theory of Island Biogeography. Princeton Univ. Press, Princeton, NJ.

MacLean, W. P., and R. D. Holt. 1979. Distributional patterns in St. Croix *Sphaerodactylus* lizards: the taxon cycle in action. Biotropica 11: 189-195.

MacLean, W. P., R. Kellner, and H. Dennis. 1977. Island lists of West Indian amphibians and reptiles. Smithsonian Herpetol. Information Service 40: 1-47.

MacLeay, W. S. 1834. A few remarks tending to illustrate the natural history of two annulose genera, namely *Urania* of Babricus and *Mygale* of Walckenäer. Proc. Zool. Soc. London 1834: 9-12.

MacPhee, R. D. E., D. C. Ford, and D. A. MacFarlane. 1989. Pre-Wisconsinan mammals from Jamaica and models of late Quarternary extinctions in the Caribbean. Anat. Res. 31: 94-106.

MacPhee, R. D. E., and R. A. Marx. 1997. The 40,000-year plague. Humans, hyperdisease and first-contact extinctions, pp. 169-217 in Natural Change and Human Impact in Madagascar, edited by S. M. Goodman, and B. D. Patterson. Smithsonian Instit. Press, Washington D. C.

Maglio, V. J. 1970. West Indian xenodontine colubrid snakes: their probable origin, phylogeny, and zoogeography. Bull. Museum Comp. Zool. 141: 1-54.

Mahon, R., and K. Aiken. 1977. The establishment of the North American bullfrog, *Rana catesbeiana* (Amphibia, Anura, Ranidae) in Jamaica. J. Herpetol. 11: 197-199.

Major, R. H. 1870. Selected Letters of Columbus. Hakluyt Soc., London.

Malfait, B. T., and M. G. Dinkelman. 1972. Circum-Caribbean tectonic and igneous activity and the evolution of the Caribbean plate. Bull. Geol. Soc. Am. 83: 251-272.

Malhotra, A., and R. S. Thorpe. 1991. Experimental detection of rapid evolutionary response in natural island populations. Nature 353: 347-348.

Malnate, E. V. 1971. A catalog of primary types in the herpetological collections of the Academy of Natural Sciences, Philadelphia (ANSP). Proc. Acad. Natl. Sci. Philadelphia 123: 345-375.

Margulis, L., and T. H. Kunz. 1984. Glimpses of biological research and education in Cuba. Bioscience 84: 634-639.

Martin. 1838. Report on a collection of snakes procured by the Euphrates Expedition. Proc. Zool. Soc. London VI: 81-84.

Martin, P. S., and R. Klein (Editors). 1984. Quaternary Extinctions: A Prehistoric Revolution. Univ. Arizona Press, Tucson.

Martin, P. S., and D. W. Steadman. 1998. Prehistoric extinctions on islands and continents, in Quaternary Extinctions edited by R. D. E. MacPhee et al., Plenum Press, New York.

Martin, R. F. 1972. Evidence from osteology pp. 37-70 in Evolution in the genus *Bufo*, edited by W. F. Blair. Univ. Texas Press, Austin, Texas.

Mateo, J., and M. Acevedo. 1989. 5. Regionalización físico-geográfica, 1:3000000. XII Paisajes. Nuevo Atlas Nacional de Cuba. Academia de Ciencias de Cuba, Cuba.

Matthew, W. D. 1915. Climate and Evolution. Ann. New York Acad. Sci. 24: 171-318.

Maurrasse, F. J.-M. R. 1982. Survey of the Geology of Haiti. Miami Geol. Soc., Miami.

McDiarmid, R. W., T. Touré, and J. M. Savage. 1996. The proper name of the neotropical tree boa often referred to as *Corallus enydris* (Serpentes: Boidae). J. Herpetol. 30: 320-326.

McDowell, S. B. 1964. Partition of the genus *Clemmys* and related problems in the taxonomy of aquatic Testudinidae. Proc. Zool. Soc. London 143: 239-279.

McDowell, S. B. 1979. A catalogue of the snakes of New Guinea and the Solomons, with special reference to those in the Bernice P. Bishop Museum. Part III. Boinae and Acrochordoidea (Reptilia, Serpentes). J. Herpetol. 13: 1-92.

McKenna, M. C. 1973. Sweepstakes, filters, corridors, Noah's arks and beached Viking funeral ships in paleobiogeography, pp. 295-308 in Implications of Continental Drift to the Earth Sciences, edited by D. H. Tarling, and S. K. Runcorn. Academic Press, New York.

McLaughlin, J. F., and J. Roughgarden. 1989. Avian predation on *Anolis* lizards in the northeastern Caribbean: an inter-island contrast. Ecology 70: 617-628.

McLaughlin, P. P., W. A. van den Bold, and P. Mann. 1991. Geology of the Azua and Enriquillo basins, Dominican Republic; 1, Neogene lithofascies, biostratigraphy, biofacies, and paleogeography. Geol. Soc. Am., Spec. Pap. 262: 337-366.

Meerwarth, H. 1901. Die westindischen Reptilien und Batrachier des naturhistorischen Museums in Hamburg. Mitt. Naturhist. Museum Hamburg 18: 1-41.

Merrem, B. 1809. Amphibiologische Beiträge. Annu. Wetterauischen Gesell. Ges. Naturk. I: 1-191.

Merrem, B. 1820. Tentamen Systematis Amphibiorum. Marburg.

Mertens, R. 1938. Amphibien und Reptilien aus Santo Domingo, gesammelt von Prof. Dr. H. Böker. Senckenberg. Biol. 20: 332-342.

Mertens, R. 1939. Herpetologische Ergebnisse einer Reise nach der Insel Hispaniola, Westindien. Abh. Senckenberg. Nat. Ges. 449: 1-84.

Mertens, R. 1940. Zoolog Reisen auf der Insel Hispaniola (from Teirleben der Tropen by R. Mertens), pp. 155-241 Verlag Dr. Waldemar Kramer, Frankfurt.

Mertens, R. 1950. Zur Taxonomie und Verbreitung der hispaniolischen Eidechsen der Gattung *Ameiva*. Senckenbergiana 31: 3-7.

Mertens, R., and H. Wermuth. 1955. Die rezenten Schildkröten, Krokodile und Brückenechsen. Zool. Jahrb. Syst. Abt. 83: 323-440.

Meshaka, W. E., Jr. 1996. Diet and the colonization of buildings by the Cuban treefrog, *Osteopilus septentrionalis* (Anura: Hylidae). Carib. J. Sci. 32.

Meshaka, W. E., Jr., B. P. Butterfield, and B. Hauge. 1994a. *Hemidactylus mabouia* as an established member of the Florida herpetofauna. Herpetol. Rev. 25: 80-81.

Meshaka, W. E., Jr., B. P. Butterfield, and B. Hauge. 1994b. Reproductive notes on the introduced gecko *Hemidactylus mabouia* in southern Florida. Herpetol. Nat. Hist. 2: 109-110.

Meshaka, W. E., Jr., B. P. Butterfield, and B. Hauge. 1994c. Geographic distribution: *Hemidactylus mabouia*. Herpetol. Rev. 25: 165.

Miles, D. B., and A. E. Dunham. 1996. The paradox of the phylogeny: character displacement of analyses of body size in island *Anolis*. Evolution 50: 594-603.

Mittermeier, R. A. 1972. Jamaica's endangered species. Oryx 11: 258-262.
Moermond, T. C. 1979. Habitat constraints on the behavior, morphology, and community structure of *Anolis* lizards. Ecology 60: 152-164.

Monroe, W. H. 1976. The karst landforms of Puerto Rico. Geol. Surv. Prof. Paper 899: iv + 69.

Moore, C. 1991. Cabaret: lithic workshop sites in Haiti. pp. 92-104 in Proceedings of the 13 International congress of Caribbean Archeology, edited by E. N. Ayubi, and J. B. Haviser. Reports Archeological-Anthropological Institute, Netherlands Antilles 9.

Moreau de Jonnès, A. 1818. Monographie du mabouia des murailles, ou Gecko Mabouia de Antilles. Bull. Sci. Soc. Philomath. Paris 3: 138-139.

Morgan, G. S., and C. A. Woods. 1986. Extinction and the zoogeography of West Indian land mammals. Biol. J. Linnean Soc. London 28: 167-203.

Morison, S. E. 1963. Journals and Other Documents on the Life and Voyages of Christopher Columbus. Heritage Press, NY.

Morrison, L. W. 1997. The insular biogeography of small Bahamian cays. J. Ecol. 85:441-454.

Moster, J. A., R. Powell, J. J. S. Parmerlee, D. D. Smith, and A. Lathrop. 1992. Natural history notes on a small population of *Anolis brevirostris* (Sauria: Polychridae) from altered habitat in the Dominican Republic. Bull. Maryland Herpetol. Soc. 28: 150-161.

Müller, L. 1923. Neue oder seltene Reptilien und Batrachier der Zoologischen Sammlung des bayr. Staates. Zool. Anz. 57: 49-61.

Müller, L. 1926. *Leptodactylus fallax* nom. nov. fr *Leptodactylus dominicensis* L. Mll. (non Cochran). Zool. Anz. 65: 200.

Murphy, J. C. 1996. Crossing Bond's Line: the herpetofaunal exchange between the eastern Caribbean and mainland South America, pp. 207-216 in Contributions to West Indian Herpetology: A Tribute to Albert Schwartz, edited by R. Powell, and R. W. Henderson. Society for the Study of Amphibians and Reptiles Contributions in Herpetology, Ithaca, NY.

Murphy, J. C., H. K. Voris, and D. R. Karns. 1994. A field guide to the snakes of the Danum Valley, a Bornean tropical forest ecosystem. Bull. Chicago Herpetol. Soc. 29: 133-151.

Murphy, R. W. The phylogenetic analysis of allozyme data: invalidity of coding alleles by presence/absence and recommended procedures. Biochem. Syst. Ecol. 21: 25-38.

Murphy, R. W., F. C. McCollum, G. C. Gorman, and R. Thomas. 1984. Genetics of hybridizing populations of Puerto Rican *Sphaerodactylus*. J. Herpetol. 18: 93-105.

Myers, C. W. 1973a. A new genus for Andean snakes related to *Lygophis boursieri* and a new species (Colubridae). Novitates 2522: 1-37.

Myers, C. W. 1973b. Anguid lizards of the genus *Diploglossus* in Panama, with the description of a new species. Novitates 2523: 1-20.

Myers, C. W., and J. A. Campbell. 1981. A new genus and species of colubrid snake from the Sierra Madre del Sur of Guerrero, Mexico. Novitates 2708: 1-20.

Myers, G. S. 1962. The American leptodactylid frog genera *Eleutherodactylus*, *Hylodes* (= *Elosia*), and *Caudiverbera* (= *Calyptocephalus*). Copeia 1962: 195-202.

Nelson, G. 1969. The problem of historical biogeography. Syst. Zool. 18: 243-246.

Nelson, G. 1973. Comments on Leon Croizat's biogeography. Syst. Zool. 22: 312-320.

Nelson, G. 1974. Historical biogeography: an alternative formalization. Syst. Zool. 23: 555-558.

Nganuma, K. H., and J. D. Roughgarden. 1990. Optimal body size in Lesser Antillean *Anolis* lizards – a mechanistic approach. Ecol. Monogr. 60: 239-256.

Noble, G. K. 1921. The bony structure and phyletic relations of *Sphaerodactylus* and allied lacertilian genera, with the description of a new genus. Novitates 4: 1-16.

Noble, G. K. 1923a. Six new batrachians from the Dominican Republic. Novitates 61: 1-6.

Noble, G. K. 1923b. Four new lizards from Beata Island, Dominican Republic. Novitates 64: 1-5.

Noble, G. K. 1927. The value of life history data in the study of the evolution of the Amphibia. Annu. New York Acad. Sci. 30:31-128.

Noble, G. K. 1933. The mating behavior of lizards; its sexual bearing on the theory of sexual selection. Annu. New York Acad. Sci. 35: 25-100.

Noble, G. K., and W. G. Hassler. 1933. Two new species of frogs, five new species and a new race of lizards from the Dominican Republic. Novitates 652: 1-17.

Noble, G. K., and G. C. Klingel. 1932. The reptiles of Great Inagua Island, British West Indies. Novitates 549: 1-25.

Norell, M. A. and K. de Queiroz. 1991. The earliest iguanine lizard (Reptilia: Squamata) and its bearing on iguanine phylogeny. Novitates 2997: 1-16.

Ober, F. A. 1880. Camps in the Caribees: the adventures of a naturalist in the Lesser Antilles. Lee & Shepard, Boston.

Oliver, J. A. 1948. The anoline lizards of Bimini, Bahamas. Novitates 1383: 1-36.

Oliver, W. L. R. 1982. The coney and the yellow snake: the distribution and status of the Jamaican hutia *Geocapromys brownii* and the Jamaican boa *Epicrates subflavus*. Dodo J. Jersey Wildl. Pres. Trust 19: 6-33.

Olson, R. E. 1995. Microhabitat in ecologically devastated areas. Bull. Maryland Herpetol. Soc. 31: 93-95.

Olson, S. L., and W. B. Hilgartner. 1982. Fossil and subfossil birds from the Bahamas, pp. 22-60 in Fossil Vertebrates from the Bahamas, edited by S. L. Olson. Smithsonian Contributions Paleobiol. 48. Smithsonian Institute, Washington, DC.

Olson, S. L., G. K. Pregill, and W. B. Hilgartner. 1990. Studies on fossil and extant vertebrates from San Salvador (Watling's) Island, Bahamas. Smitsonian Contrib. Zool. 508:1-15.

Osborn, H. F. 1931. Cope: Master Naturalist. Princeton Univ. Press, Princeton. NJ.

Ottenwalder, J. A. 1989. A summary of conservation trends in the Dominican Republic, pp. 845-850 in Biogeography of the West Indies: Past, Present, and Future, edited by C. A. Woods. Sandhill Crane Press, Gainesville.

Ottenwalder, J. A. 1996. The current status of sea turtles in Haiti, pp. 381-393 in Contributions to West Indian Herpetology: A Tribute to Albert Schwartz, edited by R. Powell, and R. W. Henderson. Society for the Study of Amphibians and Reptiles Contributions in Herpetology, Ithaca, NY.

Oviedo, G. F. de. in Stoudmere, S. A. (Trans. & Ed.). 1959. Natural History of the West Indies. Univ. North Carolina Press, Chapel Hill.

Pacala, S. W., and J. Roughgarden. 1982. Resource partitioning and intcr-specfic competition in two-species insular *Anolis* lizard communities. Science 217: 444-446.

Pacala, S. W., and J. Roughgarden. 1985. Population experiments with the *Anolis* lizards of St. Maarten and St. Eustatius. Ecology 66: 129-141.

Page, R. D. M., and C. Lydeard. 1994. Towards a cladistic biogeography of the Caribbean. Cladistics 10: 21-41.

Panton, E. S. 1952. Our ground and tree frogs – glimpses into their life and habits. Nat. Hist. Notes Nat. Hist. Soc. Jamaica 5: 87-92 + 94.

Parker, H. W. 1926a [August]. A new brachycephalid frog from Brazil. Annu. Mag. Nat. Hist. (9)18: 201-203.

Parker, H. W. 1926b. The neotropical lizards of the genera *Lepidoblepharis*, *Pseudogonatodes*, *Lathrogecko*, and *Sphaerodactylus*, with the description of a new genus. Annu. Mag. Nat. Hist. 17: 291-301.

Parker, H. W. 1933. Some amphibians and reptiles from the Lesser Antilles. Annu. Mag. Nat. Hist. 10: 155-158.

Parker, H. W. 1935. Some lizards from Venezuela and the Dutch Leeward Islands. Annu. Mag. Nat. Hist. 15: 480-484.

Parker, H. W. 1936. Some extinct snakes of the West Indies. Annu. Mag. Nat. Hist. 10: 227-233.

Parsons, K. J. 1995. *Peltophryne guentheri* (NCN). Diet. Herpetol. Rev. 26: 202.

Paull, D., E. E. Williams, and W. P. Hall. 1976. Lizard karyotypes from the Galapagos islands: chromosomes in phylogeny and evolution. Breviora Museum Comp. Zool. 441: 1-31.

Perfit, M. R., and E. E. Williams. 1989. Geological constraints and biological retrodictions in the evolution of the Caribbean sea and its islands, pp. 47-102 in Biogeography of the West Indies, Past, Present and Future, edited by C. A. Woods. Sandhill Crane Press, Gainesville.

Perkins, L. 1942. Whistling frogs in St. Annu. Nat. Hist. Notes Nat. Hist. Soc. Jamaica 1: 4-5.

Peters, G., and A. Schwartz. 1972. Ein neuer, bromelien bewohnder Kugelfingergecko (Gekkonidae, *Sphaerodactylus*) aus Oriente Cuba. Mitteilungen aus dem Zool. Museum Berlin 48: 393-399.

Peters, J. A., and R. Donoso-Barros. 1970. Catalogue of the Neotropical Squamata. Part II. Lizards and amphisbaenians. Bull. U.S. Natl. Museum 297: 1-293.

Peters, J. A., and R. Donoso-Barros. 1986. Catalogue of the Neotropical Squamata. Part II. Lizards and Amphisbaenians. Smithsonian Institute Press, Washington, DC.

Peters, J. A., and B. Orejas-Miranda. 1970. Catalogue of the Neotropical Squamata. Part I. Snakes. Bull. U.S. Natl. Museum 297: viii+347.

Peters, W. 1863. Über einige neue oder weniger bekannte Schlangenarten des zoologischen Museums zu Berlin. Monatsb. Königl. Akad. Wiss. Berlin 1863: 272-289.

Peters, W. 1869. Mittheilung über neue Gattungen und Arten von Eidechsen. Monatsb. Königl. Akad. Wiss. Berlin: 57-66.

Peters, W. 1879. Neue Amphibien des Kgl. zoologischen Museums (*Euprepes*, *Acontias*, *Typhlops*, *Zamenis*, *Spilotes*, *Oedipus*). Monatsb. Königl. Akad. Wiss. Berlin.

Peterson, J. A. 1983a. The evolution of the subdigital pad in *Anolis*. I. Comparisons among the anoline genera, pp. 245-283 in Advances in Herpetology and Evolutionary Biology, edited by A. G. J. Rhodin, and K. Miyata. Museum of Comparative Zoology, Cambridge. MA.

Peterson, J. A. 1983b. The evolution of the subdigital pad of *Anolis* 2. Comparisons among the iguanid genera related to the anolines and a view from outside the radiation. J. Herpetol. 17: 371-397.

Peterson, J. A., and R. L. Bezy. 1985. The microstructure and evolution of scale surfaces in xantusiid lizards. Herpetologica 41: 298-324.

Pfeiffer, W. J. 1996. Litter invertebrates, pp. 137-181 in The Food Web of a Tropical Rain Forest, edited by D. P. Reagan, and R. B. Waide. Univ. Chicago Press, Chicago.

Pindell, J., and S. F. Barrett. 1990. Geological evolution of the Caribbean region: a plate tectonic perspective, pp. 405-432 in The Geology of North America. Vol. H. The Caribbean Region, edited by G. Dengo, and J. E. Case. Geol. Soc. Am., Inc., New York.

Pinou, T., and H. G. Dowling. 1994. The phylogenetic relationships of the Central American snake *Tretanorhinus*: data from morphology and karyology. Amphib.-Rept. 15: 297-305.

Pitman, W. C., III, S. Cande, J. Labreque, and J. Pindell. 1993. Separation of Africa from South America, pp. 15-36 in Biological Relations between Africa and South America, edited by P. Goldblatt. Yale Univ. Press, New Haven, CT.

Poe, S. 1996. Data set incongruence and the phylogeny of crocodilians. Syst. Biol. 45: 393-414.

Poe, S. 1998. Skull characters and the cladistic relationships of the Hispaniolan dwarf twig *Anolis*. Herp. Mong. 12: 192-236.

Poey, F. 1851-55. Circulacion del cocodrilo. Mem. Sobre Hist. Nat. Cuba 1: 258-267.

Poinar, G. O., Jr. 1992. Life in Amber. Stanford Univ. Press, Stanford, CA.

Poinar, G. O., Jr., and D. C. Cannatella. 1987. An Upper Eocene frog from the Dominican Republic and its implications for Caribbean biogeography. Science 237: 1215-1216.

Pough, F. H., M. M. Stewart, and R. G. Thomas. 1977. Physiological basis of habitat partitioning in Jamaican *Eleutherodactylus*. Oecologia 27: 285-293.

Pounds, J. A. 1988. Ecomorphology, locomotion, and microhabitat structure: patterns in a tropical mainland *Anolis* community. Ecol. Monogr. 58: 299-320.

Powell, G. L., and A. P. Russell. 1992. Locomotor correlates of ecomorph designation in *Anolis*: an examination of three sympatric species from Jamaica. Canadian J. Zool. 70: 725-739.

Powell, R. 1992a. *Peltophryne guentheri*. Cat. Am. Amphib. Rept. 535: 1-4.

Powell, R. 1992b. *Amphisbaena gonavensis*. Cat. Am. Amphib. Rept. 539: 1-3.

Powell, R. 1993. Comments on the taxonomic arrangement of some Hispaniolan amphibians and reptiles. Herpetol. Rev. 24: 135-137.

Powell, R., S. S. Duer, J. J. S. Parmerlee, and D. D. Smith. 1989b. Life history notes: *Ameiva taeniura* (food habits). Herpetol. Rev. 20: 49.

Powell, R., and R. W. Henderson. 1992. Life history notes: *Anolis gingivinus* (nocturnal activity). Herpetol. Rev. 23: 117.

Powell, R., and R. W. Henderson. 1994. *Ialtris*. Cat. Am. Amphib. Rept. 590.1-590.2.

Powell, R., and R. W. Henderson. 1996a. Introduction, pp. 11-14 in Contributions to West Indian Herpetology: A Tribute to Albert Schwartz, edited by R. Powell, and R. W. Henderson. Society for the Study of Amphibians and Reptiles Contributions in Herpetology, Ithaca, NY.

Powell, R., and R. W. Henderson. 1996b. A brief history of West Indian herpetology, pp. 29-50 in Contributions to West Indian Herpetology: A Tribute to Albert Schwartz, edited by R. Powell, and R. W. Henderson. Society for the Study of Amphibians and Reptiles Contributions in Herpetology, Ithaca, NY.

Powell, R., and R. W. Henderson. 1998. *Alsophis melanichnus*. Cat. Amer. Amphib. Rept. 660:1-2.

Powell, R., R. W. Henderson, K. Adler, and H. A. Dundee. 1996a. An annotated checklist of West Indian amphibians and reptiles, pp. 51-93 in Contributions to West Indian Herpetology: A Tribute to Albert Schwartz, edited by R. Powell, and R. W. Henderson. Society for the Study of Amphibians and Reptiles Contributions in Herpetology, Ithaca, NY.

Powell, R., and J. J. S. Parmerlee. 1992. In the spotlight: *Aristelliger lar*. Dactylus 1: 37-38.

Powell, R., and J. J. S. Parmerlee. 1993. In the spotlight: *Hemidactylus haitianus*. Dactylus 2: 54-55.

Powell, R., J. J. S. Parmerlee, M. A. Rice, and D. D. Smith. 1990a. Ecological observations of *Hemidactylus brookii haetianus* Meerwarth (Sauria: Gekkonidae) from Hispaniola. Carib. J. Sci. 26: 67-70.

Powell, R., J. J. S. Parmerlee, and D. D. Smith. 1996b. Evidence of spatial niche partitioning by a Hispaniolan lizard community in a xeric habitat, pp. 317-326 in Contributions to West Indian Herpetology: A Tribute to Albert Schwartz, edited by R. Powell, and R. W. Henderson. Society for the Study of Amphibians and Reptiles Contributions in Herpetology, Ithaca, NY.

Powell, R., R. J. Passaro, and R. W. Henderson. 1992. Noteworthy herpetological records from Saint Maarten, Netherlands Antilles. Carib. J. Sci. 28: 234-235.

Powell, R., D. D. Smith, J. S. Parmerlee, C. V. Taylor, and M. L. Jolley. 1990b. Range expansion by an introduced anole: *Anolis porcatus* in the Dominican Republic. Amphib.-Rept. 11: 421-425.

Powell, R., S. G. Thornhill, and D. G. [*sic*] Smith. 1989a. Geographic distribution: *Sphaerodactylus a. altavelensis*. Herpetol. Rev. 20: 13.

Pregill, G. K. 1981a. Cranial morphology and the evolution of West Indian toads (Salientia: Bufonidae): resurrection of the genus *Peltophryne* Fitzinger. Copeia 1981: 273-285.

Pregill, G. K. 1981b. Late Pleistocene herpetofaunas from Puerto Rico. Misc. Publ. Univ. Kansas Museum Nat. Hist. 71: 1-72.

Pregill, G. K. 1981c. An appraisal of the vicariance hypothesis of Caribbean biogeography and its application to West Indian terrestrial vertebrates. Syst. Zool. 30: 147-155.

Pregill, G. K. 1984. An extinct species of *Leiocephalus* from Haiti (Sauria: Iguanidae). Proc. Biol. Soc. Washington 97: 827-833.

Pregill, G. K. 1986. Body size of insular lizards: a pattern of Holocene dwarfism. Evolution 40: 997-1008.

Pregill, G. K. 1992. Systematics of the West Indian lizard genus *Leiocephalus* (Squamata: Iguania: Tropiduridae). Univ. Kansas Museum Nat. Hist. Misc. Publ. 84: 1-69.

Pregill, G. K., R. I. Crombie, D. W. Steadman, L. K. Gordon, F. W. Davis, and W. B. Hilgartner. 1992. Living and late Holocene fossil vertebrates, and the vegetation of the Cockpit Country, Jamaica. Atoll. Res. Bull. 353: 1-19.

Pregill, G. K., and T. Dye. 1989. Prehistoric extinction of giant iguanas in Tonga. Copeia 1989: 505-508.

Pregill, G. K., and S. L. Olson. 1981. Zoogeography of the West Indies in relation to Pleistocene climatic cycles. Annu. Rev. Ecol. Syst. 12: 75-98.

Pregill, G. K., D. W. Steadman, S. L. Olson, and F. V. Grady. 1988. Late Holocene fossil vertebrates from Burma Quarry, Antigua, Lesser Antilles. Smithson. Contrib. Zool. 463: 1-27.

Pregill, G. K., D. W. Steadman, and D. R. Watters. 1994. Late Quaternary vertebrate faunas of the Lesser Antilles: historical components of Caribbean biogeography. Bull. Carnegie Museum Nat. Hist. 30: 1-51.

Presch, W. 1971. Tongue structure of the teiid lizard genera *Ameiva* and *Cnemidophorus* with a reallocation of *Ameiva vanzoi*. J. Herpetol. 5: 183-185.

Presch, W. 1988. Phylogenetic relationships of the Scincomorpha, pp. 471-487 in Phylogenetic Relationships of the Lizard Families: Essays Commemorating Charles L.Camp, edited by R. Estes, and G. Pregill. Stanford Univ. Press, Stanford, CA.

Pritchard, P. C. H. 1979. Encyclopedia of Turtles. T. F. H., Neptune, NJ.

Pritchard, P. C. H. and P. Trebbau. 1984. The Turtles of Venezuela. Soc. Stud. Amphib. Rept. Contrib. Herpetol. 3. Athens, Ohio.

Proctor, G. R. 1959. Observations on Navassa Island. Geonotes Q. J. Jamaica Group Geologists' Assoc. 2: 49-51.

Pruna Goodgall, P. M. 1994. Cronologia: Hechos historicos relacionados con la ciencia y la tecnologia acaecidos en la Habana, 1521-1988. Editorial Academia, La Habana.

Rage, J. C. 1982. La phylogénie des lépidosauriens (Reptilia): une approche cladistique. C. R. Acad. Sci. Paris 294: 563-566.

Rand, A. S. 1961. Notes on Hispaniolan herpetology 4. *Anolis koopmani*, new species, from the southwestern peninsula of Haiti. Breviora 137: 1-4.

Rand, A. S. 1962. Notes on Hispaniolan herpetology. 5. The natural history of three sympatric species of *Anolis*. Breviora 154: 1-15.

Rand, A. S. 1964a. Inverse relationship between temperature and shyness in the lizard *Anolis lineatopus*. Ecology 45: 863-864.

Rand, A. S. 1964b. Ecological distribution in anoline lizards of Puerto Rico. Ecology 45: 745-752.

Rand, A. S. 1967a. The ecological distribution of the anoline lizards around Kingston, Jamaica. Breviora 272: 1-18.

Rand, A. S. 1967b. Ecology and social organization in the iguanid lizard *Anolis lineatopus*. Proc. U.S. Natl. Museum 122: 1-79.

Rand, A. S. 1969. Competitive exclusion among anoles (Sauria: Iguanidae) on small islands in the West Indies. Breviora Museum Comp. Zool. 319: 1-16.

Rand, A. S., and S. S. Humphrey. 1968. Interspecific competition in the tropical rain forest: ecological distribution among lizards at Belém, Pará. Proc. U.S. Natl. Museum 125: 1-17.

Rand, A. S., and C. W. Myers. 1990. The herpetofauna of Barro Colorado Island, Panama: an ecological summary, pp. 386-409 in Four Neotropical Rainforests, edited by A. H. Gentry. Yale Univ. Press, New Haven, CT.

Rand, A. S., and E. E. Williams. 1969. The anoles of La Palma: aspects of their ecological relationships. Breviora 327: 1-17.

Ray, C. E. 1964. A small assemblage of vertebrate fossils from Spring Bay, Barbados. J. Barbados Museum Hist. Soc. 31: 11-22.

Reagan, D. P. 1992. Congeneric species distribution and abundance in a three-dimensional habitat: the rain forest anoles of Puerto Rico. Copeia 1992: 392-403.

Reagan, D. P. 1996a. The role of amphibians and reptiles in a West Indian rain forest food web, pp. 217-227 in Contributions to West Indian Herpetology: A Tribute to Albert Schwartz, edited by R. Powell, and R. W. Henderson. Society for the Study of Amphibians and Reptiles Contributions in Herpetology, Ithaca, NY.

Reagan, D. P. 1996b. Anoline lizards, pp. 321-345 in The Food Web of a Tropical Rain Forest, edited by D. P. Reagan, and R. B. Waide. Univ. Chicago Press, Chicago, IL.

Reagan, D. P. and R. B. Waide (eds.) 1996. The food web of a tropical rain forest. Univ. Chicago Press, Chicago, IL.

Reinhardt, J., and C. F. Lütken. 1862. Bidrag til det vestindiske Öeriges og navnligen til de dansk-vestindiske Öers Herpetologie. Vidensk. Meddel. Naturhist. Foren. Kjøbenhavn 10-18: 153-291.

Reitz, E. J., L. A. Newsome, and S. J. Scudder (Editors). 1996. Case Studies in Environmental Archaeology. Plenum, NY.

Richmond, N. D. 1964. The blind snakes (*Typhlops*) of Haiti with descriptions of three new species. Breviora 202: 1-12.

Richmond, N. D. 1966. The blind snakes, *Typhlops*, of Guadeloupe and Dominica with the description of a new species. Herpetologica 22: 129-132.

Ricklefs, R. 1970. Stage of taxon cycle and distribution of birds on Jamaica. Evolution 24: 475-477.

Ricklefs, R., and G. W. Cox. 1972. Taxon cycles in the West Indian avifauna. Am. Nat. 106: 195-219.

Ricklefs, R., and G. W. Cox. 1978. Stage of taxon cycle, habitat distribution, and population density in the avifauna of the West Indies. Am. Nat. 112: 875-895.

Rieppel, O. 1980. Green anole in Dominican amber. Nature 286: 486-487.

Rivero, J. A. 1978. Los anfibios y reptiles de Puerto Rico/The amphbians and reptiles of Puerto Rico. Univ. of Puerto Rico Press, San Juan.

Rodda, G., E. W. Campbell III, and T. H. Fritts. 1995. Extraordinary lizard densities on Guam, as revealed by complete removal sampling, Annu. Meetg. ASIH, AES, HL., Univ. of Alberta, Canada. (abstracts)

Rodriguez, G. A., and D. P. Reagan. 1984. Bat predation by the Puerto Rican boa, *Epicrates inornatus*. Copeia 1984: 219-220.

Rodríguez-Robles, J. A., and H. W. Greene. 1996. Ecological patterns in Greater Antillean macrostomatan snake assemblages, with comments on body size evolution in *Epicrates* (Boidae), pp. 339-357 in Contributions to West Indian Herpetology: A Tribute to Albert Schwartz, edited by R. Powell, and R. W. Henderson. Society for the Study of Amphibians and Reptiles Contributions in Herpetology, Ithaca, NY.

Rodríguez-Robles, J. A., and M. Leal. 1993. Effects of prey type on the feeding behavior of *Alsophis portoricensis* (Serpentes: Colubridae). J. Herpetol. 27: 163-168.

Rodríguez-Schettino, L. 1986. Algunos patrones distribucionales y ecológicos de los reptiles cubanos. Poeyana 305: 1-15.

Rodríguez-Schettino, L. 1993. Areas faunísticas de Cuba según la distribución ecogeográfica actual y el endemismo de los reptiles. Poeyana 436: 1-17.

Rodríguez-Schettino, L. 1995. Checklist and bibliography (1837-1991) of Cuban iguanid lizards. Smithsonian Herpetological Information Service 103: 1-29.

Rosen, D. E. 1976. A vicariance model of Caribbean biogeography. Syst. Zool. 24: 431-464.

Rosen, D. E. 1978. Vicariant patterns and historical explanation in biogeography. Syst. Zool. 27: 159-188.

Rosen, D. E. 1985. Geological hierarchies and biogeographic congruence in the Caribbean. Ann. Missouri Bot. Gard. 72: 636-659.

Roughgarden, J. 1990. Origin of the eastern Caribbean: data from reptiles and amphibians, pp. 10-26 in Transactions of the 12th Caribbean Geological Conference, St. Croix, USVI, edited by D. K. Larue, and G. Draper. Miami Geol. Soc., Miami.

Roughgarden, J. 1995. *Anolis* Lizards of the Caribbean: Ecology, Evolution, and Plate Tectonics. Oxford Univ. Press, New York.

Roughgarden, J., D. Heckel, and E. R. Fuentes. 1983. Coevolutionary theory and the biogeography and community structure of *Anolis*, pp. 371-410 in Lizard Ecology, edited by R. B. Huey, E. R. Pianka, and T. W. Schoener. Harvard Univ. Press, Cambridge, MA.

Roughgarden, J., and S. Pacala. 1989. Taxon cycle among *Anolis* lizard populations: review of evidence, pp. 403-432 in Speciation and Its Consequences, edited by D. Otte, and J. Endler. Sinauer, Sunderland, MA.

Roughgarden, J., W. Porter, and D. Heckel. 1981. Resource partitioning of space and its relationship to body temperature in *Anolis* lizard populations. Oecologia 50: 256-264.

Rouse, I. 1989. Peopling and repeopling of the West Indies, pp. 119-136 in Biogeography of the West Indies: Past, Present, and Future, edited by C. A. Woods. Sandhill Crane Press, Inc., Gainesville. FL.

Rouse, I. 1992. The Tainos: Rise and Decline of the People Who Greeted Columbus. Yale Univ. Press, New Haven, CT.

Ruibal, R. 1946. A new *Sphaerodactylus* from the Dominican Republic. Novitates 1308: 1-4.

Ruibal, R. 1959. *Bufo gundlachi*, a new species of Cuban toad. Breviora, Mus. Comp. Zool. 105: 1-14.

Ruibal, R. 1961. Thermal relations of five species of tropical lizards. Evolution 15: 98-111.

Ruibal, R., and R. Philibosian. 1974. The population ecology of the lizard *Anolis acutus*. Ecology 55: 525-537.

Ruibal, R. and E. E. Williams. 1961. Two sympatric anoles of the *carolinesis* group. Bull. Museum Comp. Zool. 125(7): 183-208.

Russel, A. P. 1972. The foot of gekkonid lizards: a study in comparative and functional anatomy. Ph.D. Dissertation, Univ. London, England.

Russell, A. P. 1979. Parallelism and integrated design in the foot structure of Gekkonine and Diplodactyline geckos. Copeia 1979: 1-21.

Russell, A. P., and A. M. Bauer. 1991a. *Anolis garmani*. Cat. Am. Amphib. and Rept. 513: 1-3.

Russell, A. P., and A. M. Bauer. 1991b. *Anolis grahami*. Cat. Am. Amphib. and Rept. 514: 1-4.

Russell, A. P., and A. M. Bauer. 1991c. *Anolis opalinus*. Cat. Am. Amphib. and Rept. 515: 1-2.

Sajdak, R. A., and R. W. Henderson. 1991. Status of West Indian racers in the Lesser Antilles. Oryx 25: 33-38.

Salomao, M. da G., W. Wüster, R. S. Thorpe, J.-M. Touzet, and BBBSP. 1997. DNA evolution of South American pitvipers of the genus *Bothrops* (Reptilia: Serpentes: Viperidae), pp. 89-98 in Venomous Snakes: Ecology, Evolution and Snakebite, edited by R. S. Thorpe, W. Wüster, and A. Malhorta. Clarendon Press, Oxford, UK.

Sampedro Marín, A., and O. T. Fundora. 1982. Hábitos alimentarios y actividad de *Bufo peltocephalus* Tschudi (Amphibia: Bufonidae) en el Jardín Botánico de Cienguegos. Poeyana 233: 1-14.

Sampson, L. V. 1904. A contribution to the embryology of *Hylodes martinicensis*. Am. J. Anat. 3: 473-504.

Sauer, J. D. 1969. Oceanic islands and biogeographic theory: a review. Geographr. Rev. 59: 582-593.

Savage, J. M. 1955. The lizard family Xantusiidae: an evolutionary study. Ph.D. dissertation, Stanford Univ., Stanford, California.

Savage, J. M. 1963. Studies on the lizard family Xantusiidae IV. The genera. Contrib. Sci. Los Angeles County Museum 71: 1-38.

Savage, J. M. 1964. Studies on the lizard family Xantusiidae. V. The Cuban night lizard, *Cricosaura typica* Gundlach and Peters. Copeia 1964: 536-542.

Savage, J. M. 1966. The origins and history of the Central American herpetofauna. Copeia 1966: 719-766.

Savage, J. M. 1982. The enigma of the Central American herpetofauna: dispersals or vicariance? Ann. Missouri Bot. Gard. 69: 464-547.

Savage, J. M. 1987. Systematics and distribution of the Mexican and Central American rainfrogs of the *Eleutherodactylus gollmeri* group (Amphibia: Leptodactylidae). Fieldiana Zool. 33: 1-57.

Savage, J. M., and K. R. Lips. 1993. A review of the status and biogeography of the lizard genera *Celestus* and *Diploglossus* (Squamata: Anguidae), with description of two new species from Costa Rica. Rev. Biol. Tropics 41: 817-842.

Savage, J. M., and J. Villa R. 1986. Introduction to the herpetofauna of Costa Rica. Society for the Study of Amphibians and Reptiles Contributions in Herpetology, Ithaca, NY.

Schatzinger, R. A. 1980. New species of *Palaeoxantusia* (Reptilia: Sauria) from the Uintan (Eocene) of San Diego Co., California. J. Paleontolo. 54: 460-471.

Schell, P. T., J. S. Parmerlee, Jr., and R. Powell. 1993a. *Amieva chrysolaema*. Cat. Am. Amphib. Rept. 563: 1-6.

Schell, P. T., R. Powell, J. J. S. Parmerlee, A. Lathrop, and D. D. Smith. 1993b. Notes on the natural history of *Ameiva chrysolaema* (Saura: Teiidae), from Barahona, Dominican Republic. Copeia 1993: 859-862.

Schlegel, H. 1837. Essai sur la Physionomie des Serpens. Arnz, Leiden.

Schlegel, H. 1837-1844. Abbildungen neuer oder unvollständig bekannter Amphibien, nach der Natur oder dem Leben entworfen, herausgegeben und mit einem erläuternden Texte begleitet. Arnz & Comp., Düsseldorf.

Schmidt, K. P. 1919. Descriptions of new amphibians and reptiles from Santo Domingo and Navassa. Bull. Am. Museum Nat. Hist. 41: 519-525.

Schmidt, K. P. 1920. A new *Ameiva* from Nevis Island, British West Indies. Proc. Linn. Soc. New York 33: 1-8.

Schmidt, K. P. 1921a. Notes on the herpetology of Santo Domingo. Bull. Am. Museum Nat. Hist. 44: 7-20.

Schmidt, K. P. 1921b. The herpetology of Navassa Island. Bull. Am. Museum Nat. Hist. 44: 555-559.

Schmidt, K. P. 1926. The amphibians and reptiles of Mona Island, West Indies. Field Museum Nat. Hist. Ser. 12: 149-162.

Schmidt, K. P. 1928. Amphibians and land reptiles of Porto Rico, with a list of those reported from the Virgin Islands. New York Acad. Sci. 10: 1-160.

Schmidt, K. P. 1955. Herpetology, pp. 591-627 in A Century of Progress in the Natural Sciences 1853-1953. California Academy of Sciences, San Francisco.

Schoener, A., and T. W. Schoener. 1984. Experiments on dispersal: short-term flotation of insular anoles, with a review of similar abilities in other terrestrial animals. Oecologia 63: 289-294.

Schoener, T. W. 1968. The *Anolis* of Bimini: Resource partitioning in a complex fauna. Ecology 49: 704-726.

Schoener, T. W. 1975. Presence and absence of habitat shift in some widespread lizard species. Ecol. Monogr. 45: 233-258.

Schoener, T. W. 1983. Rate of species turnover decreases from lower to higher organisms: a review of the data. Oikos 41: 372-377.

Schoener, T. W., and G. C. Gorman. 1968. Some niche differences in three Lesser Antillean lizards of the genus *Anolis*. Ecology 49: 819-830.

Schoener, T. W., and A. Schoener. 1971a. Structural habitats of West Indian *Anolis* lizards I. Lowland Jamaica. Breviora 368: 1-53.

Schoener, T. W., and A. Schoener. 1971b. Structural habitats of West Indian *Anolis* lizards II. Puerto Rico uplands. Breviora 375: 1-39.

Schoener, T. W., and A. Schoener. 1980. Densities, sex ratios, and population structure in four species of Bahamian *Anolis* lizards. J. Anim. Ecol. 49: 19-53.

Schoener, T. W., and A. Schoener. 1983a. Distribution of vertebrates on some very small islands. II. Patterns in species number. J. Anim. Ecol. 52: 237-262.

Schoener, T. W., and A. Schoener. 1983b. On the voluntary departure of lizards from very small islands, pp. 491-498 in Advances in Herpetology and Evolutionary Biology: Essays in Honor of Ernest E. Williams, edited by A. G. J. Rhodin and K. Miyata. Museum Comp. Zool., Harvard Univ. Press., Cambridge, MA.

Schoener, T. W., and A. Schoener. 1983c. Time of extinction of a colonizing propagule of lizards increases with island area. Nature 302: 332-334.

Schoener, T. W., J. B. Slade, and C. H. Stinson. 1982. Diet and sexual dimorphism in the very catholic lizard genus, *Leiocephalus* of the Bahamas. Oecologia 61: 160-169.

Schoener, T. W., and D. A. Spiller. 1987. Effect of lizards on spider populations: manipulative reconstruction of a natural experiment. Science 236: 949-952.

Schoener, T. W., and C. A. Toft. 1983. Spider populations: extraordinary densities on islands without top predators. Science 219: 1353-1355.

Schreckenberg, S. M. G. 1956. The embryonic development of the thyroid gland in the frog, *Hyla brunnea*. Growth 20: 295-313.

Schreiber, M. C., R. Powell, J. S. Parmerlee, Jr., A. Lathrop, and D. D. Smith. 1993. Natural history of a small population of *Leiocephalus schreibersi* (Sauria: Tropiduridae) from altered habitat in the Dominican Republic. Florida Sci. 56: 82-90.

Schubert, A. 1993. Conservation of biological diversity in the Dominican Republic. Oryx 27: 115-121.

Schubert, A., and G. Santana. 1996. Conservation of the American crocodile (*Crocodylus acutus*) in the Dominican Republic, pp. 425-433 in Contributions to West Indian Herpetology: A Tribute to Albert Schwartz, edited by R. Powell, and R. W. Henderson. Society for the Study of Amphibians and Reptiles Contributions in Herpetology, Ithaca, NY.

Schuchert, C. 1935. Historical Geology of the Antillean-Caribbean Region. Wiley, New York.

Schwaner, T. D., and H. C. Dessauer. 1981. Immunodiffusion evidence for the relationships of Papuan boids. J. Herpetol. 15: 250-253.

Schwartz, A. 1957. A new species of *Eleutherodactylus* (Amphibia: Leptodactylidae) from Cuba. Proc. Biol. Soc. Washington 70(23): 209-212.

Schwartz, A. 1959. A new species of toad, *Bufo cataulaciceps*, from the Isla de Pinos and Western Cuba. Proc. Biol. Soc. Washington 72(4): 109-120.

Schwartz, A. 1960. Nine new Cuban frogs of the genus *Eleutherodactylus*. Sci. Publ. Reading Public Museum and Art Gallery 11: 1-60.

Schwartz, A. 1961. A review of geckoes of the *Sphaerodactylus scaber* group of Cuba. Herpetologica 17: 19-26.

Schwartz, A. 1964 (1965). Two new species of *Eleutherodactylus* from the eastern Cordillera Central of the República Dominicana. Carib. J. Sci. 4: 473-484.

Schwartz, A. 1964a. Three new species of frogs (Leptodactylidae, *Eleutherodactylus*) from Hispaniola. Breviora 208: 1-15.

Schwartz, A. 1964b. *Diploglossus costatus* Cope (Sauria, Anguidae) and its relatives in Hispaniola. Sci. Publ. Reading Public Museum Art Gallery 13: 1-57.

Schwartz, A. 1964c. A review of *Sphaerodactylus vincenti* on the southern windward islands. Carib. J. Sci. 4: 391-409.

Schwartz, A. 1965a. Variation and natural history of *Eleutherodactylus ruthae* on Hispaniola. Bull. Museum Comp. Zool. 132: 481-508.

Schwartz, A. 1965b. Two new subspecies of the anguid lizard *Wetmorena* from Hispaniola. Proc. Biol. Soc. Washington 78: 39-48.

Schwartz, A. 1965c. Geographic variation in two species of Hispaniolan *Eleutherodactylus*, with notes on Cuban members of the *ricordi* group. Stud. Fauna Curaçao Carib. Isl. 22: 99-114.

Schwartz, A. 1965d. A new species of *Eleutherodactylus* (Amphibia: Leptodactylidae) from the Sierra de Baoruco, República Dominicana. Proc. Biol. Soc. Washington 78: 165-168.

Schwartz, A. 1965e. The *Ameiva* (Reptilia: Teiidae) of Hispaniola. I. *Ameiva lineolata* Duméril and Bibron. Carib. J. Sci. 5: 45-57.

Schwartz, A. 1965f. A new subspecies of the gecko *Sphaerodactylus microlepis*. Herpetologica 21: 261-269.

Schwartz, A. 1966a. The relationships of four small Hispaniolan *Eleutherodactylus* (Leptodactylidae). Bull. Museum Comp. Zool. 133: 371-399.

Schwartz, A. 1966b. The *Leiocephalus* (Lacertilia, Iguanidae) of Hispaniola. I. *Leiocephalus melanochlorus* Cope. J. Ohio Herpetol. Soc. 5: 39-48.

Schwartz, A. 1966c. Snakes of the genus *Alsophis* in Puerto Rico and the Virgin Islands. Studies Fauna Curaçao Carib. Isl. 23: 175-227.

Schwartz, A. 1967a. The *Leiocephalus* (Lacertilia, Iguanidae) of Hispaniola. II. The *Leiocephalus personatus* complex. Tulane Stud. Zool. 14: 1-53.

Schwartz, A. 1967b. The *Ameiva* (Lacertilia, Teiidae) of Hispaniola. III. *Ameiva taeniura* Cope. Bull. Museum Comp. Zool. 135: 345-375.

Schwartz, A. 1967c. Frogs of the genus *Eleutherodactylus* in the Lesser Antilles. Stud. Fauna Curaçao Carib. Isl. 24: 1-62.

Schwartz, A. 1968a. Two new subspecies of *Ameiva* (Lacertilia, Teiidae) from Hispaniola. Herpetologica 24: 21-28.

Schwartz, A. 1968b. The *Leiocephalus* (Lacertilia, Iguanidae) of Hispaniola. III. *Leiocephalus schreibersi*, *L. semilineatus*, and *L. pratensis*. J. Herpetol. 1: 39-63.

Schwartz, A. 1968c. Geographic variation in the Hispaniolan frog *Eleutherodactylus wetmorei* Cochran. Breviora 290: 1-13.

Schwartz, A. 1968d. Geographic variation in *Anolis distichus* Cope (Lacertilia, Iguanidae) in the Bahama Islands and Hispaniola. Bull. Museum Comp. Zool. 137: 255-309.

Schwartz, A. 1969a. A review of the Hispaniolan lizard *Anolis coelestinus* Cope. Carib. J. Sci. 9: 33-38.

Schwartz, A. 1969b. Two new subspecies of *Leiocephalus* from Hispaniola. J. Herpetol. 3: 79-85.

Schwartz, A. 1969c. The Antillean *Eleutherodactylus* of the *auriculatus* group. Stud. Fauna Curaçao Carib. Isl. 30: 99-115.

Schwartz, A. 1970a. A new species of large *Diploglossus* (Sauria: Anguidae) from Hispaniola. Proc. Biol. Soc. Washington 82: 777-788.

Schwartz, A. 1970b. A new species of gecko (Gekkonidae, *Sphaerodactylus*) from Hispaniola. J. Herpetol. 4: 63-67.

Schwartz, A. 1970c. A systematic review of *Uromacer catesbyi* Schlegel (Serpentes, Colubridae). Tulane Stud. Zool. Bot. 16: 131-149.

Schwartz, A. 1970d. *Darlingtonia haetiana* (Serpentes, Colubridae): a new subspecies. Herpetologica 26: 324-331.

Schwartz, A. 1971a. A new species of bromeliad-inhabiting galliwasp (Sauria: Anguidae) from Jamaica. Breviora 371: 1-10.

Schwartz, A. 1971b. A new species of *Eleutherodactylus* (Amphibia, Leptodactylidae) from Hispaniola. Annu. Carnegie Museum 43: 25-31.

Schwartz, A. 1971c. A systematic account of the Hispaniolan snake genus *Hypsirhynchus*. Stud. Fauna Curaçao Carib. Isl. 35: 63-94.

Schwartz, A. 1971d. A new subspecies of *Leiocephalus personatus* (Sauria, Iguanidae). Herpetologica 27: 176-182.

Schwartz, A. 1971e. Two new subspecies of *Diploglossus costatus* from Hispaniola (Sauria, Anguidae). J. Herpetol. 5: 161-165.

Schwartz, A. 1972. The native toads (Anura, Bufonidae) of Hispaniola. J. Herpetol. 6: 217-231.

Schwartz, A. 1973a. A new subspecies of *Ameiva chrysolaema* (Sauria, Teiidae) from Haiti. Herpetologica 29: 101-105.

Schwartz, A. 1973b. A third species in the Hispaniolan *shrevei* group of *Sphaerodactylus* (Sauria, Gekkonidae). Proc. Biol. Soc. Washington 86: 35-40.

Schwartz, A. 1973c. Six new species of *Eleutherodactylus* (Anura, Leptodactylidae) from Hispaniola. J. Herpetol. 7: 249-273.

Schwartz, A. 1973d. A new species of montane *Anolis* (Sauria, Iguanidae) from Hispaniola. Ann. Carnegie Museum 44: 183-195.

Schwartz, A. 1974a. A new species of primitive *Anolis* (Sauria, Iguanidae) from the Sierra de Bahoruco, Hispaniola. Breviora 423: 1-19.

Schwartz, A. 1974b. An analysis of variation in the Hispaniolan giant anole, *Anolis ricordi* Duméril and Bibron. Bull. Museum Comp. Zool. 146: 89-146.

Schwartz, A. 1975a. New subspecies of *Sphaerodactylus copei* Steindachner (Sauria, Gekkonidae) from Hispaniola. Herpetologica 31: 1-18.

Schwartz, A. 1975b. Variation in the Antillean boid snake *Tropidophis haetianus* Cope. J. Herpetol. 9: 303-311.

Schwartz, A. 1975c. A new subspecies of *Anolis baleatus* Cope (Sauria: Iguanidae) from the República Dominicana. Florida Sci. 38: 30-35.

Schwartz, A. 1976a. Two new species of Hispaniolan *Eleutherodactylus* (Leptodactylidae). Herpetologica 32: 163-171.

Schwartz, A. 1976b. Variation and relationships of some Hispaniolan frogs (Leptodactylidae, *Eleutherodactylus*) of the *ricordi* group. Bull. Florida St. Museum 21: 1-46.

Schwartz, A. 1976c. A new species of *Sphaerodactylus* (Sauria, Gekkonidae) from the República Dominicana. Florida Sci. 39: 65-70.

Schwartz, A. 1976d. Variations in the Hispaniolan colubrid snake *Uromacer frenatus* Günther (Reptilia, Serpentes, Colubridae). J. Herpetol. 10: 319-327.

Schwartz, A. 1977a. A new subspecies of *Eleutherodactylus wetmorei* Cochran (Anura: Leptodactylidae) from northern Haiti. Herpetologica 33: 66-72.

Schwartz, A. 1977b. The geckoes (Sauria, Gekkonidae) of the genus *Sphaerodactylus* of the Dominican Península de Barahona, Hispaniola. Proc. Biol. Soc. Washington 90: 243-254.

Schwartz, A. 1977c. A new subspecies of *Anolis baleatus* (Sauria: Iguanidae) from Isla Saona, República Dominicana. Florida Sci. 40: 401-405.

Schwartz, A. 1978a. Some aspects of the herpetogeography of the West Indies, pp. 31-51 in Zoogeography in the Caribbean, The 1975 Leidy Medal Symposium, edited by F. B. Gill. Acad. Nat. Sci. Philadelphia Spec. Publ., Philadelphia.

Schwartz, A. 1978b. A new species of aquatic *Anolis* (Sauria, Iguanidae) from Hispaniola. Ann. Carnegie Museum 47: 261-279.

Schwartz, A. 1978c. The Hispaniolan *Anolis* (Reptilia, Lacertilia, Iguanidae) of the *hendersoni* complex. J. Herpetol. 12: 355-370.

Schwartz, A. 1979a. A new species of cybotoid anole (Sauria, Iguanidae) form Hispaniola. Breviora 451: 1-27.

Schwartz, A. 1979b. A new species of *Eleutherodactylus* (Amphibia, Anura, Leptodactylidae) from northwestern Haiti, Hispaniola. J. Herpetol. 13: 199-202.

Schwartz, A. 1979c. A new species of *Leiocephalus* (Reptilia: Iguanidae) from Hispaniola. Proc. Biol. Soc. Washington 92: 272-279.

Schwartz, A. 1979d. The status of *Uromacer frenatus* and *U. dorsalis* (Reptilia: Serpentes: Colubridae). Herpetologica 35: 207-216.

Schwartz, A. 1979e. The herpetofauna of Ile à Cabrit, Haiti, with the descriptions of two new subspecies. Herpetologica 35: 248-255.

Schwartz, A. 1979f. The status of Greater Antillean *Phyllodactylus* (Reptilia, Gekkonidae). J. Herpetol. 13: 419-426.

Schwartz, A. 1980a. Two new species of *Eleutherodactylus* (Amphibia, Anura, Leptodactylidae) from Hispaniola. Ann. Carnegie Museum 49: 103-112.

Schwartz, A. 1980b. The herpetogeography of Hispaniola, West Indies. Stud. Fauna Curaçao Carib. Isl. 61: 86-127.

Schwartz, A. 1980c. Variation in Hispaniolan *Anolis whitemani* Williams. J. Herpetol. 14: 399-406.

Schwartz, A. 1981. Variation in Hispaniola *Anolis olssoni* Schmidt (Reptilia, Sauria, Iguanidae). Milwaukee Public Museum Contrib. Biol. Geol. 47: 1-21.

Schwartz, A. 1983. The *difficilis* complex of *Sphaerodactylus* (Sauria, Gekkonidae) of Hispaniola. Pt. 1. *Sphaerodactylus difficilis*, *S. clenchi*, and *S. lazelli*. Bull. Carnegie Museum Nat. Hist. 22: 5-30.

Schwartz, A., and M. Carey. 1977. Systematics and evolution in the West Indian iguanid genus *Cyclura*. Stud. Fauna Curaçao Carib. Is. 53: 15-97.

Schwartz, A., and R. I. Crombie. 1975. A new species of the genus *Aristelliger* (Sauria: Gekkonidae) from the Caicos Islands. Proc. Biol. Soc. Washington 88: 305-314.

Schwartz, A., and D. C. Fowler. 1973. The Anura of Jamaica: a progress report. Stud. Fauna Curaçao Carib. Is. 43: 50-142.

Schwartz, A., and R. Franz. 1976. A new species of *Sphaerodactylus* (Sauria: Gekkonidae) from Hispaniola. Proc. Biol. Soc. Washington 88: 367-372.

Schwartz, A. and O. H. Garrido. 1972. The lizards of the *Anolis equestris* complex in Cuba. Stud. Fauna Curaçao and Carib. Is. 39(134): 1-86.

Schwartz, A., and O. H. Garrido. 1975. A reconsideration of some Cuban *Tropidophis* (Serpentes, Boidae). Poc. Biol. Soc. Washington 88: 77-90.

Schwartz, A., and O. H. Garrido. 1981a. Las salamanquitas cubanos del género *Sphaerodactylus* (Sauria: Gekkonidae). 1. El grupo *copei*. Poeyana 230: 1-27.

Schwartz, A., and O. H. Garrido. 1981b. A review of the Cuban members of the genus *Arrhyton* (Reptilia, Serpentes, Colubridae). Ann. Carnegie Museum Nat Hist. 50: 207-230.

Schwartz, A., and O. H. Garrido. 1985. The Cuban lizards of the genus *Sphaerodactylus* (Sauria, Gekkonidae). Milwaukee Public Museum Contrib. Biol. Geol. 62: 1-67.

Schwartz, A., and J. E. D. Graham. 1980. The *shrevei* group of Hispaniolan *Sphaerodactylus* (Reptilia, Gekkonidae). Tulane Stud. Zool. Bot. 22: 1-15.

Schwartz, A., J. E. D. Graham, and J. J. Duval. 1979. A new species of *Diploglossus* (Sauria: Anguidae) from Hispaniola. Proc. Biol. Soc. Washington 92: 1-9.

Schwartz, A., and R. W. Henderson. 1982. *Anolis cybotes* (Reptilia, Iguanidae): the eastern Hispaniolan populations. Milwaukee Public Museum Contrib. Biol. Geol. 49: 1-8.

Schwartz, A., and R. W. Henderson. 1985. A Guide to the Identification of the Reptiles and Amphibians of the West Indies Exclusive of Hispaniola. Milwaukee Public Museum, Milwaukee, WI.

Schwartz, A., and R. W. Henderson. 1988. West Indian amphibians and reptiles: a check-list. Milwaukee Public Museum Contrib. Biol. Geol. 74: 1-264.

Schwartz, A., and R. W. Henderson. 1991. Amphibians and Reptiles of the West Indies: Descriptions, Distributions, and Natural History. Univ. Florida Press, Gainesville, FL.

Schwartz, A., and S. J. Incháustegui. 1976. A new species of *Diploglossus* (Reptilia, Lacertilia, Anguidae) from Hispaniola. J. Herpetol. 10: 241-246.

Schwartz, A., and J. F. Jacobs. 1989. A new subspecies of *Celestus costatus* (Sauria: Anguidae) from the República Dominicana. J. Herpetol. 23: 193-195.

Schwartz, A., and R. F. Klinikowski. 1966. The *Ameiva* (Lacertilia, Teiidae) of Hispaniola. II. Geographic variation in *Ameiva chrysolaema* Cope. Bull. Museum Comp. Zool. 133: 427-487.

Schwartz, A., and R. J. Marsh. 1960. A review of the *pardalis-maculatus* complex of the boid genus *Tropidophis* of the West Indies. Bull. Museum Comp. Zool. 123: 49-84.

Schwartz, A., and C. J. McCoy. 1970. A systematic review of *Ameiva auberi* Cocteau (Reptilia, Teiidae) in Cuba and the Bahamas. Ann. Carnegie Museum 41: 43-168.

Schwartz, A., and D. A. Rossman. 1976. A review of the Hispaniolan colubrid snake genus *Ialtris*. Stud. Fauna Curaçao Carib. Isl. 50: 76-102.

Schwartz, A., and R. Thomas. 1964 (1965). Subspeciation in *Sphaerodactylus copei*. Q. J. Florida Acad. Sci. 27: 316-332.

Schwartz, A., and R. Thomas. 1965. The genus *Darlingtonia* (Serpentes) in Hispaniola, including a new subspecies from the Dominican Republic. Breviora 229: 1-10.

Schwartz, A., and R. Thomas. 1975. A check-list of West Indian amphibians and reptiles. Carnegie Museum Nat. Hist. Spec. Publ. 1: 1-216.

Schwartz, A., and R. Thomas. 1977. Two new species of *Sphaerodactylus* (Reptilia, Lacertilia, Gekkonidae) from Hispaniola. J. Herpetol. 11: 61-66.

Schwartz, A., and R. Thomas. 1983. The *difficilis* complex of *Sphaerodactylus* (Sauria, Gekkonidae) of Hispaniola. Bull. Carnegie Museum Nat. Hist. 22: 1-60.

Schwartz, A., R. Thomas, and L. D. Ober. 1978. First supplement to a checklist of West Indian amphibians and reptiles. Carnegie Museum Nat. Hist. Spec. Publ. 5: 1-35.

Schwenk, K. 1988. Comparative morphology of the lepidosaur tongue and its relevance to squamate phylogeny, pp. 569-598 in Phylogenetic Relationships of the Lizard Families: Essays Commemorating Charles L. Camp, edited by R. Estes and G. Pregill. Stanford Univ. Press, Stanford, CA.

Scott, N. J. 1976. The abundance and diversity of the herpetofauna of tropical forest litter. Biotropica 8: 41-58.

Secretariá de Estado de Agricultura/Departmento de Vida Silvestre (SEA/DVS). 1990. La Diversidad Biológica en la República Dominicana: Reporte Preparado por el Departamento de Vida Silvestre para el Servicio Alemán de Cooperación Social-Técnica y Fondo Mundial para la Naturaleza (WWF-US). Secretaría de Estado de Agricultura, SURENA/DVS, Santo Domingo, República Dominicana.

Seidel, M. E. 1988a. Revision of the West Indian emydid turtles. Novitates 2918: 1-41.

Seidel, M. E. 1988b. *Trachemys decussata*. Cat. Am. Amphib. Rept., 440: 1-3.

Seidel, M. E. 1990. Growth and population characteristics of the slider turtle, *Trachemys decussata* on Grand Cayman Island. J. Herpetol., 24(2): 191-196.

Seidel, M. E. 1996. Current status of biogeography of the West Indian turtles in the genus *Trachemys* (Emydidae), pp. 169-174 in Contributions to West Indian Herpetology: A Tribute to Albert Schwartz, edited by R. Powell, and R. W. Henderson. Society for the Study of Amphibians and Reptiles Contributions in Herpetology, Ithaca, NY.

Seidel, M. E., and M. D. Adkins. 1987. Biochemical comparisons among West Indian *Trachemys* (Emydidae: Testudines). Copeia 1897: 485-489.

Seidel, M. E., and S. I. Miranda. 1984. Status of the trachemyd turtles (Testudines: Emydidae) on Hispaniola. J. Herpetol. 18: 468-479.

Seifriz, W. 1943. The plant life of Cuba. Ecol. Monographs 13: 375-426.

Sharpton, V. L., K. Burke, A. Camargo-Zanoguera, S. A. Hall, S. Lee, L. Marin, G. Suarez-Reynoso, J. M. Quezada-Muneton, P. D. Spudis, and J. Urrutia-Fucugauchi. 1993. Chicxulub multiring impact basin: size and other characteristics derived from gravity analysis. Science 261: 1564-1567.

Shaw, G. 1802. General Zoology or Systematic Natural History, 3: Amphibia. Part I & Part II. G. Kearsley, London.

Sheplan, B. R., and A. Schwartz. 1974. Hispaniolan boas of the genus *Epicrates* (Serpentes, Boidae) and their Antillean relationships. Ann. Carnegie Museum 45: 57-143.

Shochat, D., and H. C. Dessauer. 1981. Comparative immunological study of albumins of *Anolis* lizards of the Caribbean islands. Comp. Biochem. Physiol. 68A: 67-73.

Shreve, B. J. 1936. A new *Anolis* and new Amphibia from Haiti. Proc. New England Zoöl. Club 15: 93-99.

Shreve, B. J. 1968. The *notatus* group of *Sphaerodactylus* (Sauria: Gekkonidae) in Hispaniola. Breviora 280: 1-28.

Shreve, B. J., and E. E. Williams. 1963. The herpetology of the Port-au-Prince region and Gonave Island, Haiti. Part II. The frogs. Bull. Museum Comp. Zool. 129: 302-342.

Silva Lee, A. 1996. Cuba Natural. Natural Cuba. Pangaea, St. Paul.

Simberloff, D., and W. Boeklin. 1991. Patterns of extinction in the introduced Hawaiian avifauna: a reexamination of the role of competition. Am. Nat. 138: 300-327.

Simpson, G. G. 1956. Zoogeography of West Indian land mammals. Am. Museum Novit. 1759:1-28.

Singh, R. S., R. C. Lewontin, and A. A. Felton. 1976. Genic heterogeneity within electrophoretic "alleles" of xanthine dehydrogenase in *Drosophila pseudoobscura*. Genetics 84: 609-629.

Sites, J. W., S. K. Davis, T. Guerra, J. B. Iverson, and H. L. Snell. 1996. Character congruence and phylogenetic signal in molecular and morphological data sets: A case study in the living iguanas (Squamata, Iguanidae). Mol. Biol. Evol. 13: 1087-1105.

Sloane, H. 1725. A voyage to the islands Madera, Barbadoes, Nieves, St. Christophers, and Jamaica; with the natural history....of the last of these islands. 2: 335-336, London.

Slowinski, J. B. 1995. A phylogenetic analysis of the New World coral snakes (Elapidae: *Leptomicrurus*, *Micruroides*, and *Micrurus*) based on allozymic and morphological characters. J. Herpetol. 29: 325-338.

Smith, D. D., and R. Powell. 1991. Geographic distribution: *Uromacer frenatus*. Herpetol. Rev. 22: 135-136.

Smith, E. E. 1954. The forests of Cuba. Maria Moors Cabot Foundation Publ. 2: 1-98.

Smith, H. M., and C. Grant. 1958. The proper name of some Cuban snakes: an analysis of dates of publication of Ramon de la Sagra's Historia Natural de Cuba, and of Fitzinger's Systema Reptilium. Herpetologica 4: 215-222.

Smith, H. M., and E. H. Taylor. 1950. An annotated checklist and key to the reptiles of Mexico exclusive of the snakes. Bull. U.S. Natl. Museum 199: i-v + 1-253.

Smith, H. M., and V. Wallach. 1993. The date of publication of *Ialtris* Cope (Reptilia: Serpentes). Bull. Maryland Herpetol. Soc. 29: 80-84.

Smith, J. W., R. Powell, J. J. S. Parmerlee, D. D. Smith, and A. Lathrop. 1992. Natural history notes on a population of grass anoles, *Anolis olssoni* (Sauria: Polychrotidae), from the Dominican Republic. Bull. Maryland Herpetol. Soc. 30: 67-75.

Smith, M. H., D. O. Straney, and G. E. Drewery. 1981. Biochemical similarities among Puerto Rican *Eleutherodactylus*. Copeia 1981: 463-466.

Sonnini, C. S., and P.-A. Latreille. 1801. Histoire Naturelle des Reptiles. Lerrault, Paris.

Sparrman, A. 1784. *Lacerta sputator* och *Lacerta bimaculata*, tva nya Odlor fran America. Kongliga Svenska Vetenskaps-Akad Handl. 5: 164-167.

Speed, R. C. 1985. Cenozoic collision of the Lesser Antilles arc and continental South America and the origin of the El Pilar Fault. Tectonics 4: 41-69.

Speed, R. C., L. C. Gerhard, and E. H. McKee. 1979. Ages of deposition, deformation, and intrusion of Cretaceous rocks, eastern St. Croix, Virgin Islands. Bull. Geol. Soc. Am. I90: 629-632.

Spix, J. B. 1825. Animalia Nova sive Species Novae Lacertarum, quas in Itinere per Brasiliam Annis. F. S. Hübschmanni, Monachii.

Sproston, A. L., R. E. Glor, L. M. Hartley, E. J. Censky, R. Powell, and J. S. Parmerlee. 1998. Niche differences among three sympatric species of *Ameiva* (Reptilia: Teiidae) on Hispaniola. J. Herpetol. , in press.

Stamps, J. A. 1976. Rainfall, activity, and social behavior in the lizard *Anolis aeneus*. Anim. Behav. 24: 603-608.

Stamps, J. A. 1983. The relationship between ontogenetic habitat shifts, competition and predator avoidance in a juvenile lizard (*Anolis aeneus*). Behav. Ecol. Sociobiol. 12: 19-33.

Stamps, J. A. 1985. Arboreal habitat use and parachuting by a subtropical forest frog. J. Herpetol. 19: 391-401.

Stamps, J. A. 1987. The effect of familiarity with a neighborhood on territory acquisition. Behav. Ecol. Sociobiol. 21: 273-277.

Stamps, J. A. 1988. The effect of body size on habitat and territory choice in juvenile lizards. Herpetologica 44: 369-376.

Stamps, J. A. 1995. Climate driven population fluctuations in rain forest frogs. J. Herpetol. 29: 437-446.

Stamps, J. A., and G. E. Martin. 1980. Coconut husk-piles: a unique habitat for Jamaican terrestrial frogs. Biotropica 12: 107-117.

Stamps, J. A., and F. H. Pough. 1983. Population density of tropical forest frogs: relation to retreat sites. Science 221: 570-572.

Stamps, J. A., and A. S. Rand. 1991. Vocalizations and the defense of retreat sites by male and female frogs, *Eleutherodactylus coqui*. Copeia 1991: 1013-1024.

Stamps, J. A., and S. K. Tanaka. 1981. The relationship between food and social behavior in juvenile lizards (*Anolis aeneus*). Copeia 1981: 422-434.

Stamps, J. A., and L. L. Woolbright. 1998. Amphibians, in The Food Web of a Tropical Rain Forest, edited by D. P. Reagan, and R. B. Waide. Univ. Chicago Press, Chicago.

Steadman, D. W., G. K. Pregill, and S. L. Olsen. 1984a. Fossil vertebrates from Antigua, Lesser Antilles: Evidence for late Holocene human-caused extinctions in the West Indies. Proc. Natl. Acad. Sci. 81: 4448-4451.

Steadman, D. W., D. R. Watters, E. J. Reitz, and G. K. Pregill. 1984b. Vertebrates from archaeological sites on Montserrat, West Indies. Ann. Carnegie Museum 53(1):1-29.

Stearns, S. C. 1992. The evolution of life histories. Oxford Univ. Press, Oxford and New York.

Steers, J. A. 1940a. The cays and the Palisadoes, Port Royal, Jamaica. Geogr. Rev. 30: 279-296.

Steers, J. A. 1940b. The coral cays of Jamaica. Geogr. J. 95: 30-42.

Steers, J. A., V. J. Chapman, J. Colman, and J. A. Lofthouse. 1940. Sand cays and mangroves in Jamaica. Geogr. J. 96: 305-328.

Stehli, F. G., and S. D. Webb (Editors). 1985. The Great American Biotic Interchange. Plenum, New York.

Steindachner, F. 1863. Über eine neue *Epicrates*-Art aus Columbien. Densksch. Kaiserl. Akad. Wiss. Wien, math.-nat. cl. 22: 89-93.

Steindachner, F. 1867. Reptilien, pp. 1-98 in Reise der Österreichischen Fregatte Novara um die Erde in den Jahren 1857, 1858, 1859 unter den Befehlen des Commodore B. von Wüllerstorf-Urbair. K. Gerold's Sohn, Wien.

Steindachner, F. 1917. Cuban amphibians and reptiles collected for the United States National Museum from 1899 to 1902. Proc. U.S. Natl. Museum 53: 259-291.

Stejneger, L. 1899. A new name for the Great Crested Lizard of Jamaica. Am. Nat. 33: 601-602.

Stejneger, L. 1901. A new systematic name for the Yellow boa of Jamaica. Am. Nat. 23: 469-470.

Stejneger, L. 1904. The herpetology of Porto Rico. Rep. U.S. Nat. Museum 1902: 551-724.

Stejneger, L. 1916. Notes on amphisbaenid nomenclature. Proc. Biol. Soc. Washington 29: 86.

Stewart, M. M. 1979. The role of introduced species in a Jamaican frog community. Act. Symp. Int. Ecol. Trop. 1: 111-146.

Stewart, M. M., 1985. Arboreal habitat use and parachuting by a subtropical forest frog. J. Herpetol. 19(3): 391-401.

Stewart, M. M., and G. E. Martin. 1980. Coconut-husk piles – a unique habitat for Jamaican terrestrial frogs. Biotropica 12: 107-116.

Stewart, M. M. and F. H. Pough. 1983. Population density of tropical forest frogs: relation to retreat sites. Science 221: 570-572.

Stewart, M. M., and A. S. Rand. 1991. Vocalization and the defense retreat sites by male and female frogs, *Eleutherodactylus coqui*. Copeia 1991: 1013-1024.

Stewart, M. M., and L. L. Woolbright. 1996. Amphibians. in The Food Web of a Tropical Rain Forest, edited by D. P. Reagan, and R. B. Waide. Univ. Chicago Press, Chicago.

Stinnesbeck, W., G. Keller, J. de la Cruz, C. de León, N. MacLeod, and J.E. Whittaker. 1997. The Cretaceous-Tertiary transition in Guatemala: limestone breccia deposits from the South Petén basin. Geol. Rundsch 86: 686-709.

Stoudemire, S. A. 1959. Natural History of the West Indies, by Gonzalo F. de Oviedo. Univ. North Carolina Press, Chapel Hill.

Strahm, M. H., and A. Schwartz. 1977. Osteoderms in the anguid lizard subfamily Diploglossinae and their taxonomic importance. Biotropica 9: 58-72.

Stuart, L. C. 1941. Studies on the neotropical Colubrinae. VIII. A revision of the genus *Dryadophis* Stuart, 1939. Misc. Publ. Museum Zool. Univ. Michigan 49: 1-106.

Stull, O. G. 1928. A revision of the genus *Tropidophis*. Occas. Pap. Museum Zool. Univ. Michigan 195: 1-49.

Stull, O. G. 1935. A checklist of the family Boidae. Proc. Boston Soc. Nat. Hist. 40: 387-408.

Swofford, D. L. 1993. PAUP: phylogenetic analysis using parsimony, version 3.1.1 Computer program distributed by Illinios Natural History Survey, Champaign, Illinois.

Taigen, T. L., F. H. Pough, and M. M. Stewart. 1984. Water balance of terrestrial anuran (*Eleutherodactylus coqui*) eggs: importance of parental care. Ecology 65: 248-255.

Taylor, C. E., and G. C. Gorman. 1975. Population genetics of a "colonizing" lizard: natural selection for allozyme morphs in *Anolis grahami*. Heredity 35: 241-247.

Taylor, E. H. 1952. A new Panamanian tree frog. Breviora l: 1-3.

Thomas, R. 1964. The races of *Sphaerodactylus fantasticus* Duméril and Bibron in the Lesser Antilles. Carib. J. Sci. 4: 373-390.

Thomas, R. 1965a. Two new subspecies of *Amphisbaena* (Amphisbaenia, Reptilia) from the Barahona Peninsula of Hispaniola. Breviora 215: 1-14.

Thomas, R. 1965b. The genus *Leptotyphlops* in the West Indies with description of a new species from Hispaniola (Serpentes, Leptotyphlopidae). Breviora 222: 1-12.

Thomas, R. 1965c. A new species of *Typhlops* from the Barahona Peninsula of Hispaniola. Copeia 1965: 436-439.

Thomas, R. 1965d. The smaller teiid lizards (*Gymnophthalmus* and *Bachia*) of the southeastern Caribbean. Proc. Biol. Soc. Washington 78: 141-154.

Thomas, R. 1966a. A reassessment of the herpetofauna of Navassa Island. J. Ohio Herpetol. Soc. 5: 73-89.

Thomas, R. 1966b. A new Hispaniolan gecko. Breviora 253: 1-5.

Thomas, R. 1966c. Leeward Island *Typhlops* (Reptilia, Serpentes). Proc. Biol. Soc. Washington 79: 255-266.

Thomas, R. 1966d. Additional notes on the amphisbaenids of greater Puerto Rico. Breviora 249: 1-23.

Thomas, R. 1968. Notes on Antillean geckos (*Sphaerodactylus*). Herpetologica 24: 46-60.

Thomas, R. 1971. A new species of *Diploglossus* (Sauria: Anguidae) from Hispaniola. Occas. Pap. Museum Zool. Louisiana State Univ. 40: 1-9.

Thomas, R. 1974a. A new species of *Typhlops* (Serpentes: Typhlopidae) from Hispaniola. Proc. Biol. Soc. Washington 87: 11-18.

Thomas, R. 1974b. A new species of Lesser Antillean *Typhlops* (Serpentes: Typhlopidae). Occ. Pap. Museum Zool. Louisiana State Univ. 46: 1-9.

Thomas, R. 1975. The *argus* group of West Indian *Sphaerodactylus* (Sauria: Gekkonidae). Herpetologica 31: 177-195.

Thomas, R. 1976. Systematics of Antillean blind snakes of the genus *Typhlops* (Serpentes: Typhlopidae). Ph.D. dissertation, Louisiana State Univ., Baton Rouge.

Thomas, R. 1982. A new dwarf *Sphaerodactylus* from Haiti (Lacertilia: Gekkonidae). Proc. Biol. Soc. Washington 95: 81-88.

Thomas, R. 1989. The relationships of Antillean *Typhlops* (Serpentes: Typhlopidae) and the description of three new Hispaniolan species, pp. 409-432 in Biogeography of the West Indies: Past, Present, and Future, edited by C. A. Woods. Sandhill Crane Press, Gainesville, FL.

Thomas, R. 1996. Recollections of Albert Schwartz, pp. 16-18 in Contributions to West Indian Herpetology: A Tribute to Albert Schwartz, edited by R. Powell, and R. W. Henderson. Society for the Study of Amphibians and Reptiles Contributions in Herpetology, Ithaca, NY.

Thomas, R., and A. Gaa. 1998. The non-anoline reptiles of the El Verde forest, in The Food Web of a Tropical Rain Forest, edited by R. P. Reagan, and R. B. Waide. Univ. Chicago Press, Chicago.

Thomas, R., and S. B. Hedges. 1988. Two new geckos (*Sphaerodactylus*) from the Sierra Martin Garcia of Hispaniola. Herpetologica 44: 96-104.

Thomas, R., and S. B. Hedges. 1989. A new *Celestus* (Sauria: Anguidae) from the Chaine de la Selle of Haiti. Copeia 1989: 886-891.

Thomas, R., and S. B. Hedges. 1992. An unusual new *Sphaerodactylus* from Hispaniola (Squamata: Gekkonidae). J. Herpetol. 26: 289-292.

Thomas, R., and S. B. Hedges. 1993. A new banded *Sphaerodactylus* from eastern Hispaniola (Squamata: Gekkonidae). Herpetologica 49: 350-354.

Thomas, R., S. B. Hedges, and O. H. Garrido. 1992. Two new species of *Sphaerodactylus* from eastern Cuba (Squamata: Gekkonidae). Herpetologica 48: 358-367.

Thomas, R., and R. Joglar. 1996. The herpetology of Puerto Rico: Past, present and future. Annals New York Acad. Sci. 776: 181-200.

Thomas, R. and A. G. Kessler. 1996. Non-anoline reptiles. The Food Web of a Tropical Rain Forest, edited by D. P. Reagan, and R. B. Waide. Univ. Chicago Press, Chicago.

Thomas, R., R. W. McDiarmid, and F. G. Thompson. 1985. Three new species of thread snakes (Serpentes: Leptotyphlopidae) from Hispaniola. Proc. Biol. Soc. Washington 98: 204-220.

Thomas, R., and J. A. Prieto-Hernández. 1985. The use of venom by the Puerto Rican snake, *Alsophis portoricensis*. Decim. Simp. Recur. Nat. 1983: 13-22a.

Thomas, R., and A. Schwartz. 1965. Hispaniolan snakes of the genus *Dromicus* (Colubridae). Rev. Biol. Tropics. 13: 59-83.

Thomas, R., and A. Schwartz. 1966. *Sphaerodactylus* (Gekkonidae) in the Greater Puerto Rico region. Bull. Florida State Museum 10: 193-260.

Thomas, R., and A. Schwartz. 1967. The *monticola* group of the lizard genus *Anolis* in Hispaniola. Breviora 261: 1-27.

Thomas, R., and A. Schwartz. 1974. The status of *Sphaerodactylus gilvitorques* Cope and of *Sphaerodactylus nigropunctatus* Gray (Sauria: Gekkonidae). J. Herpetol. 8: 353-358.

Thomas, R., and A. Schwartz. 1977. Three new species of *Sphaerodactylus* (Sauria: Gekkonidae) from Hispaniola. Ann. Carnegie Museum 46: 33-43.

Thomas, R., and A. Schwartz. 1983a. The *difficilis* complex of *Sphaerodactylus* (Sauria, Gekkonidae) of Hispaniola. Pt. 2. *Sphaerodactylus savagei, S. cochranae, S. darlingtoni, S. armstrongi, S. streptophorus*, and conclusions. Bull. Carnegie Museum Nat. Hist. 22: 31-60.

Thomas, R., and A. Schwartz. 1983b. Variation in Hispaniolan *Sphaerodactylus* (Sauria, Gekkonidae), pp. 86-98 in Advances in Herpetology and Evolutionary Biology. Essays in Honor of Ernest E. Williams, edited by A. G. J. Rhodin, and K. Miyata. Museum Comp. Zool., Cambridge, MA.

Thompson, R. L. 1996. Larval habitat, ecology, and parental investment of *Osteopilus brunneus* (Hylidae), pp. 259-269 in Contributions to West Indian Herpetology: a Tribute to Albert Schwartz, edited by R. Powell, and R. W. Henderson. Society for the Study of Amphibians and Reptiles Contributions in Herpetology, Ithaca, NY.

Thorbjarnarason, J. B. 1988. The status and ecology of the American crocodile in Haiti. Bull. Florida State Museum, Biol. Sci. 33: 1-86.

Tihen, J. A. 1962. Osteological observations on New World *Bufo*. Am. Midl. Natur. 67:157-183.

Toft, C. A. 1980. Seasonal variation in populations of Panamanian litter frogs and their prey: a comparison of wetter and drier sites. Oecologia 47: 34-36.

Tolson, P. J. 1987. Phylogenetics of the boid snake genus *Epicrates* and Caribbean vicariance theory. Occ. Papers. Museum Zool., Univ. Michigan 715: 1-68.

Tolson, P. J. 1988. Critical habitat, predator pressures, and the management of *Epicrates monensis* (Serpentes: Boidae) on the Puerto Rico Bank: a multivariate analysis, pp. 228-238 in Management of Amphibians, Reptiles, and Small Mammals in North America, Gen. Tech. Rept. U. S. Dept. of Agriculture, Washington, DC.

Tolson, P. J. 1996. Conservation of *Epicrates monensis* on the satellite islands of Puerto Rico, pp. 407-416 in Contributions to West Indian Herpetology: A Tribute to Albert Schwartz, edited by R. Powell, and R. W. Henderson. Society for the Study of Amphibians and Reptiles Contributions in Herpetology, Ithaca, NY.

Tolson, P. J., and R. W. Henderson. 1993. The Natural History of West Indian Boas. R&A, Somerset, UK.

Townsend, D. S., and M. M. Stewart. 1986. The effect of temperature on development in a terrestrial neotropical *Eleutherodactylus*. Copeia 1986: 520-523.

Townsend, D. S. 1996. Patterns of parental care in frogs of the genus *Eleutherodactylus*, pp. 229-239 in Contributions to West Indian Herpetology: a Tribute to Albert Schwartz, edited by R. Powell, and R. W. Henderson. Society for the Study of Amphibians and Reptiles Contributions in Herpetology, Ithaca, NY.

Trivers, R. L. 1976. Sexual selection and resource-accruing abilities in *Anolis garmani*. Evolution 30: 253-269.

Trueb, L. 1970. Evolutionary relationships of casque-headed tree frogs with co-ossified skulls (family Hylidae). Univ. Kansas Pub. Museum Nat. Hist. 18: 547-716.

Trueb, L. 1972. *Hyla crucialis* Harlan, 1826 (Amphibia): proposed suppression under the plenary powers Z.N.(S.) 1982. Bull. Zool. Nomencl. 29: 39-40.

Trueb, L., and M. J. Tyler. 1974. Systematics and evolution of the Greater Antillean hylid frogs. Occas. Pap. Museum Nat. Hist., Univ. Kansas 24: 1-60.

Tschudi, J. J. 1838. Classification der Batrachier, mit Bercksichtigung der fossilen Thiere dieser Abtheilung der Reptilien. Mems. Soc. Sci. Nat. Neuchtel 2: 1-100.

Tuck, R. G., and J. D. Hardy. 1973. Status of the Ober Tobago collection, Smithsonian Institution, and the proper allocation of *Amieva suranamensis tobaganus* Cope (Sauria: Teiidae). Proc. Biol. Soc. Washington 86: 231-242.

Turner, F. B. 1960. Estimation of a Louisiana population of the skink, *Lygosoma laterale*. Ecology 41: 574-577.

Turner, F. B. 1977. The dynamics of populations of squamates, crocodilians, and rhynchocephalians, pp. 157-264 in Ecology and Behaviour, edited by C. Gans, and D. W. Tinkle. Academic Press, New York.

Turner, R. D. 1960. Land shells of Navassa Island, West Indies. Bull. Museum Comp. Zool. 122: 233-244.

Underwood, G. 1949. Notes on a hitherto unpublished Gosse illustration of Jamaican herpetology. Nat. Hist. Notes Nat. Hist. Soc. Jamaica 4: 46-48.

Underwood, G. 1950. Reptiles at Clydesdale. Nat. Hist. Notes Nat. Hist. Soc. Jamaica 4: 84.

Underwood, G. 1951. Reptilian retinas. Nature 167: 209-213.

Underwood, G. 1951-52. Introduction to the study of Jamaican reptiles (parts 1-10). Nat. Hist. Notes Nat. Hist. Soc. Jamaica 4-5: 104-107, 127-141, 144-149, 161-163, 209-213, 253-257, 15-21, 54-65, 77-82, 97-105.

Underwood, G. 1954a. Categories of adaptation. Evolution 8: 365-379.

Underwood, G. 1954b. On the classification and evolution of geckos. Proc. Zool. Soc. London 124: 469-492.

Underwood, G. 1954c. The biological resources of Jamaica (Presidential Address). Nat. Hist. Notes Nat. Hist. Soc. Jamaica 6: 163-165.

Underwood, G. 1954d. A Cuban pond-turtle in Jamaica. Nat. Hist. Notes Nat. Hist. Soc. Jamaica 6: 120.

Underwood, G. 1954e. The distribution of Antillean reptiles. Nat. Hist. Notes, Nat. Hist. Soc. Jamaica 6: 121-129.

Underwood, G. 1959a. A new Jamaican galliwasp (Sauria, Anguidae). Breviora 102: 1-13.

Underwood, G. 1959b. The anoles of the eastern Caribbean, Part III. Revisionary notes. Bull. Museum Comp. Zool. 121: 191-226.

Underwood, G. 1962. Reptiles of the eastern Caribbean. Carib. Affairs, New Series 1: 1-192.

Underwood, G. 1964. An anguid lizard from the Leeward islands. Breviora 200: 1-10.

Underwood, G. 1976. A systematic analysis of boid snakes, pp. 151-175 in Morphology and Biology of Reptiles, edited by Ad'A. Bellairs, and C. B. Cox. Linnaen Society Symposium Series 3, London.

Underwood, G. 1977. Simplification and degeneration in the course of evolution of Squamate reptiles. Colloq. Int. C.N.R.S. 266: 341-352.

Underwood, G. 1993. A new snake from St. Lucia, West Indies. Bull. Nat. Hist. Museum (Zool.) 59: 1-9.

Underwood, G., and E. E. Williams. 1959. The anoline lizards of Jamaica. Bull. Inst. of Jamaica. Sci. Ser. 9: 1-48.

Valdés, A., and F. Ruíz. 1980. Consideracione sistemáticas sobre *Bufo longinasus* (Anura: Bufonidae) de Cuba y descripción de una nueva subespecie. Poeyana 206: 1-33.

Valdés de la Osa, A., and F. N. R. García. 1980. Consideraciones sistemáticas sobre *Bufo longinasus* (Anura: Bufonidae) de Cuba y descripción de una nueva subespecie. Poeyana 206: 1-33.

Valdez Sierra, G., and J. M. M. Féliz. 1992. Sistema de Areas Protegidas de República Dominicana. Dir. Nac. Parques, Santo Domingo.

Vane-Wright, R. I., C. J. Humphries, and P. H. Williams. 1991. What to protect? - Systematics and the agony of choice. Biol. Conserv. 55: 235-254.

Vanzolini, P. E. 1951. A systematic arrangement of the family Amphisbaenidae (Sauria). Herpetologica 7: 113-123.

Vanzolini, P. E. 1961. *Bachia*: espécies brasileiras e conceito genérico (Sauria, Teiidae). Pap. Avulsos S. Paulo 14: 193-209.

Vanzolini, P. E. 1968. Geography of the South American Gekkonidae (Sauria). Archos Zool. Est. S. Paulo 17:85-112.

Vanzolini, P. E. 1978. On South American *Hemidactylus* (Sauria, Gekkonidae). Pap. Avul. Zool. 31: 307-343.

Vanzolini, P. E. 1990. Geographic Distribution: *Gymnophthalmus underwoodi*. Herpetol. Review 21: 96.

Vanzolini, P. E., and E. E. Williams. 1962. Jamaican and Hispaniolan *Gonatodes* and allied forms (Sauria, Gekkonidae). Bull. Museum Comp. Zool. 127: 479-498.

Varona, L. S. 1966. Notas sobre los cocodrilos de Cuba y descripción de una nueva especie del pleistoceno. Poeyana (serie A) 16: 1-34.

Varona, L. S. 1976. *Caiman crocodilus* (Reptilia: Alligatoridae) en Cuba. Misc. Zool. 5: 2.

Villa, J. 1984. The venomous snakes of Nicaragua. Milwaukee Public Museum Contr. Biol. Geol. 59: 1-41.

Villa, J., L. D. Wilson, and J. D. Johnson. 1988. Middle American Herpetology: A Bibliographic Checklist. Univ. Missouri Press, Columbia.

Vitt, L. J., and C. M. de Carvalho. 1995. Niche partitioning in a tropical wet season: lizards in the Lavrado area of northern Brazil. Copeia 1995: 305-329.

Vitt, L. J., and L. D. Vangilder. 1983. Ecology of a snake community in northeastern Brazil. Amphib.-Rept. 4: 273-296.

Vitt, L. J., and P. A. Zani. 1996a. Ecology of the lizard *Ameiva festiva* (Teiidae) in southeastern Nicaragua. J. Herpetol. 30: 110-117.

Vitt, L. J., and P. A. Zani. 1996b. Ecology of the South American lizard *Norops chrysolepis* (Polychrotidae). Copeia 1996: 56-68.

Vogel, P. 1983. On an ant-lizard interaction. Oecologia 58: 269-271.

Vogel, P. 1984. Seasonal hatchling recruitment and juvenile growth of the lizard *Anolis lineatopus*. Copeia 1984: 747-757.

Vogel, P., and F. v. Brockhusen-Holzer. 1984. Ants as prey of juvenile *Anolis lineatopus* (Rept., Iguanidae) in prey choice experiments. Zeitschr. Tierpsychol. 65: 66-76.

Vogel, P., and E. Curio. 1983. Response of a predator to harmful and similar harmless prey. Verh. Deutsch. Zool. Gesell. 247: 247.

Vogel, P., W. Hettrich, and K. Ricono. 1986. Weight growth of juvenile lizards, *Anolis lineatopus*, maintained on different diets. J. Herpetol. 20: 50-58.

Vogel, P., R. Nelson, and R. Kerr. 1996. Conservation strategy for the Jamaican iguana, *Cyclura collei*, pp. 395-406 in Contributions to West Indian Herpetology: A Tribute to Albert Schwartz, edited by R. Powell, and R. W. Henderson. Society for the Study of Amphibians and Reptiles Contributions in Herpetology, Ithaca, NY.

Wagler, J. G. 1830. Natürliches System der Amphibien, mit vorangehender Classification der Säugethiere und Vögel. Ein Beitrag zur vergleichenden Zoologie. J. G. Cotta, München.

Wake, D. B. 1987. Adaptive radiation of salamanders in Middle American cloud forests. Ann. Missouri Bot. Gard. 74: 242-264.

Wake, D. B., and P. Elias. 1983. New genera and a new species of Central American salamanders, with a review of the tropical genera (Amphibia, Caudata, Plethodontidae). Contrib. Sci. Nat. Hist. Museum Los Angeles County 345: 1-19.

Wake, D. B., and J. D. Johnson. 1989. A new genus and species of plethodontid salamander from Chiapas, Mexico. Contrib. Sci. Nat. Hist. Museum Los Angeles County 411: 1-10.

Wake, D. B., and J. F. Lynch. 1976. The distribution, ecology, and evolutionary history of plethodontid salamanders in tropical America. Sci. Bull. Nat. Hist. Museum Los Angeles County 25: 1-65.

Ward, W. C., A. E. Weidie, and W. Back. 1985. Geology and hydrogeology of the Yucatan and Quaternary geology of northeastern Yucatan Peninsula. New Orleans Geol. Soc. 1-19.

Weil, T. E., J. K. Black, H. I. Blutstein, K. T. Johnston, D. S. McMorris, and F. P. Munson. 1982. Dominican Republic: A Country Study. 2nd ed. Depsrtment of the Army, Washington, DC.

Weinland, D. F. 1862. Beschreibung und Abbildung von drei neuen Sauriern. (*Embryopus habichii* und *Amphisbaena innocens* von Haiti, und *Brachymeles leuckartii* von Neuholland.). Abh. Senckenb. Naturf. Ges. 4: 131-143.

Wermuth, H., and R. Mertens. 1961. Schildkröten, Krokodile, Brückenechsen. Fischer-Verlag, Jena, Germany.

Wermuth, H., and R. Mertens. 1977. Liste der rezenten Amphibien und Reptilien: Testudines, Crocodylia, Rhynchocephalia. Das Tierreich, Berlin.

Werner, F. 1909. Ueber neue oder seltene Reptilien des naturhistorischen Museums in Hamburg. 1. Schlangen. Jahrb. Wiss. Anst. Hamburg (2. Beiheft) 26: 205-247.

Westermann, J. H. 1953. Nature preservation in the Caribbean. A review of literature on the destruction and preservation of flora and fauna in the Caribbean area. Foundation Sci. Res. Surinam Netherland Antilles 9: 1-106.

Wetmore, A., and B. H. Swales. 1931. The birds of Haiti and the Dominican Republic. Bull. U.S. Natl. Museum 155: 1-483.

White, L. R., R. Powell, J. J. S. Parmerlee, A. Lathrop, and D. D. Smith. 1992. Food habits of three syntopic reptiles from the Barahona Peninsula, Hispaniola. J. Herpetol. 26: 518-520.

Wiens, J. J. 1995. Polymorphic characters in phylogenetic systematics. Syst. Biol. 44: 482-500.

Wijffels, L. C. M. 1997. *Anolis bimaculatus* als huisgenoot. Lacerta 55(5): 201-205.

Wiley, E. O. 1978. The evolutionary species concept reconsidered. Syst. Zool. 27: 17-26.

Willet, B. M. 1987. Philips' Certificate atlas for the Caribbean. Philip, London.

Williams, E. E. 1956. *Pseudemys scripta callirostris* from Venezuela with a general survey of the *scripta* series. Bull. Museum Comp. Zool. 115: 145-160.

Williams, E. E. 1959. Preface. The anoles of the eastern Caribbean. Bull. Museum comp. Zool. 121: 187-189.

Williams, E. E. 1960. Notes on Hispaniolan herpetology. 1. *Anolis christophei*, new species, from the Citadel of King Christophe, Haiti. Breviora 117: 1-7.

Williams, E. E. 1961. Notes on Hispaniolan herpetology 3. The evolution and relationships of the *Anolis semilineatus* group. Breviora 136: 1-8.

Williams, E. E. 1962a. Notes on Hispaniolan herpetology 6. The giant anoles. Breviora 155: 1-15.

Williams, E. E. 1962b. Notes on the herpetology of Hispaniola 7. New material of two poorly known anoles: *Anolis monticola* Shreve and *Anolis christophei* Williams. Breviora 164: 1-11.

Williams, E. E. 1962c. The anoles of the northern Leewards, Anguilla to Montserrat: new data and a new species. Bull. Museum Comp. Zool. 127: 453-465.

Williams, E. E. 1963a. Notes on Hispaniolan herpetology 8. The forms related to *Anolis hendersoni* Cochran. Breviora 186: 1-13.

Williams, E. E. 1963b. *Anolis whitemani*, new species from Hispaniola (Sauria, Iguanidae). Breviora 197: 1-8.

Williams, E. E. 1965a. The species of Hispaniolan green anoles (Sauria, Iguanidae). Breviora 227: 1-16.

Williams, E. E. 1965b. Hispaniolan giant anoles (Sauria, Iguanidae): new data and a new subspecies. Breviora 232: 1-7.

Williams, E. E. 1969. The ecology of colonization as seen in the zoogeography of anoline lizards on small islands. Q. Rev. Biol. 44: 345-389.

Williams, E. E. 1972. The origin of faunas. Evolution of lizard congeners in a complex island fauna: a trial analysis, pp. 47-89 in Evolutionary Biology, edited by T. Dobzhansky, M. K. Hecht, and W. C. Steere. Appleton-Century-Crofts, New York.

Williams, E. E. 1975. *Anolis marcanoi* new species: sibling to *Anolis cybotes*: description and field evidence. Breviora 430: 1-9.

Williams, E. E. 1976a. West Indian anoles: a taxonomic and evolutionary summary 1. Introduction and a species list. Breviora 440: 1-21.

Williams, E. E. 1976b. South American anoles: the species groups. Papeis Avulsos Zool., S. Paulo 29(26): 259-268.

Williams, E. E. 1977. The macrosystematics of the anoles, pp. 122-131 in The Third Anolis Newsletter, edited by E. E. Williams. Museum Comp. Zool., Cambridge, MA.

Williams, E. E. 1983. Ecomorphs, faunas, island size, and diverse end points in island radiations of *Anolis*, pp. 326-370 in Lizard Ecology, edited by R. B. Huey, E. R. Pianka, and T. W. Schoener. Harvard Univ. Press, Cambridge, MA.

Williams, E. E. 1989a. A critique of Guyer and Savage (1986): cladistic relationships among anoles (Sauria: Iguanidae): are the data available to reclassify the anoles?, pp. 433-478 in Biogeography of the West Indies: Past, Present, and Future, edited by C. A. Woods. Sandhill Crane Press, Gainesville, FL.

Williams, E. E. 1989b. Old problems and new opportunities in West Indian biogeography, pp. 1-46 in Biogeography of the West Indies: Past, Present, and Future, edited by C. A. Woods. Sandhill Crane Press, Inc., Gainesville, FL.

Williams, E. E., and S. M. Case. 1986. Interactions among members of the *Anolis distichus* complex in and near the Sierra de Baoruco, Dominican Republic. J. Herpetol. 20: 535-546.

Williams, E. E., and A. S. Rand. 1961. Notes on Hispaniolan herpetology 2. A review of the *Anolis semilineatus* group with the description of *Anolis cochranae*, new species. Breviora 135: 1-11.

Williams, E. E., and A. S. Rand. 1969. *Anolis insolitus*, a new dwarf anole of zoogeographic importance from the mountains of the Dominican Republic. Breviora 326: 1-21.

Williams, E. E., and T. P. Webster. 1974. *Anolis rupinae* new species a syntopic sibling of *A. monticola* Shreve. Breviora 429: 1-22.

Williamson, M. 1981. Island populations. Oxford Univ. Press.

Wilson, E. O. 1961. The nature of the taxon cycle in the Melanesian ant fauna. Am. Nat. 95: 169-193.

Wilson, E. O. 1991. Ants. Wings 1991: 4-13.

Wilson, L. D., J. R. McCranie, and G. A. Cruz. 1994. A new species of *Plectrohyla* (Anura: Hylidae) from a premontane rainforest in northern Honduras. Proc. Biol. Soc. Washington 107: 67-78.

Wilson, L. D., and J. R. Meyer. 1985. The Snakes of Honduras. Milwaukee Public Museum, Milwaukee, WI.

Wilson, L. D., L. Porras, and J. R. McCranie. 1986. Distributional and taxonomic comments on some members of the Honduran herpetofauna. Contributions Biol. Geol. Milwaukee Publ. Museum 66: 1-18.

Wolcott, G. N. 1923. The food of Porto Rican lizards. J. Dept. Agric. Porto Rico 7: 5-38.

Wood, G. C. 1939. The genus *Tretanorhinus* in Cuba and the Isle of Pines. Proc. New England Zool. Club 18: 5-11.

Woodley, J. D. 1971. Hellshire Hills Scientific Survey, 1970. Sci. Rep. Univ. West Indies.

Woodley, J. D. 1980. Survival of the Jamaican iguana, *Cyclura collei*. J. Herpetol. 14: 45-49.

Woods, C. A. 1989. Biogeography of the West Indies: Past, Present, and Future. Sandhill Crane Press, Gainesville, FL.

Woolbright, L. L. 1985. Patterns of nocturnal movement and calling by the tropical frog, *Eleutherodactylus coqui*. Herpetologica 41: 1-9.

Woolbright, L. L., and M. M. Stewart. 1987. Foraging success of a tropical frog *Eleutherodactylus coqui*: the cost of calling. Copeia 1987: 69-75.

Wunderle, J. M., Jr. 1981. Avian predation upon *Anolis* lizards on Grenada. Herpetologica 37: 104-108.

Wüster, W., M. G. Salomao, R. S. Thorpe, G. Puorto, M. F. D. Furtado, S. A. Hoge, R. D. G. Theakston, and D. A. Warrel. 1997. Systematics of the *Bothrops atrox* complex: new insights from multivariate analysis and mitochondrial DNA sequence information, pp. 99-113 in Venomous Snakes: Ecology, Evolution and Snakebite, edited by R. S. Thorpe, W. Wüster, and A. Malhorta. Clarendon Press, Oxford, UK.

Wyles, J. S., and G. C. Gorman. 1980. The classification of *Anolis*: Conflict between genetic and osteological interpretation as exemplified by *Anolis cybotes*. J. Herpetol. 14: 149-153.

Yang, S. Y., M. Soulé, and G. C. Gorman. 1974. *Anolis* lizards of the eastern Caribbean: A case study in evolution. I. Genetic relationships, phylogeny, and colonization sequence of the *roquet* group. Syst. Zool. 23: 387-399.

Zug, G. R., and A. Schwartz. 1958. Variation in the species of *Cadea* (Amphisbaenidae), and a record of *C. blanoides* from the Isla de Pinos. Herpetologica 14: 176-179.

Zweifel, R. G. 1967. Identity of the frog *Cornufer unicolor* and application of the generic name *Cornufe*. Copeia 1967: 117-121.

Systematic Index

A

D

F

G

H

I

J

K

L

M

N

O

P

Q

R

S

T

U

W

X

Z

The National Academies of
SCIENCES • ENGINEERING • MEDICINE

The **National Academy of Sciences** was established in 1863 by an Act of Congress, signed by President Lincoln, as a private, nongovernmental institution to advise the nation on issues related to science and technology. Members are elected by their peers for outstanding contributions to research. Dr. Marcia McNutt is president.

The **National Academy of Engineering** was established in 1964 under the charter of the National Academy of Sciences to bring the practices of engineering to advising the nation. Members are elected by their peers for extraordinary contributions to engineering. Dr. C. D. Mote, Jr., is president.

The **National Academy of Medicine** (formerly the Institute of Medicine) was established in 1970 under the charter of the National Academy of Sciences to advise the nation on medical and health issues. Members are elected by their peers for distinguished contributions to medicine and health. Dr. Victor J. Dzau is president.

The three Academies work together as the **National Academies of Sciences, Engineering, and Medicine** to provide independent, objective analysis and advice to the nation and conduct other activities to solve complex problems and inform public policy decisions. The National Academies also encourage education and research, recognize outstanding contributions to knowledge, and increase public understanding in matters of science, engineering, and medicine.

Learn more about the National Academies of Sciences, Engineering, and Medicine at **www.national-academies.org**.

The National Academies of

SCIENCES · ENGINEERING · MEDICINE

Reports document the evidence-based consensus of an authoring committee of experts. Reports typically include findings, conclusions, and recommendations based on information gathered by the committee and committee deliberations. Reports are peer reviewed and are approved by the National Academies of Sciences, Engineering, and Medicine.

Proceedings chronicle the presentations and discussions at a workshop, symposium, or other convening event. The statements and opinions contained in proceedings are those of the participants and have not been endorsed by other participants, the planning committee, or the National Academies of Sciences, Engineering, and Medicine.

For information about other products and activities of the National Academies, please visit nationalacademies.org/whatwedo.

COMMITTEE ON THE SUPPLY CHAIN FOR MIDDLE-SKILL JOBS: EDUCATION, TRAINING, AND CERTIFICATION PATHWAYS

Jeff Bingaman (*Chair*), Former U.S. Senator, New Mexico
Thomas R. Bailey (*Co-Chair*), Director, Community College Research Center, Teachers College, Columbia University
Katharine G. Frase (NAE) (*Co-Chair*), Vice President and Chief Technology Officer, IBM Public Sector (retired)
David S. C. Chu, President, Institute for Defense Analyses
Rita R. Colwell (NAS), Distinguished University Professor, Institute for Advanced Computer Studies, University of Maryland College Park, and Johns Hopkins Bloomberg School of Public Health, Baltimore, Maryland
Gary Hoachlander, President, ConnectEd: The California Center for College and Career
Harry J. Holzer, John LaFarge Jr. S.J. Professor of Public Policy, McCourt School of Public Policy, Georgetown University
Francine Lawrence, Executive Vice President, American Federation of Teachers (retired)
Jennifer McNelly, President, 180 Skills, LLC
Elizabeth Mertz, Associate Professor, Preventive & Restorative Dental Sciences, School of Dentistry, University of California, San Francisco
Paul Osterman, Nanyang Technological University Professor of Human Resources and Management, Sloan School of Management, Massachusetts Institute of Technology
Annette Parker, President, South Central College
Susan K. Sclafani, Pearson Foundation (retired)
Mark Tomkins, President and CEO, German American Chamber of Commerce of the Midwest

PROJECT STAFF

BOARD ON SCIENCE, TECHNOLOGY, AND ECONOMIC POLICY POLICY AND GLOBAL AFFAIRS

Gail Cohen, Board Director
Sujai Shivakumar, Study Director
Jeff Alstott, Christine Mirzayan Science and Technology Policy Graduate Fellow

(Continued)

David Allen Ammerman, Financial Officer
David Dierksheide, Program Officer
Fred Lestina, Senior Program Assistant
Erik Saari, Senior Program Assistant

BOARD ON HIGHER EDUCATION AND WORKFORCE POLICY AND GLOBAL AFFAIRS

Tom Rudin, Board Director

BOARD ON SCIENCE EDUCATION
DIVISION OF BEHAVIORAL AND SOCIAL SCIENCES AND EDUCATION

Heidi Schweingruber, Board Director
Margaret Hilton, Senior Program Officer

NATIONAL ACADEMY OF ENGINEERING
Ken Jarboe, Senior Program Officer

BOARD ON SCIENCE, TECHNOLOGY, AND ECONOMIC POLICY

For the National Academies of Sciences, Engineering, and Medicine, this project was overseen by the Board on Science, Technology, and Economic Policy (STEP), a standing board established by the National Academies of Sciences and Engineering and the Institute of Medicine in 1991. The mandate of the STEP Board is to advise federal, state, and local governments and inform the public about economic and related public policies to promote the creation, diffusion, and application of new scientific and technical knowledge to enhance the productivity and competitiveness of the U.S. economy and foster economic prosperity for all Americans. The STEP Board and its committees marshal research and the expertise of scholars, industrial managers, investors, and former public officials in a wide range of policy areas that affect the speed and direction of scientific and technological changes and their contributions to the growth of the U.S. and global economies. Results are communicated through reports, conferences, workshops, briefings, and electronic media subject to the procedures of the National Academies to ensure their authoritativeness, independence, and objectivity. The members of the STEP Board and staff are listed below:

Richard K. Lester (*Chair*), Massachusetts Institute of Technology
Jeff Bingaman, Former U.S. Senator, New Mexico
Ellen R. Dulberger, Dulberger Enterprises, LLC
Alan M. Garber (NAM), Harvard University
Ralph E. Gomory (NAS/NAE), New York University
Michael Greenstone, The University of Chicago
Arati Prabhakar (NAE)
Luis M. Proenza, University of Akron
Kathryn L. Shaw, Stanford University

STEP Staff

Gail Cohen, Board Director
Jeff Alstott, Christine Mirzayan Science and Technology Policy Graduate Fellow
Paul Beaton, Senior Program Officer
David Dierksheide, Program Officer
Fred Lestina, Senior Program Assistant
Erik Saari, Senior Program Assistant
Sujai Shivakumar, Senior Program Officer and Study Director

Preface

Skilled technical occupations—defined as occupations that require a high level of knowledge in a technical domain but do not require a bachelor's degree for entry—are a key component of the U.S. economy. The United States needs a workforce with the right mix of skills to remain competitive while fostering greater innovation and boosting shared prosperity.

In response to globalization and advances in science and technology, American firms are demanding workers with greater proficiency in literacy and numeracy, as well as strong interpersonal, technical, and problem-solving skills. However, employer surveys and industry and government reports have raised concerns that the nation may not have an adequate supply of skilled technical workers to achieve its competitiveness and economic growth objectives. American policy makers therefore need to consider whether they are creating the conditions that will encourage citizens to prepare for technically oriented, skilled jobs. The implementation of the recently reauthorized Workforce Innovation and Opportunity Act of 2014 represents an opportunity for federal, state, and local policy makers to make the U.S. system for workforce development more demand-driven and accountable.

In response to the broader need for policy information and advice, the National Academies of Sciences, Engineering, and Medicine convened the Committee on the Supply Chain for Middle Skilled Jobs to examine the coverage, effectiveness, flexibility, and coordination of the policies and various programs that prepare Americans for skilled technical jobs.[1] To carry out this charge, the committee undertook a review of the salient academic and evaluation literature, commissioned a series of papers on key topics, convened a national symposium, and consulted numerous experts to gather evidence and incorporate diverse viewpoints to inform its deliberations.

This consensus report, which provides the committee's conclusions and action-oriented recommendations for improving the American system of

[1]Although the committee's name includes the phrase "middle-skill," this report describes this workforce as "technically skilled," a term the committee determined to be a more appropriate designation (see Chapter 1).

technical education, training, and certification, is a result of this broad-based effort. It is the committee's hope that this report will guide policy makers and administrators, educators, employers, labor organizations, and other stakeholders in preparing Americans for well-paid technically skilled jobs that meet local and national needs to support a robust economy.

This study reflects a collaborative effort across several boards and program offices of the National Academies: the Board on Higher Education and Workforce, the Board on Science Education, and National Academy of Engineering Program Office. The study was led by the Board on Science, Technology, and Economic Policy (STEP).

ACKNOWLEDGMENTS

On behalf of the National Academies, the committee expresses its appreciation for the insights, experiences, and perspectives made available by the authors of the papers we commissioned and to the participants in the symposium and meetings we convened over the course of this study. We particularly recognize Margaret Polski for her assistance in preparing a draft of this report and in responding to the comments of reviewers. In addition, we thank Alicia Sasser Modestino for her valuable contributions to Chapter 2 of this report. We also recognize David Dierksheide, Fredric Lestina, and Erik Saari of the STEP staff, along with Rita Johnson, for their contributions to the report's preparation.

ACKNOWLEDGMENT OF REVIEWERS

This report has been reviewed in draft form by individuals chosen for their diverse perspectives and technical expertise. The purpose of this independent review is to provide candid and critical comments that will assist the institution in making its published report as sound as possible and to ensure that the report meets institutional standards for objectivity, evidence, and responsiveness to the study charge. The review comments and draft manuscript remain confidential to protect the integrity of the process.

We wish to thank the following individuals for their review of this report: Marilyn Barger, Hillsborough Community College; Crystal Bridgeman, Siemens Foundation; Ellis Cowling, North Carolina State University; Alicia Dowd, Pennsylvania State University; Miles Free, Precision Machined Products Association; Kevin Hollenbeck, W.E. Upjohn Institute for Employment Research; Melissa Kearney, University of Maryland; Livia Lam, Learning Policy Institute; Robert Lerman, American University; Daniel Marschall, AFL-CIO; Mary Alice McCarthy, New America Foundation; Darlene Miller, Permac Industries; Sanjay Rai, Montgomery College; Ursula Renolds, ETH Zürich;

James Stone, National Research Center for Career and Technical Education; Michelle Van Noy, Rutgers University; and Michael Wolf, Bureau of Labor Statistics.

Although the reviewers listed above have provided many constructive comments and suggestions, they were not asked to endorse the conclusions or recommendations, nor did they see the final draft of the report before its release. The review of this report was overseen by Cora Marrett, University of Wisconsin, and Robert Sproull, University of Massachusetts. Appointed by the National Academies, they were responsible for making certain that an independent examination of this report was carried out in accordance with institutional procedures and that all review comments were carefully considered. Responsibility for the final content of this report rests entirely with the authoring committee and the institution.

Jeff Bingaman Sujai Shivakumar

Contents

Boxes, Tables, and Figures

BOXES

TABLES

FIGURES

Abbreviations and Acronyms

AACC	American Association of Community Colleges
ACTE	Association for Career and Technical Education
ASAP	Accelerated Study in Associate Programs
ATD	Association for Talent Development
ATE	advanced technical education
BEA	Bureau of Economic Analysis
BLS	Bureau of Labor Statistics
CBO	Congressional Budget Office
CPI	Career Pathways Initiative
CPS	Current Population Survey
CSW	Corporation for a Skilled Workforce
CTE	career and technical education
CUNY	City University of New York
CWS	Credentials for Work Survey
DoD	U.S. Department of Defense
DOI	U.S. Department of the Interior
DOL	U.S. Department of Labor
ED	U.S. Department of Education
ESSA	Every Student Succeeds Act
FAFSA	Free Application for Federal Student Aid
FTE	full-time equivalent
FY	fiscal year
GAO	Government Accountability Office

HEA	Higher Education Act
HHS	U.S. Department of Health and Human Services
IT	information technology
JATC	Joint Apprenticeship and Training Committee
MIT	Massachusetts Institute of Technology
MOOC	massive open online course
NAEP	National Assessment of Educational Progress
NAF	National Academy Foundation
NASDCTEc	National Association of State Directors of Career Technical Education Consortium
NCES	National Center for Education Statistics
NGA	National Governors Association
NMI	National Manufacturing Institute
NSB	National Science Board
NSF	National Science Foundation
OECD	Organisation for Economic Co-operation and Development
PCAST	President's Council of Advisors on Science and Technology
PIACC	Programme for the International Assessment of Adult Competencies
SHRM	Society for Human Resource Management
SOC	Standard Occupational Classification (System)
STEM	science, technology, engineering, and mathematics
TAA	Trade Adjustment Assistance Act
UCX	Unemployment Compensation for Ex-servicemembers
USCM	U.S. Conference of Mayors
USMAP	United States Military Apprenticeship Program
VET	vocational education and training
WIA	Workforce Investment Act
WIOA	Workforce Innovation and Opportunity Act
WLMIS	workforce labor market information system

Summary

The United States needs a skilled technical workforce to remain competitive in the global economy and to ensure that its workers participate in the nation's economic growth. There are significant opportunities as well as major challenges in this regard. Notably, rigorous evidence indicates that the returns to investments in technical skills in the labor market are strong when students successfully complete their training and gain credentials sought by employers. At the same time, the committee found that in many instances, workers either are not taking advantage of these opportunities or are failing to complete their training programs.

To understand why, it is necessary to recognize that in the United States, the responsibility for developing and sustaining a skilled technical workforce is fragmented across many groups, including educators; students; workers; employers; the federal, state, and local governments; labor organizations; and civic associations. For the system to work well, these groups need to be able to coordinate and cooperate successfully with each other.

Unlike most other advanced economies, the United States lacks formal mechanisms that require governments, educators, labor representatives, and employers to coordinate on workforce development policies and practices at the national level. In fact, workforce development in the United States is polycentric in nature, driven by a variety of private and public investments in workforce education and training. Workers often pay for on-the-job training through lower wages. Although employers and governments share an interest in developing and maintaining a robust skilled technical workforce, their respective investments often are uncoordinated. At the same time, public investments are guided by a complex and similarly uncoordinated set of policies associated with achieving similarly divergent goals related to economic development, education, employment, health and human services, and veterans' affairs. In this polycentric system, making better use of available resources and generating better outcomes requires improving coordination between students and educational or training institutions, between secondary and postsecondary institutions, and especially between training institutions and employers through a variety of public, private, and hybrid mechanisms.

The good news is that promising experiments currently under way across the United States can provide guidance for innovation and reform, although the scalability of some of these experiments has not yet been tested. As detailed in Chapter 6, evidence suggests that integration of academic education, technical training, and hands-on work experience improves outcomes and return on investment for students in secondary and postsecondary education and for skilled technical workers in different career stages.

The findings and recommendations presented below are designed to help overcome some of the barriers identified within the current framework of federal governance, state implementation, and market incentives. They address the key elements of the statement of task for this study. To the extent possible, the recommendations call for specific actions by Congress, federal agencies, state governments, employers, and civic organizations to improve the American system of workforce development.

FINDINGS

1. The skilled technical workforce includes a range of occupations that require a high level of knowledge in a technical domain, but many of these occupations do not require a bachelor's degree for entry.
2. Although widely used to describe this segment of the workforce, the term "middle skills" fails to capture the high value and dynamism of this segment of the U.S. workforce and is seen by some as having pejorative connotations. This label can deter students and workers from these occupations at every stage of their career.
3. To remain competitive in the global economy, foster greater innovation, and provide a foundation for shared prosperity, the United States needs a workforce with the right mix of skills to meet the diverse needs of the economy. Conversely, an insufficiently skilled workforce imposes significant burdens on the U.S. economy, including higher costs to workers and employers and lower economic productivity.
4. The evidence suggests that as a nation, the United States is not adequately developing and sustaining a workforce with the skills needed to compete in the 21st century.
5. It is difficult to find rigorous evidence on how well skilled technical labor markets are functioning at aggregate levels of analysis across the nation.
6. The nation is experiencing, and will continue to experience, imbalances in the supply of and demand for skilled technical workers in certain occupations, industry sectors, and locations. The nature of the problem differs across sectors and locations. These imbalances arise from multiple sources.
7. In the United States, educators, students, workers, employers, the federal government, state and local governments, labor unions, industry and trade

associations, and other civic associations all play a role in skilled technical workforce development.

8. Skilled technical workforce development in the United States is guided and supported by a complex and often uncoordinated set of policies and funds at the local, state, and federal government levels associated with achieving goals related to economic development, education, labor and employment, health and human services, and veterans' affairs. Most resources are allocated by formulas based on demographic factors, which serve as a proxy for need, rather than on performance, outcomes, or evidence of what works best in workforce development.
9. The incentives for students and employers to invest in skill development depend on the return on these investments.
10. Policy makers and participants in workforce development lack the information they need to adjust policies and make choices. User-friendly tools that provide direct access to information about options and performance can assist labor market participants in their decision-making processes.
11. Several strategies and initiatives show promising results by improving rates of successful completion of education and training and by coordinating education and training opportunities with employer needs. These experiments, which link the different parts of the ecosystem for workforce education and training, could be instructive for designing and enforcing policy and allocating resources within appropriate contexts.
12. The Armed Services, one of the largest employers in the United States, could do a better job of assessing employment transition risks for military personnel, and designing and delivering services to mitigate these risks. They could also coordinate better with civilian policy makers, regulators, and educators to improve the transferability of military education, training, and certification.
13. Policies and governmental structures used to support education and training in such countries as Germany and Switzerland may not be directly applicable in the United States because of differences in political and economic structures. However, applicable lessons can be learned from specific programs or innovative educational techniques, and policy makers in the United States can benefit from more rigorous cross-country comparisons.

RECOMMENDATIONS

1. State and federal policy makers should support and enhance strategies that help students successfully complete their training for the skilled technical workforce. In addition, public policies should ensure that stakeholders, including students, workers, employers, and educational organizations, have

the right incentives to improve the quality of technical education and training, encourage experimentation and collaboration, and improve the collection and use of relevant information.

2. An alliance of industry, trade, academic, and civic associations and labor unions, in cooperation with the U.S. Departments of Labor and Education, should organize a nationwide public–private communication campaign to raise awareness of the value of and demand for skilled technical workers and the return on investment for individuals preparing for these careers. This campaign should be customized to recognize local variations in skilled technical workforce education, training, and labor market requirements.
3. Congress and state legislatures should improve oversight of public policies and resources, highlighting the implementation of reforms such as those called for in the Workforce Innovation and Opportunity Act (WIOA). Reforms should be accelerated through targeted incentives.
4. Congress, state legislatures, and agencies of the federal and state governments should improve the workforce labor market information system (WLMIS), labor market data, and research tools and methods, including by providing funding for such activities.
5. Federal and state agencies should remove barriers to worker mobility, such as licensing and certification requirements that are not related to public safety. They also should improve labor market information on the changing requirements for skilled technical workers to help reduce imbalances in labor markets and to align workforce development with advances in science and technology.
6. The Department of Defense should further integrate skills transition into military training rather than treating it as a separate component at the end of the service member's career.
7. While selected programs and policies from other countries have been adapted in the United States, federal agencies should further study the conditions under which particular attributes of apprenticeships or other programs can be effectively applied more broadly.

1

The Skilled Technical Workforce Development Challenge

The United States needs a skilled technical workforce to compete in the global economy and to sustain a high standard of living for its citizens. If the nation does not adequately develop and sustain this skilled workforce relative to what is needed, the consequences will be seen in lower productivity, falling revenues, and fewer job opportunities for Americans. Addressing this workforce development challenge requires understanding the scale and dynamics of the skilled technical workforce, as well as the complex and rapidly changing environments within which Americans provide and acquire skills and training for the jobs of the future.

1.1 KEY CHALLENGES FOR DEVELOPING A SKILLED TECHNICAL WORKFORCE

Globalization and scientific and technological change are affecting nearly every aspect of modern life, from how people communicate with each other to how they shop, how they make things, and how and where they work. As artificial intelligence and advanced robotics become ubiquitous, substantial changes are likely in the nature and character of work as humans find new ways to work symbiotically with the machines they have created and as machines increasingly work autonomously (see, for example, Brynjolfsson and McAfee [2014] and Frey and Osborne [2013]). As a 2017 report by McKinsey & Company (2017, p. 3) concludes, "Individuals in the workplace will need to engage more comprehensively with machines as part of their everyday activities, and acquire new skills that will be in demand in the new automation age." Adapting to this new world, American employers are demanding workers with greater proficiency in literacy and numeracy and strong interpersonal, technical, and problem-solving skills.

Many employers report experiencing difficulty finding skilled workers, although some employers may be using inappropriate screening criteria, such as requiring a 4-year degree for jobs that do not require this credential. At the same time, many students either are unaware of well-paying jobs available for skilled workers or shy away from training opportunities because of a lack of basic math or science skills. This reported shortage of skilled workers raises questions about the role of postsecondary credentials, the value of technical skills, and whether the U.S. education and workforce development systems are meeting the skills development challenge. It also raises questions about whether labor markets have the incentives and information to function properly.

While this report examines these challenges in depth, it also presents evidence in Chapter 6 that integrating academic education, technical training, and hands-on work experience produces better outcomes and return on investment for all students in elementary, high school, postsecondary, and continuing education. Moreover, a growing body of research on promising initiatives under way across the United States can be instructive for designing and implementing supportive policies and for allocating resources across organizations and programs.

1.2 THE IMPORTANCE OF SKILLS

Over the past decade, policy makers and employers have been concerned that advances in science and technology, slower population growth, and the retirement of the baby boom generation would lead to a shortage of skilled labor. These developments have important implications for the economy, society, workers, and employers.

For the economy, a skilled workforce is necessary to grow employment, output, and productivity (Lerman, 2015). If employers cannot hire workers with the right skills, they cannot produce enough of the higher-value products and services in the United States that are associated with sustained innovation and growth. Employers may then relocate or cease to operate.

For society, a skilled workforce is linked with better public and fiscal outcomes. As Figure 1-1 shows for the example of literacy skills, inequality in skills is associated with inequality in income, which creates pressure on public order and public programs. For instance, if a low-skilled workforce leads to slow economic growth and declining tax revenues, government budgets will be strained as a larger share of the population retires and draws on public programs such as Social Security and Medicare. Figure 1-1 also shows that citizens with low skill levels tend to report poorer health, lower levels of civic engagement, and less trust relative to those with higher skill levels (OECD, 2013).

For individual workers, higher skill levels are necessary to compete for increasingly skill-intensive jobs and higher wages, and are associated with a

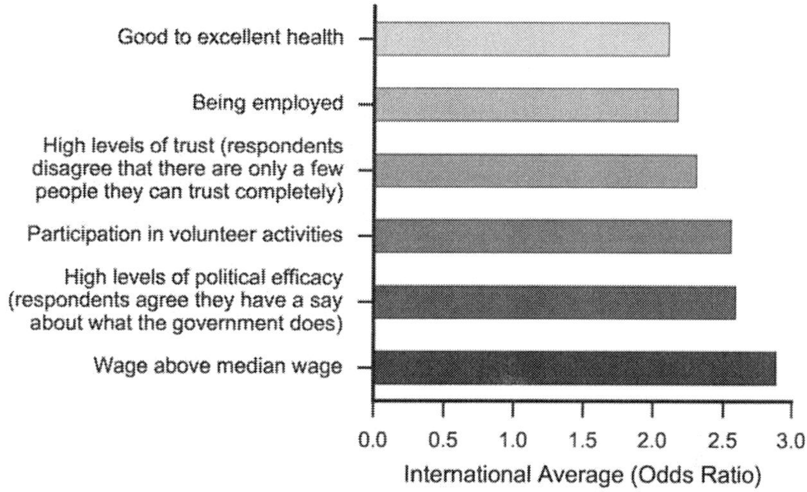

FIGURE 1-1 Likelihood of positive social and economic outcomes among highly literate adults.
NOTE: The figure shows the increased likelihood (odds ratio) of adults scoring at level 4/5 in literacy reporting high wages, high levels of political efficacy, participation in volunteer activities, high levels of trust, being employed, and having good to excellent health relative to those scoring at or below level 1 in literacy (adjusted). Odds ratios are adjusted for age, gender, educational attainment, and immigrant and language background. High wages are defined as workers' hourly earnings above the country's median.
SOURCE: OECD, 2012b, Graph II.1.1.

wide range of better outcomes. A recent cross-country survey of adult skills by the Organisation for Economic Co-operation and Development (OECD) Programme for the International Assessment of Adult Competencies found that skills have a major influence on life choices and outcomes: lower-skilled workers have fewer opportunities in many areas of life and are increasingly likely to be left behind. In an innovative economy, workers must continually upgrade their skills. Those with technical skills need to determine how best to maintain their employability, pursue career advancement, and respond to unexpected changes in demand for their current skill set. Low skill levels make it more difficult for workers to attain additional education or training when structural changes require adaptation to new methods and processes (OECD, 2013).

For employers and firms, a skilled workforce is needed to adapt to rapid technological changes. Assessment, analysis, and communication of information and action are now conducted with a wide range of software applications using personal computers, smartphones, and the Internet. As employers increasingly automate tasks and incorporate artificial intelligence in standard operating procedures and decision-making processes, the capacity of workers to use these tools to manage tasks is becoming essential in most occupations and jobs.

The demand for a skilled technical workforce is changing so rapidly that workers, employers, educators, policy makers, and civic organizations need to be highly flexible and forward looking. Likewise, the nation's approach to workforce development and skills acquisition must also adapt to changing realities at the federal, state, and local levels. Although some skilled technical jobs in such occupations as production and transportation are being eliminated through automation and outsourcing, others are in high demand. These latter jobs, which require the ability to use new technologies, can be found in growing industries associated with health care, advanced manufacturing, and information technology. Many skilled technical jobs require postsecondary credentials below the 4-year degree, yet they provide competitive salaries and benefits, as well as the opportunity to advance to even better positions. With sufficient information and resources, these jobs could be filled by people with high-quality education and training provided by community colleges, career and technical education programs, apprenticeships, and a growing number of web-based educational programs.

As American policy makers work to tackle the challenges of social and economic development, then, they must consider whether they are creating the conditions that will prepare citizens for skilled technical jobs. These jobs require problem solving and a broad range of competencies in addition to occupation-specific technical competencies. Box 1-1 lists the general set of worker competencies identified by the U.S. Department of Labor.[1]

With the implementation of the recently reauthorized Workforce Innovation and Opportunity Act of 2014, policy makers at all levels are expected to make the U.S. workforce development system more demand-driven and accountable. Tackling these challenges requires that policy makers

- understand current trends in the level and distribution of skills in the adult working-age population;
- determine the adequacy of the supply of skilled workers to meet the changing demand for skills driven by social, technological, and economic requirements;
- ensure that individuals can develop and maintain the necessary skills and adapt positively to social and economic change; and
- create the conditions that forge strong, positive linkages between skill acquisition and social and economic outcomes.

---

[1]A recent report of the National Academies of Sciences, Engineering, and Medicine (NRC, 2012) elaborates on the types of skills and learning required to be successful in the 21st century.

BOX 1-1
An Array of Skills and Competencies

The following is the general set of worker competencies identified by the U.S. Department of Labor's Competency Model Clearing House:

- **Personal skills.** These include interpersonal skills, integrity, initiative, dependability, reliability, adaptability, flexibility, and lifelong learning.
- **Academic competencies.** These include basic academic skills such as reading, writing, and mathematics; advanced knowledge and skills in science and technology relevant to the occupation; and critical and analytical thinking.
- **Workplace competencies.** These include understanding the big picture of the organization, such as inputs, throughputs, and outputs; teamwork and collaboration; customer focus; planning and organizing; problem solving and decision making; working with tools and technologies; scheduling and coordination; and checking, examining, and recording.
- **Industry-wide technical competencies.** These are the knowledge, skills, and abilities associated with a particular industry, such as health care or manufacturing. They are cross-cutting competencies common to most occupations within the industry. These competencies typically relate to comprehension, awareness, or analysis.
- **Occupation-specific technical competencies.** These are the knowledge, skills, and abilities associated with particular occupations, such as emergency medicine or automotive manufacturing technicians.

SOURCE: Competency Model Clearing House, 2015.

1.3 STUDY CHARGE, SCOPE, AND APPROACH

Recognizing the challenges outlined above, the National Academies of Sciences, Engineering, and Medicine convened the Committee on the Supply Chain for Middle-Skill Jobs to examine the coverage, effectiveness, flexibility, and coordination of programs that prepare Americans for skilled technical jobs. In carrying out its charge, the committee considered such topics as gaps in coverage and fluctuations in the labor market, the current and preferred public- and private-sector roles in financing and providing employment training and skills certification, and the incentives and information resources available for individuals to improve their work skills (see Box 1-2).

BOX 1-2
Statement of Task

An ad hoc committee will examine the coverage, effectiveness, flexibility, and coordination among the nation's programs to prepare Americans for technically oriented, skilled positions in the workforce demanding non-routine problem solving but not requiring a baccalaureate or higher degree. This system includes apprenticeship programs of schools, unions, and employers; high school Career Technical Education programs; advanced technical education and training in community colleges and for-profit colleges; employer-financed and -provided training; federal education and training programs; state learning exchanges; public-private employment training partnerships; and licensing and skills certification.

Among the topics the committee will consider are the gaps in coverage and market failures in this part of the labor market, the current and preferred public- and private-sector roles in financing and providing employment training and skills certification, and the incentives and information for individuals to improve their work skills. The committee will also consider selected employment preparation practices in other countries (e.g., European apprenticeship programs) and of foreign-headquartered firms and their relevance to the U.S. labor market.

In addition to reviewing and synthesizing the existing social science and program evaluation literature, the committee will commission papers and organize a national symposium bringing together researchers, industry representatives from diverse sectors, policy makers, and other stakeholders involved in technical workforce education and training. Based on its understanding of the evidence and experience, the committee will write a report with conclusions and action-oriented recommendations.

During its first meeting, in February 2015, the committee highlighted 11 key areas of analytical attention and identified capable researchers to prepare detailed reviews of the existing literature and practices in these areas. Box 1-3 provides a list of the papers commissioned for this study. Collectively, these papers provide comprehensive coverage of salient definitions, labor market dynamics, pathways and pipelines, the role of community colleges, the role of employers, apprenticeship programs, federal workforce development programs, pathways to skilled technical allied health care occupations, veteran unemployment benefits and transition assistance, information resources, and competing on innovation.

In addition to commissioning this research, the committee organized a national symposium, held in June 2015, that brought together researchers, representatives from various industry sectors, policy makers, and other stakeholders involved in technical workforce education and training (NASEM, 2015). The agenda for this meeting is presented in Appendix A. During the symposium, the commissioned papers were presented by their authors and

BOX 1-3
Papers Commissioned for This Study

- Jonathan Rothwell, The Brookings Institution: "Defining Skilled Technical Work"
  This paper helps frame how skilled technical workers should be defined and analyzed, with an emphasis on the skills that distinguish them from those in less advanced occupations.

- Alicia Sasser Modestino, Northeastern University: "Middle Skill Workers in Today's Labor Market"
  This paper provides a framework for analyzing the potential mismatch between the supply of and demand for skilled technical workers over the next two decades. The goal is to understand the magnitude of the problem, as well as the efficacy of alternative approaches aimed at addressing the skills gap.

- David Stern, University of California, Berkeley: "Pathways or Pipelines: Keeping High School Students' Future Options Open While Developing Technical Skills and Knowledge"
  This paper describes work-related education in high schools, and offers suggestions about the role of high school programs in the supply chain for jobs that do not require a baccalaureate or higher degree.

- Melinda Mechur Karp, Columbia University: "Community College Pathways"
  This paper examines sub-associate degrees offered by community colleges, both in terms of their stand-alone impact and in terms of their connection to a broader education and training system.

- Robert L. Lerman, Urban Institute, American University, and IZA: "Are Employers Providing Enough Training? Theory, Evidence, and Policy Implications"
  This paper examines the types and amounts of skill development, especially employer-sponsored training, that are taking place in the United States.

- Andrea Messing-Mathie, Northern Illinois University: "Apprenticeship Pathways to Skilled Technical Jobs"
  The focus of this paper is on the potential for expanding apprenticeship as part of the development of career pathway systems, particularly for young people, as they transition from secondary into postsecondary institutions and/or careers.

(Continued)

BOX 1-3
Continued

- Carolyn J. Heinrich, Vanderbilt University: "Federally Funded Workforce Development"
  This paper describes evolving administrative and organizational structures of the federal workforce development system. The paper also examines the Workforce Innovation and Opportunity Act of 2014 (WIOA) and the opportunities this law presents for improving system structures, aligning incentives, and strengthening program coordination and effectiveness.

- Bianca K. Frogner and Susan M. Skillman, University of Washington: "Pathways to Middle-Skill Allied Health Care Occupations"
  This paper explores the education and training pathways to skilled technical occupations in allied health care, which generally require less than a bachelor's degree for entry.

- Susan Payne Carter and Brian J. Miller, U.S. Military Academy: "Analysis of Army Veteran Unemployment Benefits and Transition Assistance"
  This paper analyzes how the incentives created by the Unemployment Compensation for Ex-Service Members (UCX) program affects how the program is used to facilitate ex-service members' transition back into the civilian labor market.

- Andrew Reamer, George Washington University: "Information Resources to Facilitate Middle Skills Workforce Development"
  This paper begins with a summary of the types of labor market participant decisions that require good information, follows with an overview and assessment of currently available information resources, and then offers recommendations for enhancing these information resources.

- Robert G. Sheets, George Washington University, and Jason A. Tyszko, U.S. Chamber of Commerce Foundation: "Competing on Innovation: Implications for Building the Middle-Skill Talent Pipeline"
  This paper proposes that to compete on innovation, it is necessary to rethink how high schools, community colleges, and universities could potentially partner with employers to provide more cross-functional and interdisciplinary experiences.

SOURCE: http://nas.edu/SkilledTechnicalWorkforce.

discussed by a diverse set of experts. The papers were revised based on the discussion at the symposium and later posted on the National Academies study website.[2]

The committee convened a further four times in 2015 and 2016 to gather additional evidence from experts on topics including the scope and nature of applicable federal legislation, the role of state programs and policies, the value of online learning technologies, impediments to transitioning military personnel into the civilian skilled technical workforce, and the challenges facing efforts to collect data on the market for skilled technical workers. The agendas for these meetings are presented in Appendix A. These meetings were supplemented by further research on the academic and policy literature.

As the committee learned more about current workforce development issues, its thinking about the nature of the workforce development challenges evolved. This evolution, which has important implications for the committee's findings and recommendations, included refining its definition of the workforce, and rethinking the nature of the policy problem and the feasible set of approaches to addressing related issues and challenges.

1.4 DEFINING THE WORKFORCE

As a first step in developing its findings and recommendations, the committee refined its conception of the segment of the workforce addressed by this study and decided to refer to it as the "skilled technical" rather than "middle-skill" workforce. The committee determined that the term "middle-skill," which has been widely used to describe skilled technical occupations, fails to capture the high value and dynamism of this segment of the U.S. workforce. Some participants in the June 2015 symposium also observed that the term "middle-skill" is perceived by some to have pejorative connotations. This view may make some students reluctant to obtain training for these jobs because of a stigma attached to "middle-skill" jobs and related educational pathways and careers.[3] Industry experts at the symposium noted that such

---

[2]A webcast of the symposium, research papers, and presentations can be found at http://nas.edu/SkilledTechnicalWorkforce.

[3]According to Jeff Strohl, director of research at Georgetown University's Center on Education and the Workforce, "Sub-baccalaureate education suffers the stigma of the vocational-technical high school. That's where other people's kids went. People see those programs as tracking into something that's dead end." In fact, he says, "It's very clear that that perception does not hold up" (see Marcus, 2013). The impact of social perceptions on career choices is also discussed in Holzer (2015b).

negative perceptions may limit the ability to attract workers to skilled technical jobs.[4]

After reviewing a number of different definitions and approaches to measuring the skilled technical workforce, the committee adopted the following definition: "The occupation requires a high level of knowledge in a technical domain, and it does not require a bachelor's degree *for entry*." Using this definition of skilled technical work, Rothwell (2015) estimates that in 2014, there were 16.1 million skilled technical workers in the United States, collectively representing 11.9 percent of the total U.S. workforce.

The committee noted that some analysts use slightly different definitions of and analytical approaches to measuring the skilled technical workforce, and therefore derive different estimates of its size. Box 1-4 provides an overview of the estimation challenges, which are discussed in more detail in Chapter 2.

Rothwell (2015) reports that although skilled technical workers are found in most occupation groups, 82 percent are concentrated in the areas of installation, maintenance, and repair; health care; construction and extraction; and production (see Figure 1-2). The remaining 18 percent are employed in computer and mathematical occupations (5 percent), architecture and engineering (4 percent), and 16 other occupations in a wide range of categories (9 percent). The labor markets for each of these occupations are quite different, as discussed in more detail in Chapter 2, because they are all affected by local rather than aggregate changes in demographics, technology, and educational and economic opportunities.

More detailed analyses of the nature of skilled technical work led the committee to emphasize that a simple cutoff of education or salary levels is an inadequate basis on which to define these jobs. While skilled technical jobs are often described as requiring some postsecondary education and training but less than a 4-year college degree, Rothwell (2015) found that there is considerable variation in education and training among these workers (Holzer and Lerman, 2007). He estimated that most workers in these occupations hold some form of postsecondary certification but do not hold a 4-year degree. However, his estimates indicate that 31.9 percent of workers in skilled technical occupations have a high school diploma or less, 18.7 percent have a bachelor's degree or higher, and about 15 percent have a 2-year degree. Relative to the average worker, the middle-wage occupations have a higher rate of bachelor's degree attainment and a similar rate of high school diploma or lower attainment. Finally, it appears that skilled technical workers receive significantly more on-the-job training and have more work experience relative to the average worker.

---

[4]The Manufacturing Institute has pursued the "Dream It. Do It." campaign in an effort to "overcome the negative stigma that a career in manufacturing is a dirty, low-wage job" (see The Manufacturing Institute, 2015).

BOX 1-4
Estimation Challenges

One of the committee's first tasks was to define and identify skilled technical occupations and then describe the characteristics of workers in those occupations. In a paper commissioned by the committee, Jonathan Rothwell describes the strengths and limitations of several approaches that use various proxies to estimate the skill level of a given job (Rothwell, 2015). Researchers have typically used either a relative or absolute ranking to determine skill level based on either education or wages. For example, relative rankings classify occupations by skill level from lowest to highest and then use cutoffs at specific percentiles of the distribution to determine what falls in the middle (e.g., occupations falling into the 20th to 80th percentiles).[a] By contrast, absolute rankings of skilled technical jobs use predefined skills criteria for specific jobs (e.g., education beyond high school but less than a 4-year degree) and then categorize occupations accordingly.[b] Both methods rely on the characteristics of incumbent workers (e.g., either education or wages) to categorize occupations.

Rothwell notes that it may be misleading simply to use wages or educational requirements as an indicator to identify skilled technical jobs. Workers in the middle of the wage distribution may be relatively unskilled but compensated well. Similarly, low-wage workers may be relatively skilled but subject to downward wage pressure because of changes in international competition or technology. Using educational requirements is also troublesome because there is potentially large variation in the technical skills of people who have acquired similar levels of education.

---

[a]For example, Autor and colleagues (2006) sorted (three-digit) occupations into percentiles by mean years of schooling in 1980 using data on incumbent workers in those jobs. Alternatively, Autor (2010) used the average wage of workers in 1980 as a proxy for skill to classify occupations at the three-digit level and subsequently examine employment and wage trends.

[b]For example, Holzer and Lerman (2009, p. 1) argue: "Classifying occupations into a few skill categories is awkward, given the many elements of skill required for most jobs. Under an approach that classifies jobs based on education and training levels, 'middle-skill' jobs are those that generally require some education and training beyond high school but less than a bachelor's degree. These postsecondary education or training requirements can include associate's degrees, vocational certificates, significant on-the-job training, previous work experience, or some college, but less than a bachelor's degree."

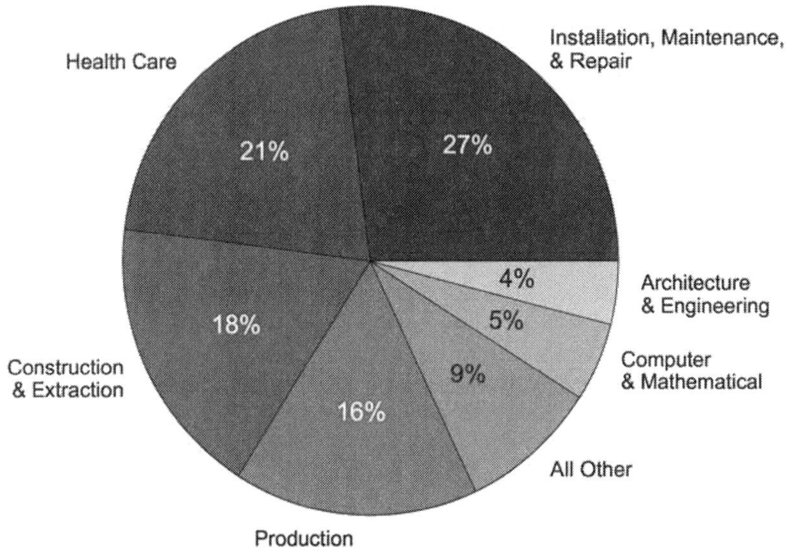

FIGURE 1-2 Skilled technical workers by major occupational group.
SOURCE: Committee's compilation based on Rothwell (2015, Table 3).

1.5 STRUCTURE OF THE SYSTEM FOR U.S. SKILLED TECHNICAL WORKFORCE DEVELOPMENT

The committee's investigation of the distributed structure and wide-ranging scale of the system for skilled technical workforce development—presented in Chapter 4—challenged assumptions that a single set of policy recommendations or reforms can address workforce development issues in the United States. As Figure 1-3 shows, the American system for skilled technical workforce development encompasses two broad components. The first is public and private academic and career and technical education and training in grades K-12, which are governed by more than 98,000 public schools in more than 18,000 public school districts, as well as more than 30,000 private schools. The second component is postsecondary skilled technical workforce development, which is equally complex. Its primary elements are degree-granting programs governed by 4,207 public and private institutions scattered across 50 states, an unknown number of certificate programs, an unknown number of certification

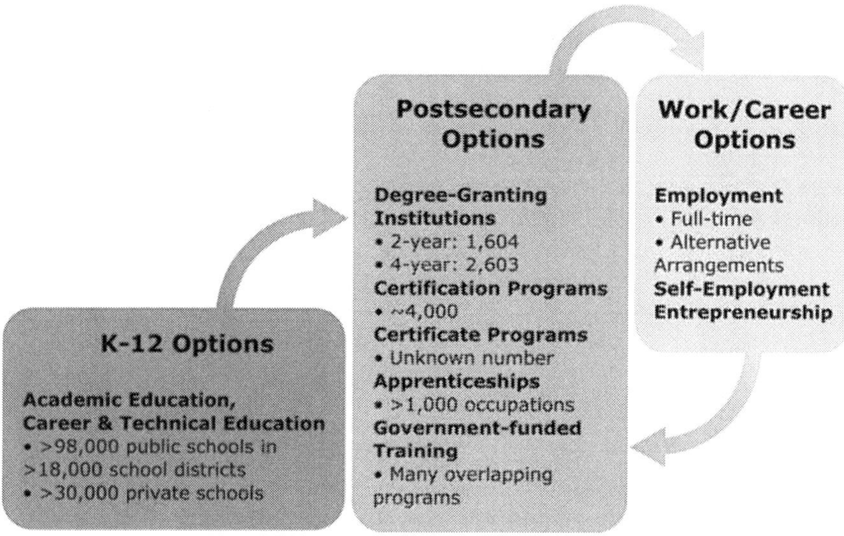

FIGURE 1-3 Overview of the components of U.S. system for skilled technical workforce development.
SOURCE: Committee's compilation.

programs governed by an estimated 4,000 certification entities, registered civilian apprenticeship programs for more than 1,000 occupations, and many overlapping government-funded training programs.

As Figure 1-3 shows, work/career options are similarly varied and include regular full-time employment, alternative and contingency work arrangements, self-employment, and entrepreneurship. In addition, as the reverse arrow in Figure 1-3 indicates, American workers can and do continue their education and training throughout their work lives, cycling back through postsecondary education and training to pursue alternative occupations, expand their knowledge and skills, maintain and expand their credentials, and retrain as advances in science and technology are disseminated into the workplace.

1.6 A NETWORK-CENTERED APPROACH

Understanding and making recommendations for improving the nation's skilled technical workforce amid a complex and rapidly changing market is a considerable challenge. In addressing this challenge, the committee sought to take advantage of the diversity of experience across the nation and the efforts of many business leaders, educators, labor organizations, and policy makers to integrate academic education, technical training, and hands-on work experience to produce better outcomes for students, workers, businesses, and states and regions (see Box 1-5 for an example). In this regard, Chapter 6 presents

BOX 1-5
Building a Network of Solutions—An Example

In his September 2015 presentation to the committee, Dale Allen of Quinsigamond Community College (QCC) in Worcester, Massachusetts, described how his institution is building robust relationships with K-16 partners, industry, universities, regional government, and other community organizations to build a robust workforce development system in his region. Key features of the strategy he described include the following:

- **Collaborating with industry.** In addition to seeking industry input to meet industry demand, QCC partners with the regional offices of the federal Manufacturing Extension Partnership (MEP) program, the state's Workforce Investment Boards (WIBs), local cluster associations, business incubators, and the National Science Foundation's (NSF's) Advanced Technical Education (ATE) programs, among others, to identify workforce development needs.
- **Redesigning education and training delivery.** This includes developing an industry-aligned curriculum and bridging courses with internships, apprenticeships, and job training to meet regional labor market demands.
- **Leveraging technology.** This includes making available customized software applications or "apps" that combine real-time data on the labor market drawn from federal, state, and local sources. Students, local firms, and educational institutions can use this information directly to weigh alternatives and guide career and investment decisions.

evidence that schools that combine high-quality career and technical education programs with a college preparatory curriculum are more effective in preparing students for both employment and postsecondary education (see, for example, Clark et al., 2007). A growing body of research on a variety of promising experiments under way across the United States—also described in Chapter 6—is instructive for designing and implementing supportive policies and for allocating resources across organizations and programs.

Accordingly, the committee has sought to draw attention to the importance of growing the rich set of networks among the various actors, as well as the significant challenges they face in making decisions and working cooperatively in the absence of sufficient information and trust. These challenges are cataloged in Chapter 5. For example, a lack of information on job opportunities available in their region may lead community college students to select courses that do not adequately equip them with the skills demanded by local firms. In turn, firms in a regional industry may need to determine how to share investments in worker training in ways that can also produce sufficient private returns. In other cases, prevailing rules and norms may lead to incentives that foster suboptimal outcomes. For example, some community colleges operate within state funding formulas that support enrollment rather than course completion (Altstadt, 2012). In another example, transition assistance provided by the military to its retiring service members to help them adjust to the civilian workforce may be misapplied as a form of unemployment benefit (Payne Carter and Miller, 2015).

In addressing its task to review the "effectiveness, flexibility, and coordination among the nation's programs," the committee has highlighted the diversity of approaches available to students, guidance counselors, workers, business owners, manufacturers, labor leaders, school administrators, and community college teachers and others for working together to create local solutions that address their workforce challenges. To support and complement these activities, the committee calls for specific actions by Congress, federal agencies, state governments, employers, and civic organizations to improve the nation's polycentric system of workforce development.

1.7 ORGANIZATION OF THIS REPORT

The committee began its work by reviewing existing efforts to standardize education and training in the United States, hearing from experts who recommended further standardization. Chapter 2 takes a closer look at the patterns and trends emerging from the current structure of the labor markets for skilled technical occupations.

A wide range of structural factors affect these trends and patterns in workforce development. They begin with child care, nutrition, and the quality of primary education, and extend to working opportunities and living conditions and the quality and quantity of postsecondary education and training. Workforce policies at the federal and state levels, which are described in Chapter 3, and the components of the workforce development system, which are described in Chapter 4, further influence the character and performance of the skilled technical workforce in significant ways. These policies and components, which are controlled and funded primarily at the local level, are in turn shaped by local social and economic structures and conditions.

A diversity of actors also is involved, including local educators, students, workers, employers, industries, associations, labor organizations, agencies of the federal government, state and local governments, and civic associations, all of which play a role in developing the nation's skilled technical workforce. These actors operate in complex environments, and they often have slightly different and sometimes competing priorities and incentives. Many of the workforce development challenges, discussed in Chapter 5, reflect significant coordination problems within and across these diverse groups of actors. In contrast, Chapter 6 provides examples of notable local strategies and initiatives aimed at overcoming these challenges to training skilled technical workers. Chapter 7 concludes with a summary of the committee's findings and recommendations for policy and program design to enhance the nation's skilled technical workforce.

2

Labor Market Patterns and Trends

In its most recent employment projections for 2014-2024, the Bureau of Labor Statistics (BLS) reports that 5 of the 15 fastest-growing occupations will require some level of postsecondary education but will not require a bachelor's degree (BLS, 2015, Table 3). Occupations that typically require postsecondary education for entry are expected, on average, to grow faster than occupations that require a high school diploma or less. However, growth rates can be misleading because some fast-growing occupations employ very few workers. For example, one of the fastest-growing occupations is biochemical engineer, but the expected increase in the number of these jobs is among the smallest of all occupations. Some analysts argue that the trend is for a "hollowing-out" of the workforce, in which high-skill occupations requiring a minimum of a 4-year degree and low-skill occupations requiring a high school diploma or less continue to grow, but skilled technical occupations shrink (see, for example, Acemoglu and Autor, 2010; Autor, 2010; Autor and Dorn, 2013; Autor and Price, 2013; Autor et al., 2003; Levy and Murnane, 2013).

This chapter turns to the question of whether the U.S. supply of skilled technical workers adequately meets demand now and in the foreseeable future. The first section looks at evidence of possible gaps and imbalances in specific skilled technical occupations, industries, and locations. The second section examines local and sectoral trends in the demand for and supply of technical skills. To understand better the policy issues affecting skilled technical labor markets, the next two sections review the dynamics in two sectors with high numbers of skilled technical occupations—health care and manufacturing. These sectors also serve as examples of how advances in science and technology can affect skilled technical labor markets. Providing an international perspective, the chapter then addresses how information and communication technologies are changing how work is organized and performed in Organisation for Economic Co-operation and Development (OECD) and partner countries. The Survey of Adult Skills conducted by OECD's Programme for the International Assessment of Adult Competencies (PIAAC) reveals that shortages of and gaps or mismatches in skills are common across the countries studied, suggesting the need for a more comprehensive analysis of the demand for and use of skills. The final section presents conclusions.

2.1 THE SUPPLY OF AND DEMAND FOR SKILLED TECHNICAL WORKERS

Recent surveys show that the demand for higher levels of workforce skills has increased over time and is likely to continue to do so. As Figure 2-1 shows, some analysts estimate that the demand for nonroutine skills in the United States rose during the period 1960-2009, while the demand for routine skills fell. However, these estimates include the demand for a wide range of skills—from those that require some postsecondary preparation but less than a baccalaureate credential to those that require doctoral and postdoctoral education and training. An increasing demand for higher-skilled workers does not necessarily provide evidence of the need for skilled technical workers.

Citing a 2010 study by Carnevale and colleagues of Georgetown University that looks specifically at the demand for skilled technical workers,

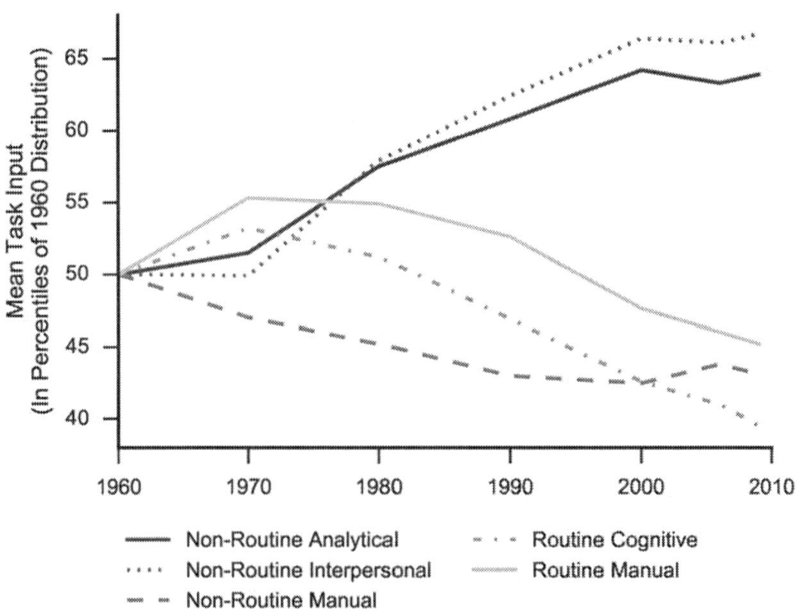

FIGURE 2-1 Change in demand for skills: Trends in routine and nonroutine tasks in occupations, United States, 1960-2009.
NOTE: The authors assigned task scores to each worker on the basis of his or her reported Census Occupation Category.
SOURCE: Autor and Price, 2013, p. 5.

the National Governors Association (NGA) Center for Best Practices warned that of the 48 million job openings projected through 2018, 63 percent will require some postsecondary education, while only 42 percent of the current workforce has an associate's or higher degree (Hoffman and Reindl, 2011). The NGA estimated that the nation will need to increase the current number of skilled technical workers by 3 million by 2018. Analyzing and projecting the demand for skills in the U.S. economy is important because these estimates are critical inputs for a wide range of workforce development decisions, including choices to pursue education and training, the design of education and training programs and infrastructure, teacher preparation, employer investments in production and innovation, and policy making with respect to economic development. However, there is considerable controversy about the nature of U.S. labor market shortages, skills gaps or mismatches, and the analytical approaches used to develop these estimates.[1]

To address questions about patterns and trends in the demand for and supply of technical skills, the committee sought data and analyses on the current supply of and demand for workers, labor market projections, and the forces likely to affect supply and demand conditions over time. As discussed in more detail in Chapters 3 and 5, the limitations of the current U.S. labor market information system make it very difficult to measure change in the skilled technical workforce precisely. Looking at recent trends and future projections across a variety of measures, however, the committee was able to develop a collage of evidence that can be used to detect broad changes in the direction of the skills gap.

For example, recent trends in job vacancies, employment, and wages can be used to suggest current labor market conditions and highlight areas in which potential imbalances may exist. In the short term, these data include persistent high vacancy rates coupled with strong employment growth and rising wages within specific occupations or regional labor markets.[2] Similarly, future projections of demographic changes and employment growth can indicate where future gaps may arise. Long-term changes in the size and composition of the available workforce can affect the cost of labor and the need for workforce development. Changes in the workforce can have important implications for the ways in which employers organize, operate, and compete: if labor is too costly in one location, employers may move jobs to another location or automate tasks that would otherwise be performed by human workers. The committee's pattern and trend analysis suggests that there may be gaps and imbalances in specific

---

[1]For additional background on the debate, see Kimmel and Martin (2015), Faberman and Mazumder (2012), and Fichen and Pellizzari (2014).

[2]Even this type of evidence can be seen as consistent with temporary imbalances due to cyclical forces. One would need to observe these trends over a period longer than the business cycle to conclude that structural forces are indeed at work.

skilled technical occupations, industries, and locations. These issues are explored at the national, local, and sectoral levels in the sections that follow.

2.1.1 National Trends in the Demand for and Supply of Skilled Technical Workers

Monitoring and measuring change in the types of skills possessed by the skilled technical workforce and the size of this workforce at the national level is difficult.[3] To derive the number of skilled technical workers in the labor force, economists typically categorize the skill level of a worker based on the person's reported educational attainment. However, these data have significant limitations (Reamer, 2015; Rothwell, 2015). Until very recently, for example, nationally representative demographic surveys of work and education experience did not ask about or draw careful distinctions between traditional formal education and other types of education and training or skills acquisition, such as alternative credentials or education and training that do not confer credit toward a credential. Thus, the use of education as an indicator of skill level can mask variation in skill levels within educational categories (Lemieux, 2006).

The National Science Board reports that data accumulated over decades reveal that in the United States, "degree is not destiny," and that only a very loose association exists between degrees and jobs at all education levels and in all degree fields (NSB, 2015). It finds that, compared with those of other nations, U.S. graduates are generally less constrained by their degree field in pursuing career options. Moreover, sorting into education and training programs is based primarily on student choices. In some nations in Europe and Asia, it is more common than in the United States to require certification for jobs and more difficult for businesses or other types of organizations to redesign jobs. By contrast, U.S. employers can be quite fluid and can acquire and use employees' skills in differing ways.

2.1.2 Exploring Imbalances across the Skilled Technical Workforce in the United States

With the foregoing caveats in mind, the committee explored imbalances in the skilled technical workforce at the national level. In a commissioned paper, Alicia Sasser Modestino of Northeastern University examines job vacancy

[3]Monitoring and measurement difficulties are identified in most of the papers commissioned for this study. For discussion of the issues regarding measurement of the entire skilled technical workforce, see Rothwell (2015), Modestino (2015), and Reamer (2015). For discussion of the issues regarding analysis of local labor markets, see GAO (2013). For discussion of the need for innovation in labor market measurement and research, see Reamer (2015).

measures to determine which major occupation groups experienced the tightest labor market conditions as of the peak of the most recent business cycle (2006), and whether these conditions persisted through the economic recovery (2012) (Modestino, 2015, 2016). She then delves more deeply into the data to examine recent wage growth (2012-2014) and future employment projections (2012-2022) by specific occupations within these major occupation groups.

Modestino finds that major occupation groups with persistently high vacancy rates employ a relatively high share of both skilled technical and higher-skilled workers (Modestino, 2015, Table 1.1). Occupation groups with critical vacancy levels include management, business, computer and mathematical science, architecture and engineering, and health care practitioners and technical occupations.[4] These occupation groups also have low unemployment relative to the number of vacancies—less than one unemployed worker per job opening—suggesting particularly tight labor markets. Many of these occupations include a significant percentage of skilled technical jobs.[5] For example, within the health care practitioner and technical occupation group, critical vacancies exist in occupations that employ a high percentage of workers with only some college or an associate's degree, such as medical and clinical laboratory technicians, surgical technologists, licensed practical and licensed vocational nurses, and medical records and health information technicians (Modestino, 2015, Table 1.2). In addition, most of the occupations that require technical skills offer high and rising wages, as well as significant projected employment growth through 2022. Among all occupations with critical vacancy levels, those that employed a larger share of skilled workers had higher job vacancy rates both before and after the recession.

Modestino also analyzed aggregate data for the supply of and demand for skilled workers by education level through 2022. She argues that the supply of skilled technical workers is likely to hold steady, but the composition of supply is not likely to keep pace with demand in the near future. Among individuals with skilled technical credentials, the percentage of those with some college is projected to increase, while the percentage of those with an associate's degree is not expected to change. Modestino attributes this forecast to low degree-

---

[4]Occupations with critical vacancy levels are defined as those having both a higher-than-average vacancy rate (number of vacancies as a percentage of total employment) and a higher-than-average vacancy share (number of vacancies as a percentage of total vacancies).

[5]In fact, skilled technical jobs are distributed throughout the broad occupation categories, suggesting that such workers are not concentrated in just a few easily identifiable sectors. Moreover, many of the jobs held by skilled technical workers appear to be complementary to those held by high-skill workers. For example, hospitals need both physicians and nursing and other support staff—jobs not easily automated or outsourced. Likewise, engineering firms need both engineers and technicians, while businesses need both systems analysts and computer support specialists.

completion rates in the nation's community colleges. She notes that these projections may overestimate the supply of technically skilled workers given that labor force participation has been declining, particularly for individuals with less than a 4-year degree.

Modestino's projections indicate that by 2022, the percentage of skilled technical job openings is likely to exceed the percentage of skilled technical workers in the labor force by roughly 1.3 percentage points, representing a shortage of about 3.4 million skilled technical workers. As the men and women of the baby boom generation continue to retire over the next decade, the absolute number of skilled technical workers will likely fall short of demand. Modestino states that her projections probably underestimate future skill imbalances because they are based on current educational requirements for jobs, which tend to increase over time.

Although Modestino's projections of future labor supply and demand may indicate where future investments in human capital may be warranted, it is important to note that future employment will be determined not only by the demands of employers and the skills of existing workers but also by future innovations that cannot fully be anticipated. For example, there is a wide-ranging debate about the impact of developments in artificial intelligence and computerization on employment and workforce development (see, for example, Arntz et al. [2016] and Frey and Osborne [2013][6]). Thus, forecasts of future labor demand are used only to place bounds on the problem and provide context rather than to pinpoint the exact number of workers that will be demanded in the future. Employers and workers are likely to adjust their behavior over the next decade if they have the incentive to do so. Workers could adjust by obtaining additional education or training or by applying their current skills in existing jobs in expanding occupations. Similarly, depending upon the structure of their incentives and the availability of skilled workers in the United States, employers could adjust by adopting new technologies, outsourcing, or restructuring jobs.

As mentioned in Chapter 1, certain conditions can create persistent labor market imbalances if demand changes rapidly and the incentives for adjustment on the supply side are weak. On the demand side, structural changes associated with advances in science and technology and competitive forces have been shown to accelerate the creation and destruction of skilled technical jobs over the course of the business cycle (Jaimovich and Siu, 2012). On the supply side, inadequate early academic preparation and weak incentives can slow the response on the part of individuals seeking to acquire technical credentials. These types of conditions could create mismatches in some specific occupational and industry sectors, such as health care. However, uncovering

[6]Whereas Frey and Osborne (2013) estimate that 47 percent of existing U.S. jobs are at risk, Arntz and colleagues (2016) estimate that this is the case for only 9 percent of existing U.S. jobs.

these conditions requires more detailed information and analyses at the organization, industry, and geographic levels. This evidence is examined in the next section.

2.2 LOCAL AND SECTORAL TRENDS IN THE DEMAND FOR AND SUPPLY OF TECHNICAL SKILLS

Many participants in the 2015 symposium convened by the committee (see Chapter 1) expressed the belief that certain regions and sectors are impeded in their ability to grow and compete because they cannot consistently meet the demand for skills.[7] Panel discussions on the current labor market (Panel 2); high school, community college, and health care pathways (Panels 3, 4, and 8); apprenticeships (Panel 6); and even federally funded programs (Panel 7) all pointed to the local and sectoral nature of skilled technical labor markets. In their keynote remarks, Senator Tim Kaine and U.S. Chamber of Commerce Foundation Senior Adviser John McKernan reinforced these comments by reflecting on their experience with and understanding of local differences.

The National Science Board argues that science, technology, engineering, and mathematics (STEM) workforces, like the skilled technical workforce, are quite heterogeneous; they consist of many different subworkforces based on degree field, occupational field, required education level, or some combination of these factors (NSB, 2015). The demand for, supply of, and career prospects for each subworkforce can vary significantly by employment sector, industry, or geographic region. The National Science Board warns that aggregate analyses and overgeneralizations that apply particular issues or challenges for a specific subworkforce to the entire workforce often lead to incorrect conclusions about the condition of the workforce. The Board argues that exploration of important questions about the STEM workforces requires a focus on disaggregated data associated with the specific subworkforce of interest.

2.2.1 Exploring the Evidence at the Local Level

A 2013 study by the Government Accountability Office (GAO) of the implementation of the federal Workforce Investment Act captures early evidence on the local and sectoral differences in skilled technical labor markets. GAO (2013) found that employers in 80 percent of the local areas surveyed in calendar year 2012 claimed some difficulty in filling skilled technical jobs and that employers' needs in some local areas were clearly not being met. Certain jobs, such as welders, machinists, and truck drivers, as well as those in health care and computer occupations, were particularly difficult to fill. Other skilled

---

[7]See the agenda for the symposium in Appendix A.

technical jobs that were difficult to fill and considered to be high-growth were in the construction and extraction and the installation, maintenance, and repair occupational groups.

The 2013 GAO study draws attention to the fact that workforce development is a local as well as a national issue. The study finds that 90 percent of local areas use state job banks and occupational projections to estimate labor market requirements. Moreover, the study highlights evidence of imbalances that arise from unique features of local labor markets, which means that a one-size-fits-all approach will not be efficient or effective. The demand for technical skills is even more likely to be specific to a given location if workers are unlikely to migrate to new jobs. Thus, workforce development activities that are guided by national trends are destined to miss their targets, and better data-driven matching and planning of current and future job openings and current and future job seekers at the local level would appear wise.

Some analysts, for example, assume that workers can and will move to obtain a job and that mobility is costless. However, a wide range of factors can affect mobility, including housing costs, the quality of education and training options, family responsibilities, social networks, and occupationally related regulation. For instance, many skilled technical jobs require occupational licensing, which imposes costs that can affect mobility. A key finding of a 2015 White House report is that licensing restricts the mobility of workers across states (The White House, 2015). Figure 2-2 illustrates the substantial differences in movement across state lines between workers in highly licensed occupations and other workers. The figure also shows that interstate migration rates for workers in the most-licensed occupations are lower by nearly 14 percent of the average migration rate compared with those in the least-licensed occupations.

2.2.2 Exploring the Evidence at the Sectoral Level

In addition to local concerns, many symposium participants in the pathways and pipelines panels (Panels 3, 4, and 8) expressed the view that workforce development efforts should focus on a more sectoral approach to address mismatches within specific industries or occupations. This concern stemmed directly from the observation that employers in certain sectors that rely on skilled technical workers often report that American workers are not adequately prepared for current and future job requirements in their industry.

It is difficult, however, to disentangle local and sectoral effects with aggregate-level analyses. For example, many information technology (IT) jobs are skilled technical positions. There is widespread consensus that the United States has a nationwide need to train many more people in the IT sector. However, recent data on the IT job market show that cities in Pennsylvania, Illinois, Oklahoma, and Alabama had higher demand for IT workers relative to their existing IT workforce compared with several other cities in other states (see Figure 2-3).

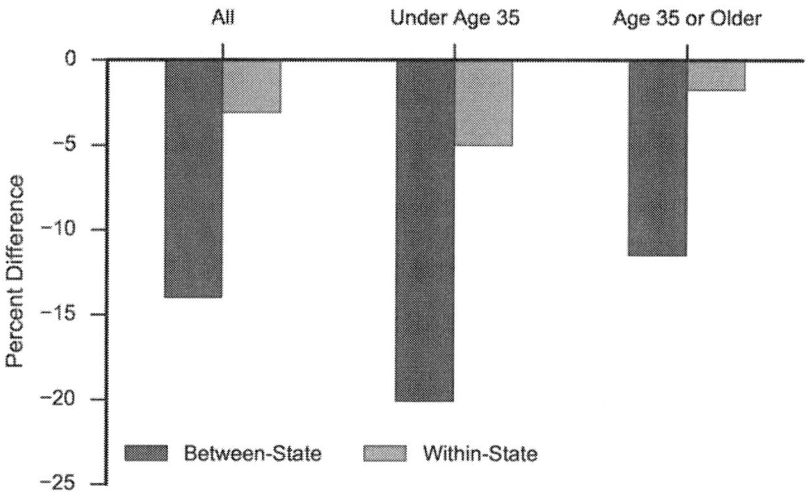

FIGURE 2-2 Difference in migration rates of workers in most- and least-licensed occupations.
NOTE: Numbers were calculated from an ordinary least squares (OLS) regression controlling for race, citizenship, sex, number of children, marital status, education, income, year, and state. Ages 25 to 65 were included.
SOURCE: The White House, 2015. Adapted from Census Bureau, American Community Survey 2010-2013; CEA calculations.

Recent employer surveys provide additional insight into the relationship between sectoral and local trends in skilled technical labor markets. In a survey of more than 800 human resources executives in February 2014, more than half of respondents reported that jobs requiring technical skills are difficult to fill, with finance, insurance, and health care companies reporting the greatest challenges (Accenture et al., 2014). Approximately 70 percent of respondent firms indicated that their inability to attract and retain individuals with technical skills frequently affected their operating performance. "Over one-third of respondents believed that inadequate availability of technically skilled workers had undermined their productivity, with manufacturing and health care the hardest hit" (Accenture et al., 2014, p. 6).

A review of Bureau of Economic Analysis (BEA) data on the sectors that contribute to growth in state economies reveals that some states and communities have larger concentrations of employers in the industries that employ high levels of skilled technical workers relative to others.[8] These

---

[8]BEA's most recent analysis of the value of the goods and services produced by states can be found in BEA (2015).

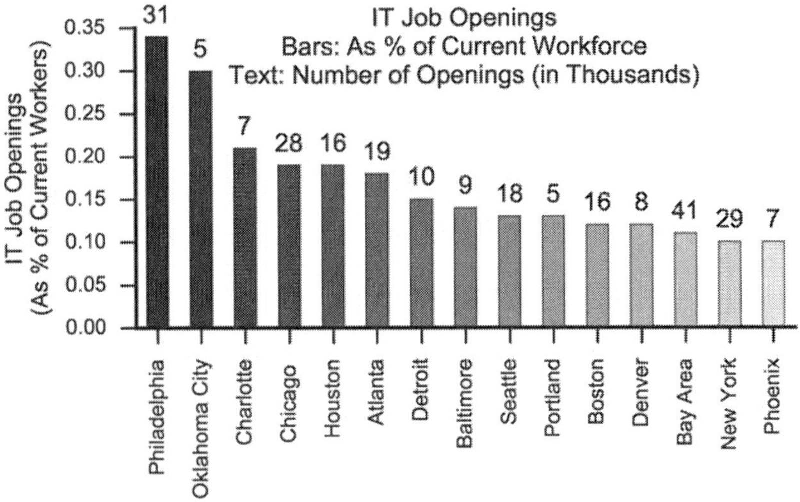

FIGURE 2-3 Cities with higher information technology (IT) demand than IT supply.
NOTE: These metropolitan areas were selected based on a "material" supply shortage. Material is defined as greater than or equal to 10 percent IT job openings as a percentage of total workforce. High vacancies can mean high turnover because jobs or employers are undesirable.
SOURCE: The White House, 2014. Adapted from CEB TalentNeuron Data.

differences are difficult to discern with analyses based on aggregate-level data. Intelligent policy responses require local labor market analyses that current labor market information systems do not adequately support.

Survey data and the preceding analysis of national occupational data on vacancies, employment, and wages suggest that certain skilled technical labor markets are experiencing tighter-than-average conditions. At the aggregate level, it does not appear that existing labor markets are currently failing to consistently provide an adequate supply of skilled technical workers. However, there may be persistent sources of imbalance in specific communities and in specific occupations and industry sectors. To understand better the policy issues affecting skilled technical labor markets, the committee looked closely at the dynamics in two sectors with high numbers of skilled technical occupations—health care and manufacturing. These sectors also serve as examples of how advances in science and technology can affect skilled technical labor markets.

2.3 TRENDS IN SUPPLY AND DEMAND IN HEALTH CARE

BLS estimates that there were about 17 million jobs in health care and social assistance in the United States in 2012 (or about 11.7 percent of total U.S.

jobs) (BLS, 2013, Table 3). Health care has been identified as an engine of job growth in the U.S. economy because of an increasing demand for health care by an aging population, as well as changes mandated by the Patient Protection and Affordable Care Act of 2010.

Bianca Frogner and Susan Skillman of the University of Washington provided the committee with an overview of labor market dynamics for health care jobs (Frogner and Skillman, 2015, 2016). They observe that many skilled technical occupations in the health care field fall within the "allied health" workforce, and while this group is not synonymous with the skilled technical workforce, there is considerable overlap.[9] Using the Occupation Finder in the BLS *Occupational Outlook Handbook*, the authors identified approximately 40 health care occupations that have prebaccalaureate entry-level requirements. They estimate that the health care field will add 3-4 million jobs over the next decade, and about 40 percent of this growth will be driven by the allied health professions, which closely resemble the committee's definition of skilled technical jobs.[10] Echoing earlier concerns, however, Frogner and Skillman argue that analyses of U.S. labor markets are very difficult to perform because models cannot predict market and population needs at the local level, and oversupply in urban areas can mask workforce gaps in rural areas.

Frogner and Skillman found several imbalances that prevent skilled technical labor markets from operating smoothly in the health care field, including the following:

- **Information deficits.** Data on the size, distribution, demographics, and work characteristics of the skilled technical workforce remain difficult to collect. Other challenges to data collection center on

---

[9]Frogner and Skillman (2016) discuss the definitional issues in the allied health field and the relationship of the workforce to the way they define skilled technical occupations. They argue that the allied health occupations are understudied, and workforce development in these occupations presents many issues similar to those identified with respect to other skilled technical occupations, such as improving data tracking and projections, understanding demand and supply, building ties between educators and employers, and identifying career trajectories. For additional background on the allied health field and related issues, see Wood (2011), Frogner et al. (2015), Ross et al. (2014), and IOM (1989, 2011).

[10]Frogner and Skillman (2015) note that although there is no single list of "allied health" occupations, it is generally accepted that allied health professions do not include physicians, dentists, or nurses. The authors further note that the Patient Protection and Affordable Care Act defines an allied health professional as "an individual who graduated with an allied health profession degree or certificate, and is employed as an allied health professional in a health care setting." Their analysis considers occupations requiring a high school degree or an associate's degree for entry, including an associate's degree for registered nursing (Frogner and Skillman, 2015, Table 1).

identifying the education and training pipeline and understanding which occupations can fill which roles. These information deficits impede policy makers' ability to estimate changes in supply and demand in skilled technical labor markets when considering workforce development policy. In addition, information deficits make it difficult and costly for labor market intermediaries, such as career centers, college placement offices, and community-based organizations, to match qualified workers with appropriate jobs.

- **Degree creep.** Occupational requirements are not always clear in the health care labor markets. For example, many health care occupations do not require a certificate or degree that is directly linked to a health care occupation. In addition, Frogner and Skillman (2015) found evidence of "degree creep," which entails setting increasingly higher educational and training requirements for practice.[11] Increasing these requirements can reduce the short-term supply of workers by delaying time of entry into the workforce, or they can reduce the long-term supply by putting education and training out of the financial reach of many workers and their employers. However, the reasons for and interpretation of "degree creep" remain unclear.
- **Accreditation and credentialing costs.** Many health care jobs require a degree from an accredited program or a specific credential, continuing education, license, or registration to practice. These requirements can provide incentives for workers to continually upgrade their skills. However, they also increase costs to invest in, prepare for, and sustain workers' qualifications. Moreover, requirements vary across states, increasing the cost of transferring from one location to another to alleviate geographic mismatches.
- **Changing needs.** The skills and training required for U.S. health care workers are changing because of shifts in the health care needs of the population; legislative changes that reshape the health care delivery system; increases in consumer safety and protection regulations; and advances in science, technology, and care modalities. These developments impose a variety of coordination costs related to education, training, and hiring of workers with the skills to meet these changing needs.[12]

---

[11]Participants in the 2015 symposium indicated that "degree creep" or "up-skilling" is a concern in many labor markets. Also see Burning Glass (2014).

[12]For a discussion of the challenges of developing training networks to support the growth of the health care workforce, see Chapter 5.

2.4 TRENDS IN SUPPLY AND DEMAND IN MANUFACTURING

BLS estimates that in 2012, the U.S. manufacturing workforce included approximately 11.9 million workers, or 8.2 percent of the total workforce (BLS, 2013, Table 3). More than 50 percent of the total manufacturing workforce involves skilled production occupations that meet the committee's definition of skilled technical jobs (Giffi et al., 2015).

Many observers have decried the loss of manufacturing jobs in the United States. A recent study by the National Academy of Engineering found that while U.S.-based companies have led high-tech manufacturing markets, they face growing competition from emerging corporations around the world (NAE, 2015). A 2014 National Academies study identified new opportunities for U.S. businesses in flexible electronics manufacturing (NRC, 2014b). As observers consistently note, however, if the United States wishes to attract and retain manufacturing facilities and compete against aggressive new competitors in developing countries, it needs to ensure that it has a qualified workforce. Robert Lerman of the Urban Institute, American University, and IZA argues in his commissioned paper that if employers cannot find the skills they need in the workforce, they may not invest or innovate (Lerman, 2015). However, the evidence with respect to this perspective is ambiguous. For example, a 2012 survey by the Massachusetts Institute of Technology found that manufacturing jobs that had moved offshore were returning to the United States (Simchi-Levi, 2012). Yet a recent A. T. Kearney (2015) report provides data that dispute this finding.

At the same time, the National Manufacturing Institute (NMI) has documented what it describes as a persistent skills gap in manufacturing.[13] Its 2015 report, prepared in collaboration with Deloitte, estimates that over the next decade, 1.5 million manufacturing jobs will go unfilled because of a skills gap in the supply of workers. According to the report, 7 of 10 manufacturing executives surveyed reported shortages of workers with adequate technology, computer, and technical training skills who can successfully pass basic screening (Giffi et al., 2015). Most of the executives surveyed indicated that the skills gap will affect their firms' ability to meet customer demand, implement new technologies, and increase productivity. Executives also reported challenges to providing effective customer service, innovating, developing new products, and expanding internationally.

On the other hand, having analyzed a nationally representative sample of manufacturing firms, Osterman and Weaver (2014b) argue that the demand for high-level skills is modest, and 75 percent of firms show no signs of hiring

---

[13]For a discussion of why manufacturing companies are not able to recruit and retain workers with the necessary skills, see Chapter 5.

difficulties, defined as long-term vacancies. Some analysts assert that there is limited upward pressure on manufacturing wages, which suggests limited evidence of a shortage of workers in these occupations. Other competitive pressures and low profit margins could explain the absence of upward wage pressure. Yet increasing skill requirements could make attracting workers more difficult unless wages rise at higher rates, creating a vicious negative spiral in manufacturing workforce development. The disagreement among analysts about the seriousness of current hiring difficulties in the manufacturing sector also highlights the need for more granular analyses of the skilled technical labor market, discussed in more detail in Chapter 4.

Some analysts in the manufacturing industry argue that structural factors or imbalances in skilled technical labor markets are likely contributors to the shortage of workers perceived by some employers. On the supply side, the number of baby boomers nearing retirement exceeds the number of younger workers who can replace them. At Boeing, for example, 28 percent of the firm's 31,000 machinists are older than 55 and eligible for retirement (Whoriskey, 2012). Yet industry leaders say that it is difficult to attract younger workers to an industry that was once considered "dying" and "dirty." They are concerned that as a result of the shift in focus of U.S. high schools toward college preparation, students are not obtaining the skills and credentials needed to enter manufacturing. In a recent poll conducted by the Foundation of Fabricators & Manufacturers Association, roughly half of all teenagers said they had no interest in a manufacturing career, and the majority of those who expressed this view agreed that a manufacturing career entailed a "dirty, dangerous place that requires little thinking or skill from its workers and offers minimal opportunity for personal growth or career advancement" (Shankel, 2010). In addition, in NMI and Deloitte's most recent study of U.S. public opinion on manufacturing, respondents aged 19 to 33 ranked manufacturing as their least-preferred career, and only a third indicated they would encourage their children to pursue a career in manufacturing (Giffi et al., 2015).

On the demand side, advances in science and technology have transformed production processes such that many skilled technical manufacturing jobs now require higher levels of science and math skills, placing manufacturing squarely in the class of STEM occupations. The recent National Academy of Engineering report *Making Value for America: Embracing the Future of Manufacturing, Technology, and Work* states that production work in the United States is shifting to require more specialized skills in such areas as robotics-controlled maintenance, advanced composites, and radio-frequency identification parts (NAE, 2015). Figure 2-4 illustrates that the number of skills required for proficiency in manufacturing occupations has consistently increased over time and is projected by leading global manufacturing firms such as Siemens to continue to do so.

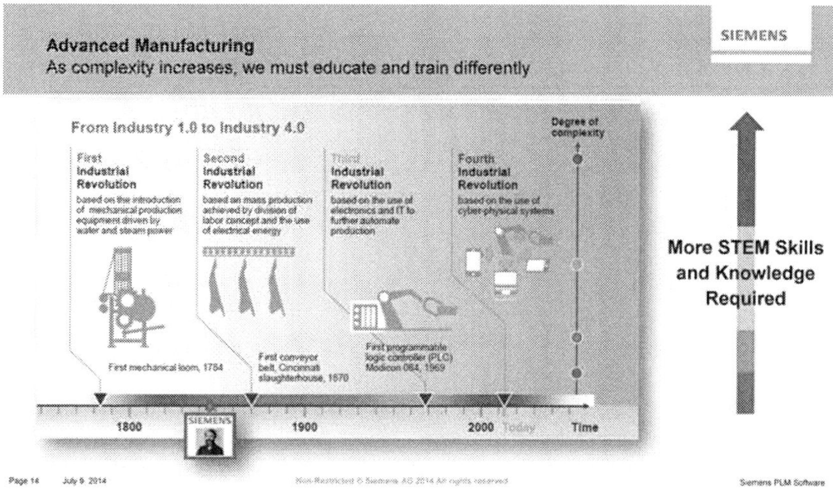

FIGURE 2-4 The evolution of skills required in production occupations.
NOTE: IT = information technology; STEM = science, technology, engineering, and mathematics.
SOURCE: Bill Boswell, Senior Director, Siemens PLM Software, Sustaining an Advanced Manufacturing Workforce presentation, September 23, 2014.

2.5 INTERNATIONAL PERSPECTIVES: OECD COUNTRIES AND PARTNERS

As the global economy continues to grow and as new markets emerge across the globe, policy makers around the world are concerned about their country's ability to trade and compete effectively (OECD, 2012a). They recognize that their workforce is a critical asset in meeting the challenges of globalization: uncompetitive firms and workers create political pressure to engage in protective measures that ultimately harm prospects for peace and prosperity at the individual, country, and global levels.

Reflecting this concern, OECD's PIAAC developed the Survey of Adult Skills to help policy makers in OECD countries assess the proficiency of adults aged 16 to 65 in key competencies that are necessary for fully integrating and participating in labor markets, education and training, and social and civic life.[14] The most recent survey identifies several trends that are influencing the development and use of technical skills in OECD and partner countries.

---

[14]The PIAAC survey does not specifically address the skilled technical workforce; therefore, its results have only general implications for this workforce.

First, access to computers and information and communication technologies is widespread and growing. More than 70 percent of households have access to computers and the Internet, and most workers have access to and use the Internet as part of their jobs (OECD, 2013). Moreover, the use of computers and information and communication technologies is changing the ways in which public and private services are provided and consumed. Many governments are increasingly delivering public services via the Internet. PIAAC estimates that in 2010, an average of 40 percent of citizens and 80 percent of businesses in OECD and partner countries interacted with public authorities via the Internet. In addition, the volume of e-commerce is growing rapidly in many countries.

The use of information and communication technologies also is changing the structure of how work is organized and performed in OECD and partner countries. Technology is enabling structural change in workplaces and markets by increasing the reach and speed of communication and reducing costs, thus facilitating the flow of goods, capital, people, and information across units, organizations, and borders. These changes mean that workplaces frequently reorganize, which compels workers to acquire new skills to adapt. According to OECD (2013, p. 48), "In most OECD countries, more than a quarter of all workers are professionals, associate professionals, or skilled technicians. Between 1998 and 2008, the number of people employed in these categories increased more rapidly than did overall employment rates in most countries." Although the evolution of employment shares for occupations with technical skills is more complex, as depicted in Figure 2-5, there is clear evidence of a significant and persistent evolution of employment in occupations requiring higher-than-average skills proficiency. Yet PIAAC estimates that imbalances between the supply of and demand for skills in OECD-country labor markets are widespread and coexist with high levels of unemployment. Skills shortages and gaps or mismatches are common, suggesting the need for "a more comprehensive account of the demand for and use of skills, including how work and organizational practices perpetuate or eliminate skills imbalances" (OECD, 2013, p. 52).

2.6 CONCLUSIONS

In the United States, the occupations that drive human development and economic growth, such as health care and manufacturing, have high proportions of jobs requiring technical skills. Such occupations as installation, maintenance, repair, and construction that develop and sustain Americans' way of life have similarly high proportions of job opportunities for workers with technical skills.

Although the ability to analyze changes in these markets with precision is limited, the evidence accumulated to date suggests that the policies and

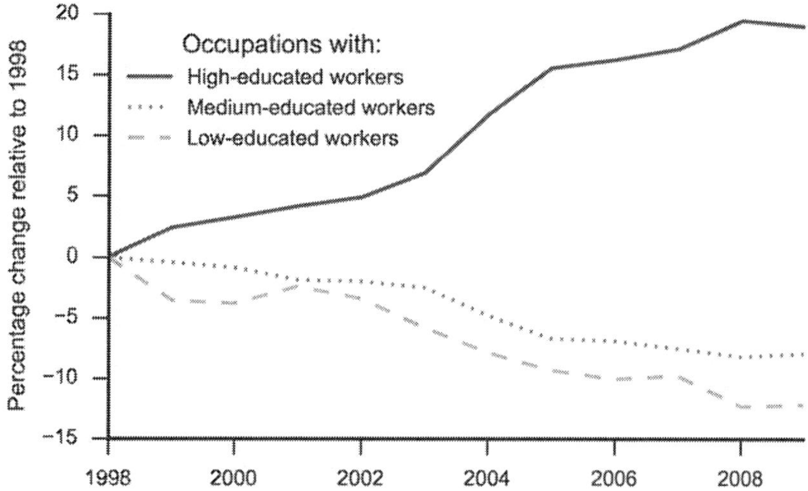

FIGURE 2-5 Evolution of employment in occupational groups defined by level of education: Percentage change in the share of employment over all employed people relative to 1998, by occupational groups defined by workers' average level of education.
NOTES: Only the 24 OECD countries available in the 1998 LFS database are included in the analysis. High level of education refers to the tertiary level or more than 15 years of schooling; medium level of education refers to no tertiary but at least upper-secondary education or around 12 years of schooling; low level of education refers to less than upper-secondary education or 11 years of schooling.
SOURCE: OECD, 2013. Adapted from Eurostat, LFS Database. Table A1.4 in Annex A.

programs governing the markets for skilled technical jobs matter. National data and industry surveys indicate a tightening of certain skilled technical labor markets. Changes in demographics, technology, and international competition create the potential for medium- and long-term mismatches in certain occupational and industry sectors. Detailed information about and analyses of how labor markets for skilled technical jobs operate at the organization, industry, sector, and local geographic levels are needed to properly understand these conditions.

A review of skilled technical occupations in health care and manufacturing suggests that markets for skilled technical jobs may not be clearing as efficiently as they could, and these imbalances are likely to be exacerbated by global patterns and trends. On the demand side, skilled technical workers face increasingly higher competency requirements as a result of structural changes associated with advances in science and technology; new business models; "degree creep"; requirements for "soft skills," including the ability to pass a drug test or meet other security-related requirements; and competitive forces.

On the supply side, factors that prevent U.S. labor markets from clearing for skilled technical jobs include a dearth of workplace and workforce information that creates barriers and raises search costs, low or stagnant wages, inadequate investment in education and training, a lack of transferability of certifications and experience that inhibits occupational and geographic mobility, and inaccurate or outdated perceptions of certain occupations that reduce incentives for individuals to invest in training. In addition, the evidence suggests that prevailing policies often affect incentives to invest in continual education and training. These incentives, discussed in more detail in subsequent chapters, have important implications for labor market clearing, where demand matches supply, in specific locations and sectors.

Developing and maintaining skills proficiency is essential to U.S. leadership in innovation, manufacturing, and competitiveness and to the creation and retention of high-wage and high-status jobs. In this report, the committee makes recommendations for how policy makers at the federal, state, and local levels can create an environment that encourages and enables adaptiveness in the skilled technical workforce. The next chapter turns to the policy landscape for workforce development.

3

The Public Policy Context

The U.S. skilled technical workforce development system is characterized by many, often competing, sources of authority and enforcement in government, business, civil society, and private life. The nation has no single, centralized system for developing a skilled technical workforce and—given the dynamics of its political and economic systems—little hope of practically achieving one. Skilled technical workforce development in the United States is the concern and shared responsibility of students; workers; labor unions and other worker associations; families; educators; businesses; industry associations; and the federal, state, and local governments.

In reviewing the public policy context, the first two sections of this chapter take note of the varied policy agenda and landscape for technical skill development. The third section summarizes key federal legislation that shapes federal policy, including the Workforce Innovation and Opportunity Act (WIOA) and the Post-9/11 Veterans Educational Assistance Act. Next is a review of the Higher Education Act (HEA) and the Carl D. Perkins Career and Technical Education Act (Perkins Act), which are under consideration for reauthorization by the U.S. Congress, followed by an overview of state policies on career and technical education. Together, these federal and state policies shape the nation's complex system of workforce education and training. Broader policy issues related to the development of a skilled technical workforce, including reforms of labor laws and rules governing occupational licensing, improvements in science, technology, engineering, and mathematics (STEM) proficiency, and upgrades to the Workforce Labor Market Information System, are then examined. The final section presents conclusions.

3.1 THE POLICY AGENDA FOR THE SKILLED TECHNICAL WORKFORCE

Educating and training a skilled technical workforce is a priority for policy makers at the federal, state, and local levels. Civic leaders, employers, industry and trade associations, labor unions, and philanthropists also have emphasized the critical role of postsecondary education and training in the development of technical skills.

Workforce development is a priority at the federal level. A bipartisan majority of Congress passed WIOA, and in 2014, President Obama signed a *Presidential Memorandum on Job Driven Training for Workers,* which tasked Vice President Biden and the secretaries of labor, commerce, and education with developing an action plan to make America's workforce development system "more job driven, integrated and effective."[1] In response to this directive, the vice president's task team reviewed what is working best nationwide and delivered a report to the President in July 2014 that includes a seven-point "Job Driven Checklist" (The White House, 2014).

The development of a skilled technical workforce is a top priority that is connected to economic competitiveness for the states' governors. Over the past 10 years, seven National Governors Association (NGA) chairs have sponsored initiatives that have included an element related to workforce development.[2] The 2013-2014 NGA chair, Governor Mary Fallin, tackled the subject in her initiative "America Works: Education and Training for Tomorrow's Jobs" (see NGA, 2015a; Simon, 2015). The NGA's current initiative, the NGA Talent Pipeline project, aims to achieve systemic change to increase the number of U.S. citizens with a postsecondary credential.[3] The focus of this initiative, which currently includes 14 states, is on aligning education and workforce development with economic development strategies and the needs of industries.

Workforce development is a concern for U.S. mayors as well. In 1977, the U.S. Conference of Mayors (USCM) created a Workforce Development Council as a USCM standing committee to ensure that workforce development would be a key focus of cities.[4] The council works to shape legislative and policy agendas related to technical skills development; provide timely information to support these efforts at the local level; and build capacity by creating networks, identifying innovative approaches, and providing technical assistance. Recent meetings reflect a concerted focus on the development of a skilled technical workforce.[5]

Policy makers' initiatives at the local, state, and federal levels all seek to engage civic leaders in the private and nonprofit sectors. In addition to employer and trade association initiatives in industries with high numbers of occupations requiring technical skills, such as health care and manufacturing, the committee noted relevant initiatives in the U.S. Chamber of Commerce, labor

---

[1] *Workforce Innovation and Opportunity Act*, Public Law 113-128 (July 22, 2014).
[2] For a historical list of NGA chair initiatives, see NGA (2017).
[3] Additional information is available at NGA (2015b).
[4] See http://www.uscmwdc.org/about-wdc.
[5] See http://www.uscmwdc.org/past meeting resources.

organizations, and philanthropic organizations across the United States. Many of these initiatives are described in more detail in the sections and chapters that follow.

3.2 THE POLYCENTRIC POLICY LANDSCAPE

Workforce development policy has a substantial history reflecting the structure of U.S. policy making: it is polycentric, having many centers of decision making and control.[6] As Figure 3-1 shows, technical skills development and regulation of labor markets are governed principally by public laws and other policies on education and labor at the federal, state, and local levels. However, they are also a concern for agencies that address health and human services, such as the Department of Health and Human Services (HHS), and the advancement of science and technology, such as the National Science Foundation (NSF). Public policies and programs that are concerned with Native Americans (Department of the Interior [DOI]) and military personnel (Department of Defense [DoD]) also address the skills development needs of their constituencies. In addition, several other initiatives, such as the Hollings Manufacturing Extension Partnership program, the America COMPETES Reauthorization Act of 2015, and small business and trade facilitation programs, provide policy direction for certain types of initiatives aimed at skilled technical workforce development.

In addition to state and local programs, several federally funded programs in the United States target specific populations such as ex-offenders, disconnected youth, and displaced or injured workers. For fiscal year (FY) 2009, for example, the Government Accountability Office (GAO) identified 47 federally funded employment and training programs administered across nine agencies (GAO, 2011a; for additional background, see GAO, 2011b,c). However, most direction and funds are provided by a relatively small set of policies and programs at the federal and state levels, and considerable effort has been made to integrate and align these policies in recent years.[7]

---

[6]For an overview of U.S. job training policy and additional context, see, for example, Decker and Berk (2011) and Holzer (2012). For an analysis of polycentricism in American political and administrative life, see the work of Ostrom (1994, 2008), for example.

[7]One of the most obvious examples of efforts to integrate and align federal and state workforce development policies occurred with the enactment of the WIOA. For an overview of these efforts, see Heinrich (2015), as well as The White House (2014). A GAO (2012) report to Congress on the Workforce Investment Act examines efforts to integrate and align state and federal activities prior to WIOA enactment. Policy alignment and integration efforts were also consistently noted in presentations and discussions during the committee's 2015 symposium and in presentations at committee meetings held

FIGURE 3-1 Federal agencies participating in the policy-making landscape for the development of a skilled technical workforce.
NOTE: ACF = Administration for Children and Families; ATE = Advanced Technological Education; CTE = career and technical education; DoD = U.S. Department of Defense; HHS = U.S. Department of Health and Human Services; HRSA = Health Resources and Services Administration; NSF = National Science Foundation; VA = U.S. Department of Veterans Affairs; WIOA = Workforce Innovation and Opportunity Act.
SOURCE: Compiled by the committee.

3.3 FEDERAL-LEVEL POLICIES AND PROGRAMS

As Table 3-1 shows, federal policy has provided scope for many federally funded employment and training programs that apply to a wide range of workforce development issues and populations. In a review of these programs for FY 2009, the GAO noted that almost all of them overlap with one or more other programs in that they provide at least one similar service to a similar population, although differences may exist in eligibility, objectives, and service

September 10, 2015, and February 17, 2016. A webcast of the symposium, research papers, and presentations can be found at http://nas.edu/SkilledTechnicalWorkforce.

TABLE 3-1 Federally Funded Employment and Training Programs by Agency, Fiscal Year 2009

Agency	Programs
Department of Labor	• Community-Based Job Training Grants • Employment Service/Wagner-Peyser Funded Activities • H-1B Job Training Grants • Homeless Veterans' Reintegration Project • Job Corps • Local Veterans' Employment Representative Program • National Farmworker Jobs Program • Native American Employment and Training • Registered Apprenticeship and Other Training • Reintegration of Ex-Offenders • Senior Community Service Employment Program • Trade Adjustment Assistance • Transition Assistance Program • Veterans' Workforce Investment Program • WIA Adult Program • WIA Youth Activities • WIA Dislocated Workers • WIA National Emergency Grants • WANTO • YouthBuild
Department of Education	• American Indian Vocational Rehabilitation Services • Career and Technical Education - Basic Grants to States • Career and Technical Education - Indian Set-aside • Grants to States for Workplace and Community Transition Training for Incarcerated Individuals • Migrant and Seasonal Farmworker Program • Native Hawaiian Career and Technical Education • Projects with Industry • Rehabilitation Services - Vocational Rehabilitation Grants to States • State-Supported Employment Services Program • Tech-Prep Education • Tribally Controlled Postsecondary Career and Technical Institutions

(Continued)

TABLE 3-1 Continued

Agency	Programs
Department of Health and Human Services	• Community Services Block Grant • Refugee and Entrant Assistance Voluntary Agency Matching Grant Program • Refugee and Entrant Assistance - Targeted Assistance Grants • Refugee and Entrant Assistance - Social Services Program • Refugee and Entrant Assistance - Targeted Assistance Discretionary Program • Temporary Assistance for Needy Families • Tribal Work Grants (Native Employment Works)
Department of the Interior	• Conservation Activities by Youth Service Organizations • Indian Employment Assistance • Indian Vocational Training - United Tribes Technical College
Department of Agriculture	• SNAP Employment and Training Program
Department of Defense	• National Guard Youth Challenge Program
Environmental Protection Agency	• Brownfield Job Training Cooperative Agreements
Department of Justice	• Second Chance Act Prisoner Reentry Initiative
Department of Veterans Affairs	• Vocational Rehabilitation for Disabled Veterans (VetSuccess Program)

NOTE: SNAP = Supplemental Nutrition Assistance Program; WANTO = Women in Apprenticeship and Nontraditional Occupations; WIA = Workforce Investment Act.
SOURCE: Testimony of Andrew Sherrill, director of education, workforce, and income security issues, before the Subcommittee on Labor, Health and Human Services, Education and Related Agencies, Committee on Appropriations, House of Representatives, GAO 11-506T.

delivery (GAO, 2011a). The GAO reports that federal employment and training programs spent approximately $18 billion on employment and training services in FY 2009. However, seven programs accounted for about three-fourths of that spending, and most participants received employment and training services through either the Employment Service/Wagner-Peyser Funded Activities (Employment Service) or the Adult Program of the Workforce Investment Act of 1998 (WIA) (GAO, 2011a, Table 1). Together, these two programs, both of which are administered by the Department of Labor (DOL), served more than

18 million individuals, or about 77 percent of the total number of participants served across all federal employment and training programs, in 2009.

Mary Alice McCarthy of the New America Foundation states that one way to understand federal policy efforts aimed at helping people gain skills for work is to analyze *all* the funding streams for programs that prepare individuals for a specific occupation, including tuition assistance (McCarthy, 2014). Looking at the sources of federal funding for postsecondary education assistance in 2013, she observes that this funding comes disproportionately from the Title IV federal student aid programs—not from the 47 federal employment and training programs identified by the GAO study and not from the 3 programs dedicated specifically to workforce development and career and technical education.[8] As Figure 3-2 shows, of the $145 billion in education assistance provided in 2013, the vast majority ($130.7 billion) was governed by Title IV of HEA, while $10 billion was governed by the Post-9/11 Veterans Educational Assistance Act of 2008, $2.5 billion by WIOA, $1.1 billion by the Perkins Act, and $575 million by the Trade Adjustment Assistance Act (TAA). In McCarthy's view, postsecondary education is influenced primarily by the goals, policies, and funding of higher education.

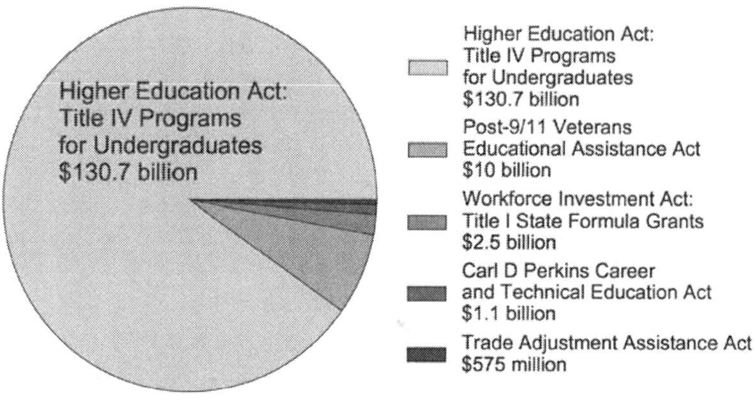

FIGURE 3-2 Annual federal funding for postsecondary education and training.
SOURCE: McCarthy, 2014.

[8]McCarthy (2014) identifies the three programs specifically dedicated to workforce development as the Perkins Act, WIOA, and the Trade Adjustment Assistance Act (TAA).

To understand the dynamics and incentives of workforce development, it is important to analyze all the policy, program, and funding mechanisms that prepare individuals for specific occupations. As described in more detail in this chapter and in Chapter 4, however, policies, programs, and funding streams for postsecondary skilled technical workforce development are largely under the purview of local- and state-level policy makers and other stakeholders.

The remainder of this section and the section that follows provide an overview of the four federal laws the committee believes are most likely to contribute to synchronizing activities focused on skilled technical workforce development across the nation and provide the greatest support for postsecondary education and training. This section continues by describing recently enacted laws that are directly related to workforce development: WIOA and the Post-9/11 Veterans Educational Assistance Act of 2008. Congress also recently passed the Every Student Succeeds Act (ESSA), summarized in Box 3-1, which indirectly affects workforce development.[9] The section that follows describes federal policy directly related to workforce development that is currently under deliberation for reauthorization: HEA and the Perkins Act. Both of these reauthorizations could have implications for policies and programs related to skilled technical workforce development.

3.3.1 Workforce Innovation and Opportunity Act

Passed on July 22, 2014, WIOA represents the first major legislative reform of the public workforce system in 15 years. It is designed to help job seekers and workers access employment, education, training, and support services that can help them succeed in the labor market and to match employers with the skilled workers they need to compete in the global economy. WIOA is administered by DOL and the U.S. Department of Education (ED). It supersedes WIA and amends the Adult Education and Family Literacy Act, the Wagner-Peyser Act, and the Rehabilitation Act of 1973.[10] Box 3-2 provides an overview of the purposes of WIOA.

In her review of federal workforce development efforts, presented at the committee's 2015 symposium, Carolyn Heinrich of Vanderbilt University states that a number of new WIOA provisions appear to be based on evidence of proven and promising strategies, such as incentivizing the development of sectoral training partnerships and requiring active coordination across agencies

---

[9]The text of ESSA can be found at https://www.congress.gov/bill/114th-congress/senate-bill/1177/text. More information about the new law is available at http://www.ed.gov/essa. Developments in career and technical education (CTE) are discussed in more detail in Chapter 4.

[10]See http://www.doleta.gov/WIOA/Overview.cfm (accessed February 28, 2017).

BOX 3-1
Every Student Succeeds Act

The Every Student Succeeds Act (ESSA), a bipartisan bill, was signed into law by President Obama on December 10, 2015. ESSA reauthorized the 50-year-old Elementary and Secondary Education Act (ESEA), replacing the prior version of the law, the No Child Left Behind Act (NCLB). The new law requires states to establish high college- and career-ready academic standards and guarantees that steps will be taken to help students and schools improve. It reduces the burden of testing that existed under NCLB while maintaining annual information for parents and students, and provides increased access to high-quality preschool.

ESSA seeks to promote local innovation and improvement in teaching and learning by authorizing various programs designed to (1) ensure that low-income and minority students are not being taught at disproportionate rates by ineffective teachers; (2) support improved teaching and learning through the implementation of human capital management systems that include sustainable performance-based compensation; and (3) support innovative and evidence-based approaches to teacher and leader recruitment, preparation, and development. To improve teaching and learning in science, technology, engineering, and mathematics (STEM), the law authorizes a STEM Master Teacher Corps program, which supports state-led efforts to elevate the status of the STEM teaching profession.

SOURCE: U.S. Department of Education, 2017.

in the development of priorities and programs (Heinrich, 2015). Recognizing the local nature of workforce development, WIOA grants states more flexibility to transfer funding between programs.

Observing that WIOA has yet to be fully implemented, Heinrich provides the following suggestions for policy makers:

- Begin training earlier for youth through approaches that blend vocational and on-the-job training, and involve employers as partners in the delivery of the training.
- Provide more funding and technical support to better link secondary, postsecondary, and workforce development data; assess the value of skills acquired in programs; and better direct job seeker training choices and American Job Center strategies.
- Improve data resources and their availability for research, and use these data to develop better performance incentive systems for guiding state and local resource allocations.

BOX 3-2
Workforce Innovation and Opportunity Act (WIOA)

Each of the act's purposes and key reforms is relevant to the development of technical skills. The purposes of WIOA are as follows:

(1) To increase, for individuals in the United States, particularly those individuals with barriers to employment, access to and opportunities for the employment, education, training, and support services they need to succeed in the labor market.
(2) To support the alignment of workforce investment, education, and economic development systems in support of a comprehensive, accessible, and high-quality workforce development system in the United States.
(3) To improve the quality and labor market relevance of workforce investment, education, and economic development efforts to provide America's workers with the skills and credentials necessary to secure and advance in employment with family-sustaining wages and to provide America's employers with the skilled workers the employers need to succeed in a global economy.
(4) To promote improvement in the structure of and delivery of services through the United States workforce development system to better address the employment and skill needs of workers, jobseekers, and employers.
(5) To increase the prosperity of workers and employers in the United States, the economic growth of communities, regions, and States, and the global competitiveness of the United States.
(6) For purposes of subtitle A and B of title I, to provide workforce investment activities, through statewide and local workforce development systems, that increase the employment, retention, and earnings of participants, and increase attainment of recognized postsecondary credentials by participants, and as a result, improve the quality of the workforce, reduce welfare dependency, increase economic self-sufficiency, meet the skill requirements of employers, and enhance the productivity and competitiveness of the Nation.

SOURCE: 29 U.S.C. § 3101, http://uscode.house.gov/view.xhtml?req=(title:29%20section:3101%20edition:prelim).

- DOL should consider making additional federal funding contingent on service levels to disadvantaged adults and youth, and on expanded training capacities in high-demand fields and for skilled technical jobs.
- DOL should lend greater financial and institutional support to cross-state research efforts and exchanges that seek to test innovative workforce development strategies.

3.3.2 Post-9/11 Veterans Educational Assistance Act of 2008

The Post-9/11 Veterans Educational Assistance Act of 2008 amends federal veterans' benefits provisions to provide updated and comprehensive educational assistance under the Montgomery GI Bill to individuals in certain length-of-service categories serving on active duty in the armed forces commencing on or after September 11, 2001.[11] Congress reports that a Congressional Budget Office (CBO) estimate for this measure has not been received.[12] The U.S. Department of Veterans Affairs reports that since the bill's implementation on August 1, 2009, it has provided more than $20 billion in educational benefits to 773,000 veterans and their family members.[13] Nearly 1 million eligible individuals participated in FY 2011, a 15 percent increase over FY 2010 and a 71 percent increase over FY 2009. Although the act requires that educational assistance be used within 15 years of an individual's discharge or release from active duty (with exceptions), it does allow individuals currently under the Montgomery GI Bill educational assistance program to elect to participate in the Post-9/11 Veterans Educational Assistance program with respect to any unused entitlement.

Congress notes in findings related to this act that the United States has a history of offering educational assistance to millions of veterans, as demonstrated by the many "GI bills" enacted since World War II. Educational assistance for veterans reduces the costs of war, assists veterans in readjusting to civilian life after wartime service, boosts the U.S. economy, and has a positive effect on recruitment for the armed forces. It is in the national interest for the United States to provide veterans who served on active duty in the armed forces after September 11, 2001, with enhanced educational assistance benefits that are worthy of such service and are commensurate with the educational assistance benefits provided to World War II veterans.

---

[11]The text of H.R.5740, the Post-9/11 Veterans Educational Assistance Act of 2008, is available at https://www.congress.gov/bill/110th-congress/house-bill/5740/text. For additional background on the bill, see Dortch (2014b).

[12]See https://www.congress.gov/bill/110th-congress/house-bill/5740/all-info.

[13]See http://www.va.gov/opa/issues/post_911_gibill.asp.

The act supports the development of technical skills in a number of ways. For example, it allows for the pursuit of (1) programs on less than a half-time basis, (2) apprenticeship or other on-the-job training, (3) correspondence courses, (4) flight training, (5) tutorial assistance, and (6) licensure and certification tests. The act also requires the secretary of the U.S. Department of Veterans Affairs to carry out the Yellow Ribbon GI Education Enhancement Program, under which colleges and universities may enter into an agreement with the secretary to cover, for certain individuals, a portion of established charges not otherwise covered and to match contributions toward such costs.

3.4 FEDERAL LEGISLATION UNDER CONSIDERATION

3.4.1 The Higher Education Act

HEA, which is administered by ED, authorizes numerous federal aid programs that provide support to individuals who pursue postsecondary education and to institutions of higher education.[14] Title IV authorizes the federal government's major student aid programs, which are the primary source of direct federal support to students who pursue postsecondary education, while Titles II, III, and V provide aid and support to educational institutions. HEA also authorizes services and support for less-advantaged students, students pursuing international education, and students and institutions engaged in certain graduate and professional degree programs. Title VIII supports the goals of improving access to postsecondary education and improving enrollment, persistence, and completion rates for specific groups of students, including veterans.

The rules and regulations governing access to HEA's financial aid programs, including institutional accreditation and student outcome reporting requirements, play a role in shaping how skilled technical education and training programs are designed and delivered and what is known about labor market outcomes for students. The nature and extent of this role need to be examined in parallel with the roles of local and state policies and programs.

Congress is considering reauthorization of HEA. The last time Congress comprehensively reauthorized the act was in 2008, when it passed the Higher Education Opportunity Act of 2008 (HEOA). HEOA authorized most HEA programs through FY 2014 but has been extended while Congress prepares for the next reauthorization (Hegji, 2014). Among the issues likely to be addressed in the final bill are affordability and college costs; access, persistence, and completion; better information for consumers; student loan programs; accreditation and oversight; innovation; and reduction of the burden of federal regulations.

---

[14]For a primer on HEA, see Hegji (2014).

On July 30, 2014, for example, Senator Kaine introduced the Jumpstart our Businesses by Supporting Students Act, which seeks to amend HEA by expanding eligibility for Pell Grants to students enrolled in short-term job-training programs.[15] The corresponding press release indicates that the initiative would authorize Pell Grants for job-training programs at community colleges and other institutions of higher education, ensure that qualifying programs align with the needs of local employers and lead to a recognized postsecondary credential, encourage eligible institutions to connect short-term credential programs to career pathways, and provide basic skills instruction to support student success (Kaine, 2015).[16]

3.4.2 The Perkins Act

The purpose of the Perkins Act is to develop the academic and career and technical skills of secondary and postsecondary education students who enroll in career and technical education (CTE) programs. The act, which is administered by ED, was last authorized on August 12, 2006 (Perkins IV).[17] The goals of Perkins IV include the attainment of rigorous academic standards by CTE students, "the integration of secondary and postsecondary CTE elements into single programs of study, and the assessment and accountability of the achievement of educational and post-educational outcomes" (Dortch, 2012). The act's original authorization period ended on June 30, 2013, and the act is currently scheduled for reauthorization.

3.5 STATE-LEVEL POLICIES

Policies, programs, and funding streams for skilled technical workforce development at the state and local levels are extensive and varied. At the K-12 level, the National Center for Education Statistics (NCES) reports that in 2013-2014, there were 98,271 operational public schools in more than 18,000 districts (NCES, 2015a, Table 2). In the fall of 2011, there were also 30,861 private schools offering kindergarten or higher grades (NCES, 2013a). NCES reports that current expenditures for public elementary and secondary schools are projected to be $634 billion for the 2015-2016 school year. State and local

---

[15]The federal Pell Grant program provides need-based grants to low-income undergraduate and certain postbaccalaureate students to promote access to postsecondary education. Financial need is determined by ED using a standard formula established by Congress. See the ED website, http://www2.ed.gov/programs/fpg/index.html.

[16]The remarks made by Senator Kaine to the committee at its 2015 symposium can be found at http://nas.edu/SkilledTechnicalWorkforce.

[17]For background on the evolution of the Perkins Act, see Dortch (2012).

government agencies are the primary source of funding for community colleges. In 2013-2014, expenditures for community colleges totaled $55.9 billion, with nearly half (49.4 percent) coming from state or local funding, 22.6 percent from federal funding, and the remainder from student tuition and fees (16.7 percent) and other sources (11.3 percent) (NCES, 2015b, Table 333.10).

Local governance and funding for K-12 education can have an important influence on postsecondary skilled technical workforce development (see Chapter 5 for more detail).[18] However, consistent with the scope of this study, the committee focused on CTE policy making, which is one of the most common ways in which states address the challenges associated with skilled technical workforce development.

CTE is offered by high schools, area CTE centers, community colleges, technical schools, and employers through classrooms, project- or work-based learning, online courses, apprenticeships, on-the-job training, and other methods (Dortch, 2014b). The goal of CTE, which must be aligned with business and industry requirements, is preparation for employment or family life. CTE occupations generally require 2 years or less of postsecondary education or training and span a wide range of technical skills and subjects in fields other than the liberal arts. For example, CTE provides preparation in homemaking, nursing, business administration, culinary arts, automotive maintenance, software programming, engineering technology, and cosmetology.

A recent paper prepared jointly by the National Association of State Directors of Career Technical Education Consortium (now known as Advance CTE and formerly known as NASDCTEc) and the Association for Career and Technical Education (ACTE) provides an overview of CTE-related state policies enacted in 2014 (NASDCTEc et al., 2015; see also Blosveren, 2015). This report shows that, for the second year in a row, a significant number of states developed and implemented new policies and programs to advance CTE at the secondary and postsecondary levels. Nearly every state had CTE-related activity: state legislatures and regulatory bodies approved approximately 150 policies across 46 states and the District of Columbia. As Table 3-2 shows, the top three issues for state action are related to (1) funding, (2) engaging employers, and (3) earning postsecondary credit in high school and articulating credit across institutions.

Advance CTE and ACTE report on several notable developments in 2014 at the state policy level. California formalized a $250 million investment in its Career Pathways Trust Grant, and South Dakota created a $50 million Build Dakota scholarship program for students entering high-demand workforce programs. Alabama and Iowa created tax incentives to encourage donations and

---

[18]The committee recognizes the important role of K-12 STEM education in fostering a skilled technical workforce, but defers to a recent National Academies study for a detailed discussion of this topic (see NRC, 2014a).

TABLE 3-2 State-Level Policy Making on Career and Technical Education in 2014

Policy Area	Number of States Addressing Policy Area in 2014	States
Funding	36	AL, AK, AZ, AR, CA, CO, CT, DE, DC, FL, GA, ID, IN, IA, KS, KY, LA, ME, MD, MA, MI, MN, MS, NE, NJ, NM, NY, OH, OR, SD, TN, UT, VT, WA, WV, WI
Industry/Partnerships/ Work-based Learning	28	AL, AK, AR, CA, CO, CT, DE, GA, IL, IN, IA, LA, ME, MD, MI, MN, MO, NH, NJ, NY, OH, OR, RI, SD, TN, VT, WV, WI
Dual and Concurrent Enrollment/Early College/Articulation	24	AL, AK, CT, DE, FL, GA, ID, IL, IN, KS, KY, LA, ME, MD, MA, MS, MO, NV, NJ, OH, OR, SD, TN, TX
Industry-recognized Credentials	19	AZ, CA, DE, FL, GA, ID, IL, IN, KS, LA, MI, MN, MS, MO, OH, RI, TN, UT, WV
Graduation Requirements	15	AZ, FL, IL, IN, LA, MI, MN, MS, MO, NY, OH, OK, SC, VA, WA
Data/Reporting/ Accountability	15	CA, CO, CT, FL, IN, IA, KY, LA, ME, MI, MN, MO, NJ, UT, VA
STEM	10	AZ, DC, IA, NH, NY, OK, OR, UT, VT, WA
CTE Standards/ Accreditation	10	CO, FL, ID, LA, MN, NJ, OK, TN, WA, WY
Technical/Employability Assessments	7	AZ, MS, MO, NV, NY, OH, SC
CTE Teacher Certification/ Development	7	IL, ME, NJ, ND, RI, TN, VA
Career/Academic Counseling	6	AL, AR, CA, KY, OH, RI
Governance	5	AL, NM, OR, RI, UT

*Note: A single bill or policy can address multiple policy areas.
NOTES: This table is not exhaustive, and therefore, not every state policy found is included. STEM = science, technology, engineering, and mathematics.
SOURCE: NASDCTEc et al., 2015, p. 2.

grants to increase business–education collaboration. Twenty-eight states approved policies to accelerate employer engagement with CTE to help align programs with labor market demands and offer work-based learning opportunities. Twenty-four states addressed earning and articulating credit issues. The Nevada State Board of Education and Board of Regents, for example, approved a new policy to develop statewide articulation agreements for all CTE programs.

Examples of policy-making activities related to business engagement and workforce preparation include measures in Kansas and Tennessee to accelerate credential attainment, Florida's measure to inform parents about the return on investment for certain credentials, and activities of Illinois and Delaware to establish grant programs that will provide training for leading industries in those states (NASDCTEc et al., 2015). In Massachusetts, the Workforce Competitiveness Trust Fund provides $5 million to support 15 active regional industry-sector partnerships of employers and education and workforce organizations that are engaged in training unemployed individuals and placing them in high-demand jobs.

3.6 CURRENT POLICY ISSUES RELATED TO DEVELOPING A SKILLED TECHNICAL WORKFORCE

3.6.1 Labor Law Reform

U.S. labor laws and enforcement agencies at the federal, state, and local levels regulate the rights, duties, costs, and benefits associated with the relationship between skilled technical workers and employers. These rules create important incentives for investing in education and training, a point made by Daniel Marschall, executive director of the AFL-CIO Working for America Institute, during the first panel discussion at the committee's 2015 symposium, and echoed in other panel discussions, papers, and presentations.[19] For example, if workers and employers can be reasonably sure that the labor market will function in a way that enables them to capture returns on their investments in education and training, they are more likely to make these investments. As discussed in more detail in Chapter 5, there is evidence that some employers may not be committed to hiring specific groups of workers, paying high enough wages, or providing sufficiently stable employment to ensure an adequate return on investment in education and training for either their workers or their organization.

[19] In his commissioned paper, Lerman (2015) explores incentive issues for employers. These issues were a particular topic of discussion in Panels 1, 2, and 5 of the committee's 2015 symposium, as well as in each of the keynote speeches.

U.S. labor rules are based principally on laws that are more than 80 years old, including the National Labor Relations Act of 1935 (Wagner Act) and the Fair Labor Standards Act of 1938. However, technological change in the U.S. economy has given rise to a proliferation of nontraditional and contingent employment relationships, which are creating pressure to review and revise these rules to ensure that incentives to promote education, training, innovation, and growth are sustained over time.[20] These developments have been referred to as the "online gig" or "on-demand" economy: work is taking place in a series of one-off or temporary assignments rather than in an ongoing relationship with a single employer. For example, technology-driven online platforms for delivering goods and services make it possible to contract for specific, on-demand services in many occupations, such as maintenance, repair, information technology (IT), and health care assistance.

Technological change presents a separate but related issue in skilled technical workforce development.[21] Most of the papers commissioned for this study, as well as discussions in the symposium panels, alluded in some way to the opportunities and challenges posed by technological change for skilled technical workers and their employers; however, these changes are an immediate and prevalent concern in the health care industry (Frogner and Skillman, 2015). As machines increasingly are integrated into the ways people think about and perform tasks, human–machine teams will proliferate in the workplace. Heterogeneous, symbiotic work teams raise entirely new issues concerning how to allocate labor rights and responsibilities, assign credit for work performed, train both human and machine workers, and allocate returns to labor.

How policy makers decide to classify and regulate these emerging work relationships has important implications for investing in the development and maintenance of a skilled technical workforce in the U.S. economy. A detailed explanation of these issues is beyond the scope of this study, but the committee believes they are important and urges policy makers across governments and locations to address them.

3.6.2 Occupational Licensing Reform

Many skilled technical jobs in maintenance, repair, construction, and health care occupations require some form of licensing or certification. In their commissioned paper, for example, Frogner and Skillman (2015) discuss the impact of licensing requirements on developing and sustaining qualifications in many health care jobs and on moving from one employer or geographic location

---

[20]For an overview of the estimates and issues, see Dokko et al. (2015).

[21]For an overview of the issues and challenges, see Brynjolfsson and McAfee (2011).

to another. And in their paper, Carter and Miller (2015) discuss the impact of licensing requirements on veterans' employment.

Occupational licensing has important benefits. It is usually justified because it improves quality and protects the public from incompetent or dangerous practitioners. Licensing helps consumers assess a provider's competency when it is difficult for them to do so before buying. In addition, licensing potentially contributes to the development of professional standards, legitimacy, and social status. It can build community and cohesiveness within an occupation and help standardize knowledge, skill, and practices. When licensing functions in these ways, it can offer benefits to providers through increased recognition of occupational work and influence.

Licensing and regulation also can increase the availability of information that helps markets function more efficiently. If consumers are unable to assess the quality of providers before purchasing a good or receiving a service, for example, low-quality providers can remain in the market without being recognized as such, reducing the average quality in the market as well as incentives for other providers to invest in quality improvements (Akerlof, 1970; Kleiner, 2000; Shapiro, 1986). If consumers are uncertain about the quality of a provider, moreover, they may not buy the service, depressing market demand for potentially valuable services. Licensing is one way to address information problems by compelling providers to meet certain quality standards and creating incentives to invest in training and skill development.

However, occupational licensing also imposes costs that can affect incentives to invest in skill development. The imposition of such requirements as additional training and education, fees, exams, and paperwork potentially reduces employment in the licensed occupation and hence competition, which can drive up the price of goods and services for consumers. This could benefit licensed providers, who might earn more than they would in an unlicensed market, or the financial benefits could flow elsewhere, such as to educational institutions or other licensing entities (Friedman and Kuznets, 1954; Kleiner, 2006; Law and Kim, 2005; Smith, 1776).

Licensing requirements also can create a barrier to entry for many jobs. If licensing places too many restrictions on workers, it can reduce the overall efficiency of the labor market. When workers cannot enter jobs that make the best use of their skills, growth and innovation may be depressed. Licensing also may affect entrepreneurship (Slivinski, 2015). Licensed workers are more likely than other workers to be self-employed. In addition, entrepreneurs in new areas that overlap with a licensed occupation may find that they are required to obtain a license because a small part of their work overlaps with that of a licensed occupation.

Finally, licensing can affect worker mobility and the availability of technical skills in labor markets. Many occupations are licensed by the states, and licensed providers typically must acquire a new license when they move to a different state. States set their own licensing requirements, which often vary with respect to education, experience, training, and testing. The resulting costs in both time and money can discourage people from entering or remaining in an occupation. Licensing requirements can be particularly costly for some workers, such as military spouses, who are very likely to move across state lines. Constraints on worker mobility can create inefficiency in labor markets if workers are unable to migrate easily to the jobs in which they are most productive.

A White House (2015) report on occupational licensing contains several noteworthy findings:

- The share of U.S. workers holding occupational licenses has grown sharply over the past several decades: more than 25 percent of U.S. workers now require a license to perform their jobs, and most of these workers are licensed by the states.
- About two-thirds of the changes in licensing stem from an increase in the number of professions that require a license, and about one-third from the changing composition of the workforce.
- Although licensing can increase incentives to invest in education and training, it also can reduce employment, lower wages for excluded workers, and increase costs for consumers if it makes entry into a profession more difficult.
- As noted above, licensing requirements can vary considerably both within and across states, potentially imposing costs on workers seeking to move across state lines, as well as inefficiencies for businesses and the economy as a whole.

These developments and issues call for DOL to work with local and state policy makers to review their licensing requirements, encourage them to limit licensing to those occupations entailing demonstrable public safety concerns, seek ways to reduce the costs of learning about and obtaining licenses, and coordinate with other in- and out-of-state authorities to harmonize requirements across occupations and sectors.

3.6.3 Science, Technology, Engineering, and Mathematics (STEM) Proficiency

STEM skills are essential components of the bundle of skills that define skilled technical occupations and qualify workers to succeed in these occupations.[22] Moreover, the STEM proficiency of the workforce is a prominent

[22]This observation is supported in each of the papers commissioned for this study, as well as in keynote speeches and discussions at the committee's 2015 symposium. See, in particular, the papers by Rothwell (2015) and Sheets and Tyszko (2015) and

concern in discussions of national competitiveness, education policy, innovation, and even immigration.

In a 2014 companion study to its annual study of STEM indicators, the National Science Board (NSB) explored STEM indicator data, examined recent STEM workforce studies and debates, and consulted numerous experts to develop insights that could facilitate more constructive discussions about the STEM workforce and inform decision makers (NSB, 2015). NSB's findings include the following:

- The STEM workforce is extensive and critical to innovation and competitiveness. As Box 3-3 shows, it is also defined in various ways and consists of many subworkforces. It includes the scientists and engineers who further scientific and technological progress through research and development (R&D) activities, workers in non-R&D jobs who use STEM knowledge and skills to devise or adopt innovations, and workers in technologically demanding jobs who need STEM capabilities to accomplish occupational tasks.
- STEM knowledge and skills allow for multiple, dynamic pathways to STEM and non-STEM occupations alike. Although many individuals with a STEM degree do not work in a STEM field, most of these workers indicate that their job is related to their STEM education. The relatively loose links between degrees and occupations are a distinctive feature of the U.S. workforce. This feature enables individuals to apply STEM skills in jobs across the economy and employers to utilize workers with STEM skills in whatever ways add the greatest value.
- Assessing, enabling, and strengthening workforce pathways are essential to the mutually reinforcing goals of individual and national prosperity and competitiveness. A well-rounded precollege education that includes significant engagement with STEM unlocks pathways into the technical STEM workforce and enables pursuit of additional STEM studies at the bachelor's, master's, and doctoral levels.

Yet a more recent NSB study suggests that policy makers still have a great deal of work to do to ensure that education and training programs effectively help students and workers build and sustain their STEM skills (NSB, 2016).

---

presentations and discussions in Panel 4, "Community College Pathways (Advanced Technical Education)," and Panel 11, "Building the Middle-Skill Talent Pipeline and Innovation." For additional insight on the topic, see NASEM (2016a,b,c) and NAE and NRC (2012). As noted earlier, a webcast of the symposium, research papers, and presentations is available at http://nas.edu/SkilledTechnicalWorkforce.

BOX 3-3
What are the STEM Capabilities?

Efforts are underway to gather data on STEM knowledge and skills. This research promises to supplement degree- and occupation-based analyses by clarifying our understanding of the interconnections among skills, education, and occupation. The U.S. Department of Labor's Occupational Information Network (O*NET) program is developing a database with detailed information on the competencies of workers in nearly 1000 occupations.[a]

Anthony Carnevale and colleagues have analyzed O*NET data to identify which competencies are highly associated with STEM occupations.[b] Among the cognitive competencies associated with STEM are knowledge of math, chemistry, and other scientific and engineering fields; STEM skills, such as complex problem solving, technology design, and programming; and STEM abilities, including deductive and inductive reasoning, mathematical reasoning, and facility with numbers. Among the non-cognitive competencies associated with STEM are preferences for investigative and independent work.

---

[a]The database, O*NET Online, can be found at http://www.onetonline.org/.

[b]Anthony Carnevale, Nicole Smith, Michelle Melton, "STEM," (Washington, DC: Georgetown University Center on Education and the Workforce, 2014). See p. 8 for a full list of the O*NET competencies associated with STEM. Retrieved from: https://cew.georgetown.edu/wp-content/uploads/2014/11/ stem-complete.pdf.

SOURCE: Excerpted from NSB, 2015.

Although 4 of 10 Americans say they are "very interested" in new scientific discoveries that contribute to innovation, growth, and national competitiveness, and 6 of 10 say they have an equally high level of interest in new medical discoveries, only 46 percent of Americans can demonstrate an understanding of scientific inquiry. Moreover, most Americans believe that other countries are doing a better job of providing STEM education and supporting lifelong education and training (NSB, 2016, Chapter 7):

- Although the percentages of 4th-, 8th-, and 12th-grade students achieving a level of proficient or higher on National Assessment of Educational Progress mathematics assessments increased from 2000 to 2013, those percentages remained well below the 50 percent mark.
- Average mathematics and science literacy scores in the United States are below those for all other developed countries, and the United States has substantially fewer high scores and more low scores than those other countries.

The development of STEM capabilities begins in K-12 education. The quality of K-12 education, particularly students' preparation in STEM subjects, has long been a concern. A 2010 report of the President's Council of Advisors on Science and Technology (PCAST) expresses concern about uneven levels of proficiency and interest in careers in many STEM fields among African Americans, Hispanics, Native Americans, and women (PCAST, 2010). The PCAST report suggests several strategies for addressing these issues, such as supporting higher standards in mathematics and science, recruiting well-prepared teachers for STEM subjects, leveraging educational technology for improvement, expanding students' opportunities to engage in rich STEM experiences, and creating more STEM-focused schools. These strategies for improvement and others are being implemented across the country. Among the most visible initiatives are the new standards in both mathematics and science—the Common Core State Standards in Mathematics (CCSS-Math) and the Next Generation Science Standards (NGSS). These standards provide a national model for states to use as they develop their own specific standards. Both the CCSS-Math and the NGSS are influencing conversations about high-quality K-12 STEM education, even in states where they have not been formally adopted.

Deep exploration of the challenges facing K-12 STEM education and the improvement initiatives across the country is beyond the scope of this study. For further information, see the portfolio of work by the Board on Science Education of the National Academies, including *A Framework for K-12 Science Education* (2012), *Successful K-12 STEM Education* (2011), *Monitoring Progress toward Successful K-12 STEM Education* (2013), *STEM Integration in K-12 Education* (2014), *Developing Assessments for the Next Generation Science Standards* (2014), *Guide to Implementing the Next Generation Science Standards* (2015), and *Science Teachers' Learning* (2015).

All citizens need STEM skills to function effectively in a modern economy that is increasingly based on advances in science and technology. The range and level of required STEM proficiency vary across occupations, including skilled technical occupations, and evolve as new discoveries and technologies are disseminated in life and work. Chapter 6 provides examples of additional approaches that some cities and states are taking to address this important issue.

3.6.4 The Workforce Labor Market Information System

The information system that supports policy making and labor market function—the workforce labor market information system (WLMIS)—is administered primarily by DOL (GAO, 2013; Reamer, 2015).[23] All stakeholders in workforce development seek high-quality data, analyses, and projections of

---

[23]The WLMIS consists of systems and data owned and controlled by each state. In addition, the Department of Commerce manages and owns data from surveys such as the census that are inputs for the WLMIS.

key segments of the U.S. workforce, such as skilled technical workers. For example, labor market data and analyses are used by policy makers and civic leaders to address education and workforce development needs; by students and workers to make choices about education, training, and credentialing; and by employers to make investment and hiring decisions. In his commissioned paper, presented at the committee's 2015 symposium during the panel on "Data Resources to Support Middle Skill Workforce Development," Andrew Reamer of Georgetown University argues that the WLMIS is less effective than it could be (Reamer, 2015).

Reamer attributes the problems in the current system to several factors. For example, funding for the WLMIS is insufficient, and coordination across local and federal agencies that contribute to and use the system could be improved. Reamer also finds a lack of understanding among analysts of how participants in labor markets at the microeconomic level rely on the WLMIS to make decisions. Moreover, many analysts are using outdated and ineffective research methods. They have not innovated in response to advances in the data and computational sciences.

Reamer notes that DOL is actively exploring how it might better use advanced, low-cost IT to improve the WLMIS. Moreover, WIOA mandates and provides a detailed framework for providing most of the information resources that labor market participants need, and aims to address many of the systemic problems. However, Reamer observes that the predecessor to WIOA (WIA of 1988) gave the secretary of labor a similar mandate to create and maintain a national employment statistics system, but succeeding secretaries did not pursue this mandate (Reamer, 2015).

Based on Reamer's detailed analysis and supporting evidence from other papers and panel discussions, the committee observes that the WLMIS does not effectively support the country's skilled technical workforce development needs.[24] Chapter 6 provides examples of additional approaches that some cities and states and are taking to address this issue.

3.7 CONCLUSIONS

In the United States, incentives to develop technical skills are influenced by a complex, polycentric system of labor and education policies and programs that operate at the federal, state, and local levels. Skilled technical workforce

[24]Every paper commissioned for this study notes the inadequacy of the existing WLMIS. It is not unusual for analysts to call for more and better data; however, some participants in Panel 10 at the 2015 symposium echoed Reamer's call for analysts to be more creative in their research methods.

development is currently a high priority for the Executive Branch, Congress, the states' governors, and the cities' mayors.

Congress recently passed WIOA, which represents the first major reform of the workforce development system in 15 years. This legislation requires extensive alignment and integration across federal, state, and local workforce development activities. Although the reforms contained in this act are appropriate to address many issues concerning the skilled technical workforce and are based on the best available evidence about effective strategies, the act has yet to be fully implemented, and ongoing research is needed. Many experts express concern about whether and how well these reforms will be fully implemented and whether the findings of salient research will be addressed. At the state level, policy makers are focusing on two interconnected types of improvements: (1) CTE, which prepares secondary and postsecondary students for skilled technical jobs and postsecondary education; and (2) better articulation and alignment of the incentives of skilled technical workers, educators, employers, trade associations, labor unions, and industry to achieve workforce development objectives. Several policy areas require high-priority attention at the federal, state, and local levels: labor law reform, occupational licensing reform, lifelong STEM skills improvement, and investment in an improved WLMIS and associated analyses.

The next chapter explores the incentives of participants who are most directly involved in producing the skilled technical workforce.

4

The Complex U.S. System of Workforce Education and Training

Education and training take place in a variety of contexts in the United States. There is no single, formal system for training workers, and Americans are responsible for finding their own path, based on their preferences, capacities, and means. Students are encouraged to begin making these choices at the secondary school level, when they have the option to pursue both academic and technical education. After earning a high school diploma, students may enter full-time employment in a technically skilled job, continue their education and training, or do both at the same time. Americans can select from a large menu of degree- and non-degree-granting postsecondary education and training programs at any time during their work lives. Unless students receive scholarships, grants, or education benefits from public or private sources, they are ultimately responsible for paying all the costs associated with their education and training by drawing on savings or current earnings or by incurring debt.

The federal, state, and local governments provide resources that support education and training both directly and indirectly. In addition, although the incentives for employers to support education and training for workers are highly variable, many employers either provide education and training directly to workers on the job or subsidize their employees' education and training through reimbursement programs. But whether employees are using these programs to their greatest advantage and to what extent employers are contributing to the development of a skilled technical workforce remain unclear.

This chapter describes the complexities of workforce development in the United States and the components of this system that are particularly relevant for preparing skilled technical workers. Section 4.1 introduces the quasi-public characteristic of education and training. Section 4.2 describes the primary components of workforce development, including career and technical education (CTE) and degree-granting postsecondary education and training programs. Section 4.3 examines other postsecondary education and training programs, including apprenticeship, certificate, and certification programs, that do not confer a degree. Section 4.4 reviews key funding sources for skilled technical workforce education and training, including federal, state, and employer sources. Section 4.5 presents conclusions.

4.1 EDUCATION AND TRAINING AS A QUASI-PUBLIC GOOD

Education and training are examples of a quasi-public good, having characteristics of both private and public goods. Society as a whole is better off when all of its members are able to develop their innate capacities. Education and training also can potentially serve public purposes by ensuring the availability of a pool of workers who can contribute to economic growth. At the same time, the benefits of private investments in education and training potentially accrue to individuals in the form of opportunities and wages. Hence both society and individuals have a role to play with respect to investment in education and training.[1]

Given that both the costs and benefits associated with investments in education and training affect individuals, employers, and the broader society, it can be difficult to separate, measure, and allocate returns on these investments.[2] For example, by observing others, their way of doing things, and the knowledge products they produce, one can educate and train oneself. In other words, knowledge and learning spill over to others, and it is difficult to fully recover the costs of production and distribution. Moreover, because some of the benefits accrue to society as a whole, individuals may undervalue and underinvest in education and training, suggesting a role for public provision or subsidies (Cornes and Sandler, 1996).

Americans have a long history of pursuing and publicly supporting elementary and secondary education and training. The colonists began to establish public schools in the early 1600s; Boston Latin School was established in 1635. By the time of the Revolution, some colonies, such as Georgia, were partially funding grammar schools (*Education News*, 2013). By 1918, attendance in elementary school was compulsory in all states, and by 1920

---

[1]Several theories aim to explain why individuals invest in education and training. For example, Becker (1964, 1993) conceptualizes the set of marketable skills of workers as a form of capital in which workers make a variety of investments to obtain better opportunities and wages. Alternatively, Spence (1973, 2002) conceptualizes education and training as a way for individuals to signal the nature of their abilities to prospective employers and thereby gain above-average wages. Theoretical perspectives are potentially important in understanding investment incentives and the structure of wages and earnings. However, exploring the range of theories and the evidence that supports them is beyond the scope of this study.

[2]For a more detailed explanation and the theoretical underpinnings of this analysis, see Cornes and Sandler (1996).

roughly one-third of Americans aged 14-17 had attended some form of secondary school. During the period 1940-2014, the percentage of U.S. adults with a high school degree increased from 24.1 to 88.3 percent (Ryan and Bauman, 2016, Table A-1).

In contrast with many other countries that are members of the Organisation for Economic Co-operation and Development (OECD), however, most federal- and state-level policies in the United States treat postsecondary education more like a private good than a public or quasi-public good. These policies often assume that workers will capture all of the returns on their investments in education and training and that no one else in society will benefit from those investments. Yet despite this fundamental misalignment of incentives, education is valued by an increasing proportion of the population. During the period 1967-2013, the percentage of Americans who completed high school and then enrolled in either a 2- or 4-year degree-granting institution increased from 25.5 to 68.4 percent (BLS, 2016a). Moreover, approximately 44 percent of individuals aged 16 or older participate in some form of formal adult education, such as basic skills training, apprenticeships, work-related courses, personal-interest courses, English as a second language classes, part-time college, or university degree programs (NCES, 2007).

Seeking to further invigorate commitment to workforce development, the OECD secretary general argued in a 2012 report that skills have become the "global currency" of 21st-century economies. Reflecting on the results of a survey conducted by the Programme for the International Assessment of Adult Competencies, he urged member countries to find ways to improve skill levels and invest continuously in the identification and acquisition of new skills so these countries can effectively organize and compete in the 21st century (OECD, 2012a).

4.2 THE PRIMARY COMPONENTS OF WORKFORCE DEVELOPMENT

As in many other aspects of public life, the United States does not have a single, formal system for educating and training workers. Americans who wish to develop technical skills and prepare themselves for success in skilled technical occupations have a wide range of education and training options that depend on several factors, including their access to funding, their social networks, the availability of programs in their geographic location, and their interests.

These specifics aside, in general Americans have the range of formal options for skilled technical education and training depicted in Figure 4-1, which they may pursue simultaneously, sequentially, or in idiosyncratic order. Beginning with secondary education, students may choose to take advantage of

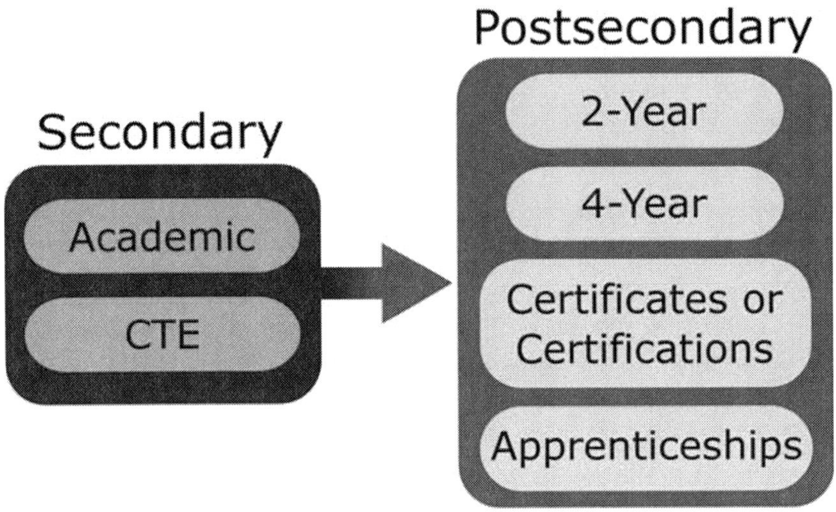

FIGURE 4-1 Decision options at the secondary and postsecondary levels for skilled technical education and training.
NOTE: CTE = career and technical education.

CTE programs offered, often in conjunction with cooperative work experience programs, in comprehensive high schools, full-time CTE schools, and area or regional CTE centers. CTE students can take work-related courses in subjects related to skilled technical occupations, acquire hands-on work experience, and prepare to enter full-time employment upon obtaining their high school diploma. Some proponents argue that secondary CTE is valuable because it serves as pedagogy for science, technology, engineering, and mathematics (STEM) subjects. Because many students learn better with a hands-on approach and skilled technical CTE requires some level of STEM proficiency, secondary CTE provides a way to reinforce and integrate academic education and technical training. Moreover, if CTE programs are overseen by effective committees whose members come from business and industry, they can advise CTE educators, administrators, and education board members on employment needs and requirements.

Concurrent with secondary studies or upon graduation from high school, a student may prepare for a skilled technical occupation through programs in an institution of higher education; an organization that certifies skill competencies, such as an industry association or professional society; or an apprenticeship. Finally, youth and adults may enroll in skilled technical job training through various programs funded by the Workforce Innovation and Opportunity Act (WIOA) through the American Job Centers or community-based organizations.[3]

To inform understanding of the education and training decisions being made by students and workers, the following subsections review enrollment and other salient quantitative data for formal skilled technical education and training programs.[4]

4.2.1 Secondary Career and Technical Education

The U.S. Department of Education reports that from 1990 to 2009, the average number of CTE credits earned by U.S. public high school graduates declined from 4.2 to 3.6, while the average number of credits earned in other subject areas increased. By contrast, average credits earned in core academic fields (i.e., English, mathematics, science, and social studies) rose during this same period. Also in this period, the percentage of graduates who reportedly earned credit in any occupational CTE area declined from 88 to 85 percent (NCES, 2013b). Within occupational CTE, however, the direction and magnitude of change differed by specific occupational area. Occupational areas with declining participation were business, manufacturing, computer and information sciences, engineering technologies, repair, and transportation. Occupational areas with increasing participation were communications and design, health care, public services, and consumer and culinary services, with communications and design being the area of greatest increase.

In the review of work-related education in high schools commissioned for this study, Stern (2015) found that CTE programs are provided in about 9,500 comprehensive high schools in the United States. In addition, he found that about 1,000 technical high schools offer both CTE and academic subjects, and about 800 area or regional technical schools offer only CTE courses.

Looking at trends in enrollment, Stern found that the percentage of high school graduates who concentrated in an occupational area by taking three or more CTE credits in the same area fell from 33 percent in 1982 to 19 percent in 2009. Overall, the percentages of CTE concentrators among high school graduates decreased among all socioeconomic, racial, and ethnic groups over the same period and are no longer skewed toward less-affluent students. As Table 4-1 shows, however, in the aggregate, CTE programs still enroll relatively large proportions of students from low-income families. Many CTE students

---

[3]These programs are relatively small in scale. See Figure 3-2 in Chapter 3.

[4]In addition to formal education and training options, some workers may elect to engage in self-study programs or to participate in employer-sponsored on-the-job training. However, it is difficult to find data on informal choices; hence, the analysis here is limited to formal programs.

TABLE 4-1 Percentages of Career and Technical Education (CTE) Concentrators from Different Demographic Groups

Group	Percentage in 2004
Lowest socioeconomic quartile	21
Middle socioeconomic quartiles	19
Highest socioeconomic quartile	12
Black	18
Hispanic	13
White	19

SOURCE: Compiled based on Stern (2015).

also have lower levels of prior academic performance. While these aggregate data may differ at the local and state levels, they suggest the need to raise awareness about the value of skilled technical work and the opportunities associated with CTE preparation. In addition, curricula need to address career awareness, exploration, and preparation in elementary as well as secondary school, and younger children need to have the opportunity to develop skills and explore technical work.

Looking at the percentage of CTE concentrators in 12th grade who state their expected level of education, Stern found that 54 percent aim to obtain a bachelor's degree or higher, and another 28 percent expect to attend some college. Stern concludes that preparing students for immediate entry into full-time employment after high school is no longer a widely espoused goal: most students who take work-related courses in high school now also complete sufficient academic coursework to enter some form of postsecondary education.

4.2.2 Postsecondary Degree-Granting Institutions

The U.S. Department of Education reports that there were 4,207 degree-granting institutions—2,603 4-year and 1,604 2-year—in the United States in 2014-2015 (NCES, 2016c, Chapter 4). Figure 4-2 shows the number of degree-granting institutions with first-year undergraduates in the academic years 2000-2001 and 2014-2015 by level (4- or 2-year) and type (public, private nonprofit, or private for-profit) of institution. Most 2-year institutions were public (919), whereas most 4-year institutions were private nonprofit (1,283). One interesting change pertains to the total number of private for-profit institutions, which almost doubled from 687 in 2000-2001 to 1,258 in 2014-2015. However, most of this growth occurred in 4-year institutions, whose number more than tripled from 207 to 656, compared with an increase in 2-year institutions from 480 to 602.

Public, private nonprofit, and private for-profit institutions are regulated in different ways. These regulations may have consequences for innovation, scale, quality, cost-effectiveness, and other factors. Investigating regulatory differences is beyond the scope of this study; however, it may be useful for policy makers to examine these differences in detail in local contexts.

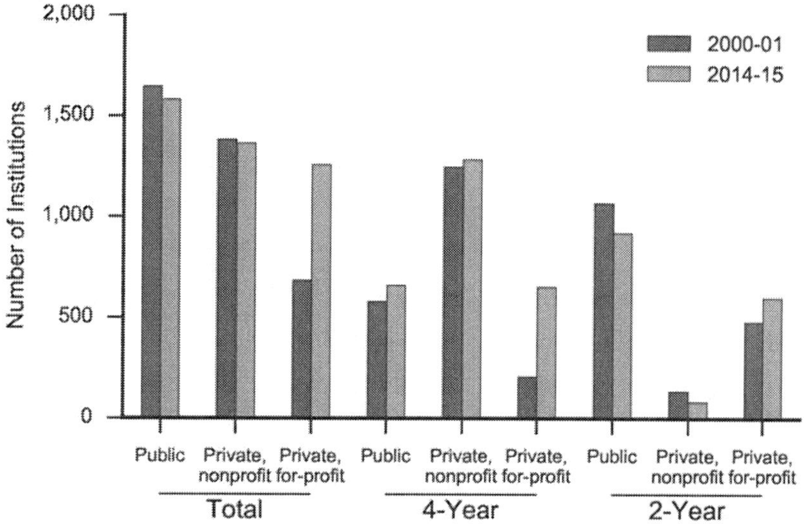

FIGURE 4-2 Number of degree-granting institutions with first-year undergraduates.
SOURCE: NCES, 2016c.

Degree-granting postsecondary institutions vary in many ways, but two dimensions in particular are frequently noted: admissions policies and completion rates.

In academic year 2014-2015, 29 percent of all 4-year institutions had open admissions policies (accepted all applicants). In that same year, a higher percentage of private for-profit 4-year institutions (66 percent) than private nonprofit (15 percent) and public (19 percent) 4-year institutions had open admissions policies. By contrast, the majority of 2-year institutions (91 percent) had open admissions policies. Open admissions policies were in operation at 98 percent of public 2-year institutions and 84 percent of private for-profit 2-year institutions, compared with 56 percent of private nonprofit 2-year institutions (NCES, 2016c).

Looking at completion rates, the majority (60 percent) of all first-time full-time undergraduates enrolled in 4-year institutions graduate within 6 years (NCES, 2016c). Private nonprofit institutions have the highest completion rates (65 percent), followed by public institutions (58 percent) and private for-profit institutions (just 27 percent). By contrast, only 28 percent of all first-time full-time undergraduates enrolled in 2-year institutions graduate within 150 percent of the normal time, and these statistics are skewed by low rates of completion in public institutions, where just 20 percent of students graduate at all.[5] The majority of students at private institutions graduate within 150 percent of the normal time, and students at for-profit institutions graduate at higher rates than those at nonprofit institutions (59 versus 51 percent).

4.2.3 Student Enrollment

In fall 2014, 17.3 million Americans were enrolled in a degree-granting institution (NCES, 2016c). Most of these students were enrolled in 4-year institutions (10.6 million), and the remainder in 2-year institutions (6.7 million). Students in 4-year institutions typically enroll full-time (76 percent), whereas students in 2-year institutions typically enroll part-time (70.4 percent). Data show rather modest participation in distance education. Students who do participate in this form of education appear to prefer combining it with residential learning: about 28 percent of students enroll in some distance education, whereas just 12 percent pursue all of their education and training through this modality (NCES, 2016c).

As Figure 4-3 shows, U.S. Department of Education statistics indicate that although most students of degree-granting institutions are young and in the early stages of their work life, a significant portion (22 percent) are age 30 or older. In contrast, just 1 percent of these students are under age 18. These data suggest that to support the achievement of skilled technical workforce objectives, institutions need to develop adept curriculum designs, pedagogy, and support services that are appropriate for students of a wide range of ages and experience levels. And given that young students will have different needs, options, and incentives from those of their older adult counterparts, policies and programs need to be designed to create incentives for both youth and older cohorts to invest in skilled technical education at all stages of their work lives.

4.2.4 Community Colleges

Most 2-year degree-granting institutions are community colleges, which are Title IV institutions that provide affordable postsecondary education and a potential pathway to a 4-year degree.[6] Community colleges are a key component

---

[5]Low completion rates impact the return on postsecondary vocational degrees and certificates. Chapter 5 reviews the evidence that the return is associated with receipt of the degree (the so-called "sheepskin effect"), not with the training alone.

[6]Title IV of the Higher Education Act of 1965 (HEA) covers the administration of the U.S. federal student financial aid programs. American colleges and universities are generally classified with regard to their inclusion under Title IV, such as under U.S. Department of Education statistics.

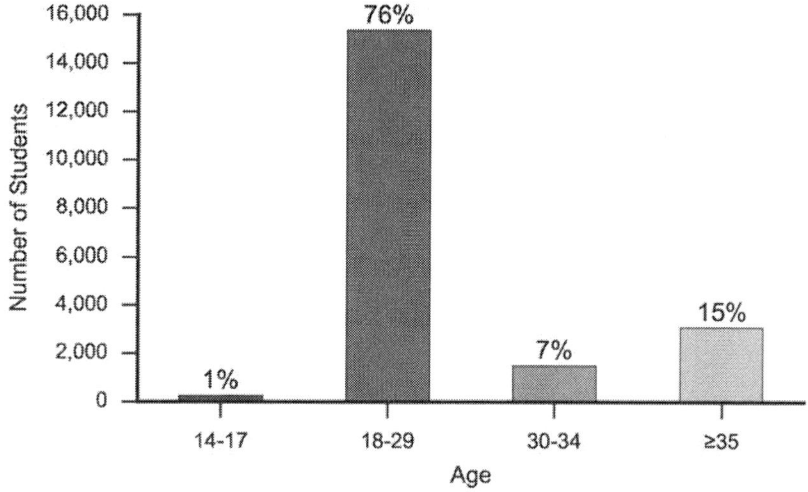

FIGURE 4-3 Postsecondary enrollment in degree-granting institutions by age group in 2015 (projected).
SOURCE: Compiled based on NCES (2014a, Table 303.40).

of skilled technical workforce development in local economies. The American Association of Community Colleges (AACC) reports that 1,108 U.S. community colleges enroll approximately 12.3 million students, or about 45 percent of all undergraduate students in the United States (AACC, 2016).

The mission of community colleges is to provide education to all individuals in a local area, and as noted above, most have an open-access admissions policy. Community colleges offer a comprehensive range of education and training to support lifelong learning that includes both credit and noncredit programs. For example, community colleges typically offer associate's degrees that are transferrable to 4-year degree programs, terminal associate's degrees, certificates, certification, apprenticeships, and contract or custom training on a noncredit basis. According to AACC, most community college students (60 percent) are enrolled in credit programs on a part-time basis (AACC, 2016). In addition, most work while they attend school: 73 percent of part-time and 62 percent of full-time students work either part- or full-time. Most community college students also receive some form of financial assistance. The U.S. Department of Education reports that 78 percent of first-time 2-year degree-/certificate-seeking students receive financial assistance to meet the costs of tuition, books, and other educational expenses (NCES, 2016c).

Aggregate data suggest that public funds, particularly state and local funds, are crucial sources of revenue for community college operations. Table 4-2 shows that most community college revenues come from tuition (29.5 percent) and from state (29.8 percent) and local (18.1 percent) sources. AACC data for

TABLE 4-2 Community College Revenues by Source (2013-2014)

Source	Revenue	%
Tuition	$17,242,025,437	29.5
Federal	8,264,032,954	14.1
State	17,442,989,183	29.8
Local	10,568,296,830	18.1
Other	4,929,830,599	8.4
Total	**$58,447,175,003**	**100.0**

SOURCE: AACC, 2016.

academic year 2011-2012 reveal that although most student financial aid was received from federal sources, state and institutional aid were also important (AACC, 2016).[7] In addition, in 2013-2014, community colleges received federal aid in the form of Pell Grants, Federal Supplemental Educational Opportunity Grants, and Federal Work-Study programs.

Researchers at Columbia University Teacher's College report that 80 percent of community college students intend to earn a bachelor's degree, but only one-quarter of them transfer to a 4-year institution within 5 years (Jenkins and Fink, 2015). However, AACC (2015) data indicate that community college students who earn an associate's degree as a first credential are more likely to transfer to a 4-year institution to earn a bachelor's degree: 41 percent of college students who earned an associate's degree in 2008-2009 as a first credential subsequently earned a bachelor's degree within 6 years. Community College Research Center analysts have shown that the completion outcomes of community college transfer students vary depending on the type of institution to which they transfer, as well as the number of credits accepted by the receiving institution (Jenkins and Fink, 2015). The analysis shows that approximately 72 percent of community college students who transferred went to public 4-year institutions. Data on bachelor's degree completion 6 years after transfer show that 65 percent of students who transferred to a public 4-year college/university earned a bachelor's degree.

4.3 OTHER POSTSECONDARY EDUCATION AND TRAINING PROGRAMS

Skilled technical workforce development encompasses education and training programs that complement degree-granting programs, including apprenticeships, work-related courses, certificate programs, and certification

[7]AACC analysis for the year 2011-2012 reveals that of students who received financial aid, 38 percent received federal grants, 19 percent received federal loans, 12 percent received state aid, and 13 percent received institutional aid.

programs. In many cases, students pursue both types of education and training either simultaneously or sequentially.

The National Center for Education Statistics (NCES) collects data on adult education that include participation in apprenticeships and work-related courses.[8] It reports that participation in all forms of adult education "among individuals aged 16 or older increased from 40 percent in 1995 to 46 percent in 2001 and then declined to 44 percent in 2005....In 2005, among the various types of adult education activities, individuals aged 16 or older participated most in work-related courses (27 percent), followed by personal interest courses (21 percent), part-time college or university degree programs (5 percent), and other activities (3 percent)" (NCES, 2007, Indicator 10).

Table 4-3 summarizes the 2005 NCES data (the most recent data) relevant to overall worker participation in skilled technical workforce development. These data suggest that aggregate participation rates vary by sex, age, and occupation: participation is higher for women than for men, for those aged 30-65 relative to other age groups, and for those working in health-related occupations relative to those working in other skilled technical occupations, although these data may vary by location.

The most common types of worker education and training programs that do not confer degrees and are particularly relevant to skilled technical workforce development are apprenticeships, certificate programs, and certification programs. Students and workers who wish to participate in such programs and earn a degree must concurrently enroll in a degree-granting program in an accredited secondary or postsecondary institution. An overview of these programs is provided below.

4.3.1 Apprenticeships

Apprenticeships are formally structured work training programs in which workers are employed in a job and earn wages while receiving on-the-job training and job-related technical instruction. In the United States, registered apprenticeship programs must meet national standards: the U.S. Department of Labor and the states' apprenticeship agencies register and administer formal apprenticeship programs. At the conclusion of the apprenticeship, the worker earns a nationally recognized credential.

The U.S. Department of Labor estimates that the U.S. registered apprenticeship network currently includes 150,000 employers and more than

---

[8]For background and links to data, see NCES (2007).

TABLE 4-3 Participation in Job-Related Courses and Apprenticeship Programs in 2005

	2005					
			Percent Participating			
Characteristic of Employed Persons	No. of Employed Persons (in thousands)		In Career- or Job-Related Courses		In Apprentice-ship Programs	
Total	**133,386**	**(1,508.1)**	**38.8**	**(0.83)**	**1.4**	**(0.24)**
Sex						
Male	71,754	(934.7)	31.7	(1.22)	2.0	(0.37)
Female	61,632	(1,219.3)	47.1	(1.43)	0.8	(0.23)
Age						
17-24	15,027	(1,030.4)	26.4	(3.01)	3.0[a]	(1.03)
25-29	14,555	(918.4)	36.1	(2.94)	3.1[a]	(1.12)
30-34	15,250	(977.2)	41.0	(3.06)	2.7[a]	(1.10)
35-39	15,286	(922.4)	41.7	(4.16)	1.0[a]	(0.46)
40-44	18,141	(946.3)	39.8	(2.73)	[b]	[c]
45-49	18,149	(842.5)	45.0	(2.15)	0.7[a]	(0.29)
50-54	14,624	(732.1)	42.6	(2.49)	0.7[a]	(0.32)
55-59	10,522	(676.0)	44.7	(2.98)	[b]	[c]
60-64	6,021	(498.8)	38.9	(3.97)	[b]	[c]
65 and over	5,812	(493.3)	21.6	(3.48)	[d]	[c]
65-69	3,385	(415.5)	19.1	(4.05)	[d]	[c]
70 and over	2,427	(282.3)	25.1	(5.81)	[d]	[c]
Occupation						
Health diagnosing and treating practitioners	978	(208.8)	78.9	(7.10)	[b]	[c]
Registered nurses, pharmacists, dieticians, therapists, and physician assistants	2,794	(238.8)	79.7	(4.60)	[b]	[c]
Health technologists and technicians	3,060	(436.7)	70.6	(7.31)	[b]	[c]
Technologists and technicians, except health	1,774	(336.5)	29.4	(8.10)	[b]	[c]
Mechanics and repairers	5,241	(521.6)	28.3	(4.47)	4.0[a]	(1.44)
Construction and extractive occupations	6,827	(647.1)	12.4	(3.04)	5.3[a]	(2.26)

TABLE 4-3 Continued

Characteristic of Employed Persons	2005 No. of Employed Persons (in thousands)		Percent Participating In Career- or Job-Related Courses		In Apprenticeship Programs	
Precision production	10,483	(839.3)	23.5	(3.79)	*b*	(*c*)
Production workers	*e*	(*c*)	*e*	(*c*)	*e*	(*c*)
Miscellaneous occupations	801	(189.4)	17.2	(6.87)	*b*	(*c*)

[a]Interpret data with caution.

[b]Reporting standards not met.

[c]Not applicable.

[d]Rounds to zero.

[e]Not available.

NOTES: Standard errors appear in parentheses. Neither this table nor the National Center for Education Statistics (NCES) data on which it is based are all-inclusive or represent a comprehensive study. The table indicates potential areas of further analysis and research.
SOURCE: Based on NCES (2014b, Table 507.30).

1,000 occupations (see DOL, n.d.).[9] Included in this total is the United Services Military Apprenticeship Program (USMAP), which accounted for nearly one in four registered apprenticeships in the United States in 2013 (Lerman et al., 2015).[10] However, NCES data (Table 4-3) suggest that fewer than 2 percent of all U.S. workers participate in apprenticeship programs. More detailed analyses of participation would likely reveal differences across groups and locations.

In a survey of apprenticeship theory and practice, Wolter and Ryan (2011) find that apprenticeships vary greatly across countries in terms of the numbers of workers who are trained and the skill content of the training.[11] Apprenticeship

---

[9]Not all registered programs are active. Interested researchers need to adopt innovative methods and tools to analyze the dynamics of U.S. apprenticeship programs and the policy implications of these dynamics.

[10]USMAP is a formal military training program implemented by the Center for Personal and Professional Development that provides active duty Coast Guard, Marine Corps, and Navy service members with the opportunity to improve their job skills and to complete their civilian apprenticeship requirements while on active duty. The U.S. Department of Labor provides the nationally recognized "Certificate of Completion" upon program completion. See Chapter 5 for further detail.

[11]For example, in a recent survey of European work-based education and apprenticeship systems, Lerman and Rein (2015) find that only one-quarter of Europe's enterprises with more than 10 staff take on apprentices. The reasons for this may be related to the costs associated with administration, apprentices' time away from the workplace, and employees' time spent training. For a comprehensive analysis of the economic issues and evidence, see Wolter and Ryan (2011).

practices also vary across sectors and occupations within countries, as do the ways in which the programs are provided and financed by employers and other stakeholders. The authors observe that no "general theory" as yet explains the full range of practices within, let alone among, countries.

Wolter and Ryan (2011) also report variation in program performance and cost. While some apprenticeships produce superior labor market outcomes, others may be poorly integrated into their local economy. Apprenticeships also are not costless, and their costs may not be symmetrically distributed.[12] If educators do not provide the appropriate curriculum to complement on-the-job training, for example, employers may end up bearing more of the apprenticeship costs, which they may or may not be able to recover. Box 4-1 reviews some of the challenges of setting up apprenticeship programs in the United States.

Despite the challenges described in Box 4-1, however, the potential benefits of apprenticeship have attracted the interest of policy makers, educators, and employers. Apprentices may produce benefits for firms that are not reflected in the value of their wages or other benefits. For program participants, benefits include the cognitive and motivational effects of combining academic learning and practice with skill learning, better integration of the content of skill learning with occupational requirements, increased youth employment rates, and better school-to-work transitions.[13]

Indeed, the past decade has seen a growing interest in apprenticeship programs in the United States. In a paper commissioned for this study, Messing-Mathie (2015) finds that interest in developing apprenticeship models in the United States is oriented around the "dual-education (or dual-study)" model, which is common in Germany, Austria, and Switzerland. This model combines classroom education in a postsecondary program of study with a direct relationship with an employer through a multiyear apprenticeship. Students apply to apprentice directly with a firm. The firm has an agreement with an institution of higher education, which in the United States is typically a community college, to provide the academic portion of the training.

U.S. apprenticeships have traditionally been limited to such sectors as construction, maintenance and repair, manufacturing, and service occupations. However, Messing-Mathie reports that since 2004, the United States has seen a

[12]Wolter and Ryan (2011) find that empirical observations show not only that there is a large degree of heterogeneity in the training costs that firms share but also that many firms reap net benefits even during the training of apprentices.

[13]In addition to the Wolter and Ryan (2011) survey of the literature, see Backes-Gellner (2014); Bolli and Hof (2014); Bolli and Renold (2015); Eggenberger et al. (2015); Muehlemann and Wolter (2014); Rupietta and Backes-Gellner (2015); Wolter and Muehlemann (2015); Ferm (2008); and Rosenfeld et al. (2014).

BOX 4-1
Apprenticeships: Challenges from Concept to Application

- **No clear definition.** A variety of different models confuse the notion of what constitutes an apprenticeship. Variations include time-based versus competency-based models and various levels of on-the-job learning versus external training. Given the lack of industry-recognized standards for apprenticeship, many short-term programs call themselves "apprenticeships" but are more like internships or coop programs.
- **High investment to start up a program.** Small businesses often find it too costly to create their own apprenticeship programs with defined curricula, to partner with community colleges or other educational institutions, and to recruit apprentices (which includes building relationships with feeder high schools and providing "train-the-trainer" courses for employees).
- **Lack of intermediaries.** One way to get around high start-up costs is to have an intermediary aggregate several companies into one program. However, funding generally is not available for such intermediaries, except for short-term grants.
- **Lack of long-term investment culture.** Given that apprenticeship programs take a minimum of 3-5 years to truly pay off, companies that offer such programs must have a culture of long-term investment. As a result, companies that offer these programs tend to be privately held, with less pressure to file quarterly or annual financial reports relative to publicly held companies.
- **Ineffective standards.** Very few U.S. apprenticeships are tied to a standardized, comprehensive, competency-based credential. Such standardization is critical in European programs and builds confidence in the ability of apprenticeships to deliver high-quality employees that meet specific standards.
- **Lack of in-company training resources.** Many companies, especially small and medium enterprises, lack the resources to develop in-house training curricula. Employees often lack the pedagogical skills and confidence to train new apprentices, human resources staff often lack the technical capability to evaluate external training experiences, and leadership for the development of a "training culture" is often absent within a company.

rise in the registration of new apprenticeship programs in such high-growth industries as health care, advanced manufacturing, information technology, and biotechnology, and that these programs—including the Kentucky Federation for Advanced Manufacturing Education (KYFAME), Apprenticeship Carolina, and the Michigan Advanced Technician Training Program (MAT2)—take a hybrid approach that deviates from the European dual-education model (Messing-Mathie, 2015).[14] Observing that apprenticeships are situated in specific social, economic, and regulatory contexts, she argues that such programs should be designed to allow for flexible responses to local requirements.

4.3.2 Alternative Credentials

Recognizing that education credentials other than academic degrees have potential labor market value, some researchers have begun to consider the role of these alternative credentials in education and training related to skilled technical workforce development (see, e.g., Carnevale et al., 2012; Kleiner and Krueger, 2010). A coordinated federal government effort to develop a better understanding of alternative credentials is currently under way as well. The Interagency Working Group on Expanded Measures of Enrollment and Attainment (GEMEnA) is developing and validating national measures of credentialing for work and is building government-wide consensus for the adoption of these measures in key federal data collections.[15] Box 4-2 lists GEMEnA's working definitions of alternative credentials. The following subsections provide an overview of issues related to educational certificates and certifications in skilled technical workforce development. Issues related to occupational licensing are described in Chapter 3.

Educational Certificates

Many postsecondary institutions offer certificate programs, which are stand-alone programs of specialized study or training that award a certificate

[14]See Chapter 6 for a discussion of the role of these intermediaries. In this hybrid approach, the educational organization acts in partnership with the employer to provide a dual-training model of apprenticeship learning, but the primary relationship is between the student (i.e., employee) and the employer (Messing-Mathie, 2015).

[15]GEMEnA consists of staff from the following federal offices: Bureau of the Census, Department of Commerce; Bureau of Labor Statistics, Department of Labor; Council of Economic Advisers; National Center for Education Statistics, Department of Education; National Center for Science and Engineering Statistics, National Science Foundation; Office of Statistical and Science Policy, Office of Management and Budget; and Office of the Under Secretary, Department of Education. To ensure the relevance of its work, GEMEnA also meets regularly with a panel of experts and collaborates with a wide range of partners and stakeholders (NCES, n.d.).

BOX 4-2
Working Definitions of Alternative Credentials

Certification

A credential awarded by a certification body based on an individual demonstrating through an examination process that he or she has acquired the designated knowledge, skills, and abilities to perform a specific job. The examination can be either written, oral, or performance-based. Certification is a time-limited credential that is renewed through a recertification process.

Educational Certificate

A credential awarded by a training provider or educational institution based on completion of all requirements for a program of study, including coursework and test or other performance evaluations. Certificates are typically awarded for life (like a degree). Certificates of attendance or participation in a short-term training (e.g., 1 day) are not in the definitional scope for educational certificates.

License

A credential awarded by a licensing agency based on predetermined criteria. The criteria may include some combination of degree attainment, certifications, certificates, assessment, apprenticeship programs, or work experience. Licenses are time-limited and must be renewed periodically.

SOURCE: GEMEnA based on Bielick et al. (2013, p. 4).

rather than a degree to students who complete the program satisfactorily. Certificate programs may be offered on a credit or noncredit basis, and usually take more than 1 month but less than 1 year to complete. NCES (2016c) estimate that public and private postsecondary institutions awarded a total of 969,353 certificates in 2013-2014. In a review of the literature commissioned for this study, Karp (2015a) defines a certificate as "a credential awarded by a postsecondary institution, recognizing completion of coursework in a discipline."[16] In general, Karp finds that individuals with a certificate are likely to earn more money in the labor market than individuals with only a high school degree or some college. Positive returns range from 7 to 24 percent after controlling for individuals' preexisting skills and abilities. However, Karp notes that recent research indicates that the return on investment for a certificate varies considerably depending on length of study, occupation, geographic region, gender, race, and ethnicity.

[16]Karp's (2015a) paper focuses on Title IV community colleges. For additional perspective and data on certificate and certification holders, see Ewert and Kominski (2014).

Karp found no studies examining the extent to which certificate holders subsequently earn an associate's degree. Her own analysis suggests that only 7 percent of first-time students who earned a certificate went on to earn an associate's degree. She also found evidence that it is uncommon for certificate completers to transfer to a 4-year institution. She concludes that the existing evidence indicates that certificates do not generally serve as pathways to degree-granting education programs, despite the aspirations expressed in the career pathways literature.

Certifications

Industry groups, professional organizations, and employers award certifications to individuals who meet program standards for knowledge, skills, and abilities. In a report for the Lumina Foundation, Ewert and Kominski (2014) estimate that more than 4,000 entities offer certifications in the United States. Governments are the most common certifying entities (71.4 percent), followed by professional associations (10.5 percent) and industries (10.3 percent) (Ewert and Kominski, 2014, Table 7). Certifications appear to be most common in health-related occupations. In one survey, the majority of individuals with certifications (76 percent) indicated that certification was required for their job, that they attended courses to prepare for certification (93 percent), and that continuing education units were required to maintain their certification (66 percent) (Ewert and Kominski, 2014).

Defining a certification as "[an award] to individuals who pass an assessment, usually sponsored by an industry or firm," Karp (2015a) finds that although the research on certifications is growing, existing studies show great variability in the involvement of community colleges in preparing students for certifications. Approaches range from aligning curricula with certification exams, to creating certification exam preparation programs, to becoming a certification-testing center. Preparation activities can include credit or noncredit courses.

4.3.3 Challenges in Assessing the Impact of Certificates and Certifications

Karp (2015a) concludes that the large volume of certification programs, the high degree of program variation, and a lack of accountability and comprehensive data make it difficult to assess the value or impact of completing these programs.

In conducting her review, Karp was unable to identify comprehensive listings of certificates or certifications in the United States or a system that collects outcome data for these programs. National datasets, for example, record certificates in very broad occupational areas, such as health and technical trades, thus precluding the more fine-grained analyses needed for large-scale workforce development initiatives. Likewise, state databases usually do not record the occupation in which certificate holders are employed; hence there is little research examining certificates as a component of workforce development programs and systems. Karp notes that the existing research also does not examine the employer perspective. These deficits make it difficult to account for all stakeholders in skilled technical workforce development and to address basic questions such as the following:

- How many certificate and certification programs exist?
- Who is in the process of acquiring a certificate or certification?
- Who has acquired a certificate or certification, and how is it associated with career trajectories and outcomes?
- What is the return on investment for certificates and certifications?

As a first step in addressing data requirements, GEMEnA is developing survey items for use in assessing (1) the attainment of nondegree credentials, including industry-recognized certifications, occupational licenses, and educational certificates; and (2) enrollment in education and training that prepare people for work. GEMEnA member NCES plans to field an in-depth survey of U.S. adults (the Credentials for Work Survey [CWS]) to provide detailed cross-sectional time series data beginning in 2016. NCES also is incorporating survey items on nondegree credentials into its post–high school longitudinal studies. Other GEMEnA agencies have started to deploy survey items on certifications and licenses in their own surveys of households and individuals, including the redesigned Survey of Income and Program Participation (Census Bureau), the National Survey of College Graduates (National Science Foundation), and the Current Population Survey (CPS) (Bureau of Labor Statistics [BLS]).

BLS recently released new data on certifications and licenses from the CPS based on the GEMEnA-developed survey items. In January 2015, questions were added to the CPS to identify persons with professional certifications and licenses. A highlight from these data is that 17.8 percent of the civilian noninstitutionalized population aged 16 and over held a currently active certification or license.[17] In addition, consistent with Karp's (2015a) analysis, the data show that certifications and licenses were more prevalent in some occupations than others: more than three-fourths of workers in health care and technical occupations held these credentials, compared with fewer than 1 of 10 in building and grounds cleaning and maintenance occupations (BLS, 2016b, Table 5).

Although promising, these data collection efforts need to be expanded, accelerated, and aligned at the local, state, and federal levels so that stakeholders

[17]The report and data table are available in BLS (2016a).

have the information necessary to make better choices about skilled technical workforce investments.

4.4 FUNDING FOR SKILLED TECHNICAL WORKFORCE EDUCATION AND TRAINING

In the United States, the costs of producing and providing postsecondary skilled technical workforce development are financed by students and workers, employers, labor unions, governments, and philanthropic organizations. In addition to investing time and effort, most Americans who wish to develop technical skills pay tuition and fees, as well as other associated costs, such as those for books, supplies, transportation, and child care, as well as opportunity costs, such as lost wages. If they are not prepared by virtue of their previous education, training, or experience, they must also invest in acquiring prerequisite skills, such as relevant STEM proficiency, or in completing a specific series of courses to qualify for admittance to a program or accreditation in the program.

On the supply side, student tuition and fees rarely cover the entire cost of building and maintaining effective skilled technical education and training centers. The remainder is supported by public revenues from federal, state, and local sources; employers; and civic-minded philanthropists.

Although co-production and co-provision are widely used strategies in the United States, they result in complex public service delivery systems that make it difficult to identify causal relations and outcomes, assign responsibility, and assess performance. "Co-production" refers to public–private partnerships to *produce* public goods and services, while "co-provision" refers to public–private partnerships to *finance or distribute* public goods and services. Education and training are examples of public goods that are co-produced and co-provided by students and their families, educators, employers, governments, and the civic sector (Davis and Ostrom, 1991). The implications of these approaches to skilled technical workforce development are discussed in more detail in Chapter 5. The following subsections describe the main components of these systems.

4.4.1 Public Funding Sources

Public sources of funding for skilled technical workforce development include federal, state, and local revenues collected from a wide range of sources. The 2016 *Condition of Education* report estimates that elementary and secondary public school revenues, which support academic and CTE programs, totaled \$618 billion in academic year 2012-2013 (NCES, 2016c). Similarly, as described in more detail in Chapter 3 and the following sections of this chapter, public funding sources offset a significant portion of the costs of employment and training services, as well as postsecondary degree-granting institutions. For

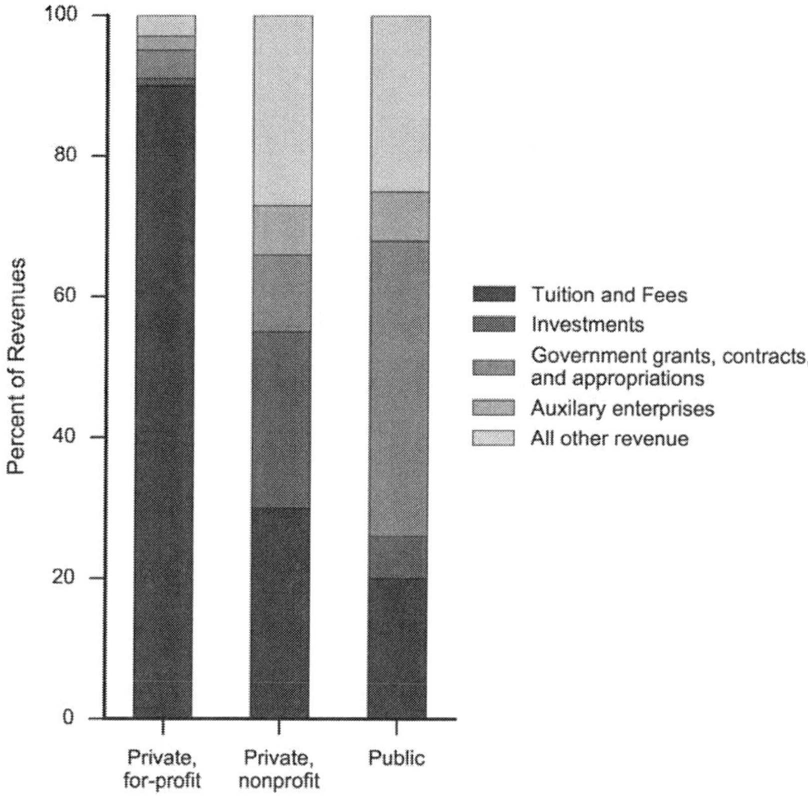

FIGURE 4-4 Sources of revenue of all postsecondary institutions.
SOURCE: NCES, 2016c.

example, Figure 4-4 shows the sources of revenue for all postsecondary institutions in 2013-2014. The primary sources of revenue for all institutions were tuition and fees, investments and government grants, contracts and appropriations, and other unspecified sources. However, the percentage shares differed significantly by type of institution. At public institutions, the largest percentage of revenue (42 percent) was from government sources, which include federal, state, and local government grants, contracts, and appropriations. At private nonprofit institutions and private for-profit institutions, by contrast, student tuition and fees constituted the largest percentage of total revenues (30 and 90 percent, respectively).[18]

[18]Note that Pell Grants are included in the federal grant revenues at public institutions but tend to be included in tuition and fees and auxiliary enterprise revenues at private

The sources of and formulas for allocating public revenues in education and training create incentives that can affect outcomes for stakeholders and society more generally.[19] However, calculating and accounting for the myriad public expenditures and the rules that govern their use in a large polycentric workforce development system such as that of the United States is a painstaking process that to the committee's knowledge has yet to be undertaken, one that also is beyond the scope of this study. To obtain a general sense of the issues, the committee focused on the largest aggregate funding sources at the federal, state, and local levels.

Federal Funding Sources

As described in more detail in Chapter 3, the federal government provides two general types of funding that are potentially relevant to skilled technical workforce development: funding for employment and training services, which is relatively small and administered principally at the local level; and funding for education assistance, which is considerable and administered through programs authorized under the Higher Education Act (HEA), the Post-9/11 Veterans Educational Assistance Act, WIOA, the Carl D. Perkins Career and Technical Education Act (Perkins Act), and the Trade Adjustment Assistance Act.[20]

Federal funds for employment and training services are provided through competitive grants, formula grants, and contracted programs, and they are subject to directives issued by the Office of Management and Budget. The majority of funding for federal employment and training services is distributed by formula grant programs that allocate the funds to state and local entities based on established noncompetitive criteria such as population, unemployment rates, or other economic conditions. In addition, greater than 90 percent of the funds appropriated under Perkins IV, which are important for skilled technical workforce development, are used to provide Basic State Grants that supplement state and local funding of CTE (Dortch, 2014a).[21] States use these funds to make

---

nonprofit and private for-profit institutions. Thus, some categories of revenue data are not directly comparable across public, nonprofit, and for-profit institutions (NCES, 2016c).

[19]This issue arose in nearly every panel discussion at the committee's 2015 symposium but was particularly prevalent in Panel 3, on "High School Pathways"; Panel 4, on "Community College Pathways"; and Panel 7, on "Federally-funded Workforce Development." Papers, presentations, and a webcast of the symposium are available at http://nas.edu/SkilledTechnicalWorkforce (accessed March 19, 2017).

[20]The Government Accountability Office (GAO) identified 47 federal employment and training services programs in fiscal year 2009, which together provided $18 billion for workforce development. The College Board estimates that the federal government funded $145 billion in postsecondary education assistance in 2013. See the section on federal policy in Chapter 3 for more detail and citations.

[21]Note that Basic State Grants are formula grants.

grants that support CTE activities at the secondary and postsecondary levels in local education agencies, CTE schools, community colleges, and American Job Centers. The funds are used to develop the career and technical skills of students, acquire instructional materials and equipment, provide professional development, offer career and academic counseling, develop CTE curriculum, and assess CTE programs.

The U.S. Department of Education estimates that 85 percent of all first-time 4-year undergraduate students and 78 percent of all first-time 2-year undergraduate students receive financial assistance from federal, state, and local governments; institutions; or private sources (NCES, 2016c). Grants and loans are the major forms of federal financial aid for undergraduates. Most of the federal funding for undergraduate education and training is authorized under Title IV of HEA and the Post-9/11 Veterans Educational Assistance Act.[22] The largest federal grant program available to undergraduates is the Pell Grant program. To qualify for a Pell Grant, a student must demonstrate financial need. Federal loans, on the other hand, are available to all students.

Although federal legislation regulates eligibility for and disbursement and repayment of financial assistance to students, it is the student-recipients who determine how they will invest in education and training. The implication is that if policy makers wish to encourage increased investment in skilled technical education and training, they need to persuade students as well as educators and employers of the potential benefits of making this investment.

Local and State Government Funding Sources

Local and state governments provide two general types of funding that are relevant to skilled technical workforce development: public education in grades K-12 and a portion of postsecondary education. On average, the states allocate about a third of their direct general expenditures to education. Most of these funds are spent on elementary education, which is intended to build the foundation for success in secondary and postsecondary academic and career and technical education, work, and life.

For example, the U.S. Department of Education reports that from 2010 to 2011, direct general expenditures of state and local governments for education averaged \$2,764 per capita, or 33.3 percent of all functions (NCES, 2013c, Table 106.50). Expenditures for elementary education averaged \$1,813 per capita, or 21.9 percent of all functions; \$810 per capita for colleges and universities, or 9.8 percent of all functions; and \$141 per capita for other types

[22]Using the GAO and College Board estimates cited in Chapter 3, total federal funding for all education and training is approximately \$163 billion annually. The total amount of education assistance under HEA and the Post-9/11 Veterans Educational Assistance Act is approximately \$141 billion annually.

of education, or 1.7 percent of all functions. The data show considerable variability in education expenditures across the states. Per capita spending on education ranged from $1,975 in Idaho, where education expenditures represented 29.9 percent of all functions, to $4,672 in Alaska, where these expenditures represented 26.4 percent of all functions.

As described in more detail in Chapter 3 and earlier sections of this chapter, in addition to funding public K-12 education and training programs, local and state governments play an important role in funding postsecondary education and training, particularly in community colleges. State and local sources provide about 48 percent of community college revenues (see Table 4-2). The implication is that state and local policy makers can use this funding as a means of exercising considerable influence over skilled technical workforce development programs offered by community colleges.

4.4.2 Private Funding Sources

Private sources of funds for skilled technical workforce development potentially include the income of students, workers, and their families and friends; the income of employers; and the income of churches, civic associations, labor unions, trade associations, and philanthropies. The latter organizations tend to fill funding gaps in the production and provision of public goods based on their specific missions and interests. In this sense, their incentives align directly with the individuals they serve and only indirectly with public policy functions. Hence, the three groups that appear to be most directly involved in producing skilled technical workers are the workers, employers, and the public (of which they are a part) in their sphere of activity.

As discussed in preceding sections of this report, U.S. policy frameworks generally place the burden of funding postsecondary education and training on individuals, who potentially earn returns on their investments in the form of increased wages, promotions, and other forms of social recognition and achievement. If properly aligned with contemporary conditions, these returns can create incentives for continual, lifelong individual development.

Recognizing that public benefits can be derived from individual education and training activities, governments subsidize individual investments by funding public schools and providing various forms of grants and financing on generous terms. However, employers also benefit from workers' self-development, as well as public subsidies for these efforts. Skilled technical workers potentially contribute to innovation, increased output, productivity, and profitability. Public policies recognize this linkage by allowing employers to deduct the cost of employee education and training from their taxable income, thereby subsidizing employer investments in workforce development.

The remainder of this section explores joint labor–management programs that support workers' participation in skilled technical workforce development and provides an overview of U.S. employers' workforce development strategies.

Joint Labor–Management Programs

Unions have historically leveraged member dues to work with employers in helping workers from a wide range of career fields develop and maintain skills to meet local needs. The AFL-CIO reports that joint labor–management skill training programs currently contribute about $1.5 billion to the American economy each year (AFL-CIO, n.d.).[23] These programs potentially offer opportunities for a diverse group of workers in skilled technical occupations in such fields as health care, building and construction, manufacturing, hospitality, and aerospace. In the building and construction industry, thousands of local Joint Apprenticeship and Training Committee (JATC) groups oversee apprenticeship and journey-level skills upgrade training in such growing occupations as energy conservation and alternative energy. The committee sees the need for additional research on the role of unions in skilled technical workforce development and on the effectiveness of joint labor–management programs.

The AFL-CIO reports that national joint training programs currently exist in the automotive, telecommunications, steel, rubber, health care, hospitality, and aerospace industries, as well as in the public sector. In the steel and rubber industries alone, 72 local joint committees in 24 states have started to offer courses in renewable energy systems and energy-efficiency technologies. The AFL-CIO estimates that there also exist 18 consortia of unions, management, universities, and health and safety organizations. Box 4-3 describes a 30-year alliance originally created through collective bargaining between the Communication Workers of America and AT&T. Additional examples of labor–management training partnerships include

- the International Association of Machinists and Aerospace Workers and Boeing,[24]
- the United Auto Workers (UAW) and Ford and UAW and GM,[25]
- the United Steelworkers,[26] and
- the Kaiser Labor Management Partnership.[27]

---

[23]An AFL-CIO program description can be found at http://www.aflcio.org/Learn-About-Unions/Training-and-Apprenticeships/Labor-Management-Partnerships (accessed March 19, 2017).

[24]See http://www.iam-boeing.com (accessed March 19, 2017).

[25]For information on the UAW-Ford partnership, see http://uawford.org (accessed March 19, 2017). For information on the UAW-GM partnership, see https://www.uawgmjas.org (accessed March 19, 2017).

[26]See http://icdlearning.org (accessed March 19, 2017).

[27]See http://www.kaiserpermanentejobs.org/labor-management-partnership.aspx (accessed March 19, 2017).

BOX 4-3
The Alliance for Employee Growth and Development, Inc.

The Alliance for Employee Growth and Development, Inc. is a private nonprofit corporation designed to provide training to individuals to enhance their employability. The Alliance was established and funded through the AT&T and Communication Workers of America (CWA) national bargaining process in 1986, and over the years has expanded to include Alcatel-Lucent, Avaya, Optical Fiber Cable and Connectivity Solutions (OFS), and the International Brotherhood of Electrical Workers (IBEW).

The Alliance has become a primary education resource for union-represented employees at its partner companies. Since its inception, it has provided more than 16.5 million training hours to more than 175,000 individuals. In the most recent fiscal year, nearly 50 percent of eligible individuals participated in more than 168,000 hours of Alliance training.

To address the needs of a diverse, nationwide workforce, the Alliance has developed an array of training delivery options that includes on-site classrooms, virtual classrooms, hands-on technical training, and online self-paced courses. Regardless of the training modality, customer satisfaction with Alliance training remains high, with a 92 percent satisfaction rating from trainees.

The Alliance remains an enduring labor–management success. Working at the local, regional, and national levels, the Alliance has joined together companies and their unions in a variety of win–win training partnerships. These cooperative efforts have had a direct impact on the workplace and technological skills of individuals and the success of the participating businesses.

SOURCE: The Alliance, http://www.employeegrowth.com/about (accessed March 19, 2017).

Employer Funding

Economists who study innovation and growth observe that labor and the STEM knowledge and skills it embodies are among the key drivers of economic growth and development.[28] The nation cannot innovate and compete without

[28]Contemporary theories of economic growth are based on Solow (1956), who argued that sustained economic growth requires capital, labor, and a residual factor associated with innovation and technological progress. North (1990) expanded on Solow's work, arguing that institutions—the rules that determine who participates in the economy and their property rights and duties—and location are additional factors. Romer (1994) further elaborated, emphasizing the endogenous role of knowledge in the standard economic production function. For a survey of the economic literature related to employer-financed skills training, see the paper by Lerman (2015) commissioned for this study. For an extensive survey of theory and evidence on apprenticeship, see Wolter and Ryan (2011).

technical skills, and if it cannot compete effectively, it will not enjoy long-term economic prosperity.

Innovation means different things to different people. In this report, the committee uses the OECD Oslo Manual definition (OECD and Eurostat, 2005):

> An innovation is the implementation of a new or significantly improved product (good or service) or process, a new marketing method, or a new organizational method in business practices, workplace organization or external relations.

The logic chain that relates innovation and workforce development suggests that employers who prioritize innovation and rely on technical skills have strong incentives to contribute to the development of a skilled technical workforce. Taking specific institutional and locational differences into account, for example, Rupietta and Backes-Gellner (2015) find that firms in developed countries that train apprentices through a combination of school- and work-based learning (a dual-education or vocational education and training [VET] approach) are more innovative than firms that do not. As discussed in Chapter 2, however, Capelli (2004, 2012, 2015) argues that employer training has declined over time, shifting the burden of skill development to individuals and governments. He attributes this trend to declining job tenure, which limits the ability of firms and workers to earn an adequate return on their training investments.

Surveying the literature on the role of employers in developing and training skilled technical workers, Lerman (2015) argues that the evidence is mixed. He finds that although employer-led training is effective for workers and firms and is widespread in the United States, data limitations make it difficult to determine the scale of training with any rigor. And as Table 4-4 shows, the survey data that have been collected vary widely. Lerman finds, for example, that government survey data provide a dated and very mixed picture of such training. Among individuals aged 25-64, the Survey of Income and Program Participation shows a decrease in employer-led training from 32.8 percent in 1996 to 21.6 percent in 2008, while the National Household Education Survey shows an increase from 31 percent in 1999 to 42.4 percent in 2005. The National Assessment of Adult Literacy survey, which reports on the incidence of training for 2003, indicates that 56.9 percent of workers in this age group were involved in employer-led training in the previous 12 months.

Employers support skilled technical workforce development in several ways, including by providing on-the-job training and apprenticeships, supporting certification and recertification, and offering tuition reimbursement programs. The Society for Human Resource Management (SHRM) (2015) estimates that 84 percent of U.S. employers provided some form of workforce development benefit in 2015, while 78 percent supported certification and

TABLE 4-4 Percentage of Workers Aged 25-64 Participating in Employer-Led Training in the Previous 12 Months

	Government Survey						
Group	Survey of Income and Program Participation				National Assessment of Adult Literacy	National Household Education Survey	
	1996	2001	2004	2008	2003	1999	2005
Total	32.8	27.9	21.8	21.6	56.9	31.0	42.4
Male	30.4	25.3	19.7	20.5	53.6	29.4	31.8
Female	35.7	31.0	24.2	22.7	60.6	21.8	46.0

SOURCE: Lerman, 2015, Table 1.

recertification (SHRM, 2015). Yet SHRM's analysis suggests that it is not clear whether employees are fully utilizing these benefits, and if not, why not (SHRM, 2015). SHRM notes, for example, that 30 percent of employees in large companies but only 1 percent of those in small companies use the benefits. SHRM speculates that employers may create more and less supportive cultures for using education and training benefits, or that the staffing requirements of some organizations make it difficult or expensive to use the benefits. More research is needed to better understand benefit utilization.

A more nuanced view emerges from examining results of a large and continuing private survey of employer training. The Association for Talent Development (ATD) regularly surveys employers about their training and development activities. ATD's 2014 *State of the Industry* report indicates that employers spent an average of $1,208 per employee on training and development in 2013, or about $167.5 billion.[29] This spending has been relatively stable over time, ranging from $1,040 per employee in 2006 to a high of $1,228 in 2010. These numbers support the idea that employees may not be fully utilizing employer-provided opportunities, such as tuition reimbursement programs, to develop their skills and credentials.

ATD finds that the average cost of workforce development typically varies according to an organization's size. In 2013, small organizations with fewer than 500 employees spent on average $1,888 per employee, whereas midsized organizations with 500 to 9,999 employees and large organizations with 10,000

[29]The ATD study is based on a survey of 340 organizations of various sizes, industries, and locations. In 2013, ATD reported that employers spent $164.2 billion on training in 2012; the committee's estimate of total spending in 2013 is based on ATD's report in 2014 that spending increased by 1 percent over the previous year. For an overview of the 2014 study, see Miller (2014).

or more employees spent about $838. The number of learning hours per year per employee also varies across the size of organizations. Although the average across all organizations in 2013 was 31.5 hours, large organizations on average reported that their employees received 36 hours of training, or approximately 4.5 days, whereas the comparable figures for midsized organizations were 27 hours, or approximately 3.5 days (Miller, 2014).

ATD also finds differences in training across industries. In 2013, for example, manufacturing organizations, which are typically large organizations in ATD's survey, reported on average spending $535 and providing 27 hours of training per employee, compared with $1,392 and 24 hours for health care and pharmaceutical organizations and $1,107 and 33 hours for the combined group of finance, insurance, and real estate organizations.

ATD finds further that organizations provide employee training in different ways, including by developing and producing courses internally, contracting with other organizations to provide the training, and reimbursing employees for tuition. ATD estimates that 10 percent of training expenses in 2013 went to tuition reimbursement programs in which employees select courses to study, subject to approval by their managers, and are reimbursed for all or some portion of their tuition costs upon demonstrating successful completion.

Finally, ATD estimates that about one-third of employer-selected training content is focused on managerial and supervisory skills, mandatory and compliance training, and profession- or industry-specific training. The remaining two-thirds of content covers such topics as processes and procedures, customer service, sales training, and executive development (Miller, 2014).

4.5 CONCLUSIONS

Skilled technical workforce development is a quasi-public good with potential benefits for individuals, employers, and society as a whole. Americans have a long-standing commitment to education and training that predates independence and the creation of a federal republic.

Mirroring the U.S. policy-making structure, the U.S. workforce development enterprise is a very large, complex system of systems: there is no single, centralized system for developing a skilled technical workforce, and the structure of the existing system cannot easily be conceptualized in terms of pathways, pipelines, or production functions or operated in this manner. Education and training programs are co-produced and co-provided by students, workers, employers, governments, and civic society. The largest funding sources for skilled technical workforce development are state and local public revenues and tuition paid by students and workers for postsecondary degree-granting programs. WIOA encourages policy makers to better integrate and align these expenditures and programs at the local level. However, assessment of the performance of systems based on polycentric co-production and co-provision is quite difficult and must be undertaken at the local level.

Americans who wish to prepare themselves for success in skilled technical occupations have a wide range of education and training options that depend on several factors, including their own socioeconomic means, the availability of programs in their preferred geographic location, and the occupation of interest. The primary vehicles for skilled technical education and training in the United States are CTE in secondary and postsecondary institutions, community college programs, and apprenticeships. However, enrollment trends indicate that while Americans may aspire to acquire career-related education and training throughout their lives, they are less likely to pursue skilled technical education and training today than they have been in the past.

Employers that wish to innovate and compete successfully also have a stake in developing a skilled technical workforce. Public tax policy subsidizes employer expenditures to encourage economic growth and development. However, data on employer-funded skilled technical workforce development are sparse, and the evidence that exists is mixed.

With this review of the highly complex U.S. system of workforce education and training as context, the next chapter examines the challenges associated with the development of a skilled technical workforce.

5

Challenges in Developing a Skilled Technical Workforce

Developing and sustaining a skilled technical workforce is a high-priority goal for employers and for policy makers at the local, state, and federal levels. The nation is making substantial public and private investments in achieving a broad range of education and training goals, spending an estimated $618 billion on public K-12 education and an additional $701.5 billion on postsecondary education and training each year.[1] This substantial and broad-based commitment has made it possible for U.S. workforce development practices to evolve so they now include a wide range of options that potentially accommodate the diverse needs of American students, workers, and employers. A key issue for the nation, then, is how effectively these substantial resources are used to develop and sustain a skilled technical workforce.

To address this issue, it is necessary to look more closely at the institutions—the rules, norms, and incentive structures—that shape the ability of students, workers, employers, educational organizations, and governments to work together to develop and sustain a skilled technical workforce.[2] The economic data reviewed in Chapter 2 capture the visible patterns created by these actors within prevailing institutions. At the microeconomic level, the data

[1]The estimate of annual spending on public K-12 education and training is based on NCES (2016c). The estimate includes federal education and training services ($17 billion), employer spending ($167.5 billion), and the total expenditures of postsecondary Title IV institutions ($517 billion). The source for federal spending on education and training services is The White House (2014). The source for employer spending is the Association for Talent Development (2015). The source for postsecondary Title IV institution expenditures is NCES (2016c).

[2]Institutions work by changing the structure of incentives individuals face in situations in which they need to cooperate and by creating shared expectations about the behavior of others (North, 1990). Ostrom (2005) observes that "a major problem in understanding institutions relates to the diversity of situations of contemporary life."

reveal a patchwork of outcomes whereby some labor markets appear to work better than others. Putting this information in context, Chapter 3 shows the great variation in policies and governance systems covering education and training across the nation's states and localities. And as seen in Chapter 4, this variation is reflected in the diversity of workforce education and training programs across the nation.

This chapter examines the various impediments and disincentives that pose challenges for fully leveraging the nation's investments in a skilled technical workforce. Section 5.1 looks at the factors that affect the perceived returns on investment in training and education for students, workers, and employers. Section 5.2 explains how some current social and policy frameworks, funding mechanisms, and data gaps impede better outcomes. Section 5.3 then explores the particular challenges of training the skilled technical workforce in the allied health and manufacturing sectors. It also takes up the unique challenges faced by about 200,000 service members who transition each year from active military duty to skilled technical occupations in civilian life. The final section presents conclusions. Chapter 6 addresses some of the policy proposals that might help overcome the challenges discussed in this chapter.

5.1 THE RETURN-ON-INVESTMENT CALCULUS

Americans share in the ideals of equal opportunity; fair competition; and the ability to capture a reasonable return on initiative, effort, and other investments. Individuals who invest in skill development should be able to generate enough income from a job or business to cover the associated costs and realize an equal or better return relative to an alternative use of the money. Similarly, enterprises that invest in training their workers should have a fair chance of generating enough revenue to realize a return on their investment. And taxpayers should receive a fair return on their tax dollars in the form of high-quality, reliable public education and training services. Yet limitations in both data and analyses make it difficult to determine whether and how individuals, employers, and the public are investing in skilled technical education and training, and whether they are realizing sufficient returns to sustain or increase their investments.

The returns on investment in acquiring technical skills are both tangible and intangible in nature. Some of these returns—such as higher wages or better benefits for workers; higher profits or greater worker productivity for firms; and additional tax revenues for local, state, and federal governments—can be estimated. Other returns, such as improved civic engagement (see Chapter 2), may be more difficult to measure but may be of equal or even greater importance. In some cases, intrinsic motivation and nonmonetary incentives that signify social recognition and respect can play a greater role in individual motivation than monetary incentives alone.[3] Potential nonmonetary rewards from education and training include the joy of doing something one likes to do or is particularly good at doing; the personal satisfaction derived from accomplishment, social recognition, or respect; and the opportunity to advance.

How students, workers, and employers estimate their return on investment depends on the local context. Education and training for skilled technical jobs are influenced primarily by local policy and practice, and there can be substantial differences across locations, institutions, and industries even within the same state.

5.1.1 Evidence on Returns on Postsecondary Education and Training

Overall, there is strong evidence that postsecondary education yields high positive returns.[4] Prior research using national longitudinal survey data has shown that individuals with an associate's degree earn more than those with only a high school diploma (Belfield and Bailey, 2011). Bailey and Belfield (2013, Chapter 6) estimate that an additional year of schooling raises yearly earnings by 5 to 10 percent, or more than 10 percent after controlling for other factors. Likewise, using data from Washington State, Dadgar and Weiss (2012) observed increases in wages with an associate's degree, although the size of the increase varied depending on the field of study. A review by the California Community Colleges system estimated average returns of 10 percent in additional earnings for short-term vocational certificates and more than 25 percent for associate's degrees (AA/AS) (Stevens et al., 2015). Similar results were seen in a study in Florida focused on minorities and disadvantaged workers (Backes et al., 2014), in which students' decisions to study general humanities and failure to complete those programs had a significant negative impact on their return on postsecondary education. Another study using data from Arkansas, Colorado, Tennessee, Texas, and Virginia found that the returns on associate's degrees or certificates for career and technical programs of study were higher than those on associate's degrees for academic programs of study. In four of these five states, moreover, average first-year earnings were higher for associate's degree holders than for bachelor's degree holders (Schneider, 2013). Interestingly, researchers using data from Kentucky found that the returns on degrees or diplomas differed for men and women, with women receiving

---

[3]For an overview of this research and an economic theory of human motivation, see Frey (1997).

[4]For evidence on positive returns to postsecondary education, see, for example, several of the papers commissioned for this study, including Heinrich (2015), Karp (2015), Rothwell (2015), and Stern (2015).

approximately $2,400 in higher quarterly earnings compared with $1,500 for men (Jepsen et al., 2012).

Finally, completing a program and receiving a degree appear to be valued more highly by employers than acquiring credits or skills (Kane and Rouse, 1995). Using data from Kentucky, Jepsen and colleagues (2012) found that the higher earnings associated with certificates or diplomas were due to completion (the so-called "sheepskin effect") rather than the number of credits taken. However, they note that the effects of completing diplomas and certificates were mixed and varied by field of study and gender (Jepsen et al., 2012).

5.1.2 Challenges in Completing Postsecondary Programs

The large returns on investment in obtaining postsecondary education despite stagnant wage premiums over the last few decades appear still to make this a good investment, for both 2- and 4-year degrees (The White House, 2016). However, college completion rates are not rising commensurately with either costs or earnings, suggesting that there is room to improve the outcomes of college investments (see, for example, Baum and Ma, 2013; Baum et al., 2013). As of October 2014, 68.4 percent of high school graduates were enrolled in colleges or universities (BLS, 2016a). As described in Chapter 4, the data suggest that a large proportion of these students, particularly those in community colleges, will drop out without earning a college credential. Although completion rates vary considerably across colleges, comprehensive analyses of the differences are not available. Researchers have shown, for example, that even after controlling for K-12 achievement, students who attend 4-year private colleges have much higher completion rates than those who attend 2-year colleges (Stern, 2015). Other research has shown that completion rates for students enrolled in 4-year colleges are higher for those enrolled in colleges with selective admission requirements (Bound et al., 2010). Low student completion rates also have been documented among for-profit colleges (Cellini, 2012; Deming et al., 2013). Some of the factors affecting college completion are reviewed below.

Inadequate Preparation

The percentages of Americans with low literacy and numeracy skills (see Table 5-1) suggest that inadequate elementary and secondary education may be a large part of the reason for students failing to complete their education.

High Costs of Remediation

Students without adequate preparation for postsecondary education and training are often required to complete remedial programs before they can

TABLE 5-1 The Nation's Report Card: Percentage of Students Who Performed at or above Proficient in Eight Subject Areas

Subject Area	Grade 4 (%)	Grade 8 (%)	Grade 12 (%)
Civics	27 (2010)	23 (2014)	24 (2010)
Economics	—	—	42 (2012)
Geography	21 (2010)	27 (2014)	20 (2010)
Mathematics	40 (2015)	33 (2015)	26 (2013)
Reading	36 (2015)	34 (2015)	38 (2013)
Science	—	32 (2011)	—
U.S. History	20 (2010)	18 (2014)	12 (2010)
Writing	—	27 (2011)	27 (2011)

SOURCE: NAEP, n.d.

participate in more advanced programs, which increases the costs of the education and training and reduces potential returns on that investment (Bettinger et al., 2013; Long et al., 2014). Remediation is costly not only for students but also for states and employers, with some estimates reaching as high as $2.3 billion per year (NCSL, n.d.). Students who are required to take remedial courses are also more likely to drop out of postsecondary programs without a degree: 9 of 10 students subject to remediation requirements when they enter their postsecondary program of study never complete the program, and only 1 in 4 students in remedial classes will eventually earn a degree from a community college.[5]

Rising Tuition Costs

Between 2008-2009 and 2013-2014, revenues from tuition and fees per full-time equivalent (FTE) student increased by 17 percent at public institutions (from $5,681 to $6,639, in constant 2014-2015 dollars) and by 6 percent at private nonprofit institutions (from $19,206 to $20,293). At private for-profit institutions, revenues from tuition and fees per FTE student were 34 percent higher in 2013-2014 than in 2008-2009 ($19,480 versus $14,515) (NCES, 2016b). Tuition increases are particularly burdensome for students with limited financial resources (Lovenheim, 2011). The high costs of tuition, as well as the opportunity costs of time spent away from work or of child care, represent another impediment to college completion.

[5] Data from the 31 Complete College America partner states indicate that roughly 22 percent of developmental education students at community colleges complete remediation requirements and associated "gatekeeper courses" within 2 years (see Complete College America, 2017).

5.1.3 Are Students and Workers Sufficiently Engaged?

Educators and policy makers rely on students to make decisions that suit their interests, capabilities, and means and to work diligently to acquire the relevant education and training. They rely on families and employers to prepare and encourage students to learn and to reinforce education and training at home and in the workplace. David Stern of the University of California, Berkeley, notes several trends regarding investment by students and workers in skilled technical education and training (Stern, 2015):

- Public high school graduates are less likely to earn career and technical education (CTE) credits than they have been in the past.
- The percentage of Americans aged 18-24 who enroll in degree-granting postsecondary education is increasing.
- Even though most bachelor's degrees awarded in the United States are in programs of study related explicitly to occupations, such as business, education, health professions, engineering, law enforcement, and agriculture, the number of students who complete a degree by age 30 is much smaller than the number of high school students who say they want to pursue a bachelor's or advanced degree.

Stern observes that enrollment and completion patterns and trends can pose a dilemma for families, educators, and policy makers. If education policies reflect students' aspirations, many students may not acquire the technical knowledge or skills needed to earn a living. However, if policies limit access to the courses required for high levels of academic success to those students who are deemed likely to succeed (an inexact assessment at best), they are likely to shortchange many talented young people, particularly those from low-income or historically underrepresented groups.

5.1.4 Are Employers Sufficiently Engaged?

Educators and policy makers rely on employers to help them understand changing skill requirements, contribute to policy and program design, assist with program delivery, and provide access to up-to-date equipment and facilities. They also rely on employers to provide graduates with jobs or business opportunities that will deliver a return on investment in education and training. Yet these expectations of employers may not be met.

Free-Ridership

Some educators and policy makers complain that employers are "free-riding" on public resources because they are not sufficiently engaged in producing and providing education or training or fail to invest their fair share in workforce development processes. As discussed in Chapter 4, the evidence on employer investments in education and training and the return on these investments is mixed, in large part because these data are missing or incomplete.

Growth of Contingent Work Arrangements

From the perspective of some students, workers, and policy makers, employers in certain industries and locations may not be committed to hiring specific groups of workers, paying high enough wages, or providing sufficiently stable employment to ensure an adequate return on investment in education and training. It appears that when making hiring decisions, companies increasingly prefer to hire workers under alternative, contingent work arrangements, such as temporary, on-call, contract, freelance, or part-time.[6]

For example, results of a 2013-2014 Harvard Business School alumni survey indicate that employers are reluctant to hire full-time workers:

> Forty-six percent of respondents agreed that their firms' U.S. operations prefer to invest in technology to perform work rather than hire or retain employees, while only 25 percent disagreed. Forty-nine percent said that their firms preferred to rely on vendors to perform work that can be outsourced, while only 29 percent reported that their firms would rather hire additional workers and keep work in-house....Companies that increased their reliance on part-time workers over the past 3 years outnumbered those that had reduced their proportion of part-time workers by two to one. (Accenture et al., 2014, p. 7)

Data show that a significant share of the workforce comprises contingent or temporary workers or those who work part-time for multiple employers, and some analysts believe that this trend is increasing at a rate that is not well captured with existing workforce measures.[7] Recent data collected by the Department of Labor (DOL) show more than 30 percent of the workforce being defined as in a contingent work arrangement (BLS, 2005).[8] Among this group,

---

[6]For an overview of current issues and data, see the discussion in Chapter 3 and Dokko et al. (2015).

[7]For an overview of data and measurement challenges associated with estimating alternative work arrangements, see Dokko et al. (2015).

[8]Contingent workers are those who do not expect their jobs to last or who report that their jobs are temporary; they do not have an implicit or explicit contract for ongoing employment. Alternative employment arrangements include people employed as independent contractors, on-call workers, temporary help agency workers, and workers provided by contract firms.

44 percent reported working part-time, and 25 percent reported being independent contractors or freelance workers. Other contingent workers are either self-employed or employed on a temporary basis through agencies or on-call relationships. Analysis of results of a 2014 survey conducted by the Freelancers Union suggests that more than 34 percent of Americans are doing at least some freelance work (Horowitz and Rosati, 2014).

Employers who are reluctant to hire full-time and long-term workers may not be sufficiently motivated to invest in skilled technical workforce development. At the same time, contingent workers may not have sufficient incentives or resources to invest in developing and maintaining their skills, a phenomenon that in turn feeds employers' reluctance and inability to hire qualified workers.

A similar dynamic exists for full-time workers. As reported by the Society for Human Resource Management (SHRM), few full-time employees utilize available tuition reimbursement programs that would permit them to enhance their skillset.[9] If employers do not unambiguously encourage and support continued education and training and provide employment opportunities that offer a return on those investments, employees may not invest in skill development—a dynamic that limits all stakeholders.

5.2 IMPEDIMENTS TO BETTER TRAINING OUTCOMES

The perceived costs and benefits of education and training are further influenced by a number of social, economic, and institutional impediments. These range from the challenges that students may face in pursuing an education while raising children and managing households, which increase perceived costs, to the lower social prestige some accord to CTE, which reduces the perceived value of its benefits. Funding formulas that incentivize the volume of enrollment over the quality of educational options, inadequate career guidance, inconsistent standards, and limited labor market data, among other factors, also make it more difficult to make informed decisions about investing in joining or growing the skilled technical workforce.

5.2.1 Limited Support for Students

As described in Chapter 4, nearly half of all Americans enrolled in postsecondary degree programs attend community colleges. Students who attend community colleges often face greater challenges than those who attend 4-year institutions. More than 50 percent of students enrolled in 2-year degree programs work either part- or full-time while pursuing their studies, and they are

[9]http://www.shrm.org (accessed March 5, 2017).

more than twice as likely as students enrolled in 4-year degree programs to be enrolled in part-time study. They also are more likely to be older, to have lower socioeconomic status, or to belong to minority groups.[10]

Support Networks

Workers who pursue continuing education and training often have a wide range of responsibilities to coordinate along with their studies, such as establishing their households, raising their families, and caring for aging family members. These challenges can affect students' return on investment in education and training and their incentives to complete postsecondary programs (Holzer and Nightingale, 2009).[11] These factors highlight the need for a wide range of services to support students who pursue skilled technical education and training and for schools to devise effective strategies for helping them access support services that meet their particular needs (Purnell and Blank, 2004). The nature of these services is locally specific and needs to be tailored to the requirements of the local population.

Most students struggle to adjust to the expectations of college coursework and campus life. However, the evidence suggests that students without strong social support networks have more difficulty accessing support services relative to those who have such networks (Karp, 2011). Some research on access to services at community colleges, for example, has found that although support services are available to all students, few low-income students find them accessible. Typically, students who take full advantage of these services have preexisting knowledge of the available resources or social relationships that help them find the information and resources they need to be successful (Karp et al., 2008).

Career Guidance

Most American students and workers who pursue postsecondary education and training have difficulty identifying appropriate programs and courses of study, funding their costs, and developing supportive educational and career networks. Many secondary and postsecondary students receive limited career guidance. Colleges typically operate under a "cafeteria-style" model of education in which students select courses from a bewildering array of choices, with limited guidance (Bailey et al., 2015). As a result, students enroll in skilled

---

[10]For a more complete discussion of community college demographics, see Chapter 4. Data sources include AACC (2016) and NCES (2016c).

[11]The authors note that "wraparound services," such as career and educational guidance, stipends, child care, and transportation, can boost program completion rates.

technical programs at relatively low rates, even after controlling for academic ability (see, for example, Holzer and Nightingale, 2009).

Recent White House and congressional reviews have sought to address these challenges. Vice President Biden's committee on workforce development reported that Americans often are not sure which education and training programs to pursue or whether jobs will be available for them when they finish (The White House, 2014). Congress also has sought to address this challenge; one of the express purposes of the 2014 Workforce Investment Opportunity Act (WIOA) is to improve coordination to increase access to opportunities and support services.[12]

5.2.2 Perceptions of Career and Technical Education

As discussed in Chapter 4, for most of the 20th century and up to the present time, federal and state policies have encouraged schools to develop curriculum for skilled technical work, known today as CTE. Prior to the 1960s, high school vocational programs provided basic job training for students that included school-based learning, work-based learning, and connections between activities and employers. An evaluation of the School to Work program of the 1990s, for example, found that these types of programs were particularly beneficial for students who were less likely to go to college, especially men (Neumark and Rothstein, 2005). However, several analysts have pointed out that as the focus in education has shifted toward preparing most students for college, the occupationally oriented CTE infrastructure in secondary schools has deteriorated (Klein and Green, 2011).

Stern (2015) observes that although policy makers have encouraged improvements in the quality of high school CTE curricula, most students and parents continue to perceive CTE as a poor substitute for college preparation rather than as a complementary or alternative pathway to a good job (see also Hoffman and Reindl, 2011; Symonds et al., 2011). This sentiment is supported by a recent report of the Independent Advisory Panel of the National Assessment of Career and Technical Education, which states that "exemplary CTE programs are seen as exceptions to mainstream options. CTE is still perceived by many as an alternative to rigorous academics—a separate track for students who are not college bound" (Independent Advisory Panel of the National Assessment of Career and Technical Education, 2014, as cited by Stern, 2015, p. 3). This diminished image of CTE and preparation for skilled technical work among students and parents inhibits efforts to improve CTE offerings, enrollment, and completion rates.

Stern notes further that experience and research over the past several decades have demonstrated the necessity and the feasibility of designing high

[12]WIOA, Public Law 113-128, Section 2.

school programs that prepare students for both college and careers.[13] In many schools in the United States, however, CTE tends to be course-specific and geared to preparation for particular jobs. This tendency may be reinforced by federal and state policies that have created bureaucratic rules and funding strategies that tend to isolate CTE from mainstream education reform.

As discussed above, perceptions about skilled technical work may lead parents and students to avoid training for these jobs. In addition, there is evidence that the additional earnings from obtaining postsecondary education are smaller for individuals seeking skilled technical credentials versus 4-year degrees. Since 2000, the wage premium for a bachelor's degree compared with a high school diploma has leveled off at approximately 80 percent, meaning that college graduates make 80 percent more over their lifetime than those with only a high school diploma (Autor, 2014). While the wage premium for an associate's degree compared with a high school diploma has increased over time, it remains much smaller, at around 30 percent (Modestino, 2010). In addition, research by Bailey and Dynarski (2011) suggests that high-income students have stronger incentives than low-income students to enroll in 4-year colleges and obtain 4-year degrees.

These considerations suggest that federal, state, and local policy makers and education professionals need to improve perceptions and beliefs about skilled technical work and better align standards for CTE with academic subjects so that students will be prepared for both work and postsecondary education that leads to a full range of credentialing options.

5.2.3 Standards for Certificate and Certification Programs

Questions exist about the overall value of alternative credentials such as certificates and certifications. Reviewing the evidence on the value of these credentials in preparing for and obtaining skilled technical work, Karp (2015) notes that this evidence is inconclusive. She calls for better data and research focused on creating coherent local systems of workforce development rather than per se developing additional credentials. Karp identifies several issues with respect to certificates and certifications.

The first is variability in value. The available research on certificates and certifications indicates that outcomes are highly context dependent. While Karp observes a positive labor market return to *certificates* in the aggregate as compared with earning only a high school diploma or completing some college without earning a certificate, different types of certificates have different value. Certificates that require a long period of study (long-term certificates) and are aligned with industry certifications provide more benefits to their holders

---

[13]For a review of the evidence, see Stern (2015). See also Hoffman and Reindl (2011) and Symonds et al. (2011).

relative to those that require short periods of study (short-term certificates). The value of certificates also varies by field and subfield: long-term certificates provide value in such fields as health care, but not in education and child care, and certificates provide differential benefits even within a single field, such as information technology (IT). It also appears that the value of certificates may vary by geographic region, gender, and race or ethnicity, and they may not be transferable across locations or related industry segments. Karp discovered few rigorous studies of *certifications*, from either an individual or systems perspective. The Lumina Foundation (2015) estimates that fewer than 10 percent of certifications are subject to third-party review or accreditation.[14] Community colleges' involvement in preparing students for certifications is highly variable and includes aligning curricula with certification exams, creating certification exam preparation programs, and/or becoming a certification testing center. Karp explains that variability may be responsive to employer needs and provide multiple access points for students, but whether it conveys value in labor markets remains an unanswered question.

The second issue identified by Karp is stackability of credentials. Many experts are promoting "stackable credentials" as a way to prepare for skilled technical jobs. A stackable credential is one of a sequence of credentials that can be accumulated over time to build qualifications, advance in one's career, and improve earning opportunities. Karp reveals some evidence that employers value this concept, but notes that certificates and certifications are potentially associated more closely with higher wages when they are stacked on a widely understood, lower-level credential, such as a general education diploma or high school diploma. Moreover, the timing of obtaining certificates and certifications may be important: they may have a positive effect on wages only when individuals earn them before obtaining an associate's or bachelor's degree. However, existing data suggest that most certificate/certification completers do not transfer to degree programs, which in turn suggests that many of these credentials serve as terminal credentials rather than stepping stones to degrees.

Finally, Karp cites the issue of inconsistent standards and limited data. Results of her survey reveal that inconsistent standards for credentialing and limited data on the role of credentials in career outcomes make it difficult to rigorously assess the value and impacts of certificates and certifications. In turn, this difficulty points to a need to adopt consistent credentialing terminology, ensure that credentials are aligned with employer and labor market requirements, and ensure that these requirements are consistently implemented and understood. This difficulty also points to a need to collect more longitudinal data on credentials and career experience and conduct more research on the

[14]In her review, Karp found only one study that examined the labor market impact of certifications—a study of Microsoft certification holders using data from 2000, which has limited generalizability.

value and impacts of particular credentials. While efforts are under way to address terminology, measurement, and data collection issues, evidence on the desirability and feasibility of articulating specific career pathways or pipelines in many occupations is difficult to find. A 2014 review of federal workforce development programs advocates using a wider range of research strategies to understand the value of credentials (DOL et al., 2014). To date, studies on credentials have used mainly nonexperimental evaluation designs. However, several experimental demonstrations are currently using random assignment designs, which will help build a more rigorous evidence-based understanding of the impact of industry-driven credentials.

5.2.4 Quality of Curricula and Pedagogy

Improving links between K-16 education and training and work is an area of further concern. Effecting these improvements includes updating curricula to reflect advances in science and technology and changing occupational requirements; focusing on an experiential learning approach that includes work-based learning rather than on a theoretical, lecture-oriented approach to teaching; growing the number of technically qualified teachers; and providing adequate and up-to-date training equipment and facilities. Training institutions, particularly community colleges, need to develop new ways to structure and deliver education and training for both traditional and nontraditional students. Although evidence on the effectiveness of some of the emerging instructional models discussed in this report is not yet available, the face validity of several new strategies appears promising for students who are also working, workers seeking to retrain for a new occupation, and low-skilled workers with little prior postsecondary education experience (DOL et al., 2014).[15] Addressing quality issues will require attention to the framework within which educators, students, workers, and other actors in the ecosystem address a number of coordination and information issues.

5.2.5 Incentives in Institutional Funding

The role of community colleges in many communities has expanded from providing relatively easy access to college studies to providing a wide range of credentials, custom education and training, programs in English as a second language, and remedial courses. Their mission, as well as the evidence on effective education strategies discussed above, requires the formation of strong employer partnerships in strategically important economic sectors in their communities. Advances in science and technology that create new occupations

[15]The role of experiential learning and work-based learning is discussed in more detail in other sections of this chapter and in Chapter 6.

with high technical and problem-solving skill requirements create additional pressure for community college faculty and administrators to develop new curricula and acquire new equipment and facilities (Autor, 2010).

There is growing concern, however, that community colleges have insufficient resources or funding incentives to provide programs of study that meet changing economic development needs and have potentially high labor market returns (Carnevale and Hanson, 2015). For example, the funding of many community colleges is based on formulas that emphasize the volume of enrollment rather than the value or quality of the services provided (Miao, 2012). This emphasis can create a financial incentive to focus investments in faculty, equipment, facilities, and support services in widely understood, high-volume education and training programs that easily generate high enrollment rather than on innovation and organizational change.

An additional challenge for community colleges is the level and stability of funding. Unlike 4-year public institutions, community colleges serve almost exclusively local students who typically continue to live and work in the community after graduation (Mellow and Heelan, 2015). This characteristic of community colleges would suggest that they potentially offer a competitive return on public investment. Yet the data displayed in Figure 5-1 indicate that in the two time periods shown (2007-2008 and 2012-2013), annual state and local revenues per FTE student were on average $3,000 lower for 2-year than for 4-year public institutions. Moreover, annual state and local resources at public 2-year institutions declined by $1,456 per FTE student between 2007-2008 and 2012-2013 (or 18 percent) (NCES, 2015c).

Similarly, even though the majority of undergraduate students are enrolled in community colleges, these 2-year Title IV institutions provide fewer resources per pupil relative to public 4-year universities. These data suggest a need to investigate two competing hypotheses: (1) community colleges are not appropriating enough money to postsecondary skilled technical education, or (2) postsecondary education funds are adequate, but they are not being allocated effectively to achieve the objectives of skilled technical workforce development.

5.2.6 Limited Data and Information

A lack of information about skill requirements makes it difficult for policy makers, educators, employers, and workers to coordinate to address quality issues in workforce development policies, processes, and programs. Policy makers and labor market participants need better data and more useful tools with which to analyze labor market dynamics, returns on investments in workforce development, and the effectiveness of alternative strategies for making adjustments to improve outcomes. Students and workers, guidance counselors, and educators need better information on skill requirements, career pathways, and potential returns on their investments in education and training. Employers

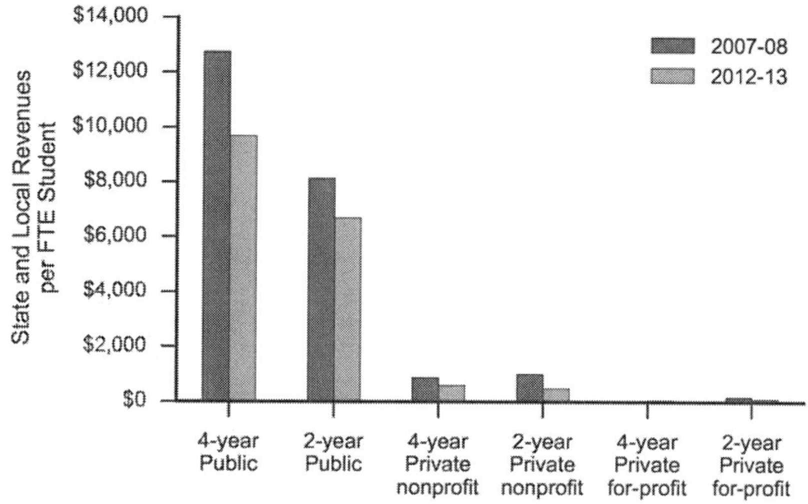

FIGURE 5-1 Revenues per full-time equivalent (FTE) student, 2007-2008 and 2012-2013.
NOTE: Revenues are from government grants, contracts, and appropriations per FTE student for degree-granting postsecondary institutions, by source of funds and institutional control and level.
SOURCE: NCES, 2015c, Figure 3, p. 220.

need better information on effective education and training resources and on viable programs and facilities for producing skilled technical workers.

The information system that supports U.S. labor market functions is funded and administered primarily by DOL. The states, which own and often collect the data, depend on DOL for supplemental data, technical support, and grants to support data capture, analysis, and research, while federal agencies depend on the states to share good-quality data in a timely and accurate manner. Reviewing the evidence on workforce labor market information issues, Reamer (2015) notes that although there are high returns to investment in the workforce labor market information system, funding has been stagnant over the past decade, and Congress has provided limited oversight. The result is a system that is less effective than it could be:

- **Funding.** Despite the potential for very high returns to investing in workforce information, Reamer observes that funding is insufficient. In his opinion, federal departments and the Office of Management and Budget tend to ask for too little funding; Congress appropriates too few funds for the President's requests; and Congress rarely approves new information budget initiatives, which typically range from $1 million to $5 million. In addition, Reamer notes a lack of federal funding for state labor market information agencies. Annual funding for Bureau of Labor Statistics grants to state agencies has remained at about $72 million for well over a decade. Similarly, the Employment Training Administration's state workforce information grant funding has been flat at $32 million. With too little money to begin with, the utility of these grants has eroded with inflation. Therefore, state agencies have been unable to provide the level of information services needed by labor market participants.

- **Data and research methods.** Advances in information and data sciences make it possible to increase significantly the volume of data collected, analyzed, and disseminated to users. However, statisticians and other labor market researchers are in many cases using out-of-date data; methods, particularly surveys; and approaches. For example, federal statistical agencies continue to rely on legacy methods for collecting and producing data, resulting in lagged data that do not reflect the true state of the labor market. New data analysis techniques offer opportunities to improve information through greater use of federal administrative records and through commercial data vendors and service providers. However, labor market information specialists have yet to determine how to work with data and statistics that fail to meet the standards of reliability defined by the Office of Management and Budget and the National Academies of Sciences, Engineering, and Medicine. Key issues include privacy and data commodification concerns related to providing data to commercial vendors and bias associated with big data analytics.[16]
- **Tools.** Current tools for sharing and understanding the meaning of labor market data are difficult to use. Reamer argues that federal economic statistics agencies lack sufficient understanding of how participants in labor markets at the local level use information to make decisions, nor do they tend to have the relationships with data users that would provide that understanding. Moreover, the federal government does not adequately assess the performance and impacts of its investments in workforce information. There is a need to develop user-friendly tools that provide direct access to information about options and performance to assist labor market participants in their decision-making processes.

WIOA mandates provision of most of the information resources that Reamer identifies as needed by labor market participants, and the law aims to address many of the systemic problems he discusses. As Reamer observes,

---

[16] For a comprehensive review of the issues, see the paper by Reamer (2015) commissioned for this study. The paper and additional material related to the topic can be found at http://nas.edu/SkilledTechnicalWorkforce.

however, the predecessor act (the Workforce Investment Act [WIA] of 1998) gave the secretary of labor a similar mandate to create and maintain a national employment statistics system, but succeeding secretaries did not pursue this mandate (Reamer, 2015). Full implementation of the information system aspects of WIOA is needed to address these data issues.

5.3 CHALLENGES IN TRAINING AND TRANSITIONING THE SKILLED TECHNICAL WORKFORCE

This section turns to the special challenges of educating and training the skilled technical workforce in two major sectors of the economy—health care and manufacturing. This is followed by a review of the challenges of transitioning the skills acquired by service members over the course of their military careers to the civilian workplace.

5.3.1 Training the Health Care Workforce

As described in Chapter 3, several changes in technology and practice, social demographics, and public law and regulations are creating the need for ongoing education and training in skilled technical health care occupations. In their paper commissioned for this study, Frogner and Skillman (2015) identify data challenges with respect to analysis of health occupations and then offer the following observations that are relevant to the coordination of workforce development in this sector:

- **Confusing education and training requirements.** Frogner and Skillman see evidence that education and training requirements for skilled technical health occupations are confusing, and that not all programs succeed in opening up pathways to well-paying careers. In their opinion, the career ladder in health care is more like a lattice because of the many options faced by entrants. Career transitions also are unclear. Credentialing through exams, certificates, and licenses is necessary to ensure quality care in some health occupations. In addition, scope-of-practice laws dictate what a worker can and cannot do. These requirements add to the confusion of occupational choice, education, and training. They also impose out-of-pocket and mobility costs for qualified workers, as credentials and scope-of-practice requirements vary by state and by occupation. In addition, obtaining credentials often requires payment of fees. Frogner and Skillman aver that, without a capable mentor or role model to serve as a guide, potential workers may find these complexities too daunting to enter the field.

- **Need for training relationships.** Public–private partnerships, including practical training partnerships between employers and educational institutions, are one way to develop necessary competencies—particularly as practice models and technologies evolve. Although clinical practicums are common in the health care industry, more formal training relationships, such as registered apprenticeships, are not, and several scale and coordination issues are associated with building sustained relationships between educators and employers.
- **Cost as a barrier to education and training.** The cost of education and training can be a barrier to entry in health care occupations for many people, and funding sources for this education and training can be unstable. Among all skilled technical jobs in allied health, nursing is the fastest-growing and offers the highest return over the span of one's career. In Frogner and Skillman's opinion, however, the cost of postsecondary education may be a barrier to entry in the nursing field for many potential workers. Other health care occupations in high demand, such as personal care assistants, have low educational entry requirements; however, they are among the lowest-paid. Many workers in these high-demand positions experience high turnover, are likely to live at the federal poverty level or below, and suffer from high rates of disability. Although the Carl D. Perkins Career and Technical Education Act's Tech-Prep programs could address funding issues, Congress has not appropriated funds for these programs since 2011, and subsidized loan programs are quickly disappearing as financial aid options shrink and the Pell Grant program struggles to meet the financial needs of eligible students amid funding cuts.

Health care goods and services are significant components of the U.S. economy in every state. Frogner and Skillman's analysis suggests that policy makers may need to create incentives for health care employers to assume a more active role in designing and funding education and training; improving job quality to recognize skill attainment; and collaborating with educators and workers to improve educational, training, and career guidance programs. In addition to better aligning workforce development incentives and programs, federal and state policy makers could work across state lines with health care industries to rationalize and harmonize health care credentialing and scope-of-practice requirements.

5.3.2 Training the Manufacturing Workforce

A recent report by Deloitte and the National Manufacturing Institute (NMI) claims that U.S. manufacturers are struggling to support their strategic, business, and production plans as a direct consequence of a shortage of skilled production workers (Deloitte and NMI, 2015). As Figure 5-2 shows, the executives surveyed by Deloitte indicated that a talent shortage affects critical aspects of business activity and expansion. Given the mixed evidence of a significant skills gap reviewed in Chapter 2, however, it is important to examine why manufacturing companies are unable to recruit and retain workers who have the necessary skills:

- **Public perception.** Public perception is a major issue for this industry: manufacturing firms often are perceived to be unstable, dirty, and dangerous workplaces with few opportunities for workers to grow and develop their skills (Panchak, 2016). Similarly, in a paper commissioned for this study, Messing-Mathie (2015) observes that students, parents, and their advisors often overlook manufacturing apprenticeship programs in favor of academic education, reflecting a bias against skilled technical education and training that tends to exist in most advanced economies. Although negative perceptions could make it difficult to attract a new generation of workers to skilled technical occupations in manufacturing, other factors also contribute to this situation.
- **Skill deficiencies.** More than half of the executives surveyed by Deloitte and NMI (2015) described difficulty in finding qualified workers and serious deficiencies in their current employees related to technology and computer, problem-solving, math, and technical knowledge. Some data show that the average manufacturing worker

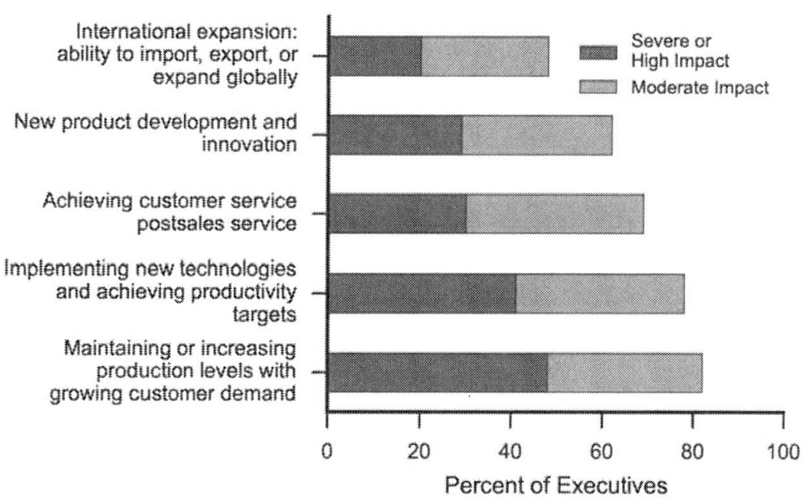

FIGURE 5-2 Business areas most affected by the talent shortage.
SOURCE: Giffi et al., 2015.

earned \$77,506 in 2013—20 percent more than what an average worker earned in other industries—and that four of five manufacturing executives are willing to pay more than the current market rate to hire and retain capable workers. However, other data suggest that skill demands have risen modestly and at a pace that makes most jobs accessible to most people. In a national survey, Osterman and Weaver (2014a) found that three-quarters of establishments can find the workforce they need but that a significant minority are experiencing difficulty. They note further that a retirement wave could exacerbate this problem. In their opinion, effective policies and institutions exist to address this problem, but they are scattered and not well supported. As discussed in more detail in Chapter 2, skill requirements for technical occupations change over time and are likely to continue to do so with advances in science and technology. These data suggest that workers who are interested in and willing to work in manufacturing are not investing in, do not have access to, or are not receiving the continuing education and training they need to be successful throughout their careers.

- **Perverse incentives.** Data collected in a recent employer survey by Accenture, Burning Glass, and the Harvard Business School suggest that manufacturing executives understand the importance of investing in employee training, development, and certification programs (Accenture et al., 2014). However, 58 percent of those surveyed reported that they use overtime to mitigate the effects of being unable to attract and retain skilled workers.[17] Yet working overtime makes it difficult for workers to find time to invest in additional education and training. In addition, loss of income due to the ability to hire a sufficiently skilled workforce makes it difficult for firms to invest in education and training, potentially creating a vicious circle of competitive decline.
- **Investments in automation and outsourcing.** Market forces also may provide an incentive for firms to invest in capital rather than labor. Over the past several decades, firms have increasingly automated or outsourced tasks and jobs previously held by skilled technical workers, resulting in job losses in specific industries and occupations. In a recent survey of Harvard Business School alumni, for example, nearly half of respondents agreed that their firms' U.S. operations preferred to invest in technology to perform work instead of hiring or retaining employees, while only 25 percent disagreed.

[17]See Accenture et al. (2014, Figure 16) for an overview of manufacturing executives' response to the question, "How important are the following techniques to mitigate the effects of existing skills shortages for a skilled production workforce?"

Forty-nine percent said their firms preferred to rely on vendors to perform work that could be outsourced, while only 29 percent reported that their firms would rather hire additional workers and keep work in house. When choosing to hire, moreover, companies also showed a distinct preference for relying on part-time workers (Accenture et al., 2014).

- **Program design.** An additional set of coordination challenges in the manufacturing sector is related to how education and training programs are designed and implemented. Messing-Mathie (2015) notes that apprenticeship programs typically require a diverse range of social partners: if key partners are excluded or if roles are not clearly specified, confusion and poor coordination will result (see also Wolter and Ryan, 2011). However, manufacturing apprenticeships and other types of education and training programs do not always offer a final certification or a portable credential upon completion, which makes them less valuable across firms and industry segments. There is a growing view that some employers, particularly small and medium-sized businesses—the predominant manufacturing firm size in the United States—are concerned about their ability to earn a return on the costs of employing and supporting apprentices. In addition, apprenticeships often pay a relatively lower wage in the early phases of the program, which may make them less competitive than other job opportunities and may create a sense of inequity that can be a disincentive for participation.

These observations suggest that the policy objectives of manufacturing-driven growth and competitiveness are more likely to be realized if policy makers better align the incentives of workers and employers to improve the returns on investment in education and training in this sector.

5.3.3 The Military Transition Challenge

The U.S. Armed Services are among the largest employers in the country (DoD, 2015).[18] Most enlisted personnel serve in the armed forces for just under

[18] The total number of military personnel in fiscal year 2014 (3.5 million) includes 1,101,939 Department of Defense (DoD) Ready Reserve and Department of Homeland Security (DHS) Coast Guard Reserve members, 214,784 Retired Reserve personnel, 13,700 Standby Reserve personnel, and 856,484 appropriated and nonappropriated fund civilian personnel. The DoD active duty and DHS Coast Guard active duty members make up the largest portion of the military forces (38.4 percent). This section focuses on skill transfer possibilities with respect to active duty military personnel in the Armed Services; therefore, the discussion does not include Reserve or civilian personnel.

7 years on average. In 1979, the United States eliminated the draft and created an all-volunteer military force. The Office of the Deputy Assistant Secretary of Defense (Military Community and Family Policy) reported that in fiscal year 2014, the military community included 1.3 million Department of Defense (DoD) active duty military personnel and 39,454 Department of Homeland Security (DHS) active duty Coast Guard members (DoD, 2015). Although the size of the armed forces is declining as the engagements in Afghanistan and Iraq wind down, Table 5-2 shows that there were still about 1.3 million women and men working in the Armed Services at the end of January 2016.

Many Americans join the armed forces as a pathway to education, training, and preparation for life and careers. As Figure 5-3 shows, 43.2 percent of all uniformed active duty personnel are aged 25 or younger. Military personnel receive excellent training in the competencies associated with skilled technical work.

Jobs in the military vary by type of work and level of responsibility. The majority of active duty jobs in the Armed Services are enlisted jobs (82.3 percent), which require a high school diploma and typically involve technical skills that could transfer to jobs in civilian occupations such as those in maintenance, health care, or IT. Some enlisted members (7.0 percent) have a bachelor's degree or higher, while most (92.1 percent) have a high school diploma and/or some college experience but less than a bachelor's degree. Officer positions, which represent 17.7 percent of active duty members of the Armed Services, involve planning, directing operations, and making critical decisions (DoD, 2015). These jobs require a minimum of a 4-year degree, and some careers, such as law and medicine, require advanced degrees. Relative to 1995, the percentage of total active duty members with an advanced degree has increased for both enlisted personnel (from 7.3 percent in 1995 to 10.1 percent in 2014) and officers (from 79.9 percent in 1995 to 86.5 percent in 2014) (DoD, 2015).

TABLE 5-2 Uniformed Armed Forces Strength Figures as of January 31, 2016

	01/31/16	12/31/15	Change from Previous Month		01/31/15
			Number	Percent	
Total Armed Services	1,347,232	1,344,258	2,974	0.22	1,360,406
Army	482,816	482,264	552	0.11	498,401
Navy	328,662	328,162	500	0.15	324,911
Marine Corps	184,418	183,197	1,221	0.66	184,781
Air Force	311,590	310,996	594	0.19	312,294
Total DoD	1,307,486	1,304,619	2,867	0.22	1,320,387
Coast Guard	39,746	39,639	107	0.27	40,019

NOTE: DoD = Department of Defense.
SOURCE: DoD, 2016.

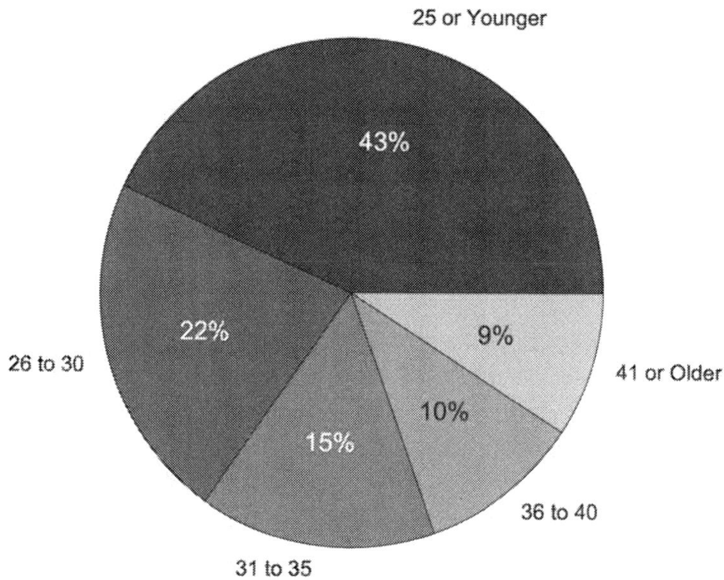

FIGURE 5-3 Age of uniformed active duty personnel in fiscal year 2014.
NOTE: Percentages may not sum to 100 because of rounding.
SOURCE: DoD, 2015. DMDC Active Duty Military Personnel Master File.

Military–Civilian Transition Issues

In 2014, 184,444 enlisted members and 20,112 officers left active military duty and returned to civilian life (DoD, 2015). The most common reason for leaving for enlisted members was voluntary separation/early release, while the most common reason for officers was non-disability-related retirement (DoD, 2015).[19] These workers have received excellent technical education and training, and they have developed the "soft skills" required to work well with others. Even though they receive comprehensive transition services, however, the Department of Veterans Affairs reports that, compared with civilians with similar demographics, veterans have an 8 percent higher total unemployment rate, and their median income is slightly below the nonveteran median of $42,317 (VA, 2015). Veterans participating in the GI Bill program complete

[19]The Pew Research Center (2011) estimated that the average length of service for enlisted personnel was 6.7 years.

programs at a rate comparable to that of traditional postsecondary students; however, they take longer to do so.

Veteran unemployment rates, earnings rates, and completion rates can vary across the services and base locations, as well as across skills, age, reason for separation, and other demographic factors. For example, the Department of Veterans Affairs reports that over the period 2005-2013, veterans aged 18-34 had a 20 percent higher unemployment rate than nonveterans, but median earnings for the same veteran age cohort were about 8 percent higher than those for nonveterans. Moreover, veterans under the age of 30 who used the GI Bill had 8 percent higher completion rates than traditional students in the general population. However, it is possible that these statistics misestimate the veteran–nonveteran difference for a wide range of reasons, including the way in which the experience of mobilized Reservists is treated.[20]

A 2014 U.S. Congressional Joint Economic Committee staff analysis indicates that more than 50 percent of veterans work in industries, such as transportation and utilities, manufacturing, construction, information, wholesale trade, and professional and business trades, that employ skilled technical workers (U.S. Congress Joint Economic Committee Democratic Staff, 2014). However, the analysis suggests that veterans are underrepresented in the fast-growing sectors of the economy, including two of the three industries that have fared the best during the past several years (in terms of job gains as a percentage of employment in the industry): the education and health services industry and the leisure and hospitality industry.[21]

To develop a better understanding of the contradictions in veteran transitions, the committee asked Susan Payne Carter and Brian J. Miller of the U.S. Military Academy to investigate the transitions of U.S. Army soldiers, who

[20]For example, the rates for the 18- to 24-year-old cohort likely reflect the experience of those who enlisted in their early 20s and left the military relatively early because it was not a good fit for them. Those who enlist at a young age and leave early are atypical because most enlistments are for 3 years or more. The proportion of those who leave early in older age cohorts may be much smaller. Additionally, the period 2005-2013 saw relatively high Reserve mobilization. Once demobilized, Reservists will appear in the data as "veterans." However, Reservists, particularly younger Reservists, may have a weaker connection to the labor force before mobilization (e.g., they may have been students or new to the workforce). Similarly, veteran–nonveteran comparative per capita income statistics may be misleading. For example, veterans who retire with an annuity or have other financial resources or support or who have fewer financial responsibilities may choose lower-paying jobs than veterans and nonveterans who have fewer financial resources or less financial support or greater financial responsibilities.

[21]The report notes that veterans represent only 4.3 percent and 3.3 percent of employees in these industries, respectively. Since December 2007, employment in the education and health services industry has expanded by 14.7 percent and in the leisure and hospitality industry by 8.9 percent.

represent the largest uniformed workforce in the Armed Services (Carter and Miller, 2015). Unlike civilian workers, departing military personnel may collect unemployment benefits even if they leave the service voluntarily. In many states, moreover, veterans may collect both GI Bill and unemployment benefits. Carter and Miller report that in 2013, DoD spent $829 million on Unemployment Compensation for Ex-servicemembers (UCX), more than one-half ($463 million) of which went to Army veterans. In 2010 and 2011, 53 percent of all transitioning veterans applied for benefits. On average, veterans drew benefits for 20 weeks during this time; 95 percent did not exhaust their benefits.

Between 2011 and 2013, 59 percent of departing Army soldiers who were eligible for UCX benefits applied, and most who did so ended up using the benefits. When data on soldiers are disaggregated across demographic and military service dimensions, however, average rates vary considerably. Surprisingly, soldiers with seemingly transferable skills, such as logisticians, apply for UCX at the highest rates, a result that holds when Carter and Miller control for education, cognitive ability, and other demographic differences. The authors speculate that some individuals in this population do not, do not wish to, or cannot directly transfer their Army skills to civilian jobs.

Similarly, Frogner and Skillman (2015) observe that military personnel have a unique career transition in health care. They point out that military skills and experience may not translate to the competencies needed or follow the path required by accreditation or licensing bodies in the civilian sector. Preparing and sitting for exams to obtain licenses despite years of experience can take more time than veterans can afford to be out of the workforce.

To explain differences in UCX application rates among demographic groups, Carter and Miller (2015) look at a wide range of demographic factors. They document evidence suggesting that a complex set of life factors can affect Army soldiers' UCX application rates (see Box 5-1). Groups more likely to apply include soldiers with 2 to 3 years of service upon separation and lower pay grades; female, black, and Hispanic soldiers relative to white male soldiers; divorced soldiers and soldiers with children relative to single soldiers; and soldiers with lower levels of educational attainment and lower cognitive ability scores on their qualification test.

Carter and Miller note further that the highest Army UCX application rates are associated primarily with service and support specialties. Veterans in the following specialties (listed in descending order) apply at higher rates: quartermaster, finance, chemical, adjutant general, air defense, and transportation. Although 62 percent of soldiers who separate because of satisfactory completion of their service contract apply for benefits, those who

BOX 5-1
Factors Affecting Unemployment Compensation for Ex-servicemembers (UCX) Applications

- Years of service
- Pay grade
- Gender and race
- Family structure
- Education and intelligence
- Military occupation
- Disability status
- Reason for leaving
- Location

separate for other reasons are more likely to apply.[22] Finally, application rates vary across base locations, suggesting differences in transition assistance programs.

Actions to Improve Veterans' Transitions

Several programs, including federal and state government programs, private-sector initiatives, and public–private partnerships, have been implemented to improve transitions for returning veterans (U.S. Congress Joint Economic Committee Democratic Staff, 2014). The United Services Military Apprenticeship Program (USMAP), for example, is a formal military training program implemented by the Center for Personal and Professional Development that gives active duty Coast Guard, Marine Corps, and Navy service members the opportunity to improve their job skills and complete their civilian apprenticeship requirements while they are on active duty. The objective of the National Apprenticeship Standards for USMAP is to provide registered certification of the training of individual service members and to achieve recognition for them equal to that of their civilian counterparts.[23] Lerman and his colleagues at the Urban Institute report that currently, about 1 in 4 enlisted

[22]The data show that soldiers with disability ratings are more likely to apply for UCX benefits. However, soldiers who have 100 percent disability ratings and are discharged for medical reasons are less likely to apply.

[23]DOL provides the nationally recognized "Certificate of Completion" upon program completion. Originally established in 1976 as a Navy program, USMAP became a single program registered with DOL (N-93063) for the three Sea Services in April 2000. USMAP allows active duty service members to complete a DOL apprenticeship program while serving their country (see DOL, 2012).

Sailors and 1 in 14 Marines participate in USMAP (Lerman et al., 2015). In their opinion, one of the program's major accomplishments is that it has registered about 100 occupations with DOL's Office of Apprenticeship that are related to civilian fields.

The VOW to Hire Heroes (VOW) Act (P.L. 112-56) strengthened and expanded tax credits for employers who hire unemployed veterans, increasing the tax credits for hiring veterans who have been seeking a job for more than 6 months, as well as those eligible for disability compensation from the Department of Veterans Affairs. Through the Hiring Our Heroes program, the U.S. Chamber of Commerce Foundation is working with businesses to obtain commitments to hire veterans and military spouses (Eversole, 2016).

Additional action is needed, however. Suggestions solicited by the Joint Economic Committee include streamlining credentialing and providing better-targeted transition assistance (U.S. Congress Joint Economic Committee Democratic Staff, 2014). Streamlining state occupational licensing, for example, helps veterans with equivalent military skills and abilities meet credentialing requirements for training and education. Veterans in jobs in health care and transportation occupations, such as emergency medical technicians, licensed practical nurses, and bus and truck drivers, often have skills that overlap with civilian occupational requirements. In addition, spouses of returning veterans often move with them across state lines and may need to obtain a certification or license in their new state. Today all 50 states and the District of Columbia allow veterans with relevant experience to receive waivers of the skills test to obtain a commercial driver's license, and dozens of states have acted in other ways to facilitate credentialing and licensing for military veterans transitioning to civilian employment and their spouses (U.S. Congress Joint Economic Committee Democratic Staff, 2014). Further legislation could facilitate the transition of additional veterans with specific skill sets into in-demand civilian occupations.

As Carter and Miller (2015) report, challenges in moving from active duty to civilian life often extend beyond simply having the right skills for available jobs. Veterans also may need assistance with more general skills that are helpful in the workplace and throughout the job search process (U.S. Congress Joint Economic Committee Democratic Staff, 2014). The VOW Act requires transitioning service members to participate in the Transition Assistance Program, which teaches them how to approach a job search, write a resume, and apply their military skills to civilian jobs (Orvis, 2016). The Veterans Employment Center, which lists private- and public-sector jobs, also helps veterans build profiles and translate their military skills into plain language. Carter and Miller argue that the Army should collect more data on soldiers' transition experiences, better analyze the efficacy of transition assistance classes, and tailor the classes to better address the challenges of transitioning from a military to a civilian career.

Other Actions

The information presented above suggests that DoD and the Department of Veterans Affairs could do a better job of assessing the transition risks for military personnel and of designing and delivering services to effectively mitigate these risks, including ensuring that the military culture encourages service members to transition without drawing UCX assistance. In addition, it appears to the committee that DoD could coordinate better with civilian policy makers, regulators, and educators to improve the transferability of military education, training, and certification.[24] A reevaluation of the Post-9/11 Veterans Educational Assistance Act may also be required.

Efforts to improve veterans' transitions into the civilian workforce have focused on actions taken immediately prior to their leaving the military at the point of transition into the civilian workforce. A seamless transition may require that planning begin earlier in—perhaps even at the start of—service members' military careers. Doing so would help ensure that service members are trained in fields in which they wish to continue after leaving the military and that these skills will be valued in both the military and civilian workforces. As discussed at a workshop convened by the committee on "The Military as a Pathway to Skilled Technical Jobs," DoD is seeking to pursue more fluid employment models that meet its workforce needs while allowing for easier transitions between the civilian and military workforces (see Box 5-2).

Given the sheer number of former service members transitioning out of the military each year (about 200,000) and the costs of UCX benefits (nearly $1 billion annually), better information on these transitions is needed, including return to military service after voluntary separation. For example, information on how often transitioning service members switch occupations and how they make occupation decisions is needed to help inform efforts of policy makers and DoD leadership to improve these transitions.

5.4 CONCLUSIONS

Policy makers, educators, employers, and workers need to better integrate economic development and workforce development objectives, undertake initiatives across and within industry sectors to align incentives, and invest in education and training. They also need to design and deliver relevant programs that meet the changing requirements of workers and employers at the local level. Addressing quality issues in skilled technical workforce development requires

---

[24]Many of these points are echoed in a recent study of USMAP. For an analysis of this program and recommendations, see Lerman et al. (2015).

BOX 5-2
Workshop on "The Military as a Pathway to Skilled Technical Jobs"

The committee convened a workshop to explore how the Department of Defense (DoD) can better integrate skills transition into military training rather than treating it as a separate component at the end of the service member's career. Participants considered how DoD can better coordinate with civilian employers, educators, and regulators to improve the transferability of military education, training, and certification to the civilian workforce. The meeting's agenda is reproduced in Appendix A.

In his keynote address, Under Secretary of Defense Peter Levine noted that education and training are at the heart of DoD's strategy both for recruiting service members and for transitioning them to civilian life. Karen Orvis of the DoD Transitions to Veterans Program Office described the Military Life Cycle (MLC) Transition Model, which starts preparation for service members' transitions early in their military careers. Service members are made aware of the Career Readiness Standards (CRS) they must meet throughout their military careers in mapping and refining their individual development to achieve their professional goals while in service and their postmilitary goals for employment, education, career technical training, or self-employment. DoD is in the process of implementing an evidence-based assessment of the Transition Assistance Program.

In his remarks, Ryan McDermott of the DoD Office for Manufacturing and Industrial Base Policy described the Skills Bridge Program, which offers service members an opportunity to participate in civilian job and employment training activities, including apprenticeships and internships. This program, he noted, also addresses the workforce needs of the manufacturing and industrial base that supports DoD's missions.

In their roundtable remarks, Debra S. Wada, Assistant Secretary of the Army (Manpower & Reserve Affairs); Brigadier General Kathleen Cook, Director of U.S. Air Force Services; and Anish Goel, Senate Armed Services Committee noted that a shift to a more integrated approach to selecting, training, and transitioning service members with technical skills is needed to facilitate the accomplishment of DoD objectives and enable workers to move back and forth between the military and civilian workforces. However, achieving this shift will require a change in DoD's mindset, authorities, and personnel management policies and procedures, which are currently focused on the more short-term goal of winning wars.

better coordination at the local, state, and federal levels and within and across economic sectors.

This chapter has considered a variety of current challenges to identifying and better aligning returns on investment in education and training on the part of students, workers, and employers. The next chapter reviews experiments under way across the United States to address many of the challenges identified in this and preceding chapters.

6

Key Lessons for Programs and Policies

Although, as discussed in the preceding chapter, significant challenges are entailed in developing and sustaining a skilled technical workforce, there is evidence for a variety of solutions that work by encouraging and facilitating collaborative partnerships across local employers; community colleges; high schools; universities; and local, state, and federal government agencies. These initiatives also can create the right incentives for employers and workers to remain engaged and vested in their local market for skilled technical work. This chapter reviews some leading programs and policies focused on achieving these ends.

Section 6.1 looks at how better to link students to skilled technical education and training opportunities and improve success rates, recognizing that students face substantial challenges in navigating, paying for, and completing such programs. Section 6.2 considers strategies for better linking secondary and postsecondary education and training, including early college schools, career academies, and dual-enrollment programs. Section 6.3 examines the links between postsecondary educational organizations and employers; the discussion encompasses strategic centers of excellence at community and technical colleges and sector-specific training programs. Section 6.4 looks at employer training programs, including the role of employers in developing apprenticeships, as well as efforts by labor unions to foster joint labor–management programs. Section 6.5 considers efforts to improve the collection, analysis, dissemination, and widespread use of data that can be used to improve the linkages discussed in the preceding section. Section 6.6 examines other policy initiatives aimed at improving career and technical education (CTE), including those focused on portable credentials and licenses, standardized credentials, licensing reforms, and competency models. Finally, Section 6.7 provides conclusions.

The polycentric nature of the skilled technical workforce development system in the United States provides the opportunity for many simultaneous policy and partnership experiments. This, in turn, makes it possible for civic leaders to examine the needs of their communities, study the experiences of others, and develop initiatives that address the identified needs.[1] Assessing these initiatives remains a challenge, however. While this report cites research pertaining to the variety of initiatives under way across the United States, the large size and complexity of the U.S. workforce development system makes it difficult to survey, evaluate, and monitor these initiatives comprehensively. The effectiveness of any particular strategy can depend not only on the validity of its concept and the rigor of its implementation but also on the presence of complementary linkages with other components of the system. Thus while this chapter highlights an array of noteworthy initiatives to develop and sustain a skilled technical workforce and presents evidence (where available) of their impacts, it is important to note that in many cases, research on these initiatives has not yet been conducted, and therefore their relative value and potential impacts have not yet been assessed.

6.1 LINKING STUDENTS TO SKILLED TECHNICAL EDUCATION AND TRAINING OPPORTUNITIES AND IMPROVING SUCCESS RATES

As discussed in Chapter 5, evidence shows high economic and social benefits of completing postsecondary education and training, but a number of factors inhibit students from either choosing or completing this course of preparation. This section reviews the evidence on the effectiveness of policy measures aimed at encouraging enrollment in skilled technical education and training and improving completion rates. These policy measures include those focused on improved counseling services, financing strategies, wraparound services, improved remediation, improved outcomes for adult learners, leveraging of online learning, and improved incentives for completion at the community college level.

Although this chapter focuses on public 2-year institutions (community colleges), many private for-profit institutions offer similar education and training programs (see Chapter 4). While for-profit institutions can help meet the demand for skilled technical education and training, the evidence suggests that completion rates, default rates, and labor market outcomes are worse for for-profit than for public 2-year institutions (Deming et al., 2013).

---

[1]The National Center on Education and the Economy and the National Conference on State Legislatures have formed a study group to investigate top-performing education jurisdictions. For information, see NCEE (2017).

6.1.1 Improved Counseling Services

There appears to be widespread consensus among education experts that policy makers, educators, and employers need to do a better job of helping students and adult workers understand the link between education and training programs and employment. The Accelerated Study in Associate Programs (ASAP) at the City University of New York (CUNY), discussed further below, provides evidence that advising, career services, and tutoring play an important role in improving completion rates (Scrivener et al., 2015).[2] Career guidance is particularly important for skilled technical workforce development to counter the common perception that the only path to lifelong occupational success is through immediate entry into a 4-year college and advanced degree programs (see Chapter 5).

Some analysts believe that without career counseling and reliable occupational information, students pay insufficient attention to labor market trends when choosing a field of study (Long et al., 2014). When industries and occupations are changing, it is particularly challenging for students and job seekers to identify viable career options and the associated education and training requirements. At the same time, data suggest that K-12 schools, postsecondary schools, and worker assistance agencies have reduced support for counselors and advisors because of budget pressures and changing priorities (Good and Strong, 2015). Inadequate occupational guidance potentially increases the costs of education and training, creates imbalances in labor markets, and reduces returns on investment in skilled technical workforce development.

Research suggests that One Stop Career Centers often provide the type of career advising that is lacking in many educational institutions, and a variety of collaborations can be encouraged, depending upon local requirements (Holzer, 2015b). Under new provisions in the Workforce Innovation and Opportunity Act (WIOA), states can rethink and expand the reach of these job centers to better serve a broad range of clients. For example, some states have purchased online job databases such as Help Wanted OnLine® to help their One Stop Career Centers better match job seekers with jobs. Nearly a quarter of One Stop Career Centers are now located on college campuses, and even more are deploying staff

---

[2]Scrivener and colleagues (2015) used randomized controlled trials to evaluate the impact of ASAP over a 3-year period compared with the usual services and courses provided at CUNY, finding that graduation rates within 3 years nearly doubled. Along with these student services, students were required to be full-time students and were provided additional financial support, such as tuition waiver, free use of textbooks, and cards for free public transportation. The ASAP program has student-advisor ratios between 60:1 and 80:1. A study of two colleges in northern Ohio found that enhanced academic counseling with an average of a 160:1 student-adviser ratio had only modest short-term effects (Scrivener and Coghlan, 2011).

to campuses on a part-time basis. States also can take steps to reform job center metrics to better reflect the quality of advising activities rather than the number of referrals.

In addition to improving completion rates, counseling and career services can increase enrollment in skilled technical education and training where the returns are higher. As discussed in Chapter 5, however, many parents, students, and workers perceive career preparation programs such as CTE to be associated with lower socioeconomic status and lower career ambitions (see, e.g., Herian, 2010). If the challenges of skilled technical workforce development are to be addressed effectively, this perception must be changed so that programs to provide job skills and training are no longer viewed as a substitute for college but as a robust way to contextualize academic learning.

The best education and training programs prepare students to enter the workforce and pursue lifelong learning by encouraging the development of general technical skills, occupational skills, and "soft skills" that make for reliable and collegial employees (Bronson, 2007). However, the current overemphasis on 4-year college-preparatory curricula may inhibit students from receiving the workplace training necessary to acquire soft skills and from learning enough technical content to become skilled workers (see Bolli and Hof, 2014; Bolli and Renold, 2015). Thus, students may need to receive counseling even before entering community college so that they realize the potential gains from pursuing this training.

Instead of offering a program solely designed to prepare students for 4-year college programs, then, another option is to prepare them by developing technical skills and then help them prepare for and choose among postsecondary education and training and career alternatives. This emphasis on a strong technical foundation can improve public perceptions of career preparation programs, enhance the employment prospects of graduates, and develop a better workforce.

6.1.2 Financing Strategies

As discussed in Chapter 5, the rising costs of community college and inadequate financial aid can be a barrier to completion for many low-income students. Some federal aid, such as Pell Grants, is limited to undergraduate students in for-credit programs and is not available for students taking continuing education classes, even when they are earning a certificate (Greenhouse, 2013). An evaluation of two colleges in New Orleans found that providing financial aid tied to academic benchmarks to supplement other financial aid increased the likelihood that students enrolled full-time and earned more credits (Scrivener and Coghlan, 2011). Although only a small fraction of students in the ASAP program received tuition waivers, it may be that the availability of such financial support increased full-time enrollment in CUNY colleges by reducing worry about future financial aid (Scrivener et al., 2015). In addition to the direct costs of community college, students have to weigh foregone earnings when in school and the need to balance college and work for many parents.

Braiding Strategies

Some states are seeking to combine public resources to support postsecondary education and training programs using "braiding strategies," which merge public dollars for a common purpose while keeping categorical funds distinct. For example, Oregon's Career Pathways Statewide Initiative braids Perkins Act, Temporary Assistance for Needy Families, Workforce Investment Act (WIA), and state general funds to finance career pathways programs in the postsecondary education system. In Arkansas, the Career Pathways Initiative (CPI) has partnered with the local division of child care and early childhood education offices to connect students to state subsidies for child care (Hoffman and Reindl, 2011). In addition, CPI students are assigned a tutor and a counselor to help them address any issues and obtain other services.

Lifelong Learning

Some states also are experimenting with strategies for ensuring that U.S. workers and employers continue to invest in education and training throughout the career span. Maine, for example, is experimenting with portable, employee-owned Lifelong Learning Accounts (LILAs) that permit employees, employers, and the state to pool funds for continual worker education and training (Hoffman and Reindl, 2011). LILAs, which are managed by the Maine Finance Authority, can be used for postsecondary education and training courses as well as books and supplies. In addition, LILA holders have access to career counselors.[3]

6.1.3 Wraparound Services

Providing education, training, and opportunity often is insufficient to ensure success for many youth and adult learners, many of whom are unfamiliar with school processes and policies or need help meeting family, work, and financial responsibilities. "Wraparound" student support services at the postsecondary level, such as those described in Box 6-1, can be particularly beneficial for low-income and first-generation students, who may not have the resources or social networks to navigate the higher education system and to persist to complete their degree programs. A recent study of the ASAP program

---

[3]For information about Maine's LILAs, see www.bos.frb.org/commdev/c&b/index.htm (accessed March 21, 2017).

BOX 6-1
Key Features of Wraparound Services

Purnell and Black (2004) identify the following key features of wraparound services:

- Academic guidance and counseling, through which students gain information on educational and course planning and graduation requirements;
- Academic supports, such as tutoring, through which students receive additional help in meeting their academic goals;
- Personal guidance and counseling, such as mental health counseling or crisis intervention;
- Career counseling, through which students receive information on careers; and
- Supplemental services, such as child care or transportation assistance.

SOURCE: Excerpted from Purnell and Blank (2004, p. 7).

at CUNY, discussed earlier, suggests that focusing on wraparound services can improve returns on investment in skilled technical education and training. The program offers full-time students intensive advising and tutoring, priority in registering for oversubscribed courses, free transportation, free textbooks, and a waiver that covers any shortfall between schooling costs and financial aid. A randomized controlled trial of the program showed significant positive outcomes. The program nearly doubled the share of students graduating within 3 years, increased the share enrolling in a 4-year college, and provided a satisfactory return on investment. The program was estimated to have increased per-student costs by 60 percent, or about $5,400 a year. However, the near doubling of graduation rates means that CUNY actually spent more than 10 percent less per college degree (Dynarski, 2015).[4]

Wraparound services may be standard and comprehensive or operate on a piecemeal or smaller scale, depending on local requirements. An example of a small-scale program is Massachusetts Bunker Hill Community College's (BHCC) emergency assistance fund.[5] This fund provides small one-time grants

---

[4]Some experts argue that wraparound services are important at the secondary as well as the postsecondary level. The committee believes this is an important issue; however, a detailed examination of secondary education is beyond the scope of this study.
[5]For information about the fund, see http://www.bhcc.mass.edu/emergencyassistancefund (accessed March 21, 2017).

to students for emergencies that occur during the semester and may cause them to drop out of college, such as a lost or stolen laptop, a medical emergency, or a car accident. Students who document such a need may receive up to $1,000 to help them get back on their feet and stay in school. In fiscal year 2012, the fund assisted 147 students and disbursed more than $100,000; the average grant was about $700. BHCC reports that the retention rate for students receiving assistance through the fund is 31 percent higher than that for the entire student population.

6.1.4 Improved Remediation

As discussed in Chapter 5, students who need to take remedial courses relative to those who do not are more likely to drop out of a postsecondary program. Nearly two-thirds of students entering community college are asked to take either remedial English or math classes (NCES, 2016a). Students who need to take two or three remedial classes in a subject are unlikely ever to complete a college-level class in that subject (Bailey and Jaggars, 2016). In addition, there is some evidence that the remediation is not effective: one study found that the vast majority (72 percent) of students who ignored the referral to a remedial class ended up completing the college-level class, compared with only 27 percent of students who complied with the referral (Bailey et al., 2010).

Reducing the Need for Remediation

As discussed in Chapter 5, many students entering college are assigned to noncredit developmental courses that increase the costs and lengthen the time for obtaining a degree. Some analysts are questioning the way students' readiness is assessed and whether the need for developmental coursework is as prevalent as some have assumed.[6] Improving readiness assessments could be a low-cost, high-benefit intervention to increase college access for many students. For example, using high school transcript information—either instead of or in addition to test scores—could significantly reduce the prevalence of assignment errors (Scott-Clayton et al., 2012).

---

[6] Many public colleges and universities use a computerized test, known as ACCUPLACER, to assess students' skills in mathematics, writing, and reading. However, recent research has found that under current test-based policies, about one in four test takers in mathematics and one in three test takers in English are assigned improperly to developmental courses. In addition to inaccurate assessments, the value of the developmental coursework has been challenged. For example, many community colleges require students to pass Algebra I before taking for-credit classes even in fields in which this mathematics track is not required.

The following subsections review evidence that integrating remediation into skills training classes, as well as using accelerated remedial coursework, can improve completion rates. Methods of reducing the need for remedial classes, including improving elementary and secondary education and better integrating secondary and postsecondary education, are then discussed.

Integrating Remediation

Schools can do a better job of assessing student readiness for postsecondary education and training and of designing more appropriate and cost-effective prerequisite requirements. Moreover, several studies indicate that it is possible to integrate academic preparation for postsecondary education and training with technical training to motivate students and to hasten readiness for skilled technical occupations (see, e.g., Hoffman and Reindl, 2011; Howington et al., 2015; Wachen et al., 2012).

Readiness can be improved by integrating developmental content into courses that are offered for credit instead of delivering it as a stand-alone course (see Bettinger et al., 2013).[7] The state of Washington's Integrated Basic Education and Skills Training (I-BEST) program, for example, offers academic instruction in the context of technical education (Hoffman and Reindl, 2011). Adult literacy and career-technical instructors co-teach courses that provide students with developmental support while they earn credit toward a certificate or degree. A multiyear evaluation of I-BEST found that when developmental instruction was combined with for-credit classes, students developed basic skills more closely tied to their field of study, were more likely to persist in their coursework, and did not waste financial resources on noncredit coursework (Wachen et al., 2012, using outcomes of I-BEST students matched to those of students with similar characteristics). The evidence also indicates that I-BEST students earn nearly twice as many college credits as non-I-BEST students and are more likely to earn occupational certifications. Moreover, the program's benefits and costs have been estimated to be approximately equal, suggesting that such programs can be designed so that no additional resources are needed to achieve better results (Wachen et al., 2012).

Several other states have launched integrated programs to help students develop more quickly the skills they need to obtain postsecondary credentials. The Minnesota Training, Resources and Credentialing for Pathways to Sustainable Employment program (FastTRAC), for example, uses a modular approach to attaining credentials that is modeled on the Washington I-BEST program (Hoffman and Reindl, 2011).

---

[7]Some of the approaches discussed in this section also serve as examples of articulating career pathways, which is discussed in more detail in a later section of this chapter.

Similarly, the state of Illinois created a plan to help adults quickly obtain a postsecondary degree valued by employers through "bridge" programs in community colleges located across the state (Hoffman and Reindl, 2011). These programs help students bridge the gap between their secondary education and the academic skills needed to succeed in postsecondary training. Students enrolled in bridge programs simultaneously receive academic and occupational instruction, which provides important context for learning.

The Florida College and Career Readiness Initiative (FCCRI) is another statewide program that assesses the college readiness of high school juniors and for those assessed as not ready for college, provides them with instruction as high school seniors (Mokher et al., 2014). A similar bridge program launched by LaGuardia Community College targets high school dropouts. This adult education program provides a bridge to a General Educational Development (GED) credential and continued postsecondary training. An evaluation by researchers at MDRC showed that bridge students at LaGuardia Community College were more likely to have passed the GED exam and enrolled in college relative to students in traditional GED preparation courses (Martin and Broadus, 2013).

Accelerating Remediation

In addition to integrating remedial classes with skills training and linking it to the students' fields of interest, providing accelerated remedial courses or co-requisite classes may reduce the incidence of students dropping out before completing required remedial courses. Acceleration strategies can include tailoring the curriculum to eliminate unnecessary topics, accelerating the pace of the remedial classes to reduce the number of dropout points, or linking college-level enrollment to a companion remedial class. Evidence from the Community College of Baltimore's Accelerated Learning Program revealed that students taking co-requisite and sequential remedial classes had the same pass rates for the college-level classes, but that students taking the co-requisite classes were able to accrue college-level credits more quickly. And Tennessee's co-requisite program showed a much higher rate of passing the college-level math and writing sequences (Bailey and Jaggars, 2016).

6.1.5 Improved Outcomes for Adult Learners

Community colleges serve adult learners in addition to students who enter immediately after graduating from high school. As discussed in Chapter 4, a significant portion of community college students are aged 30 and older. Career pathways, described below, focus on these adult learners. They entail linking two or more systems to improve education and training outcomes in a process known as "articulation" (King and Prince, 2015). There is some evidence that career pathway programs have significant impacts on employment and earnings that tend to be longer-lasting than those of typical workforce programs, including higher wages and increased employment (Smith and King, 2011). While these results are not based on rigorous experimental evidence, randomized controlled trials of these programs are ongoing.

Career pathways are "roadmaps" of the education and training required to attain credentials associated with success in specific industries, often used to guide linked learning, sector strategies, talent management, and career pipeline initiatives. This approach has been adopted in several states.

Career pathways are typically targeted to meet the demands of local labor markets. The programs and resources of local community colleges, workforce development agencies, and social service providers are integrated to create structured sequences of education, training, and on-the-job learning in strategically important occupational areas (Alssid et al., 2002). Ideally, career pathways offer "a series of connected education and training programs and support services that enable individuals to secure employment within a specific industry or occupational sector, and to advance over time to successively higher levels of education and employment in that sector" (Jenkins and Spence, 2006, p. 2). They also often involve bridge programs to address the skill deficiencies of participants and the challenge of moving them along an accelerated pathway.

Several states, including Arkansas, Montana, Oregon, Virginia, and Washington, have developed statewide career pathways initiatives that are relevant for secondary and postsecondary education (Hoffman and Reindl, 2011). One often-cited example is that of Shifting Gears, launched in 2007 with the support of the Joyce Foundation and matching funds in six states (Illinois, Indiana, Michigan, Minnesota, Ohio, and Wisconsin). This initiative prioritized industry and occupational sectors that offer skilled technical jobs; broke down longer diploma and degree requirements into shorter certificate models; and offered classes at a wider variety of places, days, and times (Strawn, 2010). In Minnesota, the Shifting Gears program evolved into the FastTRAC Adult Career Pathway partnership, which combined multiple federal, state, and philanthropic funds to serve 3,385 individuals between 2009 and 2012.[8]

Similar programs, created through the Center for Law and Social Policy (CLASP)-led Alliance for Quality Career Pathways (AQCP) Framework 1.0, focus on connecting adult learners with needed remediation and then getting them into training programs. To develop AQCP Framework 1.0, CLASP worked jointly with 10 states—Arkansas, California, Illinois, Kentucky, Massachusetts,

[8]Self-reported data indicate that 88 percent of these individuals completed industry-recognized credentials and/or credits. Among those exiting the program, 57 percent entered employment, and of those, 84.8 percent retained employment for at least 6 months. Among those with wages over the entire year, exiters earned on average $21,080 annually, a 33 percent increase over their annual earnings in the year prior to enrollment (Choitz et al., 2015).

Minnesota, Oregon, Virginia, Washington, and Wisconsin—thereby establishing a common understanding of quality career pathways and systems.[9] The framework provides criteria, indicators, and metrics for use in evaluating and creating state and local career pathways (CLASP, 2014).

6.1.6 Leveraging of Online Learning

Technology developments that support online learning make it possible for educators to create flexible learning environments that accelerate education and training, as well as facilitate the development of knowledge and practice communities that can be sustained over time using social media.

Distance education goes back decades, and many postsecondary programs are now delivering courses online. The Center for Adult Learning in Louisiana (CALL), for example, offers 13 online programs in accelerated formats for adult learners. Operated jointly by the Louisiana Board of Regents and the Southern Regional Education Board, CALL specifically targets adults and is provided exclusively online to accommodate the schedules of working students. Courses are taught in 4- and 8-week formats. Time to degree completion is accelerated through prior learning assessments, which offer students college credit for work, volunteer, or life experiences (Hoffman and Reindl, 2011). All CALL students are eligible for financial aid and are provided a range of support services. Similarly, a pilot training program funded by the Defense Advanced Research Projects Agency (DARPA) trained Navy recruits and unemployed veterans with no previous technical experience to perform at or above the competency level of a 10-year information technology (IT) expert with only 16 weeks of intensive online training (The White House, 2014). Still another example is a cross-state experiment in creating a virtual university that enrolls students from 50 states in the Western Governors University (WGU), a nonprofit online university (The White House, 2014). WGU, which costs approximately $6,000 per year and is nationally and regionally accredited, offers a competency-based curriculum that enables progress toward a degree based on demonstrated skills rather than number of credit hours.[10]

The rise of online education, particularly the recent growth of massive open online courses (MOOCs), makes it possible to increase the scale, interactivity, sophistication, and personalization of distance learning, as well as to enhance the learning experience. For example, Coursera and edX, two

---

[9]For information on the CLASP initiative, see CLASP (2014).

[10] Competency-based learning, which focuses on helping students master content, awarding credit, and permitting progress based on competency, can be considered a strategy for K-16 reform. For a definition of competency-based learning and an overview of issues and contemporary debates, see the entry on competency-based learning in *The Glossary of Education Reform* (Great Schools Partnership, 2014).

popular MOOC platforms, have more than 15 million enrollees from nearly 200 countries (Willcox et al., 2016). However, education research historically has focused on learning in the classroom and is only just beginning to consider the role and impact of online learning. A recent study by researchers affiliated with the Massachusetts Institute of Technology (MIT) Online Education Policy Initiative argues that if the proliferation of online learning is properly understood and leveraged, it can serve as a catalyst for reform in postsecondary education and training, although it cannot and should not replace teachers (Willcox et al., 2016) (see Box 6-2).

6.1.7 Improved Incentives for Completion at the Community College Level

Some states are already strengthening incentives for community colleges to respond to the skills needed in the labor market and to improve completion

BOX 6-2
Massachusetts Institute of Technology (MIT) Report on Online Learning

The committee convened a workshop to discuss online and blended learning opportunities and how they can be used to improve educational pathways. At the workshop, the authors of the 2016 report of MIT's Online Education Policy Initiative (OEPI), *Online Education: A Catalyst for Higher Education Reforms* (Wilcox et al., 2016), briefed the committee (see Appendix A).

The OEPI report provides a detailed history of online and blended education from the rise of educational games to massive open online courses (MOOCs). It highlights the use of online tools in remedial education and in reskilling of adult learners. In particular, catch-up courses offered online can help students overcome education deficiencies before starting a college program. In this way, online courses can help address some of the root causes of low community college completion rates.

The OEPI report concludes that digital technologies can play a significant role as an education enabler. However, teachers serve an essential role in learning by providing context, mentoring, and fostering reflection and discussion. Recommendations in the report include that new technologies be used to support teachers and allow them to focus on in-person interactions rather than content delivery. In the authors' opinion, "blended learning," whereby teachers and technologies cooperate across online and in-person spaces, can enhance learning but requires reorganizing the learning experience to apply the different strengths of the two modalities.

Finally, the OEPI report highlights the usefulness of learning simulators as a key online component of education. Flight simulators, for example, have transformed the training of pilots, and the report notes that training for selected skilled technical tasks can be similarly transformed using online simulation modules.

rates. Community colleges traditionally have received state-level funding based on the number of enrolled students. As of July 2015, however, 32 states had implemented funding formulas that allocate a portion of a college's funding based on such indicators as course completion, time to degree, transfer rates, the number of degrees awarded, and the number of low-income and minority graduates (NCSL, 2015). Yet it may be that reliance on these academic performance-based funding requirements will lead to more selective admissions criteria (Kelchen and Stedrak, 2016) or lowered standards for completion (Courty and Marschke, 2011). State policy makers thus might do better to attempt to create better incentives for improved completion rates by linking funding to employment metrics, such as earnings of students after graduation, or to metrics focused on increasing enrollment and completion for students who obtain credentials that are in high demand among employers or offer higher returns on investment (Holzer, 2015a).

6.2 LINKING SECONDARY AND POSTSECONDARY EDUCATION AND TRAINING

Better integration of academic and technical education and training can improve the quality of postsecondary education and training and increase the return on investment in education and training for skilled technical occupations (DOL et al., 2014). As discussed in Chapter 5, inadequate preparation in primary and secondary school may be a large part of the reason that students fail to complete their education. Strategies for reforming postsecondary education and training include creating flexible and integrated learning environments, particularly in community colleges and technical schools, and reassessing the readiness of students for postsecondary education and training and prerequisite requirements and programs.

In his paper commissioned for this study, Stern (2015) suggests that a better approach to secondary education is to prepare all high school students for both employment and a full range of postsecondary education and training options, and to avoid making assumptions about future education and training attainment or segregating students on this basis.[11] In his opinion, the coursework required for 4-year college admission can and should be available to all students willing to put in the effort. He argues that high school students should master a sequence of academic and CTE coursework that will help them earn a living

[11]This is not a new proposal. The School-to-Work Opportunities Act of 1994 (STWOA) embraced a similar philosophy and provided funds for activities classified as work-based learning, school-based learning, or connecting activities. For reviews of the challenges associated with implementing STWOA, see, for example, Hollenbeck (1997) and Neumark (2004).

regardless of whether they pursue or complete a 4-year college degree. He outlines several design principles to "make [secondary education] real and make it fair" (see Box 6-3).

Stern's findings also are relevant for state and federal policy makers. Citing a 2013 study by the National Association of State Directors of Career Technical Education Consortium and the Independent Advisory Panel for the 2014 National Assessment of Career and Technical Education, Stern calls for changing CTE standards (which now tend to be course-specific and geared toward preparing students for particular jobs) so they are aligned with a set of broader career clusters and pathways. He also makes the case that the pending reauthorization of the Perkins Act provides an opportunity for federal policy makers to eliminate bureaucratic rules that effectively maintain CTE as a silo of isolated activities and to better integrate CTE across several education reform efforts. In addition, a variety of programs, discussed in the next section, link secondary and postsecondary education and training with particular employers.

Clear linkages between secondary and postsecondary institutions can help students transition from secondary to postsecondary education and later, from education to employment. These linkages can improve student preparedness, reduce the costs of education and training, help students learn about college programs and processes, build critical social networks and supports, and improve returns on investments (Karp, 2015b). Programs designed to improve these linkages include early college schools, career academies, and dual- or concurrent-enrollment programs, discussed below.

BOX 6-3
Make It Real and Make It Fair

- Leave the choice up to the student.
- Provide support for students.
- Integrate academic and technical curriculum.
- Make the curriculum relevant with real-world linkages to jobs.
- Develop employer partnerships and work-based learning.
- Collaborate with postsecondary institutions.
- Get district support.
- Set high standards, create accountability, and let data drive decision making.
- Expand the number of career and technical education (CTE) teachers.
- Leverage the support of intermediaries.

SOURCE: Adapted from Stern (2015, p. 17).

6.2.1 Early College Schools

Early college schools integrate high school and college education and training in a rigorous yet supportive environment.[12] Each such school requires a collaborative partnership between a school district and a 2- or 4-year postsecondary school. One example is the P-Tech model, described in Box 6-4.

Most early college schools enroll a higher percentage of minority students relative to their corresponding school district and state, with an average of 60 percent of students coming from low-income families. Over the past decade, early college schools have produced significant outcomes for low-income youth, first-generation college students, and minority students. Based on a study of 100 representative early college high schools, for example, about 90 percent of early college students graduated high school, a figure 12 percentage points higher than the national average of 78 percent. Of the 2,600 early college

BOX 6-4
IBM's P-Tech Model

One of the most widely reported examples of an early college school is the Pathways in Technology Early College High School (P-Tech 9-14), launched in Brooklyn, New York, in 2011 in a partnership among the New York City Department of Education, the City University of New York, the New York College of Technology, and IBM. IBM helped develop the curriculum, which focuses on science, technology, and mathematics. P-Tech offers traditional classes, including English and history, but students also learn coding. IBM provides mentors and paid internships. Students earn their high school diploma and an associate's degree in applied science, and often have an opportunity to work for IBM.

Recently, six students from P-Tech graduated with associate's degrees in applied science and computer information systems in just 4 years instead of the expected 6 years. The students were able to finish both degrees in 4 years of secondary education because the school integrates high school and college coursework, but also includes an extended school day and academic year.

The P-Tech model has been replicated across the United States. *The Economist* (2015) reports that more than 70 small and large companies, including Microsoft, Verizon, and Lockheed Martin, have adopted or are working with schools to adopt the model. Consortia of advanced manufacturing firms and some hospitals also have partnered with school districts to replicate the model. IBM expects that more than 100 schools with 100,000 students will be operational in 2016.

---

[12]The committee's definition of early college programs embraces both 2- and 4-year postsecondary educational programs, which include "middle college" programs.

graduates who enrolled in postsecondary education and training, only 14 percent enrolled in coursework to prepare them for college-level work, compared with 23 percent nationwide. In addition, the majority of early college students earned college credit in high school, and 30 percent earned an associate's degree with their diploma. Early college school graduates thus were able to start their careers without college debt or with less debt than is typical (Webb and Gerwin, 2014).

The states of Michigan and New York have been at the forefront of the early college movement; however, a variety of programmatic models for early college schools have emerged.[13] With the help of more than $130 million in private start-up funds, the number of early college schools has grown from 3 in 2002 to 280 in 2014, serving more than 80,000 students and producing more than 5,000 graduates. Start-up costs range from $125,000 to establish an early college program within an existing high school, to $500,000 to establish a program on a college campus, to more than $3.5 million to establish a separate stand-alone early college charter school. Early college schools receive anywhere from $90,000 to $400,000 in grants from the coalition of funders that sponsors the national Early College High School Initiative. For programs established within existing high schools, these grants cover about 70 to 80 percent of start-up costs. Public school districts provide the ongoing operating budget for each early college school as they do for traditional high schools.

The Pathways to Prosperity program is another example of an initiative that links high school CTE with postsecondary education and training. In 2012, the Harvard Graduate School of Education and Jobs for the Future created the Pathways to Prosperity Network to help state, regional, and local educators, employers, and intermediary organizations build and scale up career pathways initiatives that span grades 9-14. The network's 2014 report indicates that at that time, the network included 10 states: Arizona, California, Delaware, Georgia, Illinois, Massachusetts, Missouri, New York, Ohio, and Tennessee (Jobs for the Future, 2014).

6.2.2 Career Academies

Career academies are small learning communities within schools that provide a college-preparatory curriculum with a career-related theme.[14] In some

[13]Examples of other models include middle college programs such as the Secondary Technical Education Program (STEP) in the Anoka-Hennepin School District in Minnesota and Washtenaw Technical Middle College (WTMC) in Washtenaw, Michigan. For information on the STEP program, see http://www.anoka.k12.mn.us/domain/2247 (accessed March 22, 2017). For information on WTMC, see http://www.themiddlecollege.org (accessed March 22, 2017). For an overview of early college models and student experience, see Barnett et al. (2015).

[14]See, for example, Kemple (2008) and Castellano et al. (2012). Some analysts argue that evidence of the benefits of career academies may be overstated because of bias

cases, these academies can generate large improvements in earnings for students for many years beyond graduation. In addition, many of these academies have demonstrated the ability to maintain mathematics and science instruction at high levels (Castellano et al., 2012).

Initiated in 1969 in Philadelphia with the Electric Academy at Edison High School, career academies in the United States now number more than 7,000 and cover a wide range of career areas (see, e.g., Stern et al., 2010).[15] One model is the National Academy Foundation (NAF), described in Box 6-5, which partners with educators at existing high schools, businesses, and civic leaders in high-need communities across the United States (see Orr et al., 2004, for a review of NAF's work). Other models for career academies include the Southern Regional Education Board (SREB) High Schools That Work; High Tech High; and regional technical centers in such states as Ohio, Massachusetts, and Washington.[16]

6.2.3 Dual- or Concurrent-Enrollment Programs

Dual- or concurrent-enrollment programs, also known as dual-credit technical programs, provide opportunities for high school students to take college-level courses free of charge and simultaneously earn credit toward high school completion and a college degree. Emerging from local practice in the 1980s, these programs have evolved without guidelines, regulation, or a policy framework, which has resulted in wide variation in state practices.[17] Recent years have seen a movement to learn from, accredit, and provide practice guidelines for these programs. As of the 2015-2016 school year, there were 97 accredited concurrent-enrollment programs nationwide: 59 at 2-year public colleges, 29 at 4-year public universities, and 9 at 4-year private colleges and universities (Scheffel, 2016).

One advantage of dual- or concurrent-enrollment programs is that they respond to local and state education and training needs (Scheffel, 2016). For example, if community leaders and elected officials in a specific location launch

---

introduced by self-selection, which is difficult to control for in research studies. For example, effects may be stronger for boys than for girls and for those in certain types of programs, such as information and communication technologies.

[15]For additional background and resources, see Coalition for Evidence-Based Policy (n.d.).

[16]For more information, see SREB High Schools That Work, http://www.sreb.org/about-hstw and High Tech High, http://www.hightechhigh.org/about.

[17]For a report that explores how states regulate and ensure the quality of dual-credit programs, see Taylor et al. (2015). The authors indicate that programs in Minnesota are an exception: the state was an early adopter of dual-credit programs in the 1980s and provided educators with a policy framework to guide their efforts.

BOX 6-5
A National Academy Foundation (NAF) Academy: The Academy of Finance, Northeast High School, Florida

Created in 2003, the Academy of Finance gives high school students an opportunity to learn about and prepare for college careers in business and finance. Emphasis is placed on a college preparatory curriculum that prepares students for careers in business and finance with instruction in accounting, insurance, banking and credit, financial computing, international investments, financial planning, and securities. Students are given opportunities to enter into paid internships with local financial service companies during the summer of their senior year and receive offers of part-time employment during the school year.

In academic year 2014-2015, NAF supported 667 career academies that educated 82,000 students in 38 states and the District of Columbia (NAF, 2015). NAF career academies prepare students for skilled technical jobs in finance, hospitality and tourism, information technology, engineering, and the health sciences.

an economic development initiative related to a particular industry, concurrent-enrollment courses can be designed and offered to help students learn technical skills and prepare them for a certificate or degree program that will meet the demands of local employers. Concurrent-enrollment courses also can be adapted to new programs and fields at the 4-year college level.

Of course, this localized approach also brings variation in quality. Even so, the evidence to date suggests that concurrent-enrollment programs have positive impacts for both students and society (Krueger, 2006). In her paper commissioned for this study, Karp (2015b) points out that when properly leveraged, these programs can help the United States achieve its postsecondary education goals by strategically linking high schools and colleges, stimulating educators to rethink how education is structured and delivered, and forcing operational change at the organizational level.

6.3 LINKING TRAINING AND WORK

Many employers are seeking to partner more closely with community or technical colleges. Similarly, many educators are seeking to partner with employers to develop more relevant education and training programs. These educational institution–employer partnerships have the potential to create better-integrated learning environments and meet local employers' skill requirements. The idea is that employers will provide input on courses and curricula and will contribute to ensuring that students acquire skills that will lead directly to a job upon graduation.

Flexible and integrated learning environments for skilled technical occupations involve the alignment of community and technical college education and training with economic development objectives and opportunities, as well as the creation of modular, short-term programs that can be stacked over time; delivered in a variety of ways; and integrated with skills, academic, and occupational certification programs (see Box 6-6). Effecting these types of changes requires that community and technical colleges build effective partnerships with employers, industry and trade associations, and labor unions.

These partnerships are modestly supported by the federal government. For example, the U.S. Department of Labor's Trade Adjustment Assistance Community College and Career Training (TAACCCT) grant program provides 256 grants totaling $1.9 billion over 4 years, authorized under the 2010 Health Care and Education Reconciliation Act. TAACCCT helps community colleges and other eligible institutions of higher education expand and improve their education and career training programs. The Department of Labor implements the TAACCCT program in partnership with the U.S. Department of Education.[18]

There is evidence that certain of these programs raise earnings for at-risk students, often in technical fields, when they integrate classroom training with workplace experience in a particular sector or employer. Examples of these programs include strategic centers of excellence at community and technical colleges and sector-specific training programs involving partnerships between training providers, often community colleges, and employers and trade associations in a single industry.

BOX 6-6
Community and Technical College–Employer Partnerships

Community and technical college–employer partnerships take a variety of forms:

- corporate partnerships with multiple colleges;
- college–employer partnerships facilitated by a regional workforce intermediary;
- sectoral partnerships across multiple colleges and employees;
- individual college–employer partnerships with deep employer engagement;
- career pathways partnerships that may also involve high schools; and
- partnerships facilitated by other intermediaries, including community-based organizations and vocational rehabilitation.

SOURCE: Heidkamp and Hilliard, 2014.

[18]For an overview of the evidence on federally funded education and training programs, see DOL et al. (2014).

6.3.1 Strategic Centers of Excellence at Community and Technical Colleges

Centers of excellence at community and technical colleges are designed to foster flexible and integrative learning in strategically important local industries. In 2004-2005, for example, the Washington State Board for Community and Technical Colleges created 10 centers of excellence to serve as integrators and catalysts for economic and workforce development in the state's leading industries.[19] The centers were placed on college campuses through a competitive application process and codified into state statute in 2008 (WA HB1323). They are intended to serve as a point of contact and resource hub for industry trends, best practices, innovative curriculum, and professional development. (See Box 6-7 for details on one of the centers.) Core foci of these centers include the following:

- **Economic Development Focus.** Serve as partners with various state and local agencies and regional, national, and global organizations to support economic vitality and competitiveness in Washington's driver industries.
- **Industry-Sector Strategy Focus.** Collaboratively build, expand, and leverage industry, labor, and community and technical college partnerships to support and promote responsive, rigorous, and relevant workforce education and training.
- **Education, Innovation, and Efficiency Focus.** Leverage resources and educational partnerships to create efficiencies and support development of curriculum and innovative delivery of educational strategies to build a diverse and competitive workforce.
- **Workforce Supply and Demand Focus.** Research, analyze, and disseminate information related to training capacity, skill gaps, trends, and best practices within each industry sector to support a viable new and incumbent workforce. (Washington State Centers of Excellence, 2017)

[19] These industries include aerospace and advanced manufacturing, agriculture, allied health, education, construction, clean energy, global trade, homeland security, information technology, and marine manufacturing. For more information, see Washington State Board for Community and Technical Colleges' (2017).

BOX 6-7
The Pacific Northwest Center of Excellence for Clean Energy

The Pacific Northwest Center of Excellence for Clean Energy (PNCECE) provides regional strategic coordination for workforce development in the energy industry. Integrating state, federal, and private resources, the partnership includes regional consumer- and investor-owned utilities, a federal power-marketing administration (the Bonneville Power Administration), organized labor, a national laboratory (Pacific Northwest National Laboratory), community and technical colleges and universities, and workforce and economic development councils. The center's extensive coordination has increased the number of energy programs in Washington's community and technical colleges roughly fivefold from just 5 in 2005 to 23 today, and each college has a specific technology/trade focus.

The center regularly convenes a statewide Power Generation Skill Panel that identifies, verifies, and documents precise skills, knowledge, competencies, and work functions in critical occupations in the industry. Employers use these standards to adjust job descriptions and promotion requirements. Educators and apprenticeship programs use the standards to offer training that evolves with the industry. In addition, the center offers occupational summaries, education resource guides, labor market data, white papers on topical issues, and career awareness events.

SOURCE: http://cleanenergyexcellence.org (accessed March 22, 2017).

6.3.2 Sector-Specific Strategies

Sector initiatives or partnerships (also known as sector employment programs) are regional partnerships in which a training provider agrees to prepare individuals to work in a specific industry with the expectation that the employers in that industry will hire them. In this way, sector-based strategies integrate the interests of employers, educational institutions, industries, labor, and community organizations to focus on the workforce needs of a strategically important industry within a regional labor market (see Figure 6-1). The training provider is sometimes a community college (e.g., AMTECH) and sometimes a stand-alone training institution (e.g., Per Scholas, described below).

Proponents, which include the National Governors Association (NGA) and the National Skills Coalition, argue that sector strategies can address current and emerging skill gaps; facilitate coordination across city, county, and state lines; and better align programs and resources to improve effectiveness and return on investment (NGA, 2013). Local, state, and federal government agencies that function as champions and conveners of sector-based partnerships can play a constructive role in facilitating skilled technical workforce development.

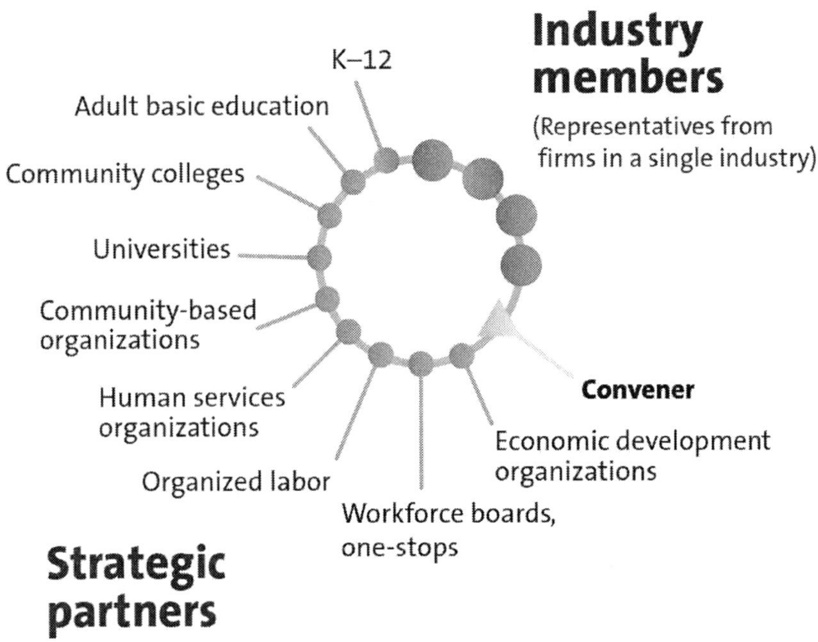

FIGURE 6-1 Sector strategy concept.
SOURCE: NGA, 2013.

Evaluations of sector partnerships suggest they have the potential to produce positive outcomes for youth and adults (Maguire et al., 2010; Roder and Elliott, 2014). Using randomized controlled trials, Maguire and colleagues (2010) found that 2 years after the sector-focused training programs they examined (Wisconsin Regional Training Partnership, JVS-Boston, and Per Scholas) ended, workers were earning about $4,500 more than those who did not participate in the sector-specific training (Maguire et al., 2010). One example, Per Scholas in the Bronx, New York, offers on-site training in IT in a 15-week program. An evaluation of Per Scholas showed large and consistent impacts on employment and earnings more than 2 years following the training (Hendra et al., 2016). However, little is known about the long-term impacts of these programs, particularly as workers change jobs and industries restructure, or what is required to build and sustain such partnerships over time. One recent study did use randomized controlled trials to track the long-term outcomes of Project

QUEST, which provides training in several in-demand occupations in San Antonio, with results to be released shortly.[20]

The NGA's Center for Best Practices estimates that 1,000 sector partnerships currently are operating across the country and that more than half the nation's states are exploring or implementing sector strategies (Maguire et al., 2010; Roder and Elliott, 2014). The center traces the origin of the concept to community-based efforts to connect workers to jobs in local industries in the 1980s. In 2010, at the conclusion of a 4-year state sector strategy project led by the NGA, the Corporation for a Skilled Workforce (CSW), and the National Network of Sector Partners (discussed below), more than 25 states were at various stages of implementing sector strategies and funding local initiatives.

In addition, over the past 10 years, the U.S. Department of Labor has funded several initiatives that are consistent with sector strategy principles, including the Sectoral Demonstration Project; the High Growth Training Initiative; the Community Based Job Training Initiative; WIRED; and several American Recovery and Reinvestment Act grants, such as State Energy Sector Partnerships, Energy Training Partnerships, Pathways out of Poverty, and other High Growth and Emerging Industry grants. And under WIOA, local workforce boards are required to use industry sector strategies to carry out training strategies.

Local and national foundations also are supporting the sector strategy concept. The National Fund for Workforce Solutions initiative, for example, involves 22 regional workforce funder collaborations, more than 80 workforce partnerships, and 200 funders. A national association, the National Network of Sector Partners, provides a way for partnerships across the country to connect and learn from each other. In June 2015, the Joyce Foundation announced that it will invest in the Business Leaders United Commitment to Action, which aims to expand the number of sector partnerships by more than 30 percent across all 50 states and to facilitate conversations between local business leaders and federal policy makers about how private, philanthropic, and public dollars can be leveraged to replicate and sustain these partnerships nationally (see Joyce Foundation, 2011).

The NGA, the CSW, and the National Skills Coalition maintain a clearinghouse for state-level sector strategy information.[21] Although all state sector strategies are based on the same basic principles, they each reflect local preferences and priorities. One way in which these differences are expressed is in who convenes the partnerships and how. In some states, for example, sector partnerships have been convened by civic leaders in business, community

[20]For a recent review of sector-based training strategies, see Holzer (2015b). See also Elliott and Roder (forthcoming).

[21]See http://www.sectorstrategies.org (accessed March 22, 2017).

service, or government, while in others, they have been convened by community colleges or economic development authorities.

6.4 EMPLOYER-BASED TRAINING PROGRAMS

This section reviews training provided by employers to workers through apprenticeship and joint labor–management programs. In many cases, this employer-sponsored training includes coursework at community colleges. American labor unions also have in many instances encouraged their members to support workplace education and training programs as a way to improve workers' employment security and help them adjust to changing workplace conditions. Connections between employers and workers can be improved as well through better communication of information about the value of credentials and licenses.

6.4.1 Apprenticeship Programs

As discussed in Chapter 4, apprenticeships combine occupational education and training with paid on-the-job work experience. Apprenticeships can be either registered or unregistered; approximately 400,000 active apprentices currently are registered with the Department of Labor. Some analysts assert that apprenticeship programs are an underutilized workforce development strategy in the United States (see, e.g., Ayres and Gurwitz, 2014). As Figure 6-2 shows, for example, the number of apprenticeship starts in the United States in 2012 was low relative to the numbers in other developed countries with well-developed apprenticeship programs, such as England, Switzerland, Germany, France, and Scotland.

A number of states have developed apprenticeship programs over the past 10 years, although the impact of such programs has not been rigorously evaluated (see, e.g., Lerman, 2010; Reed et al., 2012). In an effort to close skilled workforce gaps, for example, several programs, such as the Michigan Advanced Technician Training Program (MAT2), the Kentucky Federation for Advanced Manufacturing Education (KY FAME), and Apprenticeship Carolina in South Carolina, have created incentives to expand apprenticeships in their state's strategically important industries (see the description of Apprenticeship Carolina in Box 6-8).[22] The most interesting of these state models move beyond

[22] For information about these programs, see MAT2, http://www.mitalent.org/mat2; KY FAME, http://kyfame.com; and Apprenticeship Carolina, http://www.apprenticeship carolina.com.

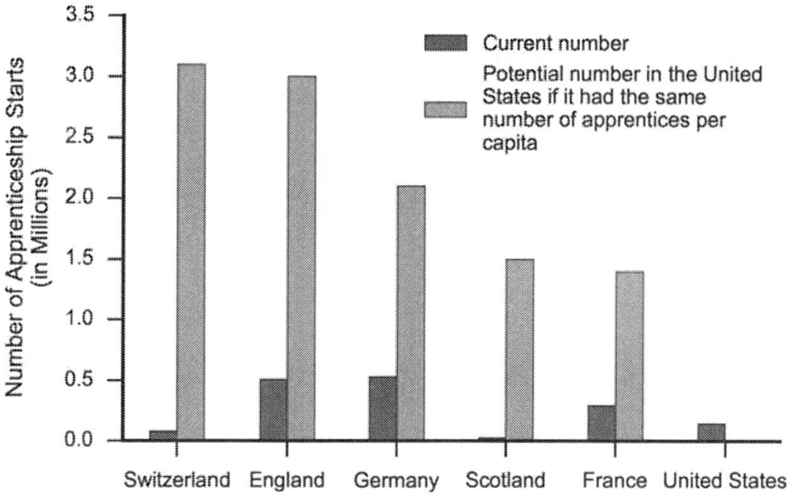

FIGURE 6-2 Apprenticeship starts by country in 2012.
SOURCE: Adapted from material published by the Center for American Progress. See Ayres, S., and E. Gurwitz. 2014. *The underuse of apprenticeships in America*. Washington, DC: The Center for American Progress (www.americanprogress.org).

BOX 6-8
Apprenticeship Carolina

This initiative expands apprenticeships in the state of South Carolina to targeted industry clusters in advanced manufacturing; construction technologies; energy; health care; information technology; tourism and service industries; and transportation, distribution, and logistics. The state increased its budget for apprenticeships to $1 million a year and offered companies a tax credit of $1,000 a year per apprentice for 4 years. The number of apprenticeship sponsors increased from 90 in 2007 to more than 600 in 2014, and the number of apprentices increased from 777 to nearly 11,000 over the same period.

time-based workplace training to integrate dual-learning principles whereby academic courses are linked to workplace training and standardized competency-based certifications.

Many newer forms of apprenticeship in the United States include coursework at a local community college, which leads to an associate's degree. Employer sponsors provide at least 2,000 hours of paid work per year in the relevant occupation and arrange for their apprentices to receive approximately 150 hours of related instruction. The average apprenticeship lasts 4 years, and apprentices earn an average of $161,000 during the course of their apprenticeship.

Given the benefits of apprenticeship and the underutilization of these programs noted above, some analysts have suggested that complementary policies and processes that make it convenient and cost-effective for employers to design and implement such programs may be necessary. Currently, it can take as long as 2 years to obtain approval for a new apprenticeship program in some states. Some have suggested offering tax subsidies to employers as a way to offset the costs of developing a new apprenticeship program (Lerman, 2010).

Yet while there is considerable enthusiasm for expanding the apprenticeship concept in the United States, including its extension to a wider range of occupations, some experts caution against directly importing apprenticeship programs from other countries into the United States at a national level.[23] They argue that significant institutional differences between the United States and other developed countries make it difficult to replicate other countries' policies and programs. In addition to the much larger size of the U.S. economy relative to those of European nations, they note the differences in national governance structures and workforce development traditions.[24] Countries such as Germany and Switzerland, for example, have stronger traditions as well as formal organizations in place that support employer involvement in education and training. Also in contrast with the policies of many other developed countries, U.S. federal and state-level policies require that most students and workers bear a significant portion of the cost of postsecondary education.

Even so, it may be that the structures of education and workforce policies and programs in some U.S. states are similar in some respects to those found in other developed countries. In such instances, there may be opportunities for state-level policy makers to learn from such countries as Denmark, Germany, Sweden, and Switzerland. Based on the review by Messing-Mathie (2015) commissioned for this study, Box 6-9 lists some key lessons that can be learned from the experience of foreign apprenticeship programs. Further comparative studies, based on a robust methodology, would increase understanding of apprenticeships, the contexts within which they are effective, and the current barriers to their expansion.

---

[23]Panel 6 at the committee's 2015 symposium covered the topic of apprenticeships. Presentations, Andrea Messing-Mathie's paper on apprenticeships, and a webcast of discussions can be found at http://nas.edu/SkilledTechnicalWorkforce.

[24]For an overview of the issues, see the section on apprenticeship in Chapter 4. For a comprehensive review of theories, evidence, and comparative issues, see Wolter and Ryan (2011).

BOX 6-9
Key Lessons from Foreign Apprenticeship Systems

- **Establish a flexible framework.** Apprenticeships "operate within the wider context of cultural traditions and aspirations of individuals and the complexity of labor market regulation" (Steedman, 2012, p. 10). Successful apprenticeships allow for flexibility within a guiding framework.
- **Set clear rights and responsibilities.** A legal framework should establish clear rights and responsibilities for all partners. In particular, legislation should secure the rights of an apprentice as an employee, but with provisions that address separate wage structures.
- **Balance incentives for employers and apprentices.** Carefully targeted tax credits and subsidies can encourage investment in the apprentice as a member of the firm and allow the apprentice to develop valuable relationships beyond his or her academic institution. These tax credits and subsidies also can encourage the apprentice to complete the program.
- **Make apprentices "employees."** The workplace, not the school, should be considered the center of the apprentice's experience. Community colleges and other educational organizations serve to complement this relationship with academic instruction.
- **Create strong intermediaries.** As Bailey (1993, p. 7) points out, "There needs to be some overarching institution trusted by employers and workers (and unions where they exist) that can coordinate individual employers' participation and that can help schools work with business."
- **Set standards and signal accomplishment.** Companies need agreement on industry standards and the competencies that apprentices should acquire in order to meet those standards. Upon graduation from their program, apprentices should receive an associate's degree or other certification that is widely recognized.

SOURCE: Messing-Mathie, 2015.

6.4.2 Joint Labor–Management Programs

Unions have a prescribed role in job training programs under WIOA, and they have a long history of designing and implementing innovative and effective education and job training programs that potentially can benefit workers, businesses, and local communities. From the earliest days of American trade unions, workers have organized to advocate for local public school systems, and they have been instrumental in the passage of federal vocational education legislation (Stacey and Charner, 1982). Craft unions developed in the 19th century and masters- and journey-level workers created apprenticeship systems that set minimum time periods for training, wage levels, and education requirements (Rorabaugh, 1986). As production and service delivery were expanded and restructured in the 20th century, distinct spheres of influence for managers and workers emerged. Enterprise owners and their professional managers sought to control strategic decision making on investments, the use of technology, the organization of work, and requirements for technical skills. Workers organized into industrial and service unions to negotiate wage and benefit levels, personnel management practices, and other procedural and process issues salient to the economic interests of their members, including education and training (Thomas and Kochan, 1992).

As the nature and organization of work have continued to evolve, the AFL-CIO, a federation of American labor unions, has encouraged its members to directly sponsor workplace education and training programs as a way of improving workers' employment security and helping them advance on the job and adjust to changing workplace conditions. In a recent report on these issues, the AFL-CIO argues that many U.S. unions have embraced this challenge and are increasingly collaborating with management to create joint programs and workplace education and training systems, and then institutionalizing these changes through collective bargaining (AFL-CIO, 2016). Box 6-10 describes one example—the Wisconsin AFL-CIO Regional Training Partnership.

More research is needed on joint labor–management strategies for establishing workplace learning systems and on community-based workforce and

BOX 6-10
Wisconsin AFL-CIO Regional Training Partnership

The Wisconsin Regional Training Partnership (WRTP) was formed by Milwaukee area business and labor leaders in 1992 in response to skill shortages created by the recovery of manufacturing, retirement of an aging workforce, and diversification of the regional economy. It is one of the largest sectoral training programs in the country, working with more than 300 firms, including small and medium-sized employers, training 1,700 workers in machining, electronics, plastics, and related sectors in 2012 alone. The WRTP has developed successful models for implementing new technologies and work processes and educating and training a diverse skilled technical workforce.

The WRTP recently collaborated with BIG STEP, an initiative led by Building Trades, and with employers and community groups to increase participation by minorities and women in the trades through training. It also launched an Industrial Manufacturing Technician apprenticeship program in partnership with the AFL-CIO Working for America Institute. The program is state certified and sponsored by the Department of Labor under its H-1B program. The program is being piloted in Milwaukee and is on pace to enroll 80 incumbent manufacturing workers. It will expand statewide as well as include Michigan and Minnesota in 2014.

The collaboration between WRTP and BIG STEP began during the Department of Labor grant term, when construction was one of the industry sectors selected for replication. With their own highly successful histories, WRTP and BIG STEP partnered to facilitate more effective coordination of employer-driven worker readiness to best ensure that individuals preparing for employment gained the skills and experiences that employers need.

SOURCE: Excerpted from AFL-CIO, 2016.

economic development initiatives. However, several innovative programs could provide inspiration for skilled technical workforce development in a wide range of areas, including health care, manufacturing, construction, and transportation, and could be adapted to meet the requirements of the nation's regions (see AFL-CIO, 2016, for an overview of some of these programs). Box 6-11 describes one example of such an innovative technical workforce development program.

BOX 6-11
The Bison Gear Example

Demonstrating that a company does not have to be large to engage in skilled technical workforce development, the Bison Gear & Engineering Corporation in St. Charles, Illinois, makes education an essential part of its work culture. According to the Aspen Institute, which has recognized Bison Gear as a model of success in its Skills for America's Future program, more than one-third of employees are involved in some type of continuing education, and Bison employees average 7.2 hours of on-the-job training per year.

The Skilled Workforce Initiative at Bison Gear provides opportunities for production-floor workers to become trained and certified and encourages all employees to attain at least an associate's degree. New and incumbent employees have the opportunity to earn a National Career Readiness Certificate (NCRC) credential, a portable credential that demonstrates achievement and a level of workplace employability skills in applied mathematics, location information, and reading for information. Bison Gear uses the NCRC as a preassessment tool: upon completion, workers become a certified production technician (CPT), which is a prerequisite for entering Bison Gear's Manufacturing Skills Standard Council (MSSC) Fast Track program.

Bison Gear's MSSC Fast Track program encourages CPT employees to attain an MSSC certificate. MSSC is a national industry-led training, assessment, and certification system that focuses on the core skills and knowledge needed in production and material handling. Interested workers complete 15-18 hours of training in modules, scheduled on their own time. The firm covers the cost of registration fees and assessments, and awards a $100 bonus to participants when they pass a module.

To encourage employees to pursue higher education degrees, Bison Gear has designed a tuition reimbursement program, Growth Education and Results (GEAR). Under GEAR, the firm has relationships with community college partners, which directly bill the company for the cost of an employee's education. Bison Gear also provides education benefits for employees' children through the Bullock Family Bison Cares Scholarship.

Bison Gear staff also contribute to workforce development efforts by working directly with local community college partners to provide information on skill requirements, project long-term hiring needs, and support the development of curriculum. They engage students through structured mentorships and host summer job training internships. Bison Gear staff regularly attend community college recruitment meetings to identify prospective job candidates, and they post job openings on each school's website.

SOURCE: Colborn and Jenkins, 2016.

6.4.3 The Push for Talent Pipelines

Complementing efforts by labor groups, the U.S. Chamber of Commerce Foundation—through its affiliation with the U.S. Chamber of Commerce and its federation of more than 3 million employers nationwide—has called upon the business community to play a leadership role in developing a workforce that will meet its strategic business needs (see Box 6-12) (U.S. Chamber of Commerce Foundation, 2014). Working with large employers such as IBM, the U.S. Chamber of Commerce Foundation promotes the use of supply chain management principles to create "talent pipelines." This approach involves linking workforce development requirements to business strategy, as well as reconceptualizing and reorganizing workforce development to nurture and sustain talent, enabling it to evolve to meet business needs over time.

Seeking to address the challenges of skilled technical workforce development, some employers have signed up to participate in the Skills for America's Future program, a federal initiative designed to foster partnerships between employers and community colleges. In addition, some employers recognize that their employees need support services to improve their skills and pursue education and training opportunities. These employers are fostering a culture that supports continual learning and are adjusting work schedules and assisting employees with child care, transportation, tuition, and other types of expenses. As discussed above and in Chapter 4, data on the extent and effectiveness of these initiatives are scarce, and more research is needed in this area.

BOX 6-12
U.S. Chamber of Commerce Foundation's Talent Management Principles

1. Link talent management to business strategy by identifying core capabilities.
2. Develop customized solutions to address core capability requirements.
3. Create shared ownership and accountability for talent management across the organization and develop sourcing strategies around the positions and capabilities that are most essential to competitiveness.
4. Establish networks of preferred providers to source and develop core capabilities.
5. Develop competency, credentialing, and other performance indicators.
6. Create balanced scorecards to align performance measures and incentives.
7. Use dashboards and predictive analytics to monitor performance.

SOURCE: Excerpted from U.S. Chamber of Commerce Foundation, 2014.

6.5 IMPROVING LINKAGES THROUGH BETTER DATA

Better data and better analyses of data are needed to make progress in understanding and addressing the challenges of skilled technical workforce development. Nearly all of the strategies discussed in this report have the potential to improve the flow of information indirectly. Efforts to improve the quality and delivery of skills education and training, for example, will inherently expose students to more information about careers and the education and training required to pursue them. Similarly, better integration across different parts of the workforce development system through dual-enrollment programs, early college schools, sector-based strategies, and career pathways will necessarily involve the sharing of information across educational institutions and employers.

6.5.1 Data Challenges

Extensive gaps exist in the data on technical workforce education and training, as pointed out by Rebecca Rust of the U.S. Bureau of Labor Statistics at the committee's 2015 symposium. Speaking at the same event, John Dorrer of Jobs for the Future argued that analysts need to improve their ability to mine newly emerging digital jobs databases, as well as existing databases, and called for additional investment in research, particularly policy-relevant research. Courtney Brown of Lumina stated that more data and analyses are needed to understand credentials, career pathways, and supply and demand dynamics in labor markets. Other participants expressed related concerns about the challenges associated with improving labor market data and research, such as privacy laws that can inhibit data sharing, the time and costs associated with information management, and the potential for false or misleading associations that can arise from poorly designed big data repositories and analyses. An additional challenge is ensuring that training programs and educational institutions can use the data collected to keep programs up to date and enable them to evolve with the changing skills required by employers.

Given the complex coordination challenges faced by policy makers, individuals, educators, employers, industry and trade associations, labor unions, and other civic organizations, improvements in the flow of information in skilled technical labor markets are clearly needed. The remainder of this section provides examples of different types of experiments currently under way to improve technical workforce development by identifying and addressing existing information deficits.

6.5.2 Using Big Data

Advances in mathematics, statistics, and the computational sciences make it possible to capture, augment, and analyze data from a broad range of very different sources. Several contributors to this study argued that applying these developments to policy analysis and research on skilled technical workforce development can provide policy makers with a stronger factual basis for decision making (see Reamer, 2015, for an overview of the issues and findings). In addition, advances in data analytics and information technologies can be used to help consumers make better decisions and create incentives to improve the quality of education and training options.

In the era of big data and new digital tools for collecting information, private firms have learned to mine the web to aggregate or crowdsource data from existing sources. Online job posting and searching, for example, which began in the early 1990s, provides a rich source of data for analysts, researchers, and labor market participants. Software and analytic tools can be designed to crawl through these data, analyze the content, and generate aggregate statistics on job vacancies more often and at finer occupational and geographic scales relative to traditional employer surveys such as the Job Opening and Labor Turnover Survey (JOLTS).[25]

Researchers at Georgetown University estimate that 60 to 70 percent of job postings are now posted online (Carnevale et al., 2014, 2015). However, they state that online job ads may not capture all job openings in all locations. Their analyses suggest that online job ads tend to be biased toward industries and occupations that seek high-skilled workers and workers with 4-year degrees. In their view, data on online job ads are useful in measuring labor demand and honing in on previously inaccessible variables, but these data have limitations. They urge users to exercise caution and utilize this tool in conjunction with traditional data sources. While noting these concerns, contributors to this study highlighted the need for researchers to use a broader set of analytic tools and develop new methodologies to take advantage of the proliferation of online databases.[26]

6.5.3 Crowdsourcing Information

A primary example of crowdsourced labor market data is the data collected by Burning Glass Technologies (BGT), one of the leading vendors of data on

---

[25] JOLTS is a monthly survey of employers that was developed to provide information on job openings, hires, and separations. Each month the JOLTS sample consists of approximately 16,000 businesses drawn from 8 million establishments represented in the Quarterly Census of Employment and Wages. The publicly available data provide a measure of labor demand across broad industry classifications at the national level or overall aggregate labor demand for four quadrants of the nation.

[26] See, for example, the paper by Reamer (2015) commissioned for this study, presentations and discussion in Panel 10 of the committee's 2015 symposium, and presentations made to the committee in February 2016 at http://nas.edu/SkilledTechnicalWorkforce.

online job ads.[27] These data include detailed information on current online job openings updated daily from such sources as job boards, newspapers, government agencies, and employer sites (Burning Glass, 2017). The data provided by BGT allow geographic analysis of occupation-level labor demand by education level and experience level over time. Moreover, the data collection process used by BGT yields a robust representation of hiring in real time.

Such crowdsourcing databases provide valuable information that can be used by educational institutions and workforce development boards. Colleges and technical schools, for example, can use these data to examine the labor market in real time and align their programs with the marketplace, giving students crucial information as they seek jobs. Similarly, labor exchanges in 14 states currently use BGT data to connect workers and employers more efficiently while providing pathways and tools that help workers identify viable careers.

These tools are valuable and increasingly used, but as discussed in the previous section, online data and analytic tools have limitations. For-profit companies own some analytic strategies, such as those offered by BGT, and their data and tools are proprietary. Therefore, maintaining publicly available data and analytic tools, which is encouraged under WIOA and requires federal, state, and local coordination, remains important.

6.5.4 Recent Initiatives

Web-Based Tools

Other web-based tools have enabled private firms to collect data from participants in a variety of markets, including the labor market. For example, Glassdoor, an online jobs and recruiting site with a crowdsourced database, has launched a new open-source map that enables job seekers to see where there are open jobs county by county across the country (The White House, 2014).[28] Workers provide information about their experience in interviewing with and working for employer organizations, and registered employers promote job opportunities to potential workers. The database is available online and via mobile applications to registered users.

Public Data Warehouses

The U.S. Department of Labor has supported state efforts to develop the data infrastructures needed to create or improve longitudinal data systems that

---

[27]Similar data are also collected by Conference Board via its Help Wanted OnLine series.

[28]For information about Glassdoor, see http://www.glassdoor.com/index.htm (accessed March 22, 2017).

link employment and earnings data with education, employment service, and training data over time (Davis et al., 2014). Integrating data from K-20 and workforce experience and then warehousing these data allows for the creation of consumer applications that estimate return on investments in education and training. In November 2010, the U.S. Department of Labor launched the Workforce Data Quality Initiative (WDQI), which has provided more than $30 million in grants to 29 states.

Consumer Report Cards

Recently, U.S. Department of Labor grants have focused on creating consumer report card systems (CRCSs).[29] As of this writing, five WDQI states—Florida, Minnesota, New Jersey, Virginia, and Washington—had existing CRCSs. Among these states, the structure of the state website and how it is used are generally similar, with each system displaying outcome data for individual programs (Davis et al., 2014).

For example, the New Jersey Department of Labor and Workforce Development partnered with the State University of New Jersey John J. Heldrich Center for Workforce Development at Rutgers to create a consumer report card website that provides information on occupational training programs in the state (Davis et al., 2014).[30] State laws require "training programs at for-profit, public 2-year, and some public 4-year schools that receive state or federal workforce funding to submit records to the state for all of their students" (The White House, 2014). The website makes it possible for students to search by occupation, location, or key word for schools and organizations that provide job training. Students can compare training providers by location, cost, outcomes, and length of training. A results section displays information about former program participants, such as employment rates, retention rates, and average earnings over time.

As another example, Florida continues to expand the longitudinal administrative data infrastructure it developed in the 1970s to integrate K-20 education and workforce data. This infrastructure supports the Florida CRCS—Smart College Choices—which makes it possible to assess and analyze participation in and outcomes of all education and training programs in the state.

---

[29] WIOA requires eligible training providers, including community colleges, to provide similar consumer/outcome data in the form of a "report card" at the program level.

[30] For additional information about the New Jersey Consumer Report Card, see http://www.nj.gov/education/cte/ppcs/students/OnliConRC.shtml. WIOA requires local workforce systems to create career pathways for in-demand industries in the local workforce area. The Department of Labor, Department of Education, and Department of Health and Human Services have agreed upon a framework for career pathways to facilitate meeting this requirement.

Florida's CRCS provides basic information about education and training programs at the state's 28 community colleges and 11 public universities. The state is currently preparing to expand the system to include additional institutions and provide more detail about existing options. Consumers can compare such institutional outcomes as completion, continuance, and employment rates and estimated earnings.

6.6 OTHER POLICY INITIATIVES

In addition to linking the key actors in the workforce training and education ecosystem, some policy initiatives help improve the portability and wider recognition of acquired technical skills. Roadmaps that link selected competencies also can help guide students and workers through the education and training they need to address the needs of local industries.

6.6.1 Portable Credentials and Licenses

The NGA's Center for Best Practices observes that up to 40 percent of community college students enroll in multiple institutions within a 6-year period (The White House, 2014). Efforts to transfer education and training accomplishments and licenses across education and training programs, employers, industries, and geographic boundaries can improve the quality of postsecondary education and training, reduce costs, and improve returns on investment (The White House, 2015, Chapter 2).

To improve the portability of credentials, some states are crafting credit-transfer agreements across their postsecondary systems. The state of Tennessee, for example, "expanded its credit-transfer agreement between 2-year colleges and the public university system. The agreement covers 100 percent of general education requirements and 75 percent of majors at the associate level" (The White House, 2015, Chapter 2). Other initiatives, discussed below, involve standardizing credentials and reducing the burden of licensing requirements.

6.6.2 Standardized Credentials

The past decade has witnessed explosive growth in the number and variety of college degrees, education certificates, industry certifications, occupational licenses, and online badges awarded by various entities and subsequently presented to employers as evidence of specific competencies. [31] This pro-

[31]Some experts observe that there is also no national standard for a high school diploma. While the committee believes this is an important issue, a detailed examination of secondary school standards is beyond the scope of this study.

liferation of credentials has created confusion on both sides of the labor market: individuals are unsure which credentials are worth investing in, and employers are questioning the quality and value of the credentials individuals present.

Efforts to standardize credentials focus on harmonizing key features related to quality, portability, and labor market value. For example, a recent partnership between The George Washington University's Institute of Public Policy and the American National Standards Institute, which is funded by the Lumina Foundation, is reviewing the credentialing activities of more than 48 major credentialing stakeholders, such as business associations; higher education associations; and federal agencies such as the U.S. Departments of Commerce, Education, Labor, Defense, Energy, and Health and Human Services. The partnership has developed definitions, coding schemes, plans for an open metadata registry for posting comparable credentialing information, and pilot projects for testing several registry applications. The implementation plan includes a public–private collaboration of stakeholders that will decide whether and how to take the system to scale and make it sustainable (Crawford and Sheets, 2015).

6.6.3 Licensing Reforms

As discussed in more detail in Chapter 3, a recent study of U.S. occupational licensing patterns and trends reveals significant growth in the number of workers holding an occupational license. When carefully designed and implemented, licensing can offer important health and safety protections to consumers, as well as benefits to workers. However, the current licensing regime in the United States also involves substantial costs, and the requirements for obtaining a license frequently do not align with the skills needed for the job. "Too often, state policy makers do not carefully weigh these costs and benefits when making decisions about whether or how to regulate a profession through licensing" (The White House, 2015, Chapter 2).

The 2015 White House study on occupational licensing argues that standards for certification and licensing are related and mutually reinforcing (The White House, 2015, Chapter 2).[32] Guidelines for occupational certifications can help reduce the need for such licensing, make it easier to transfer credentials across state lines, and improve the efficiency and effectiveness of both certification and licensing activities. The White House study includes several recommendations that are consistent with its findings:

- Limit licensing requirements to those occupations that address legitimate public health and safety concerns.
- Conduct comprehensive cost/benefit assessments of licensing laws,

[32]For related analyses in this study, see http://nas.edu/SkilledTechnicalWorkforce.

and reduce the number of licenses that are not justified by these assessments.

- Harmonize regulatory requirements across states to the extent possible, and where appropriate, enter into interstate compacts that recognize licenses from other states.
- Allow practitioners to offer services to the full extent of their current competency to ensure that all qualified workers can find services. (The White House, 2015, Chapter 2)

6.6.4 Competency Models

The Department of Labor's Employment Training Administration has developed a structured approach to articulating the competencies required for occupations in an industry or sector, from foundational competencies, to industry-wide competencies, to technical competencies within a specific subsector.[33] This competency strategy is used to identify specific employer skill requirements, define career pathways and credentials, develop competency-based curricula and training models, and create industry-defined performance indicators and certifications. It also includes the development of resources for career exploration and guidance, and is being used to encourage employers to make hiring decisions based on work-related competencies rather than credentials.[34]

Competencies can serve as a better basis for education, training, and employment decisions relative to credentials, which are imperfect proxies for occupational capacity and competency. Those who are engaged in integrated learning, sector strategies, talent management, career pathways, and career pipeline initiatives can use the competency model framework to guide their work.

6.7 CONCLUSIONS

Educators, workers, employers, industry and trade associations, labor unions, and other civic leaders face several challenges in developing a skilled technical workforce that meets local economic development requirements. Although there are no panaceas, and the available evidence has limitations, the

---

[33]For more information on competency modeling, see http://www.careeronestop.org/competencymodel (accessed March 33, 2017).

[34]See, for example, the work of the Business Roundtable (http://businessroundtable.org/issues/education-workforce) and the National Network of Business and Industry Associations (http://actfdn.org/category/national-network-of-business-and-industry-associations).

committee has identified examples of initiatives currently under way that could be instructive for policy making.

Evidence indicates that reform of secondary and postsecondary education to better integrate academic education, technical training, and hands-on work experience produces better outcomes and return on investment for all students in high school, postsecondary education, and careers (DOL et al., 2014). These approaches—which include creating flexible and integrated learning environments to meet the needs of a highly diverse student body, revising remedial programs and prerequisite requirements, creating portable credentials and licenses, and providing wraparound services—are compatible with a wide range of education and training models, including career academies, charter schools, small-themed schools, internships, and summer job programs for youth. In addition, some apprenticeship programs offer positive benefits for both participants and society as a whole.

Lifelong learning is essential for the U.S. workforce to remain globally competitive, and some policy makers, employers, industry and trade associations, and labor unions are experimenting with strategies for ensuring that U.S. workers will continue to invest in education and training throughout their work lives. In addition, they are experimenting with a number of enabling initiatives aimed at improving the flow of information, braiding financial sources, articulating career pathways, building competency models, and implementing sector strategies. Although these experiments provide ideas for policy makers who wish to innovate and implement reforms, there are no blueprints for success. Rather, experiments must be based on analyses of local requirements, and programs must be designed with these requirements in mind. Further research is needed both to support these analyses and to understand the dynamics of these highly complex social systems.

7

Findings and Recommendations

The United States needs a skilled technical workforce to remain competitive in the global economy and to ensure that its workers participate in the nation's economic growth. There are significant opportunities as well as major challenges in this regard. Notably, rigorous evidence indicates that the returns on investments in technical skills in the labor market are strong when students successfully complete their training and gain credentials sought by employers. At the same time, the committee found that in many instances, workers either are not taking advantage of these opportunities or are failing to complete their training programs.

To understand why, it is necessary to recognize that in the United States, the responsibility for developing and sustaining a skilled technical workforce is fragmented across many groups, including educators; students; workers; employers; the federal, state, and local governments; labor organizations; and civic associations. For the system to work well, these groups need to be able to coordinate and cooperate successfully with each other.

Unlike most other advanced economies, the United States lacks formal mechanisms that require governments, educators, labor representatives, and employers to coordinate on workforce development policies and practices at the national level. In fact, workforce development in the United States is polycentric in nature, driven by a variety of private and public investments in workforce education and training. Workers often pay for on-the-job training through lower wages. Although employers and governments share an interest in developing and maintaining a robust skilled technical workforce, their respective investments often are uncoordinated. At the same time, public investments are guided by a complex and similarly uncoordinated set of policies associated with achieving similarly divergent goals related to economic development, education, employment, health and human services, and veterans' affairs. In this polycentric system, making better use of available resources and generating better outcomes requires improving coordination between students and educational or training institutions, between secondary and postsecondary institutions, and especially between training institutions and employers through a variety of public, private, and hybrid mechanisms.

The good news is that promising experiments currently under way across the United States can provide guidance for innovation and reform, although the scalability of some of these experiments has not yet been tested. As detailed in Chapter 6, evidence suggests that integration of academic education, technical training, and hands-on work experience improves outcomes and return on investment for students in secondary and postsecondary education and for skilled technical workers in different career stages.

The findings and recommendations presented in this chapter are designed to help overcome some of the barriers identified within the current framework of federal governance, state implementation, and market incentives. They address the key elements of the statement of task for this study.[1] To the extent possible, the recommendations call for specific actions by Congress, federal agencies, state governments, employers, and civic organizations to improve the American system of workforce development.[2]

FINDINGS

TASK: Define the skilled technical workforce.

Finding 1: The skilled technical workforce includes a range of occupations that require a high level of knowledge in a technical domain, but many of these occupations do not require a bachelor's degree for entry.[3]

---

[1]The statement of task is presented in Box 1-2 in Chapter 1. The findings and recommendations in this chapter are cross-indexed to the tasks undertaken by the committee as follows:

- Define the skilled technical workforce: Findings 1 and 2, and Recommendation 2.
- Consider gaps and market failures: Findings 3, 4, 5, and 6, and Recommendations 3, 4, and 5.
- Consider the current public- and private-sector roles in financing and providing employment training and skills certification: Finding 7.
- Consider policy-making challenges and experiments that affect incentives and information to improve work skills: Findings 8, 9, 10, 11, and 12, and Recommendations 1 and 6.
- Consider selected employment preparation practices in other countries: Finding 13 and Recommendation 7.

[2]A summary list of selected appropriate actions for stakeholders is presented at the end of this chapter.

[3]See also the summary in Box 1-1 in Chapter 1.

a. Estimates of the number of skilled technical workers range from 11.9 percent to 23.1 percent of the total U.S. workforce, depending on how the workforce is defined.
b. Nonroutine, interpersonal, and analytic skills are becoming increasingly important for skilled technical work. It has been estimated that as of 2009, more than 60 percent of all work tasks required these skills. Competencies associated with skilled technical workers include the following:
   - **Personal skills.** These include interpersonal skills, integrity, initiative, dependability and reliability, adaptability and flexibility, and lifelong learning.
   - **Academic competencies.** These include basic academic skills such as reading, writing, and mathematics; advanced knowledge and skills in science and technology relevant to the technical occupation; and critical and analytical thinking.
   - **Workplace competencies.** These include teamwork and collaboration; customer focus; planning and organizing; problem solving and decision making; working with tools and technologies; scheduling and coordination; and checking, examining, and recording.
   - **Industry-wide technical competencies.** These are the knowledge, skills, and abilities associated with a particular industry such as health care or manufacturing. They are "cross-cutting" competencies common to most occupations within the industry. These competencies usually deal with comprehension, awareness, or analysis.
   - **Occupation-specific technical competencies.** These are the knowledge, skills, and abilities associated with specific occupations such as emergency medicine or automotive manufacturing technicians.

Finding 2: Although widely used to describe this segment of the workforce, the term "middle skills" fails to capture the high value and dynamism of this segment of the U.S. workforce and is seen by some as having pejorative connotations. This label can deter students and workers from these occupations at every stage of their career.[4]

a. **Social Perspectives.** Students, parents, and workers often fail to appreciate the value of many skilled technical occupations and wrongly believe that skilled technical jobs do not offer adequate wages, working

[4]See Section 1.4 of Chapter 1 on "Defining the Workforce."

conditions, employment security, social prestige, opportunities for advancement, or other important benefits.[5] Consequently, many students and workers overlook occupations and career opportunities that have high social value and potentially good returns to education and training, or they do not properly prepare themselves for these opportunities.

b. **High-value jobs.** In fact, the data show that skilled technical occupations are well-paying jobs with high social value and multiple points of entry that require academic preparation combined with postsecondary technical education and training and experience on the job to achieve competency.[6]

c. **Multiple pathways.** Workers in skilled technical occupations may hold any one of several different types of credentials, including associate's, bachelor's, and master's degrees; certificates; certifications; and occupational licenses.[7]

TASK: Consider gaps and market failures.

Finding 3: To remain competitive in the global economy, foster greater innovation, and provide a foundation for shared prosperity, the United States needs a workforce with the right mix of skills to meet the diverse needs of the economy. Conversely, an insufficiently skilled workforce imposes significant burdens on the U.S. economy, including higher costs to workers and employers and lower economic productivity.[8]

a. **Higher costs.** An insufficiently skilled workforce imposes costs on employers who struggle to find skilled workers, and also hurts workers who could otherwise have higher skills and earnings, thus exacerbating inequality.

b. **Lower employment and productivity.** Some economists are concerned that an insufficient number of skilled workers can limit growth rates in employment, output, and productivity. If employers cannot hire workers with the right skills locally, they cannot produce enough of the higher-value products and services that are associated with sustained innovation and growth in their location. This in turn creates pressure to relocate either within the United States or abroad. In

---

[5]See Section 1.4 of Chapter 1 on "Defining the Workforce."

[6]See Section 1.4 of Chapter 1 on "Defining the Workforce."

[7]See Section 1.5 of Chapter 1 on "Structure of the System for U.S. Skilled Technical Workforce Development."

[8]See Section 2.1 of Chapter 2 on "The Supply of and Demand for Skilled Technical Workers."

a global economy, employers often offset the lack of qualified labor in one country by seeking qualified labor in another (offshoring).

c. **Other impacts.** Citizens with lower skill proficiency tend to report poorer health, lower civic engagement, and less trust. Lower skills also make it difficult for workers to attain more education or training when structural changes require adapting to new methods and processes.

Finding 4: The evidence suggests that as a nation, the United States is not adequately developing and sustaining a workforce with the skills needed to compete in the 21st century.[9]

a. Americans perform poorly on a large international assessment of adult skills compared with citizens of other leading economies. For example:
   - Compared with the working-age population in 24 Organisation for Economic Co-operation and Development (OECD) countries, Americans rank 16th in literacy skills, 21st in numeracy skills, and 14th in problem solving in a technology-rich environment.
   - U.S. average mathematics and science literacy scores are below the average scores for all developed countries, and Americans have substantially fewer high scores and more low scores than citizens of other developed countries.
b. Americans perform poorly on U.S.-based assessments of skills:
   - Only 46 percent of Americans can demonstrate an understanding of scientific inquiry.
   - On average, American students are not proficient in mathematics, reading, or writing as measured by the National Assessment of Educational Progress (NAEP).
c. Americans are concerned about the nation's ability to develop and sustain a skilled workforce:
   - According to the National Science Board (NSB), most Americans think other countries are doing a better job of providing technical education and supporting lifelong education and training.
   - Spurred by concerns about the need for a skilled technical workforce, public officials at the federal, state, and local levels have undertaken numerous initiatives related to improving the skill of the workforce.[10]
   - Employers, industry associations, and labor unions share these

[9]See Section 3.6 of Chapter 3 on "Current Policy Issues Related to Developing a Skilled Technical Workforce."

[10]See Section 3.3 of Chapter 3 on "Federal-Level Policies and Programs" and Section 3.5 of Chapter 3 on "State-Level Policies."

concerns. Industry surveys and policy reports find evidence of current and long-term skill gaps in technical occupations.

- The available evidence suggests that efforts to create stackable credentials and career pipelines and pathways may not in themselves lead to an increase in the supply of technical workers. A recent report of the NSB cites decades of data showing that in the United States, "degree is not destiny" and that there is only a very loose association between degrees and jobs at all education levels and in all degree fields.[11]

Finding 5: It is difficult to find rigorous evidence on how well skilled technical labor markets are functioning at aggregate levels of analysis across the nation.[12]

a. Evidence indicates labor market imbalances in certain skilled technical occupations, industry sectors, and locations; however, analysts disagree as to whether there are gaps or failures in the aggregate in skilled technical labor markets, with some arguing that there is no overall evidence of market failures.
b. Labor supply and demand for skilled technical occupations vary across sectors and locations, and it is difficult to disentangle these two factors using traditional analytical methods.[13]
c. By definition, an innovative economy is likely to experience some lag in producing workers with the latest skills.[14] The ability to measure skill shortages is hampered by the dynamic nature of the labor market in that it is difficult to know whether an imbalance will be permanent or temporary. The measurement challenge arises because it is difficult to consistently identify levels of supply and demand in multiple sectors, at varying levels of aggregation, and in changing markets for skilled labor.

Finding 6: The nation is experiencing, and will continue to experience, imbalances in the supply of and demand for skilled technical workers in certain occupations, industry sectors, and locations. The nature of the

[11]See Section 2.1.1 of Chapter 2 on "National Trends in the Demand for and Supply of Skilled Technical Workers."
[12]See Section 2.1 of Chapter 2 on "The Supply of and Demand for Skilled Technical Workers."
[13]See Section 2.1 of Chapter 2 on "The Supply of and Demand for Skilled Technical Workers."
[14]See Section 1.2 of Chapter 1 on "The Importance of Skills."

problem differs across sectors and locations. These imbalances arise from multiple sources:

a. **Sectoral gaps.** Gaps are particularly evident in health care and manufacturing, industries that have undergone a shift from high-volume, low-tech jobs to dynamic, high-tech jobs and that encompass several occupations that require proficiency in science, technology, engineering, and mathematics (STEM) skills, as well as other foundational skills.[15]

b. **Demographic factors.** Changes related to population growth and the retirement of the baby boom generation create the potential for gaps and imbalances in skilled labor markets.

c. **Structural changes.** The functioning of skilled labor markets is affected by structural changes related to technological advances, changing business models, global competition, and sectoral changes associated with price movements.[16]

d. **Outmoded or misaligned policies and regulations.** Examples of misaligned policies include lengthy training requirements or licensing laws that constrain the supply of skilled technical labor for long periods of time in certain occupations, such as health care.[17]

e. **Advances in science and technology.** New technologies will impact the demand for applied technical skills: requirements will evolve over time, and continual education and training will be needed.[18]

f. **Poor incentives facing the nation's institutions.** Federal and state funding formulas can distort educators' incentives by emphasizing the volume of enrollment in institutions and programs rather than the quality of the institutions and programs or outcomes.[19] Funding formulas also can divert resources from programs that actually add value in skilled technical workforce development.

TASK: Consider current public- and private-sector roles in financing and providing employment training and skills certification.

Finding 7: In the United States, educators, students, workers, employers, the federal government, state and local governments, labor unions, industry and

---

[15]See Section 2.1 of Chapter 2 on "The Supply of and Demand for Skilled Technical Workers."

[16]See Section 2.1.2 of Chapter 2 on "Exploring Imbalances across the Skilled Technical Workforce in the United States."

[17]See Section 2.3 of Chapter 2 on "Trends in Supply and Demand in Health Care."

[18]See Section 2.3 of Chapter 2 on "Trends in Supply and Demand in Health Care."

[19]See Section 5.2 of Chapter 5 on "Impediments to Better Training Outcomes."

trade associations, and other civic associations all play a role in skilled technical workforce development.

a. **Educators** are responsible for designing and delivering education and training that will meet current and future skilled technical workforce requirements. The current education system works well to prepare some students for some skilled technical occupations.
   - **Elementary and secondary education,** which is intended to provide the foundation for qualifying for jobs in skilled occupations, is provided both publicly and privately. Public education is funded primarily with local tax revenues. Career and technical education (CTE) of varying quality and scope is provided in 9,500 comprehensive high schools, about 1,000 vocational high schools that also offer academic subjects, and about 800 area regional vocational schools that offer mainly CTE courses.
   - **Postsecondary education**, which is increasingly necessary to qualify for jobs in skilled technical occupations, is typically subsidized with state and federal funding provided to individuals and institutions. Depending on the location, options of varying quality, cost, and scope for postsecondary education and training for skilled technical jobs may include the following:
     - **Community and technical colleges** were created to provide affordable lifelong academic and technical education and training for all individuals in a local area. They offer associate's degrees that are transferrable to higher education degree programs, terminal associate's degrees, certifications, apprenticeships, and contract or custom training on a noncredit basis. Community colleges in the United States currently enroll nearly half of all students pursuing undergraduate education.
     - **Colleges and research universities** provide academic and technical education and training for skilled technical jobs for youth and adults on a competitive basis at a wide range of costs. They offer bachelor's degrees, clinical training, graduate degrees, and doctoral degrees in the arts; such vocations as business, education, government, and law; and science and technology. Slightly more than half of all students pursuing undergraduate education are currently enrolled in 4-year colleges and research universities.
     - **Apprenticeships** provide education and training in a wide range of skilled occupations and are expanding nationally. The registered apprenticeship network in the United States currently includes more than 1,000 occupations, although active programs may not exist in all of these occupations.

There are about 400,000 registered apprentices in the United States, and they are typically clustered in a few occupational sectors, such as construction, certain areas of manufacturing, and services.[20] However, there are numerous barriers to the wider adoption of apprenticeship programs, including confusion about what constitutes an apprenticeship; the high costs of starting a program; a lack of intermediating institutions needed to connect firms, workers, and educational organizations; and insufficient in-company training resources.[21]

- **Certifications** are offered by approximately 4,000 entities in the United States, including industry groups, professional organizations, and employers, often with little quality control through third-party review or accreditation.
- **Certificate programs**—typically lasting 1 year or less—are widely available at community colleges for many skilled technical fields. The limited research available suggests that the value of these certificates varies widely.

b. **Students and workers** are responsible for identifying education and training requirements, but may not be making informed choices. Higher education in public institutions in the United States is subsidized by the state governments. Even so, most postsecondary students are responsible for the up-front costs, including tuition, books, supplies, child care, and transportation.

c. **Employers** often provide on-the-job training, as well as tuition assistance for postsecondary education and training for employees. However, it is difficult to determine accurately the motivation for, the scale of, or the effectiveness of employer training because of limited independent data and analyses. The data that have been collected show wide variation across surveys. For example, estimates of the proportion of workers who participated in training in the years 2003 and 2004 vary from 22 to 57 percent. About 30 percent of employees use tuition reimbursement programs.[22]

d. **Civic organizations,** including labor unions, worker associations, industry and trade associations, religious organizations, and philanthropies, also support workforce development based on their specific missions and interests. However, there have been no

---

[20]See Section 4.3 of Chapter 4 on "Other Postsecondary Education and Training Programs."

[21]See Box 4-1 of Chapter 4 on "Apprenticeships: Challenges from Concept to Application."

[22]See Section 4.2 of Chapter 4 on "The Primary Components of Workforce Development."

comprehensive studies of the scale or effectiveness of the role of civic associations in skilled workforce development.[23]

TASK: Consider policy-making challenges and experiments that affect incentives and information to improve work skills.

Finding 8: Skilled technical workforce development in the United States is guided and supported by a complex and often uncoordinated set of policies and funds at the local, state, and federal government levels associated with achieving goals related to economic development, education, labor and employment, health and human services, and veterans' affairs. Most resources are allocated by formulas based on demographic factors, which serve as a proxy for need, rather than on performance, outcomes, or evidence of what works best in workforce development.

a. **Federal policy frameworks** include the 2015 Every Student Succeeds Act, the 2014 Workforce Innovation and Opportunity Act (WIOA), the Perkins Act, the Higher Education Act (HEA), and the Post-9/11 Veterans Educational Assistance Act of 2008.[24]
   - The majority of federal workforce training funding is distributed by formula grant programs that allocate funds to state and local entities based on established noncompetitive criteria; the rest is distributed through competitive grant programs administered by federal agencies.
   - Of the $145 billion in federal education assistance provided in 2013, the vast majority ($130.7 billion) was governed by Title IV of HEA, while $10 billion was governed by the Post-9/11 Veterans Educational Assistance Act of 2008, $2.5 billion by WIOA, $1.1 billion by the Perkins Act, and $575 million by the Trade Adjustment Assistance Act (TAA).
   - Many federal programs support workforce development; however, 95 percent of this funding is managed by four federal agencies: the Department of Labor (45.2 percent), Department of Education (29.9 percent), Department of Health and Human Services (12.5 percent), and Department of Veterans Affairs (8 percent).
   - WIOA is intended to better align federal and state workforce development efforts. However, it has not yet been fully implemented; resource allocations are uncertain; and in the past,

[23]See Section 4.4 of Chapter 4 on "Funding for Skilled Technical Workforce Education and Training."

[24]See Section 3.3 of Chapter 3 on "Federal-Level Policies and Programs."

congressional oversight of performance against policy goals has been limited.

b. **State and local policy frameworks** include those related to K-16 education, CTE, labor, and veterans' assistance.[25]
   - The National Center for Education Statistics reports that state and local expenditures for public elementary and secondary schools alone are projected to be $634 billion for the 2015-2016 school year.
   - State and local government agencies are the primary source of funding for community colleges. In 2013-2014, expenditures for community colleges totaled $55.9 billion, nearly half (49.4 percent) of which came from state or local funding, 22.6 percent from federal funding, and the remainder from student tuition and fees (16.7 percent) and other sources (11.3 percent).
   - Some forms of education and training, such as CTE, apprenticeships, certificate programs, certification programs, licensing programs, and veterans' transition assistance programs, require systematic collaboration across educators, employers, labor unions, industry and trade associations, and policy makers at the local level.
   - Although there are many examples of effective systematic coordination that prepares skilled technical workers to meet local employer needs, these models remain isolated. Many policy makers are concerned about the effectiveness of local coordination and resource allocation.

Finding 9: The incentives for students and employers to invest in skill development depend on the return on these investments.[26]

a. Overall, there is strong evidence of high positive returns to postsecondary education and training for skilled technical jobs. However, students face major hurdles in starting and completing their education and training. Students in public community colleges have lower completion rates relative to students in private technical colleges or 4-year colleges. Key factors affecting the initiation and completion of postsecondary education and training programs include
   - the rising cost of tuition;

---

[25]See Section 3.5 of Chapter 3 on "State-Level Policies."

[26]See Section 5.1 of Chapter 5 on "The Return-on-Investment Calculus."

- inadequate preparation in elementary and secondary education and the high costs of remedial programs needed to overcome this lack of preparation;
- a lack of support services to address challenges faced by many students in establishing their household, raising their children, and caring for aging family members;
- limited career guidance, which makes it difficult for students to identify appropriate courses of study; and
- the perceived low social prestige of CTE among some students and parents.

Finding 11 addresses effective strategies for improving completion rates for programs in skill training.

b. Evidence on employer investments in education and training and the return on these investments is mixed, in large part because data are missing or incomplete.
   - Some employers may look to public resources for education and training for students and workers because they fear that other employers may free-ride on their investments.
   - With the growth of contingent work arrangements, which shorten the timeframe within which employers can recoup investments in training, some employers may have insufficient motivation to invest in some types of skilled technical workforce development.

c. Investments in certification programs by students, workers, and employers depend on the perceived value of these programs. The value of certain programs may be difficult to measure or may be lower when alignment is lacking among credentialing institutions, states, or employers within an industry.
   - The available research on certificates and certifications indicates that outcomes are highly context-dependent. Certifications appear to have more labor market value in some sectors than in others, and some degree of quality control is exercised through third-party review or accreditation. Certificates that require long periods of study and are aligned with industry certifications provide more benefits to holders relative to those that require short periods of study.
   - State and local credentialing and licensing requirements can increase the costs of preparing for and entering skilled technical jobs, and make it more difficult and costly to move from one job to another or from one location to another.

Finding 10: Policy makers and participants in workforce development lack the information they need to adjust policies and make choices. User-friendly tools that provide direct access to information about options and performance can assist labor market participants in their decision-making processes.[27]

a. Limitations in data and analyses make it difficult to determine whether and how individuals, employers, and the public are investing in skilled technical education and training, and whether they are realizing a sufficient return to continue or increase their investments.[28]
b. To make adjustments that will improve workforce development outcomes, both policy makers and labor market participants need better data with which to analyze labor market dynamics, returns on investments in workforce development, and the effectiveness of alternative strategies.
   - Students and workers, guidance counselors, and educators need better information on skill requirements, career pathways, and potential returns on their investments in education and training.
   - Employers can benefit from better information on effective education and training resources and on viable pipelines for producing skilled workers.
c. Much of the data for the information system that supports U.S. labor market functions is funded and administered by the Department of Labor. However, this information system is less effective than it could be; funding has been stagnant over the past decade, and Congress has provided limited oversight.
d. Many analysts are using outdated and ineffective research methods. Investing in more research will not necessarily generate new knowledge if analysts continue to use outmoded analytical methods and tools. Advances in mathematics, statistics, and the computational sciences make it possible to capture and analyze data from a broad range of very different sources. The evidence suggests that more analysts need to apply these advances to improve workforce analysis tools and methods.
e. Local data warehouses and analytical applications, such as those funded by the Department of Labor's Workforce Data Quality Initiative, help policy makers better estimate returns on investments in education and training.

---

[27]See Section 5.2 of Chapter 5 on "Impediments to Better Training Outcomes," and Section 6.5 of Chapter 6 on "Improving Links through Better Data."
[28]See also Section 2.3 of Chapter 2 on "Trends in Supply and Demand in Health Care" and Section 5.2 of Chapter 5 on "Impediments to Better Training Outcomes."

Finding 11: Several strategies and initiatives show promising results by improving rates of successful completion of education and training and by coordinating education and training opportunities with employer needs. These experiments, which link the different parts of the ecosystem for workforce education and training, could be instructive for designing and enforcing policy and allocating resources within appropriate contexts.

a. The first set of policy measures is intended to link students to education and training and improve success rates.[29] They include the following:
   - **Counseling services.** Career guidance is particularly important to skilled technical workforce development because it can counter the common perception that the only path to lifelong occupational success is through immediate entry into a 4-year college and advanced degree programs. Without career counseling and reliable occupational information, students pay insufficient attention to labor market trends when choosing a field of study.
   - **Financial aid.** Some federal aid, such as Pell Grants, is limited to undergraduate students in for-credit programs and is not available to students taking continuing education classes, even when they earn a certificate. Evidence indicates that providing financial aid tied to academic benchmarks as a supplement to other financial aid increases the likelihood that students will enroll full-time and earn more credits.
   - **Braiding of financial sources.** Some states are finding effective ways to combine public resources to support postsecondary education and training programs by using "braiding strategies," which merge public dollars for a common purpose while keeping categorical funds distinct. More analyses of funding sources and financial flows are required to ensure that public funds can flow freely to meet changing workforce development needs.
   - **Continuing education and training.** Lifelong learning is essential for the U.S. workforce to remain globally competitive. Overall participation in all forms of adult education among Americans aged 16 and older is about 44 percent. Some states are experimenting with strategies for ensuring that U.S. workers will continue to invest in education and training throughout their work lives. Maine, for example, is experimenting with portable, employee-owned Lifelong Learning Accounts that permit employees, employers, and the state to pool funds for continual worker education and training.

---

[29]See Section 6.1 of Chapter 6 on "Linking Students to Skilled Technical Education and Training Opportunities and Improving Success Rates."

- **Wraparound services.** Providing a wide range of services, such as counseling, tutoring, child care, and transportation assistance, that support students in achieving education and training goals appears to improve student outcomes and returns on investment.
- **Remedial programs.** Students who need to take remedial classes in college are more likely to drop out. Although it has not been validated through randomized controlled trials, anecdotal evidence suggests that integrating remediation into skills training improves completion rates. Schools also can do a better job of assessing student readiness for education and training and designing more appropriate and cost-effective prerequisite requirements. Online education may be a key component of improved remediation, and the use of online simulation modules may be effective at improving training opportunities economically.
- **Career pathways.** Some evidence suggests that articulating career pathways improves education and training outcomes. Career pathways, which are roadmaps of the education and training required to attain credentials associated with success in specific industries, have been adopted in several states and often are used to guide linked learning, sector strategies, talent management, and career pipeline initiatives.
- **Online and blended learning.** New information technologies can supplement traditional educational methods. Although the use of online training has proven effective in certain professions, such as air transportation (e.g., flight simulators) and health care (e.g., surgical applications), it also offers the promise of aiding with remedial education, which may improve community college completion rates.
- **Incentives for community colleges.** Although many states have already implemented funding incentives based on such indicators as course completion and time to degree, these formulas may encourage community colleges to accept only the best students or to lower standards for completion. To boost completion rates, community colleges and other educational organizations will require incentives and funding to create more flexible and integrated programs and offer supportive services.

b. A second set of policy measures is intended to improve links between secondary and postsecondary education.[30] Better integration of academic and technical education and training can improve the quality

[30]See Section 6.2 of Chapter 6 on "Linking Secondary and Postsecondary Education and Training."

of postsecondary education and increase the return on investments in education and training for skilled technical occupations.

- **Dual-enrollment programs and early college high schools.** The existing evidence suggests that these programs, which involve formal linkages between secondary and postsecondary school credentialing programs, improve incentives for high school students to earn their high school diploma and an associate's degree in an applied science or technology and to obtain a job.
- **Career academies.** Evidence suggests that career academies, a type of small learning community within a school that provides a college preparatory curriculum with a career-related theme, improve outcomes for students during and after high school. First founded in 1969 in Philadelphia with the Electric Academy at Edison High School, career academies in the United States now number more than 7,000 and cover a wide range of career areas.

c. A third set of policy measures is intended to link training and work.[31] Many employers are seeking to partner more closely with community or technical colleges. Similarly, many educators are seeking to partner with employers to develop more relevant education and training programs. These educational institution–employer partnerships have the potential to create better-integrated learning environments and meet local employer skill requirements.

- **Strategic centers of excellence at community and technical colleges.** These centers are designed to foster flexible and integrative learning in strategically important local industries. They can serve as a point of contact and resource hub for industry trends, best practices, innovative curriculum, and professional development.
- **Sector strategies.** Evaluations of sector partnerships suggest that they have the potential to produce positive outcomes for youth and adults. Advocated by the National Governors Association (NGA), the Department of Labor, and civic associations, sector strategies are partnerships of employers within a single industry that focus on coordinating policies and resources to address the workforce needs of a strategically important industry, including through apprenticeship programs targeted to those sectors. NGA estimates that more than 1,000 sector partnerships now operate across about half of all U.S. states.

d. Apprenticeship and joint labor–management programs can serve to link employers and workers.[32]

---

[31]See Section 6.3 of Chapter 6 on "Linking Training and Work."

[32]See Section 6.4 of Chapter 6 on "Employer-Based Training Programs."

- **Apprenticeship programs.** As noted earlier, these programs combine occupational education and training with paid on-the-job work experience. A number of states have developed apprenticeship programs over the past 10 years, although the impact of such programs has not been rigorously evaluated. The most interesting of these models move beyond time-based workplace training to integrate dual-learning principles whereby academic courses are linked to workplace training and standardized competency-based certifications.
- **Joint labor–management strategies.** More evidence is needed on joint labor–management strategies for establishing workplace learning systems and community-based workforce and economic development initiatives.[33] Unions have a prescribed role in job training programs under the 2014 WIOA, and they have a long history of designing and implementing innovative and effective job training and apprenticeship programs that potentially benefit workers, businesses, and local communities.
- **Talent pipelines.** The use of supply chain management principles to create talent pipelines that link investments in workforce development with business strategy is being promoted by the U.S. Chamber of Commerce Foundation; the National Network of Business and Industry Associations; industry trade associations, such as the Precision Machined Products Association, the National Tooling and Machining Association, the Technology and Manufacturing Association, and the National Institute for Metalworking Skills; and major employers, such as IBM.

e. A final set of policy initiatives is intended to increase the portability and wider recognition of acquired technical skills.[34] Roadmaps that link selected competencies also can help guide students and workers through the education and training they need to address the needs of local industries.

   - **Portable credentials and licenses.** Given the proliferation of training programs and types of credentials, standardizing credentials would provide certainty to employers and workers about the value of the credentials. The evidence suggests that efforts to ensure that education and training accomplishments and licenses transfer across education and training programs, employers, industries, and geographic boundaries can reduce the costs of education and training and improve returns on investment.

---

[33]See Section 6.4 of Chapter 6 on "Employer-Based Training Programs."

[34]See Section 6.6 of Chapter 6 on "Other Policy Initiatives."

- **Competency models.** The Department of Labor's Employment Training Administration has developed a structured approach to articulating the competencies required in occupations in an industry or sector, from foundational competencies, to industry-wide competencies, to technical competencies within a specific subsector. The competency strategy is used to identify specific employer skill requirements, define career pathways and credentials, introduce competency-based curricula and training models, create industry-defined performance indicators and certifications, and develop resources for career exploration and guidance. The evidence suggests that assessments against competencies are a better basis for education, training, and employment decisions than credentials, which are imperfect proxies for occupational capacity and competency. Those who are engaged in linked learning, sector strategies, talent management, career pathways, and career pipeline initiatives can use the competency model to guide their work.

Finding 12. The Armed Services, one of the largest employers in the United States, could do a better job of assessing employment transition risks for military personnel, and designing and delivering services to mitigate these risks. They also could coordinate better with civilian policy makers, regulators, and educators to improve the transferability of military education, training, and certification.[35]

- Many Americans join the uniformed Armed Services as a pathway to education, training, and preparation for life and careers: 43.2 percent of all uniformed active duty personnel are aged 25 or younger.[36] Military personnel receive training in the competencies associated with skilled technical work. As of the end of November 2015, about 1.3 million individuals were working in the Armed Services.
- Each year, tens of thousands of military personnel transition back into civilian life after serving in the armed forces for just under an average of 7 years. Veterans' transition assistance programs include federal and state government programs, private-sector initiatives, and public–private partnerships. However, additional action is needed to improve

[35]See Section 5.3 of Chapter 5 on "Challenges in Training and Transitioning the Skilled Technical Workforce."

[36]See Section 5.3 of Chapter 5 on "Challenges in Training and Transitioning the Skilled Technical Workforce."

the effectiveness of these programs. Data suggest that some veterans have more difficulty with transitions than others.

- The Department of Defense (DoD) spends approximately $1 billion per year on unemployment benefits for transitioning service members but does not track how long members claim unemployment benefits or other characteristics of these benefits. DoD tracks only information on the characteristics of those who initially file for the Unemployment Compensation for Ex-servicemembers (UCX) program, regardless of whether they actually use those benefits.

TASK: Consider selected employment practices in other countries.

Finding 13. Policies and governmental structures used to support education and training in such countries as Germany and Switzerland may not be directly applicable in the United States because of differences in political and economic structures. However, applicable lessons can be learned from specific programs or innovative educational techniques, and policy makers in the United States can benefit from more rigorous cross-country comparisons.

a. Unlike most other developed countries, the United States lacks formal mechanisms that require governments, educators, labor representatives, and employers to coordinate on workforce development policies and practices at the national level.[37] Also in contrast with many other developed countries, federal and state-level policy in the United States requires most students and workers to pay for most of the cost of postsecondary education.
b. Nevertheless, state policy makers may be able to learn from the experience of other developed countries that have more coordinated employer training programs and a tradition of technical invention and innovation.[38] With regard to establishing apprenticeship programs, for example, a key lesson is the need to set clear rights and responsibilities for employers, workers, and educational organizations. Also needed are strong intermediating institutions that can coordinate individual employers' participation and help schools work with business, as well as clear industry standards and the competencies that apprentices should acquire in order to meet those standards.[39]
c. Moreover, there is a growing interest in developing apprenticeship programs in the United States, particularly in high-growth industries

[37]See Section 4.2 of Chapter 4 on "The Primary Components of Workforce Development."

[38]See Section 6.4 of Chapter 6 on "Employer-Based Training Programs."

[39]See Box 6-9 of Chapter 6 on "Key Lessons from Foreign Apprenticeship Systems."

such as health care, advanced manufacturing, information technology, and biotechnology. These programs use various approaches that often differ from the European dual-education model, and therefore direct comparisons are difficult.[40]

RECOMMENDATIONS

TASK: Consider policy-making challenges and experiments that affect incentives and information to improve work skills.

Recommendation 1: State and federal policy makers should support and enhance strategies that help students successfully complete their training for the skilled technical workforce. In addition, public policies should ensure that stakeholders, including students, workers, employers, and educational organizations, have the right incentives to improve the quality of technical education and training, encourage experimentation and collaboration, and improve the collection and use of relevant information.

a. **The Department of Education should collect and disseminate information on best practices at community colleges that facilitate timely completion and enhance employment of graduates who are being trained to be a part of the skilled technical workforce.**
    - The Department of Education should provide information to states and community colleges around the country to help them adapt and craft programs based on principles underlying successful programs.
    - Based on this information, students should be encouraged to embark on and complete education and training programs for skills that are in demand, and educational institutions should be incentivized to support them in doing so. In this regard, mandatory counseling and better student support can improve completion rates. In addition, studies in several states have shown that integrating remediation into skill training is very helpful.

b. **The Department of Labor should encourage employers to partner with industry and trade associations, labor unions, and other civic organizations; educators; workers; and policy makers to develop workforce skills.** Employers are encouraged to
    - consult regularly and partner with policy makers, educators, labor unions, industry and trade associations, and other civic associations on skilled technical workforce development issues;

[40]See Section 6.4 of Chapter 6 on "Employer-Based Training Programs."

- engage in workforce planning for future labor needs, and actively seek education and training partners to address those needs;
- create incentives and resources for employees to invest in continual technical education and training throughout their careers; and
- support programs, such as apprenticeships, that link work-based learning with contextualized coursework and competency-based standardized certifications.

c. **The Department of Labor should develop and disseminate information on the return on student and worker investments in training for skilled technical jobs.**
   - **Support state initiatives.** Congress should prioritize the provision of resources to support the states in investigating the return on investment in education and training as part of local skilled technical workforce development.
   - **Collect and disseminate information.** The Department of Labor should collect and disseminate information on investigations of return on investment, and provide technical assistance and resources for such investigations.
   - **Request studies.** State legislatures also should request studies of return on investment. These studies should include efforts to develop a better understanding of how students make decisions about education and training.

d. **The secretaries of education and labor should incentivize stakeholders to:**
   - **Develop context-based learning strategies.** Review and evaluate funding formulas and grant rules for education and training programs to ensure that they provide incentives and funding for all stakeholders to participate in the development of an equal-opportunity skilled technical workforce using integrated, work context–based learning strategies. For example, allow federal and state education and labor funds to be used for all postsecondary education and training programs at all ages and levels of career advancement.
   - **Reduce the need for remediation.** Encourage community colleges to reduce the need for remediation, integrate remediation into skills training, and improve the readiness of high school students for postsecondary or adult education programs.
   - **Reward program completion.** Implement funding formulas tied to employment metrics, such as earnings of students after graduation, or metrics focused on increasing rates of enrollment and completion for programs that are in high demand among local employers.

- **Enhance the portability of credentials.** Work with and encourage efforts of local workforce boards to harmonize key features related to the quality, portability, and labor market value of credentials.
- **Encourage participation.** Develop incentives for employers and industry and trade associations to contribute to workforce development and transition management.

The secretaries of education and labor should report to Congress or state legislatures on the results of these efforts.

e. **Innovate, experiment, and assess.** The secretaries of education and labor should encourage educators to innovate and experiment in their skilled technical education and training programs to improve the linkage between education and employment, the quality of programs, student services, information about programs, and return on investments in education and training. Examples are provided in Chapter 6.
   - Specifically, states should provide competitive funding grants to community colleges to encourage such innovation, along with accountability structures that enable assessment of what works. Principles underlying successful interventions should be identified and applied statewide and across state lines.
   - Regular internal and external assessments are essential to identify and learn the lessons of these experiments.

TASK: Define the skilled technical workforce.

Recommendation 2: An alliance of industry, trade, academic, and civic associations and labor unions, in cooperation with the U.S. Departments of Labor and Education, should organize a nationwide public–private communication campaign to raise awareness of the value of and demand for skilled technical workers and the return on investment for individuals preparing for these careers. This campaign should be customized to recognize local variations in skilled technical workforce education, training, and labor market requirements.

a. **Labor unions, industry trade associations, and other civic associations** can enhance this message by providing financial and in-kind support to the Department of Labor's communication campaign.
b. **Educational institutions** can participate by providing more transparent information about their programs and curricula that can be showcased in these campaign efforts.
c. **Employers** can help by providing salient examples of high-paying jobs and the education and training that can lead to these jobs.

TASK: Consider gaps and market failures.

Recommendation 3: Congress and state legislatures should improve oversight of public policies and resources, highlighting the implementation of reforms such as those called for in WIOA. Reforms should be accelerated through targeted incentives.

a. **Expand.** Given the gaps in data and the failure to measure important parts of the skilled workforce ecosystem, the Department of Labor should consider the need to revise and expand current data collection approaches to better monitor this ecosystem.
b. **Monitor.** The Department of Labor should continue to monitor the state of the workforce through oversight and produce data particularly relevant to the technical skill arena. These activities are important given that efforts to address the issues are spread across several stakeholders and arenas.
c. **Analyze.** Congress should require the secretaries of education and labor to prepare annual reports on progress in implementing reforms such as those called for in WIOA and on the allocation of resources and financial flows associated with implementation.
d. **Oversee.** Congress and state legislatures should conduct regular hearings on progress in implementing education reforms.
e. **Incentivize.** Community colleges and other educational organizations should be incentivized and provided the means to adopt policies and programs aimed at boosting degree or credential completion, including better integration of academic and technical education and training.

Recommendation 4: Congress, state legislatures, and agencies of the federal and state governments should improve the workforce labor market information system (WLMIS), labor market data, and research tools and methods, including by providing funding for such activities.

a. **Prepare a roadmap.** Congress should require the secretary of labor to prepare a roadmap for implementing WLMIS reforms required under WIOA and to submit an annual report on progress against this roadmap.
b. **Conduct hearings.** Congress and state legislatures should conduct regular hearings on WLMIS issues and reforms.
c. **Provide support.**
    - The Department of Labor should provide resources for improving and developing the WLMIS, as called for in WIOA.
    - Civic organizations and private industry also can provide financial and in-kind support for WLMIS initiatives.

d. **Analyze data.**
   - The National Science Foundation should develop a program to expose labor market analysts to innovations in data and information sciences, develop new user-friendly information tools and research methodologies, and train and incentivize analysts to apply these innovations in their work.
   - Labor market researchers should be required to learn about and apply advances in data science and such new tools as screen scraping, and to conduct investigations that produce useful insights for stakeholders.

e. **Disseminate results.** State departments of labor should develop an understanding of what data are required and how they are used, and provide information for students, workers, and employers that meet these requirements and help them calculate returns on investments in education and training.
   - To make well-informed decisions, high school and postsecondary students need more direct career counseling based on accurate information, as well as advice about the challenges and opportunities made available by various kinds of academic programs.
   - Educational institutions also need information about alternative pathways available to students and completion rates of various programs. For example, community college students often avoid technical fields because they plan to transfer to 4-year colleges and obtain bachelor's degrees, but most do not do so.
   - States need to include data on the labor market outcomes of students in their state student information systems so that school districts can use the data to evaluate and improve their career programs.

Recommendation 5: Federal and state agencies should remove barriers to worker mobility, such as licensing and certification requirements that are not related to public safety. They also should improve labor market information on the changing requirements for skilled technical workers to help reduce imbalances in labor markets and to align workforce development with advances in science and technology.

a. **Enforce fair employment standards.** Labor law enforcement agencies, employers, labor unions, industry and trade associations, and other civic associations should help ensure that, consistent with the laws in place, all qualified workers can participate in all occupations in the skilled technical workforce, have access to education and training opportunities throughout their careers, and be paid the regular and overtime wages to which they are entitled.

b. **Disseminate labor market information.** The National Science Foundation should develop a program to rapidly disseminate useful information to K-20 faculty and students, employers, and policy makers about changing labor market requirements and job opportunities that may result from advances in science and technology.

c. **Remove barriers.** The Department of Education should consider ways to reform financial aid that is currently limited to undergraduate students in for-credit programs so that it includes students taking continuing education classes in certificate programs.

d. **Review licensing requirements.** States should review their licensing requirements and reduce the number of licensed occupations or the number of requirements that do not pertain to public health and safety.

e. **Align training programs.** The states should consider a collective agreement with DoD that would facilitate alignment of DoD training with their 50 separate sets of certification and licensure requirements, perhaps modeled on the Military Child Education Compact.

TASK: Consider policy-making challenges and experiments that affect incentives and information to improve work skills.

Recommendation 6: DoD should further integrate skills transition into military training rather than treating it as a separate component at the end of the service member's career.

To further develop military training as a major pathway to the acquisition of technical skills, DoD should

a. investigate career pathway and transition issues in each of the Armed Services, which are among the largest employers in the United States, and use data from these investigations to better coordinate with civilian policy makers, regulators, and educators to improve the transferability of military education, training, and certification;

b. assess transition risks for military personnel, and design and deliver services to mitigate these risks, including ensuring that the military culture encourages service members to do everything they can to transition without drawing UCX assistance; and

c. collect better information on the cost and duration of unemployment benefits used by transitioning military personnel, as well as the incentives these programs create to use benefits.

TASK: Consider selected employment practices in other countries.

Recommendation 7: While selected programs and policies from other countries have been adapted in the United States, federal agencies should further study the conditions under which particular attributes of apprenticeships or other programs can be effectively applied more broadly.

a. The National Science Foundation should commission a study of how countries with more proficient workers develop their skilled technical workforces. This study should utilize concepts and methods that are common in organizational and institutional analysis, and should include the development of a rigorous methodology for conducting comparative studies and the application of insights from these studies to policy analysis and design.
b. Similarly, the Department of Education should use data from comparative organizational and institutional studies to identify ways of strengthening relationships between employers and educational institutions to improve state employment rates of graduates.

Box 7-1 organizes the committee's recommendations according to the various stakeholders to which they are targeted.

BOX 7-1
Summary of Selected Recommended Actions for Stakeholders

Congress

- Prioritize the provision of resources to support the states in investigating the return on investment in education and training as part of local skilled technical workforce development. (Recommendation 1c)
- Require the secretaries of education and labor to prepare annual reports on progress in implementing reforms and on the allocation of resources and financial flows associated with implementation. (Recommendation 3c)
- Require the secretary of labor to prepare a roadmap for implementing WLMIS reforms required under WIOA and to submit an annual report on progress. (Recommendation 4a)

Congress and State Legislatures

- Improve oversight of public policies and resources, highlighting the implementation of reforms such as those called for in WIOA. (Recommendation 3)
- Conduct regular hearings on WLMIS issues and reforms. (Recommendation 4b)

State Legislatures

- Request studies of return on investment. (Recommendation 1c)

Federal and State Departments of Labor and Education

- Review and evaluate funding formulas and grant rules for education and training programs to ensure that they provide incentives and funding for all stakeholders to participate in skilled technical workforce development. (Recommendation 1d)
- Collect and disseminate to students, workers, educators, and employers information on investigations of return on investment, and provide technical assistance and resources for such investigations. (Recommendation 1c)
- Encourage educators to innovate and experiment in their skilled technical education and training programs to improve the linkage between education and employment, the quality of programs, student services, information about programs, and return on investments in education and training. (Recommendation 1e)
- Monitor the state of the workforce and produce relevant data for students, workers, and employers. (Recommendation 3b)
- Provide resources for improving and developing the WLMIS. (Recommendation 4c)

(Continued)

BOX 7-1
Continued

Department of Defense

- Further integrate skills transition into military training rather than treating it as a separate component at the end of the service member's career. (Recommendation 6)
- Investigate career pathway and transition issues, and use data from these investigations to improve the transferability of military education, training, and certification. (Recommendation 6a)
- Collect better information on the cost and duration of unemployment benefits used by transitioning military personnel, as well as the incentives these programs create to use benefits. (Recommendation 6c)

National Science Foundation

- Expose labor market analysts to innovations in data and information sciences, develop new user-friendly information tools and research methodologies, and train and incentivize analysts to apply these innovations in their work. (Recommendation 4d)
- Commission a study of how countries with more proficient workers develop their skilled technical workforces. This study should include the development of a rigorous methodology for conducting comparative studies and the application of insights from these studies to policy analysis and design. (Recommendation 7a)

Educators (with the encouragement of the Secretaries of Education and Labor)

- Innovate and experiment in their skilled technical education and training programs to improve the linkage between education and employment, the quality of programs, student services, information about programs, and return on investments. (Recommendation 1e)

Employers, Industry and Trade Associations, Labor Unions, and Allied Training Partners (with the encouragement of the Department of Labor)

- Consult regularly and partner with policy makers, educators, and other civic associations on skilled technical workforce development issues. (Recommendation 1b)
- Create incentives and resources for employees to invest in continual technical education and training throughout their careers. (Recommendation 1b)
- Support programs, such as apprenticeships, that link work-based learning with contextualized coursework and competency-based standardized certifications. (Recommendation 1b)
- Raise awareness of the value of and demand for skilled technical workers. (Recommendation 2)

BOX 7-1
Continued

Other Civic Associations

- Consult regularly and partner with policy makers, educators, and employers on skilled technical workforce development issues. (Recommendation 1b)
- Raise awareness of the value of and demand for technical skills. (Recommendation 2)
- Provide financial and in-kind support for a skilled technical workforce communication campaign. (Recommendation 2a)
- Provide financial and in-kind support for WLMIS initiatives. (Recommendation 4c)
- Help ensure that all qualified workers can participate in all occupations in the skilled technical workforce, have access to education and training opportunities throughout their careers, and be paid the wages to which they are entitled. (Recommendation 5a)

Researchers

- Learn about and apply advances in data science, and conduct investigations that produce useful insights for stakeholders. (Recommendation 4d)

References

AACC (American Association of Community Colleges). 2015. *Completion pathways.* http://www.aacc.nche.edu/Publications/datapoints/Documents/DP_CompletPathways.pdf (accessed March 19, 2017).

AACC. 2016. *2016 fact sheet.* http://www.aacc.nche.edu/AboutCC/Documents/AACCFactSheetsR2.pdf (accessed March 5, 2017).

Accenture, Burning Glass, and Harvard Business School. 2014. *Bridge the gap: Rebuilding America's middle skills.* http://www.hbs.edu/competitiveness/Documents/bridge-the-gap.pdf (accessed February 8, 2016).

Acemoglu, D., and D. Autor. 2010. *Skills, tasks, and technologies: Implications for employment and earnings.* Working paper 16082. Cambridge, MA: National Bureau of Economic Research. http://www.nber.org/papers/w16082 (accessed February 28, 2017).

AFL-CIO (American Federation of Labor and Congress of Industrial Organizations). 2016. *Joint labor-management training and career ladder programs: Winning solutions for employers, workers and communities (in progress).* http://partners.aflcio.org/working-america-institute/workforce-intermediary-partnerships (accessed March 22, 2017).

AFL-CIO. n.d. *Labor-management partnerships.* http://www.aflcio.org/Learn-About-Unions/Training-and-Apprenticeships/Labor-Management-Partnerships (accessed March 19, 2017).

Akerlof, G. A. 1970. The market for lemons: Quality uncertainty and the market mechanism. *Quarterly Journal of Economics* 84(3):488-500.

Alssid, J., D. Gruber, D. Jenkins, C. Mazzeo, B. Roberts, and R. Stanback-Stroud. 2002. *Building a career pathways system: Promising practices in community college-centered workforce development.* Brooklyn, NY: Workforce Strategy Center. https://eric.ed.gov/?id=ED470457 (accessed March 22, 2017).

Altstadt, D. (Ed.). 2012. *Tying funding to community college outcomes: Models, tools, and recommendations for states.* Boston, MA: Jobs for the Future. http://www.jff.org/sites/default/files/publications/TyingFunding2CommColleges-042312.pdf (accessed February 28, 2017).

AMTEC (Automotive Manufacturing Technical Education Collaborative). n.d. *About AMTEC*. http://autoworkforce.org/about-amtec (accessed March 21, 2017).

Arizona Commerce Authority. n.d. *Skilled available workforce: Sector partnerships*. http://www.azcommerce.com/business-first/skilled-available-workforce/sector-partnerships (accessed March 21, 2017).

Arntz, M., T. Gregory, and U. Zierhan. 2016. *The risk for automation for jobs in the OECD countries: A comparative analysis*. http://www.oecd-ilibrary.org/social-issues-migration-health/the-risk-of-automation-for-jobs-in-oecd-countries_5jlz9h56dvq7-en (accessed February 28, 2017).

Association for Talent Development. 2015. *State of the industry report*. https://www.td.org/Publications/Magazines/TD/TD-Archive/2014-State-of-the-Industry-Report-Spending-on-Employee-Training-Training-Remains-a-Priority (accessed March 5, 2017).

Autor, D. 2010. *The polarization of job opportunities in the U.S. labor market: Implications for employment and earnings*. Washington, DC: Center for American Progress and The Hamilton Project. http://economics.mit.edu/files/5554 (accessed February 9, 2016).

Autor, D. 2014. Skills, education, and the rise of earnings inequality among the other "99 percent." *Science* 344(6186):843-851.

Autor, D., and D. Dorn. 2013. The growth of low-skill service jobs and the polarization of the U.S. labor market. *American Economic Review* 103(5):1553-1597.

Autor, D., and B. Price. 2013. *The changing task composition of the U.S. labor market: An update of Autor, Levey and Murnane (2003)*. http://economics.mit.edu/files/9758 (accessed February 28, 2017).

Autor, D., F. Levy, and R. Murnane. 2003. The skill content of recent technological change: An empirical exploration. *Journal of Quarterly Economics* 118(4):1279-1333.

Autor, D., L. Katz, and M. Kearney. 2006. Measuring and interpreting trends in economic inequality: The polarization of the U.S. labor market. *American Economic Review Papers and Proceedings* 96(2):189-194. http://economics.mit.edu/files/584 (accessed February 28, 2017).

Ayres, S., and E. Gurwitz. 2014. *The underuse of apprenticeships in America*. Washington, DC: The Center for American Progress, https://www.americanprogress.org/issues/economy/news/2014/07/22/93932/the-underuse-of-apprenticeships-in-america (accessed March 22, 2017).

Backes, B., H. Holzer, and E. Dunlop Velez. 2014. *Is it worth it? Postsecondary education and labor market outcomes for the disadvantaged*. CALDER Working Paper 117. http://www.caldercenter.org/sites/default/files/WP117.pdf (accessed March 5, 2017).

Backes-Gellner, U. 2014. *Benefits of apprenticeship training and future challenges: Empirical results and lessons from Switzerland and Germany*.

Working Paper 97. http://repec.business.uzh.ch/RePEc/iso/leadinghouse/0097_lhwpaper.pdf (accessed March 19, 2017).

Bailey, T. 1993. Can youth apprenticeship thrive in the United States? *Educational Researcher* 22(3):4-10.

Bailey, T., and C. Belfield. 2013. Community college occupational degrees: Are they worth it? In *Preparing today's students for tomorrow's jobs in metropolitan America*, edited by L. W. Perna. Philadelphia, PA: University of Pennsylvania Press. Pp. 121-148.

Bailey, M., and S. Dynarski. 2011. Inequality in postsecondary education. In *Whither opportunity? Rising inequality and the uncertain life of low-income children*, edited by G. Duncan and R. Murnane. New York: Russell Sage Foundation. Pp. 117-132.

Bailey, T., and S. Jaggars. 2016. *When college students start behind.* New York: The Century Foundation. https://tcf.org/content/report/college-students-start-behind (accessed March 22, 2017).

Bailey, T., D. Jeong, and S. Cho. 2010. Referral, enrollment, and completion in developmental education sequences in community colleges. *Economics of Education Review* 29(2):255-270.

Bailey, T., S. Jaggers, and D. Jenkins. 2015. *Redesigning America's community colleges: A clearer path to student success.* Cambridge, MA: The President and Fellows of Harvard College.

Barnett, E., E. Maclutsky, and C. Wagonlander. 2015. Emerging early college models for traditionally underserved students. *New Directions for Community Colleges* 2015(169):39-49.

Baum, S., and J. Ma. 2013. *Trends in college pricing 2013.* Trends in Higher Education Series. College Board. http://trends.collegeboard.org/sites/default/files/college-pricing-2013-full-report.pdf (accessed March 5, 2017).

Baum, S., C. Kurose, and M. McPherson. 2013. An overview of American higher education. *The Future of Children* 23(1):17-40.

BEA (Bureau of Economic Analysis). 2015. *Broad growth across states in 2014.* https://www.bea.gov/newsreleases/regional/gdp_state/gsp_news release.htm (accessed February 28, 2017).

Becker, G. S. 1964. *Human capital: A theoretical and empirical analysis, with special reference to education* (1st ed.). New York: National Bureau of Economic Research.

Becker, G. S. 1993. *Human capital: A theoretical and empirical analysis, with special reference to education* (3rd ed.). Chicago, IL: University of Chicago Press.

Belfield, C., and T. Bailey. 2011. The benefits of attending community college: A review of the evidence. *Community College Review* 39(1):46-68.

Bettinger, E., A. Boatman, and B. Long. 2013. Student supports: Developmental education and other academic programs. *The Future of Children* 23(1):93-115.

Bielick, S., S. Cronen, C. Stone, J. Montaquila, and S. Roth. 2013. *The Adult Training and Education Survey (ATES) pilot study: Technical report.* NCES 2013-190. Washington, DC: NCES.

Blosveren, K. 2015. *CTE and middle skills.* Prepared for the 3rd Committee Meeting on the Supply Chain for Middle-Skill Jobs, NASEM, September 10. http://nas.edu/SkilledTechnicalWorkforce (accessed March 2, 2017).

BLS (U.S. Bureau of Labor Statistics). 2005. *Contingent and alternative employment arrangements.* https://www.bls.gov/cps/lfcharacteristics.htm#contingent (accessed March 5, 2017).

BLS. 2013. *Employment projections—2012-2022.* News release. https://www.bls.gov/news.release/archives/ecopro_12192013.pdf (accessed February 28, 2017).

BLS. 2015. *Employment projections—2014-2024.* News release. https://www.bls.gov/news.release/pdf/ecopro.pdf (accessed February 28, 2017).

BLS. 2016a. *College enrollment and work activity of 2015 high school graduates.* Economic news release. https://www.bls.gov/news.release/hsgec.nr0.htm (accessed March 5, 2017).

BLS. 2016b. *Labor force statistics from the current population survey: Data on certifications and licenses.* http://www.bls.gov/cps/certifications-and-licenses.htm (accessed March 19, 2017).

Bolli, T., and S. Hof. 2014. *The impact of apprenticeship training on personality traits: An instrumental variable approach.* KOF Working Paper 350. Zürich, Switzerland: KOF Centre de Recherches Conjoncturelles. https://wwz.unibas.ch/fileadmin/wwz/redaktion/dekanat/Economicslunch/KOF_wp_350-2_01.pdf (accessed March 19, 2017).

Bolli, T., and U. Renold. 2015. *Comparative advantages of school and workplace environment in competence acquisition, empirical evidence from a survey among professional tertiary education and training students in Switzerland.* KOF Working Paper 389. Zürich, Switzerland: KOF Centre de Recherches Conjoncturelles. https://e-collection.library.ethz.ch/view/eth:47900 (accessed March 21, 2017).

Bound, J., M. Lovenheim, and S. Turner. 2010. Why have college completion rates declined? An analysis of changing student preparation and collegiate resources. *American Economic Journal: Applied Economics* 2(3):129-157.

Bronson, E. 2007. Helping CTE students l(earn) to their potential. *Techniques* 82(7):30-31.

Brynjolfsson, E., and A McAfee. 2011. *Race against the machine: How the digital revolution is accelerating innovation, driving productivity, and irreversibly transforming employment and the economy.* Lexington, MA: Digital Frontier Press.

Brynjolfsson, E., and A. McAfee. 2014. *The second machine age: Work, progress, and prosperity in a time of brilliant technologies.* New York: W. W. Norton & Company.

Bunker Hill Community College. 2017. *The Mary L. Fifield Endowed Student Emergency Assistance Fund.* http://www.bhcc.mass.edu/emergency assistancefund (accessed March 22, 2017).

Burning Glass. 2014. *Moving the goal posts: How demand for a bachelor's degree is reshaping the workforce.* http://burning-glass.com/research/ credentials-gap (accessed February 28, 2017).

Burning Glass. 2017. *Labor insight.* http://burning-glass.com/labor-insight (accessed March 22, 2017).

Capelli, P. 2004. Why do employers pay for college? *Journal of Econometrics* 121(1-2):213-241.

Capelli, P. 2012. *Why good people can't get jobs: The skills gap and what companies can do about it.* Philadelphia, PA: Wharton Digital Press.

Capelli, P. 2015. Skill gaps, skill shortages and skill mismatches: Evidence and arguments for the United States. *ILR Review* 68(2):251-290.

Carnevale, A., and A. Hanson. 2015. Learn and earn: Connecting education to careers in the 21st century. In *Transforming U.S. workforce development policies for the 21st century*, edited by C. Van Horn, T. Edwards, and T. Greene. Kalamazoo, MI: W.E. Upjohn Institute for Employment Research. Pp. 77-104.

Carnevale, A., S. Rose, and A. Hanson. 2012. *Certificates: Gateway to gainful employment and college degrees.* Washington, DC: Georgetown Center on Education and the Workforce.

Carnevale, A., T. Jayasundera, and D. Repnikov. 2014. *Understanding online job ads data.* Washington, DC: Center on Education and the Workforce, McCourt School of Public Policy, Georgetown University.

Carnevale, A., T. Jayasundera, D. Repnikov, and A. Gulish, A. 2015. *State online college job market: Ranking the states.* Washington, DC: Center on Education and the Workforce, McCourt School of Public Policy, Georgetown University.

Carter, S. P., and B. J. Miller. 2015. *Analysis of Army veteran unemployment benefits and transition assistance.* Prepared for the Committee on the Supply Chain for Middle-Skill Jobs, NASEM. http://nas.edu/Skilled TechnicalWorkforce (accessed March 2, 2017).

Castellano, M., K. Sundell, L. Overman, and O. Aliaga. 2012. Do career and technical education programs of study improve student achievement? Preliminary analyses from a rigorous longitudinal study. *International Journal of Educational Reform* 21(2):98-118.

Cellini, S. R. 2012. For-profit higher education: An assessment of costs and benefits. *National Tax Journal* 65(1):153-180.

Choitz, V., T. Norman, W. Smith, N. Speiser, and B. Paulson. 2015. A new way of doing business: The career pathway approach in Minnesota and beyond. In *Transforming U.S. workforce development policies for the 21st century*, edited by C. Van Horn, T. Edwards, and T. Greene. Kalamazoo, MI: W. E. Upjohn Institute for Employment Research. Pp. 265-283.

Clark, P., C. Dayton, D. Stern, S. Tidyman, and A. Weisberg. 2007. *Can combining academic and career-technical education improve high school outcomes in California?* Santa Barbara, CA: California Dropout Research Project.

CLASP. 2014. *Shared vision, strong systems: The Alliance for Quality Career Pathways framework version 1.0 summary.* http://www.clasp.org/resources-and-publications/publication-1/Alliance-WIOA-CP-Summary.pdf (accessed March 22, 2017).

Coalition for Evidence-Based Policy. n.d. *Career academies—top tier.* http://evidencebasedprograms.org/1366-2/career-academies (accessed March 22, 2017).

Colborn, J., and N. Jenkins. 2016. Bison Gear. In *Models of success: An archive of case studies done under Skills for America's Future highlighting effective partnerships that benefit community colleges, businesses, and students.* Washington, DC: The Aspen Institute. https://assets.aspeninstitute.org/content/uploads/2016/07/Models-of-Success-Archive2.pdf (accessed March 22, 2017).

Commonwealth Corporation. 2014. *Annual Report for Fiscal Year 2014.* http://commcorp.org/wp-content/uploads/2016/07/resources_2015-annual-report-fy-2014.pdf (accessed March 21, 2017).

Competency Model Clearing House. 2015. *User guide for developing competency models.* https://www.careeronestop.org/CompetencyModel/cmuserguide.aspx (accessed February 28, 2017).

Complete College America. 2017. *Resources and reports.* http://completecollege.org/resources_and_reports (accessed March 5, 2017).

Cornes, R., and T. Sandler. 1996. *The theory of externalities, public goods, and club goods.* Cambridge, UK: Cambridge University Press.

Courty, P., and G. Marschke. 2011. Measuring government performance: An overview of dysfunctional responses. In *The performance of performance standards*, edited by J. Heckman, C. Heinrich, P. Courty, G. Marschke, and J. Smith. Kalamazoo, MI: W. E. Upjohn Institute for Employment Research. Pp. 203-230.

Crawford, S., and R. Sheets. 2015. Creating and communicating critical information about workforce credentials. In *Transforming U.S. workforce development policies for the 21st century*, edited by C. Van Horn, T. Edwards, and T. Greene. Kalamazoo, MI: W. E. Upjohn Institute for Employment Research. Pp. 169-194.

Dadgar, M., and M.J. Weiss. 2012. *Labor market returns to sub-baccalaureate credentials: How much does a community college degree or certificate pay?* CCRC Working Paper 45. New York: Community College Research Center, Teachers College, Columbia University.

Davis, G., and E. Ostrom. 1991. A public economy approach to education: Choice and co-production. *International Political Science Review* 12(4):313-345.

Davis, S., L. Jacobson, and S. Wandner. 2014. *Using Workforce Data Quality Intiative databases to develop and improve consumer report card systems.* Columbia, MD: IMPAQ International.

Decker, P., and J. Berk. 2011. Ten years of the Workforce Investment Act (WIA): Interpreting the research on WIA and related programs. *Journal of Policy Analysis and Management* 30(4):906-926.

Deloitte and NMI (National Manufacturing Institute). 2015. *The skills gap in U.S. manufacturing 2015 and beyond.* http://www.themanufacturing institute.org/~/media/827DBC76533942679A15EF7067A704CD.ashx (accessed March 5, 2017).

Deming, D., C. Goldin, and L. Katz. 2013. For-profit colleges. *The Future of Children* 23(1):137-163.

DoD (U.S. Department of Defense). 2015. *2014 demographics profile of the military community.* http://download.militaryonesource.mil/12038/MOS/Reports/2014-Demographics-Report.pdf (accessed March 5, 2017).

DoD. 2016. *Current strength.* https://www.dmdc.osd.mil/appj/dwp/dwp_reports.jsp (accessed March 5, 2017).

Dokko, J., M. Mumford, and D. Schanzenbach. 2015. *Workers and the online gig economy.* Washington, DC: The Hamilton Project. http://www.hamiltonproject.org/assets/files/workers_and_the_online_gig_economy.pdf (accessed March 2, 2017).

DOL (U.S. Department of Labor). 2012. *United Services Military Apprenticeship Program (USMAP).* https://doleta.gov/OA/usmap.cfm (accessed March 5, 2017).

DOL. n.d. *Apprenticeship.* http://www.dol.gov/apprenticeship (accessed March 19, 2017).

DOL, U.S. Department of Commerce, DOE (U.S. Department of Education), and HHS (U.S. Department of Health and Human Services). 2014. *What works in job training: A synthesis of the evidence.* https://www.dol.gov/asp/evaluation/jdt/jdt.pdf (accessed March 5, 2017).

Dortch, C. 2012. *Carl D. Perkins Career and Technical Education Act of 2006: Implementation issues.* Washington, DC: Congressional Research Service.

Dortch, C. 2014a. *Career and Technical Education (CTE): A primer.* Washington, DC: Congressional Research Service.

Dortch, C. 2014b. *The Post-9/11 Veterans Educational Assistance Act of 2008 (Post-9/11 GI Bill): Primer and issues.* Washington, DC: Congressional Research Service.

Dynarski, S. 2015 How to improve graduation rates at community colleges. *The New York Times*, March 11. https://www.nytimes.com/2015/03/12/up shot/how-to-improve-graduation-rates-at-community-colleges.html (accessed April 6, 2017).

Education News. 2013. American public education: An origin story. *Education News*, April 16. http://www.educationnews.org/education-policy-and-politics/american-public-education-an-origin-story (accessed March 19, 2017).

Eggenberger, C., M. Rinawi, and U. Backes-Gellner. 2015. *Measuring the specificity of occupational training curricula and labor market flexibility: An economic perspective on the curriculum development of VET occupations*. Working Paper 106. http://repec.business.uzh.ch/RePEc/iso/leadinghouse/0106_lhwpaper.pdf (accessed March 19, 2017).

Elliott, M., and A. Roder. Forthcoming. New York: Economic Mobility Corporation, Inc.

Eversole, E. 2016. *Hiring our heroes: The private sector role.* Prepared for the Committee on the Supply Chain for Middle-Skill Jobs, NASEM. http://nas.edu/SkilledTechnicalWorkforce (accessed March 5, 2017).

Ewert, S., and R. Kominski. 2014. *Measuring alternative educational credentials: 2012.* https://www.census.gov/prod/2014pubs/p70-138.pdf (accessed March 19, 2017).

Faberman, R., and B. Mazumder. 2012. *Is there a skills mismatch in the labor market?* Chicago, IL: The Federal Reserve Bank of Chicago.

Ferm, A. 2008. *Framtidsvägen—en reformerad gymnasieskola [Road to the future—A reformed upper decondary school].* http://www.regeringen.se/49b71c/contentassets/4fbcae10413d46db91309e95d8a57a66/framtidsvagen---en-reformerad-gymnasieskola-hela-dokumentet-sou-200827 (accessed March 19, 2017).

Fichen, A., and M. Pellizzari. 2014. *A new measure of skills mismatch: Theory and evidence from the Survey of Adult Skills (PIACC).* Discussion Paper DP10280. Washington, DC: Centre for Economic Policy Research. http://cepr.org/active/publications/discussion_papers/dp.php?dpno=10280 (accessed February 28, 2017).

Frey, B. 1997. *Not just for the money: An economic theory of personal motivation.* Cheltenham, UK: Edward Elgar Publishing.

Frey, B., and M. Osborne. 2013. *The future of employment: How susceptible are jobs to computerization?* Working Paper. Oxford, UK: Oxford Martin School, University of Oxford. http://www.oxfordmartin.ox.ac.uk/publications/view/1314 (accessed February 28, 2017).

Friedman, M., and S. Kuznets. 1954. *Income from independent professional practice.* New York: National Bureau of Economic Research.

Frogner, B. K., and S. K. Skillman. 2015. *Pathways to middle-skill allied health care occupations.* Prepared for the Committee on the Supply Chain for Middle-Skill Jobs, NASEM. http://nas.edu/SkilledTechnicalWorkforce (accessed February 28, 2017).

Frogner, B. K., and S. K. Skillman. 2016. Pathways to middle-skill allied health care occupations. *Issues in Science and Technology* XXXIII(1).

http://issues.org/33-1/pathways-to-middle-skill-allied-health-care-occupations (accessed April 5, 2017).

Frogner, B. K., J. Spetz, S. Oberlin, and S. Parente. 2015. The demand for healthcare workers post-ACA. *International Journal of Health Economics and Management* 15(1):139-151.

GAO (Government Accountability Office). 2011a. *Employment and training programs: Opportunities exist for improving efficiency*. GAO 11-506T. Testimony of Andrew Sherrill, director of education, workforce and income security issues before the Subcommittee on Labor, Health and Human Services, Education and Related Agencies, Committee on Appropriations, House of Representatives. Washington, DC, April 7.

GAO. 2011b. *Multiple employment and training programs: Providing information on co-locating services and consolidating administrative structures could promote efficiencies.* GAO-11-92. Washington, DC: GAO.

GAO. 2011c. *Opportunities to reduce potential duplication in government programs, save tax dollars, and enhance revenue.* GAO-11-318SP. Washington, DC: GAO.

GAO. 2012. *Workforce Investment Act: Innovative collaborations between workforce boards and employers helped meet local needs*. GAO-12-97. Washington, DC: GAO.

GAO. 2013. *Workforce Investment Act: Local areas face challenges helping employers fill some types of skilled jobs*. GAO-14-19. Washington, DC: GAO.

Giffi, C., J. McNelly, B. Dollar, G. Carrick, M. Drew, and B. Gangula. 2015. *The skills gap in U.S. manufacturing: 2015 and beyond.* Washington, DC: The Manufacturing Institute.

Good, L., and E. Strong. 2015. Reimagining workforce policy in the United States. In *Transforming workforce development policies for the 21st century*, edited by C. Van Horn, T. Edwards, and T. Greene. Kalamazoo, MI: W. E. Upjohn Institute for Employment Research. Pp. 13-43.

Great Schools Partnership. 2014. Hidden curriculum. In *The glossary of education reform,* edited by S. Abbott. http://edglossary.org/hidden-curriculum (accessed March 22, 2017).

Greenhouse, S. 2013. The great aid gap. *The New York Times*, March 19, p. F1.

Hegji, A. 2014. *The Higher Education Act (HEA): A primer.* Washington, DC: Congressional Research Service.

Heidkamp, M., and T. Hilliard. 2014. *A review of community college-employer partnerships and initiatives: Expanding opportunities for job seekers with disabilities.* http://askearn.org/docs/AskEARN-CollegeEmployerPartnerships.pdf (accessed February 10, 2016).

Heinrich, C. J. 2015. *Federally-funded workforce development.* Prepared for the Committee on the Supply Chain for Middle-Skill Jobs, NASEM. http://nas.edu/SkilledTechnicalWorkforce (accessed March 2, 2017).

Hendra, R., D. Greenberg, G. Hamilton, A. Oppenheim, A. Pennington, K. Schaberg, and B. Tessler. 2016. *Encouraging evidence on a sector-focused advancement strategy: Two-year impacts from the WorkAdvance demonstration.* New York: MDRC. http://www.mdrc.org/sites/default/files/2016_Workadvance_Final_Web.pdf (accessed March 22, 2017).

Herian, M. 2010. *Examining public perceptions of career and technical education in Nebraska.* Paper 35. Publications of the University of Nebraska Public Policy Center. http://digitalcommons.unl.edu/publicpolicypublications/35 (accessed March 22, 2017).

Hoffman, L., and T. Reindl. 2011. *Complete to compete: Improving postsecondary attainment among adults.* Washington, DC: National Governors Association Center for Best Practices. https://www.nga.org/files/live/sites/NGA/files/pdf/1102postsecondaryattainment.pdf (accessed March 5, 2017).

Hollenbeck, K. 1997. School-to-work: Promise and effectiveness. *Employment Research* 4(2):5-7.

Holzer, H. J. 2012. *Going, going...gone? The evolution of workforce development programs for the poor since the War on Poverty.* Washington, DC: Georgetown University and American Institutes for Research. http://npc.umich.edu/news/events/war-on-poverty-june-conference/holzer.pdf (accessed March 2, 2017).

Holzer, H. J. 2015a. *Higher education and workforce policy: Creating more skilled workers (and jobs for them to fill).* Washington, DC: Brookings Institution. https://www.brookings.edu/wp-content/uploads/2016/06/higher_ed_jobs_policy_holzer.pdf (accessed March 22, 2017).

Holzer, H. J. 2015b. *Sector-based training strategies: The challenges of matching workers and their skills to well-paying jobs.* https://www.dol.gov/asp/evaluation/completed-studies/Future_of_work_sector_based_training_strategies.pdf (accessed March 22, 2017).

Holzer, H. J., and R. I. Lerman. 2007. *America's forgotten middle-skill jobs: Education and training requirements in the next decade and beyond.* Washington, DC: Urban Institute. http://www.urban.org/research/publication/americas-forgotten-middle-skill-jobs (accessed February 28, 2017).

Holzer, H. J., and R. I. Lerman. 2009. *The future of middle-skill jobs.* CCF Brief 41. Washington, DC: Brookings Institution. https://www.brookings.edu/research/the-future-of-middle-skill-jobs (accessed February 28, 2017).

Holzer, H. J., and D. Nightingale. 2009. *Strong students, strong workers: Models for student success through workforce development and community college partnerships.* Washington, DC: Center for American Progress.

Horowitz, S., and F. Rosati. 2014. *Freelancing in America: A national survey of the new workforce.* Freelancers Union and Elance-oDesk.

https://www.upwork.com/press/2014/09/03/53-million-americans-now-freelance-new-study-finds-2 (accessed March 5, 2017).

Howington, H., T. Hartfield, and C. Hillyard, C. 2015. Faculty viewpoints on teaching Quantway®. *Numeracy* 8(1).

Independent Advisory Panel of the National Assessment of Career and Technical Education. 2014. *Putting "career" in "college and career ready": The report of the Independent Advisory Panel of the National Assessment of Career and Technical Education.* https://www2.ed.gov/rschstat/eval/sectech/nacte/career-technical-education/independent-advisory-panel-report.pdf (accessed March 5, 2017).

IOM (Institute of Medicine). 1989. *Allied health services: Avoiding crisis.* Washington, DC: National Academy Press.

IOM. 2011. *Allied health workforce and services-workshop summary.* Washington, DC: The National Academies Press.

Jaimovich, N., and H. Siu. 2012. *The trend is the cycle: Job polarization and jobless recoveries.* Working paper 18334. Cambridge, MA: National Bureau of Economic Research. http://www.nber.org/papers/w18334.pdf (accessed February 28, 2017).

Jenkins, D., and J. Fink. 2015. *What we know about transfer*. New York: Columbia University, Teachers College, Community College Research Center. http://ccrc.tc.columbia.edu/media/k2/attachments/what-we-know-about-transfer.pdf (accessed March 19, 2017).

Jenkins, D., and C. Spence. 2006. *The career pathways how-to guide.* Brooklyn, NY: Workforce Strategy Center. http://files.eric.ed.gov/fulltext/ED496995.pdf (accessed March 22, 2017).

Jepsen, C., K. Troske, and P. Coomes. 2012. *The labor-market returns to community college degrees, diplomas, and certificates.* IZA Discussion Paper 6902. http://ftp.iza.org/dp6902.pdf (accessed March 5, 2017).

Jobs for the Future. 2014. *The Pathways to Prosperity Network: A state progress report, 2012-2014.* http://www.jff.org/publications/pathways-prosperity-network-state-progress-report-2012-2014 (accessed March 22, 2017).

Joyce Foundation. 2011. *Business leaders united for workforce partnerships.* http://www.joycefdn.org/news/business-leaders-united-for-workforce-partnerships (accessed March 22, 2017).

Kaine, T. 2015. *Kaine introduces JOBS Act to expand Pell grants for workforce training programs.* http://www.kaine.senate.gov/press-releases/kaine-introduces-jobs-act-to-expand-pell-grants-for-workforce-training-programs (accessed March 2, 2017).

Kaiser Permanente. n.d. *Labor management partnership.* http://www.kaiserpermanentejobs.org/labor-management-partnership.aspx (accessed March 19, 2017).

Kane, T., and C. Rouse. 1995. Labor-market returns to two- and four-year college. *The American Economic Review* 85(3):600-614.

Karp, M. 2011. *Toward a new understanding of non-academic student support: Four mechanisms encouraging positive student outcomes in the community college.* CCRC Working Paper 28. New York: Community College Research Center, Teachers College, Columbia University. http://ccrc.tc.columbia.edu/media/k2/attachments/new-understanding-non-academic-support.pdf (accessed March 5, 2017).

Karp, M. 2015a. *Community college pathways.* Prepared for the Committee on the Supply Chain for Middle-Skill Jobs, NASEM. http://nas.edu/SkilledTechnicalWorkforce (accessed March 5, 2017).

Karp, M. 2015b. Dual enrollment, structural reform, and the completion agenda. *New Directions for Community Colleges* 2015(169):103-111.

Karp, M., L. O'Gara, and K. Hughes. 2008. *Do support services at community colleges encourage success or reproduce disadvantage?* CCRC Working Paper 10. New York: Community College Research Center, Teachers College, Columbia University. http://ccrc.tc.columbia.edu/media/k2/attachments/support-services-reproduce-disadvantage.pdf (accessed March 5, 2017).

Kearney, A.T. 2015. *U.S. reshoring index.* https://www.atkearney.com/operations/ideas-insights/manufacturing-reshoring (accessed February 28, 2017).

Kelchen, R., and L. Stedrak. 2016. Does performance-based funding affect colleges' financial priorities? *Journal of Education Finance* 41(3):302-321.

Kemple, J. 2008. *Career academies: Long-term impacts on labor market outcomes, educational attainment, and transitions to adulthood.* New York: MDRC.

Kimmel, J., and S. Martin. 2015. *Sorting out the skills gap: Analyzing the evidence for a shortage of middle-skill workers in the manufacturing and healthcare industries in the Portland region.* Paper 123. Portland, OR: Institute of Portland Metropolitan Studies Publications.

King, C., and H. Prince. 2015. Moving sectoral and career pathway programs from promise to scale. In *Transforming U.S. workforce development polices for the 21st century*, edited by C. Van Horn, T. Edwards, and T. Greene. Kalamazoo, MI: W. E. Upjohn Institute for Employment Research. Pp. 195-230.

Klein, S., and K. Green. 2011. College for all: The American model for career and technical education. In *Work and Education in America*, edited by A. Barabasch and F. Rauner. *Technical and Vocational Education and Training: Issues, Concerns and Prospects* 15:123-140.

Kleiner, M. M. 2000. Occupational licensing. *Journal of Economic Perspectives* 14(4):189-202.

Kleiner, M. M. 2006. Licensing occupations: Ensuring quality or restriction competition? In *W.E. Upjohn Institute for Employment Research.* Kalamazoo, MI: Upjohn Institute Press. Pp. 1-15.

Kleiner, M. M., and A. Krueger. 2010. The prevalence and effects of occupational licensing. *British Journal of Industrial Relations* 48(4):676-687.

Krueger, C. 2006. *Dual enrollment: Policy issues confronting state policymakers.* Policy Brief. Denver, CO: Education Commission of the States.

Law, M., and S. Kim. 2005. Specialization and regulation: The rise of professionals and the emergence of occupational licensing regulation. *Journal of Economic History* 65(3):723-756.

Lemieux, T. 2006. Increasing residual wage inequality: Composition effects, noisy data, or rising demand for skill? *The American Economic Review* 96(3):461-498.

Lerman, R. 2010. *Expanding apprenticeship: A way to enhance skills and careers.* Washington, DC: Urban Institute. http://www.urban.org/research/publication/expanding-apprenticeship-way-enhance-skills-and-careers (accessed March 22, 2017).

Lerman, R. 2015. *Are employers providing enough training? Theory, evidence, and policy implications.* Prepared for the Committee on the Supply Chain for Middle-Skill Jobs, NASEM. http://nas.edu/SkilledTechnicalWorkforce (accessed February 28, 2017).

Lerman, R., and V. Rein. 2015. *Building a robust U.S. work-based education and apprenticeship system at scale: Can lessons from Europe help?* http://www.aicgs.org/publication/building-a-robust-u-s-work-based-education-and-apprenticeship-system-at-scale (accessed March 19, 2017).

Lerman, R., D. Hanson, C. Dobson, M. Tanamor, and L. Blatt. 2015. *The United Services Military Apprenticeship Program (USMAP): Implementation study and feasibility of an impact study.* Washington, DC: Urban Institute.

Levy, F., and R. J. Murnane. 2013. *Dancing with robots: Human skills for computerized work.* Washington, DC: Third Way. https://dusp.mit.edu/uis/publication/dancing-robots-human-skills-computerized-work (accessed February 28, 2017).

Long, M., D. Goldhaber, and N. Huntington-Klein. 2014. *Do students' college major choices respond to changes in wages?* Working Paper 2014-6. Seattle, WA: Center for Educational Data and Research. http://www.cedr.us/papers/working/CEDR%20WP%202014-6.pdf (accessed February 8, 2016).

Lovenheim, M. 2011. The effect of liquid housing wealth on college enrollment. *Journal of Labor Economics* 29(4):741-771.

Lumina Foundation. 2015. *Connecting credentials: A beta credentials framework: Building a system for communicating about and connecting diverse credentials.* Indianapolis, IN: Lumina Foundation. https://www.luminafoundation.org/resources/connecting-credentials (accessed March 5, 2017).

Maguire, S., J. Freely, C. Clymer, M. Conway, and D. Schwartz. 2010. *Tuning in to local labor markets: Findings from the sectoral employment impact study.* Philadelphia, PA: Public/Private Ventures. http://www.aspenwsi.org/wordpress/wp-content/uploads/TuningIntoLocalLaborMarkets.pdf (accessed March 22, 2017).

Marcus, J. 2013. Community college grads out-earn bachelor's degree holders. *CNN Money*, February 26. http://money.cnn.com/2013/02/26/pf/college/community-college-earnings (accessed February 28, 2017).

Martin, V., and J. Broadus. 2013. *Enhancing GED instruction to prepare students for college and careers: Early success in LaGuardia Community College's Bridge to Health and Business program.* New York: MDRC. http://www.mdrc.org/publication/enhancing-ged-instruction-prepare-students-college-and-careers (accessed March 22, 2017).

McCarthy, M. A. 2014. *Beyond the skills gap: Making education work for students, employers, and communities.* Washington, DC: New America Foundation. https://www.newamerica.org/education-policy/policy-papers/beyond-the-skills-gap (accessed February 28, 2017).

McKinsey & Company. 2017. *A future that works: automation, employment, and productivity.* http://www.mckinsey.com/global-themes/digital-disruption/harnessing-automation-for-a-future-that-works (accessed March 29, 2017).

Mellow, G. O., and C. M. Heelan. 2015. *Minding the dream: The process and practice of the American community college.* Lanham, MD: Rowman and Littlefield.

Messing-Mathie, A. 2015. *Building apprenticeship systems for middle-skill employment: Comparative lessons in innovation and sector-based strategies for apprenticeships.* Prepared for the Committee on the Supply Chain for Middle-Skill Jobs, NASEM. http://nas.edu/SkilledTechnicalWorkforce (accessed March 19, 2017).

Miao, K. 2012. *Performance-based funding of higher education: A detailed look at best practices in 6 states.* Washington, DC: Center for American Progress.

Miller, L. 2014. *2014 state of the industry report: Spending on employee training remains a priority.* https://www.td.org/Publications/Magazines/TD/TD-Archive/2014/11/2014-State-of-the-Industry-Report-Spending-on-Employee-Training-Remains-a-Priority (accessed March 19, 2017).

Modestino, A. S. 2010. *Mismatch in the labor market: Measuring the supply of and demand for skilled labor in New England.* New England Public Policy Center Research Report 10-2. Boston, MA: Federal Reserve Bank of Boston.

Modestino, A. S. 2015. *Middle skill workers in today's job market.* Prepared for the Committee on the Supply Chain for Middle-Skill Jobs, NASEM. http://nas.edu/SkilledTechnicalWorkforce (accessed February 28, 2017).

Modestino, A. S. 2016. The importance of middle-skill jobs. *Issues in Science and Technology* XXXIII(1).

Mokher, C., L. Jacobson, C. Ahearn, J. Atkin, M. Flory, J. Lansing, J. Rosenbaum, and C. Sun. 2014. *Assessment of the Florida College and Career Readiness Initiative: Year 2 report.* Arlington, VA: CNA Corporation. http://files.eric.ed.gov/fulltext/ED555558.pdf (accessed March 21, 2017).

Muehlemann, S., and S. C. Wolter. 2014. Return on investment of apprenticeship systems for enterprises: Evidence from cost-benefit analyses. *IZA Journal of Labor Policy* 3:25.

NAE (National Academy of Engineering). 2015. *Making value for America: Embracing the future of manufacturing, technology, and work.* Washington, DC: The National Academies Press.

NAE and NRC (National Research Council). 2012. *Community colleges in the evolving STEM education landscape: Summary of a summit.* Washington, DC: The National Academies Press.

NAEP (National Assessment of Educational Progress). n.d. *The Nation's Report Card.* http://www.nationsreportcard.gov (accessed March 5, 2017).

NAF (National Academy Foundation). 2015. *Annual report 2014-2015.* https://mis.naf.org/annualreport/2015 (accessed March 22, 2017).

NASDCTEc (National Association of State Directors of Career Technical Educational Consortium), ACTE (Association for Career and Technical Education), and CTE (Career and Technical Education). 2015. *State policies impacting CTE: 2014 year in review.* https://www.acteonline.org/uploadedFiles/Resources/Publications/2014_State_Policy_Review_FINAL.pdf (accessed February 28, 2017).

NASEM (National Academies of Sciences, Engineering, and Medicine). 2015. *The supply chain for middle-skill jobs: Education, training, and certification pathways.* Public symposium. http://nas.edu/SkilledTechnicalWorkforce (accessed March 29, 2017).

NASEM. 2016a. *Barriers and opportunities for 2-year and 4-year STEM degrees: Systemic change to support students' diverse pathways.* Washington, DC: The National Academies Press.

NASEM. 2016b. *Developing a national STEM workforce strategy: A workshop summary.* Washington, DC: The National Academies Press.

NASEM. 2016c. *Promising practices for strengthening the regional STEM workforce development ecosystem.* Washington, DC: The National Academies Press.

NCEE (National Center on Education and the Economy). 2017. *NCSL Education Benchmarking Study Group.* http://ncee.org/what-we-do/center-on-international-education-benchmarking/cieb-supported-research/ncsl-education-benchmarking-study-group (accessed March 22, 2017).

NCES (National Center for Education Statistics). 2007. *The condition of education 2007.* NCES 2007-064. https://nces.ed.gov/pubs2007/2007064.pdf (accessed March 29, 2017).

NCES. 2013a. *Characteristics of private schools in the United States: Results from the 2011-12 Private School Universe Survey*. NCES 2013-316. Washington, DC: U.S. Department of Education. https://nces.ed.gov/pubs2013/2013316.pdf (accessed April 5, 2017).

NCES. 2013b. *DATA POINT: Trends in CTE coursetaking.* NCES 2014-901. http://nces.ed.gov/fastfacts/display.asp?id=43 (accessed March 19, 2017).

NCES. 2013c. *Table 106.50. Direct general expenditures per capita of state and local governments for all functions and for education, by level of education and state: 2009-10 and 2010-11*. https://nces.ed.gov/programs/digest/d13/tables/dt13_106.50.asp (accessed March 19, 2017).

NCES. 2014a. *Table 303.40. Total fall enrollment in degree-granting postsecondary institutions, by attendance status, sex, and age: Selected years, 1970 through 2024.* https://nces.ed.gov/programs/digest/d14/tables/dt14_303.40.asp?current=yes (accessed March 19, 2017).

NCES. 2014b. *Table 507.30. Participation of employed persons, 17 years old and over, in career-related adult education during the previous 12 months, by selected characteristics of participants: 1995, 1999, and 2005.* https://nces.ed.gov/programs/digest/d14/tables/dt14_507.30.asp (accessed March 19, 2017).

NCES. 2015a. *Selected statistics from the public elementary and secondary education universe: School year 2013-14*. NCES 2015-151. Washington, DC: U.S. Department of Education.

NCES. 2015b. *Table 333.10. Revenues of public degree-granting postsecondary institutions, by source of revenue and level of institution: 2007-08 through 2013-14.* https://nces.ed.gov/programs/digest/d15/tables/dt15_333.10.asp?current=yes (accessed April 5, 2017).

NCES. 2015c. *The condition of education 2015*. NCES 2015-144. https://nces.ed.gov/pubs2015/2015144.pdf (accessed March 5, 2017).

NCES. 2016a. *Remedial coursetaking at U.S. public 2- and 4-year institutions: Scope, experiences, and outcomes*. NCES 2016-405. Washington, DC: U.S. Department of Education.

NCES. 2016b. *The condition of education: Postsecondary institution revenues*. https://nces.ed.gov/programs/coe/indicator_cud.asp (accessed March 5, 2017).

NCES. 2016c. *The condition of education 2016*. NCES 2016-144. Washington, DC: U.S. Department of Education. http://files.eric.ed.gov/fulltext/ED565888.pdf (accessed March 5, 2017).

NCES. n.d. *About GEMEnA*. https://nces.ed.gov/surveys/gemena (accessed March 19, 2017).

NCSL (National Conference of State Legislatures). 2015. *Performance-based funding for higher education.* http://www.ncsl.org/research/education/performance-funding.aspx (accessed March 21, 2017).

NCSL. n.d. *Hot topics in higher education: Reforming remedial education.* http://www.ncsl.org/research/education/improving-college-completion-reforming-remedial.aspx (accessed March 31, 2017).

Neumark, D. 2004. *The effects of school-to-career programs on postsecondary enrollment and employment.* San Francisco, CA: Public Policy Institute of California.

Neumark, D., and D. Rothstein. 2005. *Do school-to-work programs help the "forgotten half"?* Working Paper 11636. Cambridge, MA: National Bureau of Economic Research. http://www.nber.org/papers/w11636.pdf (accessed on February 9, 2016).

New Jersey Department of Education. 2014. *Online consumer report card.* http://www.nj.gov/education/cte/ppcs/students/OnliConRC.shtml (accessed March 21, 2017).

NGA (National Governors Association). 2013. *State sector strategies: Coming of age: Implications for state workforce policymakers.* https://www.nga.org/files/live/sites/NGA/files/pdf/2013/1301NGASSSReport.pdf (accessed March 21, 2017).

NGA. 2015a. *2013-14 NGA chair Gov. Mary Fallin's initiative.* https://www.nga.org/cms/home/about/initiatives-of-nga-chairs/col2-content/main-content-list/2013-14-nga-chair-gov-mary-falli.html (accessed March 2, 2017).

NGA. 2015b. *Postsecondary credential attainment by adult workers.* http://www.nga.org/cms/home/nga-center-for-best-practices/center-divisions/center-issues/page-ehsw-issues/col2-content/main-content-list/postsecondary-credential-attainm.html (accessed March 2, 2017).

NGA. 2017. *Initiatives of NGA chairs: List of NGA chairs.* www.nga.org/cms/chairs (accessed March 2, 2017).

North, D. C. 1990. *Institutions, institutional change, and economic performance.* Cambridge, UK: Cambridge University Press.

NRC (National Research Council). 2012. *Education for life and work: Developing transferable knowledge and skills in the 21st century.* Washington, DC: The National Academies Press.

NRC. 2014a. *STEM integration in K-12 education: Status, prospects, and an agenda for research.* Washington, DC: The National Academies Press.

NRC. 2014b. *The flexible electronics opportunity.* Washington, DC: The National Academies Press.

NSB (National Science Board). 2015. *Revisiting the STEM workforce: A companion to Science and Engineering Indicators 2014.* NSB-2015-10. Arlington, VA: National Science Foundation.

NSB. 2016. *Science and Engineering Indicators 2016.* NSB-2016-1. Arlington, VA: National Science Foundation.

NSF (National Science Foundation). n.d. *Advanced Technological Education (ATE).* http://www.nsf.gov/funding/pgm_summ.jsp?pims_id=5464 (accessed March 21, 2017).

OECD (Organisation for Economic Co-operation and Development). 2012a. *Better skills, better jobs, better lives: A strategic approach to skills policies.* Paris: OECD Publishing. https://skills.oecd.org/documents/OECDSkillsStrategyFINALENG.pdf (accessed February 9, 2016).

OECD. 2012b. *PISA 2012 results: Excellence through equity. Giving every student the chance to succeed (Volume II).* https://www.oecd.org/pisa/keyfindings/pisa-2012-results-volume-II.pdf (accessed February 28, 2017).

OECD. 2013. *OECD Skills Outlook 2013: First results from the Survey of Adult Skills.* Paris: OECD Publishing. http://dx.doi.org/10.1787/9789264204256-en (accessed February 28, 2017).

OECD and Eurostat. 2005. *The Oslo manual: Guidelines for collecting and interpreting innovation data* (3rd ed.). Paris, France: OECD Publishing.

Orr, M., T. Bailey, K. Hughes, M. Karp, and G. Kienzl. 2004. *The National Academy Foundation's career academies: Shaping postsecondary transitions.* IEE Working Paper 17. http://www.tc.columbia.edu/centers/iee/PAPERS/NAF_postsecondary.pdf (accessed March 21, 2017).

Orvis, K. 2016. *The Transition Assistance Program and reinventing the military life cycle.* Prepared for the Committee on the Supply-Chain for Middle-Skill Jobs, NASEM. http://nas.edu/SkilledTechnicalWorkforce (accessed March 5, 2017).

Osterman, P., and A. Weaver. 2014a. Skills and skill gaps in manufacturing. In *Production in the Innovation Economy*, edited by R. Locke and R. Wellhausen. Cambridge, MA: MIT Press. Pp. 17-50.

Osterman, P., and A. Weaver. 2014b. *Why claims of skills shortages in manufacturing are overblown.* Washington, DC: Economic Policy Institute.

Ostrom, E. 2005. *Understanding institutional diversity.* Princeton, NJ: Princeton University Press.

Ostrom, V. 1994. *The meaning of federalism: Constituting a self-governing society.* San Francisco, CA: Institute for Contemporary Studies Press.

Ostrom, V. 2008. *A political theory of a compound republic: Designing the American experiment* (3rd ed.). Lanham, MD: Lexington Books.

Panchak, P. 2016. A tale of (at least) two manufacturing worlds. *IndustryWeek*, July 6. http://www.industryweek.com/leadership/tale-least-two-manufacturing-worlds (accessed March 5, 2017).

Payne Carter, S., and B. J. Miller. 2015. *Analysis of Army veteran unemployment benefits and transition assistance.* http://sites.national

academies.org/cs/groups/pgasite/documents/webpage/pga_167648.pdf (accessed February 28, 2017).

PCAST (President's Council of Advisors on Science and Technology). 2010. *Prepare and inspire: K-12 education in science, technology, engineering, and math (STEM) for America's future.* http://www.afterschool alliance.org/documents/pcast-stemed-report.pdf (accessed February 28, 2017).

Pew Research Center. 2011. Chapter 6: A profile of the modern military. In *War and sacrifice in the post-9/11 era.* http://www.pewsocialtrends.org/2011/10/05/chapter-6-a-profile-of-the-modern-military (accessed March 5, 2017).

Purnell, R., and S. Blank, S. 2004. *Support success: Services that may help low-income students succeed in community college.* New York: MDRC. http://www.mdrc.org/sites/default/files/full_524.pdf (accessed March 5, 2017).

Reamer, A. 2015. *Information resources to facilitate middle skills workforce development.* Prepared for the Committee on the Supply-Chain for Middle-Skill Jobs, NASEM. http://nas.edu/SkilledTechnicalWorkforce (accessed March 22, 2017).

Reed, D., A. Liu, R. Kleinman, A. Mastri, D. Reed, S. Sattar, and J. Ziegler. 2012. *An effectiveness assessment and cost-benefit analysis of registered apprenticeship in 10 states.* Oakland, CA: Mathematica Policy Research. https://wdr.doleta.gov/research/FullText_Documents/ETAOP_2012_10.-pdf (accessed March 21, 2017).

Roder, A., and M. Elliott. 2014. *Sustained gains: Year Up's continued impacts on young adults' earnings.* New York: Economic Mobility Corporation.

Romer, P. 1994. The origins of endogenous growth. *Journal of Economic Perspectives* 8(1):3-22.

Rorabaugh, W. 1986. *The craft apprentice: From Franklin to the machine age in America.* New York: Oxford University Press.

Rosenfeld, S., H. Shapiro, and K. Bird. 2014. *Workplace learning in Europe and the United States.* Copenhagen, Denmark: Trans-Atlantic Technology and Training Alliance.

Ross, M., N. Svajlenka, and J. Williams. 2014. *Part of the solution: Pre-baccalaureate health care workers in a time of health system change.* Washington, DC: Brookings Institution. http://www.brookings.edu/~media/research/files/reports/2014/07/healthworkforce/healthcare-workers-report.pdf (accessed February 28, 2017).

Rothwell, J. 2015. *Defining skilled technical work.* Prepared for the Committee on the Supply-Chain for Middle-Skill Jobs, NASEM. http://nas.edu/SkilledTechnicalWorkforce (accessed February 28, 2017).

Rupietta, C., and U. Backes-Gellner. 2015. *High quality workplace training and innovation in highly developed countries.* Working Paper 74.

http://repec.business.uzh.ch/RePEc/iso/leadinghouse/0074_lhwpaper.pdf (accessed March 19, 2017).

Ryan, C. L., and Bauman, K. 2016. *Educational attainment in the United States: 2015.* U.S. Census Bureau, Current Population Reports, P20-578. https://www.census.gov/content/dam/Census/library/publications/2016/demo/p20-578.pdf (accessed March 19, 2017).

Scheffel, K. 2016. *Three views on concurrent enrollment.* http://www.nacep.org/three-views-on-concurrent-enrollment (accessed March 22, 2017).

Schneider, M. 2013. Does education pay? *Issues in Science and Technology* 30(1).

Scott-Clayton, J., P. M. Crosta, and C. R. Belfield. 2012. *Improving the targeting of treatment: Evidence from college remediation.* NBER Working Paper 18457. http://www.nber.org/papers/w18457.pdf (accessed February 9, 2016).

Scrivener, S., and E. Coghlan. 2011. *Opening doors to student success: A synthesis of findings from an evaluation at six community colleges.* New York: MDRC.

Scrivener, S., M. Weiss, A. Ratledge, T. Rudd, C. Sommo, and H. Fresques. 2015. *Doubling graduation rates: Three-year effects of CUNY's Accelerated Study in Associate Programs (ASAP) for developmental education students.* New York: MDRC.

Shankel, G. 2010. *America's most wanted: Skilled workers: A highly-skilled workforce is necessary to meet 21st century manufacturing needs.* Rockford, IL: Fabricators & Manufacturers Association, International. http://www.nutsandboltsfoundation.org/wp-content/uploads/skilled-labor-shortage2010.pdf (accessed March 29, 2017).

Shapiro, C. 1986. Investment, moral hazard, and occupational licensing. *Review of Economic Studies* 53(5):843-862.

Shedd, S. 2013. *Moving forward.* Presentation for the 2013 WIA Technical Assistance Conference. https://des.az.gov/sites/default/files/des_sector_strategy_presentation.pdf (accessed March 30, 2017).

Sheets, R., and J. Tyszko. 2015. *Competing on innovation: Implications for building the middle-skill talent pipeline.* Prepared for the Committee on the Supply-Chain for Middle-Skill Jobs, NASEM. http://nas.edu/SkilledTechnicalWorkforce (accessed March 2, 2017).

SHRM (Society for Human Resource Management). 2015. *Employee benefits: An overview of employee benefits offerings in the U.S. research report.* https://www.shrm.org/hr-today/trends-and-forecasting/research-and-surveys/Documents/2015-Employee-Benefits.pdf (accessed March 19, 2017).

Simchi-Levi, D. 2012. *U.S. re-shoring: A turning point.* Cambridge, MA: MIT Forum for Supply Chain Innovation and Supply Chain Digest.

http://supplychain.mit.edu/wp-content/uploads/2014/10/mit_forum_2012_annual_u_s_reshoring_report.pdf (accessed February 28, 2017).

Simon, M. 2015. *State models of integration and coordination of education and workforce programs and policies.* Presentation for the 3rd Committee Meeting on the Supply Chain for Middle-Skill Jobs, NASEM, September 10. http://nas.edu/SkilledTechnicalWorkforce (accessed March 2, 2017).

Slivinski, S. 2015. *Bootstraps tangled in red tape: How state occupational licensing hinders low-income entrepreneurship.* Phoenix, AZ: Goldwater Institute. https://goldwater-media.s3.amazonaws.com/cms_page_media/2015/4/15/OccLicensingKauffman.pdf (accessed March 2, 2017).

Smith, A. 1776. *An inquiry into the nature and causes of the wealth of nations*, edited by E. Cannan, 1976. Chicago, IL: University of Chicago Press.

Smith, T. C., and C. T. King. 2011. *Exploratory return-on-investment analysis of local workforce investments.* Austin, TX: Ray Marshall Center for the Study of Human Resources, Lyndon B. Johnson School of Public Affairs, The University of Texas. http://raymarshallcenter.org/files/2006/01/Capital_IDEA_ROI_Final_Aug_23_2011.pdf (accessed March 31, 2017).

Solow, R. 1956. A contribution to the theory of economic growth. *Quarterly Journal of Economics* 70(1):65-94.

Spence, M. 1973. Job market signaling. *Quarterly Journal of Economics* 87(3):355-374.

Spence, M. 2002. Signaling in retrospect and the informational structure of markets. *The American Economic Review* 92(3):434-459.

Stacey, N., and I. Charner. 1982. Unions and post-secondary education. *Education and Urban Society* 14(3)331-343.

Steedman, H. 2012. *Overview of apprenticeship systems and issues: ILO contribution to the G20 Task Force on Employment.* Geneva: International Labour Office (ILO). http://www.ilo.org/wcmsp5/groups/public/---ed_emp/---ifp_skills/documents/genericdocument/wcms_190188.pdf (accessed March 22, 2017).

Stern, D. 2015. *Pathways or pipelines: Keeping high school students' future options open while developing technical skills and knowledge.* Prepared for the Committee on the Supply-Chain for Middle-Skill Jobs, NASEM. http://nas.edu/SkilledTechnicalWorkforce (accessed March 5, 2017).

Stern, D., C. Dayton, and M. Raby. 2010. *Career academies: A proven strategy to prepare high school students for college and careers.* Berkeley, CA: Graduate School of Education, University of California, Berkeley. http://casn.berkeley.edu/resources.php?r=158 (accessed March 22, 2017).

Stevens, A., M. Kurlaender, and M. Grosz. 2015. *Career technical education and labor market outcomes: Evidence from California community colleges.* Working Paper No. 21137. Cambridge, MA: National Bureau of Economic Research. http://www.nber.org/papers/w21137.pdf (accessed March 5, 2017).

Strawn, J. 2010. *Shifting gears: State innovation to advance workers and the economy in the Midwest.* Chicago, IL: Joyce Foundation.

Symonds, W., R. Schwartz, and R. Ferguson. 2011. *Pathways to prosperity: Meeting the challenge of preparing young Americans for the 21st century.* Cambridge, MA: Pathways to Prosperity Project, Harvard University Graduate School of Education.

Taylor, J., V. Borden, and E. Park. 2015. State dual credit policy: A national perspective. *New Directions for Community Colleges* 2015(169):9-19.

The Economist. 2015. Business high school: Companies struggling to find talent are looking to teenagers. *The Economist*, July 18. http://www.economist.com/news/united-states/21657842-companies-struggling-find-talent-are-looking-teenagers-business-high-school (accessed March 29, 2017).

The Manufacturing Institute. 2015. *Inspiring youth with Dream It. Do It.* http://www.themanufacturinginstitute.org/Image/Dream-It-Do-It/Dream-It-Do-It.aspx (accessed February 28, 2017).

The White House. 2014. *Fact sheet: Ready to work at a glance: Job-driven training and American opportunity.* https://obamawhitehouse.archives.gov/the-press-office/2014/07/22/fact-sheet-ready-work-glance-job-driven-training-and-american-opportunit (accessed February 28, 2017).

The White House. 2015. *Occupational licensing: A framework for policymakers.* https://obamawhitehouse.archives.gov/sites/default/files/docs/licensing_report_final_nonembargo.pdf (accessed February 28, 2017).

The White House. 2016. *Investing in higher education: Benefits, challenges, and the state of student debt.* https://obamawhitehouse.archives.gov/sites/default/files/page/files/20160718_cea_student_debt.pdf (accessed March 5, 2017).

Thomas, R., and T. Kochan. 1992. Technology, industrial relations, and the problem of organizational transformation. In *Technology and the future of work*, edited by P. Adler. Oxford, UK: Oxford University Press. Pp. 210-231.

U.S. Chamber of Commerce Foundation. 2014. *Managing the talent pipeline: A new approach to closing the skills gap.* Washington, DC: U.S. Chamber of Commerce Foundation.

U.S. Congress Joint Economic Committee Democratic Staff. 2014. *Veterans Day 2014: An update on employment prospects for returning veterans.* http://www.jec.senate.gov/public/index.cfm/democrats/2014/11/veterans-day-2014-an-update-on-employment-prospects-for-returning-veterans (accessed March 5, 2017).

U.S. Department of Education. 2017. *Every Student Succeeds Act (ESSA).* https://www.ed.gov/essa?src=rn (accessed March 29, 2017).

VA (U.S. Department of Veterans Affairs). 2015. *2015 veteran economic opportunity report.* http://www.benefits.va.gov/benefits/docs/VeteranEconomicOpportunityReport2015.pdf (accessed March 5, 2017).

Wachen, J., D. Jenkins, C. Belfield, and M. Van Noy. 2012. *Contextualized college transition strategies for adult basic skills students: Learning from Washington State's I-BEST program model.*. New York: Teachers College, Columbia University. http://ccrc.tc.columbia.edu/publications/i-best-program-final-phase-report.html (accessed March 22, 2017).

Washington State Board for Community and Technical Colleges. 2017. *Centers of Excellence.* http://www.sbctc.edu/for-employers/centers-of-excellence.aspx (accessed March 22, 2017).

Washington State Centers of Excellence. 2017. *About us.* https://www.coewa.com/about-our-centers (accessed March 22, 2017).

Webb, M., and C. Gerwin. 2014. *Early college expansion: Propelling students to postsecondary success, at a school near you.* Washington, DC: Jobs for the Future. http://www.jff.org/sites/default/files/publications/materials/Early-College-Expansion_031714.pdf (accessed March 22, 2017).

Whoriskey, P. 2012. U.S. manufacturing sees shortage of skilled factory workers. *The Washington Post*, February 19.

Willcox, K., S. Sarma, and P. Lippel. 2016. *Online education: A catalyst for higher education reforms.* Cambridge, MA: Online Education Policy Initiative, MIT.

Wolter, S. C., and S. Muehlemann. 2015. *Apprenticeship training in Spain: A cost-effective model for firms?* https://www.fundacionbertelsmann.org/fileadmin/files/Fundacion/Publicaciones/ROI_web_EN.pdf (accessed March 19, 2017).

Wolter, S., and P. Ryan. 2011. Apprenticeship. In *Handbook of the economics of education*, Vol. 3, edited by E. Hanushek, S. Machin, and L. Woessmann. Amsterdam, Netherlands: Elsevier Science. Pp. 521-576.

Wood, C. 2011. Employment in health care: A crutch for the ailing economy during the 2007-09 recession. *Monthly Labor Review* 134(4):13-18.

A

Agendas of Committee Meetings

JUNE 24-25, 2015
SYMPOSIUM ON THE SUPPLY CHAIN FOR MIDDLE-SKILL JOBS: EDUCATION, TRAINING, AND CERTIFICATION PATHWAYS

National Academy of Sciences
Room 120
2101 Constitution Avenue, NW
Washington, DC

Day 1: Wednesday, June 24, 2015

8:30 AM **Continental Breakfast**

9:00 AM **Introduction**
Sujai Shivakumar, *The National Academies*

9:10 AM **Setting the Stage**
Senator Jeff Bingaman, *Study Chair*

9:30 AM **Keynote**
Senator Tim Kaine, *U.S Senator for Virginia, co-chair of Senate CTE Caucus*

10:00 AM **Panel 1: Defining Middle-Skill Jobs**
Moderator: Harry Holzer, *Georgetown University*
Presenter: Jonathan Rothwell, *Brookings Institution*
Discussants:

- V. Celeste Carter, *National Science Foundation*
- Brendan Lind, *LaunchCode*
- Dan Marschall, *AFL-CIO*

10:45 AM **Break**

11:00 AM **Panel 2: Middle-Skill Workers and Today's Labor Market**
Moderator: Francine Lawrence, *American Federation of Teachers*
Presenter: Alicia Sasser-Modestino, *Northeastern University*
Discussants:
- William Spriggs, *Howard University and AFL-CIO*
- Peter Cappelli, *Wharton Business School*
- Manjari Raman, *Harvard Business School*

11:45 AM **Panel 3: High School Pathways (Career Technical Education)**
Moderator: Gary Hoachlander, *ConnectEd*
Presenter: David Stern, *UC-Berkeley*
Discussants:
- Tom Volk, *Ohio Belting and Transmission*
- Grace Suh, *IBM Corporation*
- Erin Ward Bibo, *DC Public Schools*

12:30 PM **Break**

12:45 PM **Panel 4: Community College Pathways (Advanced Technical Education)**
Moderator: Annette Parker, *South Central College*
Presenter: Melinda Karp, *Community College Research Center, Columbia University*
Discussants:
- Johan Uvin, *Department of Education*
- Darlene Miller, *PERMAC*
- Marilyn Barger, *Florida Advanced Technological Education Center of Excellence*

1:30 PM **Lunch Keynote**
John McKernan, *U.S. Chamber of Commerce Foundation*

2:30 PM **Panel 5: Employer-led Training**
Moderator: Katharine Frase, *IBM*
Presenter: Robert Lerman, *American University*
Discussants:
- Dale Allen, *Quinsigamond Community College*
- Scott Ellsworth, *Business Leaders United for Workforce Partnerships*

- James Redstone, *House Education and Workforce Committee*

3:15 PM **Panel 6: Apprenticeships: Learning from Other Countries**
Moderator: Mario Kratsch, *German-American Chambers of Commerce of the Midwest*
Presenter: Andrea Messing-Mathie, *Northern Illinois University*
Discussants:

- Eric Seleznow, *Department of Labor*
- Josh Benton, *Kentucky Cabinet for Economic Development*
- Stuart Rosenfeld, *Regional Strategies Inc.*

4:00 PM **Break**

4:15 PM **Panel 7: Federally-funded Workforce Development**
Moderator: Susan Sclafani, *Pearson Foundation*
Presenter: Carolyn Heinrich, *University of Texas – Austin*
Discussants:

- Yvette Chocolaad, *National Assn. of State Workforce Agencies*
- Maureen Conway, *Aspen Institute*
- Scott Cheney, *Senate HELP Committee*

5:00 PM **Adjourn (End of Day 1)**

Day 2: Thursday, June 25, 2015

8:00 AM **Breakfast**

8:30 AM **Opening Remarks**
Tom Bailey, *Columbia University, Study Co-Chair*

8:45 AM **Panel 8: Pathways to Careers in Allied Health**
Moderator: George Zangaro, *HRSA*
Presenter: Bianca Frogner, *University of Washington*
Discussants:

- Eleni Papadakis, *Washington State Workforce Training and Education*
- Carolyn O'Daniel, *Jefferson Community and Technical College*

9:45 AM **Panel 9: Pathways for Military Credentialing & Veterans' Transitions**
Moderator: David Chu, *Institute for Defense Analyses*
Presenters: Susan Carter & Major Brian Miller, *U.S. Military Academy*
Discussants:
- Phillip Carter, *Center for a New American Security*
- Matt Notowidigdo, *Northwestern University*
- Wayne Boswell, *DoD Transitions to Veterans' Program Office*

10:45 AM **Break**

11:00 AM **Panel 10: Data Resources to Support Middle-Skill Workforce Development**
Moderator: Jennifer McNelly, *Manufacturing Institute*
Presenter: Andrew Reamer, *George Washington University*
Discussants:
- Rebecca Rust, *Bureau for Labor Statistics*
- John Dorrer, *Jobs for the Future*

12:00 PM **Panel 11: Building the Middle-Skill Talent Pipeline and Innovation**
Moderator: Rita Colwell, *University of Maryland*
Presenter: Robert Sheets, *George Washington University, and Jason Tyszko, U.S. Chamber of Commerce Foundation*
Discussants:
- Mary Alice McCarthy, *New America Foundation*
- Kimberly Green, *National Association of State Directors of Career and Technical Education Consortium*
- Byron Zuidema, *Department of Labor*

1:00 PM **Closing Remarks**
Senator Jeff Bingaman, *Study Chair*

1:15 PM **Symposium Adjourns**

SEPTEMBER 10, 2015
MEETING OF THE COMMITTEE ON THE SUPPLY CHAIN FOR MIDDLE-SKILL JOBS

Keck Center of the National Academies
Room 101
500 5th Street, NW
Washington, DC

9:00 AM	**Working Breakfast** **<u>Welcome</u>** • Senator Jeff Bingaman, *Committee Chair*
9:10 AM	**<u>Federal Legislation</u>** • Cassandria Dortch, *Congressional Research Service* • Committee Discussion
9:55 AM	**Break**
10:10 AM	**<u>Role of New Technologies in CTE Education</u>** • Dale Allen, *Quinsigamond Community College* • Committee Discussion
10:55 AM	**<u>State Programs</u>** • Martin Simon, *National Governors Association* • Kate Blosveren Kreamer, *NASDCTEc/NCTEF* • Committee Discussion
11:40 AM	**<u>Military Credentials & Transition Issues</u>** • Phillip Carter, *Center for a New American Security* • Committee Discussion
12:25 PM	**<u>Soft Skills: Briefing on *Education for Life and Work*</u>** • Margaret Hilton, *Study Staff* • Committee Discussion
1:00 PM	**CLOSED SESSION BEGINS**

FEBRUARY 17, 2016
MEETING OF THE COMMITTEE ON THE SUPPLY CHAIN FOR MIDDLE-SKILL JOBS

Keck Center of the National Academies
Room 106
500 5th Street NW
Washington, DC

9:00 AM	**Working Breakfast** **Welcome** • Senator Jeff Bingaman, *Committee Chair*
9:10 AM	**Data, Labor Market Dynamics & the Middle-Skill Workforce** • Moderator: Andrew Reamer, *George Washington University* • Kimberly Vitelli, *Department of Labor* • Michael Horrigan, *Bureau of Labor Statistics* • Sharon Boivin, *National Center for Education Statistics* • Emilda Rivers, *National Center for Science and Engineering Statistics* • Arpita Chattopadhyay, *Health Resources Services Administration*
10:30 AM	**Break**
10:45 AM	**Committee Discussion** • Tom Bailey, *Community College Research Center*
11:30 AM	**CLOSED SESSION BEGINS**

APRIL 1, 2016
WORKSHOP ON THE MIT ONLINE EDUCATION REPORT

National Academy of Sciences
The Lecture Room
2101 Constitution Avenue, NW
Washington, DC

2:00 PM **Welcome**

- Dr. Katharine Frase, co-chair of the Committee on the Supply Chain for Middle-skill Jobs

2:15 PM **Panel 1: Major Findings in MIT Online Education Report**

- Sanjay Sarma, *MIT, Dean of Digital Learning*
- Karen Willcox, *MIT, Professor of Aeronautics and Astronautics, Co-Director, Center for Computational Engineering*
- Eric Klopfer, *MIT, Professor & Director, Scheller Teacher Education Program*

3:00 PM **Panel 2: The Educators' Point of View**

Moderator: Toby Smith, *Vice President for Policy, Association of American Universities*

- Marshall "Mike" Smith, *Visiting Scholar, Carnegie Foundation for the Advancement of Teaching*
- Andrea Nixon, *Director of Educational Research, Carleton College*
- Kacy Redd, *Director of Science and Mathematics Education Policy, Association of Public and Land-grant Universities*
- Bror Saxberg, *Chief Learning Officer, Kaplan, Inc.*

4:00 PM **Break**

4:15 PM **Panel 3: Views from the Government**

Moderator: David Soo, *Senior Policy Advisor, Office of the Under Secretary, U.S. Department of Education*

- Susan Singer, *Division Director for Undergraduate Education, National Science Foundation*
- Ryan Burke, *Senior Policy Advisor, National Economic Council, The White House*

- Roberto Rodríguez, *Deputy Assistant to the President for Education, Domestic Policy Council, The White House*
- Mark Mitsui, *Deputy Assistant Secretary for Community Colleges, Office of Career, Technical, and Adult Education, Department of Education*

5:15 PM **Wrap-Up**

5:30 PM **Adjourn**

DECEMBER 6, 2016
WORKSHOP ON "THE MILITARY AS A PATHWAY TO SKILLED TECHNICAL JOBS"

National Academy of Sciences
The Lecture Room
2101 Constitution Avenue, NW
Washington, DC

12:30 PM **Welcome**
David Chu, *Committee on the Supply Chain for Middle-Skill Jobs and IDA*

12:40 PM **Keynote: The Force of the Future and the Skilled Technical Workforce**
The Hon. Peter K. Levine, *Under Secretary of Defense*

1:15 PM **Panel I: Preparing Service members for their Post-military Civilian Careers**
Moderator: David Chu, *Institute for Defense Analyses*

The Transition Assistance Program and Reinventing the Military Life Cycle
Karin Orvis, *Department of Defense*

	Signaling Skills: The DOD's Certification, Licensing, and Credentialing Efforts Diana Banks, *Department of Defense*
	Building the Skills Bridge to Post Military Success Ryan McDermott, *Department of Defense*
2:30 PM	**Panel II: Improving the Transferability of Military Training** Moderator: Susan Sclafani, *Pearson (Ret.)*
	Best Practices from a State Program for Veterans Sean Murphy, *Pacific Mountain Workforce Development Council*
	Supporting the Transition: The Federal Role Teresa Gerton, *Department of Labor*
	The Onward to Opportunity Program Mike Haynie, *Syracuse University*
3:30 PM	**Coffee Break**
3:45 PM	**Panel III: Continuum of Service and Civilian Needs** Moderator: Lieutenant General Jack Stultz, Jr., *U.S. Army Reserve (Ret.)*
	Establishing the Continuum of Service John Winkler, *RAND*
	Hiring our Heroes: The Private Sector Role Eric Eversole, *U.S. Chamber of Commerce*
	JPMorgan Chase: A Commitment to Veterans Ross Brown, *JPMorgan Chase Foundation*

4:45 PM	**Panel IV: Policy Roundtable** Moderator: Daniel Feehan, *Department of Defense* Debra S. Wada, *Assistant Secretary of the Army (Manpower & Reserve Affairs)* Brigadier General Kathleen Cook, *U.S. Air Force* Anish Goel, *Senate Armed Services Committee*
5:30 PM	**Adjourn**

B

Biographies of Committee Members

Jeff Bingaman (*Chair*) is a former U.S. senator from New Mexico, serving from 1983 to 2013. He served as chairman of the Committee on Outreach for the Senate Democratic Caucus. Senator Bingaman held several committee assignments during his tenure in the U.S. Senate, including the Committee on Energy and Natural Resources; Committee on Finance; Joint Economic Committee; Committee on Armed Services; and Committee on Health, Education, Labor, and Pensions. During much of his political career, he was involved in the immigration debate, and he has worked to protect wildlife and public lands. On the Senate Energy Committee, he contributed to every major piece of energy policy legislation in the last two decades. Prior to serving as a U.S. senator, he worked as a private practice attorney. He served as counsel to the New Mexico Constitutional Convention of 1969, and served as attorney general of New Mexico from 1979 to 1983. Senator Bingaman earned his B.A. from Harvard and his J.D. from Stanford University.

Tom Bailey (*Co-chair*) is George and Abby O'Neill professor of economics and education at Teachers College, Columbia University. He is also director of the Community College Research Center (CCRC) and two national centers funded by a grant from the Institute for Education Sciences (IES): the Center for Analysis of Postsecondary Education and Employment (CAPSEE), established in 2011, and the Center for the Analysis of Postsecondary Readiness (CAPR), established in 2014. From 2006 to 2012, Dr. Bailey directed another IES-funded center, the National Center for Postsecondary Research (NCPR). His areas of expertise include labor economics, community colleges, econometrics, education policy, developmental education, assessment, educational equity, higher education and workforce preparation, school-to-work transition, and online learning.

In 1996, with support from the Alfred P. Sloan Foundation, Dr. Bailey established the CCRC at Teachers College, where he has served since 1992 as director of the Institute on Education and the Economy (IEE). In June 2010,

U.S. Secretary of Education Arne Duncan appointed him chair of the Committee on Measures of Student Success, which developed recommendations for community colleges to comply with completion rate disclosure requirements under the Higher Education Opportunity Act. Dr. Bailey and the CCRC won the Terry O'Banion Prize for Teaching and Learning at the annual conference for the League for Innovation in the Community College in 2013 and was inducted as an American Educational Research Association (AERA) fellow in the same year. He has been a member of the National Academy of Education since 2012.

Dr. Bailey's papers have appeared in a wide variety of education, policy-oriented, and academic journals, and he has authored or co-authored several books on the employment and training of immigrants and the extent and effects of on-the-job training. Along with Shanna Smith Jaggars and Davis Jenkins, he wrote *Redesigning America's Community Colleges: A Clearer Path to Student Success*, published by Harvard University Press in 2015. Other books include *Defending the Community College Equity Agenda* (Johns Hopkins University Press, 2006), co-edited with Vanessa Morest; *Working Knowledge: Work-Based Learning and Education Reform* (Routledge, 2004), co-authored with Katherine Hughes and David Moore; *Manufacturing Advantage* (Cornell University Press, 2000), written with Eileen Appelbaum, Peter Berg, and Arne Kalleberg; and *The Double Helix of Education and the Economy* (IEE, 1992), co-authored with Sue Berryman. Dr. Bailey holds a Ph.D. in labor economics from the Massachusetts Institute of Technology (MIT).

Katharine Frase (*Co-chair*) retired from IBM after a 30-year career in 2016. Her career spanned positions in manufacturing, product and process development, strategy, research, and business development. Her most recent assignments were in support of IBM's clients and field teams in the government, cities, health care, and education industries, particularly the application of analytics and technologies such as Watson to provide actionable insights for some of the world's most important challenges. In 2006, she was elected a member of the National Academy of Engineering. Dr. Frase received an A.B. in chemistry from Bryn Mawr College and a Ph.D. in materials science and engineering from the University of Pennsylvania.

David Chu serves as president of the Institute for Defense Analyses (IDA), a nonprofit corporation operating in the public interest. Its three federally funded research and development centers provide objective analyses of national security issues and related national challenges, particularly those requiring extraordinary scientific and technical expertise. Dr. Chu served in the Department of Defense as under secretary of defense for personnel and readiness from 2001 to 2009, and earlier as assistant secretary of defense and director for program analysis and evaluation from 1981 to 1993. From 1978 to 1981 he was assistant director of the Congressional Budget Office for National Security and International

Affairs. Dr. Chu served in the U. S. Army from 1968 to 1970. He was an economist with the RAND Corporation from 1970 to 1978, director of RAND's Washington Office from 1994 to 1998, and vice president for its Army Research Division from 1998 to 2001. He earned his Ph.D. in economics, as well as a B.A. in economics and mathematics, from Yale University.

Rita Colwell is distinguished university professor at the University of Maryland, College Park and Johns Hopkins University Bloomberg School of Public Health, and chairman and chief science officer, CosmosID, Inc. Her interests are focused on genomics, biodiversity, and molecular microbial systematics and ecology. Dr. Colwell is an honorary member of the microbiological societies of the United Kingdom, Australia, France, Israel, Bangladesh, India, and the United States. She served as the 11th director of the National Science Foundation from 1998 to 2004. She has authored/co-authored 19 books and more than 800 scientific publications. She is a member of the National Academy of Sciences and has been awarded the Stockholm Water Prize; the Order of the Rising Sun, Japan; and the U.S. National Medal of Science.

Gary Hoachlander is president of ConnectEd: The California Center for College and Career. Recently established by the James Irvine Foundation, ConnectEd is dedicated to advancing practice, policy, and research designed to help young people prepare for college and career—both goals and not one or the other. ConnectEd's primary mission is supporting the development of multiple pathways by which young people can complete high school; enroll in postsecondary education; attain a formal credential; and embark on lasting success in work, community, and civic affairs.

Having begun his career in 1966 as a brakeman for the Western Maryland Railroad, Dr. Hoachlander has devoted most of his professional life to helping young people learn by doing—connecting education to the opportunities, challenges, and many different rewards to be found through work. Widely known for his expertise in career and technical education and many other aspects of elementary, secondary, and postsecondary education, Dr. Hoachlander has consulted extensively for the U.S. Department of Education, state departments of education, local school districts, foundations, and a variety of other clients. Dr. Hoachlander is also chairman of MPR Associates, Inc., an educational research and development organization closely affiliated with ConnectEd, and is also one of the country's leading policy analysts for the U.S. Department of Education, including the National Center for Education Statistics and the Office of Vocational and Adult Education. He earned a B.A. at Princeton University and holds master's and Ph.D. degrees from the Department of City and Regional Planning, University of California, Berkeley.

Harry Holzer is John LaFarge Jr. SJ Professor of Public Policy at Georgetown University. He has been at Georgetown since the fall of 2000. He served as associate dean from 2004 through 2006 and was acting dean in the fall of 2006. He is currently an institute fellow at the American Institutes for Research, a senior affiliate at the Urban Institute, a nonresident senior fellow at the Brookings Institution, and a research affiliate of the Institute for Research on Poverty at the University of Wisconsin at Madison. He has also been a faculty director of the Georgetown Center on Poverty, Inequality and Public Policy.

Prior to coming to Georgetown, Dr. Holzer served as chief economist for the U.S. Department of Labor and professor of economics at Michigan State University. He was a visiting scholar at the Russell Sage Foundation in 1995 and a faculty research fellow of the National Bureau of Economic Research.

Throughout most of Dr. Holzer's career, his research has focused primarily on the low-wage labor market, particularly the problems of minority workers in urban areas. In recent years, he has worked on the quality of jobs as well as workers in the labor market, and on how job quality affects the employment prospects of the disadvantaged, as well as worker inequality more broadly. He has also written extensively about the employment problems of disadvantaged men, advancement prospects for the working poor, welfare reform, and workforce policy. His current research focuses on disadvantaged students at college and how to improve their academic and employment outcomes. His latest book, *Making College Work* (with Sandy Baum), is forthcoming from Brookings Press in 2017.

Dr. Holzer teaches graduate courses in public policy and statistical methods for program and policy evaluation at the McCourt School, as well as courses on antipoverty policy and on labor market policy. He received his B.A. (1978) and Ph.D. (1983) from Harvard University.

Francine Lawrence was elected executive vice president of the American Federation of Teachers (AFT) by a unanimous vote of the AFT executive council and assumed the responsibilities of the office in September 2011. She ran for reelection in 2012 and served as executive vice president of AFT through July 2014. From 1997 to 2011, Ms. Lawrence was president of the 3,000-member Toledo (Ohio) Federation of Teachers, where she led contract negotiations that focused on what matters most: student achievement. She co-chaired the union and district's Intern Board of Review, which oversees Toledo's peer assistance and review plan; the Toledo plan serves as an example of school reform that works. As AFT executive vice president, she coordinated AFT's Program Policy Councils for Teachers, Healthcare, Higher Education, Paraprofessionals and School-related Personnel (PSRP), Public Employees, and Retirees. She guided the development of AFT's new LEAD initiative, focused on leadership education and development. She also led AFT's Teacher Preparation Task Force and Small Local Union Task Force.

Ms. Lawrence was a member of the Albert Shanker Institute and AFT Innovation Fund governing boards. She served as treasurer of the AFL-CIO Department for Professional Employees. She served on the United Way USA Board of Trustees and also served on and chaired the Learning First Alliance—a partnership of 17 national education associations dedicated to improving student learning in America's public schools.

Jennifer McNelly serves as president of 180 Skills, LLC, an online technical education experience supporting the nation's manufacturers. 180 Skills offers the only industry-defined, competency-based, high-quality, low-cost solution to advance the manufacturing workforce. Ms. McNelly has extensive experience in workforce development, employer engagement, and business.

Prior to joining 180 Skills, Ms. McNelly was president and executive director of The Manufacturing Institute, the nonprofit affiliate of the National Association of Manufacturers (NAM), where she advanced a national agenda to close the manufacturing skills gap and make manufacturers in America globally competitive. She is a proven leader at the Institute as the chief architect of one of the organization's flagship initiatives, the NAM-Endorsed Manufacturing Skills Certification System. Ms. McNelly is a member of the Senior Executive Service (SES) and served as an administrator for the U.S. Department of Labor's (DOL) Employment and Training Administration. Her strong private-sector experience includes serving as senior vice president of Strategic Partnerships, LLC, an international consulting firm specializing in helping Fortune 500 corporations build strategic partnerships with government agencies in support of workforce development.

In 2012, Ms. McNelly was recognized as one of the 100 inaugural Women in STEM. She is the immediate past chair of the World Economic Forum's Global Agenda Council on Advanced Manufacturing. She previously served as a member of the American National Standards Institute (ANSI) Personnel Certification Accreditation Committee, the Precision Metalforming Association (PMA) Education Foundation, and the SME Education Foundation Board.

Elizabeth Mertz is an associate professor at the University of California, San Francisco, with a joint appointment in the Department of Preventive and Restorative Dental Sciences, School of Dentistry, and in the Department of Social and Behavioral Sciences, School of Nursing. She is affiliated with Healthforce Center, the Philip R. Lee Institute for Health Policy Studies, and the Center to Address Disparities in Children's Oral Health (CANDO).

Dr. Mertz has researched, published, and lectured on a broad range of health professions workforce policy and analysis issues, including supply and demand of providers, health care regulation, state and federal workforce policy, access to care, and evolving professional practice models. She has served on a number of advisory and planning committees for organizations such as the Health Resources and Services Administration, the California HealthCare Foundation, the San Francisco Foundation, and the Institute of Medicine. She is the principal investigator on multiple research grants, including a National Institutes of Health (NIH)-funded evaluation of standardized preventive care to reduce dental disparities in children, a set of studies on workforce diversity in dentistry, and a national center for Oral Health Workforce Research.

Dr. Mertz holds a B.A. from the University of Southern California, an M.A. from the Humphrey Institute of Public Affairs at the University of Minnesota, and a Ph.D. in medical sociology from the University of California, San Francisco.

Paul Osterman is Nanyang Technological University (NTU) professor of human resources and management at the Massachusetts Institute of Technology (MIT) Sloan School of Management, as well as a member of the Department of Urban Planning at MIT. From July 2003 to June 2007, he also served as deputy dean at the Sloan School. His research concerns changes in work organization within companies, career patterns and processes within firms, economic development, urban poverty, and public policy surrounding skills training and employment programs.

Dr. Osterman has been a senior administrator of job training programs for the Commonwealth of Massachusetts and has consulted widely for government agencies, foundations, community groups, firms, and public interest organizations. His most recent book is *Good Jobs America: Making Work Better for Everyone* (Russell Sage, 2011). Other recent books include *The Truth about Middle Managers: Who They Are, How They Work, How They Matter* (Harvard Business School Press, 2009); *Gathering Power: The Future of Progressive Politics in America* (Beacon Press, 2003); *Securing Prosperity: The American Labor Market: How It Has Changed and What to Do about It* (Princeton University Press, 1999); and *Working in America: A Blueprint for the New Labor Market* (MIT Press, 2001). Dr. Osterman is also the author of *Employment Futures: Reorganization, Dislocation, and Public Policy; Getting Started: The Youth Labor Market*; *The Mutual Gains Enterprise: Forging a Winning Partnership among Labor, Management, and Government*; and *Change at Work*. He is the editor of two books, *Internal Labor Markets* and *Broken Ladders: Managerial Careers in the New Economy*. In addition, he has written numerous academic journal articles and policy issue papers on such topics as labor market policy, the organization of work within firms, careers, job training programs, economic development, and antipoverty programs.

Annette Parker is president of South Central College in North Mankato, Minnesota. In September 2013, she was selected to be on the President's Advanced Manufacturing Partnership Steering Committee. Dr. Parker was one of 19 industry, academia, and labor executives selected, including co-chairs Andrew Liveris, president, chairman, and CEO of Dow Chemical; and Rafael Reif, president of the Massachusetts Institute of Technology. Together, they were charged with identifying policies and initiatives that lead to the creation of high-quality manufacturing jobs.

Prior to moving to Minnesota, Dr. Parker worked for the Kentucky Community and Technical College System (KCTCS), where she served as system director for the National Center of Excellence in Advanced Automotive Manufacturing. The objective of this organization was to create a national, standardized curriculum that supported the needs of automotive manufacturers in 12 states. Dr. Parker brought colleges together with major automakers such as Toyota, Honda, Ford, GM, and BMW to form the standards. The Center later was recognized by the National Governors Association as a national best practice.

Prior to her position at KCTCS, Dr. Parker served as department chair of Manufacturing Engineering Technologies at Lansing Community College in Michigan, where skilled workers were in extremely high demand. GM had committed to building two new assembly plants in the area and needed workers in robotics and automation. The college responded by launching a $48 million campus dedicated to supporting advanced manufacturing.

Dr. Parker herself is a product of on-the-job training combined with higher education. After high school, she spent 11 years working at GM before pursuing an associate's degree in industrial drafting from Lansing Community College. She then continued her education, earning bachelor's and master's degrees in technical education from Ferris State University in Michigan and a Ph.D. in educational leadership from Western Kentucky University.

Susan Sclafani is retired from the Pearson Foundation, which she joined in January 2011 as vice president of programs. Previously, she served as director of state services at the National Center on Education and the Economy. Before that, she was with Chartwell Education Group, an international consulting group. She also served as assistant secretary of education for vocational and adult education from 2003 to 2005 in the U.S. Department of Education (ED). Earlier, Dr. Sclafani was counselor to U.S. Secretary of Education Rod Paige, serving as the U.S. representative to both the Organisation for Economic Co-operation and Development and Asia-Pacific Economic Cooperation.

Dr. Sclafani's passion for the task of promoting and ensuring the highest standards in education was well demonstrated during her tenure at ED. Among the highlights of her tenure there was the leadership role she played in the creation of the Mathematics and Science Initiative (MSI) to focus attention on the importance of mathematics and science in the education of all students. MSI emphasized the need for teachers knowledgeable in math and science at every level of schooling and the importance of further research in both areas.

Dr. Sclafani's international work led to her leadership of the joint E-Language Learning Project with the Chinese Ministry of Education. She also led ED's High School Redesign Initiative to better prepare students for 21st-century education and training and the workplace.

Prior to serving at ED, Dr. Sclafani was chief academic officer of the Houston Independent School District in Texas, one of the nation's largest urban school districts. In that capacity, she focused on a number of areas, including technology, curriculum development, mathematics and science initiatives, and construction management. She also has extensive state education and business experience.

Dr. Sclafani earned a bachelor's degree at Vassar College, a master's degree in Germanic languages and literature at the University of Chicago, and a master of education degree and Ph.D. in educational administration at the University of Texas, Austin.

Mark Tomkins is president and CEO of the German American Chamber of Commerce of the Midwest (GACC Midwest), where in addition to his roles in building bilateral trade and investment between the United States and Germany, he has led GACC Midwest's Skills Initiative to expand the German dual-education model of workforce development in the Midwest and throughout the United States. This has included supporting the creation and ongoing implementation of the Michigan Advanced Technician Training (MAT2) program and initiating the Industry Consortium for Advanced Technical Training (ICATT), which is currently operating in Illinois and Wisconsin. These are the first programs of their kind, where apprentices receive both an associates' degree and a German DIHK certification in various professions, such as mechatronics, industrial maintenance, and product design and visualization, while completing apprenticeships registered to U.S. Department of Labor (DOL) standards. GACC Midwest supports additional training programs at companies in more than 10 states. GACC Midwest's efforts have received recognition from the U.S. DOL, National Governors Association, *Wall Street Journal*, German Federal Government, and many others.

Mr. Tomkins is a frequent speaker in the United States and abroad on topics relating to workforce and economic development. Prior to joining GACC Midwest in 2006, he spent more than a decade in various roles of international trade and economic development.

C

Examples of Sector-Specific Strategies

MASSACHUSETTS: THE COMMONWEALTH CORPORATION

Massachusetts was the first state to formalize sector strategy work and now has more than three decades of experience. In 1981 civic leaders formed a public–private partnership, the Bay State Skills Corporation, to fund local workforce development initiatives across the state in important industries, requiring a 50 percent matching contribution from employers.

The mission of the Commonwealth Corporation is to strengthen the skills of Massachusetts' youth and adults by investing in innovative partnerships with industry, education, and workforce organizations (Commonwealth Corporation, 2014).[1] In partnership with 8 state government offices, 16 regional Workforce Investment Boards, and 10 corporate and foundation partners, it allocated \$48.6 million in 2014 to addressing the workforce training needs of businesses, unemployed workers, and youth.

The corporation's Workforce Competitiveness Trust Fund (WCTF) is an example of its partnership model in practice. In an effort to build regional capacity to address the gap between the skills held by unemployed and underemployed workers and the skills needed by employers, the WCTF made grants to 15 partnerships across the state. These partnerships involved more than 100 businesses, as well as regional stakeholders from community colleges, vocational technical schools, community-based organizations, workforce investment boards, career centers, and organized labor.

[1]In 1996, BSSC merged with the Industrial Service Program to form the Commonwealth Corporation, which is a quasi-public corporation of the Commonwealth of Massachusetts.

AUTOMOTIVE MANUFACTURING TECHNICAL EDUCATION COLLABORATIVE (AMTEC)

AMTEC is an example of a large, cross-state sector partnership that integrates federal, state, and private investments in skilled technical workforce development. Established in 2005 with a grant from the National Science Foundation's (NSF) Advanced Technical Education (ATE) program to the Kentucky Community and Technical College System, AMTEC is a collaboration of community and technical colleges and industry partners that aims to prepare technicians and engineers for work in automobile manufacturing and technology. AMTEC is one of 42 NSF ATE centers.[2]

Today AMTEC has 13 active community college and technical education partners and 15 industry partners located across eight states.[3] Industry partners include American, Asian, and German automakers and suppliers and the United Auto Workers. The sector partnership identifies worker skill needs across critical job classifications. Using a career pathway approach, AMTEC develops standardized coursework that is modular, flexible, and relevant to local contexts, and it produces "stackable" credentials. Colleges and organized labor work with manufacturers in their local area to customize training consistent with standardized curricula and credentials.

ARIZONA COMMERCE AUTHORITY

The Arizona Commerce Authority, a statewide economic development organization overseen by a public–private board, established the Workforce Arizona Council (WAC) to strengthen the workforce system and address a variety of industry workforce needs using sector strategies. WAC provides customized, intensive technical assistance to local areas, embedding sector strategies into plans, grants, and guidance for workforce plans, apprenticeships, and career pathway programs.

In 2013, Arizona had active or emerging sector partnerships in 11 counties in a wide range of sectors that depend on skilled technical occupations (see Figure C-1). For example, the Cochise Utilities Sector Partnership was established after the Sulphur Springs Valley Electric Corporation reached out to six other local utilities struggling to fill needed line worker positions. In response, the Arizona Workforce Connection worked with Cochise Community College to develop new courses and a certificate program to prepare local candidates, improve the talent pipeline, and provide long-term jobs for the

[2]For information about the NSF ATE program, see NSF (n.d.).

[3]AMTEC currently has active partners in Alabama, Kentucky, Michigan, Ohio, South Caroline, Tennessee, Texas, and West Virginia. For information on AMTEC, see AMTEC (n.d.).

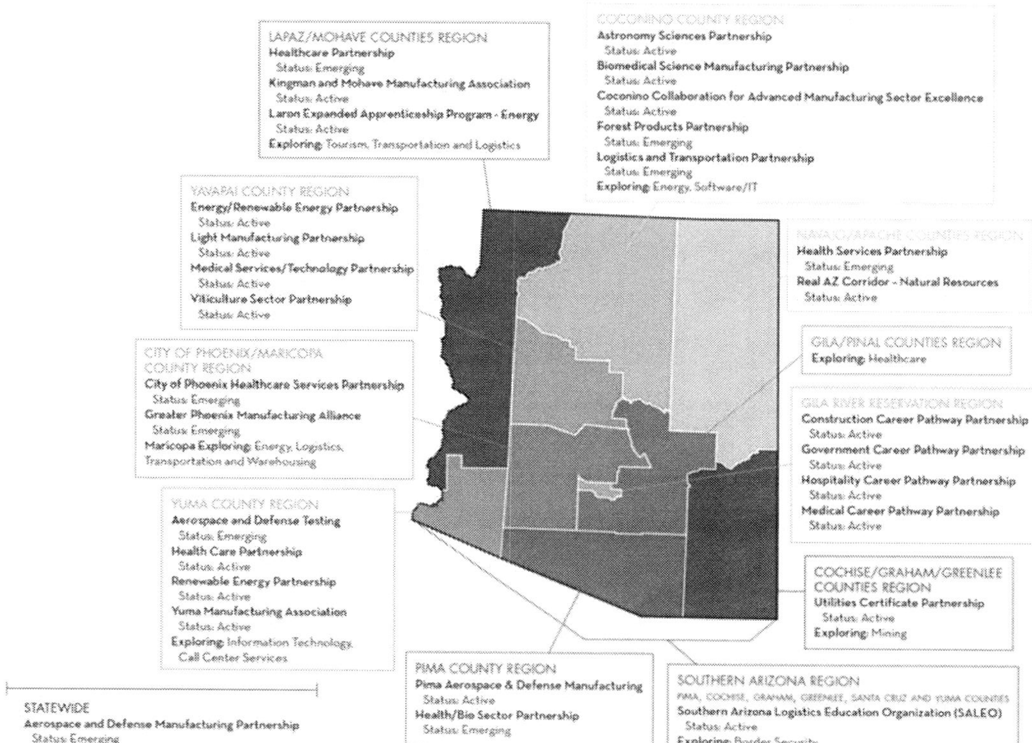

FIGURE C-1 Arizona sector partnerships in 2013.
SOURCE: Shedd, 2013.

community. Similarly, the Coconino Advanced Manufacturing Partnership is an association of 16 of the region's leading manufacturers that identified shared opportunities and requirements for growth in their sector. Working with regional partners to address workforce and training needs, the group has organized quarterly roundtables, developed incumbent worker training programs, and spearheaded a rebranding campaign to develop greater interest in careers in manufacturing (Arizona Commerce Authority, n.d.).